Contributions to Conservation Science

A Collection of Robert Feller's Published Studies on Artists' Paints, Paper, and Varnishes

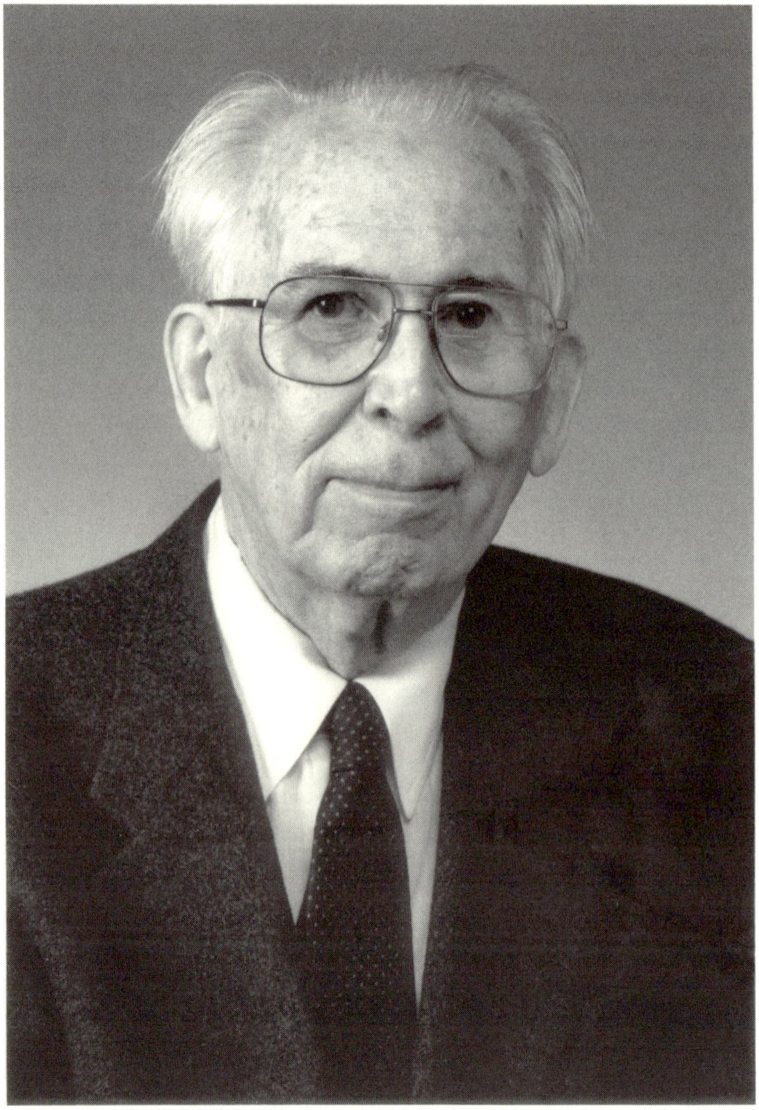

Robert Feller
2000

Contributions to Conservation Science

A Collection of Robert Feller's Published Studies on Artists' Paints, Paper, and Varnishes

edited by Paul M. Whitmore

commentaries by Robert Feller

Carnegie Mellon University Press
Pittsburgh 2002

Acknowledgments

For permission to reprint the articles by Robert Feller included in this volume, the editor and publisher gratefully acknowledge the following:

The International Institute for Conservation of Historic and Artistic Works, London, England (*Studies in Conservation*; *Recent Advances in Conservation*; *Science and Technology in the Service of Conservation*; *Application of Science in the Examination of Works of Art*); Materials Research Society, Warrendale, PA; UNESCO Publishing, Paris, France; The J. Paul Getty Trust, Los Angeles, CA, in association with the International Institute for Conservation of Historic and Artistic Works, London, England; American Chemical Society, Washington, DC; Federation of Societies for Coatings Technology, Bluebell, PA; Canadian Conservation Insitute, Ottawa, Ontario, Canada (*Conservation of Historic and Artistic Works on Paper*); American Institute for Conservation of Historic and Artistic Works, Washington, DC; Society for Imaging Science and Technology; John Wiley & Sons, Inc., New York, NY; International Council of Museums, Paris, France; American Association of Museums (*Museum News* magazine); National Gallery of Art, Washington, DC; Western Association for Art Conservation, Los Angeles, CA; Institute of Fine Arts, New York University, New York, NY; Allied Chemical and Dye Corporation, New York, NY; *The Magazine of Color Science and Technology*; and Indiana University Press, Bloomington, IN.

Carnegie Mellon University Press has attempted to obtain and has specifically received permission to use all of the material contained in *Contributions to Conservation Science*. The editor and publisher warrant that every effort has been made to ascertain and attribute the sources of initial publication.

Layout designed by Todd Sanders.

Library of Congress Control Number 2001091184
ISBN 0-88748-374-7

Printed and bound in the United States of America

10 9 8 7 6 5 4 3 2 1

Contents

Foreword

When the National Gallery of Art decided in 1950 to create a scientific research effort to advise its conservation and curatorial activities, it established a fellowship at one of the premier contract research organizations of its time, Mellon Institute in Pittsburgh. Hired to assume the role of research fellow was Dr. Robert Feller, a young chemist who went on to become one of the most accomplished conservation scientists of his generation. Dr. Feller spent his entire professional career heading the group in Pittsburgh, first directing the National Gallery of Art fellowship project, and later as the founding director of the Research Center on the Materials of the Artist and Conservator, as the research group became known when it was formed as an organization within Carnegie Mellon University in 1976. The conservation science research activity that he led continues to this day at Carnegie Mellon. It is the commemoration of these milestone events, the 50th year of Dr. Feller's career in conservation science and the 25th anniversary of the founding of the Center, that prompts the publication of this anthology of his published works.

This book is intended to bring together into a single volume the articles of lasting value that have appeared over the years in a variety of sources, some obscure or out-of-print, and which in many instances are currently available only through the reprint archives at the Center. As can be seen in the included bibliography, Dr. Feller has over 100 publications to his credit, on topics ranging from the very technical to more general treatments aimed at lay audiences. The papers in this volume were selected to capture the breadth of his interests as well as the depth to which he was able to explore these subjects over many years. Where there are substantial numbers of papers that have been published on a subject, the most significant or the retrospective summary articles have been chosen. Some subjects, such as studies of particular artists' pigments or cellulose ethers, are not included here because they are still readily available in the original published books. The papers in this collection are republished with permission from the original publishers, and they have been lightly edited to correct typographical errors and to conform to a uniform style. Apart from this, these works are unchanged from the original publications.

It is natural that this type of retrospective volume be viewed as an historical document that traces the path of the research activities and the excellence of the Pittsburgh laboratory over the years. In his section introductions Dr. Feller provides some eyewitness accounts about the laboratory activities and decisions at the time that enrich that history. While the research accomplishments are the focus of this narrative, it remains truly remarkable that such a history was possible in the first place: that a chemist was able to spend his entire career studying art conservation problems, and that a laboratory devoted to these issues was sustained for so long (from 1950 to this day) outside of a cultural institution. The secret to this longevity is revealed in the following publications, many of which conclude with an acknowledgment of the sponsorship of the Andrew W. Mellon Foundation. The history of the Center's accomplishments is due mainly to the Foundation's continuous support of conservation science in Pittsburgh. In addition to sponsoring the research, the Foundation has generously committed to the efforts to disseminate the findings of that research, and we gratefully acknowledge the Foundation's underwriting of the preparation and publication of this anthology. This book should thus serve as a lasting reminder of the Foundation's sustained commitment to art conservation and conservation science.

While more difficult to convey than the chronology of research activities, this anthology also is a testament to the significant impact Dr. Feller's work has had on a wide range of topics in conservation science. During his 38-year professional career as head of the lab, and in his continued work during the 14 years since his retirement, Dr. Feller has contributed research on fundamental subjects that has influenced subsequent studies of these topics in laboratories around the world. His research has helped shape current thinking on the chemical and physical origins of aging problems in art objects, particularly in the polymeric materials that have been incorporated into art and conservation practice over the last 50 years. Many of the insights contained in these works have become incorporated into the basic tenets of the field: of fading and polymer aging following the course of simple chemical reaction kinetics; of deterioration of materials proceeding in stages rather than as a steady gradual decline; of defining classes of stability that distinguish materials acceptable for different uses; and of using tools such as light dosimeters to monitor the risk of damage to materials in specific settings.

Beyond being a compilation of the specific findings of these studies, this book is an excellent illustration of scientific investigation in support of the conservation of art objects, for readers who may be unfamiliar with this activity. While the most publicized scientific pursuits in museums tend to be technical studies of objects, in particular those focused on authentication or attribution questions, scientific research also plays a fundamental role in the preservation and repair of objects. In this book the reader will find investigations that address specific problems in the conservation of art: identifying risks to art materials and their underlying causes, and the development of materials and tools for stabilizing and repairing the objects. One will also find insights into the more difficult issues that confront those who are charged with preserving artifacts; for example, determination of appropriate lighting conditions for museum displays that requires the balancing of exhibition and preservation needs. Similarly, the thorough understanding of the efficacy and side effects of a repair treatment is needed before one can judge the likelihood that the remedy will not jeopardize the artifact. These are all judgments that conservators and curators can best make with the information provided through scientific study. The papers in this book reflect rigorous scientific study of artifacts as material creations, but always with this implicit understanding of the aesthetic function and exhibition needs of the artifacts.

Finally, this book is offered with the deepest pleasure as a tribute to Dr. Feller from those of us fortunate to have known him personally. These pages capture, particularly in the section introductions and the more reflective papers included in the "Overview" section, some of the personality of the man who spent his entire professional career in the service of this field. His intellect is evident in the quality and breadth of the research, and the clarity with which he describes that work speaks to his skill as a teacher. Less obvious but no less important are the dedication with which he pursued this work and his generosity and eagerness to help—perhaps commonplace traits in those who are in a service-oriented profession like conservation, yet noteworthy qualities nonetheless. Most admirable, particularly for this author who continues in his stead, have been the self-effacing dignity and professionalism which he brought to the job. A more distinguished mentor for the following generation of conservation scientists is difficult to imagine.

Dr. Feller has said in more private moments that he felt genuinely privileged to do this work. He has indeed been fortunate to have led such a distinguished career, and we at the Center and in this profession are fortunate to have had his talents, enthusiasm, and generosity leave their mark on this field. Let this book be a celebration of that good fortune.

Paul M. Whitmore

Section 1:
Traditional Varnishes: Dammar and Mastic

The commercial availability of a number of durable synthetic resin-based coatings in the 1940s and 50s naturally led conservators to examine the possible uses of these new materials for restoration. In particular, colorless synthetic resins seemed attractive alternatives to natural resin picture varnishes, which discolor and degrade quickly and require frequent, and often difficult, removal and replacement.

Our laboratory began its research regarding picture varnishes in the early 1950s. At that time, little chemical and physical data were available to permit dammar and mastic resins to be compared in hardness, brittleness, and solubility with the newer materials then being considered, poly(vinyl acetate) and various methacrylic ester polymers. Simple measurements of hardness and brittleness were undertaken at first, but it quickly became apparent that determination of these basic properties would be highly dependent on the amount of solvent retained in test coatings that otherwise appeared dry to the touch. Moreover, instead of losing weight as expected owing to the evaporation of solvent, samples of dammar varnish cast from turpentine often gained in weight. This behavior was traced to the presence of peroxide in turpentine that had not been freshly distilled.

In comparing aging behaviors of natural and synthetic resin varnishes, it became necessary to develop an objective measure of deterioration that would have direct relevance to conservation practice. Conservators had long known that solvents of increasing polarity or "strength" were needed to dissolve many varnishes as they aged. Information gathered to document this behavior using solubility and removability in a series of solvent mixtures was summed up in a 1975 publication on the photochemically-induced loss of solubility of dammar, mastic, and the polycyclohexanone resins AW2 and MS2A, contrasting their behavior with that of poly(vinyl acetate) and the acrylic polymer, Acryloid B-72.

In the course of our initial investigations, it was discovered that certain authoritative texts assigned the source of dammar resin to conifers while others claimed the origin to be trees of the family Dipterocarpaceae. A search through local libraries gradually uncovered the history and reasons behind this confusion. The story was recounted in the bulletin of the Pittsburgh Section of the American Chemical Society, regrettably never enlarged upon or published elsewhere.

R. L. F.

Dammar and Mastic Varnishes — Hardness, Brittleness, and Change in Weight upon Drying

R. L. FELLER

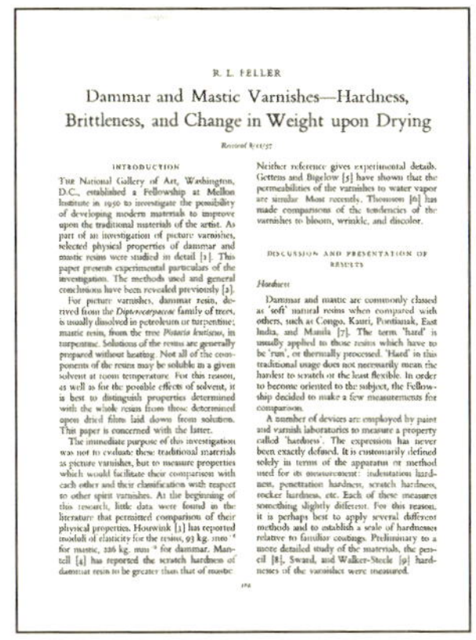

R. L. FELLER

Dammar and Mastic Varnishes—Hardness, Brittleness, and Change in Weight upon Drying

Received [illegible]

INTRODUCTION

The National Gallery of Art, Washington, D.C., established a Fellowship at Mellon Institute in 1950 to investigate the possibility of developing modern materials to improve upon the traditional materials of the artist. As part of an investigation of picture varnishes, selected physical properties of dammar and mastic resins were studied in detail [1]. This paper presents experimental particulars of the investigation. The methods used and general conclusions have been revealed previously [2].

For picture varnishes, dammar resin, derived from the *Dipterocarpaceae* family of trees, is usually dissolved in petroleum or turpentine; mastic resin, from the tree *Pistacia lentiscus*, in turpentine. Solutions of the resins are generally prepared without heating. Not all of the components of the resins may be soluble in a given solvent at room temperature. For this reason, as well as for the possible effects of solvent, it is best to distinguish properties determined with the whole resins from those determined upon dried films laid down from solution. This paper is concerned with the latter.

The immediate purpose of this investigation was not to evaluate these traditional materials as picture varnishes, but to measure properties which would facilitate their comparison with each other and their classification with respect to other spirit varnishes. At the beginning of this research, little data were found in the literature that permitted comparison of their physical properties. Houwink [3] has reported moduli of elasticity for the resins, 93 kg. mm^{-2} for mastic, 226 kg. mm^{-2} for dammar. Mantell [4] has reported the scratch hardness of dammar resin to be greater than that of mastic. Neither reference gives experimental details. Gettens and Bigelow [5] have shown that the permeabilities of the varnishes to water vapor are similar. Most recently, Thomson [6] has made comparisons of the tendencies of the varnishes to bloom, wrinkle, and discolor.

DISCUSSION AND PRESENTATION OF RESULTS

Hardness

Dammar and mastic are commonly classed as 'soft' natural resins when compared with others, such as Congo, Kauri, Pontianak, East India, and Manila [7]. The term 'hard' is usually applied to those resins which have to be 'run', or thermally processed. 'Hard' in this traditional usage does not necessarily mean the hardest to scratch or the least flexible. In order to become oriented to the subject, the Fellowship decided to make a few measurements for comparison.

A number of devices are employed by paint and varnish laboratories to measure a property called 'hardness'. The expression has never been exactly defined. It is customarily defined solely in terms of the apparatus or method used for its measurement: indentation hardness, penetration hardness, scratch hardness, rocker hardness, etc. Each of these measures something slightly different. For this reason, it is perhaps best to apply several different methods and to establish a scale of hardnesses relative to familiar coatings. Preliminary to a more detailed study of the materials, the pencil [8], Sward, and Walker-Steele [9] hardnesses of the varnishes were measured.

Originally printed in:
STUDIES IN CONSERVATION
Volume 3, Number 4 (1958), Pages 162-174.

Introduction

The National Gallery of Art, Washington, D.C., established a Fellowship at Mellon Institute in 1950 to investigate the possibility of developing modern materials to improve upon the traditional materials of the artist. As part of an investigation of picture varnishes, selected physical properties of dammar and mastic resins were studied in detail.[1] This paper presents experimental particulars of the investigation. The methods used and general conclusions have been revealed previously.[2]

For picture varnishes, dammar resin, derived from the Dipterocarpaceae family of trees, is usually dissolved in petroleum or turpentine; mastic resin, from the tree *Pistacia lentiscus*, in turpentine. Solutions of the resins are generally prepared without heating. Not all of the components of the resins may be soluble in a given solvent at room temperature. For this reason, as well as for the possible effects of solvent, it is best to distinguish properties determined with the whole resins from those determined upon dried films laid down from solution. This paper is concerned with the latter.

The immediate purpose of this investigation was not to evaluate these traditional materials as picture varnishes, but to measure properties which would facilitate their comparison with each other and their classification with respect to, other spirit varnishes. At the beginning of this research, little data were found in the literature that permitted comparison of their physical properties. Houwink[3] has reported moduli of elasticity for the resins, 93 kg. mm^{-2} for mastic, 226 kg. mm^{-2} for dammar. Mantell[4] has reported the scratch hardness of dammar resin to be greater than that of mastic. Neither reference gives experimental

1. Work by the author on the infrared spectra of these resins has been reported by him in "Dammar and Mastic Infrared Analysis," *Science*, 120 (1954), p. 1069.

2. Feller, R. L., "The Hardness and Flexibility of Natural- and Synthetic-Resin Varnishes," *The Museum News*, 29, No. 20 (1952), p. 7-8, and "Picture Varnish," *Encyclopaedia of Chemistry (Supplement)*, N.Y. (1958), p. 220-222.

3. Houwink, R., *Elasticity, Plasticity, and the Structure of Matter*, 2nd edn., Harron Press, Washington, D.C. (1953), p. 141.

details. Gettens and Bigelow[5] have shown that the permeabilities of the varnishes to water vapor are similar. Most recently, Thomson[6] has made comparisons of the tendencies of the varnishes to bloom, wrinkle, and discolor.

DISCUSSION AND PRESENTATION OF RESULTS

Hardness

Dammar and mastic are commonly classed as "soft" natural resins when compared with others, such as Congo, Kauri, Pontianak, East India, and Manila.[7] The term "hard" is usually applied to those resins which have to be "run," or thermally processed. "Hard" in this traditional usage does not necessarily mean the hardest to scratch or the least flexible. In order to become oriented to the subject, the Fellowship decided to make a few measurements for comparison.

A number of devices are employed by paint and varnish laboratories to measure a property called "hardness." The expression has never been exactly defined. It is customarily defined solely in terms of the apparatus or method used for its measurement: indentation hardness, penetration hardness, scratch hardness, rocker hardness, etc. Each of these measures something slightly different. For this reason, it is perhaps best to apply several different methods and to establish a scale of hardnesses relative to familiar coatings. Preliminary to a more detailed study of the materials, the pencil,[8] Sward, and Walker-Steele[9] hardnesses of the varnishes were measured.

The Walker-Steele and the better known Sward instruments are of the rocker type. The measurement is made by setting a rocking device in motion on the surface of a film that has been laid down on a glass or steel plate. A film is classed as softer than another if it damps the motion of the rocker more rapidly. When the motion of the rocker has been damped through a given arc, the time taken, in the case of the Walker-Steele instrument, or the number of rocks involved, in the case of the Sward instrument, is taken as a measure of hardness. The result is usually expressed as a fraction of the behavior when the instrument rocks on the uncoated glass or metal support. The higher the fraction, the harder the film.

Measurements were made at 70°F (21°C) and 50 percent relative humidity on films of the varnishes prepared with turpentine solvent. Since it was expected that retained solvent would influence the hardness, two measurements were made: the first, on films one month after having been laid down on glass plates and the second, on films that had been baked for two days at 158°F (70°C). Baking was intended to hasten drying, to drive off most of the solvent.

The results are summarized in Table 1. After one month, films of mastic-turpentine varnish were found to be harder than similar varnishes of dammar. After baking, the hardness of both varnishes proved to be identical, within the

4. Mantell, Kopf, Curtis, and Rogers, *Technology of Natural Resins*, J. Wiley & Sons, N.Y. (1942), p. 83. See also penetration-hardness measurements of Nicolardot and Coffignier in "Contribution a la Caracterization des Gommes a Vernis," *Chemie et Industrie*, 5, No. 2 (1921), p. 159, and the measurements by Stout, G. L., and Cross, H. F., in *Technical Studies*, 5 (1937), p. 241.

5. Gettens, R. J., and Bigelow, E., "The Moisture Permeability of Protective Coatings," *Technical Studies in the Field of Fine Arts*, 2 (1933-34), p. 63.

6. Thomson, G., "Some Picture Varnishes," *Studies in Conservation*, 3 (1957), p. 64.

7. Anon., *Natural Resins Handbook*, American Gum Importers Laboratories, Inc., New York, N.Y. (1939).

8. Coleman, S. D., and Smith, E. N., "The Pencil-Hardness Test for Organic Finishes," *Organic Finishing*, 10 (1949), p. 23.

9. Walker, P. H., and Steele, L. L., American Paint and Varnish Manufacturing Association (1925), Circular 229. See also Gardner, H. A., and Sward, G. G., *Physical and Chemical Examination of Paints, Varnishes, Lacquers and Colors*, 11th edn., H. A. Gardner Laboratory, Inc. (1950), Chap. 11, "Hardness, Abrasion, Impact, Adhesion."

TABLE 1
Rocker Hardness of Varnishes at 50 percent R.H. and 70°F.
Film thickness: 1.5 mil.
Precision: ± 0.03

		Walker-Steele		Sward	
Resin	Solvent	One month old (n)	Baked (n)	One month old (n)	Baked (n)
Mastic	Turpentine	0.91 (6)	0.94 (6)	0.62 (5)	0.80 (2)
Dammar	Turpentine	0.77 (4)	0.95 (6)	0.45 (5)	0.82 (2)
Polyvinylacetate AYAC	Toluene		0.74 (2)		0.63 (2)
Poly(n-butyl methacrylate) "Lucite" 44	Toluene				0.31 (4)

(n) = number of samples averaged.

limits of the measurements. On a comparative basis, the films were harder than films of polyvinylacetate and poly(n-butylmethacrylate).

In the measurement of pencil hardness, a value between HB and F was measured for the two varnishes after baking, compared to 2B for du Pont poly (n-butylmethacrylate) "Lucite" 44, F for Bakelite polyvinylacetate AYAC, and 2H for a sample of polyethylmethacrylate synthesized in the laboratory. A pencil hardness of F has been reported by Shaw[10] for dammar. Rather than a scratching of the surface, one would express more accurately the measurement on the natural resins by stating that marking by the pencil took place upon disintegration of the film of varnish owing to brittleness. The pencil hardness scale for Eagle "Turquoise" pencils in this region changes in equal steps from (soft) 3B, 2B, B, HB, F, 2H, and 4H (hard).[11] In the Fellowship's experience with spirit varnishes, a film of hardness 3B may be considered to be very soft, one of 3H, to be very hard.

10. Shaw, T. P. G., "The Properties of Some Polyvinyl Resins as Lacquer Resins," *Official Digest, Federation of Paint and Varnish Production Clubs*, 162 (1937), p. 9.

11. Private Communication, Eagle Pencil Company, 710 E. 14th Street, New York, N.Y.

Brittleness

To study further the behavior of the varnishes during drying and after baking, their ability to withstand bending was investigated. This property was measured by coating the films on 1-mil.-thick (0.001 in., 0.0254 mm.) aluminum foil with a stainless-steel paint-spreading blade and observing the diameter of mandrel around which it was necessary to bend strips of the coated foil in order to cause profuse cracking. The test was chosen for its simplicity; museum laboratories may make similar measurements without elaborate equipment. The bending test is related to the elongation-at-break of a film.[12] A brittle film is considered in this report to be one with a small percentage of elongation-at-break. The larger the diameter of mandrel, the more brittle the film.

12. Morgan, H. I., and Mathes, K. N., American Soc. Testing Materials, Special Technical Publication, *Symposium on Temperature Stability of Electrical Insulating Materials*, No. 161 (1954), p. 97.

Results are shown in Figures 1 and 2. The curves in the two figures are based on a statistical analysis of 161 individual measurements. Three different varnishes of each resin were used. Although picture varnishes are frequently applied at a thickness of 1 mil. or less, the measurements were made over a wide range of thicknesses in order to establish more firmly the characteristics of the films. A summary of the results is given in Table 2. The diameters reported in Table 2 are estimated to be precise to the extent of ± 0.1 in. for values below, and ±0.2 in. for values above 1 in. Values for several other resins are given for comparison. On the basis of the mandrel tests, the thoroughly-dried films are calculated to have an elongation-at-break of the order of 0.2 percent.

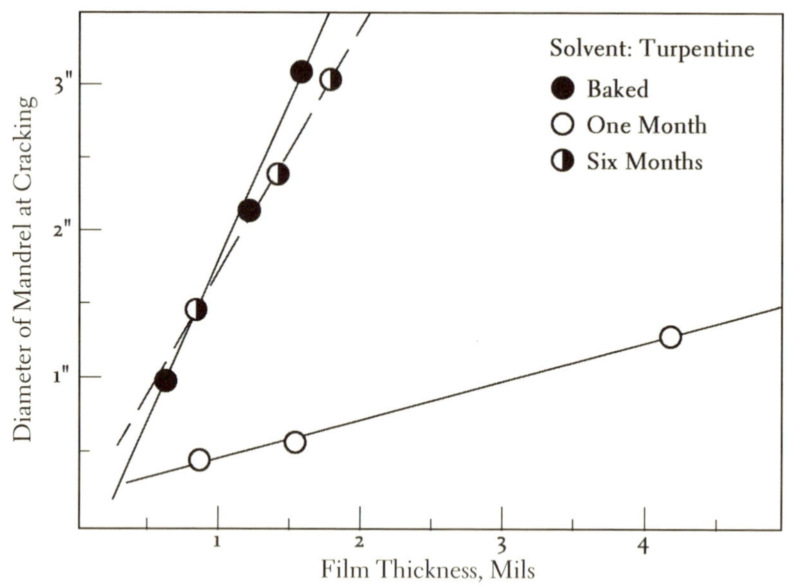

FIGURE 1
Brittleness of dammar varnish on 1.0 mil. aluminum foil at 70°F and 50% RH.

Measurements of the ability to withstand bending confirm conclusions drawn from hardness measurements. After baking, the two varnishes have essentially the same brittleness. After one month, normally-dried dammar varnish is far less brittle than mastic. When the normally-dried dammar varnish is about six months old, the brittleness is similar to that of the film dried by baking. The characteristic of mastic varnish in exhibiting greater brittleness than dammar until both are thoroughly dry is not a phenomenon dependent upon a peculiarity of turpentine. This was confirmed by repeating the tests with month-old varnishes prepared in o-chlorotoluene (b.p. 319.1°F., 159°C.) and t-butylbenzene (b.p. 334.4 - 336.2°F., 168-169°C.).

Occasionally, 5 percent stand oil by volume is added to 5 lb. cut dammar varnish. Because the oil requires time to oxidize, its addition may decrease the brittleness of the varnish during its early life. Data in Table 2 indicate, however,

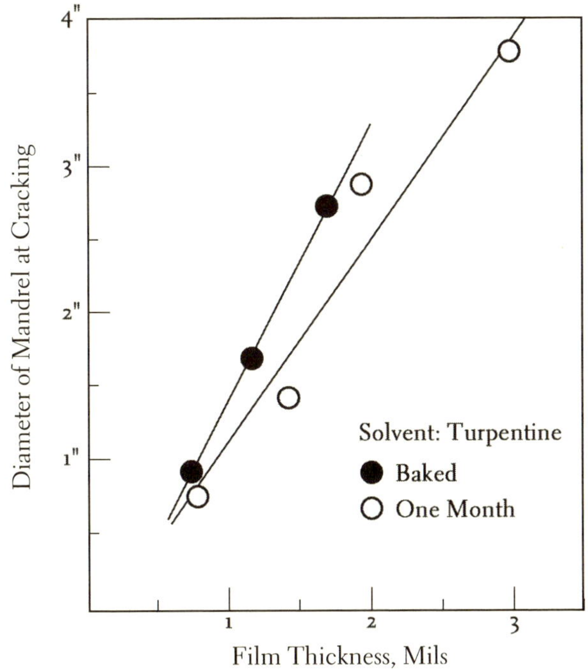

Figure 2
Brittleness of mastic varnish on 1.0 mil. aluminum foil at 70°F and 50% RH.

Table 2
Diameter of Bend Necessary to Cause Profuse Cracking of 1.5 mil. Films on 1 mil. Aluminum Foil at 70°F and 50% RH

		Diameter of Mandrel, inches	
Resin	Solvent	One Month	Baked
Mastic	Turpentine	1.9	2.5
Dammar	Turpentine	0.6	2.7
Dammar	Turpentine	2.7 *	
Dammar, 5% Stand Oil	Turpentine	2.6 *	2.4
Dammar	Toluene	1.7	
Dammar	t-Butylbenzene	0.7	
Mastic	t-Butylbenzene	2.1	
Dammar	o-Chlorotoluene	1.0	
Mastic	o-Chlorotoluene	1.7	
Badische Analin and Soda Fabrik, AW-2	Toluene		4.0
Dow Chemical, Polystyrene PS-3	Toluene		1.3
Bakelite, Polyvinylacetate AYAB	Toluene		0.6
Bakelite, Polyvinylacetate AYAA	Toluene		less than 0.1

* After six months.

that this quantity of oil is not sufficient to have a marked effect upon the brittleness of an aged film of varnish. (Recently, Thomson[6] has also pointed out that the addition of stand oil appears to have no significant effect on bloom at one month.)

When well-dried, the two varnishes may be characterized as being brittle. A 1.5 mil. film of baked dammar cracks around a 2.7 in. mandrel in this test. In comparison, unplasticized samples of polystyrene PS-3 laid down from solution in toluene, when baked dry, crack around a 1.3 in. mandrel at 1.5 mil. thickness. The Dow Chemical Company PS-3 resin is reported in the company literature to have an average molecular weight of 35,000 and to deposit what is considered to be a brittle film. The viscosity grade[13] of polystyrene PS-3, the viscosity in centipoises of a solution containing 20 percent solids by weight in toluene at 70°F., is 12.0 cp., compared with 1.4 to 1.9 cp. for dammar and mastic resins.[14]

13. Feller, R. L., "Factors Affecting the Appearance of Picture Varnish," *Science*, 125 (1957), p. 1143.

14. Viscosity grade is a means of classifying resins to give some indication regarding their average molecular weight, a factor that influences brittleness. The grade of resin AW-2, a polymer based on cyclohexanone, is 1.2 cp.[13] That of Bakelite polyvinylacetate AYAC is 9 cp., and that of AYAA is 40. The author intends to publish some results on the relationship of viscosity grade to brittleness.

Change in Weight upon Drying

The hardness and brittleness measurements on films of the well-dried varnishes proved to be similar. However, after only a month's drying under normal conditions, a difference in properties from those of the well-dried state was noted. This difference might well have been due to traces of solvent that remained in the films. It was decided to explore briefly this aspect of the problem.

A spirit varnish is expected to lose its solvent at a rate generally similar to that of curve (a) in Figure 3. It may be noted that little change in weight takes place

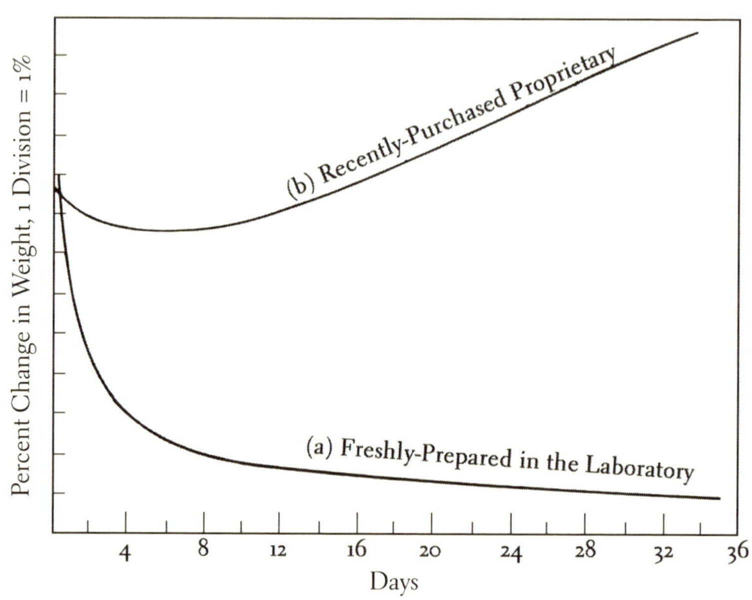

FIGURE 3
Drying of dammar-turpentine varnish at 70°F and 50% RH.

between the 15th and 30th days. Nevertheless, it is generally found that solvent still remains in a spirit varnish after one month of drying under room conditions. The shape of the curve and the presence of solvent after one month have been amply demonstrated in the laboratory with spirit varnishes formulated with solvents and synthetic resins that have little tendency to oxidize under the conditions of the experiment.

When expressed in the diagram shown in Figure 4, the brittleness data for dammar varnishes strongly suggest that, when compared to the thoroughly dry state, the ability of the varnish to withstand a bend of narrower diameter after only a month's drying is dependent upon retained solvent. For example, the rate of evaporation of toluene from the varnish is expected to be greater than that of turpentine. The presence of less solvent in a toluene-based varnish after one month would explain the greater brittleness observed. A line is drawn in Figure 4 to suggest a behavior of mastic varnishes similar to that of dammar with respect to the volatility of the solvent. From a consideration of Figure 4 one might suggest that mastic loses its solvent more rapidly than dammar and that it might be for this reason that mastic varnish is the more brittle after one month.

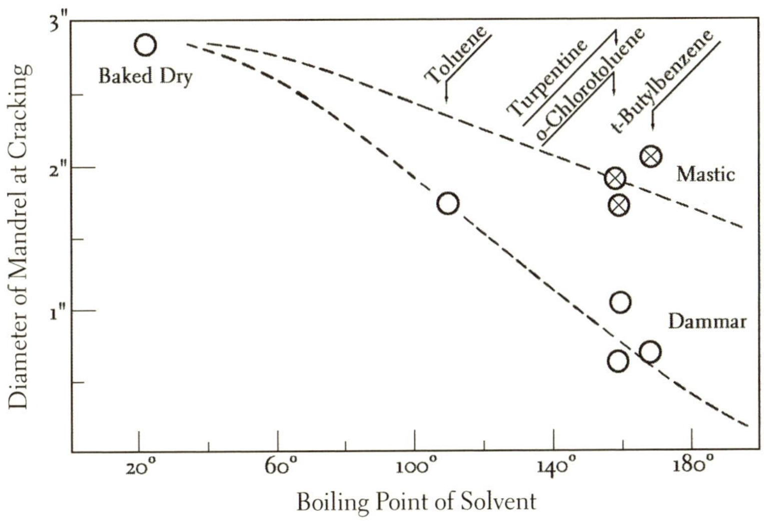

FIGURE 4
Influence of solvent upon brittleness: 1.5 mil. films of dammar and mastic after one month at 70°F and 50% RH.

An experiment was planned to demonstrate whether a wide difference in retained solvent did exist between the two varnishes after the first month. Varnishes were prepared in chlorinated solvents, allowed to dry in the laboratory, scraped off, and analyzed for chlorine. The results of four determinations appear in Table 3. Each value is the mean of samples at two different thicknesses of film, close to 2 mil. in thickness. The mean deviations averaged about 0.56 percent, largely owing to the variation in film thickness.

Table 3
Chlorinated Solvent Found in Varnishes After Drying

Solvent	Drying time in days	Solvent found, percent		Difference
		Dammar	Mastic	
Chlorobenzene	10	7.74	6.32	1.42
Chlorobenzene	6	12.09	11.76	0.35
o-Chlorotoluene	30	10.36	9.56	0.80
o-Chlorotoluene	7	12.88	11.35	1.53

The percentage of "solvent found" was calculated on the basis of the percentage chlorine found in the total weight of varnish sample taken.

The determination of solvent content in this experiment is bound to give results that are low, because the varnishes were scraped off the glass and solvent was able to evaporate from the newly exposed surfaces of the powdered varnish before the analysis took place. During scraping, the odor of the solvent was very noticeable. However, in each case a greater amount of chlorinated solvent was found in the dammar varnish. These results tend to bear out the hypothesis that mastic varnish gives up its solvent more rapidly than dammar. However, the conditions of this experiment were not sufficiently controlled that a definite conclusion could be drawn. A different method of observing the retention of solvent was next undertaken, in an attempt to confirm the results.

Table 4
Solvent Apparently Retained in Varnishes After Drying,
Based on Original Weight of Resin Taken

Varnish	Calc. wt. of resin in sample, g.	Calculated quantity of solvent retained, in percent, after			
		10 days	17 days	20 days	30 days
Dammar	0.1070	19.8		19.6	19.2
Dammar	0.0909		20.7		20.8
Mastic	0.1080	20.4		20.5	21.1
Mastic	0.0925		19.9		21.2

Solutions of dammar and mastic in t-butylbenzene were prepared. Known amounts were added to aluminum milk-bottle caps and allowed to dry in the laboratory. The film thicknesses were between 0.002 and 0.003 in. The results, each value the average of four samples, are reported in Table 4. The dammar solution was prepared at 19.9 percent solids by weighing solvent and resin. The density of the mastic solution was adjusted so that it was equal to that of the dammar, within 1 part per thousand.

The percentage of solvent reported in Table 4 is higher than that determined by scraping samples and analyzing for chlorine. The figures of about 20 percent solvent by weight in Table 4 are expected to represent more closely the actual amount of solvent present than those in Table 3 (where it is known that solvent was able to evaporate from the pulverized varnish). In fact, a similar weight of turpentine remaining after one month at 70°F. (21.1°C.) and 50 percent R.H. was observed by another method, determination of weight-loss upon baking the films: 16.5 percent for dammar and 17.1 percent for mastic varnish 0.001 in. thick, prepared in turpentine that had been freshly distilled under reduced pressure in an atmosphere of carbon dioxide. Ordinary laboratory conditions were used in drying the varnishes referred to in Tables 3 and 4, rather than the precise conditions, 70°F., 50% R.H., employed in the other experiments.

The data in Table 4 do not support the hypothesis that dammar varnish retains the greater percentage of solvent during the first month. Instead they suggest another possible explanation for the data in Table 3. Note that the mastic samples in Table 4 were gaining weight after thirty days, whereas the dammar samples continued to lose weight or remained practically unchanged. Oxidation can no doubt take place in varnishes prepared with these resins and with turpentine. If mastic varnish were to gain a greater weight by oxidation than dammar, the percentage of solvent by weight in the film of mastic resin plus oxidation products might prove to be less than that of solvent in a film of dammar plus a lesser amount of oxidation products.

Because of the unknown amount of oxidation, there is uncertainty in these first estimates regarding the percentage of solvent retained in dammar and mastic varnishes. Nevertheless, there appears to be an appreciable amount of solvent present after thirty days. A satisfactory explanation for the greater hardness and brittleness of mastic after one month must await further investigation.

The possibility of oxidation received additional consideration. Differences in behavior of the varnishes upon drying could be demonstrated with greater assurance when the change in weight was investigated with larger samples: for example, films weighing 0.5 to 1.5 g. when well dried, prepared from toluene and from turpentine solutions. Observations were made of the loss in weight under two circumstances: upon standing under controlled room conditions for a month (Figure 3), and upon baking at 158°F. (70°C.) in a Fisher "Isotemp" oven for forty-eight hours.

Upon baking varnishes that were one week old or less, it was found that nearly all samples lost weight during the first few hours, then gradually regained 1 to 3 percent above the minimum weight. The decrease in weight thus observed must have been largely through loss of solvent. Samples of pulverized resins (passing a screen of 11 mesh/cm., but not 6 mesh/cm.) from recently purchased resins and from purchases which were received over thirty years ago, did not lose or gain more than 0.8 percent in weight after seventy hours at 70°C. The loss of water is not expected to exceed this amount (see Table 6).

Analysis for solids in freshly prepared dammar varnishes of known composition gave added evidence that, at least for this resin, heating the varnish for several hours need not give rise to extensive oxidation. About 200 to 300 mg. of varnish were spread on the bottom of aluminum dishes of 2 in. diameter in making these gravimetric analyses. Upon baking the samples for four to six hours at 158°F. (70°C.) the following results were obtained: in toluene, known 19.2 and 29.9 percent solids, found 18.9 and 30.1 percent; in turpentine, known 20.0 and 30.0 percent, found 20.3 and 30.7 percent respectively.

When drying at room conditions, varnishes prepared with freshly distilled turpentine, and laid down within a day or two of their preparation, lost weight very much as expected of a spirit varnish. However, in all cases, proprietary varnishes formulated with turpentine and laboratory samples not freshly prepared behaved in a decidedly different manner. They gained considerable weight upon drying, following a loss in weight that occurred during the first few days. Figure 3 shows typical curves obtained in these experiments. Each curve in the figure is based on data from five different varnishes. Although a proprietary varnish is used in the example, curve (b) is a fair representation of the behavior of varnishes that were not freshly prepared. Similar data were obtained with mastic. A review of data on six different varnishes from each resin indicates, however, that the gain in weight usually occurred sooner in mastic varnish.

How may one explain the striking gain in weight in varnishes which were not freshly prepared? The gain probably represents oxidation, because baking films that changed weight in a manner of curve (b), rather than that of curve (a), did not cause them to lose more than 1 or 2 percent of the weight attained after one month. One would expect that if the gain had been moisture, it would have been driven off. Oxidation probably was not due to the presence of linseed or stand oil, for the hardness and the brittleness of the proprietary varnishes after one month were similar to that of the laboratory-prepared samples. Mills and Werner[15] have called attention to oxidation of the resins under the influence of ultraviolet light. However, the striking difference in behavior here, under the same conditions of illumination, appears to be associated with the fact that the solvent was no longer freshly prepared.

15. Mills, J. S., and Werner, A. E. A., "Partition Chromatography in the Examination of Natural Resins," *Journal of Oil and Colour Chemists Association*, 37 (1954), p. 131.

Since the pulverized resins and the freshly prepared varnishes did not increase much in weight, the relatively rapid gain observed with the proprietary varnishes may possibly have been enhanced by the presence of peroxides in the slightly aged turpentine.[16] Samples of turpentine were analyzed for active oxygen by the method of Siggia.[17] The following table indicates that an increased amount of active oxygen was detected in aged turpentine.

16. Mattiello, J.J., *Protective and Decorative Coatings*, Vol. 1, J. Wiley & Sons, New York (1941), p. 564.

17. Siggia, S., "Determination of Benzoyl Peroxide in Organic Media," *Industrial and Engineering Chemistry, Analytical edition*, 19 (1947), p. 872.

In all of these investigations, baking of varnishes was intended primarily to drive off solvent. No doubt some oxidation occurred. However, baking had no great effect upon the measurement of brittleness. This is shown by the similarity found between one-month-old and baked mastic varnish and between six-month-old and baked dammar.

TABLE 5
Active Oxygen in Solvents

Solvent	Milligrams of active oxygen per milliliter of solvent
Benzene, from laboratory shelf (control)	0.000
Turpentine, freshly-distilled at atmospheric pressure	0.014
Turpentine, recently purchased	0.832

Evidence was obtained that the proprietary varnishes were more readily oxidized than those freshly prepared. Further investigations may show a difference in the hardness and brittleness between these two. However, the difference, if any, in the present measurements, for either baked or air-dried films, apparently was no more than the experimental error.

To summarize these brief considerations of drying: the evidence of the lower hardness and brittleness, of the influence of solvent as suggested in Figure 4, of the presence of chlorinated solvents, and of the measurement of loss in weight, indicate that dammar and mastic varnishes retain appreciable solvent after drying for one month at room conditions. Freshly prepared varnishes of these two resins behave much as expected of a spirit varnish, with regard to the change in weight (curve (a) in Figure 3) and to the retention of solvent for a considerable time. On the other hand, varnishes of these resins in turpentine, when not freshly prepared, frequently show a striking gain in weight that gives evidence of being due to oxidation. After one month, the hardness and brittleness of mastic varnish are closer to the values of the dried state than are those of dammar; an explanation of this behavior must await further investigation.

EXPERIMENTAL

Materials

Two proprietary varnishes were tested along with a varnish that was prepared in the laboratory from a good grade of resin and freshly distilled spirits of American gum turpentine. Singapore No. 1 dammar and Chios mastic resins were purchased in 1950 from Fezandie and Sperrle, Inc., 205 Fulton Street, New York City. The melting points, determined by a modified ball-and-ring method, and water contents of these resins are reported in Table 6. Data on two older samples, used in the study of the effect of heating the pulverized resins, are also included. The content of water was determined by titration with Karl Fischer reagent.

TABLE 6
Melting Points and Water Contents of Resin

Resin	Purchase Date	m.p., °C	Water found, percent
Dammar	1950	115	0.25
Dammar	1923	109	0.57
Mastic	1950	96	0.58
Mastic	1923	111	0.80

Hardness

Varnishes were coated on 2.5 mm. thick window glass with the aid of a stainless-steel paint-spreading blade. The solutions contained 20 to 35 percent solids by weight. The dried films were scratched through to the glass at selected places and the thicknesses of the films measured to the nearest 0.1 mil. (0.0001 in., 0.00254 mm.) with a microscope having a calibrated micrometer-screw. Film thickness was plotted vs. hardness value and the hardness at 1.5 mil. thickness obtained by interpolation.

The varnishes were stored and tested in a constant temperature and humidity room maintained at 70 ± 2°F. (21.1 ± 1°C.) and 50 ± 3 percent R.H. Samples to be baked were first stored for one week under these conditions, then baked at 158 ± 2°F. (70 ± 1°C.) in a Fisher Scientific Company (Pittsburgh, Pa.) "Isotemp" oven for forty-eight hours, and returned to the conditioning room for ten days prior to testing. The laboratories at Mellon Institute are illuminated by Westinghouse 50 watt "standard cool white" fluorescent tubes at a level of about 50 footcandles on the work bench.

Tests made upon the baked films in the constant temperature and humidity room indicated that three to four days were adequate for conditioning prior to measurement. However, ten days were allowed.

Brittleness

The same varnishes that were tested for hardness were also used in these experiments. They were coated on 1 mil. thick aluminum foil and subjected to the identical schedule of conditioning. The foil was made to adhere smoothly to 8 x 8 in. glass plates with a few milliliters of toluene placed between the foil and the glass and was allowed to remain supported on the glass until the samples were tested. Tests were made by cutting 1 in. wide strips with a razor blade and bending them, varnish-side out, around aluminum mandrels. The mandrels from 0.1 in. to 0.6 in. diameter differed in 0.1 in. steps, from 0.6 in. to 2.0 in. in

0.2 in. steps, and from 2.0 in. to 4.0 in. diameter in 0.5 in. steps. Thicknesses of varnish films were determined to the nearest 0.1 mil. with an Ames gauge. Occasional cracks occurred in the films when bent at diameters close to the diameter of mandrel reported. The diameter reported at the point of profuse cracking was chosen where cracks occurred 1/8 in. apart, or closer, all along the strip as it was pulled over the mandrel.

The diameter of the mandrel, reported in Table 2, is estimated to be precise to the extent of ± 0.1 in. for values below, and ± 0.2 in. for values above 1 in. The diameters of the points in Figures 1 and 2 are equal to ± 0.1 in., ± 0.1 mil. The 161 individual observations were taken in groups that averaged about 10 observations each. The most probable values of each group of data points, determined from a plot on probability graph paper, are shown in the two figures.

To test the effect of adding 5 percent oil by volume to dammar varnish containing 35 percent solids by weight in turpentine, samples were prepared using two proprietary stand oils and two proprietary sun-thickened linseed oils. One series of samples was kept for six months in a laboratory maintained at 70°F. and 50 percent R.H. Another series was let dry for one week in the laboratory, baked for two days at 158°F. (70°C.) and allowed to stand ten days at 70°F. and 50 percent R.H. prior to testing.

Loss of Solvent

The major experiments on loss of solvent, resulting in the data given in Figure 3, were made by Mr. Stuart Raynolds in this laboratory in 1951-52.[18] In his experiments samples were coated on 8 x 16 in. aluminum foil with a 6 in. wide stainless-steel blade. As soon as the varnish resisted flow, the foil was cut so that no uncoated foil remained along the length of strip. The strip was cut along the width and the varnish removed with solvents in a manner which left about 1/2 in. of uncoated foil at each end. The final strips were about 5 x 11 in. These were rolled, varnish-side out, and fastened in the form of a cylinder by crimping the uncoated edges of foil. Their shape allowed the foils to stand upright conveniently for weighing and baking. The cylinders were hung by alligator-jawed clips to dry in the constant temperature room and were weighed periodically on an analytical balance. At the end of the experiment the varnish was removed and its weight determined by difference. Removal was accomplished by cutting away the uncoated portions of the foils, accordion-folding them and removing the resin with solvent in a Soxhlet extractor. From 0.5 to 1.5 g. of dried film was found in most of the samples. Thicknesses varied from 0.5 to 2.0 mil.

18. Present address, Department of Chemistry, University of Pittsburgh, Pittsburgh, Pa. (Ed. note: as of 1958)

Conclusions

Films of dammar and mastic resins, laid down from spirit varnishes formulated with turpentine and then well-dried, were found to be similar in brittleness and rocker hardness. They were found to be harder on the Sward and Walker-Steele hardness scale and more brittle than samples of proprietary polyvinylacetate and poly(n-butylmethacrylate). After one month of drying, mastic varnishes were harder and more brittle than dammar; this initial difference apparently becomes less in time. It was also demonstrated that the addition of 5 percent stand oil by volume to dammar varnish formulated at 35 percent solids by weight did not markedly affect the brittleness observed after 6 months' drying.

Evidence was obtained that spirit varnishes formulated with these resins retained appreciable solvent at the end of one month of drying. Evidence was also obtained that oxidation occurred during the drying of varnishes formulated in turpentine, particularly when the varnishes were not freshly prepared.

Acknowledgments

The author wishes to acknowledge the assistance of Mr. Stuart Raynolds in the preparation of the many samples used in these experiments and for his extensive investigation of the loss of solvent from varnishes during baking and natural drying. Miss M. M. Yingling of the Department of Analytical Chemistry at Mellon Institute made the determinations of the contents of water in the resins, chlorine in varnishes prepared with chlorinated solvents, and active oxygen in turpentine.

A Note on the Exposure of Dammar and Mastic Varnishes to Fluorescent Lamps

R. L. FELLER

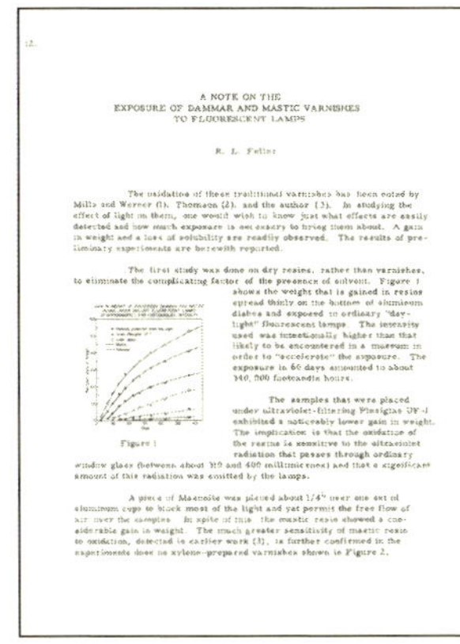

Originally printed in:
Bulletin of the American Group - IIC
Volume 4, Number 2 (1964), Pages 12-14.

The oxidation of these traditional varnishes has been noted by Mills and Werner,[1] Thomson,[2] and the author.[3] In studying the effect of light on them, one would wish to know just what effects are easily detected and how much exposure is necessary to bring them about. A gain in weight and a loss of solubility are readily observed. The results of preliminary experiments are herewith reported.

The first study was done on dry resins, rather than varnishes, to eliminate the complicating factor of the presence of solvent. Figure 1 shows the weight that is gained in resins spread thinly on the bottom of aluminum dishes and exposed to ordinary "daylight" fluorescent lamps. The intensity used was intentionally higher than that likely to be encountered in a museum in order to "accelerate" the exposure. The exposure in 60 days amounted to about 340,000 footcandle hours.

The samples that were placed under ultraviolet-filtering Plexiglas UF-1 exhibited a noticeably lower gain in weight. The implication is that the oxidation of the resins is sensitive to the ultraviolet radiation that passes through ordinary window glass (between about 310 and 400 millimicrons) and that a significant amount of this radiation was emitted by the lamps.

A piece of Masonite was placed about 1/4" over one set of aluminum cups to block most of the light and yet permit the free flow of air over the samples. In spite of this, the mastic resin showed a considerable gain in weight. The much greater sensitivity of mastic resin to oxidation, detected in earlier work,[3] is further confirmed in the experiments done on xylene-prepared varnishes shown in Figure 2.

1. Mills, J. S., and Werner, A. E. A., *J. Oil and Colour Chemists' Assoc.*, 37 (1954), p. 131.

2. Thomson, G., *Recent Advances in Conservation*, Butterworths, London (1963), p. 176.

3. Feller, R. L., *Studies in Conservation*, 3 (1958), p. 162.

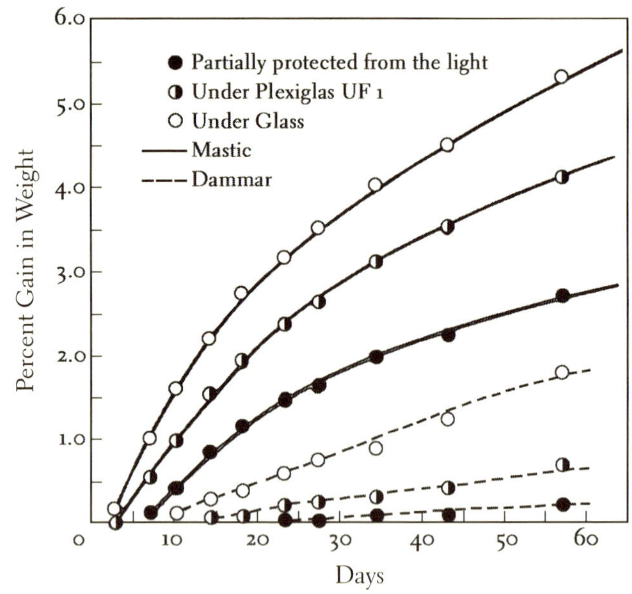

FIGURE 1
Gain in weight of pulverized dammar and mastic resins under daylight fluorescent lamps at approximately 240 footcandles intensity.

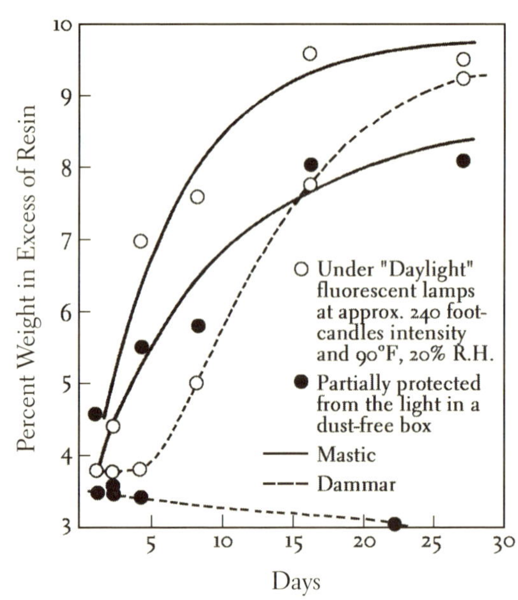

FIGURE 2
Gain in weight (oxidation) of varnishes prepared in xylene.

Xylene was used in place of turpentine to prepare the coatings in order to minimize the added factor of the oxidation of solvent. After a day or two of drying, the films still contained about 4% solvent, as indicated at the beginning

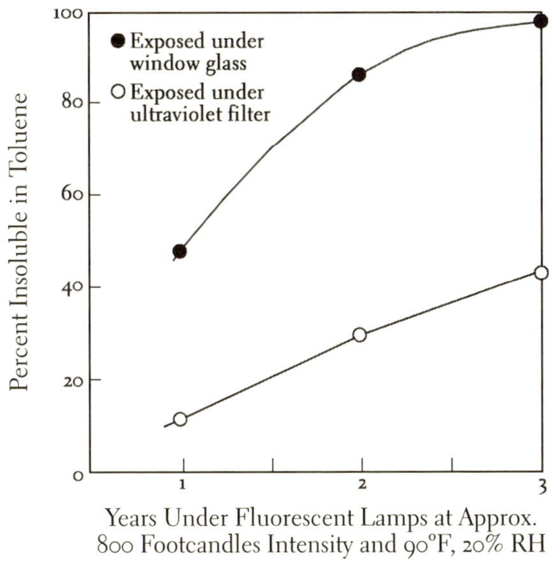

Figure 3
Rise of insoluble matter in dammar-stand oil varnishes containing 20% oil by weight.

of the curves in Figure 2. Again, the mastic gained weight even when it was partially protected from the light, whereas the dammar coating exhibited marked sensitivity to exposure to the light. The gain in weight in these films seemed to be more rapid than in the case of the dry powders.

Many films of dammar varnish contain a trace of stand oil. It is expected that gradual oxidation of the materials will make them less easily soluble in petroleum solvents. In Figure 3, films containing 20% oil (based on the weight of resin and oil present) were exposed to special "high-output" daylight fluorescent lamps . Toluene was chosen for this test and the method of analysis similar to that described earlier.[4] The results indicate that the rise of insoluble matter in toluene is not particularly rapid, but is sensitive to ultraviolet light. The matter that was insoluble in toluene was not crosslinked material. Rather, it appears to be material that had become changed in its sensitivity towards solvent through oxidation: after three years, the samples were readily dissolved in acetone, leaving only about 5% residue or less.

One should not attempt to draw elaborate conclusions from these preliminary studies. One may say, however, that (a) a gain in weight and a rise in insoluble matter can readily be detected; (b) the processes appear to be sensitive to the presence of ultraviolet light and a significant amount of ultraviolet light apparently is emitted by the ordinary fluorescent lamps used;[5] (c) the length of exposure, in terms of footcandle hours, that is necessary to cause these effects is indicated; and (d) the greater sensitivity of mastic towards oxidation is further confirmed.

4. Feller, R. L., Stolow, N., and Jones, E. H., *On Picture Varnishes and Their Solvents*, Intermuseum Conservation Association, Oberlin (1959), p. 201.

5. Sylvania Lighting Products Engineering Bulletin 0-205 states, for example, that from 1.54-7.27% of the total radiation below 700 millimicrons lies in the region of the ultraviolet below 380 millimicrons for eight lamps listed.

What's in a Name: Dammar or Serendipity in the Library

ROBERT L. FELLER

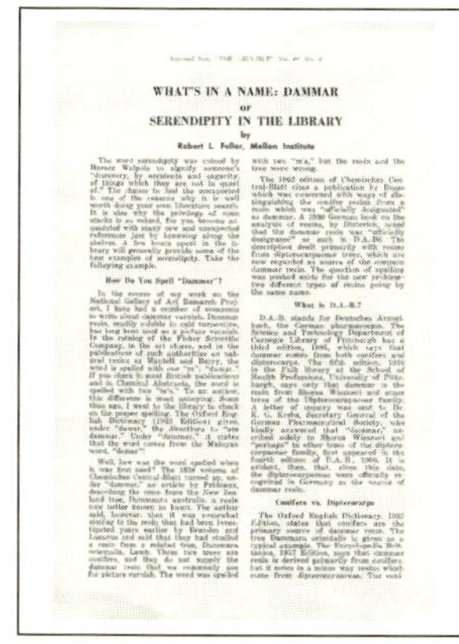

WHAT'S IN A NAME: DAMMAR
or
SERENDIPITY IN THE LIBRARY
by
Robert L. Feller, Mellon Institute

Originally printed in:
The Crucible
Volume 49, Number 8 (1964),
Pages 214, 216, 218.

The word serendipity was coined by Horace Walpole to signify someone's "discovery, by accidents and sagacity, of things which they are not in quest of." The chance to find the unexpected is one of the reasons why it is well worth doing your own literature search. It is also why the privilege of open stacks is so valued, for you become acquainted with many new and unexpected references just by browsing along the shelves. A few hours spent in the library will generally provide some of the best examples of serendipity. Take the following example.

How Do You Spell "Dammar"?

In the course of my work on the National Gallery of Art Research Project, I have had a number of occasions to write about dammar varnish. Dammar resin, readily soluble in cold turpentine, has long been used as a picture varnish. In the catalog of the Fisher Scientific Company, in the art stores, and in the publications of such authorities on natural resins as Mantell and Barry, the word is spelled with one "m": "damar." If you check in most British publications and in Chemical Abstracts, the word is spelled with two "m's." To an author, this difference is most annoying. Some time ago, I went to the library to check on the proper spelling. The Oxford English Dictionary (1933 Edition) gives, under "damar," the directions to "see dammar." Under "dammar," it states that the word comes from the Malayan word, "damar"!

Well, how was the word spelled when it was first used? The 1838 volume of *Chemisches Central-Blatt* turned up, under "dammar," an article by Prideaux, describing the resin from the New Zealand tree, *Dammara australis*, a resin now better known as kauri. The author said, however, that it was somewhat similar to the resin that had been investigated years earlier by Brandes and Lucanus and said that they had studied a resin from a related tree, *Dammara orientalis*, Lamb. These two trees are conifers, and they do not supply the dammar resin that we commonly use for picture varnish. The word was spelled with two "m's," but the resin and the tree were wrong.

The 1902 edition of *Chemisches Central-Blatt* cites a publication by Busse which was concerned with ways of distinguishing the conifer resins from a resin which was "officially designated" as dammar. A 1930 German book on the analysis of resins, by Dieterich, noted that the dammar resin was "officially designated" as such in *D. A.-B.* 6. The description dealt primarily with resins from Dipterocarpaceae trees, which are now regarded as source of the common dammar resin. The question of spelling was pushed aside for the new problem: two different types of resins going by the same name.

What is *D. A.-B.*?

D. A.-B. stands for *Deutsches Arzneibuch*, the German pharmacoepia. The Science and Technology Department of Carnegie Library of Pittsburgh has a third edition, 1895, which says that dammar comes from both conifers and dipterocarps. The fifth edition, 1910 in the Falk library at the School of Health Professions, University of Pittsburgh, says only that dammar is the resin from *Shorea Wiesneri* and other trees of the Dipterocarapaceae family. A letter of inquiry was sent to Dr. K. G. Krebs, Secretary General of the German Pharmaceutical Society, who kindly answered that "dammar," ascribed solely to *Shorea Wiesneri* and "perhaps" to other trees of the Dipterocarpaceae family, first appeared in the fourth edition of *D. A.-B.*, 1900. It is evident, then, that, since this date, the Dipterocarpaceae were officially recognized in Germany as the source of dammar resin.

Conifers vs. Dipterocarps

The Oxford English Dictionary, 1933 Edition, states that conifers are the primary source of dammar resin. The tree *Dammara orientalis* is given as a typical example. The *Encyclopedia Britannica*, 1957 Edition, says that dammar resin is derived primarily from conifers, but it notes in a minor way resins which come from Dipterocarpaceae. The conifers yield resins which are considered by the varnish trade to be of the "copal" type. They require strong thermoprocessing in order to put them into solution in turpentine. On the other hand, the dammar resin of the varnish trade dissolves in turpentine without heat, and seems to come primarily from a distinctly different family of trees, the Dipterocarpaceae. Why do so many dictionaries and textbooks say that this resin comes from both families of trees ?

An Old Mix-up

It turns out that the confusion in assigning the source of the dammar resin both to conifers and to dipterocarps is a famous one in the field of natural resins. It is still not properly appreciated even at this late date: Chatfield in his 1944 book, *Varnish Constituents*, for example, assigns dammar to *Dammara orientalis*.

Because the old mix-up has continued for so long, the thread of the tale has become much obscured. Much of the story is revealed, however, in the book by Wiesner, *Die Rohestoffe der Pflanzenreiches*. The Mellon Institute library has a 1927 edition kindly turned over to us by Dr. Leonard Cretcher not too many years ago. In the early history, many resins went under the name dammar, but Wiesner tells us that it was the dissertation of Mauch in 1898 that first showed how to distinguish the resins of conifers from those of other sources on the basis of solubility tests. (No wonder that the *D. A.-B.* was able, two years later, to make up its mind concerning the source of the turpentine-soluble dammar of commerce!)

Sources Established in 1898 and 1910

After studying specimens brought back from the East Indies, students of Wiesner decided that a dipterocarp, very similar to *Shorea Dammara selanica* Rumph was the source of the common dammar resin of commerce. They named the tree *Shorea Wiesneri*. This tree has largely been forgotten, although it is noted in the 1908 edition of *Meyers Grosses Konversations-Lexikon* and in McIntosh and Livache's, *The Manufacture of Varnishes*, Volume III (1911).

If *Dammara orientalis* does not supply the dammar of the varnish trade, what resin does it supply? Manila copal. The source of this resin was long thought to be *Vateria indica*, a basis for further errors in the old books. It was not until 1910 that the work of Foxworthy, published in the *Philippine Journal of Science*, traced the source of Manila copal to *Agathis alba* (the proper botanical name for *Dammara orientalis*). This journal can be found at the Science and Technology Department, Carnegie Library.

The Mix-up Continues

Because the research in straightening out various sources of the resins was done in Germany, many of the German authors have the story correct. Confusion persists in many English publications, however, in spite of the fact that authorities such as McIntosh (1911) and T. H. Barry (*Natural Varnish Resins*, 1932) had the essential story straight.

There are a number of reasons for the continuance and long history of this mix-up. The first is fundamental: the word "damar" simply meant resin to the natives, and many such materials were brought from the forests and given this general name. A second reason is that a number of British writers were much taken with the botanical designation "*Dammara*." In saying that a dammar resin came from a species of *Dammara*, the stress was placed on the botanical association. "Copal," on the other hand, is a word belonging to the varnish trade.

A third factor is that the artists' dammar-turpentine varnish was long called, not dammar varnish, but "crystal varnish" in England. My friends at the Winsor and Newton Company informed me that this type of varnish first appeared under the name "crystal varnish" in their 1846 catalog. It was not until 1957 that the name appeared in their catalog in the form: "crystal (dammar) varnish."

A fourth reason, and perhaps one of the original sources of trouble, is that Rudolph Brandes published a long article in the *Archiv der Pharmazie* (Volume 30 (1829), p. 1-19) on the properties of a dammar resin which was completely soluble in turpentine at room temperature. The properties he described may be recognized as those of a dipterocarp resin. Yet, for some reason, he said that this resin corresponded in properties to a sample of damar puti, from the tree *Dammara alba* Rumph (*Dammara orientalis* Lamb) brought back by M. Lesson, pharmacist on the frigate Coquille, following a trip around the world.

Sources Established in 1844 and Forgotten

It is a pity that the apparent error made by Brandes should have persisted beyond 1844. At that time, the French authority, Guibourt, reviewed the confusion in the properties of the various dammar resins (*Revue Sci. et Ind.*, Volume 16, p. 177-209). From his article it is evident that many people at the time, besides Brandes, thought that the common dammar resin came from *Dammara orientalis*. He said, moreover, that the resin that M. Lesson had furnished from this tree could not have been confused with the dammar of commerce, which he concluded corresponded with Dammar selanica, a dipterocarp resin originally described by Rumphius.

The Forgotten Footnote

It now appears that a turpentine-soluble dammar resin, probably Dammar selanica, was introduced into Europe some time around 1827. This fellow Lucanus seems to have been the first to have described a picture varnish based on a dammar resin in turpentine, in *Schweigger's Journal*, Volume LV (1829), p. 60-66. My good friend, Dr. Eric C. Hulmer, has called my attention to an

1847 *Pharmacopea Borussica* in the library of Old Economy, Ambridge, which cites this publication. Eibner, in *Malmaterialienkunde* (1909), may be the most recent author to cite the date 1828 and to remember Lucanus as the person who introduced dammar-turpentine picture varnish about that time.

In the same number of *Schweigger's Journal*, p. 242-244, Brandes briefly summarized his longer article that appeared in the *Archiv der Pharamazie*. One reason why Lucanus' publication may have been largely forgotten is that many early authors, Prideaux, R. D. Thomson, Thomas Thomson [*Chemistry or Organic Bodies: Vegetables*, J. B. Bailliere, London (1838)] and Forni [*Manuale del Pittore Restauratore*, Florence (1866)] considered that both Brandes and Lucanus were discussing the same resin, one from *Dammara orientalis*. In a footnote, however, Lucanus states definitely that his resin was different from the damar puti, the resin from this tree.

Help in the Hunt

The world-famous botanical library assembled by Rachel McMasters Miller Hunt is at nearby Carnegie Institute of Technology. The director, Dr. George Lawrence, and his associates have been most helpful in straightening out some of the botanical nomenclature in this search. One might argue that the name dammar should be associated with the resins coming from *Dammara australis* and *Dammara orientalis*. However, Dr. Lawrence has pointed out that *Dammara* is not the accepted botanical designation for these trees; the proper name for the genus is *Agathis*.

The name dammar was first used in the botanical literature by Rumphius and seems to have been intended to designate the product of the tree, the resin, and not the tree itself. The native word means resin or the tears of resin that oozed from the tree. If one wishes to read the 1741 translation of Rumphius' work, *Herbarium Amboinense*, and see what he has to say about the nearly forgotten resin, Dammar selanica, there is a copy in the Hunt Botanical Library. Dammar selanica was so common in those days that the natives simply called it "damar" whereas others were more specifically named: damar puti (white resin), damar hitam (black resin), damar batu (stone resin), etc.

How Do You Spell "Dammar"?

I prefer the native word with one "m". Various authors have used different spellings at different times, however, and it seems by far the simplest solution to continue the spelling based on the Latinization of the native word, using two "m's".

The Unexpected

So much for that. But look what was discovered: the name of the person who introduced dammar-turpentine picture varnish, the date of introduction and the initial publication. In addition, we discovered the old error in confusing the Dipterocarpaceae and conifer "dammars" still persists in the literature. Several leads have been uncovered with which the tangled thread of this story may be unravelled. We will now have to reread the literature and ask, when an author refers to a resin from *Dammara orientalis*, whether he really is describing this resin or whether he, too, is a victim of the old mix-up.

All but five of the references cited were found either in our own or in other local libraries. Four of the libraries are within a few minutes' walking distance from my desk. Without the excellence of the staffs and of the collections in these special libraries, we might not have gained an insight to this story at all. Certainly it would not have been unravelled as well as it has in so short a time!

Translation of First Description of Dammar Picture Varnish by F. Lucanus in 1829

R. L. FELLER

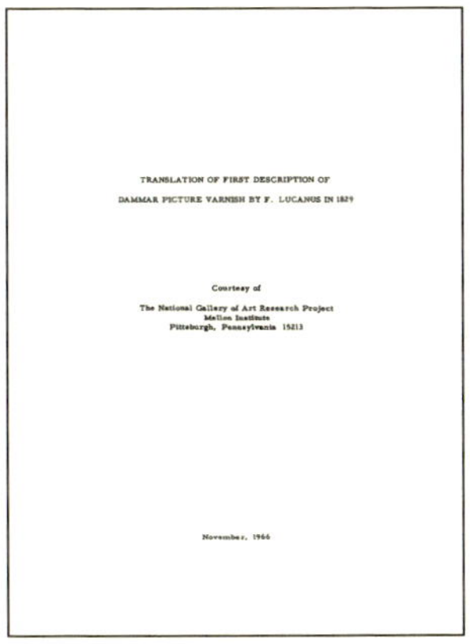
TRANSLATION OF FIRST DESCRIPTION OF
DAMMAR PICTURE VARNISH BY F. LUCANUS IN 1829

Courtesy of

The National Gallery of Art Research Project
Mellon Institute
Pittsburgh, Pennsylvania 15213

November, 1966

Originally printed in:
National Gallery of Art Research Project
(November 1966).

Dammar Resin - Concerning the Properties, Chemical Behavior, and Technical Application by F. Lucanus[1]

Resina Dammar, Motao-Cochin, or Cat's-eye resin entered the market in 1827 through Calcutta and London, and soon excited the attention of scholars and artists there.[2]

Dammar resin is to be found in Singapore[3] and usually occurs in somewhat elongated pieces weighing from a quarter to two drams; it is so colorless and transparent that it is not surpassed in these properties by either copal or mastic.

Its specific gravity is 1.060.

It is shiny when fractured, glass-like, and yields an exceedingly white powder; it can be crushed between the teeth but does not soften.

It is odorless and tasteless, melts easily, and scarcely exhibits any noticeable odor, even upon strong heating.

After moistening with strong ethyl alcohol, the surface of the resin becomes sticky.

A half dissolves in absolute alcohol, approximately a fifth in cold 80 pC R.[4] alcohol, and a quarter dissolves in hot alcohol. However, part of this precipitates out upon subsequent cooling. The residue of the resin is a white powder which is soluble in ether and in turpentine.

Dammar resin is completely soluble in turpentine and fatty oils.

Ether dissolves the resin except for a very insignificant part, which is a soft resin.

The solution in alcohol turns litmus red; I could not disclose the presence of a true acid.

1. Translation of Lucanus' publication from *Schweigger's J.*, 55 (1829), p. 60-66.

2. *Zeitschrift für Handwerker und Künstler*, No. 60 (December 1827). (Lucanus note)

3. It is not the same as Dammar puti, which is the same as *Dammara alba* Rumph. (see Thénard's *Handbuch der Chemie*, 4, translated by Fechner, Leipzig (1827), p. 1400. (Lucanus note)

4. pC. is probably percent, but the only usage found for R. is Reaumar, the old temperature scale in which the freezing point of water was set at 0° and the boiling point of water at 80°.

The resin melts in boiling concentrated acetic acid without dissolving, rather, clouding or changing only slightly.

The resin becomes brittle when heated moderately with a fourfold quantity of fuming nitric acid, is insoluble in turpentine, partially soluble in ethyl alcohol, and completely soluble in ether.

However, if one part of the resin is heated strongly with four parts of fuming nitric acid and cured until dry without ever becoming cold, then two fifths dissolves in distilled water, from which solution the resin is precipitated with ammonium as white flakes. The residue which is insoluble in distilled water is only partially soluble in ethyl alcohol, but completely soluble in ether.

Dammar resin is dissolved by fuming sulphuric acid without heating the mixture, which assumes a color similar to Orlean,[5] and is then as soluble in ethyl alcohol as in turpentine. By mixing with distilled water, the resin again precipitates as white flakes from the solution in ethyl alcohol without the addition of ammonia.

5. Orlean is a yellow dyestuff, referred to as anatto, or rocoa.

If the resin is heated considerably with sulphuric acid, then the mixture becomes dark and is soluble in ethyl alcohol.

The resin does not fulminate when ground together with iodine, even if it becomes warm.

The resin is not altered by treatment with liquid caustic ammonium.[6]

6. Ammonium hydroxide?

Thirty parts of resin, boiled with caustic potash liquor, shows a weight increment of five parts. The decanted liquor contains no dissolved resin. This potash-containing solid was boiled for some time with water, yet only a small part of it ever dissolved, and a residue of twenty parts remained which did not dissolve in water.

I dissolved only a little resin in turpentine, and boiled the whole of it in potash lye. After the turpentine was boiled off, a resin soap was formed which was completely soluble in water, and from which only the resin was precipitated in combination with the metal oxide by the metal salts. Ether and turpentine decompose this combination, since they dissolve the resin from it again. By dry distillation, sixty grains of dammar resin yield:

> a.) Three grains of aqueous acetic acid, which saturates two drops of caustic ammonium liquor.
>
> b.) forty-five grains of volatile oil, which is soluble in equal parts of ether and a fifth part of ethyl alcohol of 80 pC R.,[2] does not fulminate with iodine, and produces a deep red in litmus paper.
>
> If this oil is agitated with potash liquor, or distilled over potash, then litmus is no longer reddened by it. However, it possesses the same odor which it had before, and thus this acid[7] (probably pyro-acid) was not the principle which conferred the odor to the oil.

7. See Unverdorben in Thénard's *Chemie*, 5, No. 1, p. 434. (Lucanus note)

c.) After the oil has passed off, a yellowish solid resin sublimes in the neck of the retort. It is soluble in ether, turpentine, and ethyl alcohol, but not in potash lye, and weighs about eight grains. The solution of this resin in ethyl alcohol is yellowish and very opalescent, and displays a variety of colors in a very thin state, like the extract of horse chestnut rind. This resin appears to be the third of those species which Unverdorben prepared by dry distillation.[8]

8. See Thénard's *Chemie*, 5, No. 1, p. 435. (Lucanus note)

d.) Seven grains of carboniferous residue remain in the retort.

According to Bonastre,[9] the sub-resins, which are not usually soluble in boiling ethyl alcohol and ether, should possess the property of subliming, if they are free of volatile oils and acids.

9. See Trommsdorff's *Journal*, 8, p. 155 and 172. (Lucanus note)

Ether dissolves 59 out of 60 parts of dammar resin; absolute alcohol dissolves half. I separated a quantity of resin into two parts by digestion with absolute ethyl alcohol, and called A the part dissolved in ethyl alcohol, and B the part which was insoluble in ethyl alcohol.

The dry distillation of 40 parts of A yields 30 parts of volatile oil, including two parts of aqueous acid. Barely a trace of resin appeared in the neck of the retort opposite the distillation end. Eight parts of carboniferous residue remained.

Of 32 parts of B I obtained, 3 parts of aqueous material which did not redden litmus, then 2 parts which reddened it; further, 13 parts of volatile oil were obtained, and finally 6 parts of resin which deposited in the neck of the retort. This behaved just like the resin which I had obtained from the crude dammar.

Seven parts of carboniferous residue remained.

Since 60 parts produce eight parts of dammar resin by dry distillation, essentially only a trace coming from A and an unequally large amount from B, both together add up to 7 parts of resin product from a total of 62 parts; Bonastre's theory is indeed acceptable.

The technical aspects of dammar resin also deserve attention. Two parts of it dissolved in two and a half parts of turpentine by shaking yield an excellent varnish for paintings, drawings, lithographs, etc., whose serviceability far surpasses that of mastic. A light application of dammar varnish bestows an extraordinary clarity upon paintings, which is much more durable than that of mastic varnish, since the dry, hardened dammar varnish withstands external attack with greater certainty. Dammar varnish is not easily attacked by alcohol, so that paintings treated with alcohol during cleaning present less of a problem, particularly since dammar varnish can again be softened and removed with linseed oil and turpentine.

In order to determine whether dammar varnish might be less subject to post-yellowing than that of mastic, I set out dammar and mastic at the same temperature in different containers.

The mastic was yellowish when hardly fused; the dammar resin first assumed a trace of color after which it commenced to dry. Now, since heat and exposure to light produce post-yellowing of mastic, I believe that this bad condition need not be feared in dammar resin, inasmuch as it is sufficient to spread the dammar varnish thinly, while the thicker layers of mastic plainly will post-yellow more rapidly.

Dammar resin appears to be very good as a retouching varnish, since it forms a clear solution in all proportions with poppy-seed oil when heated, without additions of turpentine.

Halberstadt, May, 1829.

[Author's note: This article was found in the course of a search for the date of introduction of dammar picture varnish. It now turns out that there was an earlier publication by Lucanus on dammar which appeared in the *Berliner Kunst-Blatt* of February 1829, whereas this article appeared in May in another journal. A more extensive discussion on the preparation of dammar varnish was published by Lucanus in May 1836 in the *Allgemeiner Anzeiger und Nationalzeitung der Deutschen*, No. 139. A reprinting of the rarely encountered first edition of Friedrich Lucanus' important text, *Anleitung zur Restauration alter Oelgemälde* (*Introduction to the Restoration of Old Oil Paintings*), 1828, has been published by Ferdinand Enke, Stuttgart, in 1996. It is accompanied by a biography of Lucanus and an extensive discussion of the significance of the six editions of his book, based on the scholarship of Cornelia Weyer. Dammar was not discussed in Lucanus' first edition, but the three significant journal articles cited above are appended in the reprint edition.—RLF]

SOLUBILITY OF AGED COATINGS BASED ON DAMMAR, MASTIC, AND RESIN AW-2

R. L. FELLER AND CATHERINE W. BAILIE

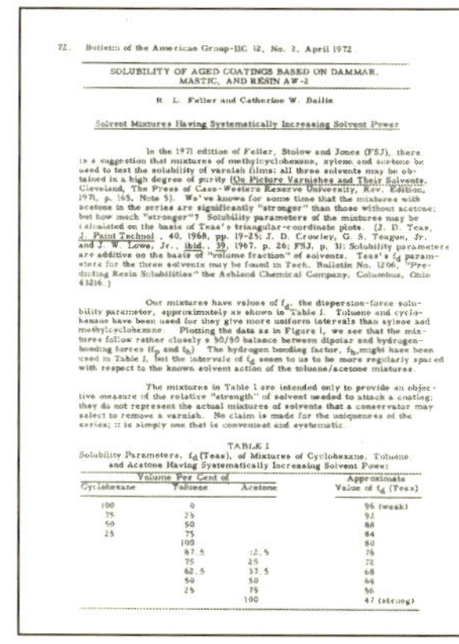
72. Bulletin of the American Group-IIC 12, No. 2, April 1972

SOLUBILITY OF AGED COATINGS BASED ON DAMMAR, MASTIC, AND RESIN AW-2

R. L. Feller and Catherine W. Bailie

Solvent Mixtures Having Systematically Increasing Solvent Power

In the 1971 edition of Feller, Stolow and Jones (FSJ), there is a suggestion that mixtures of methylcyclohexane, xylene and acetone be used to test the solubility of varnish films: all three solvents may be obtained in a high degree of purity (On Picture Varnishes and Their Solvents, Cleveland, The Press of Case-Western Reserve University, Rev. Edition, 1971, p. 165, Note 5). We've known for some time that the mixtures with acetone in the series are significantly "stronger" than those without acetone; but how much "stronger"? Solubility parameters of the mixtures may be calculated on the basis of Teas's triangular-coordinate plots. (J. D. Teas, J. Paint Technol., 40, 1968, pp. 19-25; J. D. Crowley, G. S. Teague, Jr. and J. W. Lowe, Jr., ibid., 39, 1967, p. 26; FSJ, p. 31: Solubility parameters are additive on the basis of "volume fraction" of solvents. Teas's f_d parameters for the three solvents may be found in Tech. Bulletin No. 1206, "Predicting Resin Solubilities" the Ashland Chemical Company, Columbus, Ohio 43216.)

Our mixtures have values of f_d, the dispersion-force solubility parameter, approximately as shown in Table I. Toluene and cyclohexane have been used for they give more uniform intervals than xylene and methylcyclohexane. Plotting the data as in Figure 1, we see that the mixtures follow rather closely a 50/50 balance between dipolar and hydrogen-bonding forces (f_p and f_h). The hydrogen bonding factor, f_h, might have been used in Table I, but the intervals of f_d seem to us to be more regularly spaced with respect to the known solvent action of the toluene/acetone mixtures.

The mixtures in Table I are intended only to provide an objective measure of the relative "strength" of solvent needed to attack a coating; they do not represent the actual mixtures of solvents that a conservator may select to remove a varnish. No claim is made for the uniqueness of the series; it is simply one that is convenient and systematic.

TABLE I
Solubility Parameters, f_d (Teas), of Mixtures of Cyclohexane, Toluene and Acetone Having Systematically Increasing Solvent Power

Volume Per Cent of			Approximate
Cyclohexane	Toluene	Acetone	Value of f_d (Teas)
100	0		96 (weak)
75	25		92
50	50		88
25	75		84
	100		80
	87.5	12.5	76
	75	25	72
	62.5	37.5	68
	50	50	64
	25	75	56
		100	47 (strong)

Originally printed in:
BULLETIN OF THE AMERICAN GROUP - IIC
Volume 12, Number 2 (1972), Pages 72-81.

SOLVENT MIXTURES HAVING SYSTEMATICALLY INCREASING SOLVENT POWER

In the 1971 edition of Feller, Stolow and Jones (FSJ), there is a suggestion that mixtures of methylcyclohexane, xylene and acetone be used to test the solubility of varnish films; all three solvents may be obtained in a high degree of purity.[1] We've known for some time that the mixtures with acetone in the series are significantly "stronger" than those without acetone; but how much "stronger"? Solubility parameters of the mixtures may be calculated on the basis of Teas' s triangular-coordinate plots.[2]

Our mixtures have values of f_d, the dispersion-force solubility parameter, approximately as shown in Table 1. Toluene and cyclohexane have been used for they give more uniform intervals than xylene and methylcyclohexane. Plotting the data as in Figure 1, we see that the mixtures follow rather closely a 50/50 balance between dipolar and hydrogen-bonding forces (f_p and f_h). The hydrogen bonding factor, f_h, might have been used in Table 1, but the intervals of f_d seem to us to be more regularly spaced with respect to the known solvent action of the toluene/acetone mixtures.

The mixtures in Table 1 are intended only to provide an objective measure of the relative "strength" of solvent needed to attack a coating; they do not represent the actual mixtures of solvents that a conservator may select to remove a varnish. No claim is made for the uniqueness of the series; it is simply one that is convenient and systematic.

1. Feller, R. L., Stolow, N., and Jones, E. H., *On Picture Varnishes and Their Solvents*, Rev. Edn., The Press of Case-Western Reserve University, Cleveland, OH (1971), p. 165, Note 5.

2. Teas, J. D., *J. Paint Technol.*, 40 (1968), p. 19-25; Crowley, J. D., Teague, Jr., G. S., and Lowe, J. W., Jr., *ibid.*, 39 (1967), p. 26; Feller, R. L., Stolow, N., and Jones, E. H., *op. cit.*, p. 31; Solubility parameters are additive on the basis of "volume fraction" of solvents. Teas's f_d parameters for the three solvents may be found in *Tech. Bulletin No. 1206*, "Predicting Resin Solubilities," the Ashland Chemical Company, Columbus, Ohio 43216.

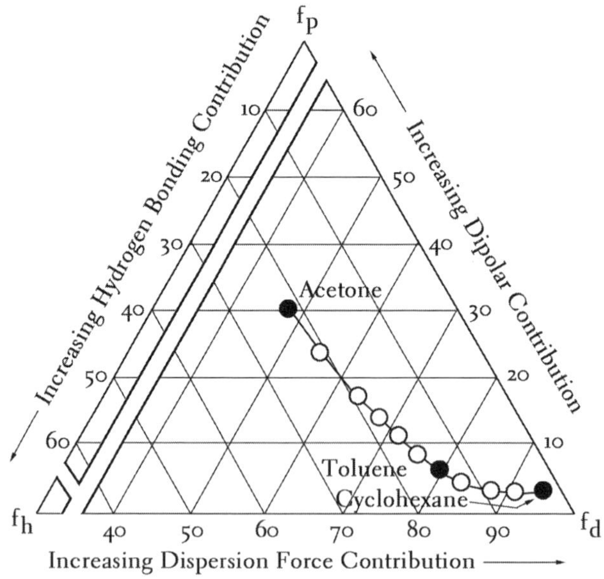

FIGURE 1
Solubility parameter plot of acetone, toluene, cyclohexane mixtures.

TABLE 1
Solubility Parameters, f_d (Teas), of Mixtures of Cyclohexane, Toluene, and Acetone Having Systematically Increasing Solvent Power

Volume Percent of			
Cyclohexane	Toluene	Acetone	Approximate Value of f_d (Teas)
100	0		96 (weak)
75	25		92
50	50		88
25	75		84
	100		80
	87.5	12.5	76
	75	25	72
	62.5	37.5	68
	50	50	64
	25	75	56
		100	47 (strong)

"STRENGTH" OF SOLVENT NEEDED FOR AGED NATURAL-RESINS

Just how difficult is it to remove aged dammar and mastic varnishes? In Feller, Stolow and Jones[3] it is said that a mixture of 25/75 acetone/xylene by volume will not remove thoroughly aged dammar. This mixture possesses an f_d solubility parameter of 75 based on Teas's values.

3. Feller, Stolow, and Jones, *op. cit.*, p. 165.

For many years Ruhemann has spoken of a "Safety Margin Test" for solvent strength, formally introduced we believe in 1940.[4] In his recent book *The Cleaning of Paintings*,[5] he cites two dozen examples of solvent mixtures used to remove varnishes; if we calculate their f_d values, the average is found to be 77.9 ± 4.8 (the mean and standard deviation for 20 values). Perhaps it would not be fair to include four additional examples in which he used pure acetone, but if this is done the average becomes 72.8 ± 12.6. Both values are remarkably close to our figure. Although Ruhemann evidently was able to use slightly milder solvents than our figure implies, this is not surprising because varnishes on paintings are not likely to be as highly oxidized as test films exposed for 200 to 300 hours on aluminum foil in the carbon-arc Fadeometer.

4. International Museums Office, *Manual On the Conservation of Paintings*, Paris (1940), p. 129.

5. Ruhemann, H., *The Cleaning of Paintings*, Praeger, N. Y., (1968).

Ruhemann describes the use of mixtures of acetone and "white spirit" in his Safety Margin Test. The locus of solubility parameters for his mixtures is likely to be more linear in Figure 1 than with our series of solvents. Nonetheless, we will continue using the convenient mixtures in Table 1 because a 10% error in mixtures such as Ruhemann's will give an error of about ±5 in f_d, whereas a 10% error among our three solvents would only be about 1.6 between cyclohexane and toluene and about 3.3 between toluene and acetone.

Ruhemann mentions the considerable resistance to turpentine and toluene exhibited by 20-year-old films of dammar and mastic varnishes.[6] After 4,300,000 foot candle hours under daylight fluorescent lamps and temperatures of about 82°F, we have found that the varnishes are only partially removed in toluene and with considerable resistance.[7]

6. *Ibid.*, p. 306.

7. In this same exposure Elvacite® 2045, poly(isobutylmethacrylate), developed 17% insolubility in toluene, but was removed in 45 sec. in xylene; crosslinking had begun, but it had not yet progressed far enough to cause difficulty in removal.

The above considerations establish that toluene will not effectively attack thoroughly aged films of the two traditional picture varnishes. We have not studied their behavior under conditions of "re-forming", but Miss Jones' data on 12-year-old films suggest that, even if re-forming is attempted, toluene will still be ineffective on thoroughly aged films.[8]

8. Feller, Stolow, and Jones, *op. cit.*, p. 176.

Solubility of Dammar and Mastic Resins

Measures of "ease of removal" and "strength" of solvent can be subject to considerable variation in execution and interpretation. We therefore decided to measure *solubility* of the resins to provide completely objective data. The results are presented in Table 2.

Procedure:

Solubility of unexposed resins was determined by analysis of the solids in saturated solutions. These were prepared in two ways: one set was shaken for 24 hours at room temperature, the other was heated on the steam bath and then allowed to return to room temperature. Preparations were centrifuged and the clear solution taken for analysis. The mean deviation in the results from the two preparations was about ± 3%.

"Aged" samples were exposed on aluminum foil for 300 hours in a carbon-arc Fadeometer at 62°F and were analyzed by shaking overnight in wire-screen baskets. The solubility measurements in Table 2 refer to solubility at room temperature, about 70-75°F.

TABLE 2
Percentages of Dammar and Mastic Resins That Dissolve at Room Temperature in Various Solvents

	Dammar		Mastic	
Solvent	As Received	Aged	As Received	Aged
n-Dodecane	84		52	
Isooctane	70		39	
Methyl Cyclohexane	100	46	80	3
Turpentine	100	–	89	–
Toluene	100	65	97	42
Isopropyl Alcohol	90	93	97	89
Acetone	–	95	–	95

Results:

We do not claim that the data are highly accurate, although the similar results with samples prepared from both hot and room temperature conditions lend confidence in the figures for the fresh resin. The unaged samples may have contained suspended colloidal material which would make the figures for solubility high. In the case of aged samples, finely divided material may pass through the screens, again giving high values for apparent solubility. Nonetheless, the fact seems to be established that mastic resin dissolves in "mild" solvents to a lesser extent than dammar.

SOLUBILITY OF AGED DAMMAR/STAND OIL VARNISHES

Drying oils are often mixed with dammar resin in picture varnishes, glazes,and paints. Because linseed oil may become at least 60 to 80% insoluble in time, it is of interest to determine the effect of added linseed or stand oil upon the solubility and ease of removal of dammar varnish. This is of particular concern in connection with the possibility of attacking glazes and paints that may contain an appreciable portion of dammar resin.

Materials and Procedure:

The experiments were conducted using five different oils. Two were obtained directly from the processors: Archer-Daniels-Midland "Alinco Z-3": density

(25°C) 0.96; Sapon. No. 190-196, Iodine No. 115-125; and Spencer Kellogg "S-70 OKO" linseed oil: density (15.5°C) 0.95, Sapon. No. 192-196, Iodine No. 115-130. Three others were from artist's materials companies: Winsor and Newton Sun-Thickened Oil, Permanent Pigments Stand Oil and Permanent Pigments Sun-Thickened Oil. The solubility of aged varnishes was determined by averaging the results of extracting films made from these oils mixed with dammar resin from the Fezandie and Sperrle Company. The varnishes were prepared in xylene. (Erratic results when turpentine was used caused experiments with that solvent to be discontinued.)

Solubility was determined by cutting out pieces of the foil, placing them in 120-mesh/in. stainless-steel wire baskets and extracting the soluble material at room temperature by shaking over night, a method previously described.[9]

9. Feller, Stolow, and Jones, *op. cit.*, p. 208.

Results: The results of measurements of the solubility of varnishes aged 250 to 300 hours on aluminum foil in an Atlas Electric Devices Co., single carbon-arc Fadeometer, Corex D filter, are shown in Figure 2.

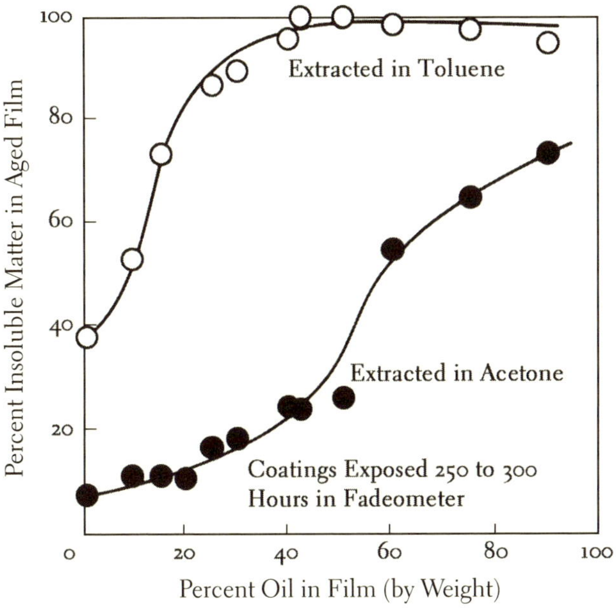

FIGURE 2
Films of dammar and stand oil.

As the amount of oil is increased, insolubility of the aged varnishes in toluene rises rapidly, from about 36% for dammar alone, to practically 100% insoluble matter for a composition of 50% stand oil. In acetone, the rise in insoluble material follows more closely the increase in the content of oil, reaching a maximum of about 80% for pure oils.

The traditional dammar/stand oil picture varnish, about 5% oil by volume added to ordinary dammar varnish, results in a film having about 13% oil by weight. A varnish of this composition may attain 62 ± 4% insolubility in toluene at room temperature; 8 ± 3% acetone. Should a varnish be formulated at a concentration of oil appreciably higher than this, it may develop 70 to 90% insoluble matter in toluene.

The data suggest that a difference may be noticed in the extent of attack of solvents upon dammar/oil glazes as opposed to traditional picture varnishes, for glazes often contain the greater amount of oil. Formulas quoted by Mayer, for example, yield films containing 32 to 70% oil.[10]

10. Mayer, R., *The Artist's Handbook*, Rev. Edn., The Viking Press, N. Y. (1957), p. 224-225.

EASE OF REMOVAL OF DAMMAR/STAND OIL VARNISHES

Knowing the amount of insoluble matter that may develop in severely exposed dammar/stand oil films, we may inquire about the ease of removal of the films in relationship to their insolubility. Determination was made of the seconds required to remove films, aged on aluminum foil 250 to 300 hours in the carbon-arc Fadeometer, when lightly rubbed with a cotton swab dipped in solvent. The results are summarized in Figures 3a and 3b. Each point is the average of two observations; six different oils were used in the study.

As seen, no special difficulty is experienced in removing the series of films until they contain more than 50 to 70% material that is insoluble in toluene. Reforming was not attempted. All exposed films were removable in less than 30 sec. with pure acetone, even those that contained up to 90% oil. Removal probably was facilitated by extensive swelling.

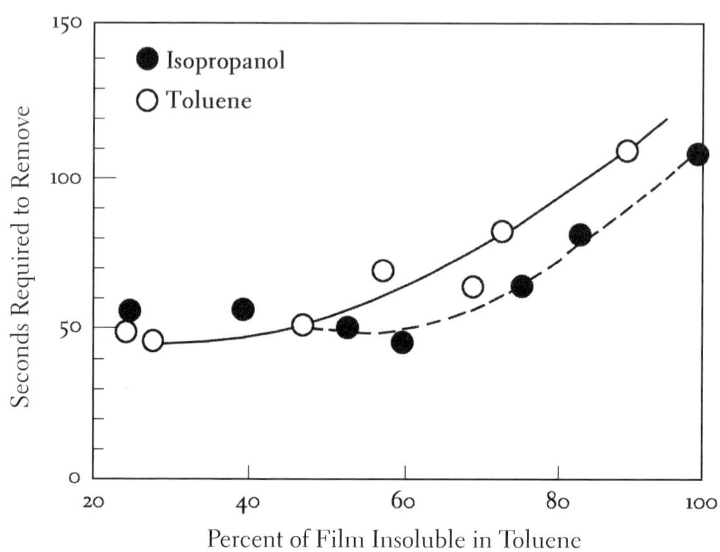

FIGURE 3A
Difficulty of removal of aged films rubbed with swabs dipped in solvent: dammar/stand oil varnishes aged 250-300 hrs. on aluminum foil in Fadeometer.

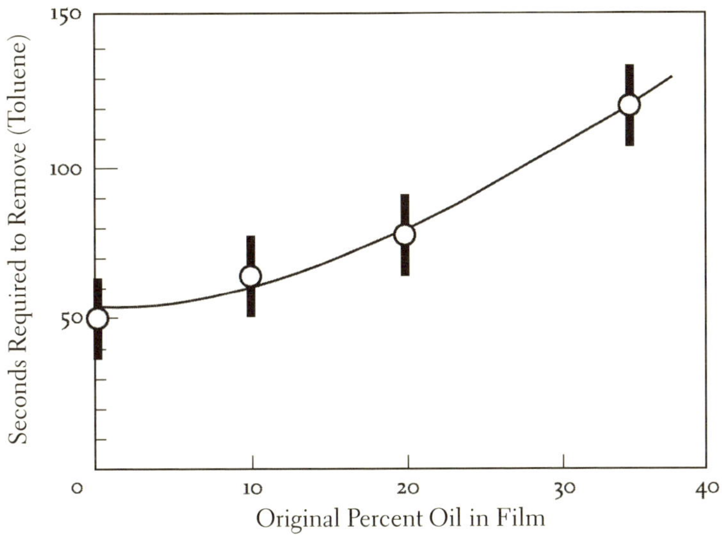

Figure 3B

Solubility of Polycyclohexanone Resin AW-2

Resin AW-2 is a proprietary polymer of Badische Anilin und Soda Fabrik, based on cyclohexanone.[11] Initially, it is soluble in heptane, cyclohexane, methylcyclohexane, xylene, and toluene, although a mixture of the resin in 140-160°C naphthenes is cloudy. As Table 3 shows, upon exposure of AW-2 in the Fadeometer, a certain amount of insolubility in toluene develops, perhaps owing to oxidation. Either a slight degree of crosslinking or a higher state of oxidation also gives rise to a small percentage of matter insoluble in acetone.

Resin AW-2 is no longer available, but we believe that the related Ketone Resin N will behave similarly; tests on this point are in progress.

11. Average molecular weight about 700; See Feller, R. L., "Characteristics of Resin AW-2," *Bulletin of the American Group-IIC*, 3, No. 1 (October 1962), p. 9-10.

Table 3
Insoluble Portion of Resin AW-2 Following Exposure on Aluminum Foil at 62°C in a Carbon-arc Fadeometer

Solvent	Insoluble Material Found After 145 hrs.	Insoluble Material Found After 395 hrs.	Time of Removal by Swab after 395 hrs.
Toluene	49%	64%	Not removed in 2 min. in toluene or in isopropanol
Acetone	2%	17%	About 45 seconds

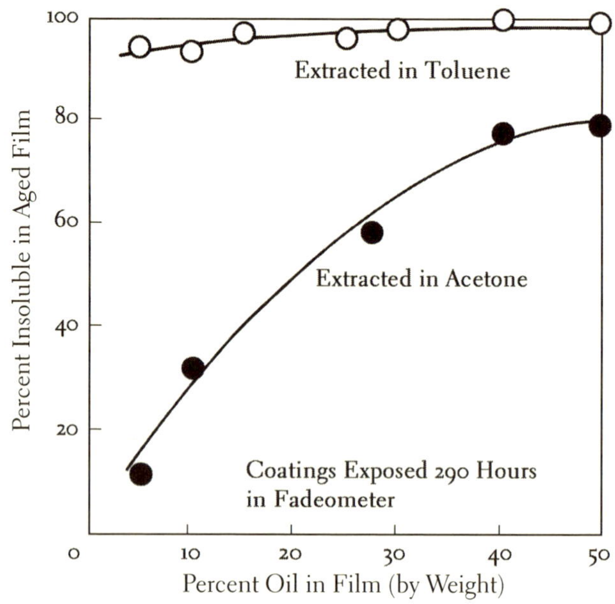

FIGURE 4
Films of resin AW-2 and stand oil.

INSOLUBILITY OF AGED FILMS OF AW-2/STAND OIL

Figure 4 summarizes similar tests on the insolubility of films on AW-2 and stand oil mixtures following 290 hours exposure on aluminum foil at 62°C in carbon-arc Fadeometer. Resin AW-2 and a trace of stand oil was at one time recommended as a picture varnish and may have been used in a number of proprietary varnishes.

Results:

The addition of "5% stand oil by volume" to varnish based on Resin AW-2 yields a varnish, when aged, more insoluble in toluene (96% vs. 30%) and in acetone (35% vs. 10%) than the traditional dammar/stand oil varnish.

The data in Figure 4 imply that AW-2 is able to crosslink with the oil, because there is an amount of acetone-insoluble matter in these films that is greater than the sum of the oil plus expected insoluble matter in the AW-2. When butylmethacrylate polymers are mixed with AW-2, we have obtained similar evidence to suggest that AW-2 can crosslink with the higher polymer.

Conclusions

How difficult is it to remove aged films of varnishes based on dammar, mastic, polycyclohexanone, and their mixtures with traditional drying oils? We have suggested a convenient series of solvents of increasing solvent power by which to test the "strength" of solvent that may be needed to attack a given coating. Aged films of the two natural resins are not completely dissolved or easily removed in toluene, and we find that a mixture of toluene containing about 25-30% acetone by volume may be necessary to attack them effectively. Whether fresh or aged, mastic resin dissolves in toluene or "milder" hydrocarbons a lesser extent than dammar.

Adding stand oil to dammar resin does not cause a noticeable increase in the difficulty of removal as long as the percentage of oil in the film is below 30% by weight. Polycyclohexanone resin AW-2 apparently develops a greater degree of insolubility when mixed with a given amount of stand oil than does the traditional combination of dammar and stand oil.

Owing to differences in exposure, extraction, and the composition of resins, oils, and solvents in the original varnish, the figures for solubility may be expected to show considerable variation should others attempt to duplicate the data; relative errors as high as ± 15% would not be unexpected, particularly when solubility is greater than 30-40%. We believe, nevertheless, that the conclusions are justified and hope that the data represent objective measures of solubility of these coatings towards toluene and acetone when they have reached the maximum possible degree of photooxidative deterioration that one may expect to encounter in practice.

Changes in Solubility and Removability of Varnish Resins with Age

R. L. FELLER AND M. CURRAN

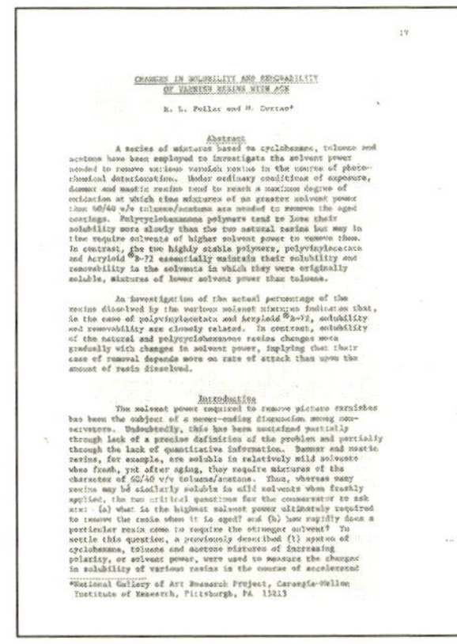

17

CHANGES IN SOLUBILITY AND REMOVABILITY OF VARNISH RESINS WITH AGE

R. L. Feller and M. Curran*

Abstract

A series of mixtures based on cyclohexane, toluene and acetone have been employed to investigate the solvent power needed to remove various varnish resins in the course of photochemical deterioration. Under ordinary conditions of exposure, dammar and mastic resins tend to reach a maximum degree of oxidation at which time mixtures of no greater solvent power than 60/40 v/v toluene/acetone are needed to remove the aged coatings. Polycyclohexanone polymers tend to lose their solubility more slowly than the two natural resins but may in time require solvents of higher solvent power to remove them. In contrast, the two highly stable polymers, polyvinylacetate and Acryloid® B-72 essentially maintain their solubility and removability in the solvents in which they were originally soluble, mixtures of lower solvent power than toluene.

An investigation of the actual percentage of the resins dissolved by the various solvent mixtures indicates that, in the case of polyvinylacetate and Acryloid® B-72, solubility and removability are closely related. In contrast, solubility of the natural and polycyclohexanone resins changes more gradually with changes in solvent power, implying that their ease of removal depends more on rate of attack than upon the amount of resin dissolved.

Introduction

The solvent power required to remove picture varnishes has been the subject of a never-ending discussion among conservators. Undoubtedly, this has been sustained partially through lack of a precise definition of the problem and partially through the lack of quantitative information. Dammar and mastic resins, for example, are soluble in relatively mild solvents when fresh, yet after aging, they require mixtures of the character of 60/40 v/v toluene/acetone. Thus, whereas many resins may be similarly soluble in mild solvents when freshly applied, the two critical questions for the conservator to ask are: (a) what is the highest solvent power ultimately required to remove the resin when it is aged? and (b) how rapidly does a particular resin come to require the stronger solvent? To settle this question, a previously described (1) system of cyclohexane, toluene and acetone mixtures of increasing polarity, or solvent power, were used to measure the changes in solubility of various resins in the course of accelerated

*National Gallery of Art Research Project, Carnegie-Mellon Institute of Research, Pittsburgh, PA 15213

Originally printed in:
Bulletin of the American Institute for Conservation of Historic and Artistic Works
Volume 15, Number 2 (1975), Pages 17-26 .

Introduction

The solvent power required to remove picture varnishes has been the subject of a never-ending discussion among conservators. Undoubtedly, this has been sustained partially through lack of a precise definition of the problem and partially through the lack of quantitative information. Dammar and mastic resins, for example, are soluble in relatively mild solvents when fresh, yet after aging, they require mixtures of the character of 60/40 v/v toluene/acetone. Thus, whereas many resins may be similarly soluble in mild solvents when freshly applied, the two critical questions for the conservator to ask are: (a) what is the highest solvent power ultimately required to remove the resin when it is aged? and (b) how rapidly does a particular resin come to require the stronger solvent? To settle this question, a previously described[1] system of cyclohexane, toluene and acetone mixtures of increasing polarity, or solvent power, were used to measure the changes in solubility of various resins in the course of accelerated aging tests involving exposure to both "daylight" fluorescent lamps and to the xenon-arc Fadeometer, with Pyrex-glass filter.

1. Feller, R. L., and Bailie, C. W., "Solubility of Aged Coatings Based on Dammar, Mastic and Resin AW-2," *Bulletin of the American Group-IIC*, 12, No. 2 (1972), p. 72-81.

Dammar and Mastic in Relation to Other Resins

Figure 1 shows the rate at which dammar and mastic come to require increasingly more polar solvent mixtures in order to remove them during exposure under "daylight" fluorescent lamps at an average intensity of about 650 footcandles. The open points refer to mastic; the several symbols indicate different samples that were placed on exposure at various times throughout the two years that these tests were in progress. The closed (black) points refer similarly to samples of dammar from various sources and from tests initiated at different times during the two-year period. ("Removability" is defined below.)

An important conclusion to be reached from these studies is that, under reasonably mild conditions of exposure, dammar and mastic resins tend to reach a level of oxidation at which point they require a solvent mixture in this series consisting of 60/40 v/v toluene/acetone to dissolve them and, for all practical purposes, they do not oxidize further. The data confirm the previously known fact that mastic tends to oxidize more rapidly than dammar, although ultimately the properties of the two resins are very similar.

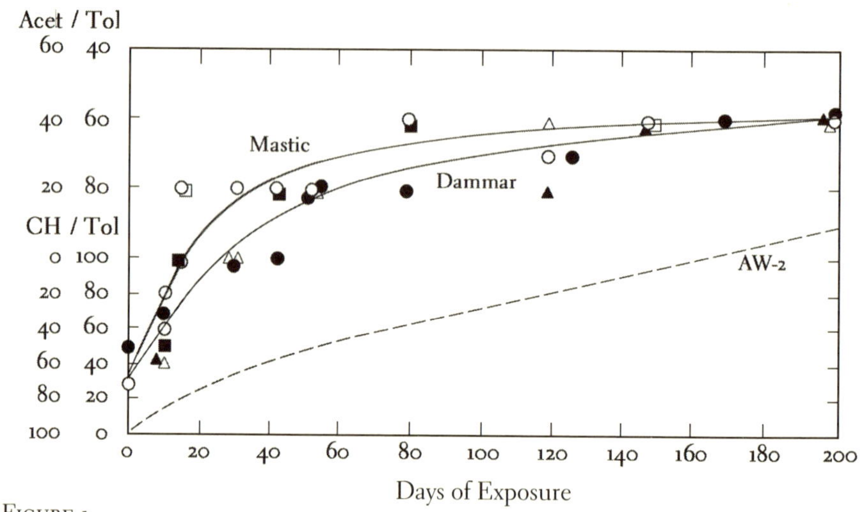

FIGURE 1
Solvent mixtures needed to remove dammar and mastic resins when exposed to daylight fluorescent lamps. (Note: CH = Cyclohexane)

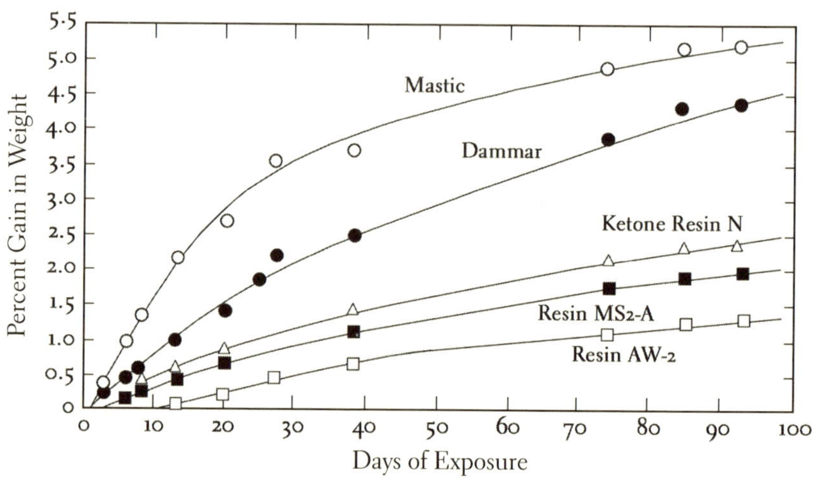

FIGURE 2
Increase in weight of various resins when exposed to daylight fluorescent lamps.

In Figure 3, the behavior of dammar and mastic resins is compared to the synthetic low-molecular-weight resins such as Ketone Resin N, resin AW-2, and resin MS2-A and two stable high-molecular-weight polymers. For convenience in referring to the two figures, a dotted line in Figure 1 refers to the curve of the experimental data for resin AW-2 in Figure 3.

POLYCYCLOHEXANONE RESINS

Data in Figure 3 indicate that the three polycyclohexanone resins tend to be removed throughout a considerable portion of their useful lifetimes in less polar solvents than required for dammar and mastic resins. It is only when they have oxidized to point B in Figure 3 that they become equally difficult to remove. Beyond that point, Ketone Resin N and resin AW-2 definitely require solvents of higher solvent power to remove them than dammar and mastic.

MS2-A, a hydrogenated variety of resin AW-2, tends to be more readily soluble in a less polar solvent than required by the others. However, the process of hydrogenating the resin may produce tertiary hydrogens in the polymer structure and, consequently, the MS2-A may oxidize more rapidly than the unhydrogenated resin, AW-2 (see Figure 2). Ketone Resin N is a resin similar to AW-2 with slightly different composition and is currently the only type available from BASF.[2]

Studies summarized in the accompanying Figure 2 clearly indicate that the rate of oxidative weight-gain of these resins closely resembles the rate at which they lose their ability to be dissolved in mild solvents. The principal difference

2. Feller, R. L., "Characteristics of Resin AW-2," *Bulletin of the American Group-IIC*, 3, No. 1 (1962), p. 9-10.

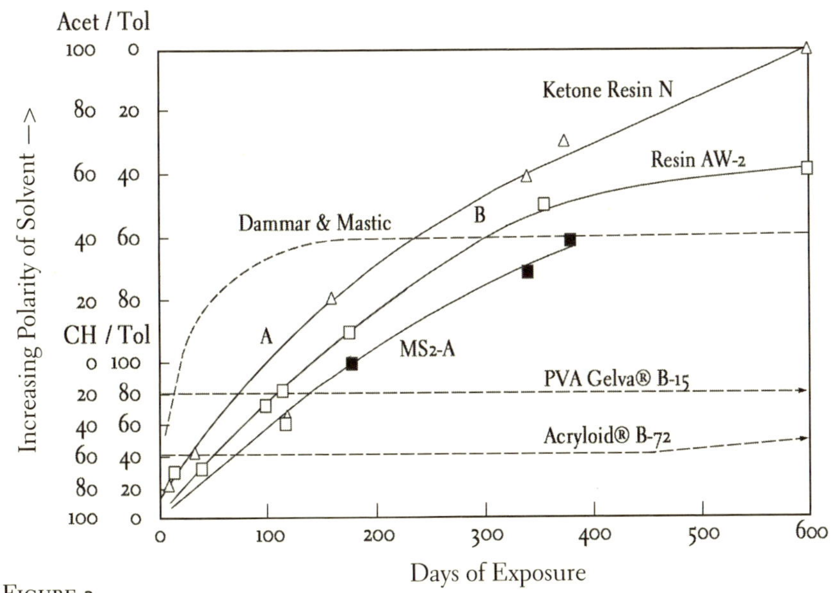

FIGURE 3
Solvent mixtures needed to remove various resins when exposed to daylight fluorescent lamps. (Note: CH = Cyclohexane)

seems to be that the order of resins AW-2 and MS2-A are reversed. The order of oxidation shown in Figure 2 has been verified in three separate experiments.

What may we conclude regarding the polycyclohexanone resins? Let us say that for a period prior to point B, Ketone Resin N, MS2-A and AW-2 are more easily soluble than the traditional natural resins, although perhaps after half this exposure (point A) they will require solvents of higher solvent power than toluene to remove them. This behavior may be contrasted with the characteristics of the following two polymers.

ACRYLOID® B-72 AND POLYVINYLACETATE

When exposed to near ultraviolet radiation, these two polymers tend, if anything, to undergo chain breaking rather than crosslinking.[3] Partially as a result of this, they tend to remain soluble in the solvents in which they were originally soluble, as indicated in Figure 3. As determined in earlier measurements of Solubility Grade, polyvinylacetate resin requires a solvent slightly more polar than that needed to dissolve Acryloid B-72.[4]

At the IIC meeting in Lisbon, we stated that these two are the most durable resins that we could suggest to the conservator of the type soluble in solvents of no greater solvent power than toluene.[5] Under museum conditions of exposure, we do not know when they would begin to change their solubility. In the xenon-arc Fadeometer with Pyrex-glass filter, polyvinylacetate coated on aluminum foil remains highly soluble in 20/80 v/v cyclohexane/toluene mixtures for at least 680 hours. As we said in 1963, slight changes in properties occur after extensive exposure in a carbon-arc Fadeometer; Acryloid® B-72 can require slightly more polar solvents to dissolve it.[6] This slight tendency to change solubility requirements is indicated in Figure 3 by the dotted line tilted upward at 600 days.

3. Feller, R. L., "Studies on the Photochemical Stability of Thermoplastic Resins," *Preprints of the Fourth Triennial Meeting, Venice*, ICOM Committee for Conservation (1975), Paper 75/22/4.

4. Feller, R. L., Stolow, N., and Jones, E. H., *On Picture Varnishes and Their Solvents*, Revised Edition, The Press of Case Western University, Cleveland, Ohio (1971), p. 34.

5. Feller, R. L., "Problems in the Investigation of Picture Varnishes," *Conservation of Paintings and the Graphic Arts*, Preprints of contributions to the Lisbon Conference, International Institute for Conservation of Historic and Artistic Works, London (1972), p. 201-209.

6. Feller, R. L., "New Solvent-type Varnishes," *Recent Advances in Conservation*, ed. Thomson, G., Butterworths, London (1963), p. 171-175.

ESTIMATE OF THE TIME SCALE IN FIGURE 3

How much time, in terms of years of exposure on a day-lighted gallery wall, are we speaking of in Figure 3? The question is not easy to answer with precision because there are factors of temperature and relative humidity to consider as well as spectral characteristics and the average intensity of light. For example, the fluorescent lamps in the present experiments were several years old and we now believe their photochemical potency was the lowest of any conditions we have employed in recent years. (Six months after the tests, the lamps began to burn out.)

The most accurate measure of the rate of deterioration can be obtained by a comparison of the behavior of butylmethacrylate polymers under these conditions of exposure with their tendency to crosslink on the walls of the Fogg Art Museum Laboratory.[7] An analysis of the full data, not given here,

7. Feller, Stolow and Jones, *On Picture Varnishes and Their Solvents*, p. 159.

indicates that point B in Figure 3, at 300 days, represents about 100 years on the day-lighted laboratory wall at the Fogg Art Museum.

A second method by which we can compare the results in Figure 3 with those that might occur on a gallery wall is by monitoring the exposure with the BS1006:1971 blue-wool fading standards.[3] We observe in Figure 4 that in 300 days, standard No. 4 has faded to a BS2662 geometric grey scale contrast of about 2 (GS 2). Years ago, the fading of these standards was observed on the wall of a gallery having an average annual exposure of about 150,000 foot-candle hours.[8] These observations indicate that standard No. 3 would fade to GS 3 in 10 years and standard No. 2 would fade to GS 1.5; an average of these two results indicates that 30 days under these lamps is equivalent to about 10 years on the day-lighted illuminated gallery wall.

Thus, on the basis of the two methods of estimation—the behavior of the standard blue cloths on the walls of a day-lighted gallery and the behavior of a number of butylmethacrylate resins on the wall of the laboratory at the Fogg Art Museum—we believe it is fair to say that point B in Figure 3 represents the properties that we may expect to observe in these resins after about 100 years on a gallery wall illuminated at about 150,000 fc. hrs. of diffuse skylight per year. The total expanse of time represented by Figure 3, therefore, represents about 200 years. The evidence speaks well for the durability of Acryloid® B-72 and the polyvinylacetate resins.

8. Feller, R. L., "Problems in Spectrophotometry," *London Conference on Museum Climatology*, revised edition, International Institute for Conservation of Historic and Artistic Works, London (1968), p. 268, Figure 11.

EASE OF REMOVAL VS. SOLUBILITY OF THE RESIN

By "removability" in these experiments, we have meant that, when a given solvent system was placed on a cotton swab, the resin came away in less than a

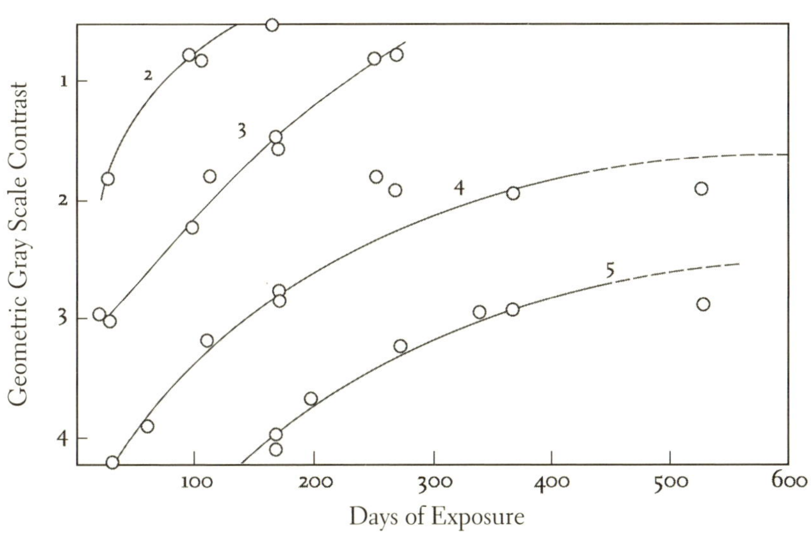

FIGURE 4
Fading of BS1006:1971 blue cloths under conditions of Figure 3.

minute of normal rubbing, about as one might wish to have occur when removing varnish conveniently from a painting. Because this is a qualitative (but reasonably reproducible) measure, we decided to investigate the actual degree of solubility of the resins in the various solvent mixtures. Particularly, we wished to measure the solubility of the resins in the solvents which were considered just able to remove the films in about one minute, the solvent mixtures indicated in Figure 3.

In these quantitative experiments, the resins were coated on aluminum foil and exposed in the xenon-arc Fadeometer (with Pyrex-glass filter) in order to speed up the results. Solubility was determined by agitating overnight in the particular solvent system cut-out sections of aluminum foil bearing the exposed coatings. To prevent undue loss of the resin in case of its separation from the foil, the samples were placed in 120-mesh-per-inch stainless steel baskets.

ACRYLOID® B-72 AND POLYVINYLACETATE

Table 1 summarizes the behavior of the acrylic and polyvinylacetate resins. In the case of the polyvinylacetate, we see that the solvent system either (a) dissolved the resin or (b) failed to do so, with a very sharp division occurring when the solvent system was sufficiently polar. Thus, a 40/60 mixture of cyclohexane/toluene dissolves only 6% of the resin while a 20/80 mixture dissolves 95%; it is the latter mixture that conveniently removes the resin throughout a long period

TABLE 1
Percentage of Polymers Insoluble in Solvents of Increasing Polarity vs. Time of Exposure in Fadeometer

Solvent Mixture, v/v		Hours of Exposure							Hours of Exposure						
		22	40	80	120	303	430	680	22	40	80	120	303	430	680
		Acryloid® B-72							PVA Vinac® B-15						
acet. / tol.															
60	40														
40	60														
20	80														
0	100				3										
cyclohex. / tol.															
20	80								• 5	•	•	•	•	•	• 13
40	60				5	4	•	• 4	94						
60	40			•	• 28	• 27		35	96						
80	20			97				94							
100	0														

• Solvent needed to remove the films in less than one minute.

of exposure. (For convenience, the mixtures were tested only in steps of 20% difference in composition; evidence indicates, however, that polyvinylacetate can be removed in a mixture containing somewhat more than 20% cyclohexane by volume.)

TABLE 2
Percentage of Polymers Insoluble in Solvents of Increasing Polarity vs. Time of Exposure in Fadeometer

Solvent Mixture, v/v		Hours of Exposure							Hours of Exposure						
		22	40	80	120	303	430	680	22	40	80	120	303	430	680
acet. / tol.		Dammar							Mastic						
60	40														
						●	●						●	●	●
40	60											2			
				●	●						●	●			
20	80					3			1			6			
			●							●					
0	100	4		14	20	30			15	23	31	35	44		
cyclohex. / tol.															
20	80			19					22		38				
40	60									36					
									●						
60	40	5							44	50					
		●													
80	20	18													
100	0	23													

Solvent Mixture, v/v		Hours of Exposure							Hours of Exposure						
		22	40	80	120	303	430	680	22	40	80	120	303	430	680
acet. / tol.		Resin MS2-A							Ketone Resin N						
								●							
60	40					4									
						●	●						●	●	●
40	60					4		9							23
				●	●						●	●			
20	80				2						2	8			35
0	100				6	33		48			18	19	32		
cyclohex. / tol.															
20	80				11							25			
40	60			5					2		24				
			●												
60	40			11		45			4						
80	20								4						
		●							●						
100	0														

● Solvent needed to remove the films in less than one minute.

The dots placed over a figure of insolubility in Tables 1 and 2 represent the solvent mixture that readily removed the resin. Thus, the dots form curves, exactly comparable to the ones in Figure 3, showing the level of solvent power or polarity necessary to remove the resin in the course of exposure.

We have mentioned that even the highly stable resins tend to become slightly oxidized and, thus may require slightly stronger solvents to dissolve them as time goes on. This is clearly illustrated in the case of the polyvinylacetate B-15 by the fact that, after 680 hours of exposure, the particular solvent system will dissolve only 87% of the resin, as opposed to 95% when the coating was initially prepared.

The behavior of Acryloid® B-72 is similar to polyvinylacetate. A 80/20 mixture of cyclohexane/toluene scarcely dissolves the resin, whereas somewhere between the 60/40 and the 40/60 solvent mixture, the resin goes completely into solution to the extent of at least 95% solubility. As we reported in 1963, Acryloid® B-72 may eventually require slightly stronger solvents to dissolve it when the photochemical exposure becomes extensive.[6] The data in Table 1 indicate that the 60/40 v/v cyclohexane/toluene mixture was not the preferred one to remove the polymer after 430 hours of exposure. The 65% solubility (35% insolubility) in the 60/40 mixture at that time also indicates that this solvent mixture will dissolve slightly less of the resin as it becomes oxidized with age.

Low-Molecular-Weight Resins

The contrasting behavior of the natural resins, dammar and mastic, and polycyclohexanone resins, MS2-A and Ketone Resin N, is reported in Table 2. Here, we find no sharp transition in solubility at a given solvent mixture. Instead, for a given degree of exposure, the use of increasingly more polar solvents in the series tends to dissolve an increasing portion of the resin. This is best illustrated in the behavior of resin MS2-A at 303 hours and in the behavior of mastic resin at 22 and 40 hours.

If we consider this behavior from another point of view, we find that solubility in a given solvent system tends to decrease more rapidly with exposure time than is the case with PVA or Acryloid® B-72. This is well illustrated in the solubility of MS2-A, dammar, and mastic resins in toluene.

In the course of aging, resins apparently become more oxidized and, as a consequence, progressively less soluble in a particular solvent system. However, the practical ease of removal in the case of the low-molecular-weight resins (both natural and polycyclohexanone) seems to depend largely upon the *rate of attack* of the solvent. The black points in Tables 1 and 2 represent the solvent mixture of lowest solvent power that will penetrate and dissolve films with sufficient speed that they may be removed in less than one minute. Thus, the points represent the solvent mixtures that attack the coating with convenient speed much more than they represent the mixtures that can dissolve a major fraction of the resin, as is apparently the case with the Acryloid® B-72 and the polyvinylacetate.

Relation to Re-forming

Because this series of experiments has suggested that the ability to remove the low-molecular-weight resins depends, within certain limits, more upon the speed of attack than on the actual amount of resin dissolved, the question naturally arises concerning the behavior of these films if one waited much longer or if one varnished them or sprayed them with solvents the day before, i.e., if the films were "re-formed".

The research that Elizabeth Jones did many years ago on this problem indicates that treatment of aged natural-resin varnishes with a solvent system that gradually evaporates over the course of 24 hours, yet which would have time to swell the resins, could effect the films sufficiently so that a milder solvent would attack them rapidly the following morning.[9] Our initial studies suggest, however, that to be effective in "re-forming" as described, the solvents applied for the purpose of re-forming must have greater solvent power than that indicated by the black dots in Tables 1 and 2. The investigation of re-forming in the light of the present information will continue.

9. Jones, E. H., "The Effect of Aging and Re-forming on the Ease of Solubility of Certain Resins," *Recent Advances in Conservation*, ed. Thomson, G., Butterworths, London (1963), p. 79-83. See also Ref. 4.

Section 2: Deteriorating Effects of Light

With the advent of the use of fluorescent lamps in museums, the architect's increased use of large expanses of window glass, and the availability of Plexiglas ultraviolet filters, considerable interest was generated within museum circles regarding the effects of lighting environments on collections. Described in reports by the National Bureau of Standards in 1951 and 1953, the Declaration of Independence had recently been put under an ultraviolet filter in a new display case. Input from the Bureau also led to the L. H. Harrison report to the Metropolitan Museum of Art in 1954 on damage factors of various light sources. There followed a report on related matters prepared by J. J. Balder for the Netherlands Museums Association in 1956. The National Gallery in London had also become interested in the color-rendering qualities of various light sources, leading to articles by Brommelle, Cooper, and Crawford. At an ICOM meeting in Barcelona, Paul Coremans and colleagues in the ICOM Committee on Conservation asked me to prepare an overview of the deteriorating effects of light, which appeared in the UNESCO magazine, *Museum*, in 1964.

As in the case of many galleries throughout the world, the National Gallery of Art had been designed with a system of skylights that would illuminate its main galleries with natural light, and the staff there also participated in the international concern regarding the numerous aspects of light exposure. With our help and advice, the Gallery arranged to install ultraviolet filters over all galleries illuminated with natural light. When the National Gallery of Art became interested in acquiring some wax models of Degas's sculptures, I was asked to offer advice regarding the heating that would occur in glass-enclosed showcases owing to the greenhouse effect. Curves showing the rise in temperature inside a test showcase appear in a review of the heating effects of incandescent lamps prepared in 1967. Our uncovering of the crosslinking phenomenon in certain methacrylate polymers served further to focus our attention on the effects of light, and that continued concern led to the subsequent studies on polymer photooxidation. Recognizing that a standard way of referring to the lightfastness of materials would be greatly useful, we suggested in the 1970s that the ISO blue wool fading reference materials could provide a readily available, reasonably priced, and internationally recognized standard. In addition to serving as metrics for lightfastness, it was found that these blue wools could also be of great value as measures of light intensities in galleries, and these materials continue to be widely used for this purpose. Several of the articles that follow served to introduce the subject to different audiences. Finally, we pointed out how several of the test procedures and materials used in our research might serve very well as standards of reference for describing the physical and chemical properties of thermoplastic resins.

R. L. F.

Control of Deteriorating Effects of Light upon Museum Objects

ROBERT L. FELLER

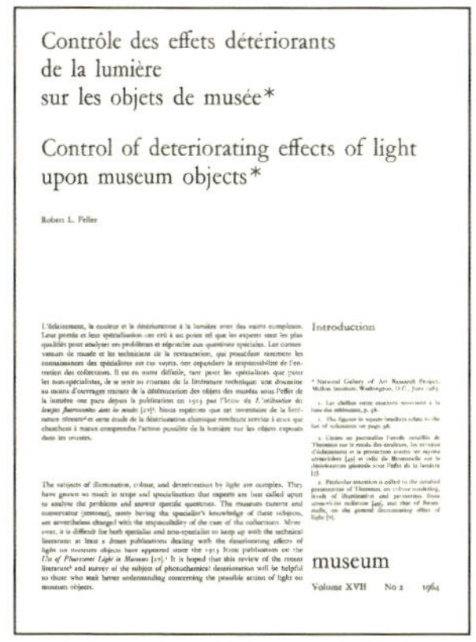
Contrôle des effets détériorants de la lumière sur les objets de musée*

Control of deteriorating effects of light upon museum objects*

Introduction

museum

Volume XVII No 2 1964

Originally printed in:
MUSEUM
Volume 17, Number 2 (1964), Pages 57-98.

Introduction

The subjects of illumination, color, and deterioration by light are complex. They have grown so much in scope and specialization that experts are best called upon to analyse the problems and answer specific questions. The museum curator and conservator (restorer), rarely having the specialist's knowledge of these subjects, are nevertheless charged with the responsibility of the care of the collections. Moreover, it is difficult for both specialist and non-specialist to keep up with the technical literature: at least a dozen publications dealing with the deteriorating effects of light on museum objects have appeared since the 1953 ICOM publication on the "Use of Fluorescent Light in Museums" [27].[1] It is hoped that this review of the recent literature[2] and survey of the subject of photochemical deterioration will be helpful to those who seek better understanding concerning the possible action of light on museum objects.

The Characteristics of Radiant Energy

Knowledge of the nature of light and sources of illumination is fundamental to an understanding of problems in illumination, color, and display. It is expected that persons interested in these problems will have knowledge of the basic aspects of the subject. Yet, a brief review is in order before a detailed discussion of the factors that control photochemical damage.

1. The figures in square brackets relate to the list of references on pages 86-87.

2. Particular attention is called to the detailed presentation of Thomson, on color rendering, levels of illumination and protection from ultraviolet radiation [49], and that of Brommelle, on the general deteriorating effect of light [7].

Light is energy

The heat one feels in sunlight, or in the direct beam of a powerful incandescent lamp, demonstrates clearly that light is energy.[3] Heat increases the general agitation of atoms and molecules and speeds up the rate of many chemical processes. But the energy of light is able to excite or "activate" the electronic systems of individual atoms and molecules in a much more specific manner, giving them an added boost of energy which makes them more reactive and liable to change. Thus "activated" or excited, a dye may fade, for example, or the long-chain molecules in fibers may become broken, leading to the embrittlement of such materials as paper and textiles; portions of other substances may be converted to colored compounds, resulting in the discoloration of paper, varnishes and drying oils. The ability of light thus to stimulate chemical change is the subject of concern here: it is called the "photochemical activity" of radiant energy.

3. It should not be necessary to discuss at length the obvious desirability of avoiding the exposure of sensitive objects to the direct rays of the sun. Neither shall we discuss the necessity of guarding against severe exposure to intense illumination from photographers' lamps, or those in television studios. The danger, principally in the latter case, is largely the temperature effect, rather than a photochemical one. Studies of the effects of intense illumination may be found in the publications of H. Kühn and C. Wolters [33] and Kureya Mityoshi [38]. Toishi Kenzo also has reviewed the arguments about general effects of artificial lighting [30].

Electromagnetic energy exhibits so many of the characteristics of wave phenomenon that the physicist is able to classify it conveniently in terms of wavelength. Figure 1 presents the physicists' well-known spectrum of electromagnetic radiation. Light is that portion of the electromagnetic spectrum that is visible to the eye, a relatively narrow range of radiation between about 400 and 760 millimicrons. (Radiations in the near-ultraviolet, 300 to 400 millimicrons, cannot be seen with the eye, yet they are sometimes called "ultraviolet light.") The eye is not uniformly sensitive to all wavelengths within the visible range; it reaches maximum sensitivity around 554 millimicrons. The variation in the sensitivity for the eye at various wavelengths, shown in Figure 2, is known as the luminous efficiency function for the CIE (Commission Internationale de l'Eclairage) standard observer.

Although electromagnetic radiation behaves in many phenomena as though it consisted of waves, in other cases it behaves as though it consisted of particles. To be sure, the effect of light is continuously manifested in passing from one wavelength to another. But the corpuscular behavior of radiation becomes particularly apparent in investigations involving radiation in the shorter

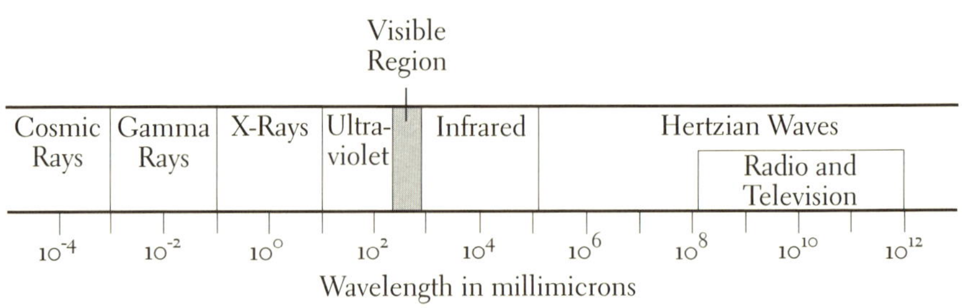

FIGURE 1
The electromagnetic spectrum.

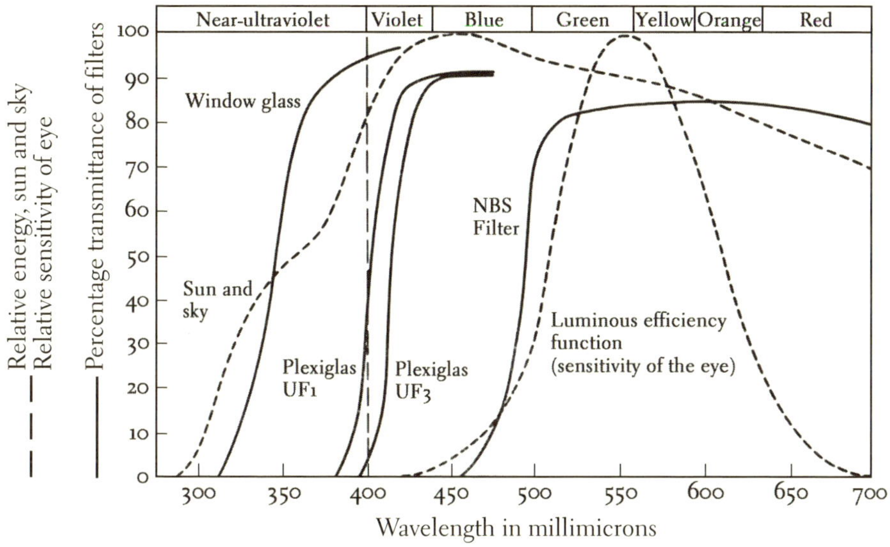

FIGURE 2
Spectral transmittance characteristics of window glass and three filters in relation to the characteristics of sunlight and the sensitivity of the eye (luminous efficiency function). The NBS filter is described in Circular 505 of the National Bureau of Standards.

wavelengths. These corpuscles, bundles or bullets of light, are called "photons." Their energy is given by the Einstein equation: $E = h\nu = hc/\lambda$. The equation relates the energy, E, of a photon to the wavelength of the radiation, λ. The frequency, ν, is inversely related to the wavelength; it is equal to the velocity of light, c, divided by the wavelength. The constant, h, known as Planck's constant, relates the units of energy to those of frequency.[4]

The corpuscular aspect of radiant energy is most important in the consideration of photochemical activity, because it reveals that the photons of radiant energy of shorter wavelength contain more energy than those of longer.[5] They are capable, therefore, of imparting greater energy to any substance that may absorb them. This concept is the basis for the often encountered statement that radiation of shorter wavelength is more energetic or potent, generally capable of greater photochemical activity.

Characteristics of radiation from various sources

In the matter of the distribution of spectral energy from various sources, the museum curator is concerned basically with two sources of illumination: incandescent and fluorescent. At very high temperatures, the atoms of a substance are in violent agitation. In this condition, they become incandescent, emitting radiant energy over a wide range of wavelengths. Light from the sun and from an ordinary incandescent electric light bulb is derived from such a source.

4. Planck's constant is equal to 6.6×10^{-27} erg-sec. The Einstein equation will yield the energy of the photon in terms of kilocalories per gram-molecular weight (kilocalories per mole) if used in the form: $E = 2.86 \times 10^4/\lambda$, where the wavelength of radiation, λ, is in millimicrons.

5. Photochemical processes are considered to be chemical reactions activated by radiation in the range of 100 to 1,000 millimicrons, having photon energies of between 29 and 290 kilocalories per mole. This can be compared with the range of energy necessary to break many chemical bonds: 35 to 100 kilocalories per mole [13]. The photochemically potent wavelengths of principal concern in the museum problem range between 310 and perhaps no more than 700 millimicrons.

The way in which the energy radiating from an object varies with its temperature can be accounted for by many of the same concepts that are used to explain the energy of a photon. A perfectly black body absorbs and radiates radiation most efficiently and is known as a "blackbody" or Planckian radiator. The distribution of radiation from such a radiator with increase in temperature is shown in Figure 3. The visible energy emitted by an incandescent substance when it is raised to higher and higher temperatures (for example, the light seen through the porthole of a furnace) passes through dull red, to yellow, white and finally blue. Figure 4 shows the change in the distribution of spectral energy as a blackbody radiator passes through this sequence of colors, plotted in the manner in which the data in the visible region of Figure 3 are more frequently presented.

It is customary in discussions of light sources to refer to the "color temperature" of the illumination. By this, it is meant that the distribution of the visible energy from the light source closely resembles the distribution of energy emitted by a blackbody radiator at the stated temperature, that is, close to some distribution of energy as shown in Figure 4. For example, most incandescent lamps give off radiation closely resembling the spectrum of a blackbody radiator at a temperature of 2,400 to 2,800°K,[6] as indicated in Figure 5B. From the distribution of energy of the sun and sky under different conditions, given in Figure 6, one can see that "daylight" is not constant, but actually varies considerably in color temperature, depending upon the angle of the sun, the conditions of the atmosphere, and the degree of scattering [41]. The color temperature of blue sky may range from 10,000 to 25,000°K. North sky light, extensively used for color matching, is considered to be between about 6,500-7,500°K.

6. The physicist's temperature scale in degrees Kelvin is equal to the degrees in Centigrade plus 273.

The visible radiation from fluorescent lamps is generated in a manner different from that of incandescent sources. In fluorescent lamps, atoms of mercury vapor are excited by the passage of current through the vapor at low pressure. The current raises the electrons in the mercury atoms to higher levels of energy. As these electrons return to their normal positions in the atom, radiant energy is emitted. The energy is in discrete bands of frequency because the electrons in the atom move only within distinct levels of energy. As a result, the spectrum is a discontinuous one, rather than a continuous one such as produced by an incandescent source.

The radiation from the mercury vapor provides little light, but it is caused to strike a fluorescent substance on the walls of the tube and excite this substance to give off visible radiation. Figure 5C shows the spectral distribution of radiation from such a lamp and compares it with that of daylight and the incandescent lamp. One can see the distinct bands of radiation from the mercury atom that occur at 578, 546.1, 435.8, 406, 365.0, 334.2, 313.2 and 302.2 millimicrons. The continuous radiation between these distinct bands is due to the fluorescence of the phosphor on the walls of the tube. Light of different color is generated in

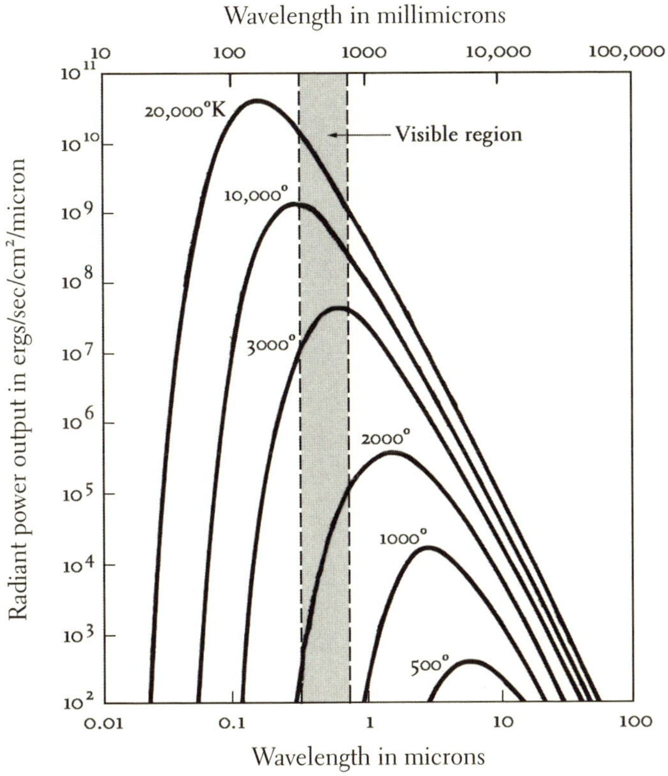

Figure 3
Distribution of radiation from a Planckian or blackbody radiator at various temperatures.

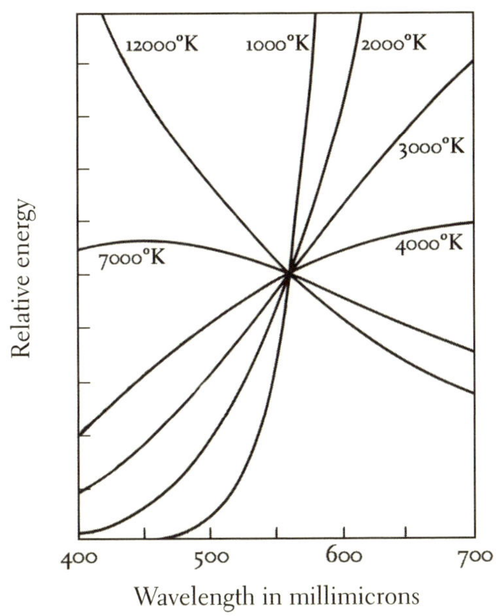

Figure 4
Relative spectral energy distribution of blackbody radiator at various temperatures (visible region).

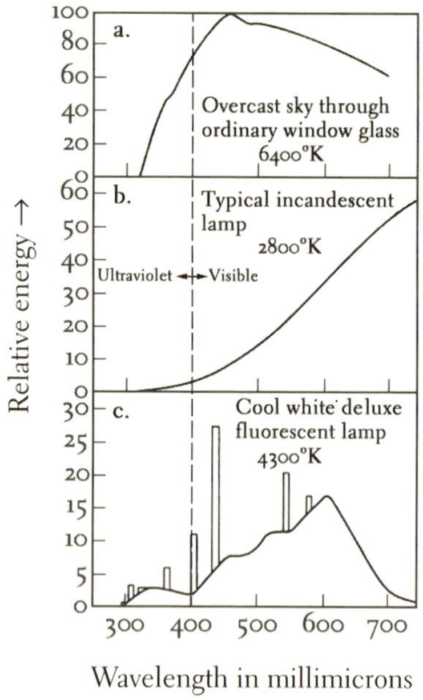

FIGURE 5
Spectral distribution of radiant energy from three principal sources of illumination [after Thomson, *Museums Journal*, 57 (1957), p. 27].

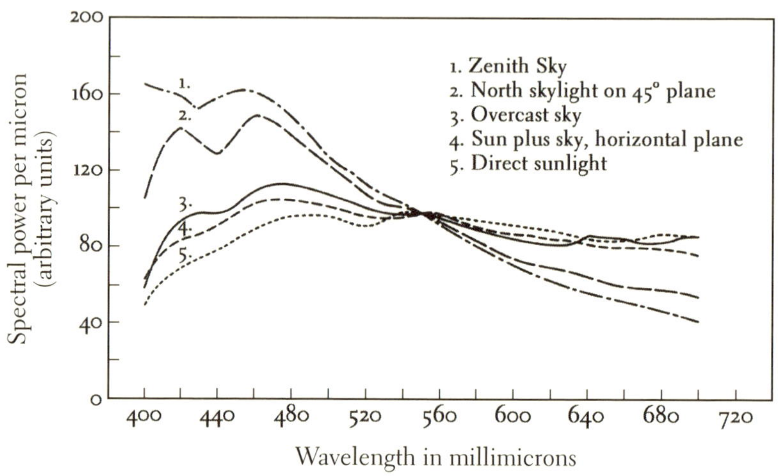

FIGURE 6
The distribution of spectral energy in skylight from different directions and under different conditions at Cleveland, Ohio.

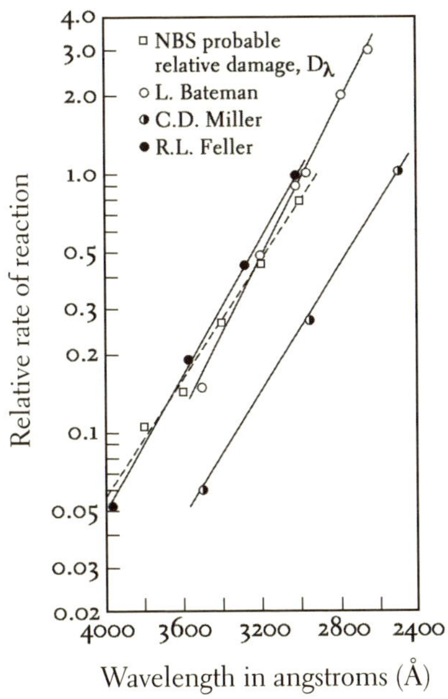

FIGURE 7
Change in the rate of several photochemical processes as a function of the wavelength of radiation.

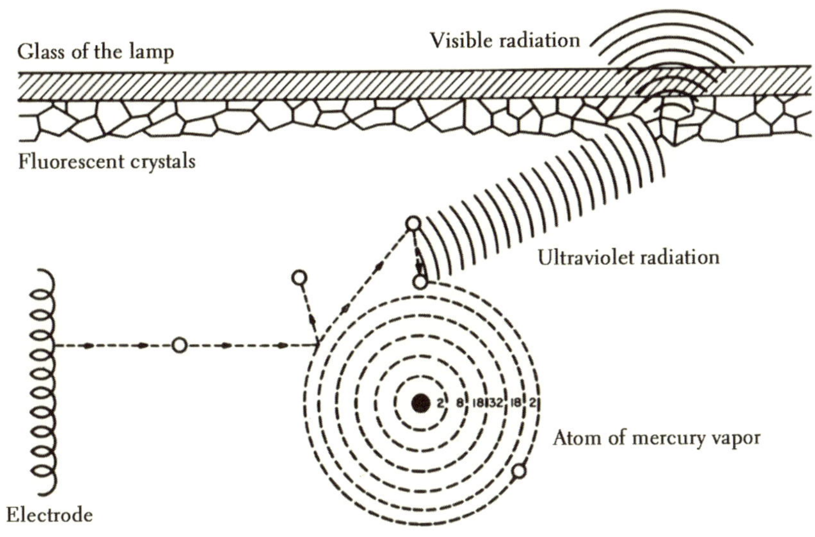

FIGURE 8
Principle of functioning of a fluorescent lamp. The electron emitted by the cathode is displaced at a great speed in the direction of the anode. In its passage, it encounters an electron in a mercury atom. Under the effect of the impact, the electron of mercury is driven from its orbit. Upon returning to the orbit, radiation is emitted (the radiation spectrum of mercury) largely situated in the ultraviolet. When the ultraviolet radiation reaches the fluorescent powder, it is absorbed and re-emitted as radiation of much greater wavelength in the visible region of the spectrum.

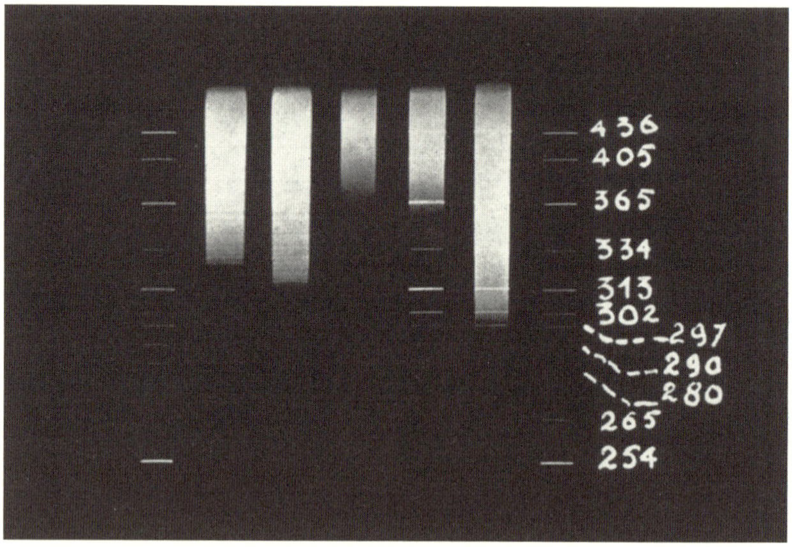

FIGURE 9
Ultraviolet radiation from various sources of light: reading from right to left, (a) electric discharge in a fluorescent lamp; (b) ultraviolet radiation from a *warm white de luxe* fluorescent lamp; (c) from a standard *warm white* lamp; (d) from an incandescent lamp; (e) ultraviolet radiation from sunlight out-of-doors; (f) indoors; (g) same as (a).

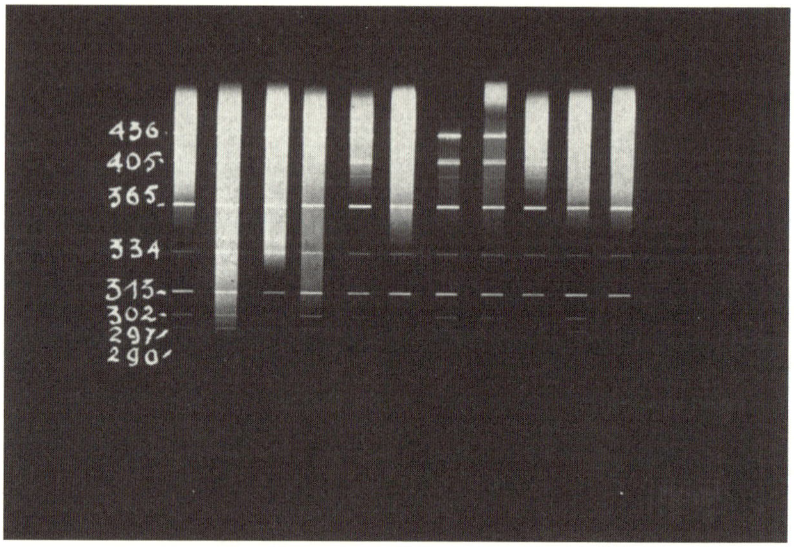

FIGURE 10
Spectra of the ultraviolet radiation from eleven different varieties of fluorescent lamps, showing differences in the relative continuous and discontinuous emission (wavelengths in millimicrons).

fluorescent lamps by using different combinations of phosphors. Harrison [23] and Brommelle [5] give the distribution of spectral energy from several such lamps.

The functioning of a fluorescent lamp is described schematically in Figure 8. Figure 9, taken from Genard [16], shows the relative amounts of continuous and discontinuous radiation emanating from various sources. The relative intensity in this case is revealed by the exposure of a photographic film, whereas that in Figure 5 is plotted graphically. For further details, the reader is referred to publications of Brommelle [5]. Genard [16] and Balder [2].

Relative photochemical potency of various sources

The shorter the wavelength, the greater the energy of the photon. Therefore, the more shorter wavelengths of visible and ultraviolet radiation that are present in illumination from a black-body radiator, the more photochemically potent the radiation is likely to be. It is easy to understand why incandescent lamplight, shown in Figure 5B, usually exhibits the least photochemical activity of the common sources of light: there is little ultraviolet present and little of the shorter wavelengths of visible radiation. On the other hand, zenith sky (11,000°K) has the greatest proportion of ultraviolet and blue light per lumen of any source of illumination under consideration. The high potency of daylight has been emphasized by Thomson [48, 49], Balder [2] and others.

Depending on their "color," fluorescent lamps may have more or less radiation in the shorter, relatively active, wavelengths of visible radiation: the violet and the blue. The relative potency of different fluorescent lamps can vary considerably, partially because the portion of ultraviolet radiation emitted can vary, as may be clearly seen in Figure 10, taken from the publication of Genard [16]. The potency of fluorescent lamplight is generally less than sky light, however. For details concerning the nature of the ultraviolet radiation emitted by fluorescent lamps, the reader is referred to the publications of Genard, Cooper [10] and the ICOM booklet [27].

In a section below, mention will be made of the measurement of intensity in terms of visible energy or total energy. Neither expression reveals much, however, about the relative potency of different sources owing to a varying proportion of the more energetic short wavelengths relative to the long. In an attempt to rate the relative potency of different sources, the National Bureau of Standards (U.S.A.) established a series of factors of probable relative damage (D_λ) which increase sharply with decreasing wavelength, as shown in Figure 7. This concept was developed to assist in the design of a filter to protect the *Constitution* and the *Declaration of Independence* of the United States [39] and later was extended to aid in the analysis of problems of illumination at the Metropolitan Museum of Art [23]. The D_λ factors are used to estimate factors of probable damage per footcandle of intensity (D/fc) for any given source, by a method of computation illustrated in Table 4 in the report by Harrison [23].

It is not considered that the D/fc factors thus calculated represent any more than one attempt to estimate the relative hazard. The concept is based on the known effect of the different wavelengths of radiation on the deterioration of low-grade paper, and although Figure 7 shows that the degradation of rubber [3], the erosion of paint [37] and the tendency for certain methacrylate polymers to become insoluble [15] behave in a similar manner, certainly this behavior represents a particular type of photosensitivity; other photochemical processes will not react precisely in this manner to changes in wavelength.

Table 1
Factors of Probable Rate of Damage per Footcandle (D/fc) and Approximate Color Temperature for Various Light Sources (largely from L. S. Harrison, Table V [23])

Source	Rated color temperature	D/fc
	°K	
Zenith sky, through window glass	11,000	1.58
Overcast sky, through window glass	6,400	0.682
Cool-white deluxe fluorescent	4,300	0.554
Warm-white deluxe fluorescent	2,900	0.444
Sun at 30-degree altitude, through window glass	5,300	0.427
Daylight fluorescent	6,500	0.402
Overcast sky through Plexiglas UF-1	—	0.243
Philips Fluorescent Lamp 34 (1955) [2]	—	0.210
Overcast sky through Plexiglas G 911 B [22]	—	0.159
Incandescent lamp	2,854	0.138
Philips Fluorescent Lamp 32 [2]	—	0.096

This limitation acknowledged, the National Bureau of Standards' factors of damage per footcandle remain a convenient method of rating the relative degree of hazard. Table 1 presents a list of the D/fc factors for a number of common sources of illumination, arranged in decreasing order of the calculated degree of hazard. The order is about as expected from our knowledge of the distribution of spectral energy in the various sources and of the potentially greater activity of radiation in the shorter wavelengths. The wide range in potency that may be encountered in fluorescent lamps is also indicated.

Susceptible Objects

Before considering any special limitation in the exposure to light, the first task of the curator is to recognize the objects that are likely to be susceptible. Because they may contain (or be covered with) fugitive pigments, dyes or inks, the following objects were considered in the ICOM booklet of 1953 as those likely to be sensitive to harm by light: textiles of all types, leather, feathers and paintings (especially watercolors).[7] Also noted were books, microscopic preparations and many natural history specimens, such as animal skins. Certain woods tend to

7. More than any special sensitivity of watercolors in themselves, it is such pigments as Vandyke brown, sepia, bistre, alizarin, indigo, yellow lake (quercitron?), carmine, gamboge and Hooker's green made with gamboge, that were found to be sensitive to light by Abney and Russell and others [42].

darken, and wood stained with dyes can fade [43, 45]. Certain colorants in pastels may also be fugitive. In addition to problems in fading, one must remember that the vehicle in paints and varnishes can yellow, shrink, be eroded on the surface and change in solubility under action of light. Textiles can become fragile and paper become yellow and embrittled.

The ICOM booklet pointed out that jewelry, objects of metal, and mineral (inorganic) substances such as ceramics, enamels, glass[8] and stone, are practically proof against deterioration by light.

8. According to Dr. Robert Brill of The Corning Museum of Glass, there is renewed interest in the possibility of discoloration of glass by light. Solarization of certain types of glass is a well-known phenomenon, for example. See also [1, 17].

Factors That Control Photochemical Damage

Present space will not permit a discussion of technical details concerning the photochemical deterioration of individual classes of materials. Nevertheless, the curator needs to know what can be done in a general way to control the possibility of damage by light, irrespective of the particular materials. On this occasion, then, the fundamental factors that influence photochemical damage will be considered, along with certain practical steps which may be taken to minimize photochemical effects. Technical details concerning the behavior of specific materials can perhaps be presented on a subsequent occasion.

The ICOM booklet of 1953 pointed out that the extent of the deterioration depended upon: (a) the intensity of the radiation, (b) the time of exposure, (c) the spectral characteristics of the radiation, and (d) the intrinsic capacity of individual materials to absorb and be affected by the radiant energy. External factors also influence the rate of deterioration, of which may be mentioned: (e) humidity, (f) temperature, and (g) the active gases in the atmosphere. High temperature, high relative humidity and the presence of oxygen usually speed up the processes of deterioration. Each of these seven will be considered individually, but they represent essentially only three fundamental factors: the characteristics of the radiation, the materials and the conditions of exposure.

Intensity of illumination

The principle of reciprocity states that the net photochemical effect is the result of the total exposure which an object receives, which is the product of the intensity multiplied by the time. In other words, 1,000 lux of radiation for 1 hour is expected to do as much damage as 10 lux for 100 hours, because the total exposure in both cases is 1,000 lux hours. The reciprocity principle has been shown to hold reasonably well in the fading of modern dyed textiles throughout a fiftyfold range of intensity [47]. In a study of more than one thousand samples of dyed textiles, Cady and Appel reported finding only about 4 percent of the cases where the rule did not hold [8].

Until laboratory tests prove the contrary in any particular case, the museum curator must assume that the reciprocity law holds true. On this basis, he may assume that the extent of photochemical damage will be reduced in direct proportion to the reduction in the intensity of illumination or the time of exposure. Reduction of the time of exposure will be discussed below.

Measurement of intensity

The eye is a poor judge of the absolute level of illumination, for it can readily adapt itself to changes in intensity. Hence, the intensity of illumination must be determined with an instrument, one that is calibrated and checked for accuracy. Two typical instruments and a chamber containing a lamp of standard intensity are shown in Figure 14. Of interest in photochemical problems is the intensity of illumination that would be reflected from a perfectly diffusing white surface in the plane of the object under consideration. In practice, the intensity of illumination on a surface is often determined by measuring the light reflected from a standard white surface of magnesium oxide, barium sulphate or a photographer's reflectance card of known percentage of reflectance.

The intensity can be expressed in terms of luminous energy or total energy. In luminous energy (the radiation that is visible), the unit is in lumens per square metre (lux), or lumens per square foot (footcandles). One footcandle is equal to 10.764 lux. The exposure in many accelerated aging tests outdoors is now reported in terms of the total energy in langleys, where the langley is equal to 1 gram-calorie per square centimeter. While the langley has not been applied to many museum problems, it or a related unit of energy may be employed more frequently in the future because it takes into account the total radiation, the ultraviolet and the infrared as well as the visible, which the measurement in terms of luminous energy alone (lux or footcandles) does not.

Measurement of neither the total energy nor the luminous (visible) energy reveals very much about the relative potency of any given source of illumination owing to the presence or absence of radiation of short wavelength. To this end, the National Bureau of Standards developed the concept of the relative factor of damage per footcandle (D/fc), already described. It was intended that this factor be multiplied by the intensity of illumination to obtain a figure of the relative "degree of hazard" of any given source.

Control of intensity

The data in Table 2 reveal that intensities of illumination ranging from about 50 lux (5 footcandles) to 300 lux for emphasis (about 30 footcandles) and averaging from 120 to 150 lux have been widely recommended for the illumination of objects on the walls of museums. It should be immediately apparent that the control of intensity of illumination is one of the most important factors for the curator to consider. It is readily possible, for example, to expose an object to two, three or more times the intensity of light than is necessary, because the eye readily adapts itself to different levels of illumination. Therefore, careful check on the intensity should be maintained with instruments.

Based on the levels of intensity in Table 2, figures are presented in Table 3 to suggest the amount of annual exposure of objects that may be encountered in a museum, calculated in terms of a 10-hour day.[9] Though expressed in terms of visible energy alone, they nevertheless provide a convenient reference and may become of increasing interest as further research better evaluates the relative hazard of different sources and determines the rate at which susceptible materials deteriorate. For example, research at Mellon Institute has indicated that a number of modern writing inks, dyed-leather book bindings, traditional textile dyes, and pastel and watercolor pigments can fade noticeably[10] in less than 1 million footcandle hours when exposed under glass at 80°F and 10 to 35 percent relative humidity (R.H.) to "daylight" fluorescent lamplight. Under these conditions, an exposure of 6 to 8 million footcandle hours has been found

9. A brief discussion of standards of exposure, particularly British Standard 1006:1961, has been published since the preparation of this manuscript [14].

10. A change in contrast of 3 on the British Standard 2662:1961 grey scale, or about 3 NBS units of color difference.

Table 2
Levels of Intensity Recommended and Used for the Illumination of Pictures in Museums*

Lux			Footcandles			Reference
Low	Medium	High	Low	Medium	High	
50			5			*Burlington Mag.*, 57 (July 1930), p. 31.
40	180		4	18		*Mouseion*, 27-28 (1934), p. 193.
30	100		3	10		*ibid.*, 33-34 (1936), p. 185.
60	100		6	10		*ibid.*, 33-34 (1936), p. 191.
	110			11		*Museum*, 5 (1952), p. 28.
	120	300		12	30	Harrison, L. S. [23].
	180			18		Balder, J. J. [2].
80		250	8		25	*Lichttechnik*, 9 (1957), p. 547.
70		200	7		20	*Couleurs*, No. 20 (1957), p. 15.
80	150	300	8	15	30	*Museum News*, 38, No. 4 (1959), p. 26.
	180			18		*Museum Journal*, 59 (1959), p. 66.
50	150	300	5	15	30	Thomson, G. [49].
	150	200		15	20	*Museum Journal*, 61 (1962), p. 259.

* For convenience, 10 lux is taken to equal 1 footcandle (1 lumen/square foot).

Table 3
Approximate Total Annual Exposure of Objects, Calculated in Terms of a 10-hour Day

Intensity		Approximate exposure	
Footcandles	Lux	Footcandle hours/year	Lux hours/year
5	53.5	18,300	195,000
15	160.5	55,000	585,000
30	321.0	110,000	1,170,000
59*	631.3	215,000	2,304,000

* Average annual value estimated for "galleries with sun louvers properly operated," L. S. Harrison [23].

to cause marked fading in a number of traditional organic pigments. One million footcandle hours is not more than about 9 years at the 30-footcandle level; 8 million, 70 years at the 30-footcandle level, and 440 years at 5 footcandles.

The intensity of daylight may be controlled by manually operated shades, louvers, and curtains.[11] The main problem in manual control is that the attendants do not always draw the blinds as soon or as often as they might. On the other hand, the use of electronic controls to operate louvers or curtains should ensure that they would be closed more effectively than if left to personal judgment. An elaborate system of electronically-controlled drapes has been designed for the New Museum of History and Technology, Smithsonian Institution, Washington, D.C. [22]. Such a system may, of course, be adapted to the operation of louvers as well as curtains.

11. If fabrics are used, Miss Strömberg called attention to the fact that their quality varied considerably, and that each lot needed to be tested with a light meter to verify that the new drapes were as effective as the previous.

The intensity of illumination can also be reduced by diffusing surfaces, materials and glasses. Or it may be reduced by a neutral-grey glass such as the Pittsburgh Plate Glass "Solargray" or the American Window Glass "Lustragray" (Solargray glass is specified in the installation described by Harrison) [22]. Plate glass that reflects radiation because of a thin deposit of metal is also available, such as the Pittsburgh Plate Glass "LHR Twindow" and "LHR Solargray Twindow."

So many excellent articles have been written on problems of display that the control of intensity of light from artificial sources will not be discussed in detail [20, 28, 34]. The eye can become well adapted to the level of illumination. Through skillful methods of display, it should be entirely possible to illuminate the objects at levels which are satisfactory both from the point of view of their conservation and adequate conditions of viewing. But, the subject of the eye's ability to adapt itself to conditions of illumination, and how this ability may be properly utilized in museum display has received little attention. Apparently, Brommelle is the first author to direct the attention of museums to technical problems concerning glare and the adaptation of the eye [6]. A variety of lighting methods is needed in a museum to avoid monotony. But Brommelle points out that the problem of viewing objects so that their details can be seen most efficiently has not been studied scientifically from the museum point of view and that a new committee has been established by the Illuminating Engineering Society, in England, under the chairmanship of Mr. J. B. Harris, one of whose tasks will be to study this aspect of display.

From the findings that will be derived from this investigation, undoubtedly will come detailed suggestions concerning the manner in which the major transitions of intensity in a museum should be made gradual, a factor pointed out by Thomson [49]. The eye needs time to adapt itself to changes in intensity. The change to a lower level of intensity upon entering rooms where susceptible objects may be displayed need not appear to be abrupt or noticeable if the visitor is led from a brightly lighted room to one of medium intensity before passing on to other areas where the intensity is at the lowest level.

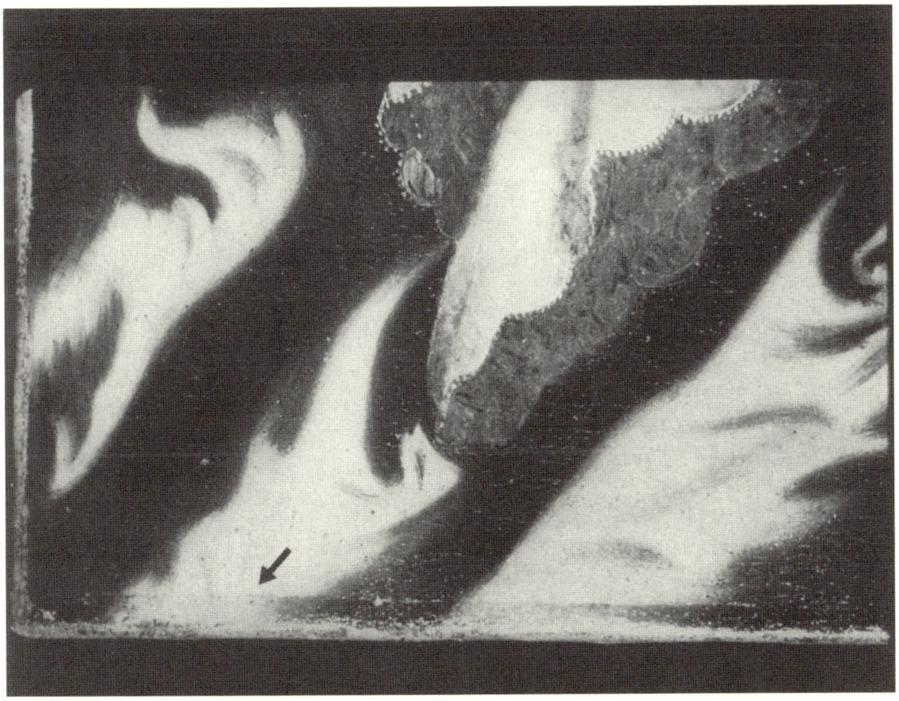

Figure 11
Darker color in the area that has been protected by rabbet of the frame shows evidence of fading in blue drapery of 18th-century oil portrait (detail).

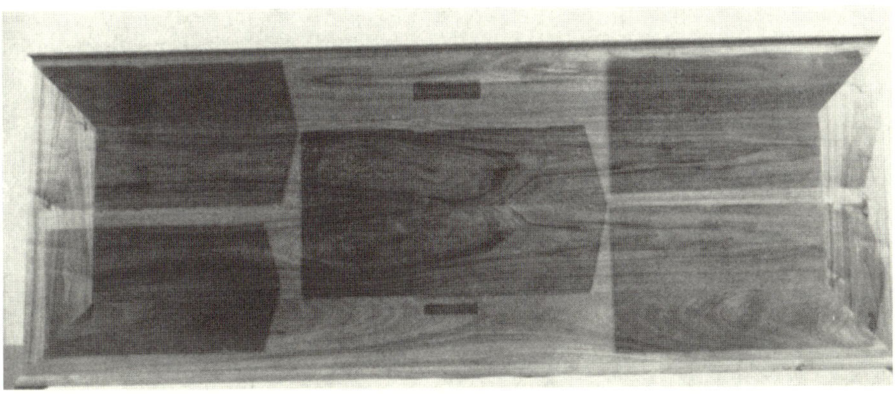

Figure 12
Panel of wood, similar to that in Figure 13a, upon which leather book covers had been exhibited.

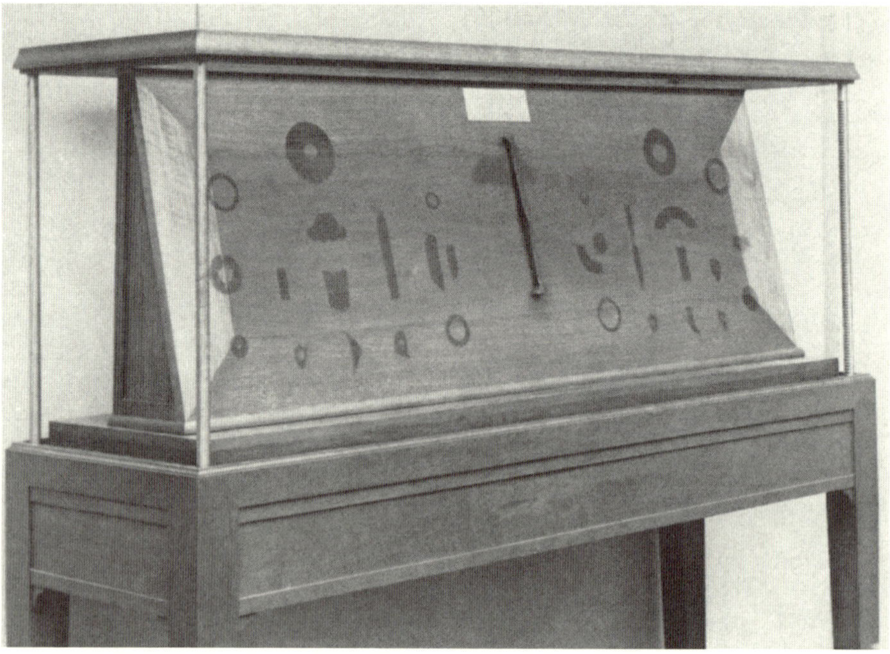

Figure 13a
Easel of wood on which artifacts of jade and bronze had been exposed in a museum case for about twenty years of natural illumination. The wood panel, expect for the areas protected by the artifacts, has been bleached.

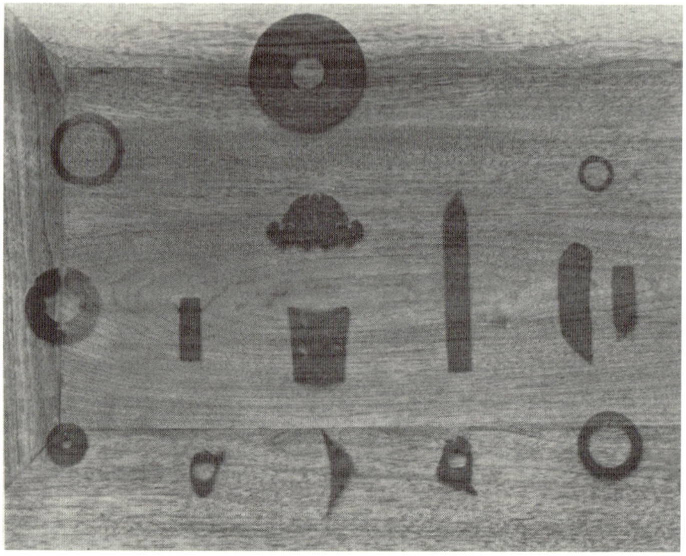

Figure 13b
Detail of Figure 13a.

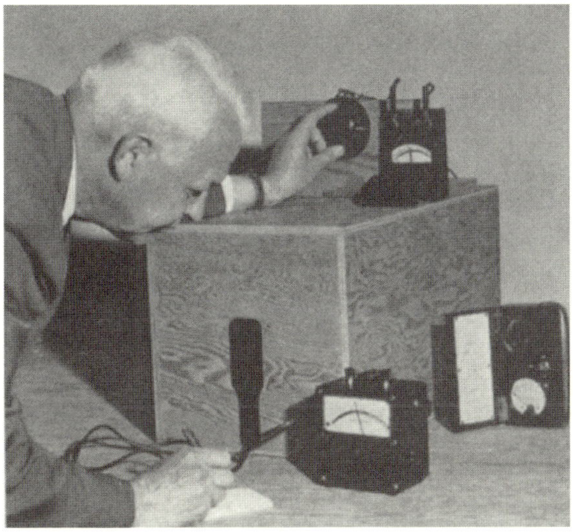

Figure 14
Checking accuracy of light meter. Black light-sensing paddle and meter in foreground is the Weston Model 657 illumination meter. To check the accuracy of the meter against a light source of known intensity, Mr. S. P. Eagleton, National Gallery of Art (Washington, D.C.) adjusts the voltage on a standard lamp which is within the plywood box. Meter at far right is the most compact, Weston Model 614.

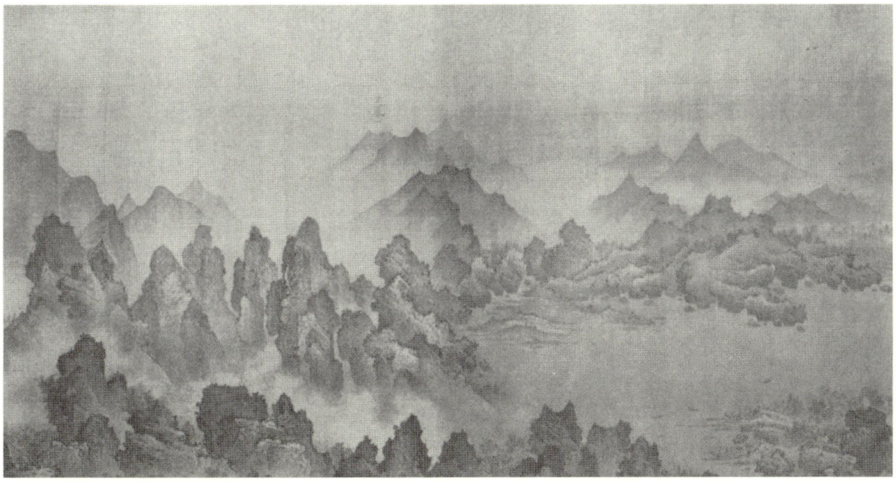

Figure 15
Part of the Chinese scroll painting on silk, Ming Dynasty, in which left half had been exposed to natural illumination for nearly forty years. Background has bleached, leaving discontinuity between this portion and that at right which had been rolled up.

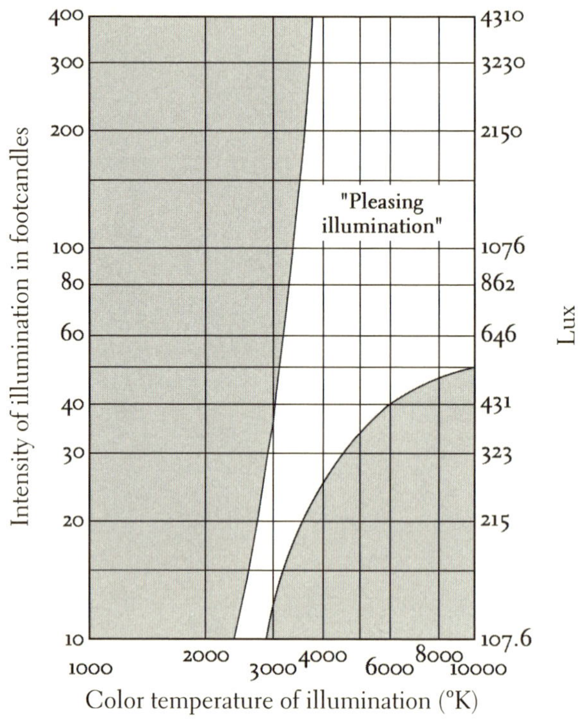

FIGURE 16
Data of Kruithof showing that for every color temperature, there exists a highest and lowest level of illumination between which the illumination is considered pleasing.

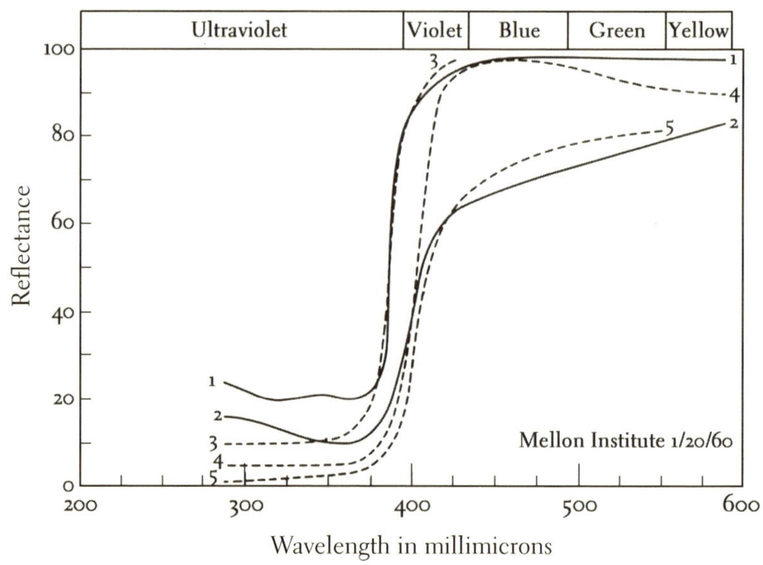

FIGURE 17
Reflection of radiant energy from a number of white paints that reflect little ultraviolet radiation.

In connection with the use of lower levels of illumination, it has been found that the observer generally prefers illumination of "warmer" hue at low intensities. Thus, on cloudy days, many museums receive complaints that the intensity in the galleries is too low. Such complaints are practically eliminated by supplementing the "bluish" natural illumination with warmer-hued incandescent. At the National Gallery of Art, Washington, D.C., the intensity of illumination is measured by the guards at key locations and the supplementary illumination is turned on when the intensity falls below a certain level. The actual level of illumination is increased only slightly in this process, but the quality of the light is so much more pleasing that few complaints are received. At the Musée du Jeu de Paume, Paris, the supplementary lights are turned on by photoelectric controls.

Figure 16 gives a chart of the preferred color temperature as a function of the intensity of the illumination, determined by investigators at the Philips Gloeilampenfabriken laboratories, Eindhoven [31]. In part owing to the phenomenon under discussion and thanks to the detailed information supplied by such authorities, color temperatures of 4,200°K [23, 28], 4,500°K [34] and 4,900°K [12] have been generally accepted as satisfactory for the illumination of museums.

Time of Exposure

Various means used to minimize the time during which the objects are exposed to light will reduce the total exposure. The custom of having curtains drawn over showcases that contain miniatures or illuminated manuscripts is an example familiar to all. Here, the objects are exposed only when the individual visitor wishes to look at them. Another method that accomplishes the same end is the use of light switches that illuminate the individual exhibit for a period of time and then automatically turn the lights off. When the objects are displayed under conditions of natural illumination, they receive a portion of their total exposure during the daylight hours when the public is not present. This has been long recognized. The British Museum of Natural History, years ago, for example, followed the custom of drawing the blinds during these hours of the day [19].

The reciprocity principle states that the total exposure is directly related to the time. Because the time of exposure is such a major factor, authorities have continually emphasized that susceptible materials should not be placed on permanent or extended periods of exhibition.

Control of spectral distribution of radiant energy

Estimated benefit to be derived from removal of ultraviolet radiation

Both visible and ultraviolet radiation can produce photochemical effects. But, though the ultraviolet radiations are not exclusively responsible for photochemical action, they are invisible and are definitely not necessary for viewing an object. Therefore, as much as possible, ultraviolet radiation should be eliminated from the illumination used to display objects that may be susceptible. If ordinary incandescent lamps are used, there is little concern, although there is a trace of ultraviolet present even in this type of illumination. But measures to eliminate ultraviolet radiation from daylight and from ordinary fluorescent lamplight should be taken wherever possible.

Can one state just how much benefit there is to be gained by elimination of ultraviolet radiation? One finds that the benefit to be derived from elimination of the ultraviolet radiations that pass through window glass, those between about 310 and 400 millimicrons, may range from practically none at all up to as much as a tenfold improvement. Figure 7 illustrates the marked and similar effect of wavelength on the deterioration of low-grade paper (D_λ factors) [39], the deterioration of rubber [3], the erosion of paint [37], and the tendency for certain methacrylate polymers to become insoluble [15]. In these four types of deterioration, elimination of radiations between 310 and 400 millimicrons leads to a decrease in the rate of deterioration to about one-tenth the value obtained when the ultraviolet radiations were present.

On the other hand, McLaren has pointed out that ultraviolet radiation generally plays a more important part in the fading of relatively light-fast dyes than in the fading of more fugitive ones [35]. Hence, in the case of the most fugitive dyes and pigments, one may not expect marked benefit from elimination of the ultraviolet radiation. Nevertheless, research at the Mellon Institute has shown that carmine, alizarin, and indigo receive noticeable protection from the removal of this radiation [36). Kuhn has also demonstrated that carmine, alizarin and gamboge benefit from such protection [32], and Thomson has shown the benefit of ultraviolet filters in reducing the fading of several dyes and water colors (including indigo and gamboge) and the yellowing of mastic resin [49].

Using the factors of damage developed by the National Bureau of Standards, one may calculate that removal of radiation below 400 millimicrons from daylight indoors under skylights would theoretically increase the estimated lifetime of objects by a factor of about three times.[12] Removing the ultraviolet radiation from a fluorescent lamplight may be estimated from the data in Harrison's Table 5 to lead to an improvement of lifetime of two- to fivefold.

Thus, one finds that the elimination of ultraviolet radiation may lead in some cases to little improvement at all, particularly with regard to the fading of fugitive

12. The ability of cotton to retain its tensile strength during outdoor exposure was found by R. A. Coleman and W. H. Peacock [9] to be improved by a factor of 4.5 times when the protection was given against ultraviolet radiations.

dyes; in other instances, it has also been observed to lead to an improvement of about tenfold. A decrease in the rate of photochemical deterioration of the order of two to three times is not to be considered unusual. It must not be forgotten that the use of filters to remove radiation in the range of 310 to 400 millimicrons rarely if ever prevents damage—it can only retard it. But, regardless of the benefit in any particular case or whether the degree of benefit is known at all, ultraviolet radiation is not needed to see objects and should be eliminated in the illumination of susceptible materials.

Filters for ultraviolet radiation

An effective and simple way to eliminate ultraviolet radiation is to reflect the light from painted surfaces containing zinc white or titanium white pigments. It has been repeatedly pointed out that such indirect lighting will contain little ultraviolet radiation [16, 27, 49]. The data in Figure 17 may perhaps dramatize the effectiveness of such a "filter" even better than Figure 75 in Genard's article [16].

Slightly-yellow colored glasses, such as Filtralux [29] and Pittsburgh Document Glass [23], have been used in a number of applications. But finding a colorless window glass that cuts off much of the ultraviolet has not been especially successful in the past. Thomson has described the colorless Cero-Crystide Glass used in some of the new rooms at the National Gallery, London, and has estimated that this glass will reduce the relative factor of damage per footcandle to approximately one-half that of ordinary plate glass [49]. Thomson has also drawn attention to a diffusing and ultraviolet-absorbing glass sold by the Shatterprufe Safety Glass Company, Ltd., Port Elizabeth, Australia (Safety Northlite UV-393). La Société Glaverbel of Brussels is reported recently to have prepared a museum glass (Verre Musée) which will filter the ultraviolet much more effectively than either of these two colorless glasses and will also filter the infra-red as well.

The best present solution to obtaining a colorless filter has been the use of plastic sheeting that contains ultraviolet-absorbing compounds.[13] These sheets are less expensive and generally more effective filters than the colorless glasses hitherto available. The first such material to be discussed in detail was Plexiglas LPC-518K (its experimental designation) described by Harrison [23] and applied in the windows of the Blumenthal Lounge in the Metropolitan Museum of Art in New York. This material at one time was sold as Plexiglas IIUF and is now known as Plexiglas UF-1 (Oroglas UF-1 in England). The spectral transmittance characteristics of an even more effective filter, the slightly yellow Plexiglas G911B (its experimental designation, now known as Plexiglas UF-3), were recently noted by Harrison [22]. The approximate curve of transmittance of this filter and that of Plexiglas UF-1 is shown in Figure 2. A Plexiglas filter having transmission characteristics intermediate between these two has recently

13. For a description of ultraviolet-absorbing compounds and materials, see [18, 46].

been placed over the laylights in the National Gallery of Art, Washington D.C. (Figure 18). Thomson has described a number of related materials varying in thickness of the sheet from 0.1 to 6 millimeters [49].

A number of varnishes and sheaths of sheeting materials are also available for the filtration of radiation from fluorescent lamps. Examples of three different types of sheaths are illustrated in Figure 19. Thomson bas described two varnishes that may be coated on glass and Kühn has described a varnish produced by Spektrafix-International J. Vincze, Bad Waldsee, Wurtt. Because of the difficulty in preparing coatings on glass that are perfectly free from distortion and also resistant to moisture, varnishes have perhaps not found as wide favor for glazing large areas as have sheet materials. However, varnishes appear to be quite satisfactory for coating fluorescent lamps.

There are a number of points to consider in the selection of ultraviolet-screening materials. Their initial specification involves a statement of the wavelengths of radiation that are filtered, the sharpness at which the transmittance falls from high transmittance of light to practically none at all, and the amount of the ultraviolet-absorbing chemical that is needed to give effective filtering. This latter property, a fundamental characteristic, is particularly important in the preparation of one's own special varnishes and sheeting, but it is of much less interest if commercially prepared materials are purchased ready to use. It is called the specific absorbance: the absorbance of the material at a concentration of one gram per liter and a thickness of one centimeter.

A fourth significant factor to be checked with such filters is the manner in which the cut-off point and the slope of the curve of transmittance changes during use. Some shift in the transmittance is expected as the material ages. It is important, however, to determine the direction of this change in time (whether there is a tendency for the filter to become more effective or less) and how rapid it is. Data of Thomson and Kühn show that a number of ultraviolet filters become less effective during exposure. On the other hand, the particular filter for fluorescent lamps in Figure 21 actually became more effective as it discolored in time.

In evaluating such materials, it is also important to consider whether the varnishes may be somewhat soft and pick up dirt. This may pose a maintenance problem if the coatings are used on laylights, for example. One would also wish to know if the plastic sheaths for fluorescent lamp tubes become brittle in time or give off volatile products that may have harmful effects on the objects. As a general policy, it is best that ultraviolet-absorbing sheets not be placed in direct contact with paper documents, textiles and similar exhibits. Upon occasion, the slightly yellow ultraviolet-absorbing compound in such sheeting has been known to migrate out of the sheeting and to stain paper, particularly in the thinner types of sheeting where a high concentration must be used.

Concerning the question of removing ultraviolet light emitted by fluorescent

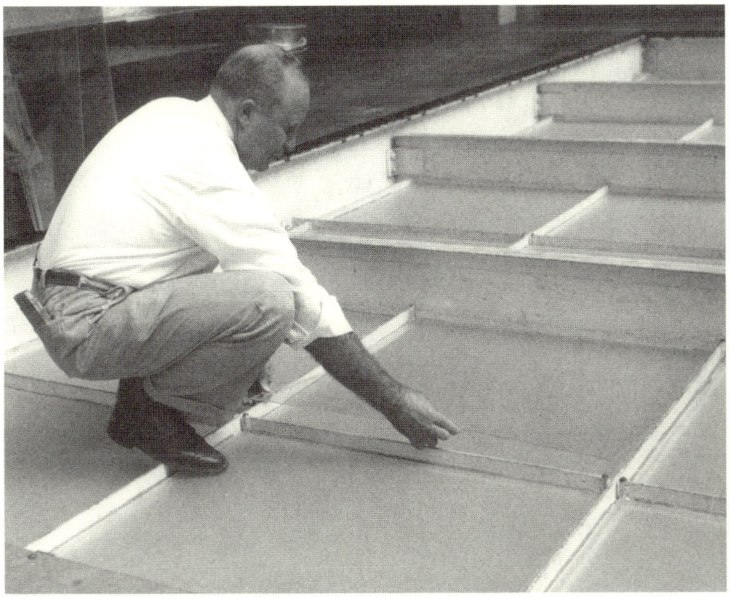

Figure 18
Placement of Plexiglas ultraviolet filter over laylights (National Gallery of Art, Washington, D.C.). (Ed. Note: alternate image used here, different from original publication)

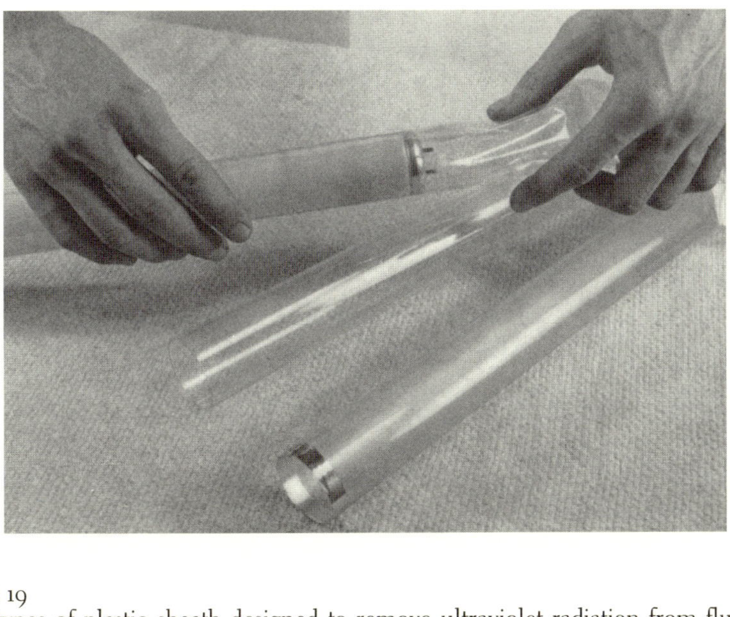

Figure 19
Three types of plastic sheath designed to remove ultraviolet radiation from fluorescent lamplight.
Flexible: Solar Screen Corp., 1023 Whitestone Blvd., Whitestone 57, NY
Rigid: Rohm and Haas Plexiglas, molded by Westlake Plastics, Lenni Mills, PA
Frosted: Infropake Corporation of America, P.O. Box 6292, Greensboro, NC

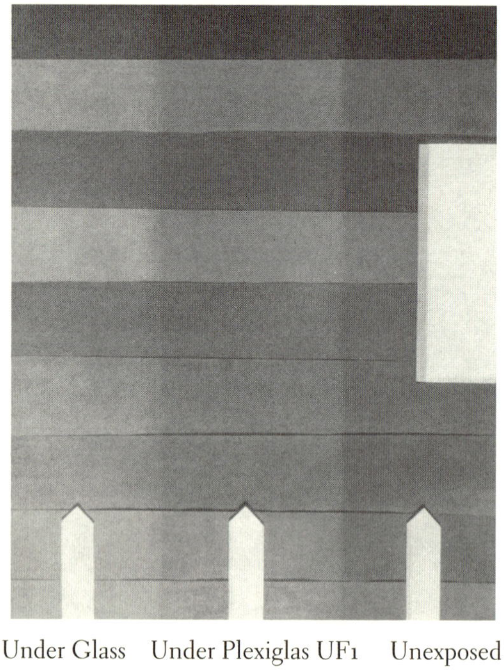

Figure 20
Modern writing inks, coated on paper strips, showing partial protection given by ultraviolet filter after 500,000 footcandle hours of exposure under ordinary fluorescent lamps.

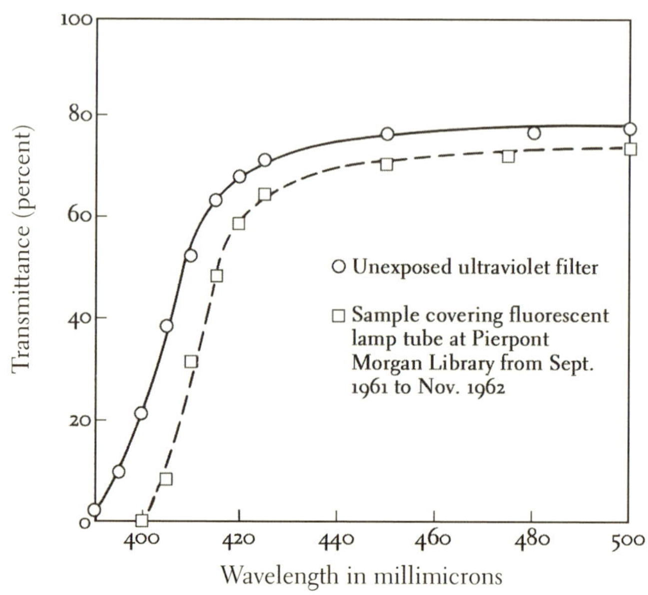

Figure 21
Change in spectral transmission characteristics of a plastic ultraviolet filter for fluorescent lamps after one year of use.

lamps: the amount emitted varies widely, but there is good evidence that as much as 5 percent of the total energy emitted below 700 millimicrons can be in the ultraviolet range. In spite of a number of research reports that apparently show that the trace of ultraviolet represents little danger to museum objects, the National Gallery of Art Research Project at the Mellon Institute has found that there is noticeable benefit from the removal of ultraviolet radiations from a number of ordinary fluorescent lamps in the protection of modern writing inks (Figure 20), alizarin, indigo, and other materials (Figure 22). Hence, whenever ordinary fluorescent lamps are used to illuminate susceptible materials, it is advisable to use ultraviolet filters or indirect reflectance of the lamplight from painted surfaces containing zinc oxide or titanium white. There are certain types of fluorescent lamps on the market, such as the Philips Color 32 and 34 lamps,[14] noted in Table 1, which emit considerably less ultraviolet radiation than the ordinary types [2]. Yet, until further research and detailed information is obtained concerning individual lamps of this special type, it is the best policy to take steps to eliminate the ultraviolet radiation whenever fluorescent lamps are used to illuminate susceptible materials.

14. Philips Color 34 lamps have been used in the illumination designed for the National Gallery, London [21].

Varying the spectral distribution of visible radiation

The possible variation in photochemical activity, owing to variations in the distribution of visible radiation, as well as ultraviolet radiation, was briefly discussed when the characteristics of various sources were described. As previously stated, radiations of shortest wavelength, the violet and blue, are likely to be the most potent. As also mentioned, there is very little of these wavelengths in the ordinary incandescent lamplight (Figure 5b); hence, the problem of filtering these wavelengths from such illumination is of little concern. In the matter of daylight, which possesses varying amounts of blue in proportion to the red, it may be generally assumed that the higher the color temperature of the illumination, the more likely it is to be of greater photochemical activity, as indicated in the data of Table 1. However, this is only a rough rule-of-thumb in the case of fluorescent lamps and can be in error if one fluorescent lamp emits a greater amount of ultraviolet radiation than another. The two "deluxe" fluorescent lamps in Table 1, for example, are more hazardous than might be expected for their rated color temperatures. To evaluate any individual source properly, the curve of spectral emission must be known and be analyzed by some method such as the calculation of the factor of damage pet footcandle.

Although the shorter wavelengths are theoretically more potent, in the last analysis, the actual extent of photochemical damage caused by the visible radiation in the blue and violet cannot be predicted and needs to be determined experimentally in any particular problem.

Where color is not important, yellow filters have been used, with the advantage that the most photochemically potent wavelengths of visible light are thus eliminated. The National Bureau of Standards, for example, chose a decidedly yellow filter for the protection of the *Declaration of Independence* and the *Constitution* of the United States [39, 40]. The curve for the transmittance of this filter is shown in Figure 2. Plexiglas UF-3, just slightly yellow, has been considered to be satisfactory for use in the windows of the new Museum of History and Technology, Smithsonian Institution, Washington D.C. [22]. Thomson has called attention to the fact that the yellowness of filters is less noticeable when they are used in this manner to control the general illumination, than when they are placed between the observer and the object to be viewed [49].

The ability of lamps to give good color rendering is an entirely separate topic which has been adequately discussed elsewhere [5, 11, 49].

Intrinsic properties of materials

Whether or not a particular material will be affected by light depends upon its intrinsic capacity to absorb radiant energy and be affected by it. The principal classes of materials that are susceptible to the action of light have been noted. Obviously, certain of these are more affected by light than others and, in each individual class, one particular type may be more susceptible than another. The stability of paper, for example, has been found to decrease in the order: new rag, sulphite, and newsprint. Of the various textiles, the order of sensitivity seems to be well established: silk, the most sensitive, followed by jute, cotton and wool [44]; the early work of Turner which supports this conclusion is shown in Figure 23. The fibers of a particular fabric are likely to be more resistant to photochemical degradation, however, the thicker they are in diameter. It is such matters that properly come under the subject of the intrinsic properties of materials and how they affect susceptibility. However, the present discussion is limited to the ways in which the action of light can be reduced and does not permit consideration of these matters. One can only point out the principal types of materials that are susceptible. On a future occasion, perhaps, consideration may be given to a discussion of the specific details which modify the sensitivity of the different classes of materials. In the meantime, Brommelle has already reviewed many of the factors that influence the fading of watercolors, the fading of dyes and the weakening of textiles [7].

Extrinsic factors also influence the susceptibility of materials. For example, the presence of weighting materials in silk may increase the susceptibility to light [26]. Giles and others have also pointed out that many dyed textiles are effectively more lightfast the greater the concentration of the dye and the greater the degree of aggregation of the dye particles [4]. Some of Giles's data that demonstrate the effect of concentration are shown in Figure 24. There are three factors, however, which are so particularly significant that special mention

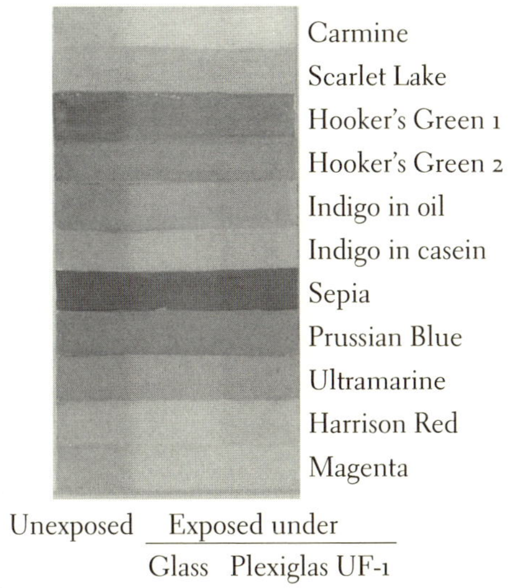

FIGURE 22
Test panel showing partial protection given by an ultraviolet filter to a number of fugitive watercolor pigments after 2.1 million footcandle hours of exposure under "daylight" fluorescent lamps.

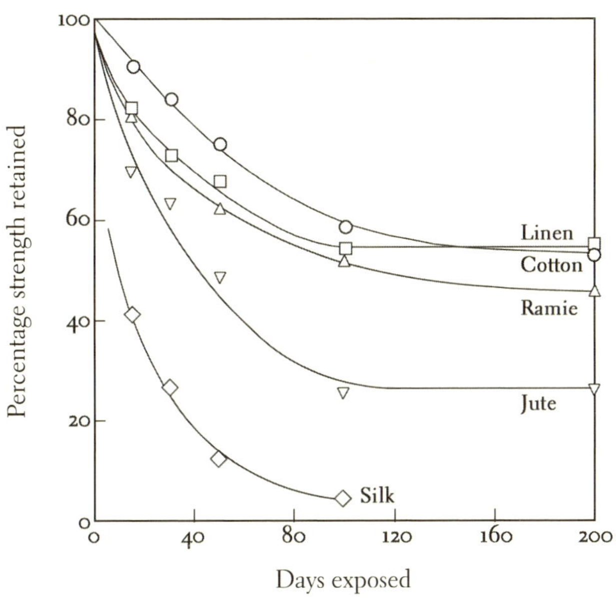

FIGURE 23
Relative rate at which various fabrics deteriorate. Data of A. J. Turner, based on weathering outdoors at Farnborough [*J. Soc. Dyers and Colourists*, 36 (1920), p. 165].

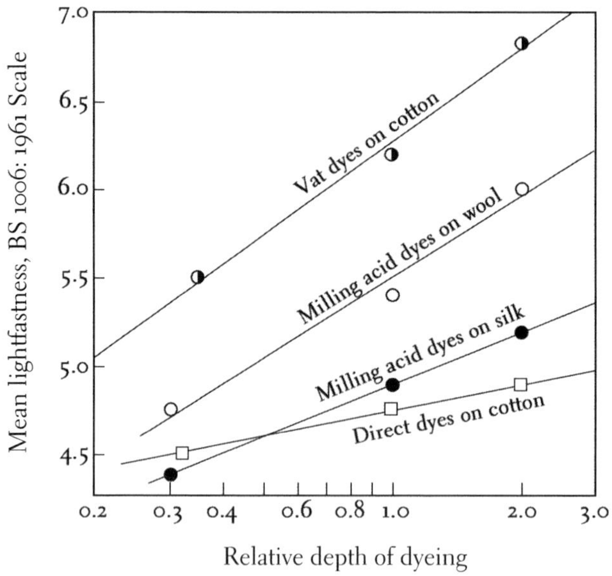

FIGURE 24
Lightfastness of dyed cloth, in relationship to the depth of dyeing [Giles, C. H., *J. Soc. Dyers and Colourists*, 73 (1957), p. 127].

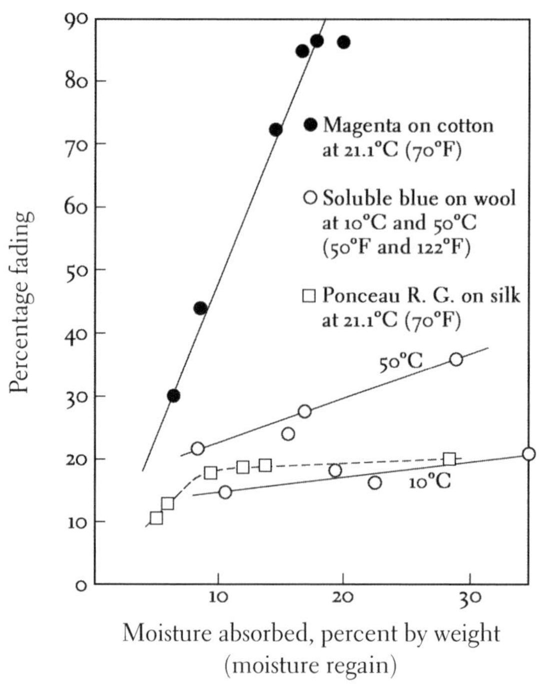

FIGURE 25
Fading of dyes under mercury arc lamp in quartz as a function of the moisture content of cotton, wool, and silk [Hedges, J. J., *J. Soc. Dyers and Colourists*, 44 (1928), p. 341].

must be made of them: the rate or speed of photochemical reactions is influenced by temperature, humidity and the oxygen in the atmosphere.

Temperature

Increasing the temperature influences the rate of chemical reactions in two ways. The first effect is to increase the general agitation of atoms and molecules, their kinetic energy. The primary process in photochemical reactions is the excitation or activation of an atom or molecule by absorption of radiant energy and the subsequent way in which this added energy is lost; this process is generally little affected by temperature. But the energy thus absorbed may be passed on to other atoms and molecules, leading to oxidation and other types of chemical reactions. Temperature frequently has a decided effect on the rate of these secondary reactions in the photochemical process. A rise in temperature will usually increase the rate of chemical reactions. There has long been a general rule of thumb, based on the Arrhenius equation,[15] that the rate of chemical reactions taking place at room temperature may be expected to double for a rise of 10°C (18°F).

In photochemical processes, the effect of temperature may be much less than this. Moreover, it is difficult to predict in advance, since the net effect depends upon the relative importance of the strictly photochemical steps (the primary process, generally independent of temperature), or the secondary reactions that are triggered off by the primary process (chemical reactions more likely to be dependent on temperature). There is a second effect of temperature, however, that is far more generally encountered and more readily explained: an increase in the temperature of an object will usually reduce its moisture content and thus reduce the rate at which many reactions proceed.

Relative humidity

Molecules of water frequently participate in chemical reactions, but they are often not directly consumed. This behavior is known as a catalytic one in which the presence of the catalyst, water in this case, speeds up a chemical reaction, but is not itself consumed. The rate of a great many chemical reactions is increased as the concentration of moisture is increased. In the fading of dyes on textiles, the data of Hedges, shown in Figure 25, demonstrate that it is the actual percentage of moisture in the fibers (moisture regain) that is the controlling factor, rather than the relative humidity of the atmosphere [24]. Certainly a sound analysis of this problem seems to be to consider the actual content of water in the object, rather than simply the relative humidity of the surroundings.

The temperature of the air has a marked effect on the relative humidity: the ratio of the actual amount of water vapor present to the maximum amount of

15. The effect of temperature in speeding up chemical reactions is well understood and is influenced by the change in the absolute, or Kelvin, temperature rather than in Centigrade or Fahrenheit. The general influence of temperature on the speed of chemical reactions is predicted by the Arrhenius equation (1889): $2.303 \log K_2/K_1 = -E/R\ (1/T_2 - 1/T_1)$, where K_2 is the rate of reaction at the higher temperature T_2, K_1 the rate at the lower temperature T_1, E the energy of activation or excitation, and R a constant. The temperature is expressed in degrees Kelvin.

water the air is able to hold at that temperature. In recognition of the major role played by temperature, some authors have used the term "object humidity" or "effective relative humidity" to signify that the actual moisture content of an object is more controlled by the temperature of the object and the atmosphere immediately surrounding it than by the relative humidity and temperature of the atmosphere at some distance away from the object, such as might be determined with the usual devices for measuring and indicating relative humidity. Owing to this interplay of temperature and humidity only on rare occasions have the individual effects of temperature and humidity been analyzed in photochemical degradations. The data of Hedges in Figure 25 for the fading of soluble blue on wool represent one of the few instances where these two effects have been separately demonstrated.

What is the importance of this matter to the museum curator? It is simply this: that, if a particular material is susceptible to photochemical degradation, the processes are likely to proceed more rapidly the higher the relative humidity. Obviously, a certain amount of humidity is necessary to keep paper, silk, and certain other materials from becoming dry and brittle. But, various investigators have shown that the total elimination of moisture practically arrests fading of many dyes and the "embrittlement" of cellulose fibers by light. Hence, although it is seldom practical and desirable to completely eliminate water vapor from the atmosphere surrounding many museum objects, efforts should be made to keep the relative humidity at a minimum wherever susceptible objects are exposed to light.

Since lamps often raise the temperature of the objects illuminated, Brommelle has suggested that a slight increase in temperature of the objects so illuminated might do much to reduce the effective relative humidity and thus reduce photochemical deterioration [7]. As Brommelle cautions, the benefit of this proposal in any particular problem must be counterbalanced against the possible deleterious effects of temperature.

The role of oxygen

The last factor to be considered is the role of reactive gases in the atmosphere. Investigators who have tested the fading of textiles outdoors are well aware that the presence of sulphur dioxide, carbon dioxide and other reactive gases in the atmosphere of industrial regions can influence their experiments. Space will not permit a discussion of the influence of such agents. Moreover, these gases are practically eliminated in air-conditioned galleries. But mention must be made of oxygen. It is in the very air about us and its role in photochemical degradation should be fully recognized.

Many of the organic substances in textiles, dyes, protective and decorative coatings, fur and feathers have a fundamental tendency to oxidize. We know that they readily burn. This represents rapid oxidation. But the slow and general process of aging, frequently hastened by the ability of light to excite the molecules, is often simply a slow process of oxidation. Investigators have demonstrated in many examples that fading and embrittlement are completely arrested when oxygen is excluded from their reaction chamber during exposure to radiation in the visible and near-ultraviolet. The knowledge that this is true may not be particularly new or of immediate and practical interest to the average curator. But it should serve to remind one again of the fundamental and pernicious way that nature tends to cause materials to disintegrate.

In special instances, museum objects can and have been placed in an inert atmosphere, free of oxygen. At the Victoria and Albert Museum in London, a painting by Turner was placed in an evacuated chamber 70 years ago [7]. A complete vacuum perhaps offers certain disadvantages and more recently the *Declaration of Independence* and the *Constitution* of the United States have been placed instead under an inert atmosphere of helium (Figure 26) [39]. Occasions undoubtedly will be found in the future in which the particular circumstances concerning the object will make a similar procedure feasible and worth while.

Conclusion

The various factors that control the photochemical deterioration of museum materials have been considered and attention called to significant publications dealing with the subject. Authorities agree that all sources of illumination represent a certain degree of hazard to materials which are susceptible. Although the shorter wavelengths of radiation are more energetic, action by light is not confined to any one region of the spectrum; both ultraviolet and visible radiation can be harmful. When compared at equal levels of intensity, radiation from the three principal sources of illumination may be rated as follows: (a) incandescent, the least hazardous, (b) fluorescent, midway, although the hazard varies somewhat, depending on the particular type of lamp, and (c) zenith skylight, the most hazardous. It is generally agreed that, whenever daylight or ordinary fluorescent lamplight is used to illuminate objects which may be susceptible, the ultraviolet radiation should be eliminated because it is potentially harmful and plays no role in the visual observation.

The three principal factors that control the effect of light are the characteristics of (a) the radiation (intensity, time of exposure, and spectral distribution), (b) the particular materials, and (c) the conditions of exposure. Levels of illumination of about 50 lux (5 footcandles) have been widely and successfully employed for the display of materials which are particularly sensitive. Less sensitive materials have been displayed advantageously at 120 to 150 lux (12 to 15 footcandles). Materials which are not sensitive allow the use of emphasis lighting of about 300 lux (30 footcandles). Each of these levels can be satisfactory because the eye is able to adjust to a wide range of intensity. It is for this reason,

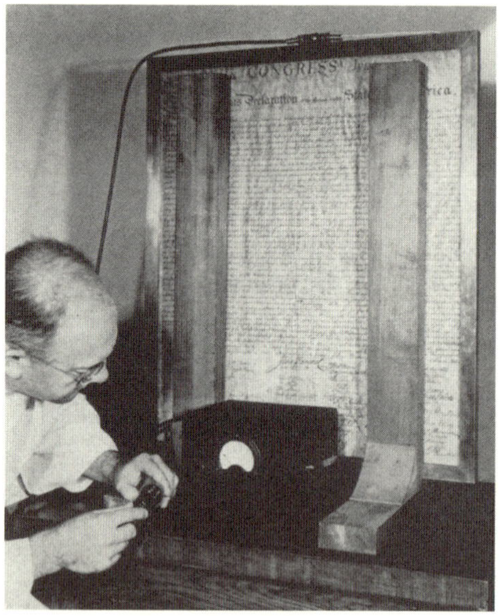

Figure 26
Mr. E. C. Creitz (National Bureau of Standards, U.S.A.) connects ammeter to the conductivity cells which are used to detect leaks in helium-filled chamber. Preliminary test being made with facsimile of *Declaration of Independence*.

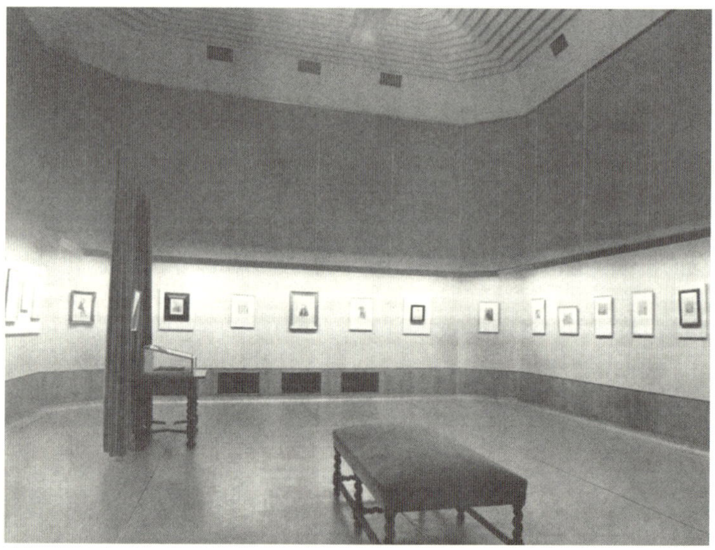

Figure 27
Exhibition of drawings by Rembrandt (1956). The lighting system was arranged in such a way that a horizontal band of light covered the drawings and their immediate surroundings while the rest of the room was kept at a suitable lower level of illumination. This arrangement permitted the drawings to be shown at good advantage at an intensity of about 180 lux (16.7 footcandles) on the drawings themselves.

moreover, that it is necessary to verify the actual level of intensity by means of suitable instruments (Figure 14).

Because time of exposure is a major factor, a decrease in the display time also reduces the hazard. Objects that are particularly sensitive should not be placed on permanent or extended display. Further protection may also be provided by the use of draw curtains or electric switches, which permits the objects to be illuminated only during the time when they are being viewed.

Special filters may be used to eliminate the unnecessary and potentially harmful ultraviolet radiation. Where color is not important, filters which are decidedly yellow in color can provide added protection. Points to be considered in the selection of such filters have been presented.

The danger in exposing metals to light and materials of inorganic nature such as minerals, ceramics, and stone, is extremely small compared to the danger to many organic substances. Textiles, low-grade paper, dyes, paint and varnish vehicles, and many pigments are likely to be affected. These materials require special attention.

Moisture generally increases the rate of photochemical deterioration. For this reason, the relative humidity of the surrounding atmosphere or, more specifically, the moisture content of the objects themselves, should not be permitted to become greater than necessary. It is important to understand also that the composition of the surrounding atmosphere plays a role. Many photochemical reactions involve the oxygen which is present in the air. In addition, certain gases present in the atmosphere in industrial areas can react with susceptible materials and may increase the over-all photochemical deterioration. Any efforts that result in the elimination of the reactive industrial gases undoubtedly will be helpful. In special cases, oxygen itself may be removed.

These are the essential facts and principles regarding the problem of photochemical deterioration. They have been known for a long time. There is a need in the future, however, to establish more precisely the list of susceptible materials and to study the specific ways in which they deteriorate, whether by fading, "embrittlement," or discoloration. It is also desirable to gain some measure of the amount of exposure necessary to cause specific amounts of these changes, and to determine more precisely the influence of temperature, humidity, and ultraviolet radiation. To find answers to some of these questions, the National Gallery of Art Research Project at the Mellon Institute has instituted an extensive program of research concerning the effects of light on paints, varnishes and other materials. Data on the effect of light upon the deterioration of varnishes have already been published and new findings concerning the rate of fading of a number of traditional pigments will shortly be available.

Suggestions concerning future investigations must not obscure the essential truth that the basic facts are already well in hand: (a) exposure of a number of materials to light can do harm if little attention is given to the problem, (b) the list of susceptible materials is reasonably well established, and (c) the principal factors of danger—high intensity, prolonged exposure, and the more potent sources—are known. Space did not permit a detailed consideration of specific materials. Nor was the occasion taken to expand upon practical solutions, although a number were presented. The author's primary intention has simply been to state the problem and the variables. The essential problem in guarding against harmful action by light is to recognize the materials likely to be susceptible, and to minimize the possible effects. It is hoped that the review will provide a basis for better understanding and more detailed study in the future.[16]

16. Since the preparation of this text, a review of the principles of photochemistry and the effect of light on paper, varnishes and paint vehicles has been published: Feller, R. L., *Museum News*, Technical Supplement No. 3 (June 1964). A review of the principles of photochemistry and of the problem of excluding oxygen has also been published by Thomson, G., *Museums Journal*, 64 (1964), p. 16.

LIST OF REFERENCES

1. Badger, A. E., and Oltoson, A. C., *J. Am. Ceram. Soc.*, 25 (1942), p. 104.

2. Balder, J. J., *The Discoloration of Coloured Objects under the Influence of Daylight, Incandescent Lamplight and Fluorescent Lamplight*, The Netherlands Museums Association, Leiden (1956).

3. Bateman, L., *J. Polymer Sci.*, 2 (1947), p. 1.

4. Baxter, G., Giles, C. H., and Lewington, W. J., *J. Soc. Dyers and Colourists*, 73 (1957), p. 386.

5. Brommelle, N. S., *Museums Journal*, 61 (1961), p. 169.

6. Brommelle, N. S., *Museums Journal*, 62 (1962), p. 178.

7. Brommelle, N. S., *Museums Journal*, 62 (1962), p. 337.

8. Cady, W. H., and Appel, W. D., *Am. Dyestuff Reptr.*, 18 (1929), p. 407.

9. Coleman, R. A., and Peacock, W. H., *Textile Res. J.*, 28 (1958), p. 784.

10. Cooper, B. S., *Museums Journal*, 53 (1953), p. 279.

11. Crawford, B. H., and Palmer, D. A., *Studies in Conservation*, 6 (1961), p.71.

12. Eeckhout, T. E., *Museum*, V (1952), p. 28.

13. *Encyclopedic Dictionary of Physics*, Pergamon Press, Oxford (1962).

14. Feller, R. L., *Bulletin of the American Group, IIC*, 4, No. 1 (1963), p. 10.

15. Feller, R. L., in *Recent Advances in Conservation*, Butterworths, London (1963), p. 171.

16. Genard, J., *Museum*, V (1952), p. 53.

17. Gibson, K. S., and Belknap, M. A., *J. Opt. Soc. Am.*, 40 (1950), p. 45.

18. Gresen, M., *Chimie des peintures*, 23 (mars 1960), p. 69.

19. Harmer, S. F., *Museums Journal*, 21 (1922), p. 205.

20. Harris, J. B., *Museums Journal*, 61 (1962), p. 259.

21. Harris, J. B., *Museums Journal*, 63 (1963), p. 36.

22. Harrison, L. S., in *Recent Advances in Conservation*, Butterworths, London (1963), p. 1.

23. Harrison, L. S., *Report on the Deteriorating Effects of Modern Light Sources*, The Metropolitan Museum of Art, New York (1954); *Illum. Engr.*, 48 (1954), p. 253.

24. Hedges, J. J., *J. Soc. Dyers and Colourists*, 44 (1928), p. 341.

25. Henderson, S. T., and Hodgkiss, D., "The spectral energy distribution of daylight," *Brit. J. Appl. Phys.*, 14 (1963), p. 125.

26. Howitt, F. O., *Bibliography on the Technical Literature on Silk*, Hutchinson, London (1945).

27. ICOM Commission for Lighting of Museum Objects, *Use of Fluorescent Lights in Museums*, International Council of Museums, Paris (1953).

28. Kalff, L. C., *Museum News*, 38, No. 4 (1959), p. 26.

29. Kantzer, D., *Musées et Collections Publiques de France et de l'Union Française*, Nouvelle série n° 16, n° 64 (1958), p. 145.

30. Kenzo, T., *Sci. Pap. Japn. Antiques*, 10 (1955), p. 19 (IIC Abstracts, No. 1295).

31. Kruithof, A. A., *Philips Technical Review*, 6 (1941), p. 65.

32. Kühn, H., *Museumskunde*, 1 (1961), p. 22.

33. Kühn, H., and Wolters, C., *Studies in Conservation*, 1 (1962), p. 1.

34. McCandless, S., *Museums News*, 37, No. 1 (1959), p. 10.

35. McLaren, K., *J. Soc. Dyers and Colourists*, 72 (1956), p. 86.

36. Mellon Institute, *Annual Report, Series No. 44* (1956-1957), p. 44.

37. Miller, C. D., *Ind. Eng. Chem.*, 50 (1958), p. 125.

38. Mityoshi, K., *Sci. Pap. Japn. Antiques*, 16 (1959), p.11 [In Japanese, see IIC Abstracts No. 2596].

39. National Bureau of Standards Circular 505, *Preservation of the Declaration of Independence and the Constitution of the United States* (1951).

40. National Bureau of Standards Circular 538, *Protective Display Lighting of Historical Documents* (1953).

41. Nayatani, Y., and Wyszecki, G., "Color of daylight from north sky," *J. Opt. Soc. Am.*, 53 (1963), p. 626.

42. Russell, W. J., and Abney, W. I., *Report to the Science and Art Department of the Committee of the Council of Education on the Action of Light on Water Colours*, H. M. Stationery Office, London (1888); *Journal of the Royal Soc. of Arts*, 37 (1889), p. 113.

43. Sandermann, W., and Schlumborn, F., *Holz als Rob-und Werkstoff*, 20 (1962), p. 245, 285.

44. Sommer, H., *Chem. Ztg.*, 51 (1927), p. 777; Greathouse, G. A., and Wessel, C. J., *Deterioration of Materials*, Reinhold Publishing Corp., New York (1954), p. 472.

45. Strömberg, E., *ICOM News/Nouvelles de l'Icom*, 3, No. 3 (1950), appendix.

46. *Synthetic Materials used in the Conservation of Cultural Property*, International Centre for the Study of Preservation and Restoration of Cultural Property (1964), p. 44.

47. Taylor, A. H., and Pracejus, W. C., *Illum. Engr.*, 45 (1950), p. 149.

48. Thomson, G., *Museums Journal*, 57 (1957), p. 27.

49. Thomson, G., *Studies in Conservation*, 6 (1961), p. 49.

Control of Deteriorating Effects of Light on Museum Objects: Heating Effects of Illumination by Incandescent Lamps

ROBERT L. FELLER

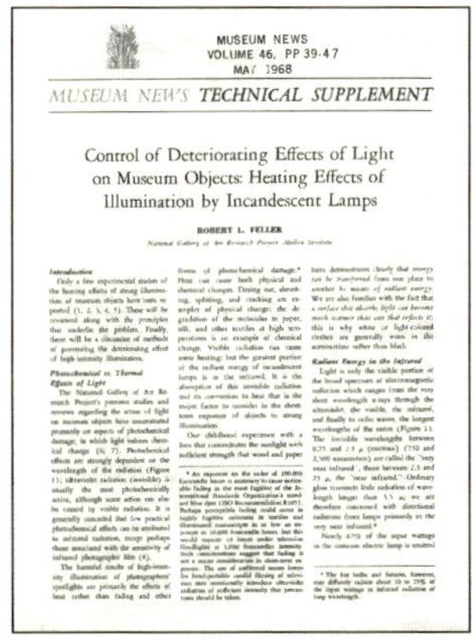
MUSEUM NEWS
VOLUME 46, PP 39-47
MUSEUM NEWS TECHNICAL SUPPLEMENT

Control of Deteriorating Effects of Light
on Museum Objects: Heating Effects of
Illumination by Incandescent Lamps

ROBERT L. FELLER

Originally printed in:
Museum News
Volume 46, Number 9 (1967), Pages 39-47.

Introduction

Only a few experimental studies of the heating effects of strong illumination of museum objects have been reported.[1-5] These will be reviewed along with the principles that underlie the problem. Finally, there will be a discussion of methods of preventing the deteriorating effect of high-intensity illumination.

Photochemical vs. Thermal Effects of Light

The National Gallery of Art Research Project's previous studies and reviews regarding the action of light on museum objects have concentrated primarily on aspects of photochemical damage, in which light induces chemical change.[6, 7] Photochemical effects are strongly dependent on the wavelength of the radiation (Figure 1); ultraviolet radiation (invisible) is usually the most photochemically active, although some action can also be caused by visible radiation. it is generally conceded that few practical photochemical effects can be attributed to infrared radiation, except perhaps those associated with the sensitivity of infrared photographic film.[8]

The harmful results of high-intensity illumination of photographers' spotlights are primarily the effects of heat rather than fading and other forms of photochemical damage.[9] Heat can cause both physical and chemical changes. Drying out, shrinking, splitting, and cracking are examples of physical change; the degradation of the molecules in paper, silk, and other textiles at high temperatures is an example of chemical change. Visible radiation can cause

1. Museum of Fine Arts (Boston) 1955 *Annual Report*, p. 43-44; see also Hartford, N. D., "Are You an 'Armchair Art Lover'?" *The Wheaton Alumnae Quarterly*, 36 (Nov. 4, 1957), p. 3-7; see IIC Abstract 1533.

2. Mityoshi, K., "An Experiment on the Influence of Strong Illumination on Some Mineral Pigments," *Sci. Papers Japan. Antiques*, 16 (1959), p. 11-18 [in Japanese; see IIC Abstract No. 2596].

3. Kollmann, F., Schneider, A., and Teichgräber, R., "Investigations on the Heating-up, Warping and Drying of Wooden Picture Panels Illuminated with High Intensities," *Holz Als Roh- und Werkstoff*, 19, No. 2 (1961), p. 4-47 [in German].

some heating; but the greatest portion of the radiant energy of incandescent lamps is in the infrared. It is the absorption of this invisible radiation and its conversion to heat that is the major factor to consider in the short-term exposure of objects to strong illumination.

Our childhood experience with a lens that concentrates the sunlight with sufficient strength that wood and paper burn demonstrates clearly that *energy can be transferred* from one place to another *by means of radiant energy*. We are also familiar with the fact that *a surface that absorbs light can become much warmer than one that reflects it*; this is why white or light-colored clothes are generally worn in the summertime rather than black.

RADIANT ENERGY IN THE INFRARED

Light is only the visible portion of the broad spectrum of electromagnetic radiation which ranges from the very short wavelength x-rays through the ultraviolet, the visible, the infrared, and finally to radio waves, the longest wavelengths of the series (Figure 1). The invisible wavelengths between 0.75 and 2.5μ (microns) (750 and 2,500 nanometers) are called the "very near infrared"; those between 2.5 and 25μ, the "near-infrared." Ordinary glass transmits little radiation of wavelength longer than 3.5μ ; we are therefore concerned with directional radiation from lamps primarily in the very near infrared.[10]

Nearly 67% of the input wattage in the common electric lamp is emitted as infrared radiation. It is the absorption of this invisible radiant energy by the illuminated object that contributes so much to the heating effect of an incandescent lamp. The infrared being invisible, much of it can be eliminated without reducing the amount of visible radiation needed to display and to photograph objects.[11]

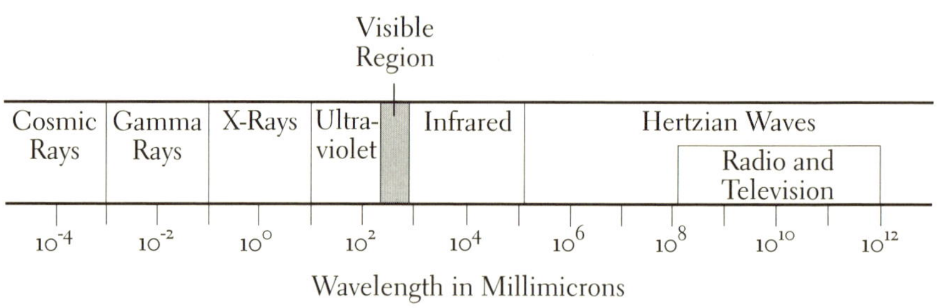

FIGURE 1
The electromagnetic spectrum.

BLACKBODY RADIATION AND COLOR TEMPERATURE

The typical electric light bulb is an example of an *incandescent* source; the light is generated by a tungsten wire heated to such a high temperature that it

4. Kühn, H., and Wolters, C., "Behaviour of Painted Wood Panels under Strong Illumination," *Studies in Conservation*, 7 (1962), p. 1-9.

5. Kühn, H., "The Use of Heat Protection Filters When Works of Art Are Filmed or Televised," *Studies in Conservation*, 12 (1967), p. 102-115.

6. Feller, R. L., "Control of Deteriorating Effects of Light upon Museum Objects," *Museum*, 17 (1964), p. 57-98.

7. Feller, R. L, "The Deteriorating Effect of Light on Museum Objects," *Museum News*, Technical Supplement No. 3 (June 1964).

8. Kistiakowsky, G. B., *Photochemical Processes*, The Chemical Catalog Company, New York (1928), p. 239.

9. An exposure on the order of 100,000 footcandle hours is necessary to cause noticeable fading in the most fugitive of the International Standards Organization's standard blue dyes (ISO Recommendation R105). Perhaps perceptible fading could occur in highly fugitive colorants in textiles and illuminated manuscripts in as low an exposure as 10,000 footcandle hours, but this would require 10 hours under television floodlights at 1,000 footcandles intensity. Such considerations suggest that fading is not a major consideration in short-term exposure. The use of unfiltered xenon lamps for hand-portable candid filming of television may occasionally introduce ultraviolet radiation of sufficient intensity that precautions should be taken.

10. The hot bulbs and fixtures, however, may diffusely radiate about 10 to 25% of the input wattage as infrared radiation of long wavelength.

glows. The manner in which the energy is emitted by such a hot wire is well understood and involves the physics of "blackbody radiation."

A "blackbody" is simply a radiating material that is, one might say, perfectly black, one that absorbs and emits all wavelengths of electromagnetic radiation with equal efficiency. The wall of a heated furnace will emit radiation close to that of the ideal "blackbody." As the temperature is raised, the radiation emitted by the hot furnace wall varies in an orderly manner (Figure 2); and the light that one would see through a small port in the wall would pass through red heat, yellow, white, and finally blue. This may be better understood by reference to Figure 3, which emphasizes the manner in which the energy in the visible region in Figure 2 changes with temperature.

Figure 4 compares the radiation emitted by three low-temperature sources. The radiation in sunlight reaching the earth is typical of an incandescent source, except that components of the earth's atmosphere absorb certain of the wavelengths more than others (Figure 5).[12]

It is customary in discussions of light sources to refer to the "color temperature" of the illumination. By this, it is meant that the distribution of the visible energy from the light source closely resembles the distribution of energy emitted by a blackbody radiator at the stated temperature, that is, close to some distribution of energy as shown in Figure 3.

The energy emitted by an incandescent lamp is closely equivalent to that of a blackbody radiator in the temperature range of 2400 to 3400° K.[13] The spectral distribution of the radiation varies somewhat with the filament temperature but most of the energy emitted is in the infrared (Figure 6).[12]

Figure 3 is a key diagram in the discussion of color and color temperature. White light is radiation made up of all visible wavelengths at practically equal intensity. Colored light has an excess of radiation at certain wavelengths more than others. Blue light (for example, radiation of color temperature 7000° K) has more of the shorter wavelengths present; red or yellow light would have more of the longer wavelengths of visible radiation. From the various illustrations, it can be seen that the color of radiation from the usual incandescent lamp is yellowish.[14]

For an object to be colored, it must absorb certain wavelengths in the visible light more strongly than others; as a result, the light reflected will contain more of certain wavelengths than others. An example can be seen in Figure 7; gold is yellow because it reflects less of the shorter (blue) wavelengths. On the other hand, steel is gray in appearance because it does not reflect the light very strongly in any one region of the spectrum more than another.

Effect of Selective Absorption

Kollmann and co-workers have measured the difference in temperature of dark, medium, and light areas of a panel painting and have shown that the dark

11. With sensitive objects, consider temperature if lamps are inside showcases. Consider temperature too if lamps outside the case are at intensities of above 12 footcandles or if lamps on objects in freely circulating air are above 30 footcandles.

12. *I.E.S. Lighting Handbook*, 4th Edn., Illuminating Engineering Society, New York (1966).

13. The physicists' Kelvin temperature scale is used in referring to blackbody radiators. This is the temperature in degrees centigrade added to 273°.

14. Particular attention must be paid to the color temperature of the light falling on a scene to be photographed in color, for films and television cameras vary in their response to the kind of light being used to illuminate the scene. For faithful color rendition, the response of the film must be carefully balanced against the character of the illumination. Under ordinary lamps, the film or video camera must be especially sensitive to blue and green. Filters used to eliminate heat may occasionally affect the visible illumination and, hence, pose problems in color photography.[5]

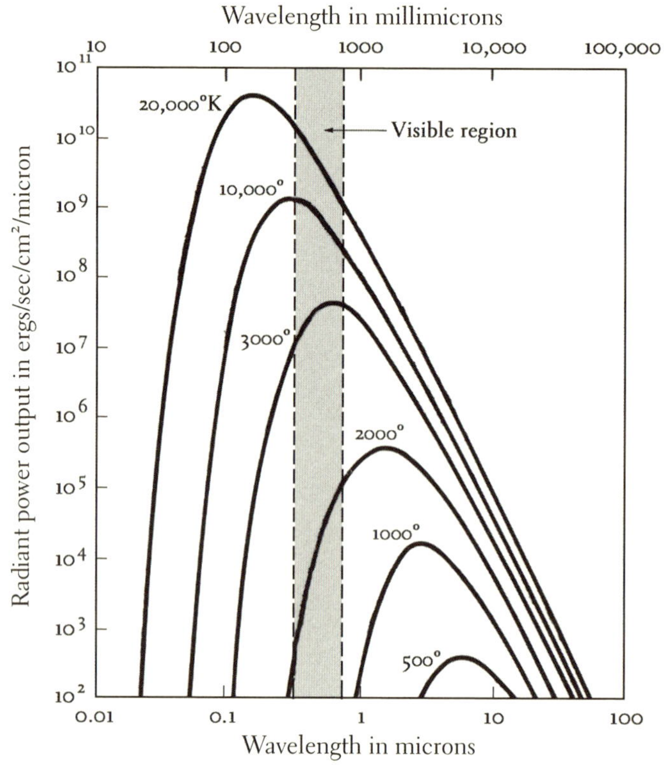

Figure 2

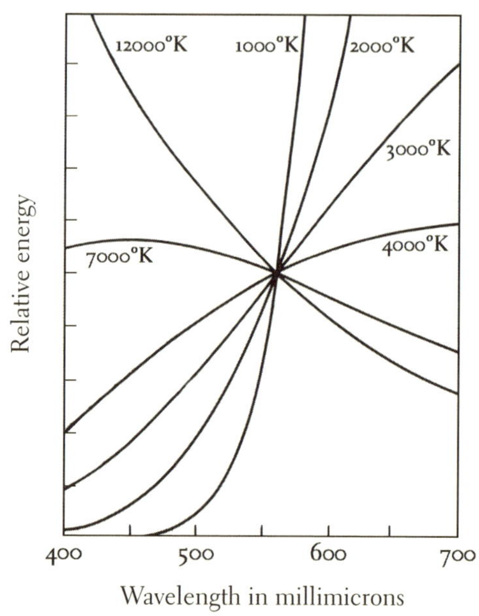

Figure 3
Relative spectral energy distributions of blackbody radiator at various temperatures.

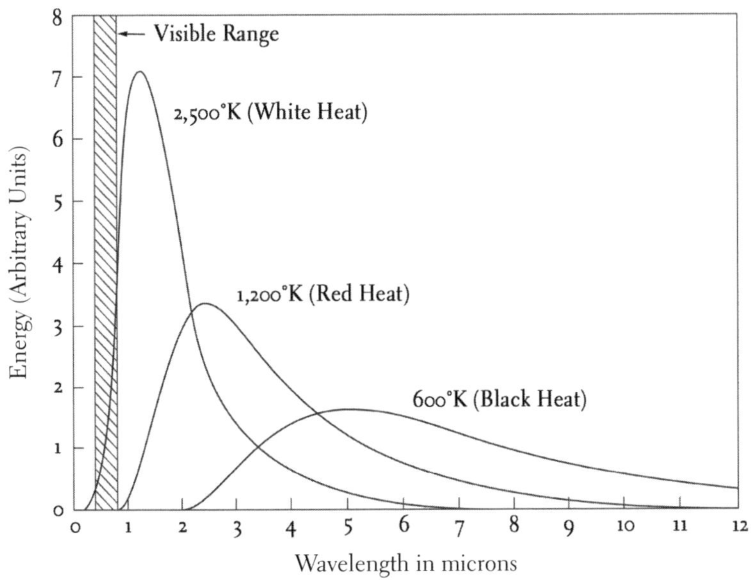

Figure 4

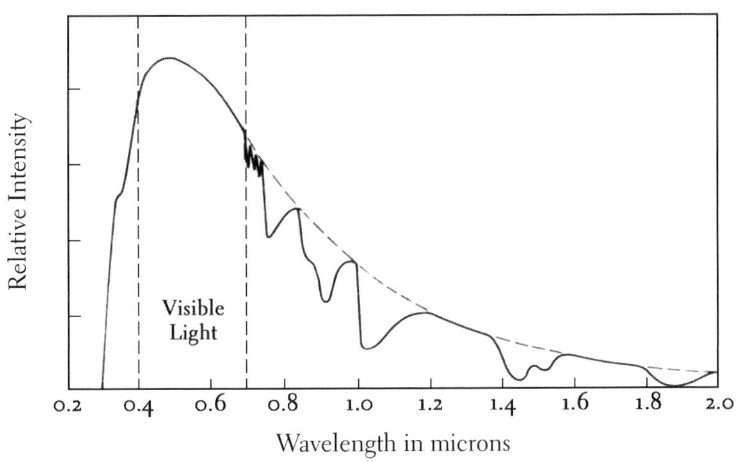

Figure 5
Distribution of radiant energy in sunlight.

areas are heated to a temperature approximately 30% higher than the light-colored areas.[3] An area covered with gold leaf reflects the infrared radiation and may not heat up as rapidly as other parts.

The way an object absorbs visible light does not necessarily indicate how it will absorb or reflect infrared radiation.[15] Figure 8 shows that the heating effect of sunlight on various painted panels is often closely related to the amount of visible radiation that is absorbed.[16] On the other hand if a black and white painted panel is illuminated by an infrared heater which emits primarily the longer wavelengths of radiation, the difference in heating rate between black and white panels will be very little.[17, 18] The reason for this is that in the 8μ region of the infrared, both the black and the white paints absorb the radiation practically to the same degree, similar to the case of the gold and steel illustrated in Figure 7.

The full and detailed study of the infrared heating of objects must take the factor of selective absorption into consideration. The difference in the absorption in the invisible wavelengths in contrast to the visible may perhaps be important in the illumination of wood. Figure 9 shows that wood varies considerably in its ability to absorb infrared radiation.[17] This may become important in the illumination of objects of inlaid wood if certain pieces heat up more than others or allow radiation to penetrate to a sufficient depth that the adhesive is affected.

15. The differences in reflectivity of objects in the infrared, as opposed to the visible region, are made use of in infrared photography for the examination of art objects and the detection of camouflage.

16. The effect of sunlight in raising the temperature of variously painted surfaces has been briefly reviewed elsewhere [Feller, R. L., *Bulletin of the American Group-IIC*, 2, No. 1 (1961), p. 15-18; *ibid.*, 2, No. 2 (1962), p. 11-12].

17. Summer, W., *Ultra-Violet and Infra-Red Engineering*, Sir Isaac Pitman and Sons, London (1962).

18. Strain, R. N. C., "Solar Reflectivity of Paints," *J. Oil and Colour Chemists' Assoc.*, 44 (1961), p. 689-712.

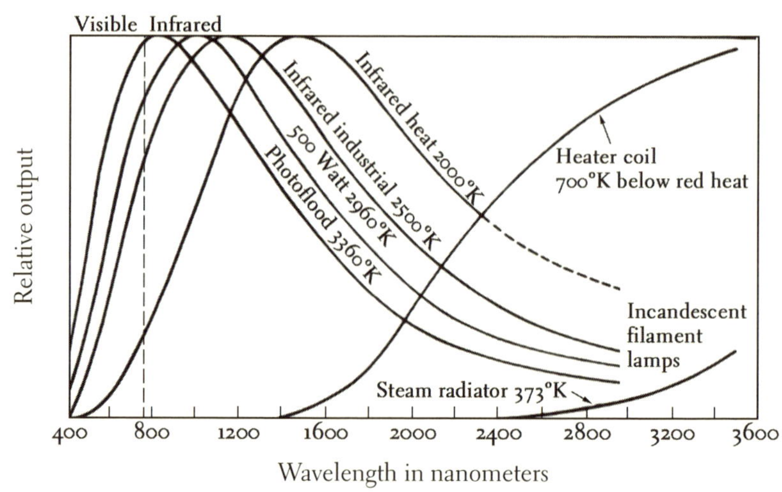

FIGURE 6

Greenhouse Effect

One of the most important practical problems associated with selective absorption and transmission of infrared radiation is the so-called greenhouse effect: when sunlight strikes objects beneath glass, they are warmed and at a low temperature give off blackbody radiation in longer wavelengths than can be transmitted by the glass (longest wavelength transmitted: about 3.5μ). As a result, the interior attains a higher temperature than it would if the radiation could escape. The earth's atmosphere similarly traps the heat of the sunlight which has been absorbed by the earth.

This same effect takes place when lamps generate heat in a museum showcase. For this reason, recommendations have been made to keep lamp fixtures either outside the case or in an especially ventilated upper compartment.[19] At the National Gallery of Art, special cases have been designed which use infrared-reflecting glass to isolate the lamp compartment at the top. In addition, dichroic-reflector lamps are employed and the cases air-conditioned.

The heating effect of incandescent lamps has been one of the major reasons why fluorescent lamps have been recommended for showcases, since they give off a lower amount of heat for the same amount of light.[20] As an added refinement, the electrical ballasts are often placed outside the case to reduce even further the heat that would be generated inside.

19. Padfield, T., "The Design of Museum Show Cases," *London Conference on Museum Climatology, Preprints*, International Institute for Conservation of Historic and Artistic Works, London (1967).

20. Olson, G., "Lighting Methods for Show Cases, The Technical Principles," *Museum*, 4 (1951), p. 203-205.

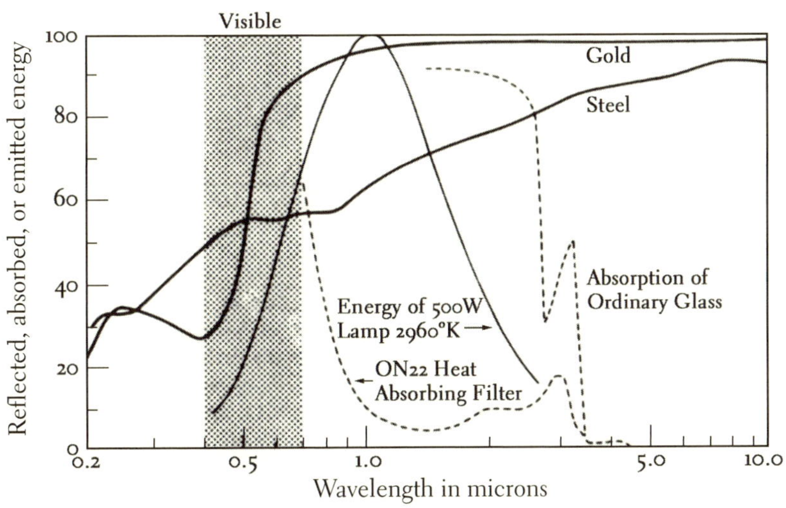

Figure 7

Surface	Temperature of Panel
Black	212°F
Unpainted Iron	156°F
Aluminum (Paint)	144°F
Yellow	106°F
White	82°F

Figure 8
Rise in temperature of painted panels in sunlight. [From Cottony and Dill, *Bureau of Standards Materials and Structures Report* No. 64 (1941), p. 1]

English Name	Thickness, mm
Larch	5-7
White Fir	6-7
Spruce	6
Pine	3-4
Poplar, Maple	4
Oak	2
White Walnut	0.2
Teak, Ebony	0?

Figure 9
Thickness of wood permeable to infrared radiation. [From Summer, W., *Ultra-Violet and Infra-Red Engineering*, Sir Isaac Pitman and Sons, London (1962)]

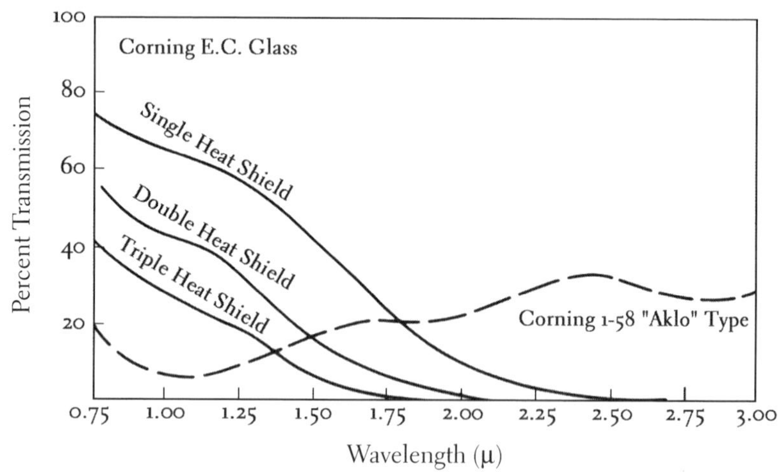

Figure 10
Characteristics of infrared-reflecting glass.

An illuminated object framed and sealed under glass may occasionally build up sufficient temperature that special ventilation of the exterior or interior of the frame will be advisable.

IRR Glass and Dichroic-Reflector Lamps

Surface-treated glasses constitute a special topic concerning selective absorption and transmission. A metal oxide-coated type transmits the visible wavelengths and reflects the infrared. The transmission curve of one such product, the Corning Electrically-Conducting (EC), Infrared Reflecting (IRR) glass, is given in Figure 10. It can be seen that this type of filter is most efficient in the longer wavelengths and that its effectiveness can be improved by employing several

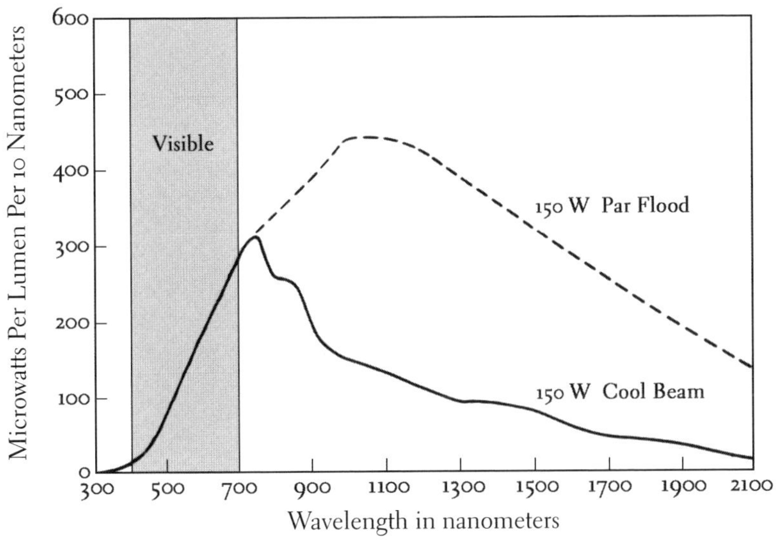

FIGURE 11
Approximate initial spectral energy distribution. (Data courtesy of General Electric Company)

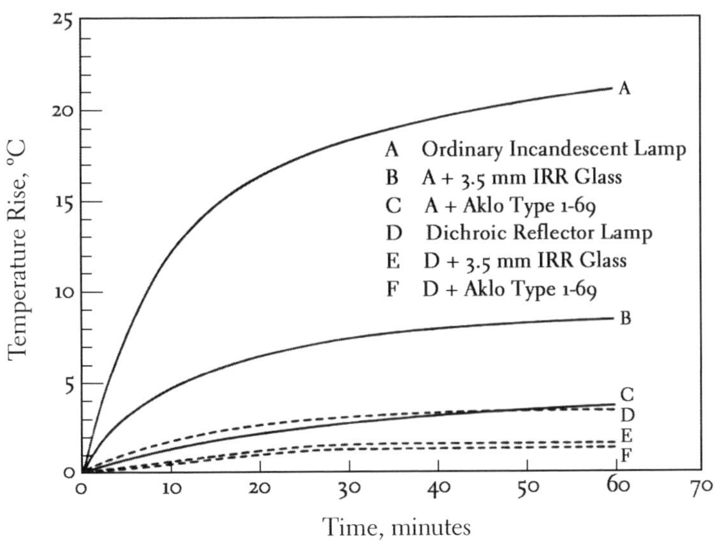

FIGURE 12
Temperature rise of a black aluminum panel at 200 footcandles in a glass-enclosed case.

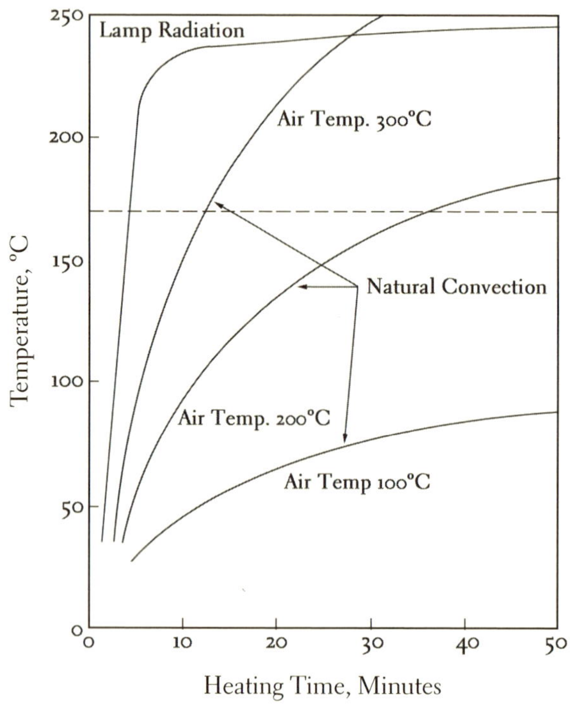

FIGURE 13
Temperature build-up by convection vs. radiant transfer. (from Ref. 17)

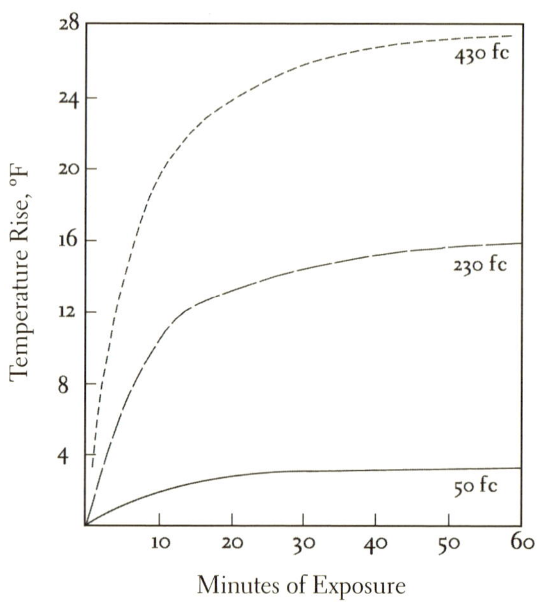

FIGURE 14
Temperature increase of blackened pine panel 0.9 x 55 x 42 cm. (from Ref. 3)

layers of the glass. The Kliegl Brothers' "No Heet" filter for theatrical spotlights is of the IRR type and is said not to affect the color temperature of the light significantly.[21]

Spotlights are also available that have especially coated reflectors which permit much of the heat to be transmitted through the rear of the lamp.[22] The spectral emission of one such *dichroic-reflector* lamp, the General Electric Cool-Beam®, is shown in Figure 11.[23, 24]

The effect of IRR glass in reducing the temperature buildup both by an ordinary lamp and by a dichroic reflector-type spotlight can be seen in Figure 12. The effectiveness of a pale green Aklo-type glass, Corning 1-69, is also shown. Kühn has published similar data accompanied by the IR transmission characteristics of various filters and measurements of their effect on the color temperature of the illumination.[5]

21. More, H., Kliegl Brothers Universal Electric State Lighting Company, private communication.

22. Carlson, F. E., Howard, G. T., Turner, A. F., and Schroeder, H. H., "Temperature Reduction on Motion-Picture and Television Sets Using Heat-Controlling Coatings," *J. Soc. Motion Picture and Television Engineers*, 65 (1956), p. 136-139.

23. Beesley, E. M., Makulec, A., and Schroeder, H. H., "Lamps with Multilayer Interference-Film Reflectors," *Illum. Engineering*, 53 (1963), p. 380-385.

24. I am indebted to Mr. H. G. Williams, of the General Electric Company, Nela Park, Cleveland, Ohio, for these data and for calling my attention to footnote 23.

Transfer of Energy

The transfer of energy by radiation is particularly efficient. Figure 13 shows that radiant heating can increase the temperature of an object much faster than can be done by simply heating up the air around the object (the latter process is known as convection heating).[17] It must be kept in mind, then, that heat buildup in objects under strong illumination can be far more rapid than the buildup of the temperature of the air in the room. The data in Figures 12 and 14 show that the major temperature rise takes place in about 20 minutes.[3-5] The rapidity of heating can be important, because objects often suffer greater damage if the changes in temperature and humidity take place rapidly.

If the radiant energy falling on an object is entirely converted to heat, the amount converted at a given time is proportional to the intensity (I) and the overall absorption (A).[25] This absorbed energy, kIAdt, is shared between the heat gained by the object, cDLdT, and the heat lost by convection, $h(T - T_a)dt$.[26] The rate of change of temperature at any stage in the heating up process, therefore, is given by the equation

$$\frac{dT}{dt} = \frac{kAI}{cDL} - \frac{h(T - T_a)}{cDL} \qquad (1)$$

The temperature will reach a maximum value when the heat being absorbed by an object is balanced by that being lost to the surroundings. Since the rate of change in temperature

$$\left(\frac{dT}{dt}\right) \qquad (2)$$

25. Tiller, F. M., and Garber, H. J., "Infrared Radiant Heating," *Ind. Eng. Chem.*, 34 (1942), p. 773-781; *ibid.*, 42 (1950), p. 456-463.

26. $cDLdT = kAIdt - h(T - T_a)dt$, where c is the heat capacity; D, the density; L, the thickness of the object; dT, the change in temperature; h, the coefficient of convection heat loss; k, a proportionality constant; T, the temperature of the object; T_a, the ambient temperature; and dt, the time interval during which the temperature changes by the amount dT.

at this point is zero, the equation can readily be solved to yield

$$T_{max} = T_a + \frac{kAI}{h} \qquad (3)$$

This important equation shows that the maximum attainable temperature is directly affected by the intensity of the radiation (I), the absorption (A), and the temperature of the ambient air (T_a); but it is independent of the heat capacity,[27] density, and thickness of the object (except as they may affect h). The maximum temperatures in Figure 14, for example, are closely related to the intensity of illumination. Forced circulation of air past the object will increase the heat loss to the surroundings (increases h), but the equation states that an irradiated object always will be at a temperature above the surrounding air.

27. The heat capacity of an object may be said to be the number of calories that must be absorbed in order to raise its temperature 1°C. One calorie is required to raise 1 g of water 1°C; most materials have lower specific heats (calories per gram per degree).

The equation demonstrates that a higher maximum temperature will be attained by the illuminated object if the temperature of the air in a room or showcase increases. In order to maintain the temperature of an irradiated object at room temperature or below, the temperature of the ambient air obviously must be lowered. This was done in the special showcases mentioned above which were designed by Mr. Sterling Eagleton at the National Gallery of Art to exhibit wax objects. To reduce the heating effect even further, dichroic filter lamps were placed in an air-cooled duct above the case and the illumination passed through a filter of IRR and diffusing glass. The lights do not turn on until the air temperature is below 65°F; and for safety, they turn off and sound an alarm if it gets above 69°F. The curves in Figure 15 show the changes in the temperature at various levels inside a test object in these showcases. It may be noted that the temperature of the object remained above that of the surrounding air and that the temperature near the outer surface rose about 3° when the lights were turned on. This is the direct result of the absorption of the radiation.

The *rate* of temperature rise is dependent on the heat capacity, thermal conductivity, and thickness of the object, even though the maximum attainable temperature is not. Tiller and Garber give equations which relate these properties to the temperature at any given time.[25] They point out that a plot of the object temperature vs. the difference in temperature for a fixed interval of time should give a straight line, the slope of which is equal to

$$\frac{-cDL}{h} \qquad (4)$$

When this is done with the experimental data in Figure 16, it can be seen that many of the materials exhibit similar values of

$$\frac{cDL}{h} \qquad (5)$$

(the same slope) in the initial stages of heating. The distinctly higher rate of heating (lower slope) of the lead plate, however, may have been due to the fact

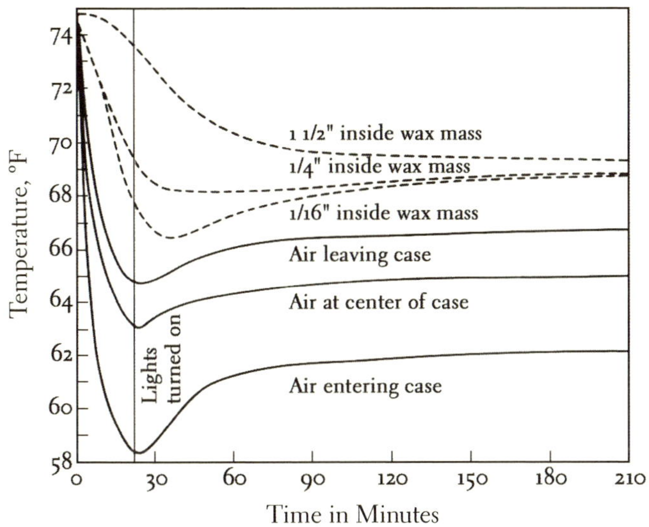

Figure 15
Temperature of experimental object of wax in air-conditioned showcase.

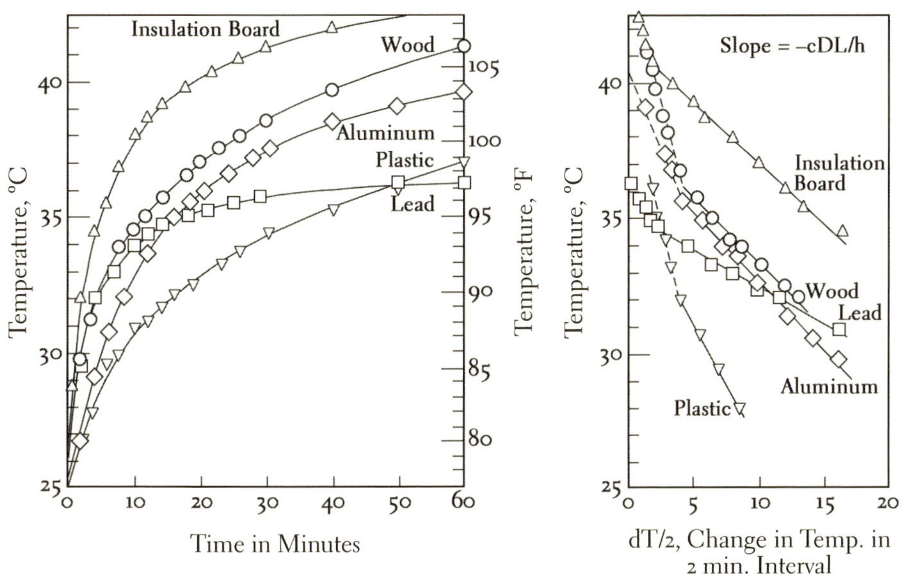

Note: Rectification of data in accord with the equation

$$T = T_m - \frac{cDL}{h} \times \frac{dT}{dt} \qquad (6)$$

Figure 16
Temperature rise, 200g black panels of various materials at 200 footcandles intensity incandescent lamplight in a glass-enclosed case.

that its heat capacity is only about one-tenth that of the other materials. A break occurs in the curves after about 20 minutes (when the temperature difference, dT, becomes small). This suggests a change in the ease of heat loss from the object (a change in h), perhaps owing to the increased temperature of the walls and the ambient air within the experimental glass chamber.

The absorption of radiation (or, to put it another way, the reflectance of the surface) apparently is the principal property of the object that influences the rise in temperature. Further tests will be necessary to determine if any museum objects heat up at a significantly different rate owing to differences in heat capacity. Thermal conductivity can become an important factor in thick materials that have poor conductivity, causing the surface to heat up more rapidly than the interior. In the experiments illustrated in Figure 16, higher surface temperatures were attained with insulation board and wood than with the samples of metal. Tiller and Garber have presented data on the variation of temperature within a 0.75-inch-thick block of oak and the effect of the thickness of metal stock on the rate of heating.[25]

POSSIBLE DAMAGE

Objects that are sensitive to changes in humidity and temperature, particularly those that markedly change dimension, are most likely to be harmed by exposure to high-intensity illumination.[28] Among the most critical are objects of wood, especially paintings on wood and polychrome sculpture, which are likely to develop flakes and cracks in the paint. Serious problems can also be encountered with musical instruments. The possibility of problems with inlaid wood or veneered furniture has already been mentioned. One might also encounter shrinkage and cracking where metal and ivory are combined with wood, primarily because of the difference in expansion and contraction between the various materials. Barnard has called attention to an object in which the inlays were held in place by wax.[29]

Other objects particularly sensitive to changes in humidity are parchment, leather, and paintings on canvas. Besides being sensitive to changes in humidity, paper and textiles, particularly silk, can become permanently embrittled owing to the deteriorating effects of temperature. Ivory will crack.[5]

Kühn has recently published data to show that a rise in temperature of 10°C (18°F) will increase the fading of a number of fugitive lake pigments by a factor of about 1.3.[30]

The question is often asked whether the effects of exposure are accumulative. The harmful effects of short-term exposures to the heat of lamps are primarily because of the movement that is induced. To the extent that the dimensions of the object return to the original state when it is cooled down again, there may be relatively little accumulative effect in one cycle of change; on the other hand, when an object is subjected to a number of cycles of heating and cooling

28. Materials affected by moisture are usually sensitive to the relative humidity of the air, which is a function of the temperature of the air. For a more detailed discussion see footnote 6 and *Museum*, 8, No. 4 (1960), p. 202-289.

29. Barnard, P., "Problems and Techniques in Producing Museum Programs (Television)," *Museum News*, 45 (March 1967), p. 35-37.

30. Kühn, H., "The Effect of Oxygen, Relative Humidity, and Temperature on the Fading Rate of Watercolors," *London Conference on Museum Climatology, Preprints*, International Institute for Conservation of Historic and Artistic Works, London (1967), p. 79-85.

then the effects induced by repeated changes in dimension will in a sense be accumulative. The embrittlement of paper and textiles induced by heat, owing to the deterioration of the polymer molecules, is obviously accumulative.

Minimization of Possible Damage During Filming

When a museum prepares to exhibit objects on television, the staff conservator should be consulted regarding the general safety of the objects and the particular circumstances of exposure. He is the person ultimately responsible for the conditions under which objects are exposed, whether for exhibition, photography, or television. The National Gallery of Canada has a special form to be signed by the person responsible for the photography, giving precautionary measures for the operation of temporary lighting equipment used for still photography, motion pictures, and television.[31] The form states that it is the responsibility of the camera operator to ensure that the lamps do not increase the surface temperature of the painting more than 5°C and that the maximum interval for illumination is no more than 15 minutes.[32]

The use of infrared filters is one of the simplest and most direct steps that can be recommended to reduce the heating effect of lamps. As previously mentioned, there are a number of lamps on the market which are constructed with special lenses, reflectors, and filters to reflect much of the heat to the back and side. These are perhaps not widely used as yet in showcases, photographic studios, and the usual television stage. For the special purposes of museum work, however, the utility of such lamps should be fully considered. The simplest precaution is to place an infrared filter in front of the lamps, although attention must be given to the effect of the filter on color temperature if the illumination is to be used for color photography.[5] Electrically conducting IRR glass should have a negligible effect on color temperature.

In special problems, a current of air may be circulated past an object to increase the conduction of heat away from it. The data in Table 1 show that a forced draft will reduce the rise in temperature of a panel painting to less than one-half that without forced draft.[3] If the object is to remain at room temperature, it will be necessary to cool the ambient air as well as to circulate it.

With especially sensitive objects, as low an intensity as possible should be used. In the past, color television often has required higher intensities of light than black and white. There are now video cameras in the industry which permit the use of intensities of light as low as 100 footcandles, however.[12]

The utilization of substitutes and props on television in place of especially sensitive objects should not be overlooked. Photographs, drawings, reproductions, and diagrams of objects can sometimes be used to advantage with far less risk than showing the actual object.

31. Stolow, N., National Conservation Research Laboratory, The National Gallery of Canada, Ottawa, private communication.

32. When television or photographers' spots are being placed in position, they are often operated at a low intensity and only brought up to full intensity and color temperature just before the actual "shooting" begins. This does much to minimize the heating effects of the illumination.

TABLE 1
Influence of Forced-Draft Cooling of Painting on Panel Illuminated 60 Minutes at 1000 Footcandles (Ref. 3)

Air Circulation	Free	Forced-Draft	
Velocity of Air Flow, meters/sec	About 0.2	1.7	3.2
Temperature Increase, °C (°F)			
Darkest Area	27 (55)	14.1 (26.8)	12.1 (23)
Lightest Area	19.6 (37.3)	9.0 (17.11)	8.0 (15.2)

The use of fluorescent lamps, infrared-reflecting spotlights and filters, and special ventilation in showcases has been mentioned.

MEASUREMENT OF TEMPERATURE OF OBJECTS

In closing, a word may be said about instruments that can be used to measure the temperature of objects on display. One of the most interesting is the Barnes Engineering Company's IT3 Infrared Sensing Thermometer, shown in Figure 17. It costs about $2,000. This instrument is simply pointed at an object and its temperature is measured by the infrared radiation that it gives off. (If an object should be behind glass, however, the instrument will measure the temperature of the glass.)

The conventional instruments for measuring the temperature of surfaces and specific areas of objects involve thermocouple wires or thermistors attached to a potentiometer. Thermocouples consist simply of wires of two different metals joined together in a tiny bead-like junction which is attached to the place to be measured (Figure 18). The junction produces an amount of electric current in proportion to the temperature. The data reported in Figures 12 and 14 were obtained with thermocouples.

A simple arrangement is to have two matched thermometers mounted side-by-side, one inserted in a close-fitting hole in a block of metal painted white (about 2" x 2" x 5/16") and the other inserted in a similar block painted flat black; a drop of melted wax or a liquid in the holes will ensure better thermal contact. A difference in temperature between the two will largely be due to the greater absorption of radiation by the black.

Such measurements cannot be made effectively without the operator having some training and experience with the instruments, but the staff consultant on conservation can arrange to have measurements taken or to take them himself if he is called upon to assist.

Conclusion

High levels of illumination by incandescent lamps can induce rapid and uneven heating of objects. It is primarily objects of wood, ivory, silk, parchment, leather, and paper that require the greatest attention. With the aid of infrared filters, forced air ventilation of objects, and cooling of the air, it is possible to maintain the temperature of objects within a few degrees of room temperature. In all cases, the advice of a conservator should be sought concerning the sensitivity of objects to heat, their general condition, and the suitability of studio and display arrangements for their safety.

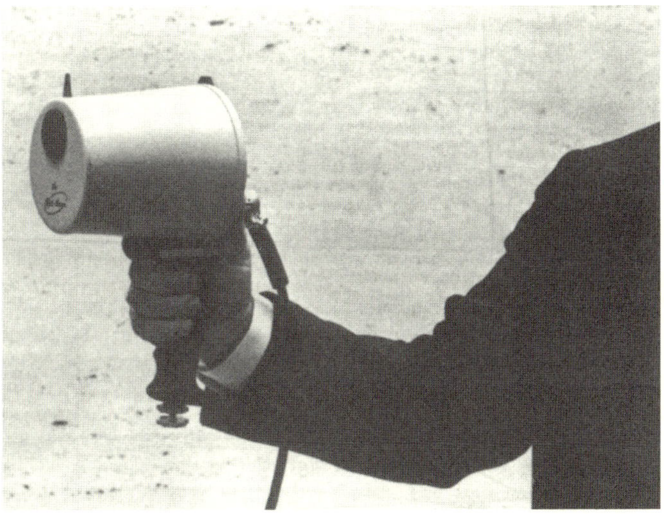

Figure 17

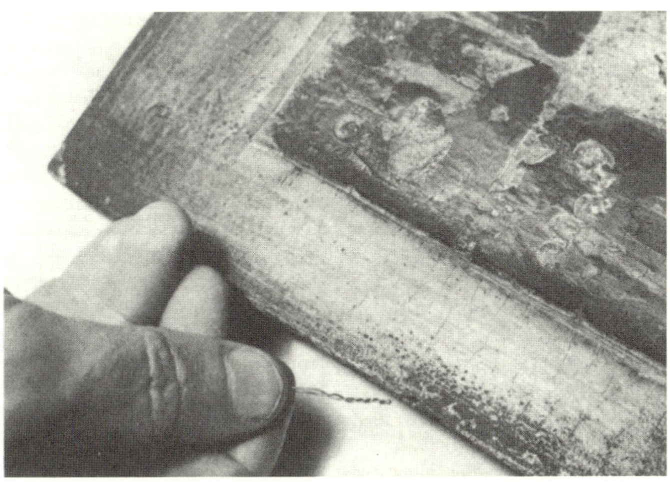

Figure 18

Speeding Up Photochemical Deterioration

ROBERT L. FELLER

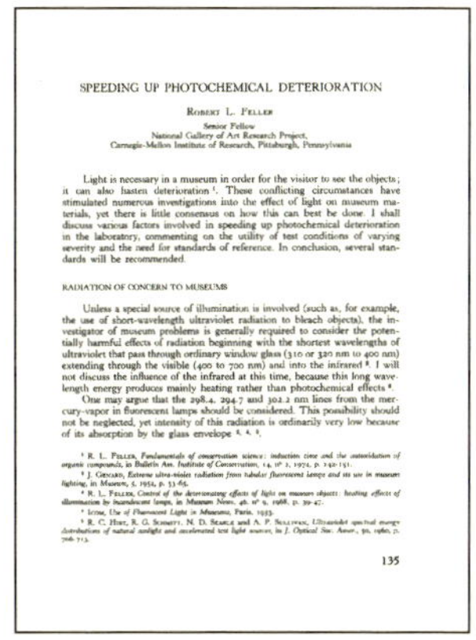

SPEEDING UP PHOTOCHEMICAL DETERIORATION

ROBERT L. FELLER

Senior Fellow
National Gallery of Art Research Project,
Carnegie-Mellon Institute of Research, Pittsburgh, Pennsylvania

Light is necessary in a museum in order for the visitor to see the objects; it can also hasten deterioration [1]. These conflicting circumstances have stimulated numerous investigations into the effect of light on museum materials, yet there is little consensus on how this can best be done. I shall discuss various factors involved in speeding up photochemical deterioration in the laboratory, commenting on the utility of test conditions of varying severity and the need for standards of reference. In conclusion, several standards will be recommended.

RADIATION OF CONCERN TO MUSEUMS

Unless a special source of illumination is involved (such as, for example, the use of short-wavelength ultraviolet radiation to bleach objects), the investigator of museum problems is generally required to consider the potentially harmful effects of radiation beginning with the shortest wavelengths of ultraviolet that pass through ordinary window glass (310 or 320 nm to 400 nm) extending through the visible (400 to 700 nm) and into the infrared [2]. I will not discuss the influence of the infrared at this time, because this long wavelength energy produces mainly heating rather than photochemical effects [3].

One may argue that the 298.4, 294.7 and 302.2 nm lines from the mercury-vapor in fluorescent lamps should be considered. This possibility should not be neglected, yet intensity of this radiation is ordinarily very low because of its absorption by the glass envelope [2, 4, 5].

[1] R. L. Feller, *Fundamentals of conservation science: induction time and the autoxidation of organic compounds*, in *Bulletin Am. Institute of Conservation*, 14, nº 2, 1974, p. 142-151.

[2] J. Genard, *Extreme ultra-violet radiation from tubular fluorescent lamps and its use in museum lighting*, in *Museum*, 5, 1952, p. 53-65.

[3] R. L. Feller, *Control of the deteriorating effects of light on museum objects: heating effects of illumination by incandescent lamps*, in *Museum News*, 46, nº 9, 1968, p. 39-47.

[4] Icom, *Use of Fluorescent Light in Museums*, Paris, 1953.

[5] R. C. Hirt, R. G. Schmitt, N. D. Searle and A. P. Sullivan, *Ultraviolet spectral energy distributions of natural sunlight and accelerated test light sources*, in *J. Optical Soc. Amer.*, 50, 1960, p. 706-713.

135

Originally printed in:
BULLETIN DE L'INSTITUT ROYAL DU PATRIMOINE ARTISTIQUE
Volume 15 (1975), Pages 135-150.

Light is necessary in a museum in order for the visitor to see the objects; it can also hasten deterioration.[1] These conflicting circumstances have stimulated numerous investigations into the effect of light on materials, yet there is little consensus on how this can best be done. I shall discuss various factors involved in speeding up photochemical deterioration in the laboratory, commenting on the utility of test conditions of varying severity and the need for standards of reference. In conclusion, several standards will be recommended.

Radiation of Concern to Museums

Unless a special source of illumination is involved (such as, for example, the use of short-wavelength ultraviolet radiation to bleach objects), the investigator of museum problems is generally required to consider the potentially harmful effects of radiation beginning with the shortest wavelengths of ultraviolet that pass through ordinary window glass (310 or 320 nm to 400 nm) extending through the visible (400 to 700 nm) and into the infrared.[2] I will not discuss the influence of the infrared at this time, because this long wavelength energy produces mainly heating rather than photochemical effects.[3]

One may argue that the 298.4, 294.7 and 302.2 nm lines from the mercury-vapor in fluorescent lamps should be considered. This possibility should not be neglected, yet intensity of this radiation is ordinarily very low because of its absorption by the glass envelope.[2, 4, 5]

1. Feller, R.L., "Fundamentals of conservation science: induction time and the autoxidation of organic compounds," *Bulletin Am. Institute of Conservation*, 14, No. 2 (1974), p. 142-151.

2. Genard, J., "Extreme ultra-violet radiation from tubular fluorescent lamps and its use in museum lighting," *Museum*, 5 (1952), p. 53-65.

3. Feller, R.L., "Control of the deteriorating effects of light on museum objects: heating effects of illumination by incandescent lamps," *Museum News*, 46, No. 9 (1968), p. 39-47.

MEANS OF SPEEDING UP THE EFFECTS OF LIGHT

The following means can be employed to speed up the effects of light: (a) increase the intensity of illumination, (b) apply the radiation continuously, in contrast to the rise and fall of natural illumination, (c) increase the intensity of the near-ultraviolet over the amount normally encountered, (d) increase the energy of the ultraviolet radiation, employing wavelengths shorter than 310 nm (I disapprove of this method and will point out the reasons), (e) increase the temperature of the samples, (f) increase the moisture content in the samples and, (g) increase the oxygen in the surrounding atmosphere. These will be discussed individually in reverse order.

Increased Concentration of Oxygen

Oxygen is a major factor in deterioration, yet it occupies only about 21% by volume of air. An effective way, then, to increase the rate of deterioration is to increase the concentration of oxygen. This is feasible when reasonably small samples are used so that the danger of conflagration is at a minimum. The test results can present certain computational problems if one wishes to predict the rate under normal atmospheric conditions, but these are not insurmountable.

Moisture Content

The concentration of water in a substance frequently plays an important role in photochemical deterioration. The fact that fading can sometimes be reduced markedly when moisture is excluded was demonstrated in some of the earliest studies in fading of museum materials.[6] Because samples are generally heated in accelerated-aging tests,[7] either by ovens or by absorption of the intense radiation, moisture content is frequently an uncontrolled factor.

If the investigator does not heat an object to more than 90°F (32°C), significant moisture should generally remain in materials such as wool, silk, paper and wood. Nevertheless, the importance of relative humidity should be checked when moisture-sensitive materials are tested. By carrying out fading experiments in a climate-controlled chamber, Stolow has been able to maintain the samples at room temperature, or even below if desired.[8]

The International Standards Organization's blue-dyed-wool fading standards were especially chosen because their rates of fading are not particularly sensitive to relative humidity or temperature.[9-11]

Temperature

Temperature has little influence on the primary step in photochemical reactions, but it does affect the relative importance of various secondary

4. ICOM, *Use of Fluorescent Light in Museums*, Paris (1953).

5. Hirt, R. C., Schmitt, R. G., Searle, N. D., and Sullivan, A. P., "Ultraviolet spectral energy distributions of natural sunlight and accelerated test light sources," *J. Optical Soc. Amer.*, 50 (1960), p. 706-713.

6. Brommelle, N. S., "The Russell and Abney report on the action of light on water colours," *Studies in Conservation*, 9 (1964), p. 140-152.

7. Hoey, C. E., and Hipwood, H. A., "An appraisal of artificial weathering methods for assessment of the durability of paint films," *J. Oil and Colour Chemist's Assoc.*, 57 (1974), p. 151-160.

8. Stolow, N., private communication.

9. McLaren, K., "The use of fading lamps for determining light fastness," *J. Soc. Dyers and Colourists*, 70 (1954), p. 553-55; "A critical appraisal of the international daylight fastness test," *J. Soc. Dyers and Colourists*, 73 (1957), p. 121-127.

10. British standard 1006 : 1971, *Methods for the determination of the colour fastness of textiles to light and weathering*, British Standards Institution, 2 Park St., London WIA 2BS; International Organization for Standardization, ISO Recommendation R105. The geometric grey scale used is British Standard 2662.

11. Jaeckel, S. M., Ward, C. D., and Hutchings, D. M., "Variations in assessment of light fastness and in rates of fading and spacing of the blue standards," *J. Soc.Dyers and Colourists*, 79 (1963), p. 702-722.

chemical reactions.[12] Temperature also influences physical properties such as flexibility, fluidity of the thermoplastic materials, and the tendency to pick up dirt, in addition to causing changes in the water-vapor content of moisture-sensitive materials. When testing unstable materials (Class C, described below), every effort should be made to keep the temperature as close to normal room conditions as possible. However, higher temperatures can often be tolerated when testing highly stable materials.

12. Feller, R. L., "The deteriorating effect of light on museum objects: principles of photochemistry, the effect on varnishes and paint vehicles and on paper," *Museum News, Technical Supplement*, No. 3 (June 1964).

Effect of Shorter Wavelengths

The energy of photons, the "bullets" of electromagnetic energy, increases as the wavelength diminishes. This photochemical principle is expressed by the equation $E = h\nu$, in which E is the energy, h is Planck's constant, and ν, the frequency of the radiation (the latter is the inverse of the wavelength multiplied by the speed of light). The U. S. National Bureau of Standards has shown that the deterioration of low-grade paper increases logarithmically as a function of decreasing wavelength. Other photochemically activated deterioration processes have also been shown to exhibit similar sensitivity with respect to wavelength.[12]

It does not follow that photochemical reactions will invariably obey this same rule, increasing in rate as radiation of shorter and shorter wavelengths is employed. As the wavelength decreases, the predominant path of the photochemical processes may change and may increasingly involve photolytic dissociation—that is, the direct breaking apart of chemical bonds as the result of the absorption of the radiant energy.

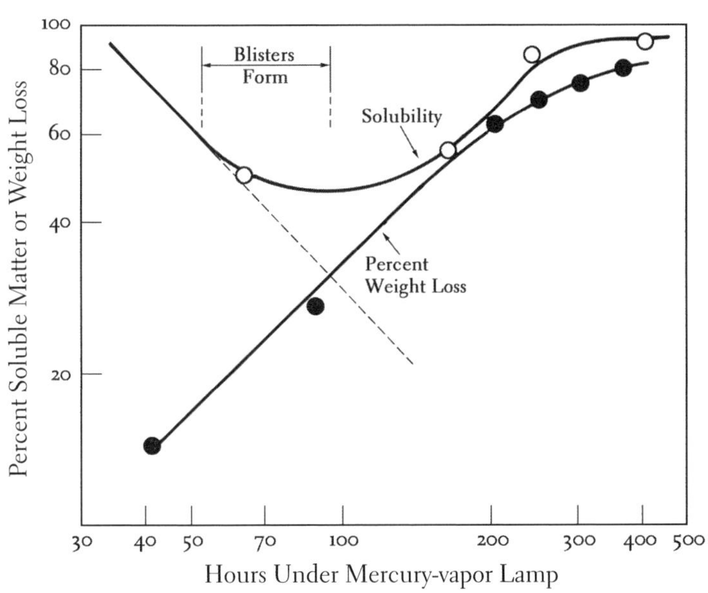

FIGURE 1
Changes in solubility and weight of Elvacite® 2046 butyl methacrylate copolymer exposed to germicidal mercury-vapor lamp.

One of the best known examples of the difference in the effects of near-ultraviolet and the middle ultraviolet wavelengths is the landmark investigation of Launer and Wilson on the deterioration of paper.[13] Here it was found that water vapor *retarded* the rate of deterioration under 254 nm radiation and the presence of oxygen was without effect, whereas under near-ultraviolet, water vapor and oxygen *increased* the rate of deterioration.

13. Launer, H. F., and Wilson, W. K., "The photochemistry of cellulose. Effects of water vapor and oxygen in the far and near ultraviolet regions," *J. Amer. Chem. Soc.*, 71 (1949), p. 958-962.

The literature on polymers and textiles abounds with studies regarding the effects of 254 nm ultraviolet. Many acrylic resins tend to depolymerize almost completely when subjected to this energetic radiation,[14, 15] whereas under near-ultraviolet certain of these polymers exhibit a marked tendency to crosslink. Figure 1 shows the result of exposure of a copolymer of approximately 50/50 normal butyl/isobutyl methacrylate, DuPont Elvacite® 2046, to the radiation of a "germicidal" (quartz enclosed) mercury-vapor lamp. Under ordinary fluorescent lamps, RS-type sunlamps, and carbon- and xenon-arcs, the major photochemical effect is that of crosslinking, with the result that the films become insoluble in organic solvents. Insolubility can reach 90% in about 300 hours, indicated by the dotted line in Figure 1 (see also Figure 4). In contrast, under the 254 nm radiation emitted by the germicidal lamp, the same copolymer begins to form blisters after it has become approximately 50% insoluble and, at this point, the amount of insoluble material begins to decrease (Figure 1 and Table 1). As determined in a separate experiment, the copolymer film is almost completely volatilized if exposed long enough to the germicidal lamp, apparently because of depolymerization of the copolymer. Thus, the reactions of the copolymer to the two different wavelength regions are considerably different.[16]

14. Jellineck, H. H. G., *Degradation of Vinyl Polymers*, Academic Press, New York (1955).

15. Fox, R. B., "Photodegradation of high polymers," *Progress in Polymer Science*, 1, ed. Jenkins, A. D., Pergamon Press, London (1969), p. 45-89.

16. Renz, R., Topfel, H., and Vesper, W., "Einfluss des UV-Lichtes auf die Zerstörung von Lack Filmen," *Farb und Lack*, 80 (1974), p. 513-517.

TABLE 1
Analysis of material in Films of Elvacite® 2046
Exposed to Germicidal Mercury-Vapor Lamps

Time, Hours	Per Cent Original Material Remaining	Portion Insoluble	Portion Soluble
40	86	20.6	65.4
60	80	41.6	38.4
100	67	30.1	36.9
150	50	17.5	32.5
200	37	11.1	25.9
250	30	3.0	27.0

An additional example of the inappropriateness of 254 nm radiation for investigating deterioration "under normal museum conditions" is the fact that BS1006 : 1971 standard blue-wool cloths turn yellow brown and develop the odor of degraded protein, failing to show the usual stepwise rate of fading in which the standards take approximately twice as long to fade to grey scale contrast 3 as they do to fade to contrast 4, etc.

Increased Intensity of the Near-Ultraviolet

A number of test methods have been recommended in which fluorescent "black" lamps are used either alone or in conjunction with other sources to increase the intensity of near-ultraviolet radiation.[9, 17, 18] This is an excellent suggestion, for fluorescent lamps contribute little heat. The spectral emission of a typical fluorescent blacklight lamp is seen in Figure 2.

17. Cipriani, L. P., Giesecke, P., and Kinmouth, R., "New way to peg plastics aging," *Plastics Technology, PLTEA*, 12 (May 1964), p. 34.

18. Wyzsecki, G., and Stiles, W. S., *Color Science*, J. Wiley & Sons, New York (1967). This text provides a readily available survey of the spectral emission characteristics of various lamps.

The usual carbon-arc emits strongly in the near-ultraviolet, but is not especially intense in the visible.[5] We have found the carbon-arc with ordinary glass filter useful in evaluating highly stable resins, but less satisfactory in fading tests in which a significant portion of fading is caused by visible radiation.

I am told that the Purple-X-Blacklight (Figure 2) has too short a life to be practical. A mercury-tungsten lamp has recently been shown to be useful in the testing of dyed textiles but I have no experience with this type of lamp.[19]

19. Giles, C. H., Shah, C. D., and Bailee, D., "Economical and efficient fading lamps," *J. Soc. Dyers and Colourists*, 85 (1969), p. 410-417.

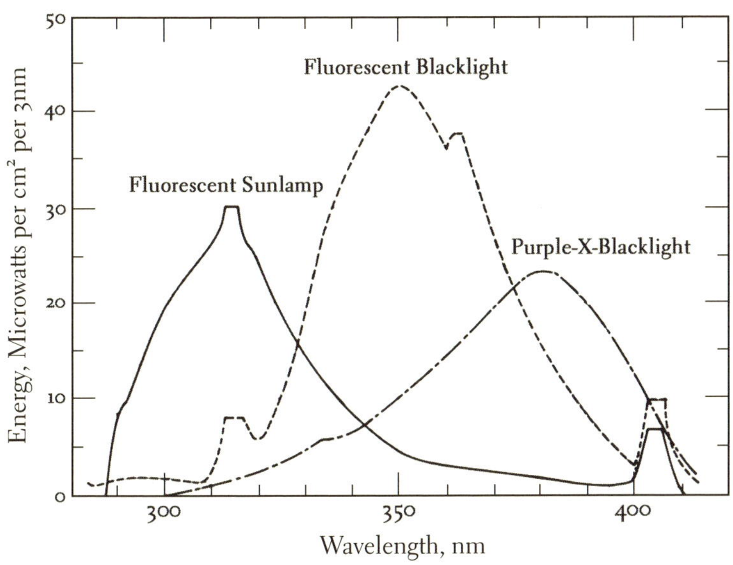

FIGURE 2
Spectral radiant flux distributions of various fluorescent lamps in the ultraviolet. (After Wyszecki and Stiles, Ref. 18)

Continuous vs. Discontinuous Exposure

Exposure of materials to continuous illumination is practical and convenient. There is an aspect of normal exposure, however, that should not be overlooked: illumination in museums, whether natural or artificial, is generally diminished if not completely eliminated at night, resulting in discontinuous exposure. It has been found that deterioration sometimes proceeds faster under alternating light and dark than under continuous illumination.[20] This is particularly true when the source of illumination warms the objects, thereby reducing their

20. Launer, H. F., "Light-sensitive papers as controls for testing textile colorfastness and stability of materials under arc lamp exposure," *National Bureau of Standards J. Research*, 41 (1948), p. 169-177.

moisture content and causing stresses within systems that have layers of different materials. Alizarin, for example, fades more extensively under a given number of footcandle hours of exposure to alternating light and dark than it does under the same number of footcandle hours of continuous illumination during which the panels attain a higher-than-normal temperature.[21]

There is another general problem associated with continuous illumination: the quantity and quality of light emitted often changes as a lamp ages. The radiation is thus continuous but not *constant*. The intensity and spectral distribution of the radiation from xenon-arc lamps changes so much that it is customary to replace them after about 2000 hours.[22, 23] As the data in Table 2 indicate, older lamps can result in longer induction periods and significant changes in the ratio of crosslinking to chain-breaking. The fall-off of intensity, shown in Figure 3, may be largely responsible for the change in introduction time, but a decrease in the shorter wavelengths of ultraviolet is probably the cause of the higher ratio of chain-breaking to crosslinking.

21. Levison, H., former president of Permanent Pigments, Inc. private communication.

22. Clark, J. E., and Harrison, C. W., "Accelerated weathering of polymers: radiation," in *Weatherability of Plastic Materials*, ed. Kamal, M. R., Wiley-Interscience, New York (1967), p. 97-110.

23. Allen, E., "A monitoring device for ultraviolet energy in the Atlas xenon weather-ometer," *Applied Optics*, 4 (1965), p. 835-838.

TABLE 2
Effect of Age of Xenon Lamp Upon Induction Time and Ratio of Chain Breaking to Crosslinking, β/α

Polymer	Age of Lamp, Hours	Observed Relative Induction Time	Calculated Increase in Induction Time*	Ratio β/α
poly(n-buytl methacrylate)	0	1.0	–	0.0
poly(n-buytl methacrylate)	700	1.1	1.3	0.5
poly(isoamyl methacrylate)	0	1.0	–	0.25
poly(isoamyl methacrylate)	700	1.0	1.3	0.4
poly(isoamyl methacrylate)	1414	2.0	1.5	0.5
50/50 copolymer of n-butyl/isobutyl methacrylate	0	1.0	–	0.1
50/50 copolymer of n-butyl/isobutyl methacrylate	700	1.25	1.3	0.15

* Calculated from average decrease in intensity, Figure 3.

Increased Intensity

What degree of intensification can the investigator expect in photochemical degradation studies? It is relatively easy to obtain 600 to 1000 footcandles of intensity by placing samples a few inches away from banks of ordinary fluorescent lamps. One year of continuous exposure under these conditions, about 8,000,000 footcandle hours, would be similar to the total exposure encountered in about 70 years in a museum at a level of 30 footcandles and about 400 years at a level of 5 footcandles.

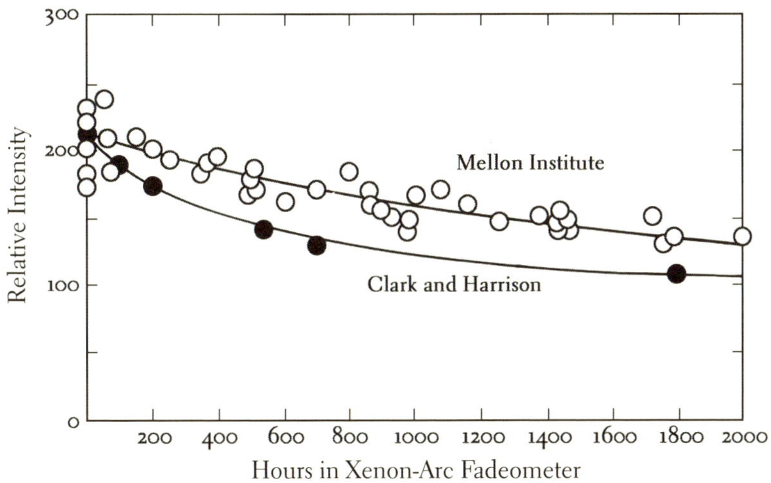

FIGURE 3
Decrease in intensity of water-cooled xenon arc with time.

With an Atlas Electric Devices 600 WRC Xenon-Arc Fadeometer, we have obtained the same amount of crosslinking of certain acrylic resins in 50 hours of exposure as observed in specimens that have been on the walls of the laboratory at the Fogg Art Museum for a little more than 11 years.[24] This indicates that the process of crosslinking was speeded up about 2000 times. Through control of the concentration of water vapor and oxygen, perhaps an increase in velocity of deterioration could be achieved as much as 4000 times greater than that occurring in a museum.

24. Feller, R. L., Stolow, N., and Jones, E. H., *On Picture Varnishes and Their Solvents*, Press of Case Western Reserve University, Cleveland (1971), Table 7-5, p. 159.

AVAILABLE LIGHT SOURCES

In view of the variables cited, my recommendation is that the investigator should not rely solely on a single method of testing. The National Gallery of Art Research Project regularly employs a variety of exposure methods. Test panels under glass outdoors, facing south, on the roof in Pittsburgh provide inexpensive data concerning the lightfastness of new dyes and pigments. Samples placed under the skylights of the National Gallery of Art in Washington provide conditions that are less severe; Thomson has used such an exposure area at the National Gallery, London.[25]

Carbon-arc machines are useful in the testing of highly durable coatings, but we have reduced our use of this equipment because the samples are heated to a temperature of about 63 ± 5°C and the radiation does not contain a high portion of visible wavelengths.

As the illustration in Figure 4 indicates, the rate of crosslinking of a standard test resin is about the same under both the carbon- and xenon-arc Fadeometers.[26] In our air-cooled xenon-arc apparatus, the radiation is filtered through Pyrex glass rather than quartz in order to achieve a distribution of

25. Thomson, G., "A new look at colour rendering, level of illumination and protection from ultraviolet radiation in museum lighting," *Studies in Conservation*, 6 (1961), p. 49-70.

26. Feller, R. L., "Problems in the investigation of picture varnishes," in *Conservation of Paintings and the Graphic Arts, Preprints of the Lisbon Congress*, International Institute for Conservation, London (1972), p. 201-209.

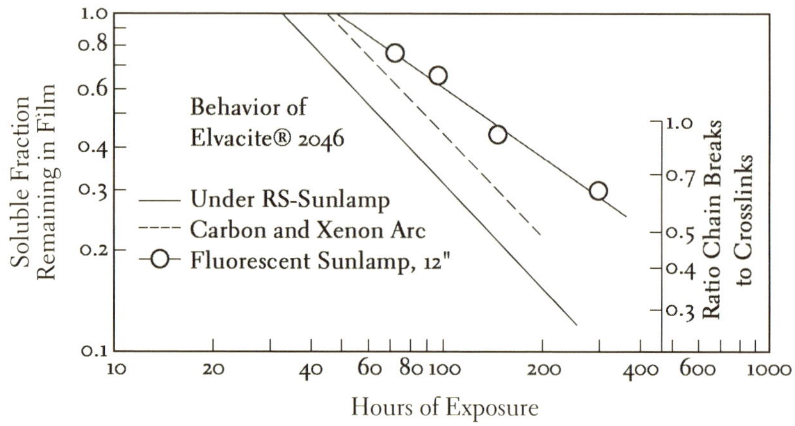

FIGURE 4
Rate of formation of insoluble matter in Elvacite® 2046 butyl methacrylate copolymer under various sources of illumination. Samples coated on aluminum foil.

radiation closer to that encountered in museums; to compensate for the resultant decrease in activity, the samples have been placed closer to the source. Except for the slight changes in the intensity and spectral distribution of the radiation with time, xenon-arc machines provide the highest intensity of both visible and near-ultraviolet radiation under the best controlled and defined conditions that we regularly employ.

Mercury-vapor-discharge sunlamps, of the RS-type were proposed a number of years ago as a test system by the American Society for Testing Materials (ASTM).[27] The rate of crosslinking of samples placed 10 inches away from an RS sunlamp is indicated in Figure 4. This equipment has uses, particularly when one considers that the cost of the xenon- or carbon-arc Fadeometers is about 1000 times greater. However, excessive heat is a major drawback; samples attain a temperature of at least 55°C. The short life of the lamp is also a nuisance; lamps are customarily replaced after about 1000 hours.

In recent years, the ASTM has recommended fluorescent sunlamps to replace the RS-type lamps (see Figure 2).[28] Figure 4 includes recent results obtained with a Westinghouse F-20T12 fluorescent sunlamp placed 12 inches away from the samples. At this distance, the polymer begins to crosslink just about as fast as it does under the other test methods but, perhaps because of the lower temperature or the shorter wavelengths involved, there is a certain degree of chain-breaking. Without further study, I would not recommend unfiltered fluorescent sunlamps for the investigation of museum deterioration problems because much of the radiation is below 320 nm.

Test conditions milder than those described are provided by a set of ordinary fluorescent lamps in a room maintained at 21°C (70°F) and 50 to 65% relative humidity. This system was suggested to me by Garry Thomson, although I believe he has not particularly stressed his own use of it. A similar rack was used by Padfield and Landi.[29, 30] We have reported the results of exposures of

27. "Accelerated weathering of plastics using S-1 bulb and fog chamber," *ASTM Standard*, D795-57T (1965).

28. Reinhart, F. W., and Mutchler, M. K., "Fluorescent sunlamps in laboratory testing for plastics," in *ASTM Bulletin*, ASTBA, No. 212 (1956), p. 45-51; *ASTM Standard*, D1501-65T.

29. Padfield, T., and Landi, S., "The light-fastness of the natural dyes," *Studies in Conservation*, 11 (1966), p. 181-193.

30. Feller, R. L., "Control of the deteriorating effects of light on museum objects," *Museum*, 17, No. 2 (1964), p. 57-98.

up to three years on these racks[12] and find them an excellent low-cost method of testing, providing that controls accompany each series of exposures.

Lastly in our range of test methods, we have been fortunate in the last fifteen years to have had the use of the north wall of a gallery at the National Gallery of Art that is not open to the public.

CONTROLS

In view of the variability in temperature, humidity, and spectral distribution of the radiation, exposure tests should always be accompanied by "controls." The blue-wool cloths of the International Standards Organization provide inexpensive standards. These have admitted faults, but they have been extensively studied, are readily available and internationally accepted.[9-11] We regularly accompany each set of exposures with one of the little cards sold by the Swiss Standards Organization, shown in Figure 5.[31, 32] The fading of the eight standards in terms of Langleys has been published,[33] also their rate of fading under fluorescent lamps[11, 34] and on a gallery wall.[34] They provided valuable reference samples in Padfield and Landi's tests on fugitive textile dyes[29] and in Hanlan's test on the effects of an electronic flash.[35]

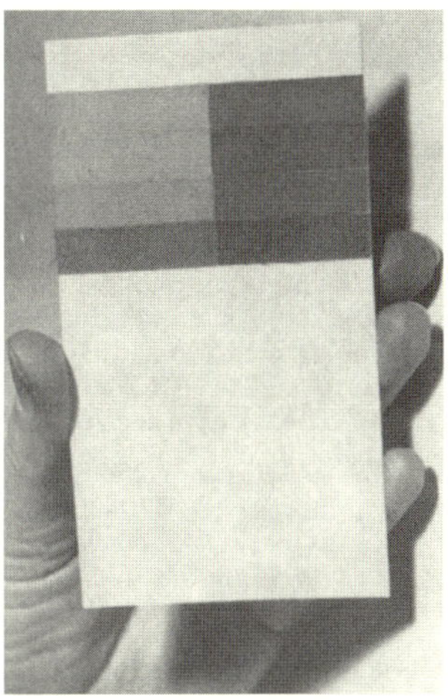

FIGURE 5
The International Standards Organization blue-wool fading standards (ISO Recommendation R105/I, part II; BS1006:1971) as supplied by the Schweizerische Normen-Vereinigung.

The highly durable acrylic resin Acryloid® B-72 and the DuPont rutile pigment Ti-Pure® R-610 (more recently, Ti-Pure® R-902) have been used at our laboratory to prepare the reference panels of alizarin at a tint of Munsell value about 7.[36, 37]

As a control for crosslinking tests, we have recommended polymers of normal butyl or isobutyl methacrylate in the range of intrinsic viscosity of the DuPont

31. Feller, R. L., "Standards of exposure to light," *Bulletin American Group-IIC*, 4, No. 1 (1963), p. 10-12; *ibid.*, 7, No. 2 (1967), p. 8, 32.

32. Schweizerischer Verband für die Materialprüfungen der Technik, Unterstrasse 11, St. Gallen, SNV 95809, 1961. I am grateful to Dr. Bruno Mühlethaler for calling my attention to the test cards issued by this organization.

33. Schmidt , C. H. A., "Daylight fastness testing by the Langley system," *Am. Dyestuff Reporter*, 51, No. 18 (1962), p. 31-42. Also, Rowland, O., "Fading of the Bntish dyed wool light fastness standards in the U.K.," *J. Soc. Dyers and Colourists*, 79 (1963), p. 697-701.

34. Feller, R. L., "Problems in spectrophotometry," *London Conference on Museum Climatology*, International Institute for Conservation of Historic and Artistic Works, London (1968), p. 257-269.

35. Hanlan, J. F., "The effect of electronic photographic lamps on the materials of works of art," *Museum News*, 48, No. 10 (June 1970), p. 33-41.

36. Feller, R. L., "Felt-tipped markers and the need for standards of lightfastness for artists' colorants," *Bulletin Am. Group-IIC*, 8, No. 1 (1967), p. 24-26; reprinted in *Inter-Society Color Council Newsletter*, 192 (Jan.-Feb., 1968), p. 10-11.

37. Feller, R. L., Keisch, B., and Curran, M., "Notes on modern pigments," *Bulletin American Group-IIC*, 12, No. 1 (1971), p. 60-62.

resins familiar in America, Elvacite® 2044 and 2045.[24, 26] Presently, we prefer the copolymer of the two, Elvacite® 2046, for this purpose. Data of the type shown in Figure 4 represent a convenient and rather precise monitor of photochemical exposure.

The Rohm and Haas acrylic resin Acryloid® B-72 (Paraloid® B-72 in Europe) can serve as a reference standard of an excellent thermoplastic of the type soluble in a solvent having a solubility parameter no higher than xylene: it does not seriously crosslink when exposed on aluminum foil for 2000 hours in the carbon-arc Fadeometer. Paints formulated with B-72 have retained their integrity and gloss when exposed under glass on the roof in Pittsburgh for two years.

When seeking merely to "improve" methods and materials, not to meet specific standards, the investigator should accompany his exposure tests with control samples of the particular materials which he wishes to improve upon.

CLASSES OF USE AND STABILITY

What degree of exposure can a "highly durable" material be expected to withstand? What is a "fugitive" pigment in these terms, or an "unstable" resin? To facilitate discussion, let me define four classes of materials to be used in conservation, one defining use, three defining stability, summarized in Table 3.

Use - Class T : Materials to be used only in temporary physical or vapor-phase contact with objects.

"Temporary contact" would extend from a few hours to perhaps something

TABLE 3

Standards of Intended Use and Photochemical Stability for Materials in Conservation

Class	Classification	Intended Useful Lifetime	Approximate Equivalent Standard of Photochemical Stability
T	Materials in Temporary Contact	less than 6 months?	–
C	Unstable or Fugitive	less than 20 years	BS1006 Class 3 or less
B	Intermediate	(20-100 years)	(3 to 6 years)
A	Excellent	(A2?) greater than 100 years	greater than BS1006 Class 6
		(A1?) greater than 500 years.	?

under six months. These would be materials such as moistureproof covers that might be used to store materials temporarily, Mylar and silicone-treated papers used to prevent sticking, or adhesives used for facings. Such materials do not have to remain stable for long periods of time, but each material would have to be tested for its suitability under the particular circumstances proposed for its use.

Stability - Class C: Unstable or Fugitive Materials.

These would be materials that would be expected to last more than six months but no more than twenty years of exposure in a museum without serious changes in their properties. Whereas there may be little role for many of these in preservation methods, this class is introduced so that such materials can be clearly recognized and labeled when they are revealed in the course of a testing program.

Several years ago, we suggested three classes of lightfastness for artist's colorants: fugitive, intermediate, and lightfast.[36] For the fugitive class, materials that would have a lightfastness equivalent to BS1006 blue cloth No. 3 or less were suggested. According to our previously published data on the fading of these standards on the wall of a daylighted gallery, blue cloth No. 3 should fade considerably (that is, to a British Standard 2662 gray scale contrast of about 2.5) in about 20 years in such circumstances.[34] This specification could be applied more generally; class C materials in photochemical stability could be defined as those that discolor or fade to geometric gray scale contrast 2.5, or that decrease in tensile strength about 20%, in an exposure to a mixture of visible and near-ultraviolet radiant energy that is sufficient to cause BS1006 standard blue cloth No. 3 to fade to a geometric gray scale contrast of 2.5. (Let us say that energy in the near-ultraviolet should be no more than 20% of the total visible plus near-ultraviolet used in the exposure tests.)

Stability - Class B: Intermediate Materials.

This would correspond to the intermediate class of lightfastness proposed for artists' colorants, intended to represent materials that would experience appreciable fading, or that would change in their properties sufficiently that we would wish to remove or replace them, in less than hundred years but more than twenty years.

Stability - Class A: Excellent Materials.

This would represent substances that possess excellent resistance to deterioration; the intention being that the materials would be expected to last for better than hundred years under normal conditions. With respect to photochemical stability, such materials should not seriously change in properties, let us say less than 20%, during an exposure longer than that which corresponds to BS1006 class 6.0.

I believe that it is possible today to select organic polymers that can be expected to fulfill their desired functions for hundred. years. One might consider these as class A2; others, expected to last five hundred years, class A1. Acryloid® B-72 is certainly a class A material.[26]

The physical scientist undoubtedly opens himself to criticism if he attempts to predict useful lifetimes longer than hundred years; certainly if he speculates that the life of polymers and colorants will be longer than five hundred years. Yet, we are already working with a number of materials that promise to maintain their useful properties within the museum circumstances for just such lengths of time. Tests of Permalife-type paper, for example, have led to the predication that this material will maintain its useful physical properties for two hundred years.[38, 39]

Many industrial coatings today are designed to withstand 5, 7 or more years of outdoor exposure.[40, 41] At Carnegie-Mellon Institute of Research, we have tested a number of pigments and thermoplastics that withstand outdoor exposure under glass on the roof for a period of at least one year; certain paints have remained in good condition after two years under such exposure. Let us estimate what the latter exposure means: if the annual exposure outdoors in the latitude of Pittsburgh is on the order of 15 million footcandle hours of daylight per year and the annual exposure indoors in an art gallery or museum at an average daytime intensity of 30 footcandles amounts to about 100,000 footcandle hours per year,[42, 43] then we may roughly assume that the annual exposure on the roof is equivalent to the same amount of visible radiation as encountered in at least 150 years of the exposure in an art gallery. The added effects of ultraviolet, temperature and humidity would make the roof-exposure equivalent to a period much longer than 150 years indoors.[44]

38. Hobbs, R. C., "Estimating the life expectancy of book papers," *Permanent/Durable Book Paper*, The Virginia State Library, Richmond (1960), p. 47-53.

39. If we follow the usual rule of thumb that the changes in paper heated at 100°C in an oven for 72 hours (3 days) correspond roughly to 25 years of aging under normal library and archival conditions, then in this sense, a paper that might last 100 years could be detected in about two weeks of testing.

40. Newland, G. C., "The role of ultraviolet absorbing agents in the production of weather resistant polymers," *Solar Energy*, 12 (1968), p. 65-73.

41. Berg, C. J., Jarosz, W. R., and Salathe, G. F., "Performance of polymers in pigmented systems," *J. Paint Technology*, 39 (1967), p. 436-453.

42. Ellis, C., and Wells, A. A., *The Chemical Action of Ultraviolet Rays*, Reinhold, New York (1941), p. 202.

43. Thomson, G., "Annual exposure to light within museums," *Studies in Conservation*, 12 (1967), p. 26-35.

Conclusion

Because museum conditions are rather mild with respect to temperature and the intensity of ultraviolet radiation, the investigator concerned with new methods and materials for conservation must give special attention to the conditions of photochemical "accelerated-aging" tests if he is to avoid misleading results. In the first part of my discussion, I reviewed various factors that must be considered, drawing particular attention to the general inappropriateness of 254 nm radiation. For general tests of photochemical deterioration, I would limit the short wavelengths to the near-ultraviolet (320 to 400 nm). If the middle ultraviolet is employed (280 to 320 nm), tests should be especially conducted to demonstrate the effect of including and excluding these wavelengths.

Three classes of photochemical stability have been described and practical standards suggested by which three may be differentiated. Other standards will be needed to define similar classes with respect to thermochemical stability.

The investigator should not attempt to rely on one method of exposure for all classes of materials. Class C materials should be tested under relatively mild conditions but, in order to shorten the necessary testing time, Class A materials are customarily exposed in xenon- and carbon-arc machines. The accompanying higher temperatures can in many cases be tolerated, but if a material should fall slightly short of Class A in such a test, it is advisable to repeat the exposure under milder conditions. "Controls" should accompany each set of exposures.

Acknowledgments

The experimental measurements referred to in Figures 1, 3 and 4 were carried out by Mrs. R. A. Curran.

Additional Bibliography

AATCC Committee on Colorfastness to Light, "A study of variables encountered in natural lightfastness testing," *Am. Dyestuff Reptr.*, 46, No. 23 (Nov. 18, 1957), p. 861-883.

Bush, C. F., "New artificial weathering methods," *Am. Dyestuff Reptr.*, 49, No. 3 (1960), p. 33-39.

McLaren, K., "Use of grey scale for assigning effects on patterns," *J. Soc. Dyers and Colourists*, 68 (1952), p. 205-6; "The importance of temperature and relative humidity in light fastness testing," *ibidem*, 72 (Nov. 1956), p. 527-537; "A critical appraisal of the international daylight fastness test," *ibidem*, 73 (Apr. 1957), p. 121-127.

Neville, G. H. J., "Some observations on the weathering of titanium dioxide pigmented paint films," *J. Oil and Colour Chemists' Assoc.*, 46 (1963), p. 753-781.

Nordhammer, G., "Conditions of temperature and humidity in the Fadeometer," *Am. Dyestuff Reptr.*, 38, No. 16 (Aug. 8, 1949), p. 571-574, 593.

44. For convenience in the present discussion I have referred to footcandles to give the reader a feeling for the magnitude of intensity, fully realizing that this only refers to the amount of visible energy in a given source. Such an expression is not without error and ambiguity if we attempt to express the radiation of mercury vapor lamps, fluorescent lamps and incandescent lamps in these terms. Citation of exposure in terms of Langleys also leads to ambiguity because the Langley can include infrared radiation as well as visible and ultraviolet. The most basic practice would be to report the spectral distribution of a given light source in terms of the microwatts per square centimeter in a particular region of the spectrum. One Langley is 1 g cal/cm^2.

Use of the International Standards Organization's Blue-Wool Standards for Exposure to Light. I. Use as an Integrating Light Monitor for Illumination Under Museum Conditions

Robert L. Feller and Ruth M. Johnston-Feller

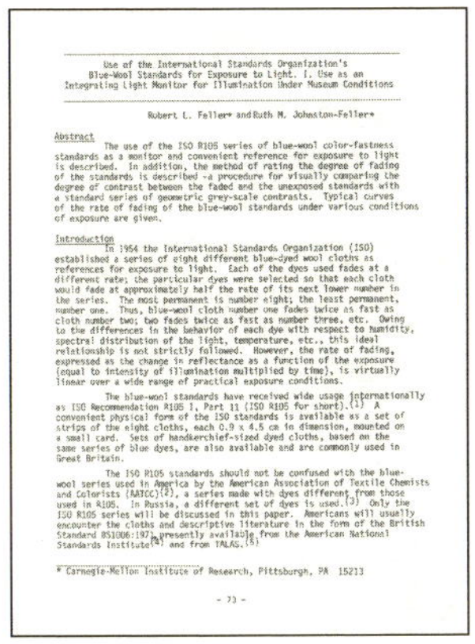

Use of the International Standards Organization's Blue-Wool Standards for Exposure to Light. I. Use as an Integrating Light Monitor for Illumination Under Museum Conditions

Robert L. Feller* and Ruth M. Johnston-Feller*

Abstract

The use of the ISO R105 series of blue-wool color-fastness standards as a monitor and convenient reference for exposure to light is described. In addition, the method of rating the degree of fading of the standards is described -a procedure for visually comparing the degree of contrast between the faded and the unexposed standards with a standard series of geometric grey-scale contrasts. Typical curves of the rate of fading of the blue-wool standards under various conditions of exposure are given.

Introduction

In 1954 the International Standards Organization (ISO) established a series of eight different blue-dyed wool cloths as references for exposure to light. Each of the dyes used fades at a different rate; the particular dyes were selected so that each cloth would fade at approximately half the rate of its next lower number in the series. The most permanent is number eight; the least permanent, number one. Thus, blue-wool cloth number one fades twice as fast as cloth number two; two fades twice as fast as number three, etc. Owing to the differences in the behavior of each dye with respect to humidity, spectral distribution of the light, temperature, etc., this ideal relationship is not strictly followed. However, the rate of fading, expressed as the change in reflectance as a function of the exposure (equal to intensity of illumination multiplied by time), is virtually linear over a wide range of practical exposure conditions.

The blue-wool standards have received wide usage internationally as ISO Recommendation R105 I, Part II (ISO R105 for short).(1) A convenient physical form of the ISO standards is available as a set of strips of the eight cloths, each 0.9 x 4.5 cm in dimension, mounted on a small card. Sets of handkerchief-sized dyed cloths, based on the same series of blue dyes, are also available and are commonly used in Great Britain.

The ISO R105 standards should not be confused with the blue-wool series used in America by the American Association of Textile Chemists and Colorists (AATCC)(2), a series made with dyes different from those used in R105. In Russia, a different set of dyes is used.(3) Only the ISO R105 series will be discussed in this paper. Americans will usually encounter the cloths and descriptive literature in the form of the British Standard BS1006:1971, presently available from the American National Standards Institute(4) and from TALAS.(5)

* Carnegie-Mellon Institute of Research, Pittsburgh, PA 15213

- 73 -

Originally printed in:
Preprints of Papers Presented at the Sixth Annual Meeting, Fort Worth, Texas
American Institute for Conservation, Washington, DC (1978), Pages 73-80.

Introduction

In 1954 the International Standards Organization (ISO) established a series of eight different blue-dyed wool cloths as references for exposure to light. Each of the dyes used fades at a different rate; the particular dyes were selected so that each cloth would fade at approximately half the rate of its next lower number in the series. The most permanent is number eight; the least permanent, number one. Thus, blue-wool cloth number one fades twice as fast as cloth number two; two fades twice as fast as number three, etc. Owing to the differences in the behavior of each dye with respect to humidity, spectral distribution of the light, temperature, etc., this ideal relationship is not strictly followed. However, the rate of fading, expressed as the change in reflectance as a function of the exposure (equal to intensity of illumination multiplied by time), is virtually linear over a wide range of practical exposure conditions.

The blue-wool standards have received wide usage internationally as ISO Recommendation R105 I, Part 2 (ISO R105 for short).[1] A convenient physical form of the ISO standards is available as a set of strips of the eight cloths, each 0.9 x 4.5 cm in dimension, mounted on a small card. Sets of handkerchief-sized dyed cloths, based on the same series of blue dyes, are also available and are commonly used in Great Britain.

The ISO R105 standards should not be confused with the blue-wool series used in America by the American Association of Textile Chemists and Colorists (AATCC),[2] a series made with dyes different from those used in R105. In Russia, a different set of dyes is used.[3] Only the ISO R105 series will be discussed in this paper. Americans will usually encounter the cloths and descriptive literature

1. British Standard 1006:1971, *Methods for the Determination of the Colour Fastness of Textiles to Light and Weathering*. This 37-page pamphlet essentially represents the principal aspects of International Organization for Standardization Recommendation R105: 1, Part 2.

2. *AATCC Technical Manual* (1976), p. 132.

3. Holzer, P. J., "Die Objective Auswertung von Prüfungsergebnissen, Farbechtheit betreffend," *Melliand Textilber.* (1961), p. 811-815.

in the form of the British Standard BS1006:1971, presently available from the American National Standards Institute[4] and from TALAS.[5]

Two major applications for the R105 blue-wool cloths in museum conservation investigations are (1) use as an "effective exposure" light monitor to measure the integrated exposure on museum walls and in illuminated showcases; and (2) as a reference for specification of the relative photochemical stability of particular materials—for textiles, as a reference for the stability of dyestuffs and fibers—for paint or varnishes, as a reference for the stability of pigments and vehicles. Standards of photochemical stability based on the ISO R105 cloths have recently been proposed.[6, 7]

The standard procedure designed for reporting the fading of the blue-wool cloths is a convenient visual method—visually comparing the degree of fading with a standard geometric-grey-scale. Although precise measurements are possible by measuring the reflectance changes spectrophotometrically,[8, 9] the present description of their use as monitors of museum exposure conditions will deal with the method of visual evaluation. In a later publication we will describe the instrumental method for measuring reflectance.

Method of Use

The small cards containing mounted strips of the eight blue-wool cloths are available from Schweizerische Normen-Vereinigung[10] and in this country from TALAS.[5] These cards, illustrated in a previous publication,[6] are probably the most convenient form for use in museums, because of their size and the prior assembly of the series of cloths. However, if so desired, the handkerchief-size blue cloths may be obtained from the American National Standards Institute.[4] Suitably-sized specimens can be cut from these larger cloths and mounted as needed, as also illustrated earlier.[11]

To use the blue-wool series as monitors of effective exposure to light, a reference set of cloths must be kept from exposure. This may be done either by covering a portion of the standards with an opaque material such as aluminum foil or, as we prefer, simply by cutting off a portion of the cardboard-mounted set and preserving it in the dark. The rest of the set is placed in position and left to fade where the monitoring of exposure to light is desired. Periodic inspection and comparison of the exposed with the unexposed cloths can be made to determine when sufficient exposure has occurred to result in visible changes in the most fugitive members of the series, or the set may simply be left exposed for an appropriate period of time before inspection, such as one year on a gallery wall.

The degree of fading between the exposed and unexposed portions is estimated visually by comparison to a geometric five-step neutral grey scale. The scale used with BS1006:1971 is British Standard BS2662, available from the British Standards Institution.[12] This geometric grey scale (GGS) has five

4. American National Standards Institute, 1430 Broadway, New York, NY 10018.

5. Pamphlets describing the standard practice (BS1006:1971), the geometric grey scale (see reference 12) and the small cards comprised of strips of all eight cloths are available from TALAS, 104 Fifth Avenue, New York, NY 10011.

6. Feller, R. L., "Speeding Up Photochemical Deterioration," *Bulletin de l'Institut Royal du Patrimoine Artistique*, 15 (1975), p. 135-150.

7. Feller, R. L., "Studies on Photochemical Deterioration," *Preprints of the 4th Triennial Meeting, Venice*, ICOM Committee for Conservation (1975), Paper 75/19/4.

8. Friele, L. F. C., "A Comparative Study of Natural and Xenotest Exposure Conditions for Measuring Fading and Degradation," *J. Society of Dyers and Colourists*, 79 (1963), p. 623-631; see also Friele, L. F. C., *Textilveredlung*, 5 (1970), p. 899-911.

9. Feller, R. L., "Problems in Spectrophotometry," *London Conference on Museum Climatology*, International Institute for Conservation of Historic and Artistic Works, London (1968), p. 257-269.

10. Schweizerische Normen-Vereinigung, SNV 95 8 10, c/o EMPA-C Unterstrasse 11, 9001 St. Gallen, Switzerland. According to this document the standard blue-wool cloths are also available from BeuthVertriebs-GmbH, Berlin W 15 und Köln; and from Association pour la Determination de la Solidite des Teintures sur Textiles, Paris 8me, 12 Rue d'Anjou, France.

pairs of grey-painted paper panels which differ in contrast from zero (no contrast) in step 5 (established at a Munsell neutral value of 4.0) to a marked contrast in step 1. Step 4 in the geometric grey scale presents a small contrast (1.5 NBS units of color difference), step 3 twice as much (3 NBS units), step 2 twice as much as step 3 (6.0 NBS units) and step 1 twice as much as step 2 (12 NBS units). The five pairs of grey-painted papers are mounted on a black-covered slide-rule-like device with a window through which one pair at a time can be observed.

11. Feller, R. L., "Standards of Exposure to Light," *Bulletin of the American Group-IIC*, 4, No. 1 (1963), p. 10-12; see also *ibid.*, 7, No. 2 (1967), p. 8, 32.

12. British Standard BS2662:1961, available from British Standards Institution, Two Park Street, London W1, England and TALAS (see footnote 5).

To rate the degree of fading of the exposed blue-wool standards, each faded cloth in the series is placed side by side with the corresponding unexposed standard and the degree of contrast (degree of fading) of the exposed and unexposed blue cloths is then compared visually with the pair of greys closest in contrast. Employment of a grey mask that reveals only the exposed and adjacent unexposed blue cloth will make evaluation of the contrast relative to the geometric grey scale more reliable.

A typical degree of fade will be reported as follows: "a particular blue-wool standard, after a certain exposure time under specified conditions, has faded to a geometric grey-scale contrast of 2 (GGS2) or perhaps between 2 and 3 (GGS2-3, or GGS2.5)." The authors recommend that the fading of the blue cloths be rated by two to three observers, that the grey-scale contrast be estimated in terms of 1/4 steps, such as 2, 2.25, 2.5, 2.75 or 3 and that the average of the individual observer's ratings be taken. Because of the uncertainties of visual rating, the authors do not consider the use of more than three observers necessary.

The precision in an individual rating on the five-step grey scale is admittedly no more than ±0.5 grey scale units; however, the authors have found it useful to attempt to rate the fading in terms of the 1/4 steps. Moreover, we believe that the precision with which the contrasts can be estimated can be improved by use of a nine-step grey scale described in ASTM method D 2616-67.[13] This scale, in which the half steps between the five-step scale are provided, is available from the Munsell Color Company.[14] Use of the nine-step scale should more nearly improve the standard deviation in visual evaluations from ±0.5 to about ±0.25 GGS units.

13. American Society for Testing Materials ASTM D2616-67 (Reapproved 1972), Standard Method for Evaluating Change in Color with a Grey Scale.

14. Munsell Color Company, 2441 N. Calvert Street, Baltimore, MD 21218. (As of 1975—Ed.)

An illustration of the use of the blue-wool series as light exposure monitors is presented in Figure 1, reprinted from a previous publication.[9] The data are based on exposure of blue-wool cloths 1, 2 and 3 on the same north wall of a daylighted gallery illuminated at about 150,000 footcandle hours per year. The individual data points are based on the average visual rating by two experienced observers and represents the exposures of three sets of the cloths begun at different times over a ten-year period. The graph, therefore represents the average of replicate determinations made during the period. The uncertainty in the visual rating, ±0.4 GGS units, is indicated by the length of the lines drawn vertically through the individual points.

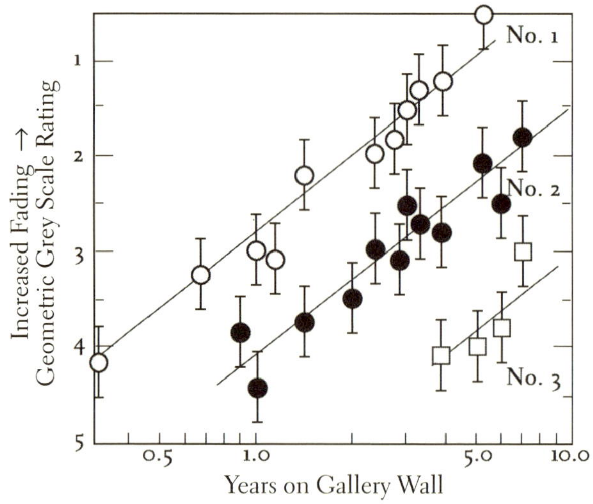

Figure 1
Fading of British Standard 1006:1971 blue-cloth standards 1, 2 and 3 on north wall of daylighted gallery.

Graphical Plotting of the Rates of Fading

Figure 2 gives a reference series of curves for change in the blue-wool cloths for "daylight" fluorescent lamps at an average intensity of about 1000 footcandles. Figures 3 and 4 summarize the results of fading on the roof and under an RS-type sunlamp. These curves, previously reported to the ICOM Committee for Conservation,[7] must be regarded as only general indications of the type of results to be expected; many more experiments will have to be done before the degree of precision can be stated with complete confidence. The relative spacing

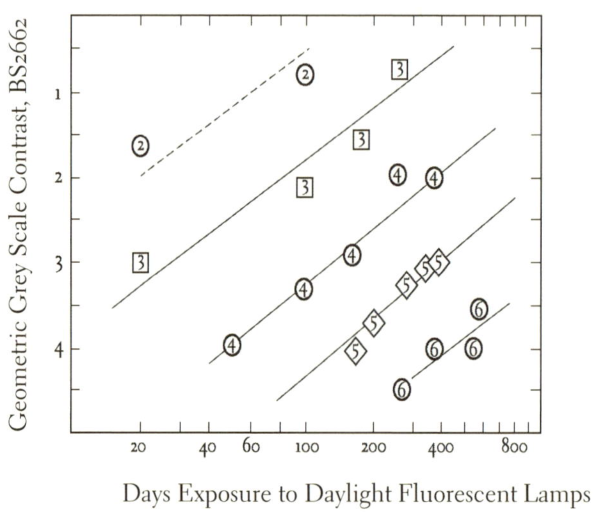

Figure 2
Fading of BS1006:1971 blue-wool standards under fluorescent lamps.

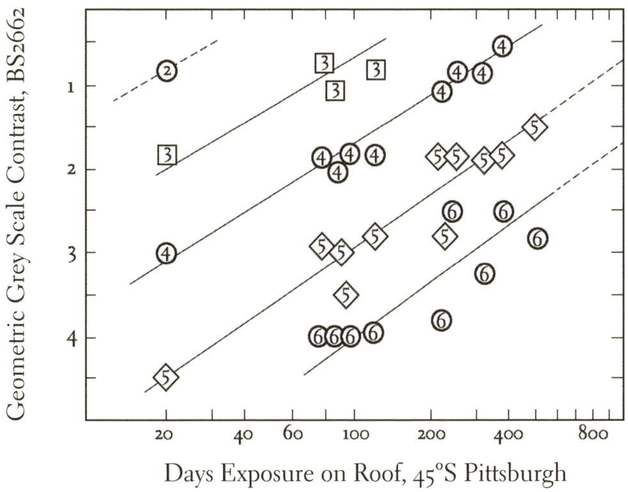

FIGURE 3
Fading of BS1006:1971 blue-wool standards on roof.

between the various standards and the slope of the curves must be further investigated. The slope varies from about 0.6 in Figures 2 and 3 to about 1.0 in Figures 1 and 4 although the reasons for these differences are not presently understood.[15]

The important point to be gained from these initial results is that the rate of change in grey-scale contrast is virtually linear when plotted against the log of exposure. The authors recommend that the observations of fading be plotted in this manner. The authors also recommend that two to three sets of the blue cloths be placed on exposure in a given location, each beginning after a

15. Jaeckel, S. M., Ward, C. D., and Hutchings, D. M., "Variations in Assessment of Lightfastness and in Rates of Fading and Spacing of the Blue Standards," *J. Society of Dyers and Colourists*, 79 (1963), p. 702-722.

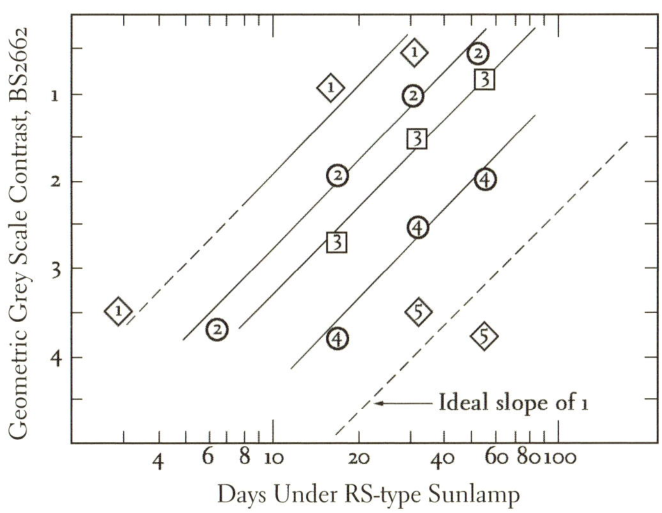

FIGURE 4
Fading of BS1006:1971 blue-wool standards.

convenient interval of time such as 6 months, in order to obtain multiple checks on the exposure results. Use of semilogarithmic paper for plotting makes the results easy to interpret: the grey scale rating is plotted linearly on the vertical axis and time (or intensity times time, if available) is plotted on the logarithmic, horizontal axis (thus making this direction a "geometric" scale also). For graph papers in which the equal grey-scale steps are one inch in dimension and a doubling on the log scale is also one inch, an "ideal slope" of 1.0 (direct proportionality) will be a 45° angle.

The stage of fading at which the standard or an unknown has faded to a geometric grey-scale contrast of 3 (GGS3) is a standard degree of fading that is recommended in ISO R105 as a general basis for comparison. This can be found with improved precision, we believe, from an "interpolation" of the graph of the observations.

Darkening of some of the blue-wool standards may occur initially before fading (lightening) occurs. When seeking to determine the onset of change, the region of "just perceptible difference" (GGS4), the darkening phenomenon may cause some confusion.[16] This is further reason for making multiple visual evaluations over a range of time and plotting curves for the purpose of interpolation.

16. Padfield, T., and Landi, S., "The Light Fastness of the Natural Dyes," *Studies in Conservation*, 11 (1966), p. 181-193.

Conclusions

Experience with a number of museums as well as with private collectors has shown the value of the visual evidence of light exposure provided by the blue-wool standards. The authors hope that the standards will become more widely used in the future for monitoring effective light exposure levels and that results will be reported in these terms for the purpose of intercomparison.

The ISO R105 standards are admittedly far from perfect and much is yet to be learned about their response to variations in humidity, temperature and spectral distribution of the illumination; research with these objectives is currently in progress. Nonetheless, the graphical method of plotting the changes occurring over a period of time and the use of multiple sets of the blue-cloth standards in any given location is recommended as a simple method of monitoring exposure that offers a practical, inexpensive and internationally recognized standard of reference.

Further Studies on the International Blue-Wool Standards for Exposure to Light

R. L. FELLER

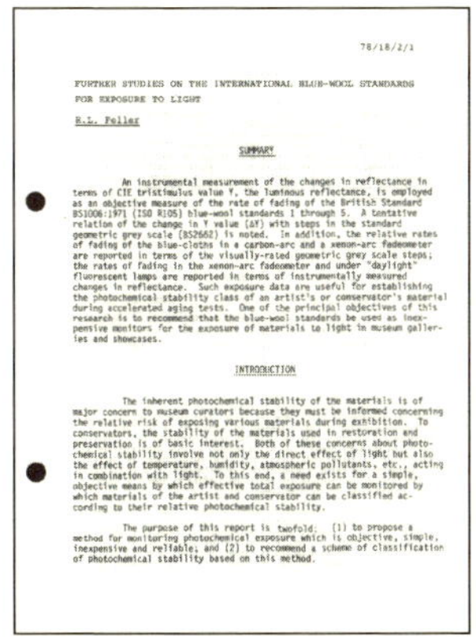
78/18/2/1

FURTHER STUDIES ON THE INTERNATIONAL BLUE-WOOL STANDARDS FOR EXPOSURE TO LIGHT

R.L. Feller

SUMMARY

An instrumental measurement of the changes in reflectance in terms of CIE tristimulus value Y, the luminous reflectance, is employed as an objective measure of the rate of fading of the British Standard BS1006:1971 (ISO R105) blue-wool standards 1 through 5. A tentative relation of the change in Y value (ΔY) with steps in the standard geometric grey scale (BS2662) is noted. In addition, the relative rates of fading of the blue-cloths in a carbon-arc and a xenon-arc fadeometer are reported in terms of the visually-rated geometric grey scale steps; the rates of fading in the xenon-arc fadeometer and under "daylight" fluorescent lamps are reported in terms of instrumentally measured changes in reflectance. Such exposure data are useful for establishing the photochemical stability class of an artist's or conservator's material during accelerated aging tests. One of the principal objectives of this research is to recommend that the blue-wool standards be used as inexpensive monitors for the exposure of materials to light in museum galleries and showcases.

INTRODUCTION

The inherent photochemical stability of the materials is of major concern to museum curators because they must be informed concerning the relative risk of exposing various materials during exhibition. To conservators, the stability of the materials used in restoration and preservation is of basic interest. Both of these concerns about photochemical stability involve not only the direct effect of light but also the effect of temperature, humidity, atmospheric pollutants, etc., acting in combination with light. To this end, a need exists for a simple, objective means by which effective total exposure can be monitored by which materials of the artist and conservator can be classified according to their relative photochemical stability.

The purpose of this report is twofold: (1) to propose a method for monitoring photochemical exposure which is objective, simple, inexpensive and reliable; and (2) to recommend a scheme of classification of photochemical stability based on this method.

Originally printed in:
Preprints of the 5th Triennial Meeting, Zagreb
ICOM Committee for Conservation (1978),
Paper 78/18/2.

Introduction

The inherent photochemical stability of the materials is of major concern to museum curators because they must be informed concerning the relative risk of exposing various materials during exhibition. To conservators, the stability of the materials used in restoration and preservation is of basic interest. Both of these concerns about photochemical stability involve not only the direct effect of light but also the effect of temperature, humidity, atmospheric pollutants, etc., acting in combination with light. To this end, a need exists for a simple, objective means by which effective total exposure can be monitored by which materials of the artist and conservator can be classified according to their relative photochemical stability.

The purpose of this report is twofold: (1) to propose a method for monitoring photochemical exposure which is objective, simple, inexpensive and reliable; and (2) to recommend a scheme of classification of photochemical stability based on this method.

Background

In the report made at the last triennial meeting (Paper 75/19/4)[1], the author proposed three classes of photochemical stability based on British Standard BS1006:1971.[2] (The International Standards Organization's (ISO) Recommendation R105, which we will refer to throughout this report, is the basis for British Standard BS1006:1971, the Swiss SNV 95 8 10 and similar documents issued by individual countries.) These proposed classes are

1. Feller, R. L., "Studies on Photochemical Deterioration," *Preprints of the 4th Triennial Meeting, Venice,* ICOM Committee for Conservation (1975), Paper 75/19/4.

2. British Standard 1006:1971, *Methods for the Determination of the Colour Fastness of Textiles to Light and Weathering,* International Organization for Standardization Recommendation R105.

summarized once again in Table 1. The intended Useful Lifetime in Table 1 is based on exposure on a well-illuminated museum gallery wall at about 150,000 footcandle hours of exposure per year.[1]

TABLE 1
Standards of Photochemical Stability for Materials in Conservation

Class	Classification	Intended Useful Lifetime	Approximate Equivalent Standard of Photochemical Stability
C	Unstable or Fugitive	less than 20 years	BS1006 Class 3 or less
B	Intermediate	(20-100 years)	(3 to 6 years)
A	Excellent	greater than 100 years	greater than BS1006 Class 6

The internationally recognized ISO R105 fading standards consist of eight different blue dyes applied to wool, each dye of which fades at a different rate. The individual dyes were selected so that each one in the series would fade at half the rate of its lower neighbor in the scale starting from the least permanent (most fugitive), No. 1, to the most stable, No. 8. Depending on the individual dye and the specific exposure conditions, this ideal relationship in the rate of fading is not strictly adhered to. Nevertheless, rate of fading as a function of the exposure (intensity multiplied by time) is linear over a wide range of exposure conditions.

The degree of fading of the blue-wool cloths can be reported by comparing them to a standard grey scale such as British Standard BS2662:1961.[3] (Other countries recognize and issue this same grey scale because it is also part of the International Standard Organization's R105.) This standard of comparison consists of five pairs of grey panels in which one half of each pair is the reference, set at a Munsell value of 4. The other pairs each contain chips of grey that are progressively lighter in the series. Thus the geometric grey scale contrast of panel 4 (GGS4) is equivalent to 1.5 NBS units of color difference, panel 3 (GGS3) has a pair of grey cards that differ by 3.0 NBS units of color difference; GGS2 is equivalent to 6.0 NBS units of color difference and GGS1 to 12 NBS units. To rate the degree of fading, the exposed (faded) blue cloth is compared to unexposed cloth and the degree of contrast (degree of fade) is compared visually with the degree of contrast in one of the four standard contrast steps in the geometric grey scale. Thus a typical degree of fade might be stated as follows: "a given blue-wool standard, after so many hours of exposure under specified conditions, has faded to a geometric grey scale contrast of between 2 and 3

3. British Standard BS2662:1961, available from British Standards Institution, Two Park Street, London W1, England.

(GGS2-3)." A plot of the grey-scale change vs. the logarithm of the exposure, as shown in Figures 1 and 2, is essentially linear, although the proportionality may require a power function other than 1.0 (in other words, the slope may not be ideally 1).

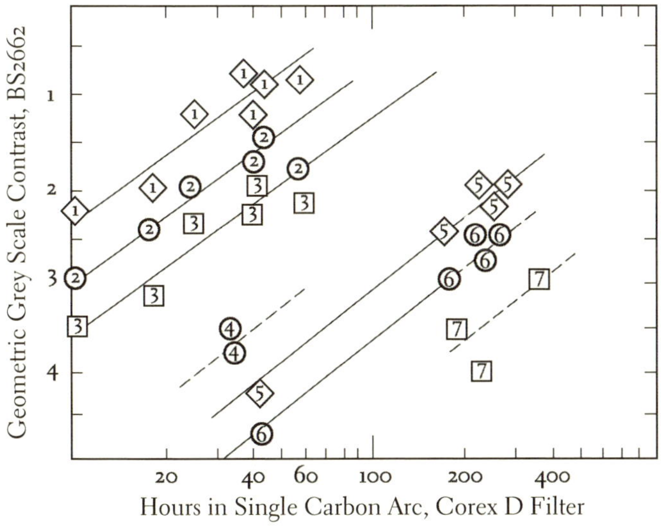

Figure 1
Fading of BS1006:1971 blue-wool standards in Fadeometer.

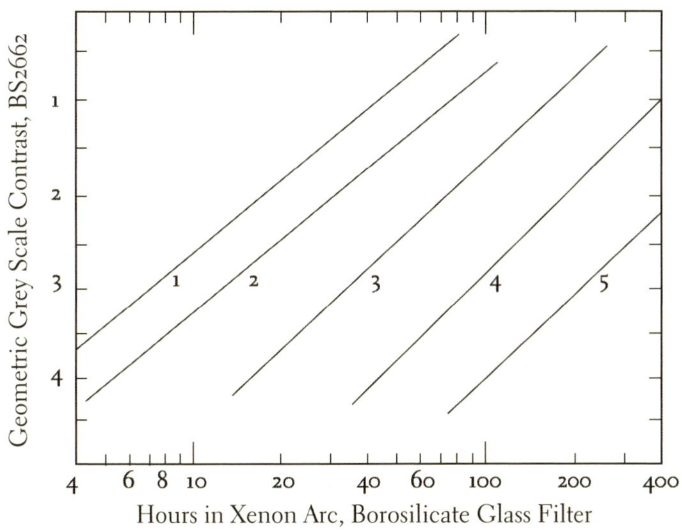

Figure 2
Fading of BS1006:1971 blue-wool standards in Fadeometer.

A convenient form of these ISO standards for museums is the small card available from Schweizerische Normen-Vereinigung.[4] One of these was illustrated in a recent publication.[5] The card contains rows of the eight standards, each lightly glued adjacent to one another and in order. Each small piece of the standard cloths measures 0.9 x 4.5 cm. The total size of the card is 12 x 4.5 cm., leaving room for labeling and attachment. During exposure, one half of the set of standards is either cut off and set aside for an unexposed "control" or is covered with an opaque material such as aluminum foil; the other half is left exposed to the light. The amount of fading can be periodically evaluated relative to the four-step geometric grey scale.

4. Schweizerische Normen-Vereinigung SNV 95 8 9-10. According to this document the standard blue-wool cloths are also available from Beuth-Vertriebs-GmbH, Berlin W 15 und Köln; and from Association pour la Determination de la Solidite des Teintures sur Textiles, Paris 8me, 12 Rue d'Anjou, France.

5. Feller, R. L., "Speeding Up Photochemical Deterioration," *Bulletin de L'Institut Royal du Partimoine Artistique*, 15 (1975), p. 135-150.

Rates of Fading Relative to Geometric Grey Scale

It is frequently useful to be able to compare relative speed of fading or other photochemical changes under different conditions of exposure in a museum showcase, on a gallery wall, or under accelerated-aging conditions. Our last report (75/19/4) presented curves of the rate of fading of blue-wool standards under four different conditions: (a) a north wall in a gallery at about 150,000 footcandle hours per year (about 1.5 million lux hours/year), (b) an RS-type sunlamp, (c) "daylight" fluorescent lamps and (d) under glass on the roof in Pittsburgh facing south. The degree of fading in these curves was visually evaluated by comparison with the Geometric Grey Scale, British Standard BS2662. At this time we wish to report similar curves for exposure in a xenon-arc Fadeometer (borosilicate-glass filters) (Figure 2) and in an old model Atlas single-carbon-arc Fadeometer ("Sunshine" carbons, Corex D glass) (Figure 1). The xenon-arc machine is an Atlas Electric Devices 600 WRC Weatherometer with the samples carried on a drum of 60 cm.-diameter instead of the standard 96 cm. The slopes of these curves vary from 0.8 to 0.9 for the xenon-arc to 0.7 to 0.8 for the carbon arc.

Increased Precision in Measuring Fading of Blue Cloths

Admittedly, the precision in visually rating the degree of fading in terms of the grey scale is limited. To increase the precision of such fading curves, an objective instrumental measurement of fading is needed. Several years ago we reported some preliminary results of using spectrophotometric measurements to increase the precision of reporting fading rather than visual comparison with the geometric grey scale.[6] In the past year The Materials Center of Carnegie Mellon University has completed extensive measurements of the rate of fading of the blue-wool standards using the Small-Area-View (SAV) Color-Eye instrument[6] and the small cards supplied by the Schweizerische Normen-Vereinigung. The SAV Color-Eye is able to measure the reflectance of samples that are only 1/4" in diameter. Typical results are shown in Figures 3 and 4. The curves in Figures

6. Feller, R. L., "Problems in Spectrophotometry," *London Conference on Museum Climatology*, International Institute for Conservation of Historic and Artistic Works, London (1968), p. 257-269.

2 and 3—one based on visual ratings and the other on instrumental—are derived from the same exposed standards. No individual observations are reported in Figure 2 because the extensive numbers of data points are too numerous to be placed in this one diagram; they will be reported in a subsequent publication.

In Figure 4, data are reported in terms of footcandle hours under a bank of "daylight" fluorescent lamps in a room maintained at 50% R.H. and at sample

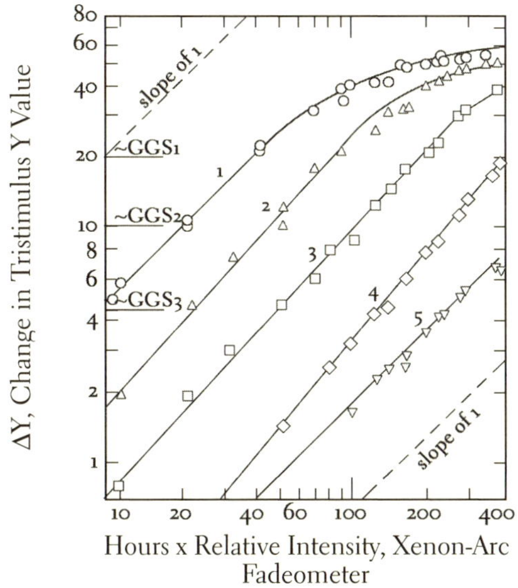

Figure 3
Fading of BS1006:1971 Blue-Wool Standards 1 to 5 in Xenon-Arc Fadeometer, Borosilicate-Glass Filter.

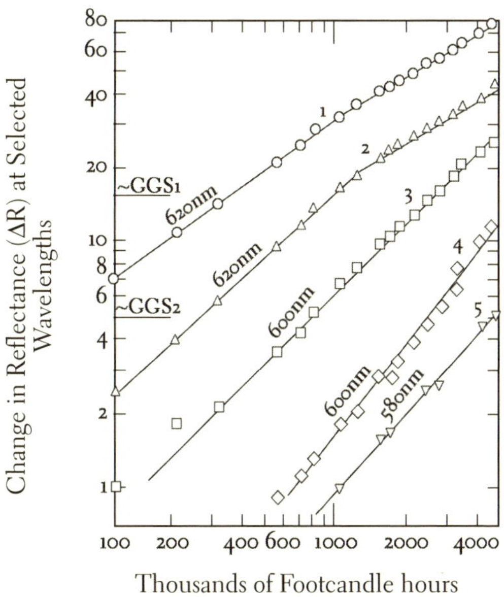

Figure 4
Fading of BS1006:1971 Blue-Wool Standards 1 to 5 under "Daylight" Fluorescent Lamps.

temperatures of about 80°F. These data, by Dr. Jonathan Arney of the Center's staff, are based on the reflectance at various wavelengths selected to be close to the point of maximum absorption of the particular blue-wool standard. In 1963 Friele reported similar results in terms of millions of lux hours in the Xenotest apparatus although he simply used the reflectance at 620 for all standards.[7]

7. Friele, L. F. C., "A Comparative Study of Natural and Xenotest Exposure Conditions for Measuring Fading and Degradation," *J. Society of Dyers and Colourists*, 79 (1963), p. 623-631; see also Friele, L. F. C., *Textilveredlung*, 5 (1970), p. 899-911.

We have analyzed the results of the xenon-arc exposures (Figure 3) both using the luminous reflectance (Y) and the wavelength of maximum absorption and find equally good correlation either with the Y value or with the reflectance values at the approximate wavelength of maximum absorption, 620 nm, the measurement Friele used. Correlation coefficients of Y values versus exposure (relative intensity multiplied by the time in the xenon-arc Fadeometer with borosilicate glass filters) were as follows: 0.994 for blue cloth No. 1 and 0.995 for blue cloth No. 2 (using data only up to geometric grey scale contrast of 1), 0.992 for No. 3, 0.978 for No. 4, 0.974 for No. 5 and 0.973 for No. 6. We are convinced that the use of objectively measured values in terms of Y (which we prefer to use at present), or 620 nm or the wavelength of principal absorption are all more satisfactory to follow the rate of fading than plots of color difference (ΔE) as has been employed by Berger and Brockes and others.[8]

8. Berger, A., and Brockes, A., "Zur Beurteilung des Ausbleichens von Lichtechtheitstypen mit Farbdifferenzformeln," *Die Farbe*, 11 (1962), p. 263-274.

In Figure 4, one can observe that the spacing between the five standards and the steeper slope for standard number 4 when exposed to fluorescent lamps are similar to the behavior under the xenon-arc shown in Figure 3. Data for blue cloth No. 6 are not shown in three of the curves presently reported because its low rate of fading is so slow that insufficient data were available at the time of the preparation of this report.

RELATIONSHIP OF CHANGE IN REFLECTANCE AND GEOMETRIC GREY SCALE CONTRAST

It should be possible to relate a given degree of change in instrumentally measured luminous reflectance or spectral reflectance of the samples to a given visually rated grey scale contrast between faded and unfaded samples. Berger and Brockes determined, for example, the color difference (ΔE) that was related to the geometric grey scale 3.[8] Our initial results, shown in Figure 3, indicate the changes in luminous reflectance (ΔY) that are approximately equivalent to geometric grey scale contrasts 3, 2, and 1. This work will be confirmed more thoroughly through additional experiments. The principal purpose of these initial observations, however, was to show that the plots of the logarithm of time vs. change in reflectance in Figures 3 and 4 was linear throughout the range of visual change in which the blue cloths are normally used—that is, over a range in grey scale contrast from 4 to 1. The equivalent grey scale changes using the reflectance at the wavelength of maximum absorption in Figure 4 resemble those in Figure 3 but do not occur at precisely the same intervals.

The reason for the marked nonlinearity of the change in Y value at high degrees of fading, seen especially in blue cloths Nos. 1 and 2, is that the dye has almost completely faded at this stage and the wool itself has begun to yellow.

Limitations to Present Data

The importance of the four curves reported in 1975 and two presented now is to demonstrate the principle that such curves can be developed with practice and little expense, simply by visually comparing the degree of fading of the blue-wool standards with the steps on the geometric grey scale, very much as the recommended procedure in the International Standards Organization's R105. The plot of grey scale change vs. the logarithm of the exposure tends to be linear. ("Exposure" is time multiplied by intensity of illumination, or simply the time if the intensity is practically constant.)

It is recognized that the uncertainty in such visual ratings is considerable, likely to be no better than ±0.5 grey scale units, even if three to five persons rate the degree of fading in terms of the grey scale and an average of their ratings is taken.[9] This degree of uncertainty is indicated in our data for fading on a gallery wall (Figure 1 in Report 75/19/4)[1] and by the scatter of the observations in the present Figure 1. It is for this reason that all of the visually rated curves that the author has reported have been based on the average of at least three separate sets of blue-cloth exposures rated by two observers. (Figures 1, 2 and 3 of this present report are based on the average of six separate exposures and the average visual rating by two observers.)

9. Jaeckel, S. M., Ward, C. D., and Hutchings, D. M., "Variations in Assessment of Lightfastness and in Rates of Fading and Spacing of the Blue Standards," *J. Society of Dyers and Colourists*, 79 (1963), p. 702-722.

The six sets of visually rated curves and the instrumentally measured curves in Figures 3 and 4 must be regarded as preliminary findings, giving only approximate rates of fading. Much more work will have to be done to determine the variations that will be encountered when different investigators attempt to duplicate the measurements under similar but obviously not identical conditions of exposure.

Conclusion

The author's purpose in reporting the sets of fading curves under six different exposure conditions has been first of all to demonstrate the principle of how to obtain such plots and, secondly, to establish an initial set of curves in which the relative rates of fading of the blue-wool standards under various exposure conditions can be further studied and compared. These preliminary measurements, for example, have provided us with initial estimates of how long it would take for accelerated aging tests to determine if a material is in class C or class A of the proposed standards of photochemical stability (Table 1).

Another potential use of these standards is as an inexpensive and unobtrusive monitor for exposure. Light intensity on museum walls and showcases usually varies with the time of day and season of the year. In a sense, the blue-wool standards "integrate" the varying intensity, providing an indication of the net amount of exposure (intensity vs. time, along with a certain influence of temperature and humidity). Thus, if we determine their rates of fading—that is, the slope of the curves—then by observation of the rate of fading of the most fugitive standards, we can predict how long it will take to accumulate a given number of lux hours (shall we say) in a particular location. For example, we know that alizarin tints are equivalent to BS1006:1971 standards 4-5 on lightfastness, and gamboge washes are equivalent to standard cloths 2-3.[10] Many natural dyes on textiles fade at a rate equivalent to BS1006 blue cloths 2 and 3.[11] Thus, after brief exposure of a set of the blue cloths in a given location in a museum, and observation of the degree of fading of standards 1 or 2, one can predict with some assurance how long it would take materials equivalent to standards 3 and 4, such as alizarin or gamboge, to fade in the same location either to a geometric grey scale contrast of 4 (GGS4, just perceptible difference) or to a more severe degree of fading such as GGS3, 2 or 1.

Additional measurements will be carried out in the future to increase the precision and reliability of these initial findings. We intend to repeat the measurements many times over in the coming three-year period, using spectrophotometric measurements to increase the precision in which the rates of fading can be followed.[12]

10. Feller, R. L., "Standards of Exposure to Light," *Bulletin of the American Group - IIC*, 4, No. 1 (1963), p. 10-12.

11. Padfield, T., and Landi, S., "The Light Fastness of the Natural Dyes," *Studies in Conservation*, 11 (1966), p. 181-193.

12. Friele recommends the use of $\Delta R/R$ at 620 nm instead of ΔY or ΔR. Full consideration will be given to his views in this matter.

Acknowledgment

The data in Figures 1, 2 and 3 were obtained by Mrs. R. A. Curran, who also determined the correlation coefficients and slopes for the curves.

Use of the International Standards Organization's Blue-Wool Standards for Exposure to Light. II. Instrumental Measurement of Fading

Robert L. Feller and Ruth Johnston-Feller

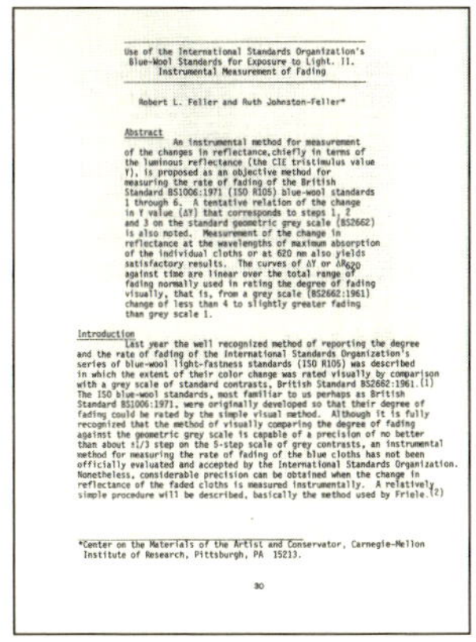

Use of the International Standards Organization's Blue-Wool Standards for Exposure to Light. II. Instrumental Measurement of Fading

Robert L. Feller and Ruth Johnston-Feller*

Abstract

An instrumental method for measurement of the changes in reflectance, chiefly in terms of the luminous reflectance (the CIE tristimulus value Y), is proposed as an objective method for measuring the rate of fading of the British Standard BS1006:1971 (ISO R105) blue-wool standards 1 through 6. A tentative relation of the change in Y value (ΔY) that corresponds to steps 1, 2 and 3 on the standard geometric grey scale (BS2662) is also noted. Measurement of the change in reflectance at the wavelengths of maximum absorption of the individual cloths or at 620 nm also yields satisfactory results. The curves of ΔY or ΔR_{620} against time are linear over the total range of fading normally used in rating the degree of fading visually, that is, from a grey scale (BS2662:1961) change of less than 4 to slightly greater fading than grey scale 1.

Introduction

Last year the well recognized method of reporting the degree and the rate of fading of the International Standards Organization's series of blue-wool light-fastness standards (ISO R105) was described in which the extent of their color change was rated visually by comparison with a grey scale of standard contrasts, British Standard BS2662:1961.(1) The ISO blue-wool standards, most familiar to us perhaps as British Standard BS1006:1971, were originally developed so that their degree of fading could be rated by the simple visual method. Although it is fully recognized that the method of visually comparing the degree of fading against the geometric grey scale is capable of a precision of no better than about ±1/3 step on the 5-step scale of grey contrasts, an instrumental method for measuring the rate of fading of the blue cloths has not been officially evaluated and accepted by the International Standards Organization. Nonetheless, considerable precision can be obtained when the change in reflectance of the faded cloths is measured instrumentally. A relatively simple procedure will be described, basically the method used by Friele.(2)

*Center on the Materials of the Artist and Conservator, Carnegie-Mellon Institute of Research, Pittsburgh, PA 15213.

30

Originally printed in:
Preprints of Papers Presented at the Seventh Annual Meeting, Toronto, Canada
American Institute for Conservation, Washington, DC (1979), Pages 30-36.

Introduction

Last year the well-recognized method of reporting the degree and the rate of fading of the International Standards Organizations series of blue-wool lightfastness standards (ISO R105) was described in which the extent of their color change was rated visually by comparison with a grey scale of standard contrasts, British Standard BS2662:1961.[1] The ISO blue-wool standards, most familiar to us perhaps as British Standard BS1006:1971, were originally developed so that their degree of fading could be rated by the simple visual method. Although it is fully recognized that the method of visually comparing the degree of fading against the geometric grey scale is capable of a precision of no better than about 1/3 step on the 5-step scale of grey contrasts, an instrumental method for measuring the rate of fading of the blue cloths has not been officially evaluated and accepted by the International Standards Organization. Nonetheless, considerable precision can be obtained when the change in reflectance of the faded cloths is measured instrumentally. A relatively simple procedure will be described, basically the method used by Friele.[2]

Methods of Reporting Changes in Reflectance

It is natural to want a scale of numbers, a "yard-stick", with which to report changes. Yet color is three dimensional and the attempt to find a suitable scale of units with which to express the degree of fading in colored materials continues to occupy the attention of authorities in the paint, textile and plastic industries. Various "color-difference equations" are usually employed. These equations

1. Feller, R. L., and Johnston-Feller, R. M., "Use of the International Standards Organization's Blue-Wool Standards for Exposure to Light. I. Use as an Integrating Light Monitor for Illumination Under Museum Conditions," *Preprints of Papers Presented at the Sixth Annual Meeting, Fort Worth, Texas*, American Institute for Conservation, Washington, DC(1978), p. 73-80.

2. Friele, L. F. C., "A Comparative Study of Natural and Xenotest Exposure Conditions for Measuring Fading and Degradation," *J. Soc. Dyers and Colourists*, 79 (1963), p. 623-631; see also Friele, L. F. C., *Textilveredlung*, 5 (1970), p. 899-911.

pose many problems, partially because they have been designed generally to correlate with the visual estimates of small color differences and never were expected to be especially appropriate in expressing very large differences that can occur during extensive discoloration and fading of paints and textiles. Moreover, the usual color-difference calculations fail to take into consideration changes in gloss and the development of chalking in the case of paints.[3]

An extensive study of the use of color difference equations in the rating of the ISO blue-wool cloths was made more than fifteen years ago by Berger and Brockes.[4] However, these authors were concerned with following the fading of the blue cloths only up to grey scale contrast 3. This corresponded to a color difference, ΔE, by the Nickerson Stultz equation of no more than 20,[5] or 22 to 28 based on the MacAdam color difference equation.[6] The exposures outdoors under glass at a 45° angle facing south ran from 50,000 footcandle hours for blue cloth No. 1 to about 100x that for blue cloth No. 6.

Last year, Vaeck in Belgium proposed a new color difference formula:

$$\Delta E = [\Delta L^2 + \Delta A,B]^{1/2} \qquad (1)$$

where
$$L = 25(100Y/Y_o)^{1/3} - 16$$
$$A = 500[(X/X_o)^{1/3} - (Y/Y_o)^{1/3}]$$
$$B = 200[(Y/Y_o)^{1/3} - (Z/Z_o)^{1/3}]$$
$$\Delta A,B = (\Delta A^2 + \Delta B^2)^{1/2}$$

and X, Y and Z are the tristimulus values of the color in the standard CIE system. Illuminant D65 and the 10° observer were used.

The principal difference between this and the usual ANLAB or CIELAB formulas is that ΔA,B is not squared. Vaeck claims that this equation gives excellent correlation between visually rated grey-scale steps and values of ΔE based on instrumental measurements of the tristimulus values X, Y and Z.[7] His suggestion will undoubtedly receive careful consideration by ourselves and others.

For the moment we have rejected the use of color difference equations in favor of measurement of changes in luminous reflectance (ΔY) or in the reflectance at a wavelength near or at the point of maximum absorption (ΔR). These measurements were employed by Friele[2] but apparently his practice aroused little interest. We have found, in the results to be reported here, that curves of change in reflectance versus exposure (footcandle hours or intensity x time), for cloths 1 through 6, have proven to be linear throughout a total change that corresponds to a degree of fading less than grey scale contrast 4 to slightly more than that which corresponds to grey scale contrast 1. Figure 1 shows the data obtained by Mrs. R. A. Curran of this laboratory using changes in luminous reflectance (ΔY) vs. total exposure (time x relative intensity of illumination) in a xenon-arc Fadeometer equipped with a Pyrex-glass filter.

3. Johnston-Feller, R. M., and Osmer, D., "Exposure Evaluation: Quantification of Changes in Appearance of Pigmented Materials," *J. Coatings Technology*, 49 (1977), p. 25-36.

4. Berger, A., and Brockes, A., "Zur Beurteilung des Ausbleichens von Lichtechtheitstypen mit Farbdifferenzformeln," *Die Farbe*, 11 (1962), p. 263-274.

5. Nickerson, D., and Stultz, K. F., "Color Tolerance Specification," *J. Optical Soc. Amer.*, 34 (1944), p. 550-570.

6. MacAdam, D. L., "Specification of small color differences," *J. Optical Soc. Amer.*, 33 (1943), p. 18-26. Using values of f (ΔY) of Davidson, H. R., and Hanlon, J. J., *J. Optical Soc. Amer.*, 45 (1955), p. 617-620.

7. Vaeck, S. V., "An improved color-difference formula for use in lightfastness testing," *J. Soc. Dyers and Colourists*, 94 (1978), p. 301-306.

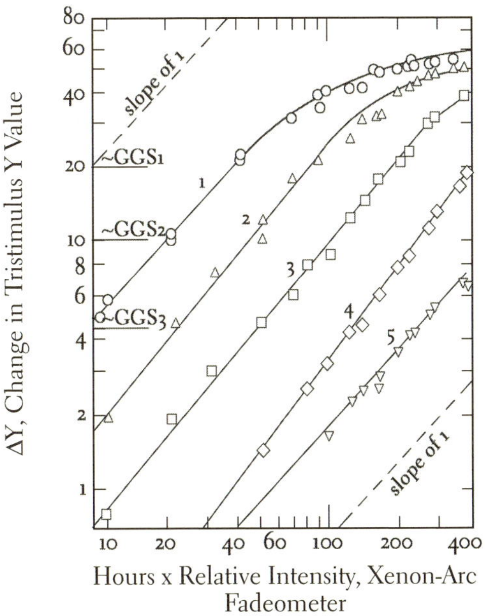

FIGURE 1
Fading of BS1006:1971 blue-wool standards 1 to 5 in xenon-arc Fadeometer, borosilicate-glass filter.

Figure 2 shows the data obtained by Dr. J. Arney, also of this laboratory, based on measurements of the change in reflectance at the wavelength of maximum absorption of each standard cloth in the series. Marked nonlinearity at high

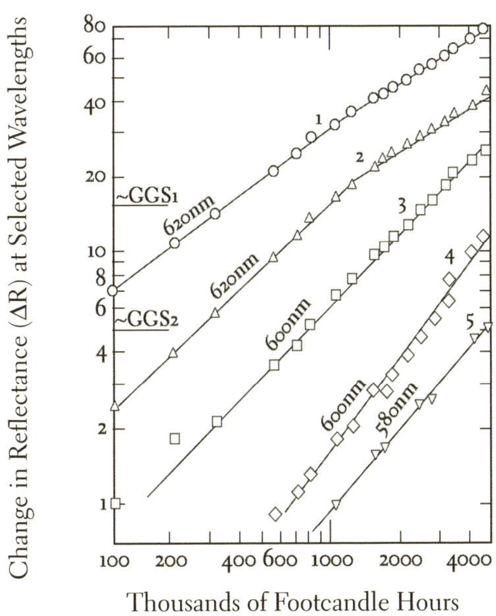

FIGURE 2
Fading of BS1006:1971 blue-wool standards 1 to 5 under "daylight" fluorescent lamps.

degrees of fading, seen particularly in blue cloths Nos. 1 and 2, is to be expected because the dye has almost completely faded at this stage and the discoloration of the wool affects the results.

METHOD OF MEASUREMENT

The data in Figures 1 and 2 were reported at the ICOM meeting in Zagreb, October 1978.[8] Reflectances were measured using the Kollmorgan Small-Area-View (SAV) Color-Eye® instrument.[9] The ISO R105 blue cloths were those mounted on small cards as supplied by the Schweizerische Normen-Vereinigung and other standards organizations on the continent.[1] The dimensions of the individual strips of cloth are about 0.9 x 4.5 cm. The SAV Color-Eye® is able to measure the reflectance of the small strips of the standards in an area that is only 1/4" in diameter. The data in Figure 2, obtained by exposure of the blue-wool standards under a bank of "daylight" fluorescent lamps in a room maintained at 50% relative humidity (sample temperature about 80°F), are based on the reflectance at wavelengths selected to be close to the initial point of maximum absorption of the particular blue-wool standard. In 1970 Friele reported similar results in terms of millions of lux hours in the Xenotest apparatus although he simply used the reflectance at 620 nm for all standards.[2] We have obtained much the same results as those in Figure 2 when the reflectance at 620 nm was used throughout.

8. Feller, R. L., "Further Studies on the International Blue-Wool Standards for Exposure to Light," *Preprints of the 5th Triennial Meeting, Zagreb*, ICOM Committee for Conservation (1978), Paper 78/18/2.

9. Feller, R. L., "Problems in Spectrophotometry," *London Conference on Museum Climatology*, International Institute for Conservation of Historic and Artistic Works, London (1968), p. 257-269.

CORRELATION COEFFICIENTS OF REFLECTANCE CHANGE VS. EXPOSURE

We have analyzed the results of the xenon-arc exposures (Figure 1) using the change in luminous reflectance (ΔY) as well as in reflectance at 620 nm and have found equally good correlation either with the ΔY value or with ΔR, the reflectance values at the approximate wavelength of maximum absorption, 620 nm (the measurement Friele used). Correlation coefficients of ΔY values versus exposure (relative intensity multiplied by time) were as follows: 0.994 for blue cloth No. 1 and 0.995 for blue cloth No. 2 (using data only up to geometric grey scale contrast of 1), 0.992 for No. 3, 0.978 for No. 4, 0.974 for No. 5 and 0.973 for No. 6. The lower correlation coefficients for cloths 4, 5 and 6 arose in part because the total extent of fading was less than that for cloths 1, 2 and 3 at the end of the experiment. The data were plotted on a log-log graph paper for convenience.

We are convinced that the use of objectively measured values in terms of ΔY (which we prefer to use at present), or of the change in reflectance at 620 nm (or at the wavelength of principal absorption), are all more satisfactory for following the rate of fading than determinations based on recognized color-difference equations. However, the equation of Vaeck will be considered in future investigations.

Relationship of Change in Reflectance and Geometric Grey Scale Contrast

It should be possible to relate a given difference in measured reflectance between faded and unfaded samples to a visually rated grey-scale contrast. Berger and Brockes determined, for example, the color-difference (ΔE) that was related to the geometric grey scale 3.[4] Vaeck has compared all four grey-scale steps with values of ΔE calculated from his modified equation. Our initial results, shown in Figures 1 and 2, indicate the changes in luminous reflectance (ΔY), and (ΔR), that are approximately equivalent to geometric grey scale contrasts (GGS) 3, 2 and 1. The purpose of noting the approximate location of the grey-scale steps was to show that the log-log plots of exposure vs. change in reflectance in Figures 1 and 2 are linear throughout the range in which the blue cloths are normally used in the visual rating method—that is, over a range in grey scale contrasts from about 4 to 1.

In Figure 2, one can observe that both the general spacing between the five standards, and the steeper slope for standard number 4, found during exposure to fluorescent lamps, are similar to the behavior under the xenon-arc shown in Figure 1. Such evidence would not be as convincingly presented if one had to rely on visual ratings of the cloths.[1]

Suggested Practice to Obtain Objective (Instrumentally Determined) Measurements of the Degree of Change in the Blue-wool Fading Standards

Because many instruments have the capability of measuring luminous reflectance, we suggest that the change in luminous reflectance (ΔY) be measured for each blue cloth as an objective measure of the extent of fading. If so desired, the reflectance at 620 nm (ΔR_{620}) can be used instead. Friele suggested the calculation of ΔR/R (ΔY/Y can also be used) to allow for differences in the depth of dyeing of the standards prepared on different occasions, a problem known to occur in varying degree. For any given set of dyed standards in one's possession, this does not seem to be of major importance. However, it is recommended that the initial reflectance values of each standard be reported for the record in case a calculation of the change relative to initial values should later prove of interest.

The type of instrument used for making the measurements should always be reported. If the change in the Y tristimulus value is used, as recommended here, the method of measurement needs to be described so that intelligent comparison can be made between laboratories. Is the tristimulus value obtained from a tristimulus colorimeter or by integration of a spectrophotometric curve from a spectrophotometer? What is the geometry of illumination and viewing? What is the size of the area of sample measured? Description in terms of the instrument manufacturer and the model number will eliminate questions.

The method of sample presentation for measurement is vital because one layer of cloth does not completely hide the color of the substrate. If only one layer of cloth is used, the color of the backing becomes important and must be specified. When the individual pieces of cloth are used, three layers are recommended for measurement, eliminating any effect from the substrate (backing). (Only the top layer need be the exposed cloth.) However, when the cards are used which have only a single layer of cloth over a white paperboard substrate, measured absolute results will be different from those measured with multiple layers, so the technique to be used must be constant for any series and specified for precise communication between laboratories. The results described here were obtained using the latter type of standards, a single layer mounted on cards. As in all measurements, the reflectance standard used should also be specified. In the measurements reported here, $BaSO_4$ was the white reference standard. The reflectance of the white cardboard on which the set of blue cloths were mounted, measured on the back of the cards, was 85.7% relative to barium sulfate.

Summary and Conclusion

The purpose of this report is first of all to emphasize the excellent precision that can be obtained by instrumental measurement of the degree of fading of the ISO blue-wool fading cloths and secondly to suggest that measurements of luminous reflectance (CIE "Y") values can be used for this purpose rather than calculations of color difference, ΔE. It is recommended that the information be plotted in terms of ΔY vs. time on linear or logarithmic graph paper. If desired, measurements of the reflectance at 620 nm can be used with similar results. It is further recommended that the initial values of reflectance be recorded so that the data may be further expressed in terms of $\Delta Y/Y$ or $\Delta R/R$ if so desired.

Measurement of the rate of fading in this manner will improve the precision in the work of any given laboratory. Because of expected variations owing to different instruments and different ways of presenting the samples for measurement, it is expected that the resulting curves will differ in detail between investigators using different methods. To document the potential sources of these expected differences it is suggested that the instrument used be described and that the reflectance values of the unexposed cloths be reported. The cloths should be backed up with white (the present method) or by a black background (the reflectance of which white or black is also reported) or backed by at least two layers of the identical cloth standards of which the top one only would be the exposed sample; the two behind, unexposed.

It is further suggested that the cloths be rated visually in the usual manner[1] at the same time that they are instrumentally measured. This procedure will provide a recognized reference to the extent of exposure and change being monitored.

Acknowledgment

Measurements of six different sets of the standards during exposure in the xenon-arc Fadeometer and calculations of the correlation coefficients for these data were carried out by Mrs. Curran.

Standards in the Evaluation of Thermoplastic Resins

ROBERT. L. FELLER

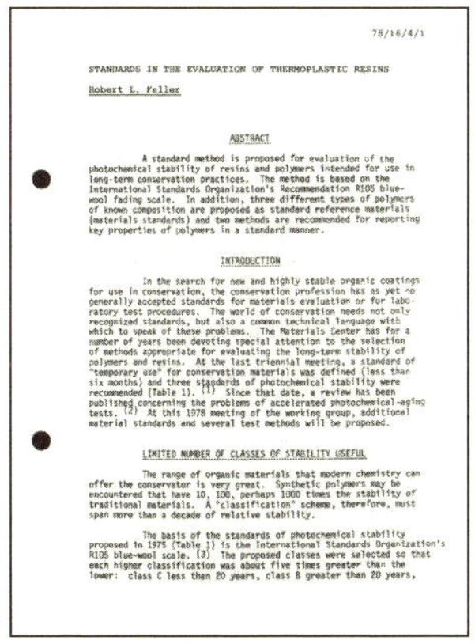

78/16/4/1

STANDARDS IN THE EVALUATION OF THERMOPLASTIC RESINS

Robert L. Feller

ABSTRACT

A standard method is proposed for evaluation of the photochemical stability of resins and polymers intended for use in long-term conservation practices. The method is based on the International Standards Organization's Recommendation R105 blue-wool fading scale. In addition, three different types of polymers of known composition are proposed as standard reference materials (materials standards) and two methods are recommended for reporting key properties of polymers in a standard manner.

INTRODUCTION

In the search for new and highly stable organic coatings for use in conservation, the conservation profession has as yet no generally accepted standards for materials evaluation or for laboratory test procedures. The world of conservation needs not only recognized standards, but also a common technical language with which to speak of these problems. The Materials Center has for a number of years been devoting special attention to the selection of methods appropriate for evaluating the long-term stability of polymers and resins. At the last triennial meeting, a standard of "temporary use" for conservation materials was defined (less than six months) and three standards of photochemical stability were recommended (Table 1). (1) Since that date, a review has been published concerning the problems of accelerated photochemical-aging tests. (2) At this 1978 meeting of the working group, additional material standards and several test methods will be proposed.

LIMITED NUMBER OF CLASSES OF STABILITY USEFUL

The range of organic materials that modern chemistry can offer the conservator is very great. Synthetic polymers may be encountered that have 10, 100, perhaps 1000 times the stability of traditional materials. A "classification" scheme, therefore, must span more than a decade of relative stability.

The basis of the standards of photochemical stability proposed in 1975 (Table 1) is the International Standards Organization's R105 blue-wool scale. (3) The proposed classes were selected so that each higher classification was about five times greater than the lower: class C less than 20 years, class B greater than 20 years,

Originally printed in:
Preprints of the 5th Triennial Meeting, Zagreb
ICOM Committee for Conservation (1978), Paper No. 78/16/4.

Introduction

In the search for new and highly stable organic coatings for use in conservation, the conservation profession has as yet no generally accepted standards for materials evaluation or for laboratory test procedures. The world of conservation needs not only recognized standards, but also a common technical language with which to speak of these problems. The Materials Center has for a number of years been devoting special attention to the selection of methods appropriate for evaluating the long-term stability of polymers and resins. At the last triennial meeting, a standard of "temporary use" for conservation materials was defined (less than six months) and three standards of photochemical stability were recommended (Table 1).[1] Since that date, a review has been published concerning the problems of accelerated photochemical-aging tests.[2] At this 1978 meeting of the working group, additional material standards and several test methods will be proposed.

Limited Number of Classes of Stability Useful

The range of organic materials that modern chemistry can offer the conservator is very great. Synthetic polymers may be encountered that have 10, 100, perhaps 1000 times the stability of traditional materials. A "classification" scheme, therefore, must span more than a decade of relative stability.

The basis of the standards of photochemical stability proposed in 1975 (Table 1) is the International Standards Organization's R105 blue-wool scale.[3] The proposed classes were selected so that each higher classification was about five

1. Feller, R. L., "Studies on the Photochemical Stability of Thermoplastic Resins," *Preprints of the Fourth Triennial Meeting, Venice*, ICOM Committee for Conservation (1975), Paper 75/22/4.

2. Feller, R. L., "Speeding Up Photochemical Deterioration," *Bulletin de l'Institut Royal du Patrimoine Artistique*, 15 (1975), p. 135-150.

3. British Standard 1006:1971, "Methods for the Determination of Colour Fastness of Textiles to Light and Weathering," (International Standards Organization ISO Recommendation R105); Schweizerischer Verband Materialprüfungen der Technik, SNV 95809, 1961 and similar documents issued by other countries that recognize ISO R105.

times greater than the lower: class C less than 20 years, class B greater than 20 years, class A2 greater than 100 years, and class A1 greater than 500 years.

TABLE 1
Standards of Intended Use and Photochemical Stability for Materials in Conservation

Class	Classification	Intended Useful Lifetime	Approximate Equivalent Standard of Photochemical Stability
T	Materials in Temporary Contact	less than 6 months	–
C	Unstable or Fugitive	less than 20 years	BS1006* Class 3 or less
B	Intermediate	(20-100 years)	(Between Class 3 and 6)
A	Excellent	(A2?) greater than 100 years (A1?) greater than 500 years	greater than BS1006* Class 6 5x more stable than A2

*British Standard BS1006:1971 is equivalent to the International Standards Organization (ISO) Recommendation R105.

A larger number of "classes" of stability could be chosen—there are, after all, *eight* blue-wool standards. However, three basic classes should prove adequate and should do much to minimize disputes regarding "borderline" cases of stability.

MINIMUM PHOTOCHEMICAL STABILITY OF MATERIALS SUITABLE FOR CONSERVATION

Modern chemistry has given us polymeric materials that possess sufficient photochemical and thermal stability that they can be expected to maintain their essential physical and chemical properties for many hundreds of years under the relatively mild conditions represented by the indoor climate of museums. The proposed general class "A", thus, in a sense, represents only a minimum standard because of the difficulty of predicting a much longer lifetime with assurance. If one wished to refer to materials that might meet a higher standard, the author has suggested "A1" to be used to designate materials that promise a useful lifetime of more than 500 years; "A2," for those likely to be stable between 100 and 500 years. Indeed, once the field of conservation begins to consider the most stable types of modern materials, fluoropolymers, for example, a high classification such as A1 may prove to be necessary.

The conservator's usual requirements for (a) "reversibility" (b) applicability at low temperatures and (c) removability in relatively "mild" solvent systems have generally served to limit the number of materials, particularly thermoplastic resins, that can meet the proposed class A standard. Thus, although many fluoropolymers possess outstanding durability, their application and removal present problems that have thus far limited their use by conservators.

What, then, is the definition of a material that is of "sufficient photochemical stability that it can be used in conservation"? In its search for new organic coatings and adhesives to be used in conservation, the Materials Center at Carnegie-Mellon Institute has considered only class A materials (Table 1). The basic objective has been to select materials that could be expected to suffer a loss of no more than about 20% of their essential properties in 100 years exposure under normal museum conditions.

The ability of a substance to meet such a requirement cannot be estimated with complete assurance on the basis of accelerated aging tests. However, the defined standard based on ISO blue-wool standard No. 6 provides a readily available, well recognized standard that will allow the conservator to select materials that should come close to meeting, if not exceeding, the objective of a minimum of 100 years of stability. It can be demonstrated that standard No. 6 will only fade to Geometric Grey Scale contrast of 3 (British Standard BS2662) after about 213 million lux hours (20 million footcandle hours) of "daylight" fluorescent lamplight under average room conditions. Assuming annual exposures on museum walls to be 1 to 1.5 million lux hours,[4] it seems reasonable to propose that a class A material should be at least as photochemically resistant as an ISO R105 blue-wool standard No. 6.

4. Thomson, G., "Annual Exposure to Light Within Museums," *Studies in Conservation*, 12 (1967) p. 26-35.

Method of Using Standards of Photochemical Stability

The use of the standards proposed in Table 1 is as follows: If one wishes to evaluate a resin, paint or pigment as to (a) whether it might be expected to maintain its desirable properties for no more than a decade or two on a well-illuminated museum wall or (b) whether it may be expected to maintain its integrity for perhaps more than a century, the test specimen can be placed on accelerated photochemical exposure along with a set of the ISO R105 blue-wool cloths and compared to their behavior. The blue-cloths act as an exposure monitor. Whatever physical property is of interest and is known to be affected by exposure to the normal light sources in museums and archives, it can be monitored in this manner. Thus, whether the physical property be brittleness, loss of solubility, discoloration, cracking or fading—if unsuitable deterioration fails to appear in the test specimen before blue-cloth No. 6 noticeably fades—then the assumption can be made that the material under test has excellent photochemical resistance with respect to the particular physical property under consideration. In the case of Paraloid® B-72, for example, neither yellowing, cracking nor marked loss of solubility occurs in the usual xenon-arc Fadeometer or "daylight" fluorescent lamp exposures by the time ISO R105 blue-cloth No. 6 has significantly faded.

Problems Posed by Materials of Intermediate Stability

It is relatively easy to determine those materials that are highly unstable. Class A materials, on the other hand, usually require many months of "accelerated aging" before deterioration can be demonstrated. Thus, the more stable a substance is, the longer the test period that is generally needed to demonstrate its resistance to deterioration.

The usual alternative to waiting a long time is to increase the severity of the exposure. However, increased temperature and intensity of light often give false indications of what will occur under natural conditions of aging. The author has reported in one example that the high temperature of specimens in the usual carbon-arc Fadeometer favors crosslinking of isobutylmethacrylate polymers, whereas at a lower temperature, chain breaking is the major change that occurs instead of crosslinking.[5]

5. Feller, R. L., "Stages in the Deterioration of Organic Materials," *Preservation of Paper and Textiles of Historic and Artistic Value*, American Chemical Society, Advances in Chemistry Series No. 164 (1976), p. 314-355.

The investigator usually attempts to circumvent the many doubtful factors involved in the application of accelerated test conditions by comparing new products to the old—the old serving as a "control" sample. This procedure is most successful when the new materials far outlast the old; that is, when the new is obviously better than the old.

Suppose, however, that a material fails under accelerated photochemical testing, but proves to be almost as good as the control. This poses a problem for the investigator because he has no assurance that it would not "pass"—would not be better than the control—under milder conditions. Consider, further, the following results from an accelerated test: product X, previously shown to be a class C material, fails in 10 ± 2 hours, product Y (under study) fails in 70 ± 5 hours, and Paraloid® B-72, the "control", a class A material, fails in 110 ± 5 hours. From previous studies, it is known that Paraloid® B-72 is considerably better than the minimum for a class A material. The question arises: should one reject product Y as "not being class A"? While product Y clearly exhibited lower stability under accelerated aging than the control, it is seven times more stable than product X and may, therefore, be in photochemical class A (Table 1) even though it is significantly less stable than the control. An example of such a borderline case, difficult to classify with assurance, is the evaluation of the stability of ethylenevinylacetate copolymers reported to this working group at the Venice meeting in 1975.

Factors of Safety in Selection of Materials of Class A Photochemical Stability

Although product Y in our example obviously is not as stable as the control under accelerated test conditions, it may be of sufficient stability for use in conservation, particularly under the mild conditions of exposure encountered in a museum. Coming back to the question, then, of what materials can be judged to be "sufficiently stable for use in long-term conservation practices," several factors are evident. First of all, it is relatively easy to determine those materials that are unstable. Secondly, materials of outstanding stability can be selected with reasonable confidence; the Materials Center has selected Paraloid® B-72 for a standard of an excellent (class A) thermoplastic polymer that is initially soluble in solvents equivalent to or milder than xylene and which retains a high degree of solubility during long periods of exposure. However, the determination of "acceptable" materials exhibiting stability less than Paraloid® B-72—that is, the establishment of the "lower limit of acceptable stability"—is open to question.[6]

The problem of establishing minimum performance characteristics of a material that still represents sufficient stability to be safe for use in long-term conservation treatments will remain a challenging one. The Materials Center has sought to circumvent this problem by selecting a few standard materials that are considered to be significantly more stable than the absolute minimum. Such a decision in choosing a practical working lower limit of acceptability for a material in conservation can be compared to the engineer's procedure of employing a factor of safety in designing mechanical devices: an attempt is made to provide a degree of strength (photochemical resistance or other vital physical or chemical characteristic) in excess of what is estimated to be the absolute minimum requirement. Helmut Ruhemann expressed very much the same principle many years ago in his effort to specify a "factor of safety" in the selection of solvents for the removal of varnishes.[7]

Conservation specialists should be able to agree upon a few materials that are of excellent stability and also to identify a few that are obviously of intermediate stability. Hopefully, it should not be necessary to argue at length about the acceptability of products of intermediate stability—materials that in effect are "good enough". In conservation practice, "good enough" should not be the final goal; the search must be for the most stable, the very best, materials available.

6. The author has previously pointed out that the maximum permissible brittleness (minimum elongation-at-break) for a picture varnish is open to question: "Problems in the Investigation of Picture Varnishes," *Conservation of Paintings and the Graphic Arts*, International Institute for Conservation of Historic and Artistic Works, London (1972), p. 201-209.

7. Ruhemann, H., *The Cleaning of Paintings*, Praeger, New York (1968), p. 192; see also *Manual on the Conservation and Restoration of Paintings*, International Museums Office, Paris (1940), p. 129.

Controls: Materials Standards

Because accelerated-aging tests involve uncertainties when attempts are made to "extrapolate" the results to predict behavior at room temperatures, reference standards, or "controls" are usually employed. Several materials have been suggested by this laboratory in the past as controls[2] and the reasons for suggesting them will be reviewed here again.

Suggested Principal Standards: (1) the ISO R105 blue-wool fading standards represent internationally recognized, readily available and reasonably priced standards of photochemical sensitivity. These are proposed as principal control samples to measure photochemical exposure.[1, 2]

(2) Paraloid (Acryloid)® B-72 is a class A material of excellent photochemical stability. The very slow change in solubility of this polymer in a carbon-arc Fadeometer has been reported;[8] its ease of removal relative to dammar, mastic and polycyclohexanone resins measured;[1, 9] its tendency to chain break relative to polyvinylacetate demonstrated;[1, 5] and its lack of brittleness relative to polymers of lower viscosity grade pointed out.[10] The viscosity grade of Paraloid® B-72 is about 29 centipoises (for definition of viscosity grade, see below).[10]

Suggested Secondary Standards: (3) Polyvinylacetate homopolymers have long been made in similar molecular-weight grades throughout the world.[10] For convenience, one grade—that of viscosity grade 80, corresponding to Union Carbide AYAF, Airco Vinac® B-15, Mowolith 50, Vinavil K50 and Vinnapas UW1—could be selected as a principal control polymer if this is considered desirable. Whenever any particular polyvinylacetate or similar thermoplastic resin is used, however, its viscosity grade should be identified. It is even possible to specify the viscosity grade of the resin in a polymer emulsion, although more effort is involved to obtain this information than in the case of the lower molecular-weight resins supplied as 100% solids. To the best of the author's knowledge, polyvinylacetate homopolymer is a class A material.

(4) Poly(butylmethacrylate)—iso, normal or copolymers thereof—can be prepared by refluxing overnight in toluene with about 0.5% benzoyl peroxide

8. Feller, R. L., "New Solvent-Type Varnishes," in *Recent Advances in Conservation*, Butterworths, London (1963), p. 171-175.

9. Feller, R. L., and Curran, M., "Changes in Solubility and Removability of Varnish Resins with Age," *Bulletin of The American Institute for Conservation*, 15, No. 2 (1975), p. 17-26.

10. Feller, R. L., Stolow, N. and Jones, E. H., *On Picture Varnishes and Their Solvents*, The Press of Case-Western Reserve University, Cleveland (1971).

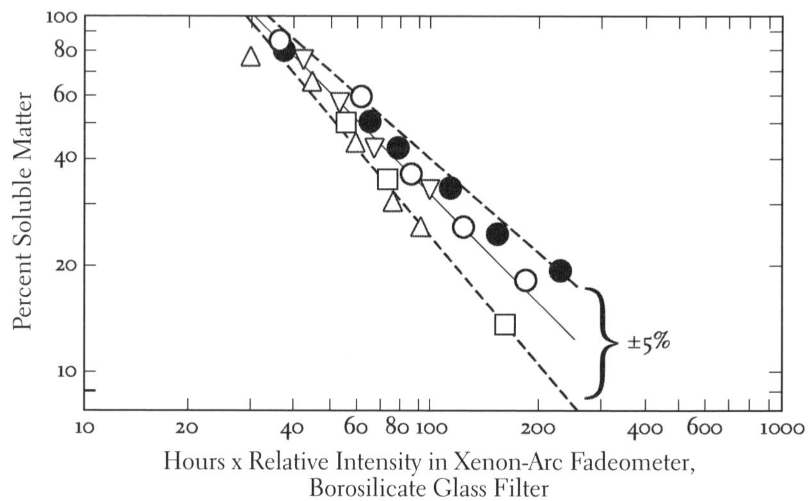

FIGURE 1
Behavior of six "control" samples of Elvacite® 2046 (butyl methacrylate copolymer) over a seven-year period.

as a catalyst based on the weight of monomer. The product that is precipitated by adding alcohol, subsequently vacuum-dried or redissolved in a suitable solvent, corresponds closely in photochemical stability to the long known commercial products, Elvacite® 2044, 2045 and 2046 (du Pont). These polymers crosslink and lose their solubility at a known rate under near-ultraviolet radiation. The quantitative measurement of their loss of solubility, as shown in Figure 1, can serve as a useful standard of photochemical stability for a class B material. The method illustrated in Figure 1 has previously been proposed as a standard procedure for reporting the loss of solubility of polymers.[6]

One of the major reasons for proposing these three types of polymers as materials standards is that their composition is known and, if required, they can be prepared from the monomers in the laboratory in viscosity grades that correspond as closely as desired to the commercially available products.

A Word of Caution: Materials Standard vs. Suitability in Specific Applications

The above standards represent reasonably well-characterized materials of tested stability; in themselves they exhibit class A or B photochemical stability. Such a classification does not imply that for a particular use—as an adhesive, paint, consolidant or clear protective coating—they may fail more rapidly in a given situation than some other material. No one substance is universally applicable to every conservation problem. The emphasis is on the stability of the material, not the *system*. These particular polymers and the blue-wool standards are recommended as an initial set of reference materials that are reasonably easy to obtain and that can be expected to have reasonably well-defined properties. Other materials undoubtedly will be proposed in the future and investigators working in the field will want to evaluate such new proposals as to their suitability and appropriateness. However, it is hoped that the several materials referred to here will prove to be a useful initial set of reference "control" standards in testing and evaluation of conservation materials and methods.

Standard Laboratory Test Methods

The viscosity grade of a polymer and the "mildest" solvent that will dissolve it are useful properties to specify. The following two methods of measuring these key properties are simple to employ.

Viscosity Grade of Polymers: The average molecular-weight grade of many polymers can be conveniently indicated by reporting their "viscosity grade," the viscosity of a 20% solution of the resin in toluene at 70°F (21.1°C).[10] For very high molecular-weight materials, the viscosity of a 2% solution in toluene, water, or if necessary, other suitable solvent can be used.

Series of Solvents of Increasing "Strength": A key property of a thermoplastic

TABLE 2
Solubility Parameters, f_d (Teas), of Mixtures of Cyclohexane, Toluene, and Acetone Having Systematically Increasing Solvent Power

Mixtures: Volume Percent of			
Cyclohexane	Toluene	Acetone	Approximate Value of f_d (teas)
100	0		96 (weak)
75	25		92
50	50		88
25	75		84
	100		80
	87.5	12.5	76
	75	25	72
	62.5	37.5	68
	50	50	64
	25	75	56
		100	47 (strong)

or lightly crosslinked polymer is the "mildest" solvent that will either dissolve or swell the coating. In 1972, mixtures of cyclohexane, toluene and acetone were proposed that would provide a series of solvents of regularly changing "strength" based on Teas' triangular-coordinate plots of three parameters.[11, 12] This series of solvent-mixtures, given in Table 2, has been used to compare the rate of loss of solubility of the following natural and synthetic resins: dammar, mastic, Resin AW-2®, Ketone Resin®, MS-2A®, polyvinylacetate Gelva® B-15 and Paraloid® B-72.[9] The essential results of the testing program on these resins were summarized in Figure 4 in the author's 1975 report to this working group of the ICOM Committee for Conservation.[1]

11. Feller, R. L., and Bailie, C. W., "Solubility of Aged Coatings Based on Dammar, Mastic and Resin AW-2," *Bulletin of the American Group - IIC*, 12, No. 2 (1972), p. 72-78.

12. Feller, R. L., "The Relative Solvent Power Needed to Remove Various Aged Solvent-Type Coatings," in *Conservation and Restoration of Pictorial Art*, ed. Brommelle, N., and Smith, P., Butterworths, London (1976), p. 158-61.

Solvents having both very high or very low solubility parameters relative to a given resin tend not to dissolve it. The solvent mixtures listed in Table 2 will usually be used to determine the lower or "milder" limit of a solvent system that has the ability to dissolve a resin or film. This is illustrated in Figure 2; the open circles are solvent systems that will not dissolve polyvinylacetate; the closed circles are within the area of solvent power that will dissolve this polymer. The solvent mixture that has just enough toluene (in this case) to dissolve the resin can thus be determined with considerable precision.

It is proposed that the solvent mixtures given in Table 2 be used as a convenient standard series by which to specify the mildest solvent system that will dissolve or remove a fresh or aged resin. For convenience, the stepped series of about ten solvents in Table 2 can be used. However, for precise measurements, the exact composition that just succeeds in dissolving the resin, or in swelling it to a given degree, can be determined. The principal standard for reporting solubility would be a diagram of the complete solubility characteristics on the Teas' diagram.

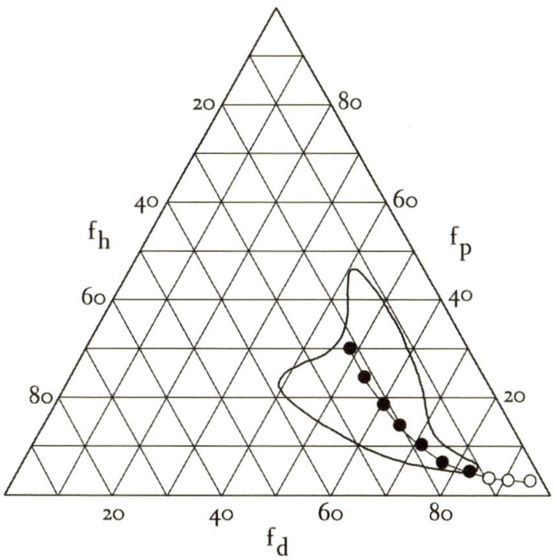

FIGURE 2
Solubility characteristics of polyvinylacetate.

Conclusion: Summary of Suggested Standards

The author proposes that, during the next three-year period, the members of the working group on protective coatings concentrate on the problem of establishing standards of materials and of test procedures that can be used to coordinate the international effort to develop durable new materials for conservation. The following are suggested as material standards:

Principal Standards
ISO R105 Blue-Wool Fading Standards
Paraloid (Acryloid)® B-72 (a class A material)

Secondary Standards
Polyvinylacetate (a class A material)
Polybutylmethacrylate (iso, normal or copolymer) (a class B material)

The following are suggested as standard laboratory test methods:

Specification of photochemical stability: the three classes given in Table 1.
Specification of viscosity grade: viscosity of a solution at 70°F (21.1°C) preferably at 20% concentration in toluene.
Minimum strength of solvent to dissolve or remove resins: Use of mixtures of cyclohexane, toluene and acetone.

As the number of technical laboratories devoted to the study of conservation materials grows, there is a corresponding increasing need for standard methods for use in describing the characteristics of the many polymers and resins which might be considered.

It is hoped that members of the working group will give immediate consideration to these proposed standards or to others which may be suggested.

Section 3: Light-induced Color Changes of Pigments

Another long-standing concern of the National Gallery of Art was the harmful effects of light exposure on colored materials. In addressing these issues in a laboratory program, it soon became evident that we would need a tool to measure color and changes in color. An instrument salesman suggested that we consult with Ruth Johnston at the nearby laboratories of the Coatings and Resins Division, PPG Industries. Ruth had helped draw up specifications for a portable abridged spectrophotometer, then used throughout the company. The instrument, known as the "Color-Eye," was designed to be capable of measuring the color of paint samples of many sizes, including even that of a typical easel painting. The laboratory under her direction was also regularly analyzing pigments with the use of the Davidson and Hemmendinger Colorant Mixture Computer (COMIC). This instrument allowed one to subtract out the spectral reflectance characteristics of individual colorants that might be present in a sample, enabling determination of both the kind and quantity of each colored constituent. A joint publication in 1963 drew attention to the usefulness of this technique of differential spectral curve analysis, so widely used in industry.

At about that time an incident occurred in which examples of in-painting executed with the mercuric sulfide pigment, vermilion, turned black during exhibition on the walls of a daylit gallery. While looking into the problem, I discovered among the books of Edward Forbes being kept at Oberlin College a booklet of 1914 by Alexander Eibner: *Uber Lichtechte Zinnober* (On Lightfast Cinnabar). By 1967, it was possible to present an extensive report on the darkening phenomenon in vermilion.

In the 1980s Keiko Keyes and her husband, Roger, had been commissioned to study and catalog the Mary A. Ainsworth Collection of Japanese prints at Oberlin College. During a visit with them, the possibility was discussed that the Color-Eye could provide a non-destructive way of identifying colorants in the prints and verifying the presence of those that were particularly fugitive. The project that resulted was funded in part by a small grant from Keiko. Several of the pigments present were identified as having poor lightfastness, and it was found that some of these, such as safflower red and dayflower blue, could be distinguished on the basis of their distinctive reflectance spectra. Moreover, it was demonstrated that a number of them would receive little benefit from removal of ultraviolet radiation.

Ruth and I had long believed that fading must be a process governed by basic laws of optics and principles of photochemistry. Using those optics laws, particularly the Kubelka-Munk formulas relating reflectance changes to changes in the concentration of colorant, it was possible to analyze fading as a photochemical reaction obeying well-known laws of chemical kinetics. This was demonstrated in 1984 using the example of the artist's pigment, alizarin lake. Later, that fading reaction was examined in more detail, showing that the production and subsequent fading of a yellow intermediate product explained the hue shift that occurred when alizarin lake paint faded. Finally, a brief study was performed to examine the influence of humidity on fading rates.

R. L. F.

The Use of Differential Spectral Curve Analysis in the Study of Museum Objects

Ruth M. Johnston and Robert L. Feller

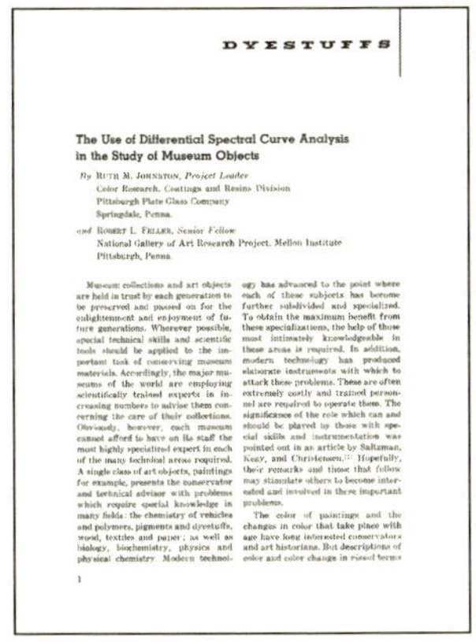
DYESTUFFS

The Use of Differential Spectral Curve Analysis in the Study of Museum Objects

By Ruth M. Johnston, *Project Leader*
Color Research, Coatings and Resins Division
Pittsburgh Plate Glass Company
Springdale, Penna.

and Robert L. Feller, *Senior Fellow*
National Gallery of Art Research Project, Mellon Institute
Pittsburgh, Penna.

Museum collections and art objects are held in trust by each generation to be preserved and passed on for the enlightenment and enjoyment of future generations. Wherever possible, special technical skills and scientific tools should be applied to the important task of conserving museum materials. Accordingly, the major museums of the world are employing scientifically trained experts in increasing numbers to advise them concerning the care of their collections. Obviously, however, each museum cannot afford to have on its staff the most highly specialized expert in each of the many technical areas required. A single class of art objects, paintings for example, presents the conservator and technical advisor with problems which require special knowledge in many fields: the chemistry of vehicles and polymers, pigments and dyestuffs, wood, textiles and paper; as well as biology, biochemistry, physics and physical chemistry. Modern technology has advanced to the point where each of these subjects has become further subdivided and specialized. To obtain the maximum benefit from these specializations, the help of those most intimately knowledgeable in these areas is required. In addition, modern technology has produced elaborate instruments with which to attack these problems. These are often extremely costly and trained personnel are required to operate them. The significance of the role which can and should be played by those with special skills and instrumentation was pointed out in an article by Saltzman, Keay, and Christensen.[1] Hopefully, their remarks and those that follow may stimulate others to become interested and involved in these important problems.

The color of paintings and the changes in color that take place with age have long interested conservators and art historians. But descriptions of color and color change in visual terms

1

Originally printed in:
Dyestuffs
Volume 44, Number 9 (1963), Pages 1-10.

Museum collections and art objects are held in trust by each generation to be preserved and passed on for the enlightenment and enjoyment of future generations. Wherever possible, special technical skills and scientific tools should be applied to the important task of conserving museum materials. Accordingly, the major museums of the world are employing scientifically trained experts in increasing numbers to advise them concerning the care of their collections. Obviously, however, each museum cannot afford to have on its staff the most highly specialized expert in each of the many technical areas required. A single class of art objects, paintings for example, presents the conservator and technical advisor with problems which require special knowledge in many fields: the chemistry of vehicles and polymers, pigments and dyestuffs, wood, textiles and paper; as well as biology, biochemistry, physics and physical chemistry. Modern technology has advanced to the point where each of these subjects has become further subdivided and specialized. To obtain the maximum benefit from these specializations, the help of those most intimately knowledgeable in these areas is required. In addition, modern technology has produced elaborate instruments with which to attack these problems. These are often extremely costly and trained personnel are required to operate them. The significance of the role which can and should be played by those with special skills and instrumentation was pointed out in an article by Saltzman, Keay, and Christensen.[1] Hopefully, their remarks and those that follow may stimulate others to become interested and involved in these important problems.

1. Saltzman, M., Keay, A. M., and Christensen, J., "The Identification of Colorants in Ancient Textiles," *Dyestuffs*, 44, No. 8 (1963), p. 241-251.

The color of paintings and the changes in color that take place with age have long interested conservators and art historians. But descriptions of color and color change in *visual* terms are not very useful in determining the nature of the chemical and physical changes that are taking place. Colored objects exhibit spectral curves characteristic of the materials used in making them, depending primarily on the chemical structure of these materials. A physical-analytical approach, based on an analysis of the character of the spectral curves, can yield much information, besides a description of the color, to the specialist conversant with the language. For example, the spectral reflectance curves shown in Figure 1 are characteristic of different blue pigments. It can be seen that each is distinctively different. Red pigments, yellows, greens, etc., are likewise different from one another.

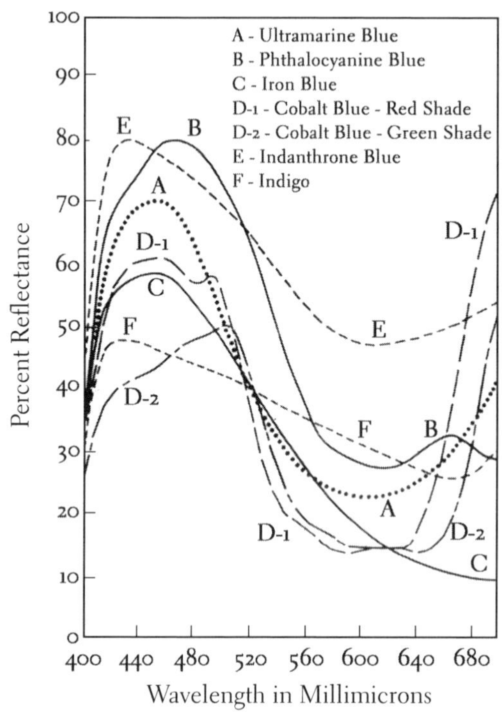

FIGURE 1
Spectral reflectance curves of different blue pigments.

Differential spectral curve analysis is a technique for analyzing the spectral curve of an object into its constituents. Much can be learned about the colorants present in a mixture through the application of this nondestructive analytical technique. Exploitation of the technique naturally depends upon the availability of the characteristic spectra of known materials. Unfortunately, there have been few measurements made on the traditional artists' pigments, except for the early work of Abney and Russell in 1888 and that of Barnes about 25 years ago.[2, 3] It seems vital, therefore, to accumulate this information for as many of the colorants used in the history of painting as possible. It is particularly

2. Russell, W. J., and Abney, W. de W., *Report to the Science and Art Department of the Committee of the Council on Education on the Action of Light on Water Colours*, H. M. Stationery Office, London (1888); Abney, W. de W., *Colour Measurement and Mixture*, Soc. for Promoting Christian Knowledge, London, F. & J. G. Young & Co., New York (1891); see curves in Luckiesh, M., *Color and Its Applications*, Van Nostrand Co., New York (1915).

3. Barnes, N. F., *Technical Studies in the Field of Fine Arts*, 7 (1939), p. 120-138; "Color Characteristics of Artists' Pigments," *J. Opt. Soc. Am.*, 29 (1939), p. 208-214.

important to seek out those that are no longer used and are becoming increasingly rare. The authors are in the process of doing this. When the task has been accomplished, it will be possible to gain additional information concerning the pigments and pigment mixtures used in paintings. Even without an extensive library of curves, however, the method can be useful, as will be indicated in the examples that follow.

Differential spectral curve analysis is a well-established technique in the spectrophotometry of solutions. More than ten years ago, Stearns described this "Method of Absorbancy Coefficient Differences" in detail[4] and designed a computing aid to apply the technique to the analysis of reflectance curves.[5] Saunderson,[6] Duntley,[7] Duncan[8] and many others have shown how the equation stated by Kubelka-Munk[9] can be used to predict the spectral reflectance, and hence the color, of pigment mixtures. The simple form of the Kubelka-Munk equation, relating reflectance to the absorption coefficient and to the scattering coefficient, is now used widely for industrial color control:

$$\frac{(1-R)^2}{2R} = K/S \qquad (1)$$

where K is the absorption coefficient and S is the scattering coefficient.

The reflectance for a mixture is related to the sum of the fractional contributions of the absorption and scattering coefficients for the particular materials present. Because these coefficients are different at each wavelength for each colored pigment, it is necessary to solve many simultaneous equations. Consequently, practical development of the method of differential spectral curve analysis depended upon the manufacture of high-speed computing aids. A special analog computer is now available from Davidson and Hemmendinger, Easton, Pennsylvania. This computer has the advantage of displaying the absorption curve in the form of a series of sixteen dots on an oscilloscope screen. Their recent publication describes the theory and technique in detail.[10]

Our purpose here is to describe a few applications of the technique of differential spectral curve analysis to the study of paintings and art materials.

COMPOSITION OF ARTISTS' COLORS

In research concerning the behavior of different pigments, validation of the pigment composition used in the experiments is vital. While it is desirable to use other chemical and physical techniques for positive identification of the pigment in question, an analysis of the spectral reflectance curve is an immediate guide. For example, four tubes of artists' watercolors labeled "indigo" were examined prior to planning experiments on fading. The spectral curve of each, painted in a translucent wash on watercolor paper, is shown in Figure 2. Only curve # 4 shows the characteristic reflectance curve of indigo. Analysis showed that the materials used to make the other colors were as follows: curve

4. Stearns, E. I., in *Analytical Absorption Spectroscopy*, ed. Mellon, M. G., John Wiley & Sons, Inc., New York (1950), p. 364-380.

5. Stearns, E. I., "Color Predictor for Pigments," U. S. Patent No. 2,540,798, February 6, 1951.

6. Saunderson, J. L., "Calculation of the Colors of Pigmented Plastics," *J. Opt. Soc. Am.*, 32 (1942), p. 727-736.

7. Duntley, S. Q., "The Optical Properties of Diffusing Materials," *J. Opt. Soc. Am.*, 32 (1942), p. 61-70.

8. Duncan, D. R., "The Colour of Pigment Mixtures," *Proc. Phys. Soc.*, 52 (1940), p. 390-401; "The Colour of Pigment Mixtures," *J. Oil and Colour Chem. Assoc.*, 32 (1949), p. 296-321; "The Identification and Estimation of Pigments in Pigmented Compositions by Reflectance Spectrophotometry," *ibid.*, 45 (1962), p. 300-324.

9. Kubelka, D., and Munk, F., "Ein Beitrag zur Optik der Farbanstriche," *Zeits. f. tech. Physik.*, 12 (1931), p. 593-601.

10. Davidson, H. R., Hemmendinger, H., and Landry, J. L. R., Jr., "A System of Instrumental Color Control for the Textile Industry," *J. Soc. Dyers and Colourists*, 79 (1963), p. 577-589.

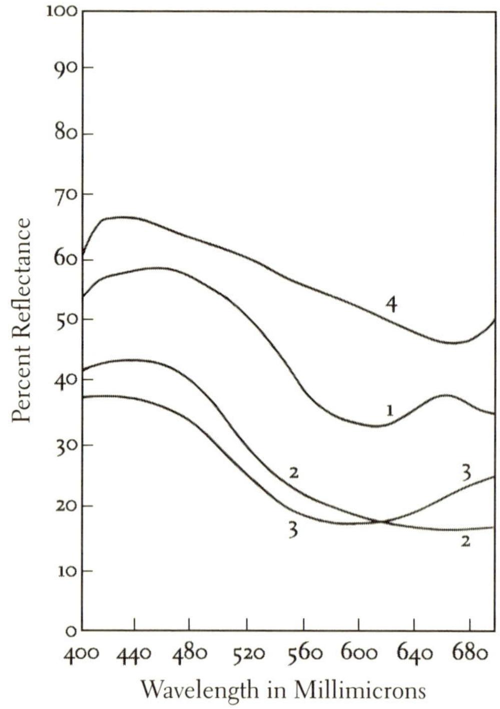

FIGURE 2
Spectral reflectance curves of four artists' watercolors labeled "Indigo."

#1, phthalocyanine blue and black; curve #2, iron blue, ultramarine blue, and black; and curve #3 ultramarine blue, phthalocyanine green, phthalocyanine blue, pyrazolone red, and black. Certainly, each of these "indigo" formulations would behave differently on exposure to light. The different characteristics that would be exhibited to the artist are: different color behavior with changes in concentration and in mixture with other pigments.

Thus, the measurement and analysis of the spectral curve aid in the identification of pigment composition.

STUDY OF COLOR CHANGES

Of major concern to the conservator is the magnitude of the visual color change of pigments following exposure to light: how much will a painting change in ten years of exposure in a museum illuminated with a certain level and type of illumination? The change in visual appearance with exposure may be computed from measurements or may be rated by comparison with visual standards. But to the research investigator, who wishes to learn what happens and why, such results are not sufficient. He wishes to learn the answers to such questions as: what component in a mixture is changing and how much; does the pigment change to a colorless form; what is the nature of the darkening that may occur; what is the rate of pigment change; how much does a filter which absorbs the ultraviolet energy alter the rate and nature of the change?

Information which can be helpful in answering such questions can be obtained by analyzing the differences in the spectral curves. First, however, calibration curves must be made that relate the optical behavior to the changes in the concentration of the materials in question. From such data, it is then possible to interpret the nature of some of the changes that occur.

Consider, for example, the case in which each of two panels of pigmented plastic was observed to have undergone the same visual change, 2.2Δe (MacAdam), following three months of exposure in Florida. One panel was pigmented with cobalt blue and the other with cadmium yellow. Analysis of the spectral reflectance curves before and after exposure showed that, although the change in visual color was of the same magnitude, the cobalt blue pigment had not changed—all of the change in the blue panel was caused by yellowing of the vehicle. On the other hand, the cadmium yellow had faded. Since the vehicle had also yellowed, the color change in the pigment was thus partially compensated for. The analysis also showed that the amount of color change due to the vehicle in the blue panel was four times that which occurred in the yellow. Apparently the yellow pigment protected the vehicle, perhaps because it absorbed more of the more energetic short wavelength radiation.

These examples of the information that may be gained concerning the type and magnitude of changes that take place on exposure to light indicate the usefulness of the analytical technique. By this means, there also can be derived rates of specific changes, effects of specific materials and conditions on the rates, and perhaps greater understanding regarding photochemical effects.

The importance of studying the effect of light on the stability of artists' materials is heightened by the current trend towards the use of higher levels of illumination in museums and other public buildings. In the old days, when museums were dark and dingy, the dangers from exposure to light were considerably less. But the modern point of view considers that art and other museum objects should be displayed attractively and enjoyed by a large public. If not properly employed, however, the increased levels of illumination can create added problems for the museum staff, hastening the deterioration of the very objects that should be preserved for public enjoyment—and the objects are not replaceable—they are fragile.

It is not always a simple matter to demonstrate, in any particular painting, that exposure to light presents a problem or that fading has taken place. Occasionally, however, an opportunity presents itself to demonstrate clearly that light represents a hazard. Such a case may be seen in Figure 3, a photograph of a detail of a 19th-century painting that has been removed from its frame. Here, the extent of the fading may be seen by comparing the general color of the drapery with the narrow band of more intense color along the bottom edge. Along this edge, the rabbet of the frame has protected the pigments from the light.

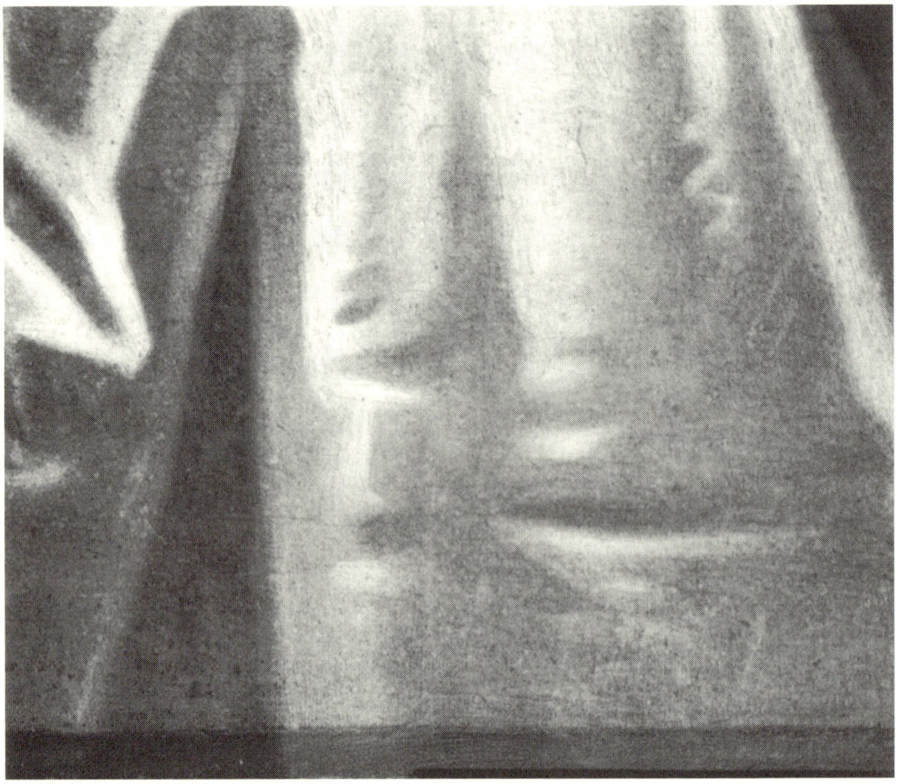

FIGURE 3
Detail of 19th-century painting. Note the intense blue of the drapery along the bottom edge where the paint was protected by the rabbet of the picture frame. (Ed. note: original illustration appeared in color.)

The technique of spectral curve analysis is currently being used to determine the extent and type of changes that may occur in museum materials under controlled conditions of exposure. The results of these investigations will add considerably to the information needed for the preservation of our art treasures. As well, they will assist the modern artist in the choice of new materials.

ANALYSIS OF GLAZES

Many artists add color to their paintings by applying coats of tinted translucent varnishes, called *glazes*. This method of painting presents a number of problems to the "restorer" of fine paintings. Among these is the need to know whether the protective varnish has merely become dirty and yellow or brown with age or whether there was a colorant used in the varnish beneath. Santini, appreciating that it might be helpful to analyze the spectral reflectance curves before and after the removal of varnish, published curves in 1955 made with a modified Beckman DU Spectrophotometer.[11] There is a change in the surface characteristics of the object, however, when the protective varnish is removed

11. Santini, M., *Bollettino Dell' Istituto Centrale del Restauro*, 23-24 (1955), p. 95-129; see also *Il Restauro della "Maesta" di Duccio*, Istituto Poligrafice dello State (1959).

and, for this reason, it is essential that the reflectance measurements be made incorporating the *total reflectance*. Unfortunately, it is not possible to gain this required information with the Beckman DU or Santini's modification.

To illustrate the application of the technique to the analysis of glazes, consider the curves shown in Figure 4. These were made on a red area of paint before and after removal of the varnish that covered the paint. The large increase in the reflectance at the short wavelength, blue, region indicates at once that the removed varnish was a dirty yellow. It is not apparent, however, whether there was a colorant removed as well. From the shape of the resultant curve after the removal of the varnish, the red pigment in the paint is recognizable as alizarin mixed with white (probably zinc white). The red color was considerably more orange before the varnish had been removed. In removing this varnish, the conservator would wish to know: was the bluer red that resulted after the removal of the varnish the color that the artist originally intended? Or did the artist intend that the red color should be more orange, using a colorant in the varnish to get this effect or did he intend something else?

The measured reflectance values for the two curves in Figure 4 were expressed as absorption function, K/S, and the difference in absorption between these two curves is shown as curve #1 in Figure 5. The absorption curves of a typical yellow varnish, #2, and of alizarin, #3, are also shown. It is readily discernible

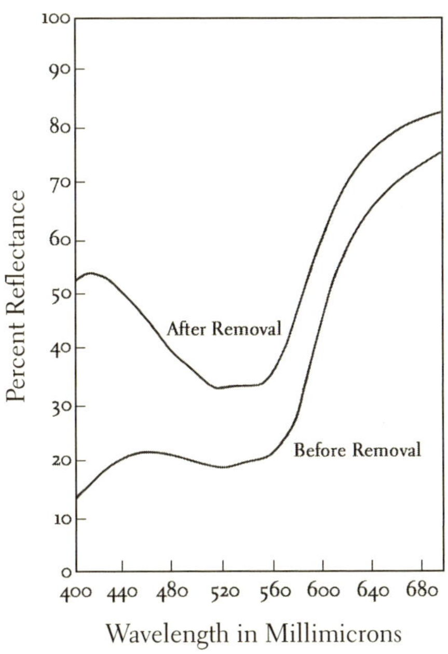

Figure 4
Spectral reflectance curves related to the removal of varnish covering an area of red paint.

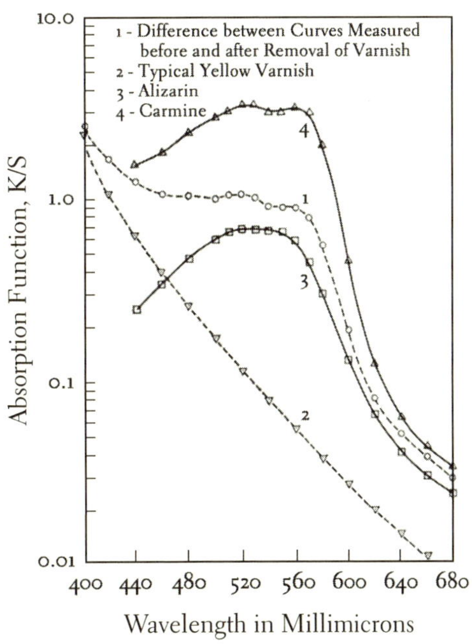

Figure 5
Spectral absorption curve (1) showing difference between curves in Figure 4. The curves of the color of alizarin, carmine, and a typical varnish are also given.

that the coating that was removed, curve #1, contains absorption which represents more than yellow varnish. From the location of the two maxima, it is apparent that the absorption is not that of alizarin. In this particular case, the curve shape is recognized as that characteristic of carmine, shown as curve #4. The artist had used a carmine glaze over alizarin.

The technique of measuring and analyzing the changes in reflectance as the layers of the painting are removed can be extremely helpful, not only in analyzing varnishes and glazes, but also in studying the overpainting used in "restoring" paintings. The characterization of pigment layers and the study of film formation, drying effects, and the efficacy of solvents are other areas of application where the technique may be helpful.

Conclusions

The technique of differential spectral reflectance curve analysis can be a useful tool in the study of art objects and their conservation. Further useful applications can be in research designed to create better materials and techniques for the contemporary artist.

It must be borne in mind, however, that the interpretation of the spectral curves requires the knowledge and experience of the trained spectrophotometrist working with an extensive background experience and a library of reference curves. Therefore, the technique probably cannot be applied to the everyday problems in the museum, but only to those of special importance. Nevertheless, it offers the museum scientist a powerful tool in the analysis of color and color change. The applications now being made to aid the National Gallery of Art Research Project at Mellon Institute serve as a prime example of an instance where the expensive instruments and highly specialized techniques used in modern industrial laboratories can be called upon to aid the scientist associated with museum problems.

Acknowledgments

The National Gallery of Art Research Project is most grateful to the Pittsburgh Plate Glass Company for its kind cooperation in the current studies of color and color change in artists' pigments. Their interest and assistance in these problems have done much to advance our knowledge of pigment identification, fading, and related matters in the care of works of art. Miss Johnston particularly wishes to thank Dr. H. L. Gerhart, Director of Research and Development of the Coatings and Resins Division of the Pittsburgh Plate Glass Company, for his interest and encouragement in these investigations, and to thank the Pittsburgh Plate Glass Company for the privilege of publishing these remarks.

Studies on the Darkening of Vermilion by Light

ROBERT L. FELLER

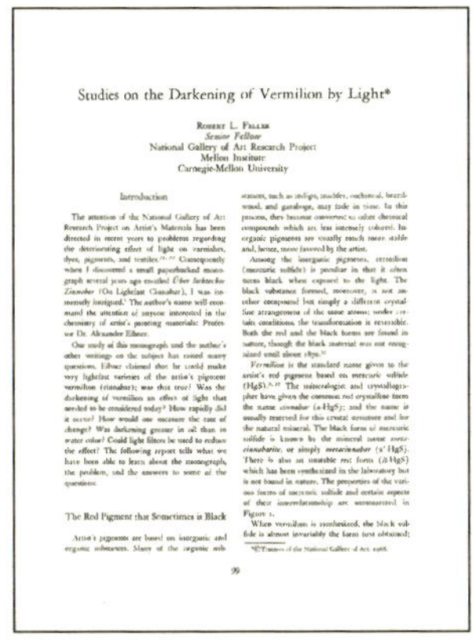
Studies on the Darkening of Vermilion by Light*

ROBERT L. FELLER
Senior Fellow
National Gallery of Art Research Project
Mellon Institute
Carnegie-Mellon University

Introduction

Originally printed in:
REPORT AND STUDIES IN THE HISTORY OF ART
National Gallery of Art, Washington, DC (1967), Pages 99-111.

Introduction

The attention of the National Gallery of Art Research Project on Artist's Materials has been directed in recent years to problems regarding the deteriorating effect of light on varnishes, dyes, pigments, and textiles.[1, 2] Consequently when I discovered a small paperbacked monograph several years ago entitled *Über lichtechte Zinnober* (On Lightfast Cinnabar), I was immensely intrigued.[3] The author's name will command the attention of anyone interested in the chemistry of artist's painting materials: Professor Dr. Alexander Eibner.

Our study of this monograph and the author's other writings on the subject has raised many questions. Eibner claimed that he could make very lightfast varieties of the artist's pigment vermilion (cinnabar); was that true? Was the darkening of vermilion an effect of light that needed to be considered today? How rapidly did it occur? How would one measure the rate of change? Was darkening greater in oil than in water color? Could light filters be used to reduce the effect? The following report tells what we have been able to learn about the monograph, the problem, and the answers to some of the questions.

1. Feller, R. L., "Control of the Deteriorating Effects of Light on Museum Objects," *Museum*, 17, No. 2 (1964), p. 57-98.

2. *The National Gallery of Art, a Twenty-five Year Report*, Washington, D.C. (1966), p. 28-40.

3. Eibner, A., *Über lichtechte Zinnober*, Munich (1914), anonymous review in *Farbenzeitung*, 19 (1914), p. 2296-2299.

The Red Pigment that Sometimes is Black

Artist's pigments are based on inorganic and organic substances. Many of the organic substances, such as indigo, madder, cochineal, brazilwood, and gamboge, may fade in time. In this process, they become converted to other chemical compounds which are less intensely colored. Inorganic pigments are usually much more stable and, hence, more favored by the artist.

Among the inorganic pigments, vermilion (mercuric sulfide) is peculiar in that it often turns black when exposed to the light. The black substance formed, moreover, is not another compound but simply a different crystalline arrangement of the same atoms; under certain conditions, the transformation is reversible. Both the red and the black forms are found in nature, though the black material was not recognized until about 1870.[4]

Vermilion is the standard name given to the artist's red pigment based on mercuric sulfide (HgS).[5,6] The mineralogist and crystallographer have given the common red crystalline form the name *cinnabar* (α-HgS); and the name is usually reserved for this crystal structure and for the natural mineral. The black form of mercuric sulfide is known by the mineral name *metacinnabarite*, or simply *metacinnabar* (α´-HgS). There is also an unstable red form (β-HgS) which has been synthesized in the laboratory but is not found in nature. The properties of the various forms of mercuric sulfide and certain aspects of their interrelationship are summarized in Figure 1.

When vermilion is synthesized, the black sulfide is almost invariably the form first obtained; this is subsequently converted to the reddish-orange artist's pigment through appropriate control of the conditions of manufacture.

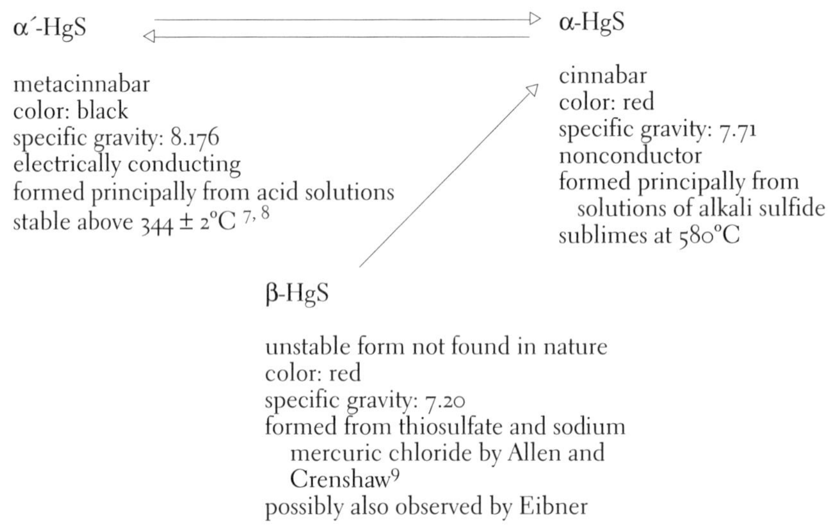

FIGURE 1
Summary of the properties and relationship of the various crystalline forms of mercuric sulfide.

The reasons why the red form may subsequently turn black are still poorly understood, although the phenomenon has been known since ancient times. The oldest story frequently quoted is that of the Roman author Vitruvius, who said that the notary Faberius had ordered the walls of his house on the Aventine to be painted with cinnabar; yet after only thirty days, the walls became so dark and ugly that it was necessary to repaint them with another pigment. Authorities

4. Moore, G. E., "Über das Vorkommen des amporphen Quecksilbersulfids in der Natur," *Journal für Praktische Chemie*, 2 (1870), p. 319-330.

5. British Standard 2876, 1957, "Powder Pigments for Artists' Use."

6. U.S. Department of Commerce, Commodity Standards Division, *Artists' Oil Paints*, Commercial Standard CS98-62, Superintendent of Documents, U.S. Government Printing Office, Washington, D.C. (1962).

7. Curtis, O. L., Jr., "Effect of a Phase Transformation on the Vapor Phase Growth of Single-Crystal HgS," *Journal of Applied Physics*, 33 (1962), p. 2461-2463.

8. Dickson, F. W., and Tunell, G., "The Stability Relations of Cinnabar and Metacinnabar," *American Mineralogist*, 44 (1959), p. 471-487.

9. Allen, E. T., and Crenshaw, J. L., "The Sulphides of Zinc, Cadmium and Mercury; Their Crystalline Forms and Genetic Conditions," *American Journal of Science*, 34 (1912), p. 341-396.

agree that Faberius must have been using natural mineral and that its source was probably the famous mines at Sisopa in Spain, which Pliny said supplied nearly all of Rome's mercury.

Synthesis by the Dutch and German Methods

Although the natural mineral was pulverized and used as a pigment in ancient times, methods of synthesizing vermilion have been known at least since the thirteenth century. The synthetic product gradually displaced the natural, and about 1880 the latter finally disappeared altogether from the shops of the artist's colormen.[3]

The earliest method of synthesis used is known as the *dry* or *Dutch* method, whereby mercury and sulfur were simply ground together until the black sulfide was formed. Subsequently, the black material, sometimes called "ethiops mineral," "mohr," or "moor," was sublimed.[10] When the process was completed, the red crystalline material was found on the underside of the lids of the containers in which the ethiops had been heated. Mellor says that this method was described by Geber and by Albertus Magnus in about the thirteenth century.[11] A similar method, used by the Chinese, yielded what was perhaps the finest vermilion available, although for a long time their manufacturing procedure was unknown. Anonymously written descriptions of the Chinese process appeared in the English literature about 1829.[13]

The *wet* or *German* process, introduced by Gottfried Schultz in 1687, converted the black ethiops to vermilion by heating it in a solution of ammonium or potassium sulfide rather than by sublimation.[14] In the western world, the dry method was gradually displaced by the wet process, partly because the latter produced more brilliant pigments, although Eibner later showed that they tended to blacken more readily than those made by the dry process.[3]

The Inconsistent Lightfastness of Vermilion

Vermilion is classed in the U.S. commercial standard codes as an artist's pigment of reasonable fastness to light.[6] Nevertheless, few authorities consider it to be a "permanent" pigment because of its capricious tendency to darken. A number of years ago, for example, vermilion was used in retouching a group of paintings that were later hung in a prominent museum under bright natural illumination; in a few years' time, the vermilion turned black and the work had to be redone completely. This case prompted the National Gallery of Art Research Project to expose test panels of vermilion oil paints on a typical gallery wall. Half of the panels showed perceptible darkening within five years; their total exposure amounted to about 650,000 footcandle-hours of diffused daylight illumination.

Although modern vermilion may often exhibit poor fastness to light, many paintings are known in which the pigment has maintained its brightness for centuries. Apparently part of the secret is in the way that the pigment is prepared.

10. A substance that sublimes passes directly from the solid to the gaseous state without melting and becoming a liquid; the gas also condenses directly to a solid without first liquefying.

11. Eibner has assigned Geber's description to the eighth century in one reference[3] and to the eleventh century in another.[12]

12. Eibner, A., "Über Lichtwirkungen auf Malerfarbstoffe, III. Erhöhung der Lichtechtheit der auf nassem Wege hergestellten Zinnober," *Chemiker-Zeitung*, 41, No. 55 (1917), p. 385-386.

13. Mellor, J. W., A *Comprehensive Treatise on Inorganic and Theoretical Chemistry*, vol. IV, London (1923).

14. Eibner, A., "Über Lichtwirkungen auf Malerfarbstoffe, III. Erhöhung der Lichtechtheit der auf nassern Wege hergestellten Zinnober," *Chemiker-Zeitung*, 41, No. 58 (1917), p. 402-404.

15. Wagner, H., *Die Körperfarben*, Stuttgart (1928).

16. Eibner, A., Kaiserliches Patentamt Patentschrift Nr. 263472, July 4, 1912, "Verfahren zur Herstellung von lichtechten hellen und dunklen Zinnobern auf nassem Wege."

17. Eibner, A., Kaiserliches Patentamt Patentschrift Nr. 281483, August 12, 1913, "Verfahren zur direkten Darstellung von hellen und dunklen lichtechten Zinnobern aus Zinnobererzen."

18. Eibner, A., U.S. Patent No. 1,137,467, April 27, 1915, "Producing a Light Pigment from Cinnabar or Other Ores Containing Mercury Sulphide."

EIBNER'S DIRECTIONS FOR MAKING LIGHTFAST VERMILION

It is somewhat surprising that vermilion pigments of poor lightfastness continue to be manufactured when fifty-five years ago the famous German authority on artist's materials, Alexander Eibner, received two patents for processes that produced vermilion in a variety of hues able to resist darkening for at least ten years' exposure to direct sunlight.[15] His basic patent was granted in Germany, Austria, and the United States.[16-19]

Eibner's extensive investigation concerning vermilion was one of a number of important research contributions that he made in association with the German Society for the Promotion of Rational Painting Techniques.[20] In the course of the research, he discovered some of the basic reasons for the poor lightfastness of the commercial preparations of vermilion and, later, developed improved methods of manufacture.

The German patents on methods for preparing the pigment were granted to Eibner in 1912 and 1913. A year later, he recounted the story of his investigations in the fifty-three page monograph mentioned earlier, *Über lichtechte Zinnober.*[3] Figure 2 shows the title page of this rare publication, the second in a series of seven monographs published from about 1914 to 1921 by the German Society for the Promotion of Rational Painting Techniques.[21] It was reviewed in *Farbenzeitung*, 1914.[3] Three years later, Eibner described his research more fully in a series of six articles, completing the description with a seventh in 1918.[12, 14, 22-26]

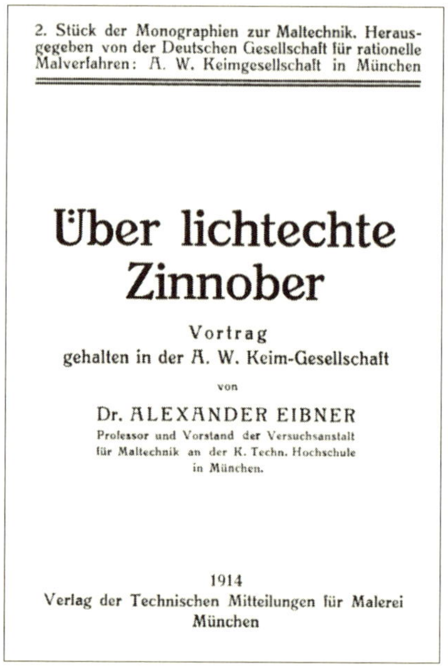
2. Stück der Monographien zur Maltechnik. Herausgegeben von der Deutschen Gesellschaft für rationelle Malverfahren: A. W. Keimgesellschaft in München

Über lichtechte Zinnober

Vortrag
gehalten in der A. W. Keim-Gesellschaft
von
Dr. ALEXANDER EIBNER
Professor und Vorstand der Versuchsanstalt für Maltechnik an der K. Techn. Hochschule in München.

1914
Verlag der Technischen Mitteilungen für Malerei
München

FIGURE 2
Title page of Eibner's monograph.

19. Eibner, A., "Verfahren zur direkten Darstellung von hellen und dunklen lichtechten Zinnobern aus den Zinnobererzen und sonstigen quecksilbersulfidhaltigen Mineralien auf nassem Wege," Kais. Königl. Patentamt Österreichische Patentschrift Nr. 71443 (October 1, 1915).

20. Part of the history of this society and its influence on improving the knowledge of artists' materials may be found in A. W. Keim's book, *Über Mal-Technik*, Leipzig (1903).

21. A list of their titles appears on the back of vols. 6 and 7; five were by Eibner.

22. Eibner, A., "Über Lichtwirkungen auf Malerfarbstoffe, III. Erhöhung der Lichtechtheit der auf nassem Wege hergestellten Zinnober," *Chemiker-Zeitung*, 41, No. 61 (1917), p. 423-425.

23. Eibner, A., "Über Lichtwirkungen auf Malerfarbstoffe, III. Erhöhung der Lichtechtheit der auf nassem Wege hergestellten Zinnober," *Chemiker-Zeitung*, 41, Nos. 65/66 (1917), p. 447-450.

24. Eibner, A., "Über Lichtwirkungen auf Malerfarbstoffe, III. Erhöhung der Lichtechtheit der auf nassem Wege hergestellten Zinnober," *Chemiker-Zeitung*, 41, No. 70 (1917), p. 482-483.

25. Eibner, A., "Über Lichtwirkungen auf Malerfarbstoffe, III. Erhöhung der Lichtechtheit der auf nassem Wege hergestellten Zinnober," *Chemiker-Zeitung*, 41, Nos. 74/75 (1917), p. 510-513.

In order to evaluate Eibner's contributions and to be able to share the information with colleagues, the Research Project had complete translations made of his monograph, patents, and publications. In addition, samples of vermilion were prepared according to his directions; one was found to be at least ten times more resistant to blackening than recently purchased proprietary pigments.

Only Eibner's first patent is mentioned in Mellor's treatise on theoretical and inorganic chemistry.[13] His patents and publications on vermilion are not mentioned at all by such authorities on artist's materials as Gettens and Stout,[27] Laurie,[28] or de Wild.[29] Perhaps one of the reasons for the lack of attention given to this work is that it appeared at the time of the First World War, when German publications were not widely available.[30]

Key Factors in the Preparation

Eibner began his investigations in 1909. In the course of the research, a thorough survey was made of the vermilion pigments being manufactured in the period 1910 to 1912. In all, about fifty-five pigments were tested. Their lightfastness was found to vary considerably and to bear little relation to a given manufacturing process or trade designation. Samples of the natural mineral were also tested.

Several of the early experiments were designed to demonstrate that the blackening was indeed a physical change rather than a chemical one; in other words, that both the black and red materials were mercuric sulfide. He further showed that the presence of sulfur or mercury impurities in the final product did not hasten the darkening.

The key facts that helped unravel the problem were: (a) that the specific gravity of the red form is lower than that of the black; and (b) that the red is a nonconductor of electricity, while the black is a conductor (see Figure 1).[31] Eibner's technique for producing a lightfast variety, then, was to employ conditions that favored the formation of a dense, low-conducting mercuric sulfide. He noted, for example, that the usual wet process employed an excess of mercury over sulfur. He reasoned that an excess of mercury might favor the product with the higher electrical conductance and recommended that sulfur be kept in excess instead. He also suggested, in an alternative wet procedure, that polysulfides of calcium and barium be used rather than those of the alkali metals, for the former elements have a lower metallic (conducting) character. This line of reasoning might be considered old-fashioned today, yet it led this gifted chemist to the development of vermilion pigments of remarkable fastness to light.

Perhaps Eibner's principal preparative innovation was the conversion of the black to the red by digestion in the presence of *freshly made* potassium pentasulfide. Usually the excess sulfur was removed in the wet process by

26. Eibner, A., "Lichtwirkungen auf Malerfarbstoffe, IV. Darstellung von Zinnobern mit erhöter Lichtechtheit aus den Erzen durch Auslaugen," *Chemiker-Zeitung*, 42, Nos. 116/117 (1918), p. 469-470.

27. Gettens, R. J., and Stout, G. L., *Painting Materials*, New York (1966).

28. Laurie, A. P., *The Painter's Methods and Materials*, Philadelphia (1926).

29. de Wild, A. M., *The Scientific Examination of Pictures*, London (1929).

30. Another example of a long forgotten German contribution on artists' materials from the same period is that of Wiesner in 1914 concerning dammar resin. See Feller, R. L., "What's in a Name: Dammar," *The Crucible*, 49, No. 8 (1964), p. 214, 216, 218.

31. The differences in density and in specific heat clearly suggest that the atoms are arranged differently in the two forms. In Eibner's day, however, X-ray diffraction had not yet been used to disclose the differences in their crystal structure.

heating the raw product in a potassium hydroxide solution. However, thiosulfate can form in this solution, and Eibner's careful investigations revealed that vermilion prepared in the presence of thiosulfate is particularly sensitive to darkening.[32] A solution of sodium or potassium sulfide is also able to dissolve the contaminating sulfur and may be readily converted to the pentasulfide used to transform the raw product into vermilion. Hence, he recommended that the alkali be abandoned and that a sulfide solution be used instead. It must be freshly prepared, however; otherwise thiosulfate may accumulate and impair the lightfastness of the vermilion.

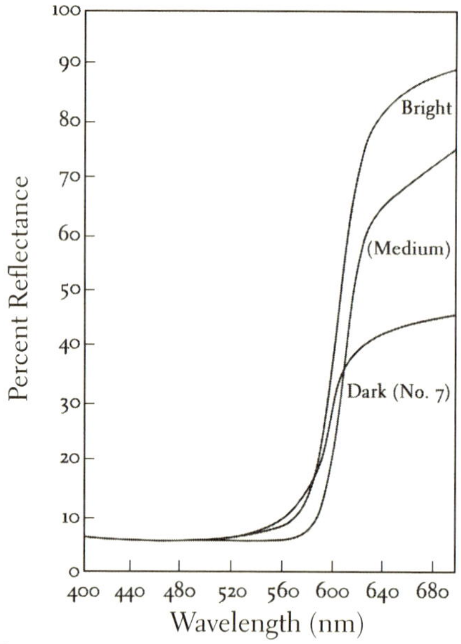

FIGURE 3
Reflectance of vermilion pigments (masstone in oil).

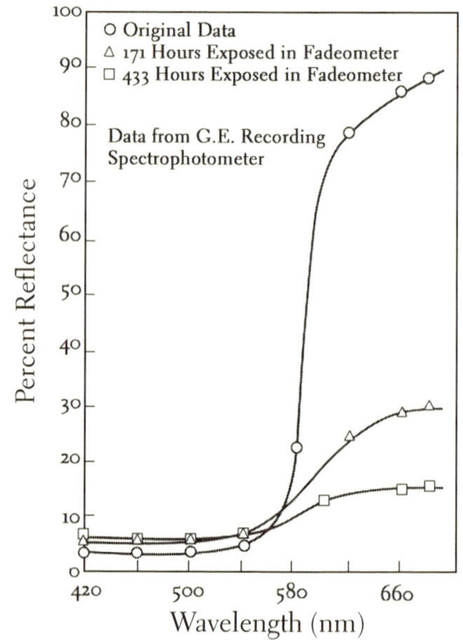

FIGURE 4
Reflectance of vermilion in oil.

THE SPECTRAL REFLECTANCE AND COLOR OF VERMILION

Vermilion pigments vary in masstone color from red to vivid reddish-orange of high chroma.[33] The spectral reflectance curve of a bright high-chroma vermilion is shown in Figure 3; wavelengths up to about 560 nanometers (nm) are absorbed strongly, but the longer wavelengths are reflected (orange and red).[35] The pigment may also be obtained in darker varieties of lower chroma; reflectance curves of these types are also shown.

Figure 4 illustrates the change that occurs in the reflectance curve of a bright vermilion following exposure to light. The drop in the reflectance in the long wavelength (red) region is a measure of the darkening. Curves illustrating this change have also been published by Toishi.[36]

The reflectance curves for the darkened states of the bright vermilion (Figures 4 and 12) are similar to those of the dark and medium varieties (Figure 3).[37] It

32. Throughout the 1914 monograph, Eibner refers to salts of "unterschwefeligsäure." These could be translated as "hyposulfites." However, he was apparently employing an obsolete word for "thiosulfates," as the latter term is used in his patents and later publications.

33. Masstone is the color of a single pigment not mixed with any other. The three attributes of color according to the Munsell system are hue, value (lightness), and chroma (purity or saturation). Vermilion is such a high-chroma pigment that D. V. Thompson has suggested that vivid pigments such as azurite (blue) and malachite (green) were necessary to balance it in the medieval artist's palette.[34]

34. Thompson, D. V., Jr., *The Materials and Techniques of Medieval Painting*, New York (1956).

35. The ordinate scale represents the relative amount of light reflected. Black will be zero on this scale, and perfect white, such as freshly smoked magnesium oxide, will be 100. The wavelength of light falling on the sample is plotted on the abscissa, wavelength being expressed in nanometers (nm). (A nanometer is 10^{-6} millimeters.) Wavelengths between 400 and 700 nanometers constitute the region of the electromagnetic spectrum visible to the human eye, varying from violet at 400 nm to red at 700 nm.

can be seen, therefore, that it is difficult, if not impossible, to distinguish, either visually or spectrophotometrically, whether the dark color of a vermilion paint is its original or a darkened state.

A reflectance curve for the masstone of metacinnabar is not given, because it would simply be a line across the bottom of the figure at low reflectance. The curves of mixtures of the black metacinnabar and white in Figure 5, however, demonstrate that the spectral reflectance is uniform, causing the mixture to form neutral rather than bluish or brownish grays.[38] (The sharp decrease in reflectance at wavelengths below 420 nm is due to the lower reflectance of the titanium white in this region.)

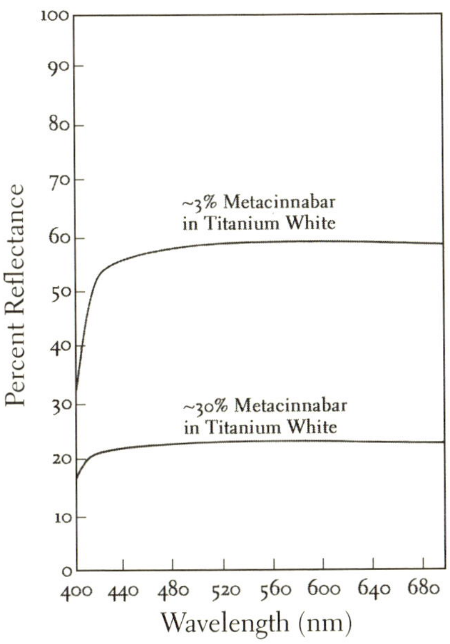

Figure 5
Reflectance of mixtures of metacinnabar with rutile titanium white at 35%PVC in oil.

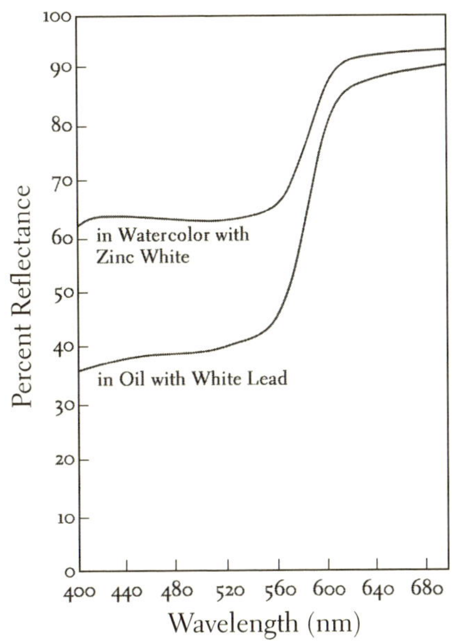

Figure 6
Reflectance of vermilion mixed with white pigments.

36. Toishi, K., and Kenjo,T., in *Scientific Papers on Japanese Antiques and Art Crafts*, 17 (1963), p. 6-22, in Japanese; abstracted as "Colour Fading in Mineral Pigments," IIC abstract 4145. See also Liberti, S., "Ricerche sulla natura e sulla origine delle alterazioni del cinabro," *Bolletino dell'Istituto Centrale del Restauro*, 3-4 (1950), p. 45-64.

37. It cannot be assumed that the low-chroma type of vermilion is dark because it contains more of the metacinnabar. Other factors can affect the chroma of the pigment, such as the average particle size and the distribution of particle sizes in the pigment. Methods other than analysis of the spectrophotometric curves must be used to determine the relative amounts of the red and the black forms in a given pigment.

38. Johnston, R. M., and Feller, R. L., "Optics of Paint Films: Glazes and Chalking," in *Application of Science in Examination of Works of Art*, Museum of Fine Arts, Boston (1967).

Reflectance curves of mixtures of bright vermilion and white are given in Figure 6. This combination is widely used for flesh color in painting and will turn towards gray as the vermilion darkens.

What is the color of vermilion and its mixtures with white? Measurements made with a General Electric recording spectrophotometer were used to describe the colors of six artist's oil paints. Their masstone colors were found to fall in the range of Munsell 5R to 7.5R (red).[39] According to the standard nomenclature in the ISCC-NBS method of designating color names, these may be properly called strong red and vivid reddish-orange, depending on the particular pigment.[40] Mixtures of vermilion with white fall in the standard classification of pink and yellowish-pink. The changes in color, chromaticity, and brightness that take place upon darkening will be discussed in detail in a subsequent publication.

39. *Munsell Book of Color*, Munsell Color Company, Inc., Baltimore, MD (1929-42).

40. U.S. Department of Commerce, National Bureau of Standards, "The ISCC-NBS Method of Designating Colors and a Dictionary of Color Names," Circular 553, Superintendent of Documents, U.S. Government Printing Office, Washington, D.C. (1952).

MEASUREMENT OF THE RATE OF DARKENING

When the Research Project began the study of the rate of darkening of vermilion, the assumption was made that the darkening was caused principally by the change of the red pigment to the black metacinnabar. This is a reasonable assumption, yet it must be admitted that incontestable proof of the formation of metacinnabar has not yet been obtained in the present investigation, for this requires analytical methods other than spectrophotometry.[41] There is a good indirect evidence, however, that red pigment is indeed being converted into the black: as the absorption in the long wavelengths increases (reflectance decreases), we observe in Figure 4 that absorption in the shorter wavelengths decreases (reflectance at wavelengths from 420 to about 540 nm increases). The red pigment strongly absorbs the short wavelengths of light; an increase in reflectance in this region therefore implies that the concentration of the red pigment is decreasing at the same time that the darkening occurs. This is just as one would expect if the red pigment were being converted into the black.

41. X-ray diffraction is not particularly sensitive in detecting the presence of metacinnabar. An X-ray pattern is determined predominantly by the bulk of the pigment crystal. If the darkening occurs primarily on the surface of the pigment particles, the relatively small amount of black on the surface of the red may not be detected. On the other hand, electron diffraction should be sensitive to changes on the surface. Research on analytical methods is in progress.

We may logically assume, then, that the decrease in reflectance at the long wavelengths is influenced by the amount of metacinnabar formed. The change in the reflectance of paints, however, is not *directly* related to changes in the concentration of the pigments. Instead, the variation of reflectance in relation to pigment changes can be described on the basis of the Kubelka-Munk equation, which associates the reflectance with the absorption and scattering coefficients of the individual pigments present:

$$\frac{(1-R_m)^2}{2R_m} = \frac{C_1K_1 + C_2K_2 + \ldots}{C_1S_1 + C_2S_2 + \ldots} \qquad (1)$$

where R_m is the fractional reflectance of the mixture at a single wavelength; K, the absorption coefficient; S, the scattering coefficient; and C_1, C_2 the concentration of the individual pigments.[38, 42] Hence for the mixture,

42. Judd, D. B., *Color in Business, Science and Industry*, New York (1952).

$$\frac{(1-R_m)^2}{2R_m} = \left(\frac{K}{S}\right)_{\text{mixture}} \qquad (2)$$

These equations suggest that it would be most useful to follow the change in (K/S) with darkening rather than simply to observe the change in reflectance. It is the value of (K/S) for the mixture, not the reflectance, that should be most directly affected by the change in the concentration of the metacinnabar. It may be seen from the last equation that as the value of (K/S) increases, the reflectance, R, decreases; that is, the color gets darker.

It was considered that measurement of the change in (K/S) at a wavelength where the vermilion was highly reflecting would provide an indication of the amount of the highly absorbing black that had formed. This would be in the orange and red wavelengths, where there was little absorption by the vermilion; the change in absorption in this region of the spectrum should be primarily due to the black pigment present. To confirm that this would be the case, oil paints containing known mixtures of vermilion and metacinnabar were prepared and their reflectance measured at 660 nm. As Figure 7 shows, the concentration of the metacinnabar in these mixtures is linearly related to a function of (K/S) at 660 nm.[43]

A plot of (K/S) at 660 nm vs. time of exposure is shown in Figure 8 for the darkening of samples of vermilion under fluorescent lamp light and in Figure 9 for the darkening in a carbon-arc Fadeometer. The fact that the experimental

43. The absorption curve for black mercury sulfide is rather "flat" (Figure 5); its absorption at 660 nm is much the same as that at any other wavelength which might have been selected between 440 and 700 nm.

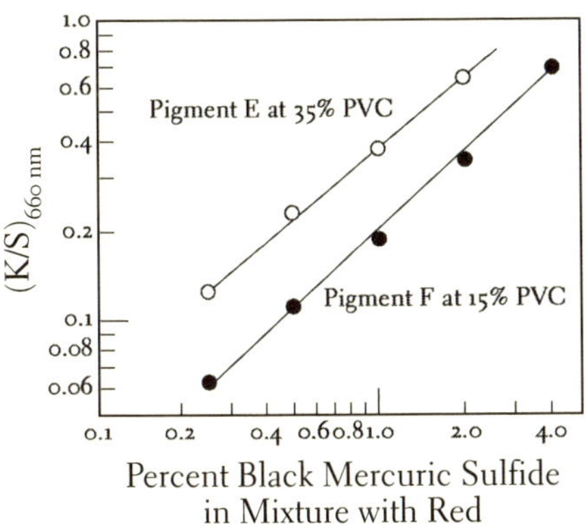

FIGURE 7
Relation of Kubelka-Munk function (K/S) at 660 nm to the percentage of metacinnabar pigment (black, α′-HgS) in mixtures with vermilion (red, α-HgS) in oil.

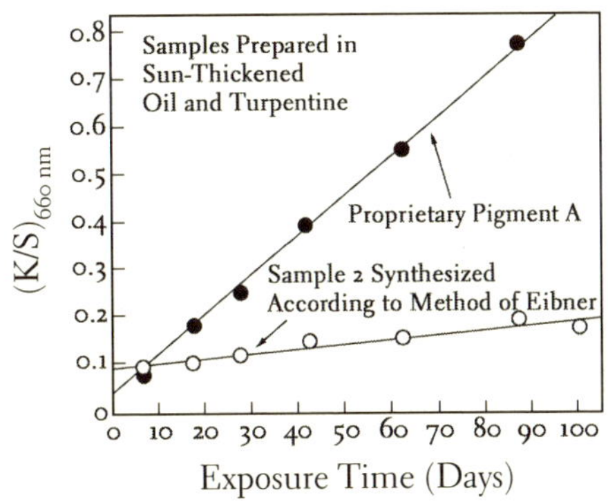

FIGURE 8
The darkening of vermilion as a function of exposure time under daylight fluorescent lamps.

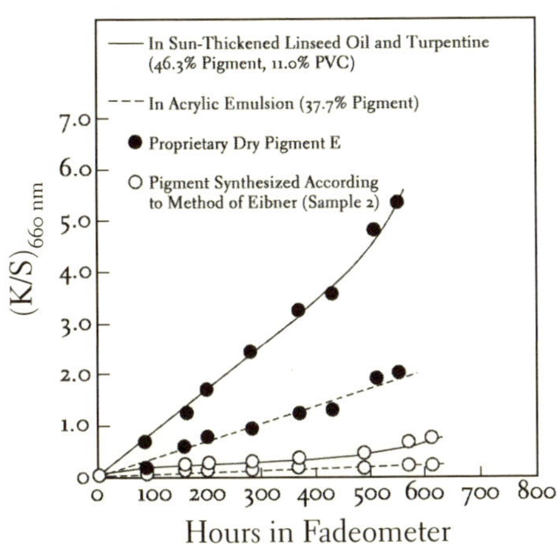

FIGURE 9
The darkening of vermilion as a function of exposure time in a Fadeometer.

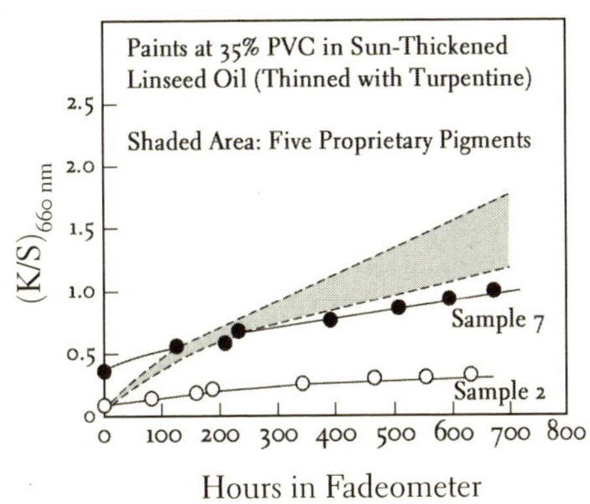

FIGURE 10
The darkening of vermilion as a function of exposure time in a Fadeometer.

points fall practically in a straight line implies that the amount of the black variety formed is directly related to the time of exposure. These results, considered in relation with the data in Figure 7, indicate that analysis of the darkening of vermilion in terms of the (K/S) function is an excellent technique by which the process may be studied in a rigorous manner. The fading of pigments can also be analyzed on the same basis.[44]

44. Feller, R. L., "Fading of Paints and Textiles," unpublished lecture on the use of the Kubelka-Munk equation, presented at the Annual Meeting of IIC American Group, New York (June 6, 1963).

Sample no. 2, prepared at Mellon Institute according to one of the methods described by Eibner, is seen in Figures 8 and 9 to be about ten times more lightfast than recently purchased vermilion pigments. In Figure 10, both sample no. 2 and a second laboratory-prepared sample, no. 7, are shown to darken less readily than five proprietary pigments. Vermilion no. 7 may not at first appear to be especially stable, but it should be noted that the curve begins at a higher (K/S) value than the others (it is a dark pigment, as may be seen by its reflectance curve, shown in Figure 3). In Table 1, the data illustrated in Figure 10 are expressed in terms of changes in reflectance; the manner in which pigments no. 2 and no. 7 were prepared is also briefly noted.

TABLE 1
Effect of Exposure of Various Pigments for 540 Hours in the Carbon-arc Fadeometer
Formulation: 35% PVC in Sun-thickened Linseed Oil

	Reflectance at 600 nm		Percent Loss of Reflectance at 660 nm
	Initial	Final Darkened State	
Proprietary Vermilions:			
Brand A	81.9	22.3	72.8
Brand B	77.8	24.5	68.5
Brand C	78.6	19.8	74.8
Brand D	70.0	20.6	70.6
Vermilion Pigments Synthesized According to Eibner's Directions:			
Sample No. 2 (from $Hg(NH_3)_2Cl_2$ and Barium Polysulfide)	62.5	47.1	24.6
Sample No. 7 (from Red HgO and Potassium Polysulfide)	44.1	28.4	35.6

There are a number of reasons why the experimentally determined curves of (K/S) vs. exposure may not be perfectly linear even though metacinnabar may be formed in direct relation to the exposure to light. Changes in the vehicle and in the roughness of the surface of the paint can affect the color independently of the formation of metacinnabar. For example, if a film or its surface deteriorates in the course of exposure, scattering of the light may increase. As a result, the increase in the value of (K/S) may not continue at the same rate that it did initially; and the paint may not become as dark as it would have if the surface had remained glossy. An apparent slowing down of the rate of darkening does occur in the Fadeometer exposures illustrated in Figure 10; further research will be necessary, however, before it can be said positively that film deterioration is the precise cause.

Darkening Influenced by Pigment Volume Concentration

Eibner pointed out that darkening would be greatest in vehicles that allowed the light to penetrate most deeply into the paint layer. Therefore, the darkening of paints based on oil, water color, and polymer emulsions may not be compared to one another readily because the absorption and scattering coefficients can differ considerably in these various media. Moreover, a matte paint may never appear as dark as a glossy one. Eibner demonstrated that vermilion changed more rapidly in linseed oil than in water color; our results show greater darkening in oil than in an acrylic emulsion (Figure 9).[14]

It is obvious from these considerations that, since the ratio of pigment to vehicle influences the depth to which the light can penetrate into the film, comparative tests of paints and varnish glazes must be conducted at nearly the same pigment volume concentration (PVC).[45] Experiments illustrated in Figure 11 clearly demonstrate that the darkening of vermilion in oil varies with the PVC.

Analysis of five artists' vermilion oil paints revealed that they were formulated at PVC's between about 21% and 44%, with an average of about 35%. The pigments made in the laboratory were therefore formulated at 35% PVC in preparation for the tests illustrated in Figure 10. The proprietary pigments were also extracted from the paints and reformulated in the same drying oil at exactly 35% PVC in order to place them under comparable conditions. Supplementary tests proved that the reformulation did not significantly alter the rate of darkening of the proprietary pigments.

45. Feller, R. L., and Matous, J. J., "Critical Pigment Volume Concentration and Chalking in Paints," *Bulletin of the American Group- IIC*, 5, No. 1 (1964), p. 25-26.

Protection by Filters

Ultraviolet filters are often used in museums to provide partial protection from the deteriorating effects of daylight or fluorescent lamplight.[1] Eibner suggested that paintings that contained much vermilion would benefit even under glass.[14] Preliminary tests indicated, however, that removal of the near-ultraviolet radiation by filters or glass would do little, if anything, to reduce the darkening of vermilion. Accordingly, it was decided to test the effect of a series of filters that would remove progressively more of the short wavelengths of radiation; perhaps a yellow or red filter would be effective where a colorless one might not.

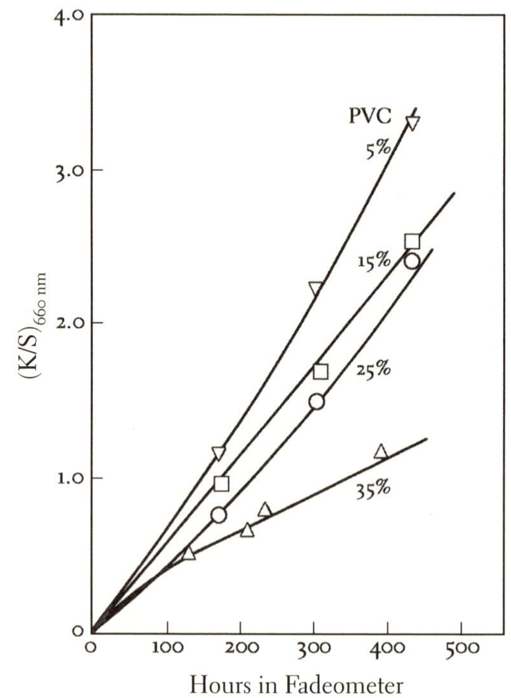

FIGURE 11
Darkening of proprietary vermilion pigment E in oil at various pigment volume concentrations (PVC).

The transmission characteristics of the filters and the results of the exposure are shown in Figure 12. It can be seen that the darkening of vermilion was progressively less under the colorless ultraviolet (UV), yellow, orange, and red filters. However, as well as removing an increasingly greater portion of the short-wavelength end of the spectrum, the filters also let through a decreased amount of light; the red filter was quite dark in color. The results, therefore, should be plotted with regard to the amount of radiant energy transmitted by each filter, not the range of wavelengths removed. When this is done (Figure 13), it can be seen that the reduction in the darkening may largely be accounted for by the amount of light transmitted by the particular filter.

These experiments are of interest in the history of painting technique; it was long the custom to place glazes of madder or cochineal over vermilion.[46] Toch and Eibner have said that the red glazes would partially protect vermilion from darkening; Eibner even pointed out that a yellow glaze was less effective.[14, 47] The present results substantiate their claim that such glazes can protect vermilion underpainting from darkening.

46. Zechmeister, L., "Die Erzeugung des Zinnobers in China, seine Eigenschaften und dessen Anwendung für die Malerei," *Technische Mitteilungen für Malerei*, 27 (1911), p. 201-203, 213-215.

47. Toch, M., *Materials for Permanent Painting*, New York (1911).

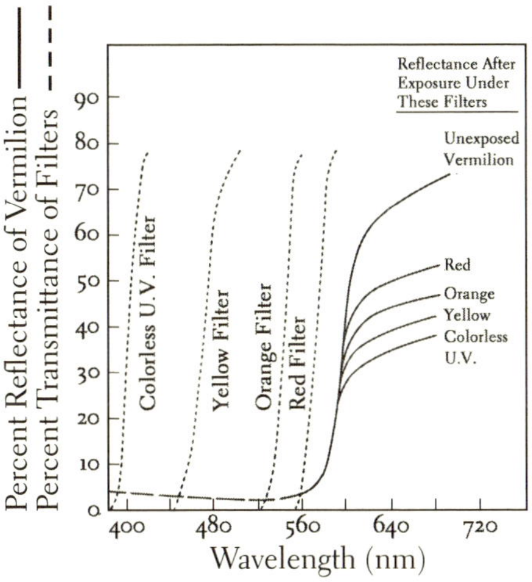

FIGURE 12
Darkening of vermilion oil paint after exposure to daylight fluorescent lamplight under various sharp-cut-off filters.

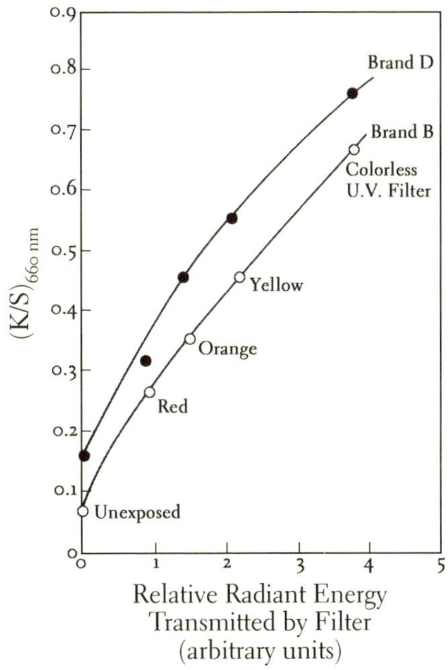

FIGURE 13
Darkening of proprietary artists' vermilion oil paints exposed to daylight fluorescent lamplight under various sharp-cut-off filters.

WHY HASN'T THE DARKENING ATTRACTED MORE ATTENTION?

William Holman Hunt complained about the poor quality of vermilion in 1875.[46] German investigators were concerned with the problem of darkening in 1896 and again in Eibner's day.[48-50] Why hasn't the phenomenon attracted more attention in recent years?

Perhaps the major reason is that dark varieties of the pigment exist, and, as has been stated, it is difficult to tell whether the color of the vermilion in a painting is the original or a darkened state.[51] Another reason may be that the darkening of vermilion is reversible under certain circumstances.

In addition to these considerations, the fact remains that some varieties possess excellent stability to light. Eibner found that vermilions made by the dry method tended to be more resistant to change than those made by the wet; the excellent reputation of vermilion of Chinese manufacture has long been known. Apparently, the varieties that are made with special care, either by Eibner's or related methods, or by sublimation, possess considerable resistance to darkening.

48. [Backenhus, G.], "Der Zinnober, seine Herstellung und sein Gebrauch in der Malerei der letzten acht Jahrhunderte," *Technishce Mitteiltingen für Malerei*, 28 (1911), p. 28-31, 38-40, 44-46.

49. Keim, A. W., "Zum Kapitel der Lichtbeständigkeit unserer Mal- und Anstrichfarben," *Technische Mitteilungen für Malerei*, 23 (1907), p. 212-217.

50. Keim, A. W., "Über die Eigenschaften und die Verwendbarkeit des Zinnobers," *Technische Mitteilungen für Malerei*, 28 (1911), p. 7-9, 17-19.

51. Unless a sharp contrast in color is seen under the rabbet of a frame, or unless a spot of dark varnish or overpaint has covered an area of a painting, it has not been easy in the past to be certain of fading or of darkening caused by exposure to light. For this reason, both problems have received less notice than perhaps they should. Modern methods of measuring color in paintings may help detect these changes more effectively in the future.[52]

52. Feller, R. L., "Problems in Spectrophotometry," *London Conference on Museum Climatology*, International Institute for Conservation of Historic and Artistic Works, London (1968), p. 257-269.

PROBLEMS FOR FURTHER STUDY

One of the first pigments prepared at Mellon Institute according to Eibner's directions was found to be about ten times more resistant to darkening than recently purchased vermilion. His pigments that remained unchanged after ten years outdoors were probably even more stable than this.[53, 15] The preparation of vermilion according to Eibner's directions therefore must be continued in order to evaluate his procedures more completely and to determine the maximum degree of stability that can be achieved.

53. Eibner, A., "Arbeitsumfang der Versuchsanstalt für Maltechnik an der Technischen Hochschule zu München," *Technische Mitteilungen für Malerei,* 42 (1926), p. 4-12.

It would be highly desirable to learn more about the reversibility of the blackening and to determine the threshold of exposure beyond which reversibility may be lost. Further studies of the effect of scattering and pigment volume concentration are important, for they relate to, practical differences in behavior between water colors, oils, and emulsion paints. Sensitive methods of analysis should be developed to determine the lightfastness of a pigment before it is exposed and to measure the change in vermilion that may already have taken place in paintings.

Theoretical chemistry today has the means to explain the transformations of vermilion more precisely than Eibner could. This should be done, for increased understanding of the phenomenon is bound to lead to further advances.

CONCLUSION

Let us go back to the original questions. On the basis of these preliminary tests, we have good evidence that Eibner's explanations for the darkening of vermilion and directions for the preparation of lightfast types were basically sound; pigments made according to his methods proved to be more resistant to darkening than proprietary vermilions currently available. The initial success in substantiating many of his statements will prompt us to reconsider his contributions on the subject in greater detail.

Tests on currently available vermilions have shown that significant darkening may occur in five to ten years of normal gallery exposure, depending on the specific pigments and conditions. In general, the degree of darkening is affected by the paint structure, particularly the ratio of pigment to vehicle. Vermilion in water color or polymer emulsion, for example, darkens less than in linseed oil. The results of tests with yellow, orange, and red filters indicate that the traditional practice of overglazing with madder or cochineal provided protection from darkening, principally by reducing the intensity of the light transmitted.

Spectrophotometric analysis of the changes that occur in the reflectance of vermilion has shown that absorption by the red pigment is lost in the course of darkening and that the darkened states of bright vermilion are similar in appearance to varieties of the pigment that were originally dark. Analysis of the changes in reflectance in terms of the Kubelka-Munk function (K/S) has provided evidence indicating that the amount of metacinnabar formed is directly related to the quantity of light that falls upon the sample. Application of the Kubelka-Munk concept thus, once again, has proven to be an excellent tool by which darkening and fading processes can be studied in a rigorous manner.

The information gained from these investigations is being shared with manufacturers and suppliers in the hope that continued advancements can be made in the materials available to the conservator of paintings and the creative artist.

Acknowledgments

The author wishes to acknowledge the contributions of Mrs. Derrick J. Page and Mrs. Roy A. Curran in the laboratory investigations. Dr. Hermann Kühn, of the Doerner Institut, Munich, kindly provided copies of the publications from the rare journal *Technische Mitteilungen für Malerei*. The Research Project is also most grateful to Miss Ruth M. Johnston, who made many of the spectrophotometric measurements on the samples and who gave valuable advice and criticism while she was Project Leader in Color Research, Coatings and Resins Division, PPG Industries, Springdale, Pennsylvania. We also wish to express special appreciation to Dr. H. L. Gerhart, Director of Research and Development, Coatings and Resins Division, PPG Industries, for his interest and encouragement of the cooperation between the Color Research Laboratory and the Research Project.

IDENTIFICATION OF TRADITIONAL ORGANIC COLORANTS EMPLOYED IN JAPANESE PRINTS AND DETERMINATION OF THEIR RATES OF FADING

ROBERT L. FELLER, MARY CURRAN AND CATHERINE BAILIE

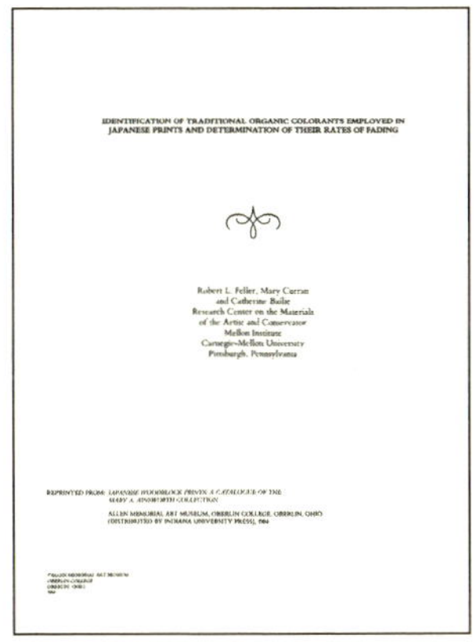

Originally printed in:
JAPANESE WOODBLOCK PRINTS: A CATALOGUE OF THE MARY A. AINSWORTH COLLECTION
Roger S. Keyes, Allen Memorial Art Museum, Oberlin College, Oberlin, Ohio (1984), Pages 253-266.

INTRODUCTION

The investigation that will be described here was undertaken on behalf of the Allen Memorial Art Museum to provide information concerning the potential rate of fading of traditional organic colorants found in Japanese prints. This undertaking was supported in part by a portion of the funds provided by the National Endowment for the Arts to enable the museum to inspect and catalogue its collection of prints and to survey the needs for their conservation. Because of the long-term interest that the Research Center on the Materials of the Artist and Conservator has had concerning the potentially damaging effects of light on museum collections, the project was expanded further than originally commissioned.

Roger and Keiko Keyes (consultants), who in 1979 undertook the cataloguing and technical examination of the Ainsworth collection of Japanese prints in the Allen Art Museum, had long been aware that many Japanese prints in collections throughout the world are now in seriously faded states. Keiko Keyes, particularly, urged both the curatorial staff of the Museum and the scientists at the Research Center to seek information on the rate of fading of traditional colorants used in the prints. Among the questions that needed to be answered were: What colorants used in these prints are especially sensitive to light? How rapidly will they fade? Finally, are such pigments to be found in the prints in the collection of the Allen Art Museum? At the beginning of this project, the Keyes knew of no decisive information available which would answer these questions. Accordingly, a portion of the grant to the Museum was assigned to obtain this information, vital to the care and preservation of prints executed in this time-honored art form.

PLAN OF INVESTIGATION

In order to obtain the desired information concerning the sensitivity of the traditional Japanese colorants to light, the authors first had to obtain a number of authentic samples of these materials. Two—*ai* (indigo) and *beni* (safflower red)—were obtained from Keiko Keyes in the form of dry pigments as well as strips of paper on which they had been painted. Two others, *aigami* (dayflower blue) and *enju* (Japanese pagoda tree), were provided through the kindness of the distinguished collectors, Dr. and Mrs. James B. Austin of Pittsburgh. Other colorants of interest to this work were available from the collection of the National Gallery of Art Pigment Bank maintained at the Center; the lightfastness of pigments such as carmine, natural and synthetic alizarin, Prussian blue, turmeric and gamboge, for example, had already been investigated on a number of occasions in the past, some more extensively than others.

It was necessary to determine whether such pigments could be identified in a print and, if so, whether they could be proven to exist in prints of the Ainsworth collection. The spectrophotometric reflectance curves of the authentic samples of both *aigami* (dayflower) and *ai* (indigo) proved to be sufficiently characteristic that they can be detected and distinguished when these blue pigments are used alone or in mixtures with a yellow or red. *Beni* (safflower) fluoresces under near-ultraviolet radiation and has a sufficiently characteristic spectrophotometric reflectance curve that it, too, can be identified with reasonable confidence. With this information, we were able to show that these three colorants were indeed present in representative prints from the collection that were sent to the Center for examination. The measurement of spectral reflectance curves is a "non-destructive" method of analysis; no sample needs to be taken. A detailed description of the methods used to measure the spectral reflectance curves of various areas on the prints, by simply placing the particular colored area at the viewing port of the instrument, is given in the addendum section.

Finally, using samples of dyeings and paint-outs of traditional colorants, as well as small areas of two prints of the 1800s kindly provided to the laboratory for this purpose, fading tests were conducted under a bank of fluorescent lamps and the observed rate of fading of the samples compared with the International Standards Organization's blue-wool fading standards. The results showed that at least seven of the traditionally employed colorants are indeed fugitive. *Beni* (safflower) and *aigami* (dayflower blue) were found to be especially sensitive to light.

A formal report on the initial investigation was presented to the Allen Art Museum on Dec. 18, 1980.[1] The basic findings on lightfastness were also published in 1981 in the preprints of a meeting of the International Council of Museums' (ICOM) Committee for Conservation.[2] Since then, the work has been extended to include studies on (a) the fading of some of the same colorants

1. Feller, R. L., and Bailie, C. W., *Identification and Rate of Fading of Some Traditional Organic Colorants Employed in Japanese Prints*, Final Technical Report to the Allen Art Museum from the [Research] Center on the Materials of the Artist and Conservator, Mellon Institute Pittsburgh, PA (1980).

2. Feller, R. L., and Johnston-Feller, R. M., "Continued Investigations Involving the ISO Blue-Wool Standards of Exposure," *Preprints of the Sixth Triennial Meeting, Ottawa, Canada*, ICOM Committee for Conservation (1981), Paper 81/18/1.

applied as textile dyes, (b) the general ineffectiveness of ultraviolet filters in preventing fading, and (c) the direct measurement of fading in two inexpensive prints of the 1800s.

Concerning the Traditional Colorants

Pigments

The coloring materials that artists employ fall into two principal chemical classes: inorganic (mineral) and organic compounds. The latter are usually complex substances based on carbon along with hydrogen, oxygen, and perhaps nitrogen or sulfur. The compounds can be subclassified further, if so desired, into those of natural or of synthetic origin. With regard to long-term durability, it is generally recognized that the mineral pigments tend to be highly stable. In contrast, the organic-based coloring materials, particularly those traditionally derived from various natural vegetable or animal sources, tend to exhibit moderate to poor lightfastness. The Japanese printmaker employed both inorganic and organic colorants, but our chief concern here will be with the traditional natural organic materials, for these have long been considered to be the least stable of the various colorants used.

A coloring agent is particularly useful to the artist if it stays where he puts it (does not "run") and does not tend to wash away or to dissolve into the paint vehicle or a protective varnish. In order to immobilize a colorant based on soluble dyestuffs, the dyer of cloth has traditionally reacted his coloring matter with mineral-based "mordanting" agents which will precipitate or otherwise insolubilize the colored chemical substance. The painter traditionally has insolubilized dyestuffs by a process called "laking": the soluble colorant is absorbed on a relatively colorless but porous mineral, or is coprecipitated with an otherwise colorless mineral. Colorants attached to inorganic colorless minerals in this way are known as "lakes." There are also organic-based coloring agents that are, or can become, insoluble in themselves and thus do not require laking; indigo is such an example.

Because the authors cannot be certain of the exact nature of many of the colored materials traditionally used by the Japanese printmaker, we will employ the general term "colorant" in this text without attempting to define more precisely whether the coloring matter is based on a lake, an insoluble organic substance, or perhaps a mixture of either a mineral pigment and an organic-based colorant or two organic-based colorants. Mixtures were often used—a yellow with a blue to make green—or a red with a blue to make purple, for example. In such cases, one of the two colorants may fade more rapidly than the other, leaving a highly distorted color balance in the picture. The reader may recall that the color in the foliage of European tapestries is frequently found as blue today; it is probable that it was originally green but that the yellow dye used in mixture with the blue has completely faded away. One may expect greens and purples in many Japanese prints to suffer similar changes in hue as one component fades in time more rapidly than the other.

A further factor affecting the color of Japanese prints is the color of the paper. Inks are more or less transparent, depending in part on the concentration and in part on the light-absorption characteristics of the colorants used. The result is that the hue of the paper can affect the printed color. Papers frequently yellow with age, even in the dark. Moreover, as the colorants themselves fade, more and more of the color of the paper is able to show through, further affecting a change in appearance. When it was possible to do so, we have measured the color of the unprinted paper in order to aid in our analysis and identification of the colorants.

A paint or printing ink is based on a colorant dispersed in a binder or vehicle combined with a solvent. Unfortunately, we cannot say much about the binder at this time, for that used in traditional Japanese prints is not well documented. It was frequently a water-dispersible material, possibly based on a rice-starch paste.

For the sake of clarity throughout this report, we shall use the term "color" to apply to the perceived color; it is the visual effect, that which the eye sees. The term "colorant" applies to the coloring matter or material used to create the color perceived.

The authors are not familiar with the history of Japanese printmaking but can refer to the comments of Arthur Davison Ficke in order to provide some idea of the dates involved.[3] Ficke says that the black and white period of Japanese printmaking changed after about 1700 with the introduction of red lead ("tan"). He further states that, about 1710, citrine and yellow were introduced and, by 1715, "beni," the red made from Carthamus tinctorius L., the Beniband plant. Hand-coloring prevailed up to about 1742; then green and red (*beni*) colors printed from blocks appeared in the period of 1742-64. After that date multi-colored prints began to appear, extending up to the present time. Ficke considered a fifth period, 1806-1858, to mark the close of the great era of figure-prints, although it was the time of the outstanding landscape prints of Hokusai and Hiroshige. From this brief summary, one learns that the inquiry concerning lightfastness will be concerned primarily with colorants available in the eighteenth century.

Table 1 presents a list of some of the principal coloring agents that one may expect to find in traditional Japanese prints.[4, 5] To the best of our knowledge, all the organic colorants in this list tend to be fugitive, with alizarin, indigo and Prussian blue the most stable of the series, perhaps even possessing what might be considered "intermediate" lightfastness.[6] Information concerning the lightfastness of eleven in this list of organic-based colorants has been obtained as a result of the present investigations, summarized principally in Tables 4 and 5.

3. Ficke, A. D., *Chats on Japanese Prints*, Unwin, London (1915).

4. Yoshida, T., and Yuki, R., *Japanese Print Making*, Charles E. Tuttle, Rutland, Vermont (1966).

5. Schaubstadter, C., *Care and Repair of Japanese Prints*, Idlewild Press, Cornwall-on-Hudson, New York (1948).

6. The characterization of colorants with respect to their being in a class of lightfastness designatable as "fugitive," "intermediate" or "excellent" is discussed in the section entitled "Determination of Lightfastness."

TABLE 1
Some Traditional Japanese Colorants

Japanese Name	Typical or Nominal Color	Source (Common Name)
Organic Colorants		
Beni*	red or pink	Carthamus tinctorius (Safflower)
Ukon*	yellow	Curcuma longo (Turmeric)
Ai*	blue	Polygonum tinctorium (Indigo)
Aigami*	blue	Commelina communis (Dayflower, lily family)
Zumi	mustard yellow	Malus Sieboldii or Pyrus Toringo
Shio	yellow	Garcinia morella (Gamboge)
"Yen Chih"	red	Mirabilis Jalapa L.
Suo	red	Caesalpina sappan L. (Red bud)
Kariyasu	yellow	Miscanthus tinctorius (Miscanthus)
Enju	yellow	Styphnolobium japonicus (Jap. pagoda tree)
Kihada	yellow	Phellodendron amurense (Jap. yellow wood)
Yama-momo	yellow	Myrica rubia
Kusa-ai	green	Combination of yellow with indigo
Bero-ai	blue	Prussian blue
Enji	crimson lake	Traditionally, a carminic acid colorant from cochineal insect
Akane	red	Rubia akane and tinctorum; pseudopurpurin- and alizarin-based colorants
Inorganic Colorants		
Enpaku**	white	Lead white
Gofun	white	Clam-shell white
Shu**	vermilion	Mercuric sulfide
Tan**	orange	Red lead
Benigara	iron oxide red	"Red Shell"
Kio or Sekio	yellow	Orpiment, "Stone Yellow"

* Mrs. Keyes considers these four to be the organic colorants of principal concern with respect to exposure to light. The present study did not include *ukon* (turmeric), but prior experience with this material indicates that it is fugitive. Cloth dyed with ukon proved to be very fugitive.

** As noted in the text these can darken.

Although inorganic pigments tend to be stable, brief mention should be made of the fact that some of these can also readily change in appearance. Three deserve comment. Red lead and white lead can discolor owing to the chemical reaction of these pigments with sulfides in the atmosphere, causing them to darken owing to the formation of black lead sulfide. Ficke mentions, for example, that the white—this would probably have been lead white—had turned black on some of Hiroshige's works.[3]

In what may be a reference to the darkening of red lead by sulfides but which also may be a reference to the darkening of vermilion upon exposure to light, Ficke states in another passage, "The rich orange color of Koriusai and Shunshō, and the delicate pink used by Harunobu, Kiyonaga and Shunshō, are transformed in the course of time to a rusty black . . ." Vermilion (mercury sulfide) is able to darken owing to its transformation into a black form, a physical rather than chemical alteration. Articles by Gettens[7] and Feller[8] discuss in considerable detail the phenomenon of the blackening of vermilion caused by exposure to light. In 1964, the authors noted a set of Japanese-prepared vermilion pigments that had been kept in glass jars in a showcase in the hallway outside the conservation laboratory of the Fogg Art Museum of Harvard, perhaps placed there by Edward Forbes in the 1930s; all had turned black at the outer wall of the jars where they had been exposed to the light. The samples were labeled: *Shu* (*akaguchi*, deep), *Shu* (*kiguchi*, orange) and *Shu* (*kodai*, burnt).

7. Gettens, R. J., Feller, R. L., and Chase, W. J., "Vermilion and Cinnabar," *Studies in Conservation*, 17, No. 2 (1972), p. 45-69.

8. Feller, R. L., "Studies of the Darkening of Vermilion by Light," *Report and Studies in the History of Art*, National Gallery of Art, Washington, D.C. (1967), p. 99-111.

Traditional Dyes

Although our specific concern is with the colorants in prints, it is of some interest to review the natural coloring agents traditionally used in the dyeing of cloth. Schraubstadter[5] notes that the dyeing of cloth was a fine art when the first prints were made and, hence, the colorants used in treating cloth were likely to have been employed initially in printmaking, both the mineral and vegetable. Later on, cochineal, imported from China, was substituted for *beni*. Sometime after its discovery date of 1704, Prussian blue was imported from Holland and used in many cases in place of indigo.

At an international symposium on the conservation and restoration of cultural property held in Tokyo and Tsukuba, Japan, in November of 1978, Hayashi presented a paper on the plant dyes used in ancient Japanese textiles. Therein the sources and chemical composition of common dyestuffs were described.[9] Professor Hayashi's list of yellow dyestuffs include:

Kihada, Phellodendron amurense var. suberosum (Rutaceae). The dye constituent is berberine, which has an intense yellow fluorescence under near-ultraviolet radiation.
Woren, Coptis japonica (Ranunculaceae). The dye constituent here is also berberine.
Kariyasu, Miscanthus tinctorius and Arthraxon hispidus (both belonging to Graminae). The dye constituents are anthraxin and related flavonoids (mainly luteolin glycosides).

9. Hayashi, K., "Chemical Procedure for the Determination of Plant Dyes in Ancient Japanese Textiles," *International Symposium on the Conservation of Cultural Property*, Tokyo National Research Institute of Cultural Properties (1979), p. 39-50. Through the courtesy of Dr. John Winter of the Freer Gallery of Art the following publications have come to our attention since this project was completed: Kashiwagi, M., and Yamazaki, S., "The Lightfastness of Traditional Vegetable Dyes," *Scientific Papers on Japanese Antiques and Art Crafts*, 27, Tokyo National Research Institute of Cultural Properties (1982), p. 54-65, in Japanese; Hiraoka, A., "Spectrophotometric Characteristics of Ancient Yellow Papers and their Fading Behaviors," *ibid.*, 27 (1982), p. 66-72.

Ukon (Japanese turmeric), Curcuma domestica (Zingiberaceae). The dye constituent is curcumin, the only yellow dye not requiring a mordant to fix it on cloth. This apparently fluoresces as well as berberine-based colorants.
Kuchinashi, Gardenia jasminoides (Rubiaceae). The dye constituent is crocin.
Enju, Sophora japonica (Leguminosae). The dye constituent is rutin (quercetin glycoside).
Haze or *Haji* (Japanese sumack), Rhus succedanea (Anacardiaceae). The dye constituent is fustin.

The red dyestuffs are:

Alkane, Rubia akane (Rubiaceae), Japanese Madder. The dye constituent is pseudopurpurin, which is yellow-orange but becomes red when mordanted with alkali.
Seiyo-akane, Rubia tinctorum (Rubiaceae). The dye constituent is alizarin.
Benibana, safflower, Carthamus tinctorius (compositae). The two dye constituents are carthamin and carthamon.
Suwo (Brazil wood, originated in Malaya, India), Caesalpinia sappan (Leguminosae). The dye constituent is brazilein, which is a yellowish brown color but becomes scarlet red when mordanted by alum.

The only violet dye reported was:

Murasaki or *Shikon*, Lithospermum erythrorhizon (Borraginaceae). The dye constituent is shikonin, a derivative of naphthoquinone.

The major blue dye is:

Ai, Polygonum tinctorium (Polygonaceae). The dye constituent is indigotin (indigo). Besides this plant, there are Strobilanthes flaccidifolium (Acanthaceae) and Isatis sp. (Cruciferae), the main dye constituent of which is also indigotin. *Aigami* does not appear in Hayashi's list of dyes used on cloth.

The major brown to black dyes are tannins. There are several plant sources for the tannins, but it is impossible to distinguish them at this time.

DETERMINATION OF CHARACTERISTIC SPECTRAL REFLECTANCE CURVES AND MEASUREMENT OF RATE OF FADING WITH KNOWN SAMPLES OF TRADITIONAL JAPANESE ORGANIC COLORANTS

Preparation of Samples of Known Colorants

Mrs. Keyes kindly provided the initial samples. *Beni* (safflower red) was received as a brushed-out specimen on a strip of paper; a small piece was cut out for exposure tests. The sample of *aigami* (dayflower blue) was received as a paper pad saturated with the colorant, a time-honored way in which dyes and lake pigments have been supplied to the artist. A small piece was cut off, and suspended in distilled water; the extracted blue coloring matter was then brushed on filter paper. The *ai* (indigo) was received in the form of a solid cake; a fragment was broken off, ground and dispersed in water and applied to filter paper.

A second sample of *aigami* (dayflower blue) was prepared by Dr. and Mrs. James Austin to demonstrate how the colorant could be readily obtained from the plants growing in their garden simply by crushing the flower petals directly onto a piece of filter paper. The result was a beautiful intense blue.

Dr. Austin also brought to the laboratory a spire of flowers from a tree blossoming nearby which he said was known as the Japanese pagoda tree. We extracted the petals and stamens separately in hot distilled water, brushing each resulting solution onto filter paper. Although we are not certain that this is the traditional technique for preparation of the colorants, the stamen extract yielded a bright clean yellow, and that from the petals, a somewhat darker yellow. The Japanese name for the yellow colorant from the pagoda tree is *enju*.

The test samples on filter papers were exposed 3 1/2 inches below a bank of fluorescent lamps in a room maintained at 70°F (21.1°C) and at 65% relative humidity. The International Standards Organization's blue-wool standards (ISO R105, BS1006:1971) were exposed along with the colorant samples to monitor the rate and extent of fading. (Use of these fading reference standards is discussed in the section entitled "Determination of Lightfastness".) Initially, and at intervals during the exposures, the samples were removed for spectrophotometric reflectance measurements. The results are illustrated in Figures 1 to 7. As fading occurred, the reflectance increased.

Spectral Curve Characteristics and Changes on Fading

Figures 1 and 2 illustrate the changes that took place in the spectral reflectance curves of the two samples of *aigami* (dayflower blue) as fading occurred. The crushing of the petals had deposited the greatest amount of colorant on the filter paper (Figure 2); this is indicated by the lower reflectance than that of

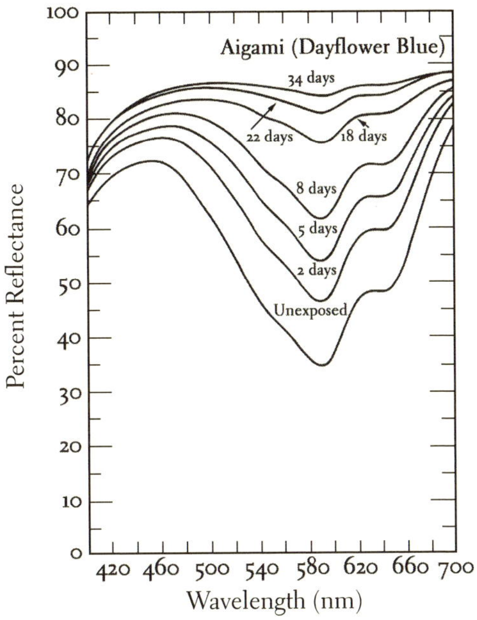

Figure 1
Spectral reflectance curves of pale wash of *Aigami* (dayflower blue) exposed for a total of 34 days (9 million lux hours) under "soft white" fluorescent lamps.

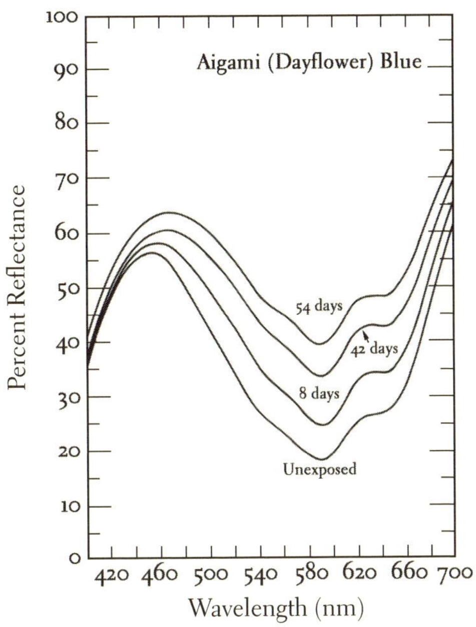

Figure 2
Spectral reflectance curves of higher concentration of *Aigami* (dayflower blue) than in Figure 1 exposed for a total of 54 days (14.3 million lux hours) under "soft white" fluorescent lamps.

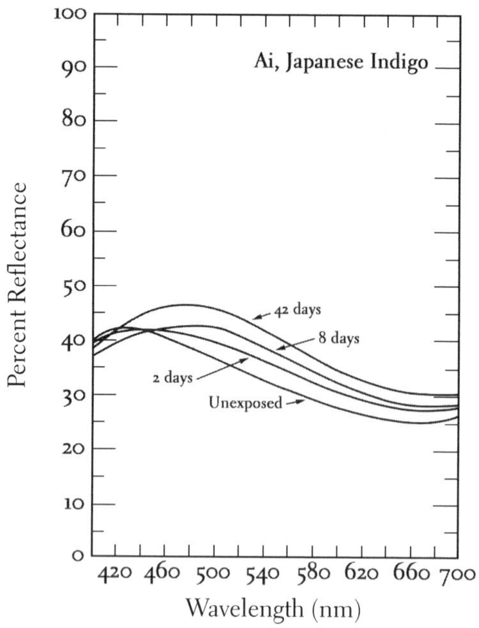

FIGURE 3
Spectral reflectance curves of *ai*, Japanese indigo, exposed for a total of 42 days (12 million lux hours) under "soft white" fluorescent lamps.

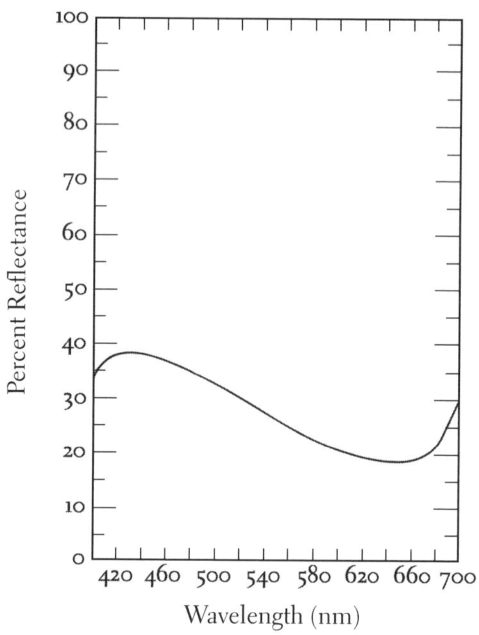

FIGURE 4
Typical spectral reflectance curve of indigo.

the colorant extracted from the paper pad (Figure 1). Both samples exhibit a characteristic absorption pattern in the region of low reflectance: a major absorption peak at 585 nanometers (nm) and a major shoulder at 630 nm and minor shoulder at about 550 nm. Because absorptions are additive, these curve characteristics will appear regardless of whether the colorant is mixed with other colorants such as a yellow or a red and will provide a unique fingerprint for identifying the presence of this colorant.

Figure 3 shows the changes that took place in the spectral reflectance curve of *ai* (indigo). The shape of the curves in the absorption region of 660-700 nm remains much the same even though the short-wavelength reflectance region tends to shift towards the green in the course of fading. Figure 4 illustrates a curve which the authors regard as typical of indigo, exhibiting a somewhat higher reflectance at 700 nm than found in Figure 3. However, the differences seen here are not uncommon in samples of a natural dyestuff such as this. The reader can readily see by comparing Figures 1 and 2 with Figures 3 and 4 that there would be no difficulty in distinguishing between these two blue colorants.

The spectral reflectance curve of safflower red is not as unmistakable, although the general curve shape, the location of the absorption maximum, and the fact that the colorant fluoresces under near-ultraviolet radiation can make its identification a reasonably certain matter. If one were to measure the spectral sensitivity of emission and the spectral emission curve of the fluorescence of safflower red under near-ultraviolet radiation, as we have reported in the case of madder and of Indian yellow,[10] the characterization of *beni* could be firmly established with this additional objective measurement, one again that does not require the taking of a sample. Figure 5 indicates that, in the course of fading, the pigment loses some of the sharpness of the slight reflectance peak at 450 nm. The curve shape of this pigment is not sufficiently distinctive that one would have a high degree of confidence in using the curve as a sole means of identifying beni in a print. However, as we have said, the fact that the safflower-based colorant also fluoresces under 365 nm ultraviolet radiation is a useful supplementary characteristic that warrants a high degree of confidence in our current identifications in the Ainsworth collection.

10. Feller, R. L., "Problems in Spectrophotometry," *London Conference on Museum Climatology*, International Institute for Conservation of Historic and Artistic Works, London (1968), p. 257-269.

The curves of the two extracts from the flowers of the Japanese pagoda tree, *enju*, shown in Figures 6 and 7, are not sufficiently distinct to be used for identification purposes. The coloring matter proved to be very fugitive indeed.

The above described investigations, carried out on samples of known composition, convinced us that, if the spectral reflectance curves of colorants in actual prints could be measured, these three colorants—safflower red and particularly the *aigami* and indigo blues—could be identified with considerable confidence even in a faded condition. These preliminary studies also convinced us that the three colorants were markedly sensitive to fading upon exposure to light.

IDENTIFICATION OF *AIGAMI* (DAYFLOWER), INDIGO AND SAFFLOWER RED IN SPECIFIC PRINTS

Having shown that three very fugitive and commonly encountered colorants in traditional Japanese prints can be identified with considerable confidence simply by measuring their spectral reflectance curves and by observing the tendency of the red to fluoresce, attention was turned to the possibility of identifying these particular colorants in the prints of the Ainsworth collection. The effort to measure the spectral reflectance curves of various reds, blues, violets and greens in the prints was successful; we are able to report the presence of these three colorants, which readily tend to fade upon exposure to light, in a representative set of ten prints in the collection. The measured spectral reflectance curves, some examples of which can be seen in Figures 8, 9, and 10, permit the colorants to be identified by reference to the curves of known specimens, shown in Figures 1, 3 and 5, and by observation of the fluorescence of the red.

Details of the methods used to measure the spectral reflectance curves on the prints themselves are described in the addendum. Because of the generally small size of areas of uninterrupted color that were to be found in a typical print, the Small-Area-View (SAV) Color-Eye® was usually used to obtain the reflectance data; the circular measurement-port on this instrument is only 1/4" in diameter. The reflectance was measured every 20 nm across the visible spectral region of wavelengths (400-700 nm) relative to a barium sulfate (white) standard. Owing to the fact that most of the colored areas were not completely opaque, a sheet of Millipore® filter paper was placed behind each print to provide a standard white-reflecting background behind the prints. This material has a reflectance of 100.5 relative to usual barium sulfate standard.

Because the prints are generally on an off-white background (perhaps the paper has discolored in many cases owing to age), the curves were frequently altered by the yellowness of the paper. Nevertheless, this was not sufficient to obscure the characteristic absorption peaks of the two blue pigments and certain mixtures thereof. Figure 9, for example, shows a typical result in which area b2 in the sky of print 410, now an olive grey owing to the discoloration of the paper, can still be identified as having once been a blue or bluish-grey made with *aigami*.

Table 2 presents a summary of the pigment identifications in ten prints in the collection. Complete data on all measurements are reported in the Final Technical Report submitted to the Allen Art Museum.[1]

Aside from the identification of the various colorants, it is hoped, through the availability of these records of spectral reflectance, that someone at a future date might be able to detect a change in the reflectance in these particular

TABLE 2

Summary of the Results from the Examination of Prints in the Collection

Prints of the 1820s –

Print 416 – (print in excellent condition; possibly unfaded?)–the blue of the water is based on indigo. Area (b)*

Print 410 – the blue in sky and the water is based on *aigami* (b; b1, 2). See Figure 9.

Print 411 – the blue of the water is *aigami* but the color is more pale than in 410 (b).

Prints of the 1850s –

Print 1113 – (print essentially unfaded?)–red in both calligraphic panels at top right is probably *beni* (r1, 2). See Figure 8.

Print 1111 – (print considered by Mrs. Keyes perhaps to have undergone an intermediate degree of fading)–red signature panels at top right: basically the same spectrophotometric curves as 1113 but paler hues; also more pale than red cloud which fluoresces strongly under near-ultraviolet radiation. Probably *beni* (r1, 2, 3).

Print 323 – (a print which has, in the opinion of Mrs. Keyes, faded somewhat since 1972)–the purples of the kimono appear to be red plus *aigami*; a paler hue has the same curve shape as the darker one (with little less blue and more red in it). The darker one is a slightly more blue–purple (p1, 2).

Print 479 – the red curves are not distinctive enough to learn much, but fluorescence in the red kimono at the far left figure suggests *beni* (r). See Figure 8.

Print 194 – the blue in the left-most figure is *aigami* and the purple in the figure on the extreme right is based on a mixture of red and *aigami* (b; p1, 2).

Print 288 – the pale blue seems to be *aigami*; the red fluoresces strongly and therefore may be *beni* (safflower) (see Figures 8, 9) (b1, 2, 3; r1, 2). The purple appears to be a mixture of red and *aigami* (p1, 2). The green apparently is a yellow plus indigo, as seen in Figure 10 (g1, 2).

Print 290 – the blue of the kimono at far left appears to be indigo (see Figure 10); the purple showing at the neck of the lady at the left, a mixture of red and indigo (b1, 2; p1, 2).

* Locations of the areas measured are noted in the illustrations that accompanied the Final Technical Report submitted to the Allen Art Museum on Dec. 18, 1980 (Reference 1). The particular area is designated by a letter or a number: (b2) is blue area 2; (r3) is red area 3; p, purple. See example of the marked copies of the prints in Figure 14.

prints, especially if fading should occur to an appreciable extent. Accurate and confident measurement of the extent of a color change between these present measurements and a time five to ten years hence is not an easy task, but we believe that the achievement of such an objective is possible with reasonable success.[10, 11]

11. Johnston, R. M., and Feller, R. L., "The Use of Differential Spectral Curve Analysis in the Study of Museum Objects," *Dyestuffs*, 44, No. 9 (1963), p. 1-10.

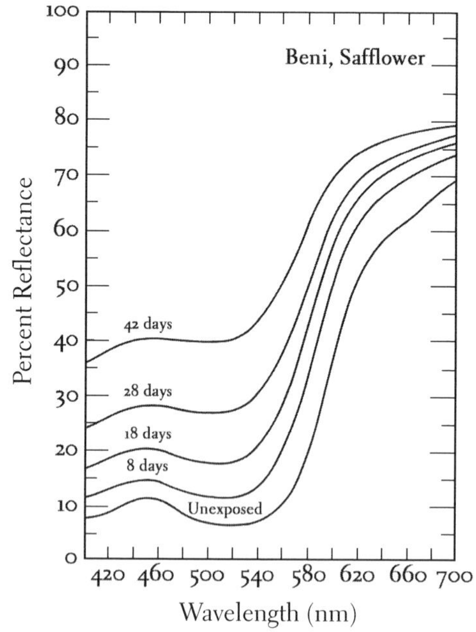

FIGURE 5
Spectral reflectance curves of *beni* (safflower) exposed for a total of 42 days (12 million lux hours) to "soft white" fluorescent lamps.

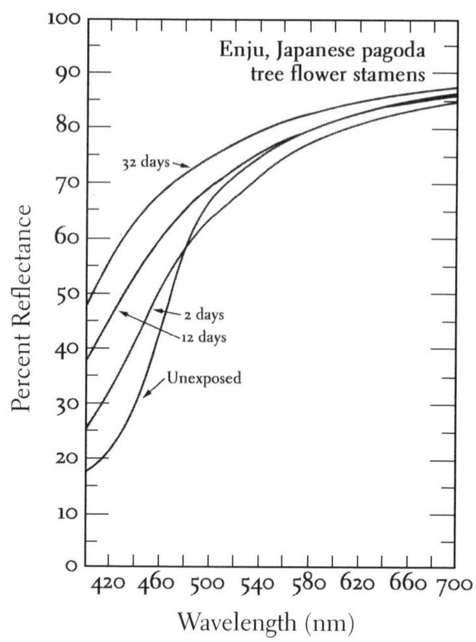

FIGURE 6
Spectral reflectance curves of *enju*, from flower stamens of Japanese pagoda tree, exposed for a total of 32 days (9 million lux hours) to "soft white" fluorescent lamps.

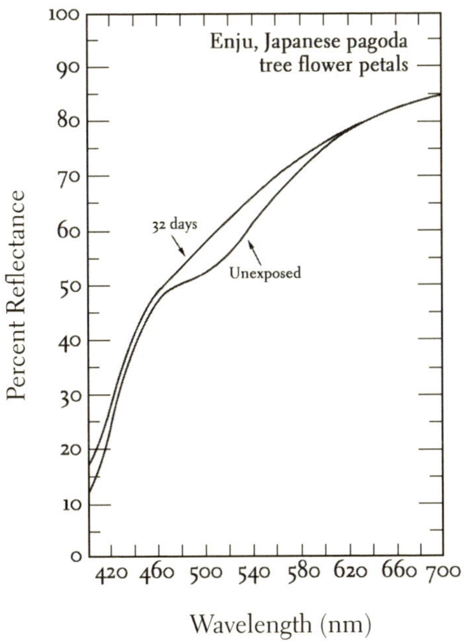

FIGURE 7
Spectral reflectance curves of *enju*, from petals of Japanese pagoda tree, exposed for a total of 32 days (9 million lux hours) to "soft white" fluorescent lamps.

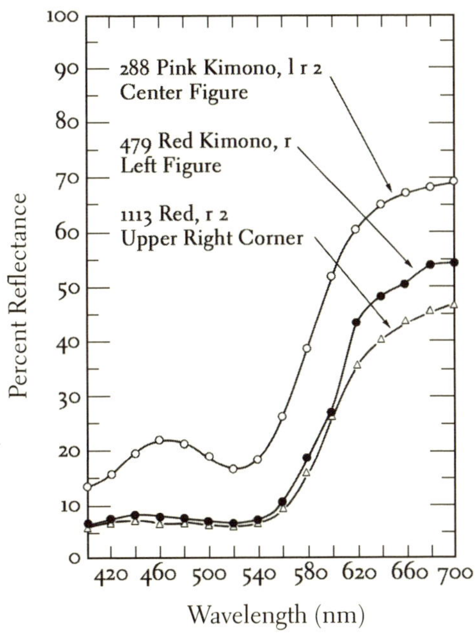

FIGURE 8
Spectral reflectance curves of a red colorant that fluoresces under near-ultraviolet radiation (*beni*?). Measurements made with the Model D-1 Color-Eye® spectrophotometer.

DETERMINATION OF LIGHTFASTNESS

Concerning the Specification of Lightfastness

The International Standards Organization's Recommendation R105 on Lightfastness: While we are all reasonably familiar with materials that fade and realize, moreover, that some textiles and paints tend to fade very rapidly and that others are highly resistant to change, the characterization of the relative lightfastness of materials remains a challenging task even for the scientist. The difficulties arise because the rate at which a material fades depends not only on the particular coloring matter involved but also on whether it is associated with other substances such as mordants and whether it is mixed with ultraviolet-absorbing or photochemically-active pigments. Fading also depends on the concentration of the coloring matter and on the vehicle or binder with which it may be associated.

The complexities of this situation have been somewhat circumvented by the textile chemists who have developed a series of standard dyeings that fade at a known rate. These lightfastness standards are simply placed alongside of an unknown material to be tested and, after a period of exposure to daylight or to a source of artificial illumination, one simply states whether the material under test has faded faster, equal to, or less than the rate of one of the standard dyeings. The International Standards Organization (ISO), for example, supplies a series of eight blue-dyed wool cloths, each one of which has been designed to fade at about half the rate of the preceding standard in the series. These are known as the blue-wool fading (or lightfastness) standards and are described in ISO Standard R105. English-speaking colleagues can most readily obtain a copy of this standard in the form of British Standard BS1006:1971.[12-14]

The simplest way to specify the lightfastness of various materials is to take a sample of a dyed cloth or test panel of paint, place it side by side with a set of the ISO blue cloths, expose them to a light source and, after a period of time, report whether the particular colored sample under test has faded at much the same rate as one of the eight ISO standards. Standard No. 1 is the most fugitive; No. 8, the most stable.

If one wishes to be a bit more objective in describing the extent of fading, various standards organizations have issued sets of five pairs of grey patches, each pair of which differs in contrast throughout a series of five major steps. The first pair is identical (geometric grey scale contrast 5; unchanged) and the last (contrast No. 1) shows considerable difference in the lightness between the two patches. Known as the geometric grey scale, this simple device is readily available as British Standard BS2662.[15] The American Association of Textile Chemists and Colorists and the American Society for Testing Materials also specify virtually the same standard. Using this series of grey patches, one can objectively state, as the result of a visual comparison between the exposed and

12. British Standard 1006:1971, *Methods for the Determination of the Colour Fastness of Textiles to Light and Weathering*. This 37-page pamphlet essentially represents the principal aspects of International Organization for Standardization Recommendation R105: 1, Part II. Pamphlets describing the standard practice (BS1006:1971), the geometric grey scale (see footnote 15) and the small cards comprised of strips of all eight cloths were available from TALAS, 104 Fifth Avenue, New York, NY, 10011. If presently unavailable from this source, write to the American National Standards Institute 1430 Broadway, New York, NY, 10018.

13. Feller, R. L., "Further Studies on the International Blue-Wool Standards for Exposure to Light," *Preprints of the Fifth Triennial Meeting, Zagreb*, ICOM Committee for Conservation (1978), Paper 78/18/2. Available from International Centre for Conservation, Via di San Michele, 13, 00153 Rome, Italy.

14.a. Feller, R. L., and Johnston-Feller, R. M., "Use of the International Standards Organization's Blue-Wool Standards for Exposure to Light. I. Use as an Integrating Light Monitor for Illumination Under Museum Conditions," *AIC Preprints of Papers Presented at the Sixth Annual Meeting, Fort Worth, Texas* (1978), p. 73-80.

b. Feller, R. L., and Johnston-Feller, R. M., "Use of the International Standards Organization's Blue-Wool Standards for Exposure to Light. II. Instrumental Measurement of Fading," *AIC Preprints of Papers Presented at the Seventh Annual Meeting, Toronto, Ontario, Canada* (1979), p. 30-36.

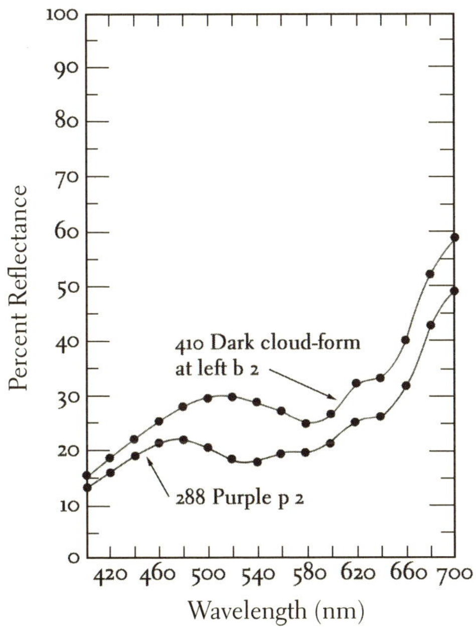

FIGURE 9
Spectral reflectance curves, computed from the MASK program, for a blue cloud form in the sky at left and a purple in kimono, in prints 410 and 288, colorants based on or containing aigami. See Figure 14 for areas that were measured in print 288. The formerly blue area b2 is now, as recorded here, an olive grey owing to the discoloration of the paper.

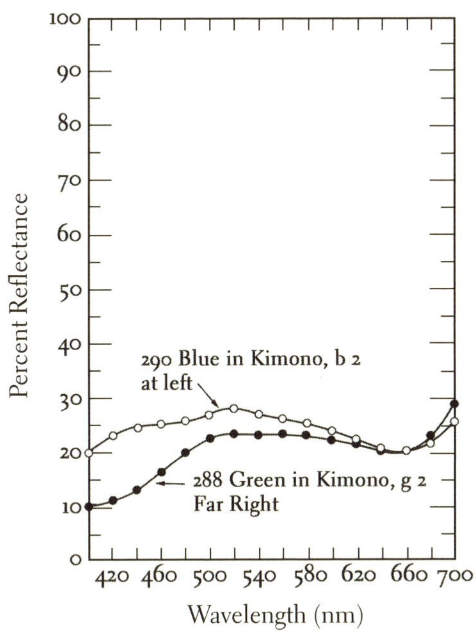

FIGURE 10
Spectral reflectance curves for areas of prints 288 and 290 based on indigo or mixtures with indigo. Measurements taken with the Kollmorgen-Leres "Trilac" instrument and computed from the MASK program.

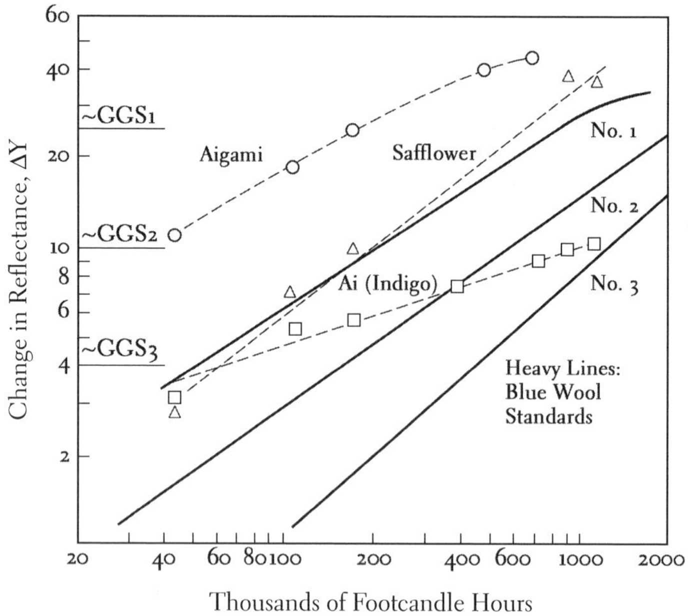

FIGURE 11
Rate of fading, measured by the change in luminous reflectance (ΔY), of samples of *aigami* (dayflower), *beni* (safflower) and *ai* (indigo) under "daylight" fluorescent lamps, compared with the equivalent change in ISO R105 blue-cloths Nos. 1, 2 and 3.

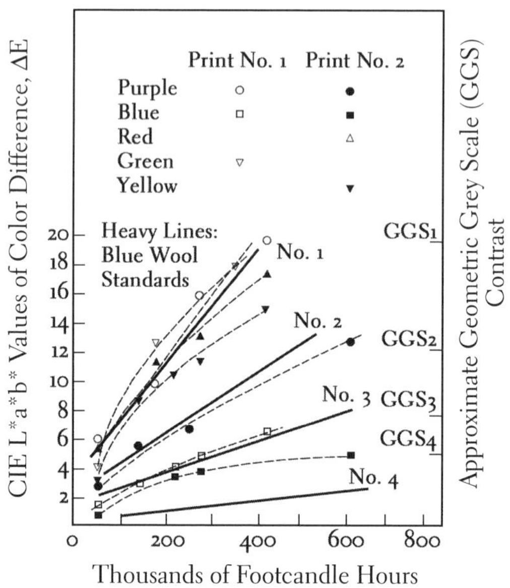

FIGURE 12
Rate of fading, measured by the CIE L*a*b* units of color difference (ΔE), of seven colorants in two prints, taken from an illustrated Japanese book of about 1800, kindly donated for these studies by Mrs. Keyes. Illumination: "daylight" fluorescent lamps.

unexposed portions of an unknown material and an inspection of both exposed and unexposed portions of one of the blue cloths, that the unknown colored material being tested and a particular blue cloth have both faded to the same geometric grey scale (GGS) contrast. The contrast between faded and unfaded portions in a particular case may have been equivalent to GGS step 4, let us say. If such were the case, this information would convey the fact that very little fading had taken place because grey scale step 4 (GGS 4) is considered to be a "just perceptible change." If, on the other hand, one stated that the standard and test piece had changed to GGS 1, this information would establish the fact that considerable fading had taken place.

15. British Standard BS2662:1961, Available from British Standards Institution, Two Park Street, London, W1, England.

For those workers who do not possess elaborate color-measuring equipment, the International Standards Organization intended that the degree of fading of test samples would be reported by visually comparing the extent of fading that took place in any given test with one of the steps in the geometric grey scale. If color-measuring equipment is available, one can report the degree of fading in terms of the total color difference, a mathematically calculated unit usually designated ΔE.[16] Because there are many different equations that have been devised to try to represent the visual evaluations of color difference (none is perfect), one must always specify the particular equation being used. This laboratory has regularly used an equation recommended by the Commission Internationale de l'Eclairage known as the CIE $L^*a^*b^*$ equation for color difference. However, we have also found that measurement of the change in luminous reflectance, signified by ΔY, will give results nearly as useful for our purposes.[14b]

16. Judd, D. B., and Wyszecki, G., *Color in Business, Science, and Industry*, 3rd ed., J. Wiley, New York (1975), p. 313-330.

In view of the existence of these internationally-recognized standards, we have chosen to report the lightfastness of various samples of the dyed or painted-out colorants in terms of the ISO R105 blue cloth that fades at approximately the same rate. Thus, we report in Figure 11 that the beni (safflower) fades at the same rate as blue cloth No. 1 as measured by the change in the luminous reflectance, ΔY. In Figure 12, measurements of ΔE for blue cloths Nos. 1 to 4 and the particular colorant samples are compared, again allowing us to report the lightfastness of the various colored areas of the prints in relationship to the rate of color change in one of the standard blue cloths. For convenience, the approximate location of the geometric grey scale changes is also indicated in the two figures.

A Practical Standard of Lightfastness for Conservators: Because variations in the concentration of the colorant, the kind of mordant or laking material, and the type of vehicle or substrate have such a profound effect on the rate at which a colorant apparently fades, the Center has recommended that the lightfastness of artists' colorants be referred to simply in terms of three broad classes, specified in Table 3. The first classification would be considered *fugitive*; such materials would be expected to exhibit noticeable color change in no more than 20 years of normal exposure to light under museum or household conditions. Materials that might be expected to last more than 100 years without much change (more specifically: 100 years of exposure at an annual level of about 100,000 footcandle hours, or approximately 1 million lux hours per year) could be considered to be of *excellent* lightfastness; those in between would be considered to exhibit *intermediate* lightfastness.[17] One can, of course, classify the lightfastness of various test pieces as being equivalent to one of the eight ISO blue-wool fading standards, but we consider that there would be far less disagreement over the precision and reliability of such measurements if the test material were simply placed in one of three classifications: fugitive (unstable), intermediate or excellent (stable).

17. Feller, R. L., "Speeding Up Photochemical Deterioration," *Bulletin de l'Institut Royal du Patrimoine Artistique*, 15 (1975), p. 135-150.

TABLE 3
Proposed Standard Classifications of Photochemical Stability for Materials in Conservation[17]

Class	Classification	Intended Useful Lifetime	Approximate Equivalent Standard of Photochemical Stability
C	Unstable or Fugitive	less than 20 years	ISO R105 Class 3 or less
B	Intermediate	(20-100 years)	(between Class 3 and 6)
A	Stable or Excellent	greater than 100 years	greater than ISO R105 Class 6

Details of the determination of the lightfastness of representative organic colorants that might be found in Japanese prints will be reported in the following sections. In nearly every case there is no question that the colorants described are fugitive.

The Traditional Organic Colorants

Based on the spectrophotometric data from the experiments shown in Figures 1, 2, 3 and 5, the rate of fading of the brushed-out traditional colorants under a bank of "soft white" fluorescent lamps was compared to the rate of fading of the first few ISO R105 blue-wool cloths, measured as the change in the luminous reflectance of the specimens, ΔY. The rated color temperature of the "soft white" lamps is 3025°K; ultraviolet emission, 3-7% of the total radiation in the ultraviolet-plus-visible range.

The results are shown in Figure 11. The Y value reported here (capital Y or "cap Y") is the colorimetric measurement known as the CIE tristimulus Y value. The reader will recall that colors can be specified in terms of the amount of red-, green- and blue-colored lights that can be mixed to duplicate the color. This is essentially what the tristimulus X, Y, Z values in the CIE system of colorimetric specifications represent. The net color change, usually designated as ΔE, can be based on any number of mathematical equations involving the changes in X, Y, and Z taken together. However, in the case of monitoring the changes in the blue cloths, the authors have found that the change in the luminous reflectance, (ΔY) gives satisfactory results.[14b] We have monitored the fading of the colorants in the present investigation both ways, using ΔY in Figure 11 and ΔE in Figure 12. The results will not be precisely the same by these two methods, but when comparing materials of widely different degrees of lightfastness, one can demonstrate with considerable confidence whether a given test sample possesses either fugitive (having a lightfastness equal to or less than ISO R105 blue-wool standard 3), intermediate, or excellent lightfastness (greater than ISO R105 standard 6).

Over the years this laboratory has investigated the lightfastness of a number of other pigments that are related to the present question. Most extensively studied have been alizarin and carmine lakes, but we also have prior data on the behavior of indigo, gamboge, turmeric and quercitron. Table 4 sums up the information that we have on the principal colorants that one can expect to find in Japanese prints, essentially as reported in a brief communication in 1981.[2]

Table 4
Lightfastness of Some Traditional Colorants

Pigment	Approximate ISO R105 (BS1006:1971) Lightfastness Grade	Protection Afforded by Ultraviolet Filter
Synthetic Alizarin Lake (Control)	4 to 5	some
Indigo	3 to 4	some
Gamboge, Garcinia Morella	2 to 3	some
Carmine Lake (Carminic Acid Lake)	1 to 2	little if any
Safflower Red, Carthamus Tinctorius (*Beni*)	1*	little if any
Tumeric, Curcuma Longo	1	little if any
Dayflower Blue, Commelina Communis (*Aigami*)	<1*	some

* Test exposures were made on samples, dispersed in water and painted out on Whatman filter paper, placed 3 1/2" from General Electric HO "daylight" fluorescent lamps (rated color temperature 6250°K) in a room maintained at 70°F (21°C) and 50%RH. Of the total irradiance between 320 and 700 nm, about 3 to 4% was determined to be in the near-ultraviolet (320-400 nm). Those marked with an asterisk (*) were exposed under "soft white" fluorescent lamps (rated color temperature 3025°K; ultraviolet 3 to 7%).

One can see that most of the traditional colorants are very fugitive indeed and will be little protected under ordinary gallery conditions of exposure through the use of an ultraviolet filter. Under slightly different light sources, conditions of temperature and humidity, and concentrations or type of colorant, the lightfastness classifications given in Table 4 may change by one grade, or possibly two, in the ISO series of eight standards. Nonetheless, the general conclusion is that nearly all of these colorants are very sensitive to light, tending to fade markedly (to a geometric grey scale contrast greater than 2) in less than two million footcandle hours (about 20 million lux hours) of illumination similar to diffuse daylight through window glass.

Fading of Cloth Dyed with Traditional Materials

Several samples of silk dyed by Mr. Seiju Yamazaki, Director of the Institute of Vegetable Dyeing, Takasaki City, Gumma Prefective, were kindly provided by Professor Kôzô Hayashi. Small specimens were used for tests of lightfastness. The dyes, their mordants, and their classification with respect to lightfastness are listed in Table 5.

TABLE 5
Fading of Representative Dyed Cloths

Color	Raw Plant Dyes	Mordanted by	ISO R105 Lightfastness Classification
Yellow	*Ukon* (Curcuma root stock)	citric acid	<1
Yellow	*Enju* (Sophora flower bud)	aluminum acetate	3*
Red	*Beni* (Carthamus flowers)	citric acid	1
Red	*Suwo* (Caesalpinia wood)	aluminum acetate	3*
Brown	*Suwo* (Caesalpinia wood)	aluminum acetate and ferric acetate	4
Violet	*Shikon* (root of Lythospermum erythrorhyzon)	aluminum acetate	3
Violet	*Shikon* (root of Lythospermum erythrorhyzon)	aluminum acetate and lime	3
Brown (?)	*Yama-momo***	alum	2*
Yellow	*Kariyasu***	alum	2*

* Scarcely any protection to these dyeings was provided by an ultraviolet filter when fading carried out under "daylight" fluorescent lamps.

** Most of the samples in this table were kindly supplied by Prof. Hayashi, but these two dyeings were carried out at Pittsburgh with materials and directions supplied by Mrs. Keiko Keyes. Perhaps if they had been dyed by an experienced Japanese craftsman, the lightfastness would be one or two grades higher.

The important conclusions to these studies is that *ukon* and *beni* proved to be very sensitive to light when applied in the traditional manner of dyeing cloth. All samples, save a brown-colored *suwo*, fell in the "fugitive" classification, exhibiting a lightfastness no better than ISO R105 blue-cloth No. 3.

Reviewing Padfield and Landi's work on a number of dyes that can be related to the present question, one finds turmeric (curcuma tinctoria) being rated as equivalent to an ISO R105 class 1 dye; likewise, orchil (red from lichens), sanders wood, young fustic (wood of Rhus cotinus), annatto (seed pods from Bixa orellana), gardenia (G. Florida), tesu (flowers of Butea frondosa).[18] Of the dyeings that exhibited reasonable lightfastness, quercitron was rated between ISO class 3 and 4; cutch about the same. Indigo exhibited a lightfastness between 4 and 5 on cotton in their tests; higher, when dyed on wool. The point to be made in citing this work is to provide further evidence that the traditional natural dyestuffs are well known to be fugitive.

18. Padfield, T., and Landi, S. "The Light Fastness of the Natural Dyes," *Studies in Conservation*, 11 (1966), p. 181-193.

Fading of Colorants in Actual Prints

Two small prints, taken from a Japanese illustrated book published about 1800, were kindly donated to the Center by Mrs. Keiko Keyes to be used to determine the fugitive character of colorants in actual prints of the period.

Each print was wrapped completely in household aluminum foil. An area of about 3 square inches was cut out of the foil where an exposed color could be visually compared to an unexposed area. In order to fade only a small portion of this total area of the print, a flap of aluminum foil was placed over the cut-out area, overlapping onto the large piece of foil that covered the print. A small window was cut in this flap to expose just a small area of color to the light, about 1/4" x 1/4". Because the flap was firmly taped in place (hinged) to the overall covering foil, this procedure made it possible to examine the faded versus the adjoining unfaded portion periodically, simply by lifting up the hinged flap with its 1/4" x 1/4" opening, taping it shut again after inspection, and thereby not losing the location of the area being exposed.

Small areas of various colors in the prints were thus selected and exposed to high-output "daylight" fluorescent lamps in a room held at 50% RH and about 70°F (21.1°C). A set of blue-wool standards (BS1006:1971) was placed beside the prints and exposed at the same time. The progress of fading was measured on the Small-Area-View SAV Color-Eye® and, from the tristimulus values thus determined, the CIE $L^*a^*b^*$ color differences (ΔE) were calculated.

Figure 12 shows the increase in the color differences (ΔE) with exposure time for each colorant area in the prints and compares them with the color change experienced by the blue-wool standards. It can be seen that the colorant areas selected are indeed fugitive. Only one was more lightfast than blue-wool No. 3 and at that only slightly better.

Analysis of the spectrophotometric reflectance curve indicated that the blue in print No. 1 contained *aigami* (dayflower). By 400 hours, the sample was considerably faded. This exposure is equivalent to no more than 550,000 footcandle hours. If this print were to be exposed on a gallery wall illuminated for eight hours a day at 25 footcandles (about 250 lux), the blue would be seriously faded in less than ten years. As seen in Figure 12, this blue—perhaps *aigami* mixed with carbon black—is one of the more stable colorants in the two prints.

Following these initial tests, additional unexposed areas of the prints were selected and the prints re-exposed, this time under an efficient ultraviolet filter that removed most of the radiation of wavelength below 400 nm. The results, summarized in Table 6, indicate that little if any protection was gained by placing the prints under an ultraviolet filter. This important conclusion is further discussed in the following section.

TABLE 6
Lightfastness of Colorant Areas in Two Prints Taken from Japanese Book of About 1800. Ratings as to Blue-Wool Standard Fading Classes 1 to 4 (ISO R105; BS1006:1971)

	Print #1		Print #2	
	No Protection	Under Ultraviolet Filter*	No Protection	Under Ultraviolet Filter*
Purple Area	< 1	1**	2	2
Blue Area	3	3	3-4	3
Red Area			1	1-2
Green Area	< 1	1-2		
Yellow Area			1-2	2-3

* Efficient ultraviolet filter that transmits less than 70% at a wavelength of 400 nm and 5% at 390 nm.

** The rating number means that this colorant area of the print faded at the same rate as ISO blue-wool standard No. 1.

This investigation has shown that six different areas of color in these two prints contained fugitive colorants. A seventh, apparently based on indigo, was scarcely more stable. Moreover, the rate of fading of these under normal exhibition conditions is not likely to be significantly improved by the use of filters that remove most of the ultraviolet radiation.

Because very small areas of the prints were only partially faded, the prints were returned to Mrs. Keyes in relatively good condition; the change in some of the test areas was scarcely noticeable.

Ineffectiveness of Ultraviolet Filters

Daylight entering a museum through window glass or the illumination from ordinary fluorescent lamps frequently contains a portion of the total radiation which is in the ultraviolet, usually the near-ultraviolet, region. This radiation, which cannot be seen by the human eye, is very energetic with respect to potential photochemical harm. Of the total radiation emitted by such sources in both the visible and ultraviolet regions of the spectrum (310 to 700 nm [nanometers]), sometimes as much as 7% will be in the ultraviolet, that is, radiation below 400 nanometers. In the fluorescent lamps used in the fading experiments described herein, the percentage fell between 3 and 7%. Because of this fact, it has become the custom of many institutions to exhibit prints, drawings, textiles and similar materials under plastic ultraviolet filters.[19] It is well known, however, that materials that are particularly susceptible to exposure are frequently highly sensitive to radiation in the visible region and are not primarily sensitive to the ultraviolet.[20] Thus, as has been stated many times, the use of an ultraviolet filter may result in very little protection whatsoever. This seems to be particularly true of many of the most sensitive colorants found in the traditional Japanese print.

19. Feller, R. L., "Control of the Deteriorating Effects of Light Upon Museum Objects," *Museum*, 17, No. 2 (1964), p. 57-98.

20. McLaren, K., "The Spectral Regions of Daylight Which Cause Fading," *J. Society Dyers and Colourists*, 72 (1956), p. 86-99.

As mentioned in the previous section, Table 6, none of the seven colored areas of the two prints received significant protection from the use of an ultraviolet filter. The dyed cloths reported in Table 5 also were scarcely protected. In another series of tests, a paint based on cochineal carmine was found to fade practically at the same rate regardless of whether it was protected by an ultraviolet filter or not. The same was found true of carmine oil paints and water colors exposed under diffuse daylight-through-window-glass on the wall of an art gallery for more than three years: practically no protection at all was given to this pigment by the use of an ultraviolet filter.

Colorants of particular interest in this work that may receive some benefit from ultraviolet filters are, first of all, indigo, which tests indicated might fade at a rate only about a third as much if it is protected by a filter. We have also observed that Prussian blue and alizarin will usually receive some benefit from the use of ultraviolet filters, although we do not have quantitative data to offer at this time.

The tests to verify these findings have not been elaborate. However, the limited number of measurements that have been made thus far confirm an opinion that has long been held by conservators and conservation scientists: in the case of very fugitive colorants, the use of ultraviolet filters tends to offer little protection against fading under normal museum conditions of illumination by diffuse daylight and fluorescent lamplight. This is not to say that the precaution of using ultraviolet filters should not be taken in order to do everything possible to protect the paper and vehicles as well as the colorants from potential damage. Nonetheless, one must not be lulled into thinking that the use of ultraviolet filters will all but eliminate the danger of fading. Visible light can also cause a significant degree of fading in many materials; this is expected to be especially true in those that exhibit the greatest tendency to fade.

There are a number of plastic sheets and foils available on the market designed to filter out ultraviolet radiation; most are very satisfactory. The accepted definition of the ultraviolet region refers to those radiations of shorter wavelength than 400 nm; a more or less ideal ultraviolet filter would thus be one that removed all radiation below 400 nm. The absorbing substances used to filter out radiation do not possess a sharp on-again, off-again ability to transmit certain wavelengths and to remove others, however. Instead, most filters remove a few percent of the radiation on the edge of the visible region; their transmission then sharply falls to a point of very low percentage at, let us say, about 390 nm, perhaps further tapering off, transmitting ever decreasing amounts of the ultraviolet radiation, down to possibly 380 nm. The precise shape of the curve of spectral transmission of such filters will vary with the thickness of the filter, the concentration of the filtering compound that is in the transparent sheeting and the particular chemical substance that has been used to filter the ultraviolet radiation. One might say that an "ideal" ultraviolet filter would cut off very sharply at about 400 nm, transmitting better than 90% of the visible radiation greater than 410 nm and less than 1% of the radiation below 390 nm. This describes more or less the characteristics of the practically colorless filters manufactured for this purpose by many companies.

There is another filtering material that many museums find convenient to use: this type removes some of the blue visible radiation, with the result that the plastic is ever-so-slightly yellow. In a sense, this type removes just a bit more of the potentially damaging radiation, that is the shorter wavelengths of the visible, blue, region of the spectrum. Whereas it might be considered somewhat more protective to use this just-slightly-yellow material, the point that we wish to make in the connection with the care of Japanese prints can be demonstrated with the type of filter that removes primarily radiations below 400 nm. In discussing the effects of attempting to use an ultraviolet filter, we have been referring to the effectiveness of the first type, the practically colorless filters, rather than the slightly yellow kind. Nonetheless, if either type had been used in such studies, we are confident that the essential findings would have been much the same.

CONCLUSIONS

This investigation has shown that *aigami* (dayflower) and *ai* (indigo) can be identified positively in Japanese prints without taking samples, simply by measuring the spectral reflectance curves of specific areas of the prints. Because of its fluorescence and reasonably distinctive spectral reflectance curve, safflower red (*beni*) can also be identified with considerable assurance. In the future, the identification of *beni* can be made with even greater confidence if its spectral-fluorescence characteristics were to be measured.

Fading tests on known samples of these three colorants have shown that they fall in the fugitive class of lightfastness, or very close to it, characterized as having a tendency to fade equivalent to or less than ISO R105 blue-wool cloth No. 3. *Aigami* fades more rapidly than the most fugitive of the blue-wool standards (cloth No. 1) and *beni* fades just as rapidly as blue-wool cloth No. 1. The lightfastness of indigo is just slightly better, equivalent to somewhere between blue-wools 3 and 4. *Aigami* and *beni* will fade noticeably in about

50,000 footcandle hours (500,000 lux hours) and fade drastically in less than 200,000 footcandle hours (2 million lux hours) of exposure to illumination similar to that emitted by "soft white" fluorescent lamps (3025°K).

These particular colorants were detected in every one of the ten prints examined from the collection of the Allen Art Museum. Five prints were shown to contain the very fugitive blue, *aigami*; four apparently contained safflower (*beni*) red and three cases involved indigo. Five examples of a purple made with one of these fugitive blues were found and two greens made with indigo and a yellow.

Even if the exposure standard recommended by the Museum Exchange subcommittee of the United States-Japan Conference on Cultural and Educational Interchange (CULCON)[21] is followed, in which it is recommended that materials highly sensitive to light be exhibited to the extent of no more than 5,000 footcandle hours of exposure per year, prints that contain the traditional colorants *aigami* and *beni* can be expected to fade noticeably in ten years under these reasonably cautious restrictions. If such prints were to be exhibited every museum day throughout the year at levels of illumination of only 5 footcandles, it is predicted that these areas of color may noticeably fade in no more than 200 weeks of exhibition. This will certainly disrupt the harmony and balance of colors, for it is expected that other areas of the print, based on more stable colorants, will not fade as readily.

21. Japan Society, *Report of the Study Group on Care of Works of Art in Traveling Exhibitions of the Museum Exchange Subcommittee of the United States-Japan Conference on Cultural and Education Interchange*, New York (1980).

In the course of this investigation, the lightfastness of more than a dozen natural-dye colorants traditionally used in Japan was determined: seven pigments, reported in Table 4 and nine dyed cloths, reported in Table 5. Only two were significantly more lightfast than the ISO R105 blue-cloth No. 3, and these scarcely more than one step better. The basic fact of the high sensitivity of these materials to light seems well established, although other investigators will undoubtedly wish to verify, confirm and refine specific details. The findings place the traditional Japanese print among the most light-sensitive of materials that a museum or gallery may have in its collection. Certainly works containing such fugitive colorants should be displayed only for a limited time at the lowest possible levels of illumination.

Acknowledgments

Special appreciation goes to Keiko Keyes for her encouragement of these studies and to her and to Dr. James B. Austin for samples of authentic materials. Technical advice on color measurement was kindly supplied by Ruth Johnston-Feller who also arranged for use of the computational program MASK through the courtesy of the Ciba-Geigy Corporation, Ardsley, N.Y. David Encke measured the light intensities and assisted in setting up the easel system on which the prints were mounted for spectrophotometric measurements.

The authors particularly wish to thank Dr. Kôzô Hayashi of the Research

Institute of Evolutionary Biology, Tokyo and Mr. Seiju Yamazaki, Director of the Institute for Vegetable Dyeing, Takasaki, for their kindness in providing authentic samples of several traditional dyes and dyeings. Through Mrs. Keyes, Mr. Iwataro Oka also supplied the project with several plant materials from which dye extracts were made.

The project designed to identify the presence of *aigami*, *ai* and *beni* in representative prints of the collection was supported by a grant from the Oberlin cataloguing project sponsored by the National Endowment for the Arts, using matching funds arranged through the efforts of Mrs. Keyes. The extensive work of characterizing known colorants and measurement of their rates of fading was carried out as part of the general research program of the Research Center on the Materials of the Artist and Conservator, an undertaking that receives its basic support through the generosity of the Andrew W. Mellon Foundation.

METHODS USED TO MEASURE THE SPECTRAL REFLECTANCE OF THE PRINTS

Measuring Large Matted Japanese Prints with the Color-Eye®

To facilitate easy movement and handling of the color-measuring instrument, the Kollmorgen Small-Area-View (SAV) Model D-1 Color-Eye® was placed on a typewriter table equipped with locking wheels. After the instrument was standardized in the usual manner in the specular-component-included (SCI) mode, the sample holder and its spring support were removed from the sample-viewing port and a piece of black construction paper, with a hole cut in it for the light beam to pass through, was taped over the port area to prevent any sharp edges of metal on the instrument from abrading the print.

An artist's easel was used to support the prints. Weights were taped to the supporting legs of the easel to keep it firm and free from tipping forward. Several layers of large filter-paper sheets were strapped to a pressed-wood support using bands of chromatography filter-paper strips. The sheets of filter paper served to cushion the back of the matted print. A two-inch-deep trough of glassine paper was likewise strapped around the pressed-wood support near the bottom to hold and protect the bottom of the hinged mat and its print. The support panel thus prepared was securely clamped upon the easel.

The SAV Color-Eye® requires only a 1/4"-diameter area for color measurement. Before measuring an area of color in a print, a Xerox® copy was made. By measuring from the top, left-hand border, it was possible to specify the exact area of the print that was measured. The spot was also marked on the photocopy. Before attaching the print and its mat to the support, six pieces of Millipore® filter paper were placed between the print and the backing mat board. Because the prints tended to be translucent (at incomplete hiding), all measurements were made with the print having this white background, which has a measured reflectance of 100.5 relative to the usual barium sulfate

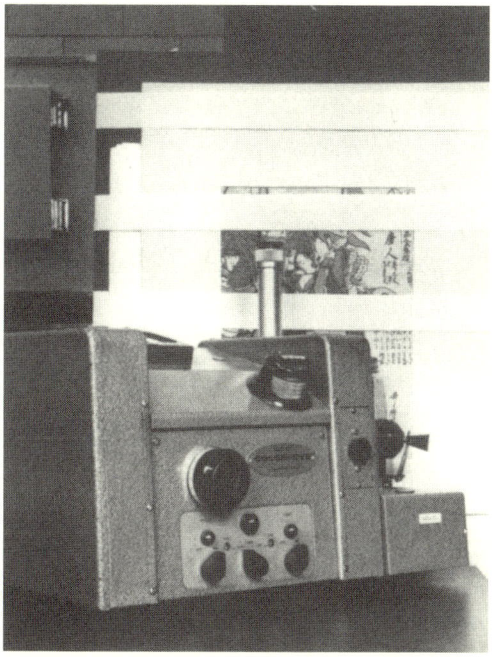

Figure 13
The Model D-1 SAV Color-Eye® with a print on its supporting easel in place for measurement.

standard. The cover of the matted print was folded back and the print and its matte-board support were then tucked into the glassine trough; the print and its Millipore® backing was held in place on the support with filter-paper strips, as shown in Figure 13. The print thus supported on the easel could be adjusted up and down, and the Color-Eye® on its moveable table could be moved sideways, to align the print with the port of the instrument. When everything was in proper position, the wheels on the table were locked, and a soft black cloth draped around the print and port area in order to reduce stray light in case the print was not perfectly tight against the measurement port.

The measurements thus carried out resulted in no damage to the prints. At some date in the future, they can be remeasured in the same locations. Copies of all prints showing the measured areas are available in the Final Technical Report submitted to the Allen Art Museum in 1980.[1] An example of such a record, marked where both the small-area-view (SAV) and the larger ("mask") areas have been measured, is shown in Figure 14.

Measuring Large Matted Japanese Prints with the Trilac® Spectrophotometer

The Center's Kollmorgen-Leres Trilac® recording spectrophotometer has an especially-modified removable metal section which holds spring supports for both the standard and sample. After standardizing the instrument in the usual

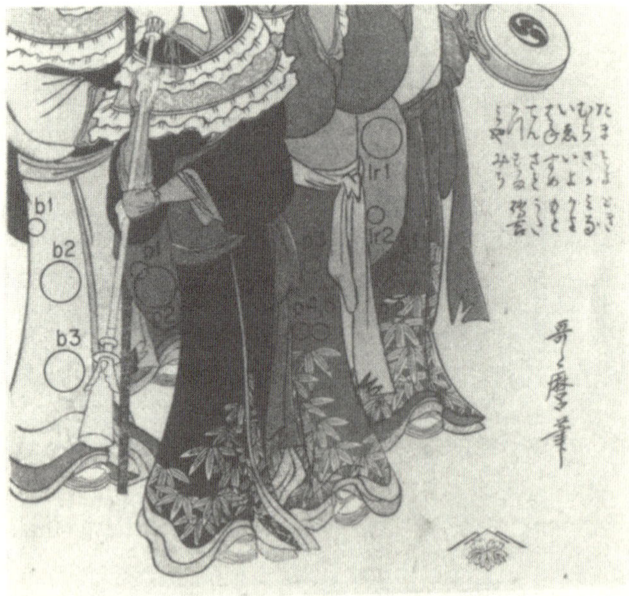

Figure 14
Illustration of the photocopied detail of print 288 with areas marked, for the record, where measurements were taken with the small-area-view (SAV) Color-Eye® and larger Leres "Trilac" instrument fitted with 1/2"-diameter "mask." Similar records were made of all prints in which the spectral reflectance measurements were taken.

manner, this section was removed so that the print in its supporting mat could be placed in a vertical position firmly against the measuring port of the instrument and adjusted in position so that a specific area of the print could be measured. The specular-component-adjusting wedge, which is screwed onto the sphere, is normally held in place with a special screw that has sharp knurled edges. Because this was considered potentially dangerous to the prints, the wedge was removed and measurements made with specular component excluded (SCE). Rather than replacing the screw in the hole, the opening was covered with black electrical tape. The sliding cover, which usually protects the standard and sample ports from light, was also removed and a soft black cloth draped over the sample and port area to eliminate stray illumination. Because the prints have practically no gloss, it was considered that measurements in the SCE mode would be much the same as measurements made with the specular component included (SCI).

Areas of relatively uniform color in the prints that one might wish to measure are generally smaller than the two-inch-diameter port of the Trilac® instrument. For this reason, a black mask with a 1/2"-diameter opening in it, made from flexible paperbacked aluminum foil painted with black optical paint, was taped over the sample port. With the barium sulfate reference-white plaque taped to the "standard" port, the Trilac® was ready to measure the spectral reflectance of specific areas of the prints. The instrument bench has a pressed-wood board securely fastened to it, which extends out beyond the viewing ports. This served as a table upon which to rest the support on which the prints were placed for measurement purposes.

Unlike the Color-Eye®, the Trilac® has no periscopic peephole with which to align the sample in the proper position. The back of the print and its support, not the front, is all the operator could see. Because of this, a large 16" x 10" piece of glass, bound with tape to cover the edges, was used to support the print. First, a Xerox® copy on clear plastic was made of the print. This transparent copy was taped to the back of the glass and, thus supported, was positioned at the sample port so that the area to be measured could be lined up with the 1/2" port. By using a bench jack, the bottom edge of the glass was raised or lowered to the desired height; for alignment in the horizontal direction, marks were made at the bottom edge of the glass to show the position of the bench jack. Its position thus aligned, the glass support could be returned to exactly the same position even though the operator would no longer be able to see the sample port. The print was carefully placed on the front of the glass panel so that the bottom and one edge of the mat would permit it to be exactly lined up with its Xerox® transparency on the other side of the glass. The edges of the mat assembly were gently held in place with strips of filter paper. Millipore® filter paper was inserted behind the print for a standard white background. The glass support with the print now taped to its front was placed back where it had been during the alignment and held firmly there for the measurement. A soft black cloth, draped over the easel and other paraphernalia at the end of the instrument, excluded stray light. The use of a Xerox®-copy transparency mounted on the rear removed the problem of guesswork concerning the area of the print being measured. Figures 9 and 10 show the results of a measurement using the MASK computer program; this computational program gives the reflectance values one would expect to measure if the normal 2" ports had been employed rather than the black mask with its 1/2"-diameter opening.

Details of the exact areas measured on ten prints in the collection and the data relating to thirty-eight spectrophotometric measurements can be found in the Final Technical Report, "Identification and Rate of Fading of Some Traditional Organic Colorants Employed in Japanese Prints" on file at the Allen Art Museum.

The Kinetics of Fading: Opaque Paint Films Pigmented with Alizarin Lake and Titanium Dioxide

RUTH JOHNSTON-FELLER, ROBERT L. FELLER,
CATHERINE W. BAILIE AND MARY CURRAN

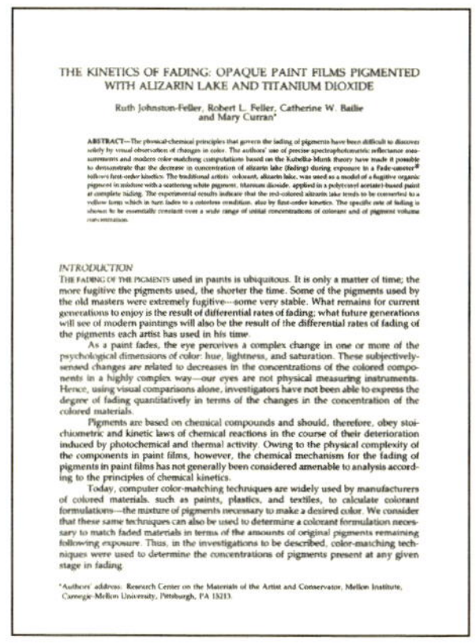

THE KINETICS OF FADING: OPAQUE PAINT FILMS PIGMENTED WITH ALIZARIN LAKE AND TITANIUM DIOXIDE

Ruth Johnston-Feller, Robert L. Feller, Catherine W. Bailie and Mary Curran*

ABSTRACT—The physical-chemical principles that govern the fading of pigments have been difficult to discover solely by visual observation of changes in color. The authors' use of precise spectrophotometric reflectance measurements and modern color-matching computations based on the Kubelka-Munk theory have made it possible to demonstrate that the decrease in concentration of alizarin lake (fading) during exposure in a Fade-ometer® follows first-order kinetics. The traditional artists' colorant, alizarin lake, was used as a model of a fugitive organic pigment in mixture with a scattering white pigment, titanium dioxide, applied in a poly(vinyl acetate)-based paint at complete hiding. The experimental results indicate that the red-colored alizarin lake tends to be converted to a yellow form which in turn fades to a colorless condition, also by first-order kinetics. The specific rate of fading is shown to be essentially constant over a wide range of initial concentrations of colorant and of pigment volume concentration.

INTRODUCTION

THE FADING OF THE PIGMENTS used in paints is ubiquitous. It is only a matter of time; the more fugitive the pigments used, the shorter the time. Some of the pigments used by the old masters were extremely fugitive—some very stable. What remains for current generations to enjoy is the result of differential rates of fading; what future generations will see of modern paintings will also be the result of the differential rates of fading of the pigments each artist has used in his time.

As a paint fades, the eye perceives a complex change in one or more of the psychological dimensions of color: hue, lightness, and saturation. These subjectively-sensed changes are related to decreases in the concentrations of the colored components in a highly complex way—our eyes are not physical measuring instruments. Hence, using visual comparisons alone, investigators have not been able to express the degree of fading quantitatively in terms of the changes in the concentration of the colored materials.

Pigments are based on chemical compounds and should, therefore, obey stoichiometric and kinetic laws of chemical reactions in the course of their deterioration induced by photochemical and thermal activity. Owing to the physical complexity of the components in paint films, however, the chemical mechanism for the fading of pigments in paint films has not generally been considered amenable to analysis according to the principles of chemical kinetics.

Today, computer color-matching techniques are widely used by manufacturers of colored materials, such as paints, plastics, and textiles, to calculate colorant formulations—the mixture of pigments necessary to make a desired color. We consider that these same techniques can also be used to determine a colorant formulation necessary to match faded materials in terms of the amounts of original pigments remaining following exposure. Thus, in the investigations to be described, color-matching techniques were used to determine the concentrations of pigments present at any given stage in fading.

*Authors' address: Research Center on the Materials of the Artist and Conservator, Mellon Institute, Carnegie-Mellon University, Pittsburgh, PA 15213.

Originally printed in:
Journal of the American Institute for Conservation
Volume 23, Number 2 (1984), Pages 114-129.

Introduction

The fading of the pigments used in paints is ubiquitous. It is only a matter of time; the more fugitive the pigments used, the shorter the time. Some of the pigments used by the old masters were extremely fugitive—some very stable. What remains for current generations to enjoy is the result of differential rates of fading; what future generations will see of modern paintings will also be the result of the differential rates of fading of the pigments each artist has used in his time.

As a paint fades, the eye perceives a complex change in one or more of the psychological dimensions of color: hue, lightness, and saturation. These subjectively sensed changes are related to decreases in the concentrations of the colored components in a highly complex way—our eyes are not physical measuring instruments. Hence, using visual comparisons alone, investigators have not been able to express the degree of fading quantitatively in terms of the changes in the concentration of the colored materials.

Pigments are based on chemical compounds and should, therefore, obey stoichiometric and kinetic laws of chemical reactions in the course of their deterioration induced by photochemical and thermal activity. Owing to the physical complexity of the components in paint films, however, the chemical mechanism for the fading of pigments in paint films has not generally been considered amenable to analysis according to the principles of chemical kinetics.

Today, computer color-matching techniques are widely used by manufacturers of colored materials, such as paints, plastics, and textiles, to calculate colorant formulations – the mixture of pigments necessary to make a desired color. We consider that these same techniques can also be used to determine a colorant formulation necessary to match faded materials in terms of the amounts of original pigments remaining following exposure. Thus, in the investigations to be described, color-matching techniques were used to determine the concentrations of pigments present at any given stage in fading.

BACKGROUND

These investigations were first undertaken in 1962, and we have been further convinced, since that time, that many organic colorants in paint systems tend to fade according to first-order kinetics. In 1976, with the availability of modern computational techniques for colorant formulation, we designed a fifth series of experiments to study the kinetics of the fading of the widely used artists' pigment, alizarin lake, incorporating in this study many of the variables in the formulation of paint films. A few of the principal experimental results will be reported here for the case of opaque paint films made with a mixture of alizarin lake and titanium dioxide white pigment. Principally, we shall demonstrate that the process of fading of these opaque films can be described on the basis of first-order kinetics for the loss of alizarin lake as a function of exposure time.

The statement that the fading of a colorant system takes place according to "first-order kinetic behavior" means that the rate at which pigment is lost at any given moment is a constant fraction of the quantity of pigment present at that point in time. In other words, the rate of disappearance of the colored compound is proportional to its concentration. This is expressed mathematically as:[1]

$$-\frac{dC}{dt} = k_1 C \qquad (1)$$

where $-dC/dt$ is the rate of change in concentration, C, and k_1 is the proportionality constant, also known as the specific rate constant.

Equation 1 may be expressed in the integrated form:

$$\ln C = \ln C_o - k_1 t \qquad (2a)$$

where ln refers to the natural logarithm to the base e, C is the concentration remaining at time, t, and C_o is the initial concentration. The equation may also be expressed in logarithms to the base 10:

$$\log C = \log C_o - k_1 t / 2.303 \qquad (2b)$$

A review of the literature revealed only a few articles describing studies of the fading of pigmented or dyed materials in terms of chemical kinetics. In

1. See, for example, Moore, J. W., and Pearson, R. G., *Kinetics and Mechanism*, John Wiley & Sons, NY (1981).

1954, Haug suggested that paints often faded according to the first-order kinetic law, but his experimental data were not convincing and his work has rarely been cited.[2] Baxter, Giles, and coworkers in the late 1950's, investigating films of molecularly-dispersed dyes in gelatin, methyl ethyl cellulose, and collodion, and measuring the changes in the concentration in terms of the logarithm of the absorbance of the dyes in the transparent films, showed that the rate of fading frequently followed first-order behavior.[3, 4] Several years later, Kortum used the Kubelka-Munk K/S function (Equation 4) to follow the thermally-induced fading of photo-darkened dyes absorbed on various substrates.[5] Subsequently, the Kubelka-Munk function (Equation 4) was used to demonstrate that light-induced fading of dyes in pharmaceutical tablets followed the first-order law.[6]

In the results to be described, the relative amount of alizarin lake remaining in the light-exposed paint panels was computed by means of the Kubelka-Munk equations for opaque pigment mixtures:[7]

$$(K/S)_{mixture} = \frac{C_1K_1 + C_2K_2 + \ldots C_iK_i}{C_1S_1 + C_2S_2 + \ldots C_iS_i} \qquad (3)$$

where

$$(K/S)_{mixture} = \frac{(1 - R_\infty)^2}{2R_\infty} \qquad (4)$$

and R_∞ is the internal reflectance of a paint film at complete hiding—that is, a film sufficiently thick to obscure completely any layer below. Concentrations were expressed as relative dry pigment concentration—the percentage of alizarin lake relative to the total weight of pigment present (alizarin lake plus titanium dioxide).

In order to solve Equations 3 and 4, it is necessary to use the internal reflectance as R_∞; this refers to the reflectance that would occur if there were no surface reflectance as a result of the refractive index differences between the paint medium and air. This is computed from the following equation:[8]

$$R_i = \frac{R_T - \rho_1}{1 - (\rho_1 + \rho_2) + R_T\,\rho_2} \qquad (5)$$

where R_i is the internal reflectance, ρ_1 is the specular reflectance of the light entering the film, ρ_2 is the reflectance at the under surface of the film for the exiting reflected light, and R_T is the total measured reflectance. The value used for ρ_1 was 0.04 (4%) and for ρ_2, 0.6 (60%); these are the theoretical values for films of refractive index 1.5 and for perpendicular or near-perpendicular incident light and diffuse exiting light. The actual values are not critical provided they are constant.[9]

2. Haug, R., "Über den Zusammenhang der Lichtechtheit organischer Farbkörper mit ihrer Konzentration in Mischungen mit Substraten," *Farbe und Lacke*, 58 (1952), p. 340-344; "Untersuchungen über die Lichtechtheit organischer Pigmente," *Deutsche Farben Zeitung*, 15 (1961), p. 373-378.

3. Baxter, G., Giles, C. H., McKee, M. N., and Macaulay, N., "The Influence of the Physical State of Dyes Upon Their Light Fastness," *J. Soc. Dyers and Colourists*, 71 (1955), p. 218-235.

4. Baxter, G., Giles, C. H., and Lewington, W. J., "Relation Between Physical State and Rate of Fading of Dyes," *J. Soc. Dyers and Colourists*, 73 (1957), p. 386-392.

5. Kortum, G., Kortum-Seilev, M., and Bailey, S. D., "Optical Properties of 2-(2'-4'-Dinitrobenzyl)-Pyridine in the Absorbed State," J. *Physical Chem.*, 66 (1962), p. 2439-2442.

6. Everhard, M. E., and Goodhart, F. W., "Stability of Certified Dyes in Tablets 1. Fading of FD&C Red No. 3 in Tablets as a Function of Concentration, Time and Light Intensity," *J. Pharm. Sci.*, 52 (1963), p. 281-283.

7. See, for example, Kuehni, R. G., *Computer Colorant Formulation*, Lexington Books, D. C. Heath and Company, Lexington, MA (1975).

8. Judd, D. B., and Wysecki, G., *Color in Business, Science and Industry*, 3rd Edition, John Wiley & Sons, NY (1975), p. 417.

9. Brockes, A., "The Influence of Glossy Surfaces on Reflectance Measurements", *Die Farbe*, 9, No. 13 (1960), p. 53.

The K's and S's in Equation 3 are the original absorption (K) and scattering (S) constants determined on the unexposed samples.[7] From these characterized values, the concentrations (the C's) of the coloring matter in any mixture can be computed, using Equation 3 and the measured reflectances of the exposed samples converted to K/S by Equation 4. The equation applies to a single wavelength of light only; hence, the reflectances at many wavelengths across the spectrum are necessary for the evaluation of the C's, solved simultaneously to fit the measured spectral data.

In a previous paper dealing with alizarin glazes (transparent or translucent films containing only alizarin-lake pigment), it was shown that use of the Kubelka-Munk equations appropriate to the case of films at *incomplete hiding* would permit prediction of the color changes occurring during the fading of glazes.[10] The results to be reported here are those obtained with alizarin lake mixed with a scattering white pigment, rutile titanium dioxide (TiO_2), for the case of films made at *complete hiding*, that is, at a thickness so great that no light penetrates through the paint film to the substrate. The effect of a number of characteristic variables in paint formulation were investigated for both glazes and opaque films, but details of our findings with other variables must await subsequent publication.

10. Johnston-Feller, R., and Bailie, C., "Analysis of the Optics of Paint Glazes: Fading," in *Science and Technology in the Service of Conservation*, ed. Brommelle, N. S., and Thomson, G., International Institute for Conservation of Historic and Artistic Works, London (1982), p. 180-185.

It was our belief at the inception of this research that if the spectrophotometric curves of the original and of the faded states of a paint could be analyzed by color-matching computational techniques, it would then be possible to determine, for each faded state, the concentration of the original colored material that remained in a faded zone near the surface of the paints. In addition, if a new colorant of different hue were generated, the relative amount of the new colored substance could also be calculated. Thus, in order to obtain the concentrations of the colored components present at any given time during the course of the exposure of an alizarin lake-titanium white paint, we simply used a color-matching program to determine the percentage composition of pigments needed to "match" the color of the paint at each stage in the course of fading.

EXPERIMENTAL

Pigments and Vehicle

The vehicle used for these studies was a solvent-dry system based on a moderate molecular weight poly(vinyl acetate), Vinac® B-7 (Air Products Chemicals, Allentown, Pa.), a polymer of excellent photochemical stability similar to those sometimes used by conservators as a vehicle for retouching. The pigments used were alizarin lake (Ciba-Geigy X-686) and rutile titanium dioxide (du Pont Ti-Pure® R-960). The latter exhibits low photochemical activity.[11]

11. "Ti-Pure® Titanium Dioxide for Coatings," Technical Brochure E-42003, duPont Chemicals and Pigments Department, Wilmington, Del. (1981).

Preparation of the Panels

Separate dispersions of the alizarin lake and of the titanium dioxide were made by ball-milling in a toluene solution of the poly(vinyl acetate) vehicle. These dispersions were then blended together to achieve the desired pigment and vehicle composition. The pigment ratios, alizarin lake/TiO_2 on a weight basis, and the pigment volume concentrations (PVC) used in this study are described in Table 1.

TABLE 1
Compositions of Test Paints in Poly(vinyl acetate)

	Percent Relative Dry Pigment			
Sample No.	Alizarin X-686	Titanium White R-960	Total PVC* in Dried Film	PVC* of TiO_2 in Dried Film
F23	2.59	97.41	2.66	2.50
F18	2.47	97.53	7.98	7.50
F19	2.50	97.50	11.79	11.07
F15	2.50	97.50	20.90	19.64
F20	18.25	81.75	3.87	2.48
F21	17.58	82.42	11.44	7.45
F22	17.41	82.59	17.38	11.36
F25	50.11	49.89	8.82	2.50
F72	17.73	82.27	17.01	11.03
F74	48.20	51.80	8.76	2.62

* PVC = Pigment volume concentration

The paints were applied to Eastman Kodak glass slides, 3 1/4" x 4", by draw-down, using bars with a clearance of not greater than three mils. The draw-downs were made in a jig containing nine slides, of which only the center slide, uniform in appearance, was used. The glass was always weighed before coating so that the film thickness could be calculated from the weight and density of the solids in the paint. The purposes of applying the paint to glass slides were (a) to determine the completeness of hiding by measuring the reflectance over a white and over a black substrate and (b) to achieve a surface as smooth and free of irregularities as possible. The freshly coated panels were allowed to dry at ambient conditions for about an hour and were then baked overnight at 130°C to remove solvent. When necessary to achieve complete hiding (a condition indicated by the identity of the measurements over white and black substrates), additional coats of paint were applied and the same drying procedure followed.

Measurements

Spectrophotometric curves of the paint panels initially and at intervals during exposure were measured using the integrating-sphere mode on a "Trilac" recording spectrophotometer (Leres, Alfortville, France) equipped with a small digital computer for 0% and 100% line corrections of the measured data at 10 nm intervals. Simultaneously, as the reflectance curves were being measured, computations at 10 nm intervals were performed to yield the CIE tristimulus values for Illuminant C, the chromaticity coordinates, and color differences all based on corrected reflectances. The CIE 1976 L*a*b* color-difference equation was used.[12] Measurements were made with the specular (gloss) reflectance included (designated SCI or total reflectance, R_T) and excluded (designated SCE) to determine changes in specular reflectance. Measurements were also made over a white and over a black substrate to make sure that the paint had remained at complete hiding throughout the exposure. Pressed barium sulfate was used as the white reference standard.

12. Billmeyer, F. W., Jr., and Saltzman, M., *Principles of Color Technology*, Second Edition, John Wiley & Sons, NY (1981), p. 63.

Exposure Technique

The panels were exposed in an Atlas Electric Devices (Chicago, Ill.) Model 65-WRC xenon-arc Fadeometer equipped with Pyrex®-glass filters which effectively eliminated radiation below 310 nm, closely simulating the range of spectral energy that passes through window glass. The air temperature in the chamber averaged 30.5° to 35°C; the black-panel temperature between 59° and 60°C. The ambient air in the chamber measured 20% R.H., with little deviation. Each time the panels were returned after measurement their position on the rotating rack in the Fadeometer was randomized with respect to the upper and lower positions. The painted glass panels were backed with a glass slide covered half with a stack of six white Millipore® filters (Millipore Corp., Bedford, Mass.) and half with black matte paint applied on aluminum foil. A pressed-wood support was arranged to leave an air space of 1/8" between the samples and the black and white background. The purpose in mounting the samples over the black and white panels was to determine if there were any differences in the results owing to heating of the panels even though hiding in the visible spectral range was complete; none was found for these opaque samples.

Because the intensity of the polychromatic irradiance in the Fadeometer declines with the age of the xenon-arc lamps, the rate calculations are expressed in terms of "exposure" rather than time—that is, the product of light intensity times clock hours. The Fadeometer is equipped with a light-monitoring device that measures the intensity of radiation on the samples at a selected wavelength, in this case 420 nm. Exposure, therefore, is expressed in kilojoules/meter^2 (kJ/m^2) as measured at 420 nm. During these experiments, 14.8 kJ/m^2 was equivalent to about one hour of exposure. In the course of only a few hundred hours time, the distribution of the spectral power was not expected to change significantly. The fact that the rate constant determined in a second series of experiments at a later date (Figure 5) was identical within experimental error supports this expectation.

Results

Kinetics

One of the first effects noted in our original study conducted in 1962-3—observed initially with alizarin added to a commercial flat alkyd-base paint and subsequently in a series of panels made with the acrylic emulsion, Rhoplex® AC-33 (Rohm and Haas Company, Philadelphia, Pa.)—was that the alizarin lake did not fade directly to a colorless form. Instead, a yellow intermediate was formed which, in turn, apparently faded to a colorless state. This behavior has been observed in all of our studies on the exposure of alizarin lake in mixture with a white pigment when the paint film was opaque (at complete hiding). Figure 1 illustrates some of the spectral curves measured for one of the samples, drawn in solid lines, and the curves that would be obtained if no yellow intermediate were formed, plotted as dotted lines. It can be seen that the curves as measured on the exposed samples reflected less violet and blue light than did the series of curves calculated for the case in which no yellow form is present, indicating absorption by a yellow component. The development of a yellow intermediate is also evidenced when the CIE chromaticity coordinates are calculated and plotted as in Figure 2; here, the solid circles and connecting lines represent the data obtained for the exposed samples and the open circles and dotted connecting lines represent those which would have resulted if no yellow form had been present and the red alizarin had faded directly to a colorless form instead.

Kineticists frequently describe behavior such as this in terms of a series of consecutive chemical reactions as indicated by Equation 6,

$$A \xrightarrow{k_1} Y \xrightarrow{k_2} C \qquad (6)$$

where A is the initial concentration of alizarin, Y is the concentration of the yellow intermediate, and C is the concentration of the colorless form. The constant k_1 is the same as in Equation 1 and k_2 is the specific reaction rate for the fading of the yellow form. Equation 6 can be used to describe the fading of alizarin lake through a yellow intermediate to a colorless form in which each step proceeds according to first-order kinetics. Details of the method for determining k_2, however, will require a separate publication.

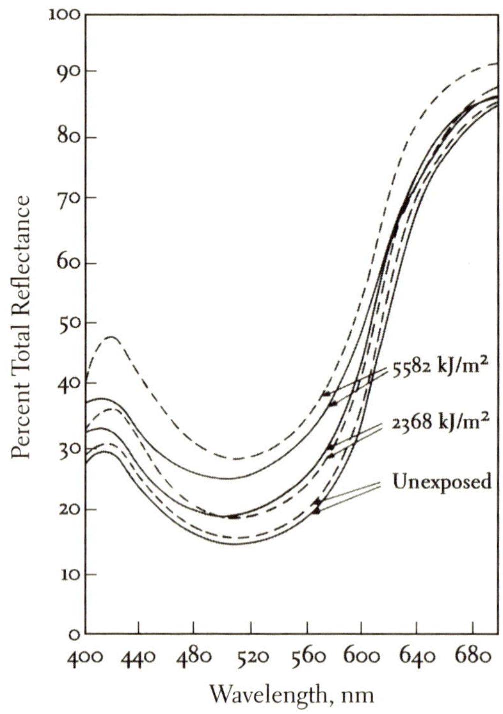

FIGURE 1
Changes in spectral reflectance curves of an alizarin lake-rutile titanium white-poly (vinyl acetate) (Vinac® B-7) paint at complete hiding during exposure in a Fadeometer, showing experimentally determined curves (solid), and the curves calculated for the case in which no yellow component is formed (dashed). Exposure monitored in terms of kJ/m² at 420 nm.

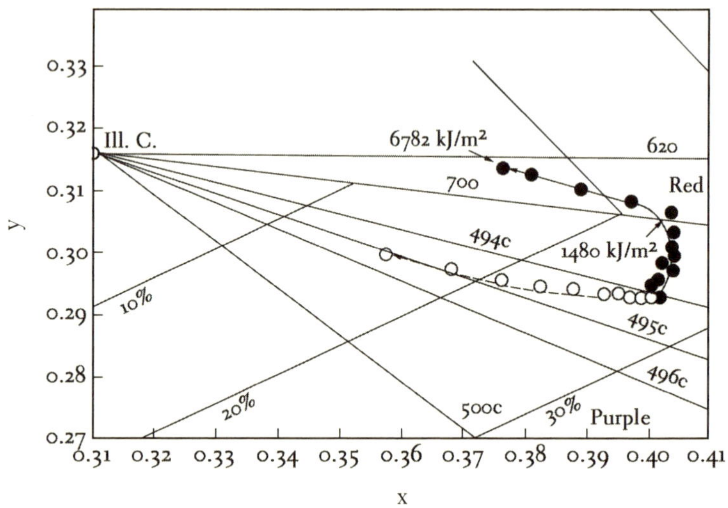

FIGURE 2
Change in CIE chromaticity coordinates (Illuminant C) of alizarin lake-rutile titanium white-poly(vinyl acetate) paint throughout 540 hours of Fadeometer exposure. Experimental data shown, in which yellowness developed (closed circles), together with calculated values for the change that would have occurred if no yellow had developed (open circles). Paint at complete hiding.

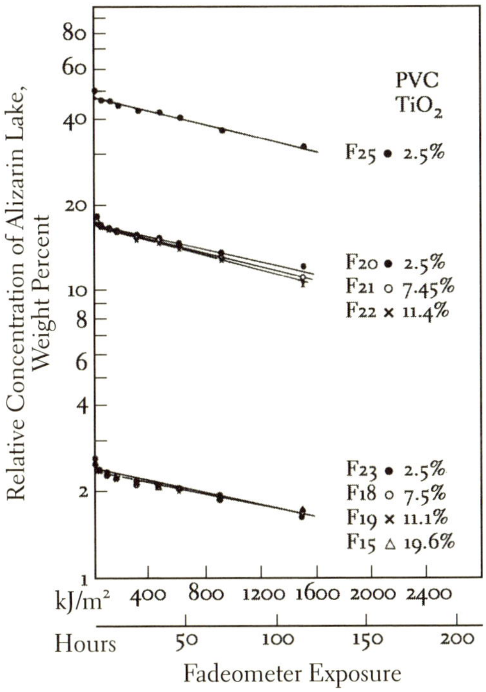

FIGURE 3
Concentrations of alizarin lake remaining in alizarin lake-rutile titanium white-poly(vinyl acetate) paints throughout 120 hours of Fadeometer exposure. Concentrations are expressed on a logarithmic scale as weight percent of alizarin lake relative to the amount of alizarin lake plus titanium white. Exposure monitored in terms of kJ/m^2 at 420 nm. Paints at complete hiding.

If the rate of fading of a colorant (alizarin lake in this case) follows a first-order rate law (Equation 1), the experimenter may verify this by plotting the experimental data according to the integrated form of the equation (2a or 2b); a graph of the logarithm of the concentration of the colorant versus the exposure will yield a straight line, the slope representing the specific rate constant, k_1. Figure 3 shows the relative concentration of alizarin lake measured at various intervals of exposure, placed on a logarithmic scale, versus the exposure up to 1500 kJ/m^2; the specific rate constants (the slopes) together with the 95% confidence interval, calculated from Equation 2a by means of the least-squares solution, are given in Table 2. Two facts become clear: first, regardless of the relative initial concentration of alizarin lake and regardless of the pigment volume concentration in the paints, the relationship is linear; second, within experimental error, the specific rate, k_1, is constant over the considerable range of concentrations and PVC's employed. During the 1500 kJ/m^2 period of exposure about one third of the alizarin colorant had faded. In this length of time a control sample of the ISO R105 blue-wool fading standard No. 4, placed in the Fadeometer along with the alizarin panels, had faded to the extent of Geometric Grey Scale Contrast 3.[13]

13. Feller, R. L., and Johnston-Feller, R. M., "Use of the International Standards Organization's Blue-Wool Standards for Exposure to Light. I. Use as an Integrating Light Monitor for Illumination Under Museum Conditions," *Preprints of Papers Presented at the Sixth Annual Meeting, Fort Worth, Texas*, American Institute for Conservation of Historic and Artistic Works, Washington, DC (1978), p. 73-80.

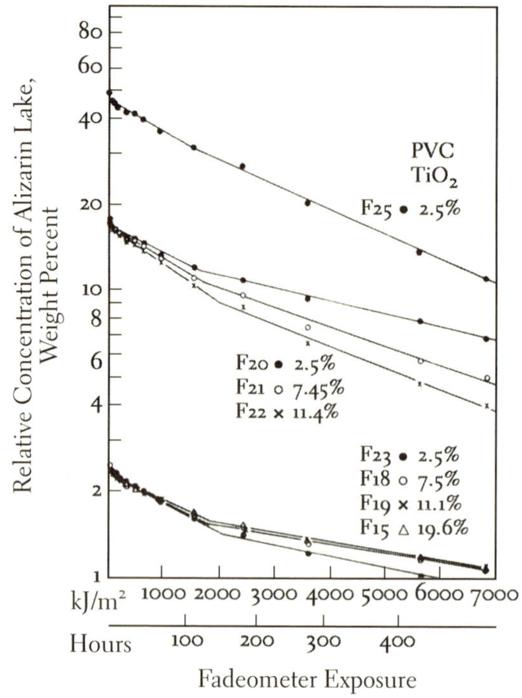

FIGURE 4
Concentrations of alizarin lake remaining in the same alizarin lake-rutile titanium white-poly(vinyl acetate) paints shown in Figure 3 throughout 540 hours of Fadeometer exposure. Concentrations are expressed on a logarithmic scale as weight percent of alizarin lake relative to the amount of alizarin lake plus titanium white. Exposure monitored in terms of kJ/m^2 at 420 nm. Paints at complete hiding.

TABLE 2
First Order Rate Constant, k_1, for Fading of Alizarin in Vinac® B-7 at Complete Hiding

Sample No.	Rel. % Alizarin (wt.)	PVC TiO_2	$k_1 \times 10^4/kJ/m^2$*
F23	2.59	2.50	− 2.4 ± 0.1
F18	2.47	7.50	− 2.2 ± 0.2
F19	2.50	11.07	− 2.1 ± 0.2
F15	2.50	19.64	− 2.0 ± 0.5
F20	18.25	2.48	− 2.2 ± 0.2
F21	17.58	7.45	− 2.7 ± 0.2
F22	17.41	11.36	− 3.1 ± 0.1
F25	50.11	2.50	− 2.6 ± 0.2
		Average ±95% confidence interval:	− 2.6 ± 0.4
F72	17.73	11.03	− 2.4 ± 0.1
F74	48.20	2.62	− 2.3 ± 0.2
			or − 2.6 ± 0.4

* From data in Figures 3 and 5 with initial (unexposed) data point omitted in the calculation of k_1.

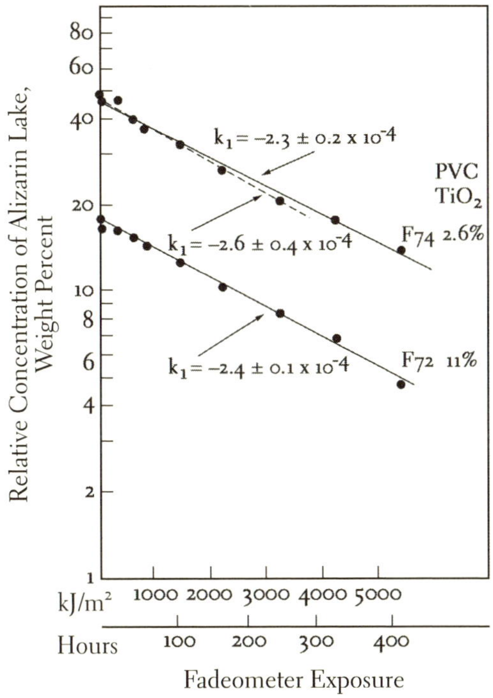

FIGURE 5
Concentrations of alizarin lake remaining in alizarin lake-rutile titanium white-poly(vinyl acetate) paint throughout 420 hours of Fadeometer exposure. Two samples from a commercial lot and paint dispersion different from those samples in Figures 3 and 4. Concentrations are expressed on a logarithmic scale as weight percent of alizarin lake relative to the amount of alizarin lake plus titanium white. Exposure monitored in terms of kJ/m^2 at 420 nm. Paints at complete hiding.

The exposure of this set of panels was continued for almost 7000 kJ/m^2. The extended data are shown in Figure 4. The curves appear to have two first-order stages: that up to 1500 kJ/m^2 and a slower stage after this period. In addition, an abrupt change took place initially in the first few hours of exposure. These two characteristics have been observed by previous authors.[4, 5, 14, 15] The break at 1500 kJ/m^2 may be due to the more rapid fading of the smaller pigment particles, a fact long observed in the paint industry.[16] Fractionation of the X-686 alizarin lake pigment by sedimentation confirmed this—paints made with the very fine particles faded at the initial rate observed here, much faster than paint made with the larger-sized particles that had rapidly settled out of the suspension.

Further investigation will be necessary before a fully confident explanation of the apparent break in the curves can be provided. In many samples of this same paint system that we have prepared since these initial investigations, we have found no such sharp break. Figure 5 illustrates this behavior in a set of samples prepared several years later. These paint films were made using a dispersion prepared in a different laboratory with a different commercial lot of

14. Giles, C. H., Johari, D. P., and Shah, C. D., "Some Observations on the Kinetics of Dye Fading," *Textile Research J.*, 38 (1968), p. 1048-1056.

15. Lachman, L., Swartz, C. J., Urbanyi, T., and Cooper, J., "Color Stability of Tablet Formulations II: Influence of Light Intensity on the Fading of Several Water-Soluble Dyes," *J. Am. Pharmaceutical Assoc.*, 49 (1960), p. 165-169.

16. Hafner, O., "Lightfastness of Organic Pigments as a Function of their Particle Size Distribution," *J. Paint Technology*, 47 (1975), p. 64-69.

alizarin lake, X-686. In the case of sample F-72, the least-squares determination of the slope yielded a value for k_1 of -2.4×10^{-4} with a 95% confidence limit of $\pm 0.1 \times 10^{-4}$. The linear relationship held over 5400 kJ/m² of exposure (monitored at 420 nm), at which point 70% of the original concentration of alizarin lake had apparently been lost. In the case of sample F-74, the least-squares solution of the slope was $-2.3 \pm 0.2 \times 10^{-4}$ over the same extent of exposure and loss of alizarin. If one chose to use only the first eight data points, the least squares solution would yield a specific rate constant of $-2.60 \pm 0.4 \times 10^{-4}$ up to about 3400 kJ/m² exposure. At this point about 67% of the original concentration of alizarin lake has been lost. These constants compare to the average value of $-2.60 \pm 0.4 \times 10^{-4}$ determined with the earlier samples (Table 2).

The experimental results show that the specific rate of fading of the alizarin lake is, within experimental error and over a loss of between 30 and 70% of the initial concentration of lake, essentially independent of the relative concentration of alizarin lake, and of the pigment volume concentration (PVC). Thus, the pale tints did not lose their coloring matter faster than darker colors; the alizarin lake was converted at virtually the same specific rate regardless of its initial concentration.

The specific reaction rate for the fading of the yellow form, the k_2 in Equation 6, was also determined for all samples; however, description of the complex mathematical technique used for calculating k_2 will not be included here. The important conclusions are that k_2 was found also to relate to a first-order reaction and that k_2 was always larger than k_1—that is, the specific rate of fading of the yellow intermediate is faster than the specific rate of concentration change of the red alizarin lake.

Visual Color Changes

Munsell notations were calculated for all the exposed samples in order that the color changes might be visualized by reference to the Munsell Book of Color. The Munsell data for one of the samples, mid-way in terms of relative alizarin lake-titanium dioxide concentration and pigment volume concentration, are presented in Table 3.

Color differences as the panels progressively faded from the original color were also calculated using the CIE 1976 L*a*b* equation. This equation expresses the total color difference, ΔE, as the square root of the sum of the squares for the lightness-darkness difference, ΔL^*, the redness-greenness difference, Δa^*, and the yellowness-blueness difference, Δb^*.[12] One unit of ΔE (L*a*b*) represents a good commercial color match; one-half unit of ΔE (L*a*b*) is, on the average, a minimum perceptible difference. The calculated values of ΔL^*, Δa^*, and Δb^* approximate the direction of the color shift and the ΔE represents the total color difference that the eye perceives.

Figure 6 shows the total color difference, ΔE, plotted against the exposure

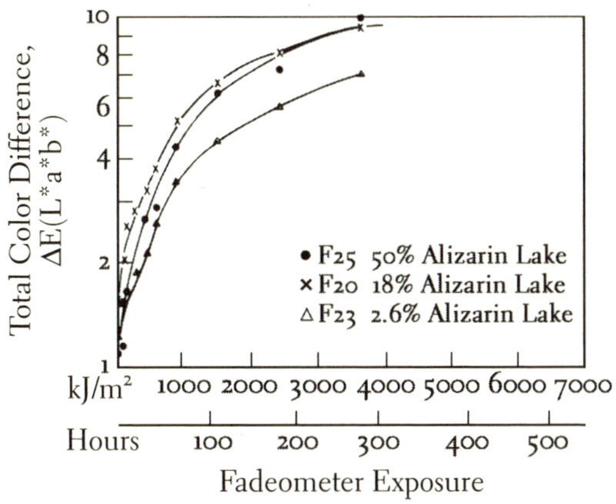

FIGURE 6
CIE 1976 color differences, $\Delta E(L^*a^*b^*)$, plotted against length of exposure. Three paint samples are shown, each made with titanium dioxide at 2.5% pigment volume concentration but at different ratios of alizarin lake to total dry pigment concentration. All films were made with poly(vinyl acetate) vehicle and were applied as complete hiding.

TABLE 3
Summary of Total Reflectance Measurements
and Calculated Munsell Notations on Exposed Sample F21

Exposure kJ/m^2	x	y	Y	Munsell Hue	Munsell Value	Munsell Chroma	Calculated Percent Alizarin Remaining*
0	.4014	.2926	24.58	9.84 RP	5.50	8.79	100.0
30	.4001	.2947	24.52	9.81 RP	5.50	8.49	95.5
90	.4010	.2956	24.56	0.81 R	5.50	8.50	93.8
150	.4036	.2972	24.60	0.48 R	5.50	8.53	90.9
296	.4017	.2980	25.02	0.44 R	5.54	8.40	86.9
444	.4036	.2991	24.91	0.21 R	5.53	8.41	85.8
592	.4033	.3007	25.10	0.14 R	5.55	8.27	81.8
888	.4036	.3031	25.69	0.41 R	5.61	8.18	75.0
1480	.4038	.3065	27.15	1.02 R	5.75	8.12	64.2
2368	.3966	.3078	28.68	0.94 R	5.88	7.70	55.9
3552	.3888	.3102	31.79	1.15 R	6.15	7.09	43.4
5582	.3807	.3124	35.39	1.36 R	6.44	6.53	32.9
6782	.3765	.3134	37.17	1.54 R	6.58	6.24	28.7

* Corrected for surface reflectance.

for three samples, all at 2.5% PVC titanium dioxide, which contained 2.5%, 18%, and 50% alizarin lake by weight relative to the total dry pigment concentration. One would not be able to recognize first-order fading behavior from such data. Moreover, the mid-concentration range, 18%, exhibited the highest color change, ΔE, up to about 3000 kJ/m^2 exposure; the lowest net color change occurred in the case of the palest tint, 2.5% alizarin. These data additionally demonstrate the fact that for every pigment there is a range of relative pigment-white concentration for which the *visual* differences in color will be highest for a particular percentage loss in the concentration of colorant. Experience and theory reveals that the concentration of colorant which will exhibit the maximum ΔE (visual change) is one that will yield a reflectance of about 40% at the wavelength of absorption maximum.

Discussion

The formation of a yellow intermediate compound found in these studies contrasts with the results obtained previously with glazes (alizarin lake, without titanium dioxide, applied at *incomplete hiding*).[10] In that investigation we found that the color of the dark samples also made an abrupt shift towards yellow during fading. In the case of the dark glazes, however, the shift towards the yellow was calculable on the basis of the changes taking place in the spectrophotometric curve shape as the concentration of alizarin diminished; no yellow compound was ever needed to match the color of the faded glazes.

Included in our investigations were a number of samples of alizarin lake mixed with titanium dioxide at *incomplete hiding*, not reported in the paper on glazes nor in this article. In brief, we found that, whenever hiding of the alizarin lake-titanium white mixtures was incomplete, no yellow compound was found. This is to be expected considering that the specific reaction rate for the fading of the yellow intermediate, k_2 in Equation 6, was found to be greater than the rate of disappearance of the alizarin lake, k_1. Thus, when hiding is incomplete, and radiant energy can penetrate all the way through the film, there is no underlying layer in the paint in which the yellow compound can be detected in the process of being formed. The controlling reaction rate in successive reactions is the slowest one, in this case k_1. The result is that, when hiding is incomplete, the yellow form is rapidly converted to the colorless form. We will report full details of these results at a later date.

In the publication on the glazes, we described a simplified version of the Kubelka-Munk equation for incomplete hiding, a "single-constant" equation, useful but not as accurate as the two-constant equation in which both K and S are variables. The scattering coefficient, S, was assumed to be constant, and only the absorption coefficient, K, was assumed to vary during the exposure. We can also describe a simplified equation applicable to a relatively transparent pigment, such as alizarin lake, in mixture with a highly scattering white pigment in a paint film applied at complete hiding. Again the assumption is made that the scattering of the paint film is constant. As in the case of the glazes, we may then assume that the only change in color is caused by a change in the absorption coefficient, K. On this basis we can calculate a color match to the faded samples using a simplified "single-constant" equation applicable to the case of complete hiding:

$$(K/S)_M = C_1(K_1/S) + C_2(K_2/S) + \ldots C_w(K_w/S) \qquad (7)$$

where the subscript w refers to the white pigment. This is the equation used in our early, unreported, studies. It was sufficiently accurate to lead us in the direction of applying first-order kinetics to the study of fading.

If a pigment fades directly to a colorless form—that is, if no intermediate compound is formed which absorbs light of the same wavelength as that absorbed at the absorption maximum of the pigment being tested—a further simplification can be made. If both assumptions mentioned are true—S does not change, and K at the absorption maximum is not affected by the formation of another species absorbing at this wavelength—then we can apply kinetic analysis by using the (K/S) value (Equation 4) at the wavelength of maximum absorption as the measure of concentration of the absorbing pigment remaining. Thus, to analyze the kinetics of the fading process, the log (K/S) at the absorption maximum can be plotted against time or net exposure. In two of the publications cited[5, 6] the authors used this abbreviated method to demonstrate the first-order behavior of the colorants.

Conclusion

Precise spectrophotometric reflectance measurements, made before and during the exposure of paint films to radiant energy, coupled with computer color-matching calculations based on the Kubelka-Munk equations, can be used to compute the percentages of pigment remaining after exposures for various lengths of time. With such quantitative data regarding the concentration of colorants at any particular time in the course of fading, the principles of chemical kinetic analysis may be applied. Using the artists' pigment, alizarin lake, as a model chromatic pigment in mixture with rutile titanium dioxide white, we have shown that the fading of alizarin lake, in an opaque, high-gloss paint film made with poly(vinyl acetate) as the vehicle, approximates first-order behavior: the amount of colorant faded in any given interval of time is a constant percentage of the amount of alizarin lake pigment present. Within experimental error, this relationship is independent of the relative alizarin lake-titanium white concentration and the pigment volume concentration. Exposure was carried out in a xenon-arc Fadeometer, filtered to approximate solar radiation through window glass in the near ultraviolet and visible spectral regions.

Spectrophotometric curve analysis also revealed that the red alizarin lake pigment tends to fade to a yellow intermediate before fading to a colorless form. By means of Munsell Notation and CIE color-difference calculations, curves were developed to show how the orderly change in concentration relates in a non-linear fashion to the perceived color of the paint. The inadequacy of visual evaluations for quantifying the changes in pigment concentrations is thus illustrated.

The techniques described are equally applicable to the study of the fading of other pigments, alone or in mixture.

Determination of the Specific Rate Constant for the Loss of a Yellow Intermediate During the Fading of Alizarin Lake

ROBERT L. FELLER, RUTH M. JOHNSTON-FELLER
AND CATHERINE BAILIE

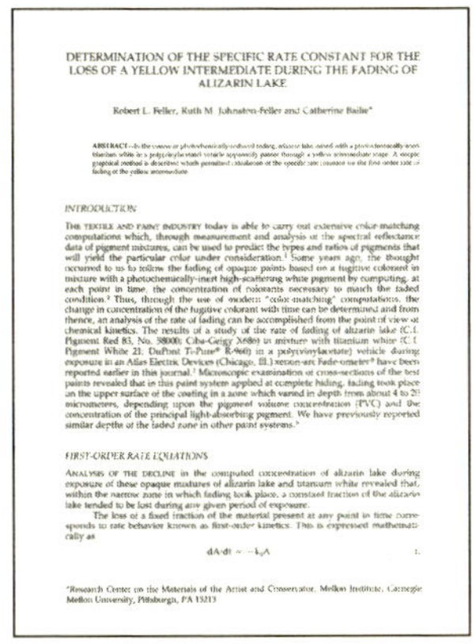
DETERMINATION OF THE SPECIFIC RATE CONSTANT FOR THE LOSS OF A YELLOW INTERMEDIATE DURING THE FADING OF ALIZARIN LAKE

Robert L. Feller, Ruth M. Johnston-Feller and Catherine Bailie*

INTRODUCTION

THE TEXTILE AND PAINT INDUSTRY today is able to carry out extensive color-matching computations which, through measurement and analysis of the spectral reflectance data of pigment mixtures, can be used to predict the types and ratios of pigments that will yield the particular color under consideration.[1] Some years ago, the thought occurred to us to follow the fading of opaque paints based on a fugitive colorant in mixture with a photochemically-inert high-scattering white pigment by computing, at each point in time, the concentration of colorants necessary to match the faded condition.[2] Thus, through the use of modern "color-matching" computations, the change in concentration of the fugitive colorant with time can be determined and from thence, an analysis of the rate of fading can be accomplished from the point of view of chemical kinetics. The results of a study of the rate of fading of alizarin lake (C.I. Pigment Red 83, No. 58000; Ciba-Geigy X686) in mixture with titanium white (C.I. Pigment White 21; DuPont Ti-Pure® R-960) in a poly(vinylacetate) vehicle during exposure in an Atlas Electric Devices (Chicago, Ill.) xenon-arc Fade-ometer® have been reported earlier in this journal.[2] Microscopic examination of cross-sections of the test paints revealed that in this paint system applied at complete hiding, fading took place on the upper surface of the coating in a zone which varied in depth from about 4 to 20 micrometers, depending upon the pigment volume concentration (PVC) and the concentration of the principal light-absorbing pigment. We have previously reported similar depths of the faded zone in other paint systems.[3]

FIRST-ORDER RATE EQUATIONS

ANALYSIS OF THE DECLINE in the computed concentration of alizarin lake during exposure of these opaque mixtures of alizarin lake and titanium white revealed that, within the narrow zone in which fading took place, a constant fraction of the alizarin lake tended to be lost during any given period of exposure.

The loss of a fixed fraction of the material present at any point in time corresponds to rate behavior known as first-order kinetics. This is expressed mathematically as

$$dA/dt = -k_1A \qquad 1.$$

*Research Center on the Materials of the Artist and Conservator, Mellon Institute, Carnegie Mellon University, Pittsburgh, PA 15213

Originally printed in:
Journal of the American Institute for Conservation
Volume 25, No. 2 (1986), Pages 65-72.

Introduction

The textile and paint industry today is able to carry out extensive color-matching computations which, through measurement and analysis of the spectral reflectance data of pigment mixtures, can be used to predict the types and ratios of pigments that will yield the particular color under consideration.[1] Some years ago, the thought occurred to us to follow the fading of opaque paints based on a fugitive colorant in mixture with a photochemically-inert high-scattering white pigment by computing, at each point in time, the concentration of colorants necessary to match the faded condition.[2] Thus, through the use of modern "color-matching" computations, the change in concentration of the fugitive colorant with time can be determined and from thence, an analysis of the rate of fading can be accomplished from the point of view of chemical kinetics. The results of a study of the rate of fading of alizarin lake (C.I. Pigment Red 83, No. 58000; Ciba-Geigy X686) in mixture with titanium white (C.I. Pigment White 21; DuPont Ti-Pure® R-960) in a poly(vinylacetate) vehicle during exposure in an Atlas Electric Devices (Chicago, Ill.) xenon-arc Fadeometer have been reported earlier in this journal.[2] Microscopic examination of cross-sections of the test paints revealed that in this paint system applied at complete hiding, fading took place on the upper surface of the coating in a zone which varied in depth from about 4 to 20 micrometers, depending upon the pigment volume concentration (PVC) and the concentration of the principal light-absorbing pigment. We have previously reported similar depths of the faded zone in other paint systems.[3]

1. See, for example, Kuehni, R. G., *Computer Colorant Formulation*, Lexington Books, D. C. Heath and Company, Lexington, MA (1975).

2. Johnston-Feller, R. M., Feller, R. L., Bailie, C. W., and Curran, M., "The Kinetics of Fading: Opaque Paint Films Pigmented with Alizarin Lake and Titanium Dioxide," *J. American Institute for Conservation*, 23 (1984), p. 114-129. Since this date a more extensive treatment of the subject has been published: Johnston-Feller, R. M., "Reflections on the Phenomenon of Fading," *J. Coatings Technology*, 58, No. 736 (1986), p. 32-50.

3. Feller, R. L., "Studies on Photochemical Deterioration," *Preprints of the 4th Triennial Meeting, Venice*, ICOM Committee on Conservation (1975), Paper 75/19/4.

FIRST-ORDER RATE EQUATIONS

Analysis of the decline in the computed concentration of alizarin lake during exposure of these opaque mixtures of alizarin lake and titanium white revealed that, within the narrow zone in which fading took place, a constant fraction of the alizarin lake tended to be lost during any given period of exposure.

The loss of a fixed fraction of the material present at any point in time corresponds to rate behavior known as first-order kinetics. This is expressed mathematically as

$$dA/dt = -\ k_1A \qquad (1)$$

In the integrated form this becomes

$$\ln A = \ln A_0 \ - \ k_1t \qquad (2)$$

In these equations, A represents the weight-percent concentration of alizarin at any time, t [alizarin + white + yellow (see below) = 100], dA/dt is the instantaneous change in concentration of A with time, and A_0 the initial weight concentration of alizarin; ln A is the logarithm of A to the base e.[4]

Because the intensity of illumination diminishes with the age of the xenon lamps, instead of *time* we used the product of intensity times time, or *exposure*, for t. Thus, instead of reporting the fraction of alizarin lake lost per hour, the rate, in a sense, is expressed in terms of the fraction lost per footcandle hour or, in our case, per kiloJoule/cm^2. The irradiance was monitored at a wavelength of 420 nm.

As reported,[2] the specific rate constant, k_1, was found to be essentially constant throughout a range of initial concentrations of alizarin lake pigment of from 50 to 2.5% relative to the titanium white and a range of pigment volume concentration (PVC) of titanium white from 19 to 2.5%. This is indicated by the identical slopes of the plots of log A (equivalent to ln A/2.303) vs. exposure in Figure 1, regardless of initial concentration.

4. In our first presentation of these concepts (Footnote 2), we used the symbol C (for concentration) to describe the first-order rate equations, but the different set of symbols A, Y, and C seem particularly appropriate here to represent the concentration of the three chemical components expressed in equation 3. The authors also consider that the minus sign is an essential part of the specific rate constant, although this is not the usual convention found in textbooks of physical chemistry. Nonetheless, equations 2 and 5 are the form in which the equations are usually found.

FORMATION OF A YELLOW INTERMEDIATE

Upon calculation of the concentration of pigments necessary to match the spectrophotometric curves of the faded states of the test panels of alizarin lake and titanium white formulated in poly(vinyl acetate), applied at complete hiding, the color-matching computations indicated that the addition of a yellow component was usually required. Figure 1 shows a typical result: along with the expected decrease in the amount of alizarin lake with increased exposure, there was also at first a rising, then a declining, concentration of a yellow component present in many of the exposed panels. Such behavior can be expected as the result of a sequence of chemical steps that may be represented by the equation

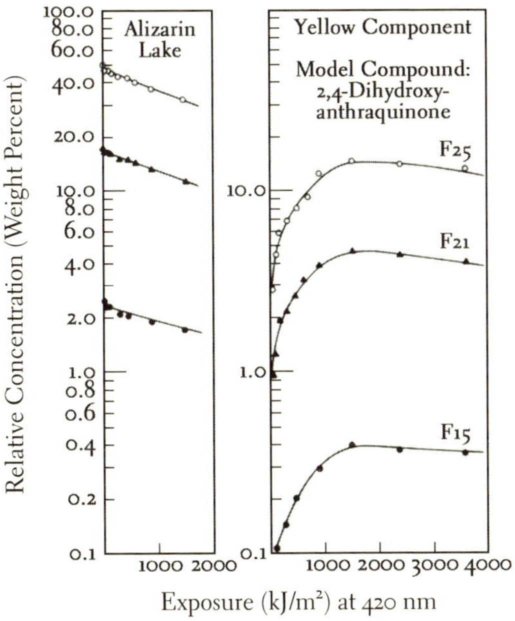

FIGURE 1
Semilog "first-order" plot of the loss of concentration of alizarin lake formulated at three different initial concentrations, showing also the accompanying rise and fall of a yellow component. Exposure carried out in a xenon-arc Fadeometer monitored in terms of kJ/m^2 at 420 nm.

$$A \xrightarrow{k_1} Y \xrightarrow{k_2} C \qquad (3)$$

in which initial colorant A (in our case, alizarin lake) forms component Y (in our case, a yellow substance for which the yellow, unlaked, form of alizarin was used for calculation purposes). This in turn yields a further product, C. In our situation, C must have been colorless because no pigments other than alizarin lake, titanium white, and a yellow were needed to match the color of the faded states of the test panels.[4] Other organic colorants also are known to go through an intermediate colored form before passing on to a colorless, completely faded, condition.

In our initial report,[2] we demonstrated that the apparent rate of disappearance of the alizarin lake tended to follow equation 2. We further reported that the fading of alizarin and its yellow intermediate appeared to follow a sequence of events such as shown in equation 3 and cited a specific first-order rate constant, k_2, for the disappearance of the yellow. However, the method used to obtain the rate constant for this second step was not explained. We wish now to describe the technique used to obtain values of k_2 by a simple graphical procedure.

EQUATIONS FOR THE RATE OF FORMATION AND LOSS OF THE INTERMEDIATE

In view of equation 3, the concentration of intermediate Y at any time, t, can be accounted for by the following equation:

$$\frac{dY}{dt} = \frac{-dA}{dt} - k_2Y \qquad (4)$$

This equation states that the observed change in the amount of Y at any time (dY/dt) is equal to the opposite from the rate of loss of A (this is because Y is produced from A; hence dY/dt = – dA/dt in step k_1) minus the rate of disappearance of Y in step k_2, which we have here assumed also to follow a decay law such as expressed for component A in equation 1. Taking into consideration 1, this equation may be written:

$$\frac{dY}{dt} = k_1A - k_2Y \qquad (5)$$

Dividing by Y and rearranging, the equation yields an expression for k_2:

$$k_2 = \left(k_1A - \frac{dY}{dt}\right) / Y \qquad (6)$$

We chose to obtain k_2 by rearranging 6 further to conform to the general equation for a straight line, y = b + mx, where m is the slope and b is the intercept on the y ordinate. Thus:

$$\left(\frac{dY}{dt}\right) / Y = -k_2 + k_1\left(\frac{A}{Y}\right) \qquad (7)$$

Thus, if equation 3 indeed represents the sequence of chemical steps involved, each of which proceeds by first-order kinetics, a graph of the values of (dY/dt)/Y versus A/Y should yield a straight line with the slope equal to k_1 and the intercept, k_2. A typical graph is shown in Figure 2.

GRAPHICAL SOLUTIONS TO EQUATIONS

In order to utilize equation 7, values of (dY/dt)/Y must be obtained. A graphical method can be used to provide this information. To accomplish this, the concentrations of the yellow component in the exposed panels, which the color-matching computations showed was required to match the faded state, were plotted on linear graph paper versus the exposure (t), expressed in terms of kiloJoules per square meter, kJ/m^2, monitored at 420 nm. These data resulted in a series of points through which a smooth curve could be drawn showing the rise and subsequent fall in concentration of the yellow component, illustrated by the heavy-lined curve of Figure 3. By inspection, lines are drawn tangent to the curve at various points of tangency. The slope of these lines is dY/dt, the change in the concentration of Y per unit of exposure at that particular exposure, t, and at that particular concentration of yellow, Y. The tangents in our example are illustrated by light lines numbered 1 to 6 in Figure 3. The value of Y at the points of tangency, similarly numbered, can be read from the graph. The concentration of alizarin lake at this same time is obtained from the experimental data, seen here in the graphs of concentration versus exposure, such as in Figure 1. From this information, the value of A/Y at exposure stage, t, may be calculated. From the slopes of the tangents, values of (dY/dt)/Y may be determined and plotted against values of A/Y, as done in Figure 2.

An example of the data necessary for preparing a graph such as that in Figure 3, is presented in Table 1. Typical data, which we usually employed only for the calculated concentrations of yellow before it had reached its maximum concentration (at which point, incidentally, the slope is zero), are shown in Figure 2. According to equation 6, the intercept on the ordinate axis of the line connecting these points is the value for k_2 and should be negative. While the example here is for a sample made at complete hiding, the same procedure for obtaining k_2 has been successfully applied to the data for samples at incomplete hiding.

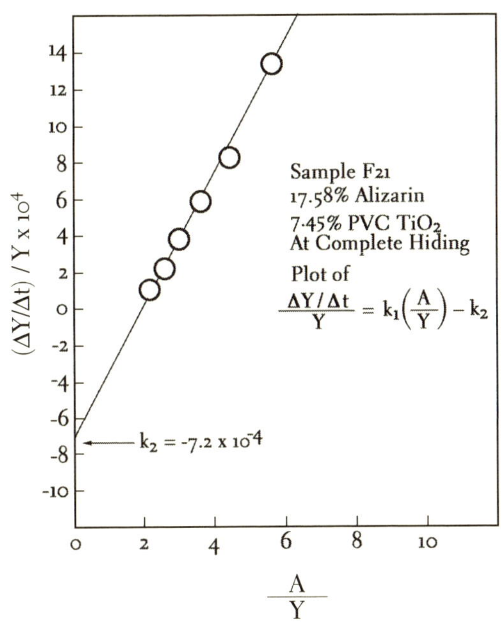

FIGURE 2
Graphical solution of equation 7, $(\Delta Y/\Delta t)/Y$ on y axis versus A/Y on x axis. Data taken from Table 1 and Figure 3.

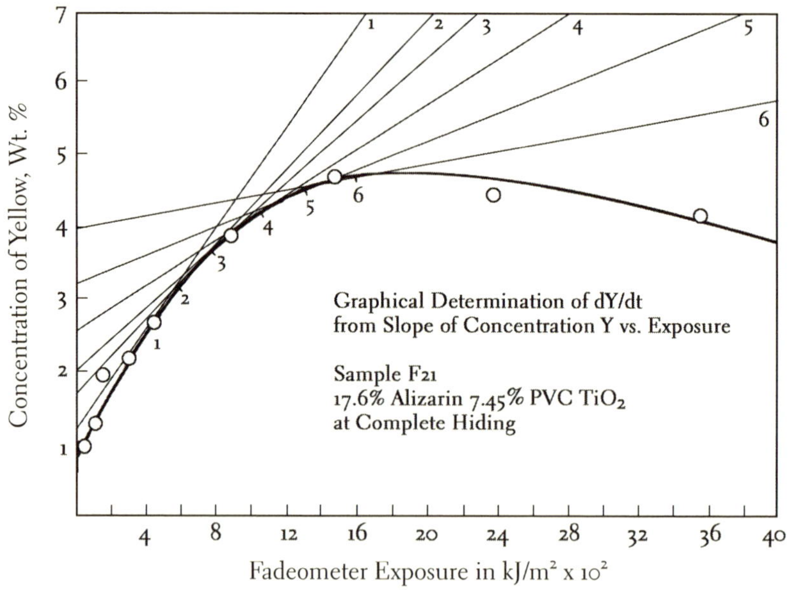

FIGURE 3
Concentration of Y versus exposure, t, on linear graph paper with heavy-line curve connecting the data points and various tangents drawn in light lines. The slopes of the tangent lines represent dY/dt.

TABLE 1
Sample F21: Example of Data Used to Calculate k_2 From Curve of Percent Yellow Versus Exposure

Data Point in Figure	Concentration of Yellow*	Concentration of A at This Value of Y (from Figure 1)	$\frac{A}{Y}$	Slope of Tangent at Y: $\frac{\Delta Y^{**}}{\Delta t}$	$\frac{\Delta Y / \Delta t}{Y}$
1	2.63	14.99	5.70	35.30×10^{-4}	13.42×10^{-4}
2	3.20	14.20	4.43	26.10	8.16
3	3.66	13.30	3.63	21.80	5.96
4	4.17	12.60	3.02	15.75	3.78
5	4.49	11.70	2.60	9.97	2.22
6	4.68	10.70	2.29	4.45	0.95

* Calculated as though this component were 1,2-dihydroxyanthraquinone (unlaked form of alizarin).
** $\Delta t = t_2 - t_1$, $\Delta Y = Y_2 - Y_1$

VALUE OF k_2 INDEPENDENT OF TRUE CONCENTRATION OF YELLOW COMPONENT

The precise concentration of component Y was unknown since we do not know the chemical nature of the yellow intermediate compound and, hence, we could not know the precise spectral reflectance data to employ in the color-

TABLE 2
Specific Rate Constant for Fading of Yellow-Form, k_2, Using the Spectral Reflectance Characteristics of Three Different Yellow Pigments in the Color-Matching Calculations When Computing the Concentration of the Yellow Component Vehicle: Vinac® B-7; Paint at Complete Hiding

Sample No.	Computed Values of $-k_2 \times 10^4$/kJ/cm² 2,4-Dihydroxy-anthraquinone (Yellow Alizarin)	Yellow Iron Oxide (Pfizer Y-06087)	Permacal Yellow 6664 (Hoechst) C.I. Pigment Yellow 97
F15	7.5	7.0	7.2
F20	7.2	7.3	7.3
F21	7.2	8.0	7.8
Avg.	7.3	7.4	7.4

matching computations. Arbitrarily, the spectral reflectance curve or the yellow (unlaked) form of alizarin, 2,4-dihydroxyanthraquinone, was selected for our initial calculations, but we do not wish to imply that this is indeed the intermediate, Y. As the results in Table 2 indicate, however, the use of any one of three different yellow colorants in the color-matching program, even though each had slightly different spectral reflectance characteristics, had little effect

TABLE 3
Specific Rate Constants k_1 and k_2 for Fading of Alizarin Lake in Mixture With Titanium White in Xenon-Arc Fadeometer
(Pyrex/Pyrex Filters)
Vehicle: Poly(vinyl acetate) Vinac® B-7; Complete Hiding

Sample No.	Rel. % Alizarin (wt.)	Pigment Volume Concentration TiO_2	$-k_1 \times 10^4$/kJ/cm² For Fading of Alizarin	$-k_2 \times 10^4$/kJ/cm² For Fading of Yellow Intermediate
F23	2.59	2.50	2.4	8.3
F18	2.47	7.50	2.2	7.7
F19	2.50	11.07	3.2	7.5
F15	2.50	19.64	2.0	7.5
F20	18.25	2.48	2.2	7.2
F21	17.58	7.45	2.7	7.2
F22	17.41	11.36	3.1	8.4
F25	50.11	2.50	2.6	8.4
	mean and standard deviation:		2.6 ± 0.4 (±15%)	7.8 ± 0.5 (±6%)

on the values of k_2. Consideration of equation 7 will show why this is so: the intercept, k_2 [that is, (dY/dt)/Y, when A/Y = 0] is not sensitive to uncertainties in Y. This is because any correction factor necessary to yield the true value of Y will appear in both the denominator (dY/dt) and numerator (Y). Thus, the effect on the intercept k_2 of a systematic factor of error in Y tends to cancel out.

CONCLUSION

Table 3 summarizes the values of k_1 previously reported (obtained from equation 2) and of k_2 obtained from the graphical solution of equation 7. For the particular batch and dispersion of alizarin lake/titanium white paints used in this investigation, the specific rate of disappearance of the yellow intermediate indeed seems to be a constant, suggesting that the fading of the yellow intermediate also proceeds according to first-order kinetics.

These initial interpretations concerning the apparent rate of fading of alizarin lake confirm the potential usefulness of modern color-matching computations in facilitating the analysis of the phenomenon of fading in paints according to well-known principles of chemical kinetics.

The Fading of Some Traditional Pigments as a Function of Relative Humidity

C. W. Bailie, R. M. Johnston-Feller and R. L. Feller

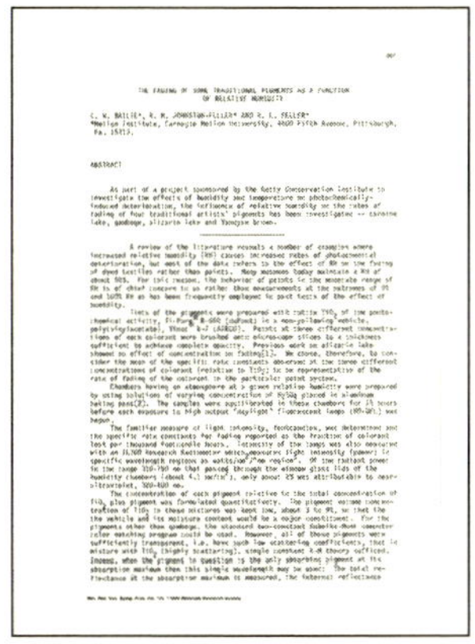

Originally printed in:
Materials Issues in Art and Archaeology
Materials Research Society Symposium Proceedings, Volume 123, ed. Sayre, E. V., Vandiver, P. B., Druzik, J., and Stevenson, C., Materials Research Society, Pittsburgh, PA (1988), Pages 287-292.

A review of the literature reveals a number of examples where increased relative humidity (RH) causes increased rates of photochemical deterioration, but most of the data refers to the effect of RH on the fading of dyed textiles rather than paints. Many museums today maintain a RH of about 50%. For this reason, the behavior of paints in the moderate range of RH is of chief concern to us rather than measurements at the extremes of 0% and 100% RH as has been frequently employed in past tests of the effect of humidity.

Tints of the pigments were prepared with rutile TiO_2 of low photochemical activity, Ti-Pure® R-960 (duPont) in a non-yellowing vehicle poly(vinyl acetate), Vinac® B-7 (AIRCO). Paints at three different concentrations of each colorant were brushed onto microscope slides to a thickness sufficient to achieve complete opacity. Previous work on alizarin lake showed no effect of concentration on fading.[1] We chose, therefore, to consider the mean of the specific rate constants observed at the three different concentrations of colorant (relative to TiO_2) to be representative of the rate of fading of the colorant in the particular paint system.

Chambers having an atmosphere at a given relative humidity were prepared by using solutions of varying concentration of H_2SO_4 placed in aluminum baking pans.[2] The samples were equilibrated in these chambers for 15 hours before each exposure to high output "daylight" fluorescent lamps (HO-DFL) was begun.

The familiar measure of light intensity, footcandles, was determined and the

1. Johnston-Feller, R. M., *Journal of Coatings Technology*, 58, No. 736 (1986), p. 33.

2. Lange, N. A., *Handbook of Chemistry*, 9th edn., Handbook Publishers, Inc., Sandusky, Ohio (1956), p. 1423.

specific rate constants for fading reported as the fraction of colorant lost per thousand footcandle hours. Intensity of the lamps was also measured with an IL700 Research Radiometer which measures light intensity (power) in specific wavelength regions as watts/cm^2/"nm region." Of the radiant power in the range 310-750 nm that passed through the window glass lids of the humidity chambers (about 4.1 mW/cm^2), only about 2% was attributable to near-ultraviolet, 320-400 nm.

The concentration of each pigment relative to the total concentration of TiO_2 plus pigment was formulated quantitatively. The pigment volume concentration of TiO_2 in these mixtures was kept low, about 3 to 9%, so that the vehicle and its moisture content would be a major constituent. For the pigments other than gamboge, the standard two-constant Kubelka-Munk computer color matching program could be used. However, all of these pigments were sufficiently transparent, i.e. have such low scattering coefficients, that in mixture with TiO_2 (highly scattering), single constant K-M theory sufficed. Indeed, when the pigment in question is the only absorbing pigment at its absorption maximum then this single wavelength may be used: The total reflectance at the absorption maximum is measured, the internal reflectance calculated and from this, the internal $(K/S)_i$ ratio which is proportional to the concentration of pigment.[3] No difference in results was obtained between this simplified form and the more complex complete-wavelength two-constant theory for the three pigments.

3. Johnston-Feller, R.M., in *Pigment Handbook*, Vol. III, ed. Patton, T. C., John Wiley & Sons, Inc., New York (1973), p. 264-267.

The equation for a first-order kinetic reaction is:

$$-\frac{dC}{dt} = k_1 C \qquad (1)$$

Integrated:

$$\ln C = \ln C_o - k_1 t \qquad (2)$$

where C = concentration at any time t (in this case, at any exposure expressed in footcandles), C_o = initial concentration and k_1 = the specific reaction rate constant. A summary of the results is given in Table 1. The specific first-order rate constants were calculated by the least-squares solution of Equation (2).

Figure 1 is a graphic presentation of the results. The absolute values of the rate constants reflect the fact that carmine lake and gamboge are much more fugitive pigments than alizarin lake and Vandyke brown. With the exception of gamboge, an increase in RH caused faster fading. Carmine lake and Vandyke brown faded almost three times as fast at 80% RH as at 5%, and alizarin lake five and a half times as fast. Of particular interest is the moderate range of 40-60% RH which one might encounter under exhibition conditions. Thus, Table 2 also expresses the data in terms of what may be called the humidity sensitivity index (HSI), the percentage increase in rate for a rise of 10% RH. This is

TABLE 1
The Effect of RH on the Specific Rate Constants of Fading of Four Traditional Pigments Mixed with TiO_2 in Poly(vinyl acetate) Vinac® B-7 Vehicle
Mean and Standard Deviation of Rates of Fading (Fraction Lost per Thousand Footcandle Hours) of Three Concentrations of Colorant at Each Relative Humidity

	Relative Humidity				
	5%	20%	40%	60%	80%
Carmine Lake	$5.3 \pm .6 \times 10^{-3}$	$8.6 \pm 1.0 \times 10^{-3}$	$10.5 \pm 1.5 \times 10^{-3}$	$13.1 \pm 2.0 \times 10^{-3}$	$14.4 \pm 2.2 \times 10^{-3}$
Gamboge*	$1.5 \pm .3 \times 10^{-3}$	$1.7 \pm .6 \times 10^{-3}$	$1.3 \pm .4 \times 10^{-3}$	$1.6 \pm .6 \times 10^{-3}$	$1.7 \pm .6 \times 10^{-3}$
Alizarin Lake	$0.6 \pm .1 \times 10^{-4}$	$0.9 \pm .1 \times 10^{-4}$	$1.2 \pm .1 \times 10^{-4}$	$1.9 \pm .3 \times 10^{-4}$	$3.3 \pm .3 \times 10^{-4}$
Vandyke Brown	$1.1 \pm .2 \times 10^{-4}$	$1.3 \pm .2 \times 10^{-4}$	$1.8 \pm .2 \times 10^{-4}$	$2.3 \pm .3 \times 10^{-4}$	$2.9 \pm .6 \times 10^{-4}$
Mean Temp. °C	28.0	28.5	28.2	27.9	28.9

* Rate calculations based on decrease of $(K/S)_i$ at 440 nm.

TABLE 2
Summary of Data for Fading Rates at Different RH in Poly(vinyl acetate) Vinac® B-7

Tint	"Overall Increase": Increase in Rate of Fading From 5-80% RH	Humidity Sensitivity Index (HSI): Approximate Increase in Rate of Fading for a Change of 10% RH in 40-60% RH Range
Carmine Lake	2.7 x	11%
Gamboge	none	–
Alizarin Lake	5.5 x	22%
Vandyke Brown	2.6 x	12%

calculated as one half the difference in the rates of fading over a range of 20% RH divided by the average of the two rates, times 100.

Because poly(vinyl acetate) is a relatively hydrophobic vehicle, dispersions were also prepared in two more hydrophilic vehicles: carboxymethylcellulose (CMC)® and hydroxypropylcellulose (Klucel L)®. The reflectance changes of the paints were measured during the course of exposure and changes in $(K/S)_i$ at the wavelength of maximum absorption calculated. The concentrations were not computed by the two-constant color-matching program in this series of experiments because the necessary standard data for the colorants in these vehicles were not determined.

A summary of the results is presented in Table 3. One reason for the apparent poor precision in the results for gamboge is that a measurable but as yet

TABLE 3
The Effect of RH on the Specific Rate Constants of Fading of Two Traditional Pigments Mixed with TiO_2 in Cellulose Ether Vehicles
Mean and Standard Deviation of Rates of Fading (Fraction Lost per Thousand Footcandle Hours) of Three Concentrations of Colorant at Each Relative Humidity

Colorant in Hydroxypropylcellulose (Klucel L)	% RH 20	40	60	80
Carmine Lake*	$10.1 \pm .7 \times 10^{-3}$	$12.0 \pm 1.4 \times 10^{-3}$	$12.0 \pm 1.4 \times 10^{-3}$ $13.2 \pm 1.1 \times 10^{-3}$***	$14.9 \pm 2.1 \times 10^{-3}$
Gamboge**	$7.5 \pm 1.4 \times 10^{-4}$	$7.3 \pm 1.8 \times 10^{-4}$	$8.6 \pm 3.1 \times 10^{-4}$	$12.9 \pm 4.7 \times 10^{-4}$
Colorant in Carboxymethylcellulose (CMC)				
Carmine Lake*	$11.9 \pm 0.3 \times 10^{-3}$	$14.5 \pm 0.4 \times 10^{-3}$	$16.7 \pm 0.9 \times 10^{-3}$	$20.8 \pm 1.6 \times 10^{-3}$
Gamboge**	$14.3 \pm 2.8 \times 10^{-4}$	$17.2 \pm 3.5 \times 10^{-4}$	$16.4 \pm 3.8 \times 10^{-4}$	$18.0 \pm 4.3 \times 10^{-4}$
Mean Temp. °C	27.6	28.2	28.5	29.0

* Rate calculations based on decrease of $(K/S)_i$ at 520 nm.
** Rate calculations based on decrease of $(K/S)_i$ at 440 nm.
*** A repeat run

unexplained effect of colorant concentration occurred in these formulations; fading took place more rapidly in the paler tints. A summary of the results for carmine lake in Figure 2 also includes the data on carmine from Figure 1. In general the order of fading rates of carmine lake in the three vehicles is in the order of the equilibrium moisture contents determined at 50% RH: 1.7% for Vinac® B-7; 7.7% for Klucel ® L; and 26.2% for carboxymethylcellulose. However, a subsequent series carried out in poly(vinyl pyrrolidone), having an equilibrium moisture content similar to CMC, faded at rates even lower than in poly(vinyl acetate). Apparently the chemical nature of the vehicle also influences the rate of fading and many more studies will be necessary before attempts can be made to draw general conclusions.

It was our intention to study alizarin lake and Vandyke brown in the more hydrophilic vehicles, but good dispersions of these pigments could not be made without major modification of the formulations.

CONCLUSIONS

These studies have alerted us to the fact that organic pigments in paint systems often fade at an increased rate as the relative humidity is increased, even in a relatively hydrophobic vehicle such as poly(vinyl acetate). This tendency must be considered in carrying out precise accelerated fading tests and in attempts

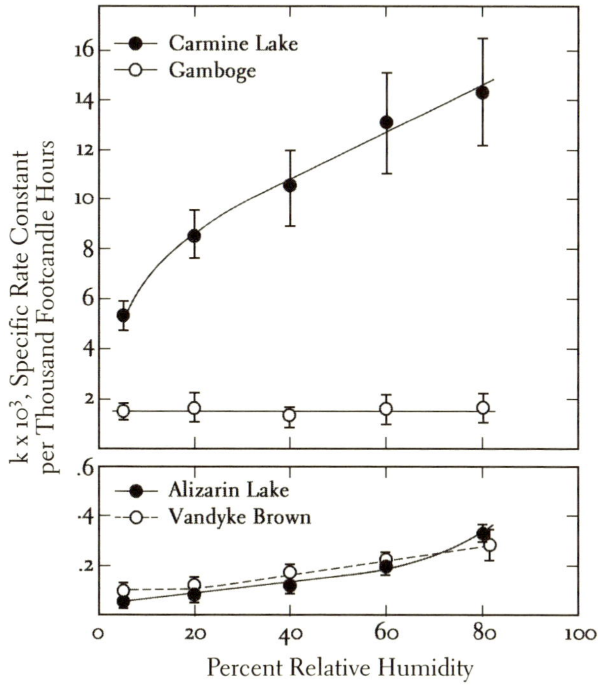

FIGURE 1
Effect of RH on rates of fading of various poly(vinyl acetate)-based paints.

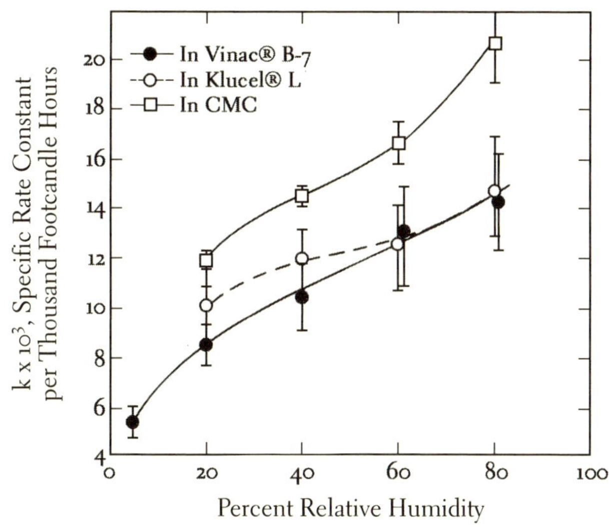

FIGURE 2
Rates of fading of carmine lake/TiO_2 tints in three different vehicles.

to evaluate the "characteristic" rate of fading of particular pigment/vehicle systems. In museums, the influence of relative humidity on the rate of fading of objects on exhibition is a factor that must be considered in instances where every effort is to be made to provide the ultimate in care and preservation of artifacts that may contain fugitive dyes and pigments.

Section 4: Pigment Volume Concentration and Chalking

The selective absorption of certain wavelengths of light underlies the phenomenon of color. Pigments also vary in their ability to scatter particular wavelengths. Hence, to account for the color of paints, the degree of scattering of the component pigments must be taken into consideration as well as their characteristic absorption. However, should the body of a paint or varnish or its surface physically disintegrate, the resultant increase in the scattering of light would lead to an increased lightness of the perceived color. Without a thorough analysis and understanding of the change thus brought about in the spectral reflectance data, the increased scattering could be mistaken for the loss of colorant molecules, the case of true fading. Over the years, a number of articles were prepared to draw attention to the effect on color arising from surface powdering (chalking) and other types of increased scattering in paint systems, and to the related concept of critical pigment volume concentration (CPVC). In the course of an internship at the Research Center, Noel Kunz, a student in the conservation training program at New York University, demonstrated that a time-honored observation, that paints formulated with animal glue or casein (distempers) tend to be lighter in color than those formulated in oil, was a phenomenon clearly explainable on the basis of pigment volume concentration and CPVC, and not upon differences in the refractive index of the vehicle.

R. L. F.

Critical Pigment Volume Concentration and Chalking in Paints

R. L. FELLER AND J. J. MATOUS

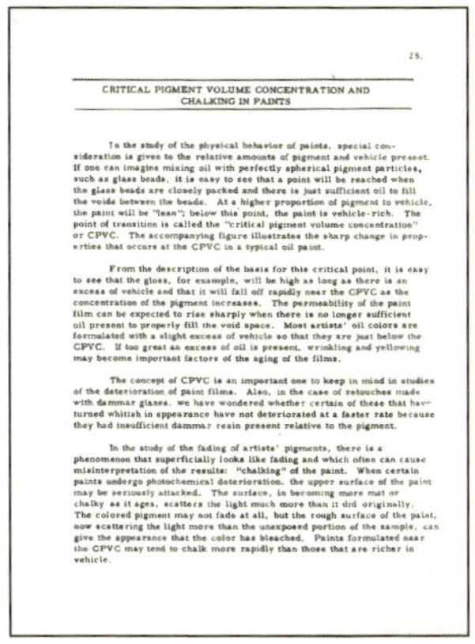

25.

CRITICAL PIGMENT VOLUME CONCENTRATION AND CHALKING IN PAINTS

In the study of the physical behavior of paints, special consideration is given to the relative amounts of pigment and vehicle present. If one can imagine mixing oil with perfectly spherical pigment particles, such as glass beads, it is easy to see that a point will be reached when the glass beads are closely packed and there is just sufficient oil to fill the voids between the beads. At a higher proportion of pigment to vehicle, the paint will be "lean"; below this point, the paint is vehicle-rich. The point of transition is called the "critical pigment volume concentration" or CPVC. The accompanying figure illustrates the sharp change in properties that occurs at the CPVC in a typical oil paint.

From the description of the basis for this critical point, it is easy to see that the gloss, for example, will be high as long as there is an excess of vehicle and that it will fall off rapidly near the CPVC as the concentration of the pigment increases. The permeability of the paint film can be expected to rise sharply when there is no longer sufficient oil present to properly fill the void space. Most artists' oil colors are formulated with a slight excess of vehicle so that they are just below the CPVC. If too great an excess of oil is present, wrinkling and yellowing may become important factors of the aging of the films.

The concept of CPVC is an important one to keep in mind in studies of the deterioration of paint films. Also, in the case of retouches made with dammar glazes, we have wondered whether certain of these that have turned whitish in appearance have not deteriorated at a faster rate because they had insufficient dammar resin present relative to the pigment.

In the study of the fading of artists' pigments, there is a phenomenon that superficially looks like fading and which often can cause misinterpretation of the results: "chalking" of the paint. When certain paints undergo photochemical deterioration, the upper surface of the paint may be seriously attacked. The surface, in becoming more mat or chalky as it ages, scatters the light much more than it did originally. The colored pigment may not fade at all, but the rough surface of the paint, now scattering the light more than the unexposed portion of the sample, can give the appearance that the color has bleached. Paints formulated near the CPVC may tend to chalk more rapidly than those that are richer in vehicle.

Originally printed in:
BULLETIN OF THE AMERICAN GROUP - IIC
Volume 5, Number 1 (1964), Pages 25-26.

In the study of the physical behavior of paints, special consideration is given to the relative amounts of pigment and vehicle present. If one can imagine mixing oil with perfectly spherical pigment particles, such as glass beads, it is easy to see that a point will be reached when the glass beads are closely packed and there is just sufficient oil to fill the voids between the beads. At a higher proportion of pigment to vehicle, the paint will be "lean"; below this point, the paint is vehicle-rich. The point of transition is called the "critical pigment volume concentration" or CPVC. The accompanying figure illustrates the sharp change in properties that occurs at the CPVC in a typical oil paint.

From the description of the basis for this critical point, it is easy to see that the gloss, for example, will be high as long as there is an excess of vehicle and that it will fall off rapidly near the CPVC as the concentration of the pigment increases. The permeability of the paint film can be expected to rise sharply when there is no longer sufficient oil present to properly fill the void space. Most artists' oil colors are formulated with a slight excess of vehicle so that they are just below the CPVC. If too great an excess of oil is present, wrinkling and yellowing may become important factors of the aging of the films.

The concept of CPVC is an important one to keep in mind in studies of the deterioration of paint films. Also, in the case of retouches made with dammar glazes, we have wondered whether certain of these that have turned whitish in appearance have not deteriorated at a faster rate because they had insufficient dammar resin present relative to the pigment.

In the study of the fading of artists' pigments, there is a phenomenon that superficially looks like fading and which often can cause misinterpretation of the results: "chalking" of the paint. When certain paints undergo photochemical deterioration, the upper surface of the paint may be seriously attacked. The surface, in becoming more matte or chalky as it ages, scatters the light much more than it did originally. The colored pigment may not fade at all, but the rough surface of the paint, now scattering the light more than the unexposed portion of the sample, can give the appearance that the color has bleached. Paints formulated near the CPVC may tend to chalk more rapidly than those that are richer in vehicle.

In studies on the fading of artists' pigments, chalking confused some of the early results. As an example, in Figure 22 of the article on the "Control of Deteriorating Effects of Light upon Museum Objects,"[1] there is a sample of ultramarine mixed with white casein paint. The figure is labeled as containing a number of fugitive watercolor pigments. We would not ordinarily consider ultramarine to be fugitive, but it happened to be used in this particular panel. The fact also may at first appear surprising that the ultramarine shows a considerable effect of light, particularly the wavelengths of ultraviolet radiation that are able to pass through window glass, those from about 310 to 400 millimicrons. This apparently is an example where the vehicle has chalked.

We have also noticed the "chalking" type of fading in a number of pigments mixed with white lead oil paint. White lead does not require much oil to make a paste.[2] For this reason, perhaps, paints mixed with white lead may be formulated close to the CPVC and may, therefore, not require much photochemical deterioration of the oil before chalking occurs. The chalking of paints prepared with certain types of anatase titanium white pigments, of

1. Feller, R. L., *Museum*, 17 (1964), p. 57-98.

2. White lead also screens out little ultraviolet radiation in comparison to zinc and titanium white. See Stutz, G. F. A., *J. Franklin Inst.*, 200 (1925), p. 87.

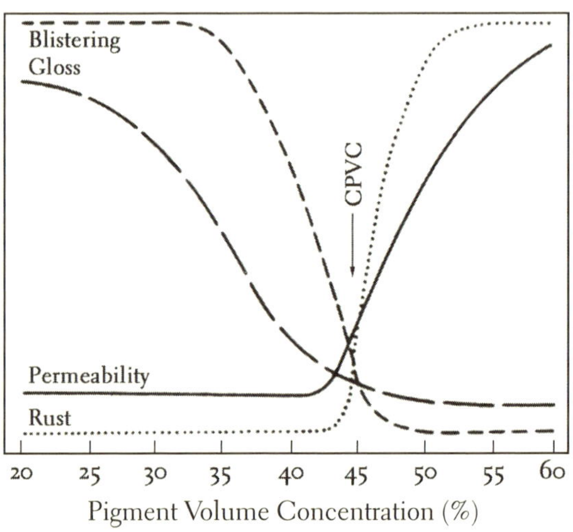

Figure 1
Appearance and performance as function of Pigment Volume Concentration.[3]

3. Asbeck, W. K., and Van Loo, M., "Critical Pigment Volume Relationships," *Ind. Eng. Chem.*, 41 (1949), p. 1470-1475.

course, is a phenomenon well-known to the field. The ultraviolet region of the spectrum seems to be particularly potent in causing chalking. Chalking can be distinguished from fading by an analysis of the change in the spectrophotometric reflectance curve before and after exposure.[4]

The concept of CPVC has been briefly discussed because it is a fundamental one to be considered in the behavior of paint. There has been considerable interest among colleagues in the appearance of various types of "blanching" in both modern oil paintings and those of historic interest. The possibility should not be overlooked that some of these difficulties are due to chalking of the vehicle. It is essential to consider this type of change in the study of the apparent and actual fading of artists' pigments.

4. Technique suggested by Johnston, R. M., and Feller, R. L., "The Use of Differential Spectral Curve Analysis in the Study of Museum Objects," *Dyestuffs*, 44, No. 9 (1963), p. 1-10.

Problems in Retouching: Chalking of Intermediate Layers

ROBERT L. FELLER

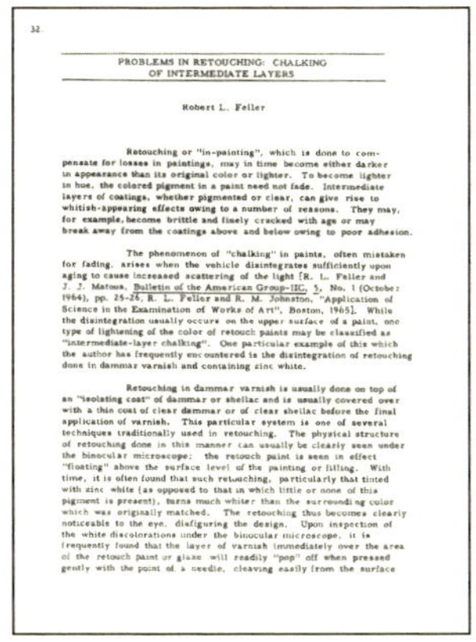
32

PROBLEMS IN RETOUCHING: CHALKING OF INTERMEDIATE LAYERS

Robert L. Feller

Originally printed in:
BULLETIN OF THE AMERICAN GROUP - IIC
Volume 7, Number 1 (1966), Pages 32-34.

Retouching or "in-painting", which is done to compensate for losses in paintings, may in time become either darker in appearance than its original color or lighter. To become lighter in hue, the colored pigment in a paint need not fade. Intermediate layers of coatings, whether pigmented or clear, can give rise to whitish-appearing effects owing to a number of reasons. They may, for example, become brittle and finely cracked with age or may break away from the coatings above and below owing to poor adhesion.

The phenomenon of "chalking" in paints, often mistaken for fading, arises when the vehicle disintegrates sufficiently upon aging to cause increased scattering of the light.[1] While the disintegration usually occurs on the upper surface of a paint, one type of lightening of the color of retouch paints may be classified as "intermediate-layer chalking." One particular example of this which the author has frequently encountered is the disintegration of retouching done in dammar varnish and containing zinc white.

Retouching in dammar varnish is usually done on top of an "isolating coat" of dammar or shellac and is usually covered over with a thin coat of clear dammar or of clear shellac before the final application of varnish. This particular system is one of several techniques traditionally used in retouching. The physical structure of retouching done in this manner can usually be clearly seen under the binocular microscope: the retouch paint is seen in effect "floating" above the surface level of the painting or filling. With time, it is often found that such retouching, particularly that tinted with zinc white (as opposed to that in which little or none of this pigment is present), turns much whiter than the

1. Feller , R. L., and Matous, J. J., *Bulletin of the American Group-IIC*, 5, No. 1 (October 1964), p. 25-26; Feller, R. L., and Johnston, R. M., in *Application of Science in the Examination of Works of Art*, Museum of Fine Arts, Boston, MA (1967), p. 86-95.

surrounding color which was originally matched. The retouching thus becomes clearly noticeable to the eye, disfiguring the design. Upon inspection of the white discolorations under the binocular microscope, it is frequently found that the layer of varnish immediately over the area of the retouch paint or glaze will readily "pop" off when pressed gently with the point of a needle, cleaving easily from the surface of the retouch paint.[2] In many cases, the retouch paint, thus exposed, is found no longer to be a continuous film but a fine powder that can be moved about with the point of a needle (see Figure 1).

2. "Paint" consists of pigment plus vehicle; in this case, dammar varnish is the vehicle.

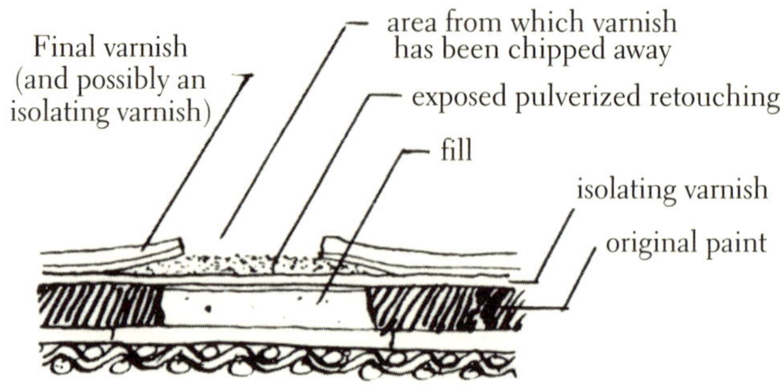

Figure 1

The whitish appearance in such a case is due to the fact that the retouch paint has broken up into a finely fractured layer which scatters the light. The whitishness is not due to fading. This can be readily demonstrated by the fact that the color will return almost to an exact match with the surrounding area if the powdered retouch, exposed by the chipping off of the upper layer of varnish, is wet with a drop of hexane or related petroleum hydrocarbon. Since this will usually not dissolve the pulverized retouch paint, the whitish appearance quickly returns when the hydrocarbon evaporates.

In an extensive series of test panels prepared under the direction of Mr. Mario Modestini some years ago, it was clearly demonstrated (a) that exposure to light brought about the whitish discolorations, (b) that near-ultraviolet radiation had a particularly potent effect, and (c) that the combination of dammar varnish and zinc white was especially prone to deteriorate in this manner. To a lesser degree, titanium white may lead to a similar disintegration. The trouble did not occur, however, when white lead was used. When whitish-discolored retouches were examined in a number of paintings, the presence of zinc white was found in nearly all cases.

An explanation of this particular type of deterioration may be based partly on the fact that dammar is a readily oxidizable and fundamentally brittle material, so much so that the retouch paint hardens and eventually fractures as it ages, leaving a virtual cloud of disintegrated retouch paint "floating" between the final varnish and the coatings below. Zinc white may also hasten this disintegration by speeding up the oxidation of the varnish. It is known, for example that zinc white can absorb near-ultraviolet radiation and can exhibit a chalking effect on oil paints.[3] Zinc oxide is basic and may perhaps interact with the acids in dammar in some way that further enhances the tendency of the films to disintegrate.[4]

Our study and solution of the problem of whitened retouches has been referred to elsewhere.[5] This particular form of deterioration can be avoided by eliminating the use of zinc white and the essentially brittle and easily oxidizable dammar varnish in retouching. When polyvinylacetate was substituted for dammar in Mr. Modestini's series of test panels, the retouches showed practically no tendency to undergo this type of deterioration, undoubtedly because this vehicle provided flexible oxidation-resistant films.

The "whitening" discoloration described, largely associated with retouching done in dammar varnish in the presence of zinc white, represents a particular type of "intermediate-layer chalking." Near-ultraviolet radiation hastens the deterioration; therefore, it undoubtedly can occur under certain types of fluorescent lamplight as well as daylight. This interesting form of chalking provides one more example of photochemical deterioration that may occur on the walls of a museum.

3. Ruben et al., *J. Am. Chem. Soc.*, 75 (1953), p. 2850; Winter, G., and Whittem, R. N., *J. Oil and Colour Chemists' Assoc.*, 33 (1950), p. 477 (see also p. 471, 484); Clay, C. F., *ibid.*, 40 (1957), p. 935.

4. Daniels, R. G., *Nitrocellulose Lacquer Manufacture*, Leonard Hill, London (1933), p. 37, mentions that dammar varnish can "feed" on zinc oxide when the two are ground together.

5. Walker, J., *National Gallery of Art, Washington, D. C.*, Harry M. Abrams, Incorporated, New York (1964), p. 55.

Optics of Paint Films: Glazes and Chalking

RUTH M. JOHNSTON AND ROBERT L. FELLER

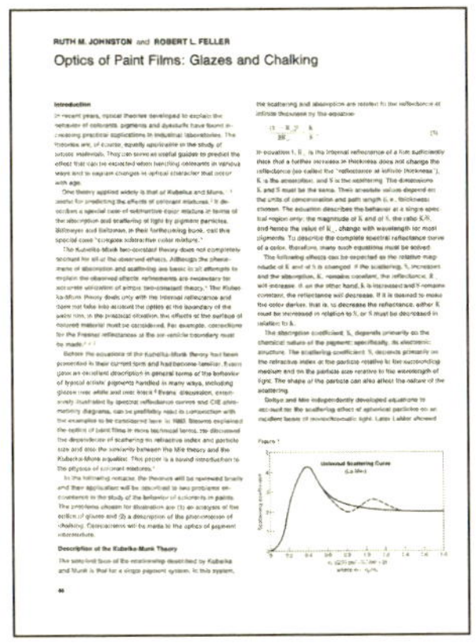
RUTH M. JOHNSTON and ROBERT L. FELLER

Optics of Paint Films: Glazes and Chalking

Originally printed in:
Application of Science in the Examination of Works of Art
Museum of Fine Arts, Boston, MA (1967), Pages 86-95.

Introduction

In recent years, optical theories developed to explain the behavior of colorants, pigments and dyestuffs have found increasing practical applications in industrial laboratories. The theories are, of course, equally applicable in the study of artists' materials. They can serve as useful guides to predict the effect that can be expected when handling colorants in various ways and to explain changes in optical character that occur with age.

One theory applied widely is that of Kubelka and Munk,[1, 2] useful for predicting the effects of colorant mixtures.[3] It describes a special case of subtractive color mixture in terms of the absorption and scattering of light by pigment particles. Billmeyer and Saltzman, in their forthcoming book, call this special case "complex subtractive color mixture."

The Kubelka-Munk two-constant theory does not completely account for all of the observed effects. Although the phenomena of absorption and scattering are basic in all attempts to explain the observed effects, refinements are necessary for accurate utilization of simple two-constant theory.[4] The Kubelka-Munk theory deals only with the internal reflectance and does not take into account the optics at the boundary of the paint film. In the practical situation, the effects of the surface of colored material must be considered. For example, corrections for the Fresnel reflectances at the air-vehicle boundary must be made.[5-7]

1. Kubelka, P., and Munk, F., "Ein Beitrag zur Optik der Farbanstriche," *Z. Tech. Physik.*, 12 (1931), p. 593.

2. Kubelka, P., "New Contributions to the Optics of Intensely Light-Scattering Materials, Part 1," *J. Optical Soc. Am.*, 38 (1948), p. 38.

3. Duncan, D. R., "The Identification and Estimation of Pigments in Pigmented Compositions by Reflectance Spectrophotometry," *J. Oil and Colour Chemists' Assoc.*, 45 (1962), p. 300-324.

Ed. Note: reference in text to Billmeyer, F. W., Jr., and Saltzman, M., *Principles of Color Technology*, Second Edition, John Wiley & Sons, NY (1981).

Before the equations of the Kubelka-Munk theory had been presented in their current form and had become familiar, Evans gave an excellent description in general terms of the behavior of typical artists' pigments handled in many ways, including glazes over white and over black.[8] Evans' discussion, extensively illustrated by spectral reflectance curves and CIE chromaticity diagrams, can be profitably read in conjunction with the examples to be considered here. In 1953, Stearns explained the optics of paint films in more technical terms. He discussed the dependence of scattering on refractive index and particle size and also the similarity between the Mie theory and the Kubelka-Munk equation. This paper is a sound introduction to the physics of colorant mixtures.[9]

In the following remarks, the theories will be reviewed briefly and their application will be described to two problems encountered in the study of the behavior of colorants in paints. The problems chosen for illustration are (1) an analysis of the optics of glazes and (2) a description of the phenomenon of chalking. Comparisons will be made to the optics of pigment intermixture.

DESCRIPTION OF THE KUBELKA-MUNK THEORY

The simplest form of the relationship described by Kubelka and Munk is that for a single pigment system. In this system, the scattering and absorption are related to the reflectance at infinite thickness by the equation:

$$\frac{(1 - R_\infty)^2}{2R_\infty} = \frac{K}{S} \qquad (1)$$

In equation 1, R_∞ is the internal reflectance of a film sufficiently thick that a further increase in thickness does not change the reflectance (so-called the "reflectance at infinite thickness"), K is the absorption, and S is the scattering. The dimensions K and S must be the same. Their absolute values depend on the units of concentration and path length (i. e., thickness) chosen. The equation describes the behavior at a single spectral region only; the magnitude of K and of S, the ratio K/S, and hence the value of R_∞, change with wavelength for most pigments. To describe the complete spectral reflectance curve of a color, therefore, many such equations must be solved.

The following effects can be expected as the relative magnitude of K and of S is changed: if the scattering, S, increases and the absorption, K, remains constant, the reflectance, R_∞, will increase. If, on the other hand, K is increased and S remains constant, the reflectance will decrease. If it is desired to make the color darker, that is, to decrease the reflectance, either K must be increased in relation to S, or S must be decreased in relation to K .

The absorption coefficient, K, depends primarily on the chemical nature of the pigment; specifically, its electronic structure. The scattering coefficient, S, depends primarily on the refractive index of the particle relative to the surrounding medium and on the particle size relative to the wavelength of

4. Duntley, S. Q., "The Optical Properties of Diffusing Materials," *J. Opt, Soc. Am.*, 32 (1942), p. 61-70.

5. Ryde, J. W., "Scattering of Light by Turbid Media - Part 1," *Proc. Roy. Soc.*, A 131 (1931), p. 451-464.

6. Saunderson, J. L., "Calculation of the Color of Pigmented Plastics, " *J. Opt. Soc. Am.*, 32 (1942), p. 727-736.

7. Walsh, J. W. T., "The Reflection Factor of a Polished Glass Surface for Diffused Light," appendix of Taylor, A. K., and Grieveson, C. J. W., "The Transmission Factor of Commercial Window Glasses," *Department of Scientific and Industrial Research, Illumination Research Technical Paper No.* 2 (1926).

8. Evans, R. M., *An Introduction to Color*, J. Wiley & Sons, New York (1948), p. 278-294.

9. Stearns, E. I., "Principles Underlying the Color and Appearance of Coatings," *Off. Dig. Federation of Soc. Paint Technol.*, No. 336 (1953), p. 55-69.

light. The shape of the particle can also affect the nature of the scattering.

Debye and Mie independently developed equations to account for the scattering effect of spherical particles on an incident beam of monochromatic light. Later LaMer showed that the scattering coefficients of a number of materials could be plotted on a single curve, in which the coefficient is expressed as a function of the diameter of the particles, D, the wavelength of light, λ, and the refractive indices of the particles (n_p) and the medium in which they are dispersed (n_v).[10, 11] This "universal scattering curve" for spherical particles is shown in Figure 1. Based on this concept, Stearns calculated a curve to predict the optimum size of a spherical pigment that would give the maximum scattering when immersed in a vehicle of refractive index 1.53.[9] It is this line of reasoning that may be used to explain why traditional pigments of low refractive index, such as smalt, develop their maximum effectiveness in oil paints if they are coarsely ground.

10. Stratton, J. A., and Houghton, H. G., "A Theoretical Investigation of the Transmission of Light through Fog," *Phys. Rev.*, 38 (1931), p. 159-165.

11. LaMer, V. K., et al., *OSRD Report* 4904, PB 32,208, U.S. Dept. of Commerce (1945).

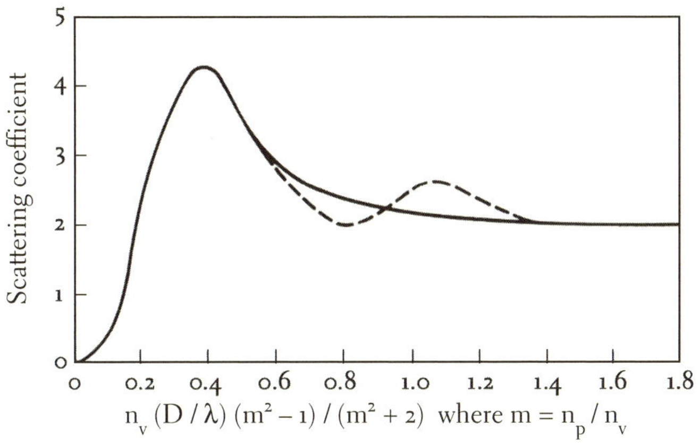

FIGURE 1
Universal scattering curve. (LaMer, Ref. 11)

When mixtures of pigments are made, the resulting reflectance at infinite thickness is described by the contribution of the sum of the individual absorption terms divided by the sum of the individual scattering terms:

$$\frac{(1 - R_\infty)^2}{2R_\infty} = \frac{C_1K_1 + C_2K_2 + \ldots}{C_1S_1 + C_2S_2 + \ldots} \qquad (2)$$

In this equation, the C's refer to the concentration of individual pigments, identified by subscripts. Scattering depends on the size of a particle relative to the wavelength of light and on the number of particles encountered by the beam of light. Therefore, strictly speaking, the concentrations must be given in volume fractions. For approximation, weight fractions are frequently used, but it must be remembered that this may introduce errors, particularly when high-density pigments are involved.

When the film thickness is not sufficiently great to hide the substrate, or when a transparent or translucent layer is applied over a substrate, the reflectance characteristics of the substrate or the "ground" which shows through must be taken into account. The use of a transparent or translucent layer over an opaque substrate is basically the technique of the painter known as "glazing." A form of the Kubelka-Munk equation that describes the relationship of the various factors under these circumstances is:

$$R = \frac{1 - R_g\,(a - b\ \mathrm{ctgh}\ bSX)}{a - R_g + b\ \mathrm{ctgh}\ bSX} \qquad (3)$$

In equation 3, R is the resulting internal reflectance of the combination of the upper layer and the substrate, R_g is the reflectance of the substrate, X is the factor relating to the thickness of the upper layer and the colorant concentration, "a" is a constant derived from the relationship of K and S in the upper layer (a is $\frac{KX + SX}{SX}$), b is a constant related to "a" ($b = \sqrt{a^2 - 1}$), and ctgh is the abbreviation for hyperbolic cotangent. Again, the equation applies to only one wavelength.

The solution of many such equations in order to describe the spectral reflectance curve of a color may seem formidable, but it can be done. Modern computing aids are, of course, a great help.

The effects obtained in glazes may be described in general terms. If the substrate reflects the light falling on it (if it is white, for example, where reflectance R_g is large), most of the light that falls on the substrate is reflected back with the exception of that which is absorbed by the colorant in the upper or glaze layer above the substrate. Under these circumstances, if the scattering in the upper layer is small, that is if the layer is transparent, the absorption will not be offset by scattering and the absorption curve of the upper layer will have the predominant effect; the resultant color will be brilliant, a color high in chroma.[12] On the other hand, if the substrate absorbs the light striking it (is black, for example), the only light reflected will be that which is scattered back by the upper or "glaze" layer before the light reaches the substrate. The scattering properties of the pigment in the upper layer thus become apparent, and the color is frequently milky. These effects will be illustrated in detail below. In order to use the Kubelka-Munk equations to predict the behavior of pigments, it is necessary to determine the constants K and S. This can be done by preparing films of known composition, measuring the suitable reflectances, and solving for K and S at a sufficient number of wavelengths to describe the curve. From these values, then, the reflectance curves of masstones, of pigment mixtures, and of glazes may be predicted. The equations given here, and a number of others suitable for determining K and S, may be found in Judd.[13]

12. The three attributes of color according to the terminology of the Munsell system are hue, value and chroma. Other designations of the attribute of chroma are "saturation" or "purity."

13. Judd, D. B., *Color in Business, Science and Industry*, J. Wiley & Sons, New York (1952); Judd, D. B., and Wyszecki, G. (second edition) (1963).

Description of Glazes

Glazes may be of two basic types. They may be transparent, in which case the scattering is very low in relation to the absorption—the result is similar in effect to a dyed filter—or they may be translucent, in which case the scattering is large in relation to the absorption—equivalent to placing a diffusing filter over the substrate. A completely transparent film of any color applied over black appears black. A nonabsorbing translucent, or diffusing, film applied over white appears white.

Colors of high chroma result when one region of the spectral reflectance curve has very low reflectance, and another region has very high reflectance. Such an effect is obtained with glazes over white. The chroma or brilliance achieved by this means is greater than that achievable by a mixture of the colored pigment with white pigment.[8] The striking effects of glazes, particularly this aspect of the optics of the films, were well understood and appreciated by the painters of the past who explored the technique.

In contrast, when a colorant is *mixed* with white pigment, the high scattering of the white pigment over all of the spectrum tends to increase the reflectance in the region of low reflectance of the colored pigment, i.e., in the region of high absorption. The mixture with white therefore decreases the contrast between the regions of low and high reflectance and results in decreased chroma. It may be noted in equation 2 that the introduction of a considerable scattering term in the denominator will increase the value of the reflectance, R_∞.

When pigments of high scattering ability are applied as glazes over dark substrates, the effect is the opposite of that of low-scattering glazes over white—low chroma colors result, often of surprising spectral character, eerie and foggy.

All of these effects that have been described in a general way are predictable in terms of the Kubelka-Munk two-constant theory relating the absorption and the scattering to the reflectance.

To illustrate the effects of glazes, four chromatic pigments were studied over white and over black substrates. Pairs of yellows and reds were selected to illustrate different degrees of transparency—light chrome yellow, relatively opaque; flavanthrone yellow, illustrating a transparent yellow; cadmium red, relatively opaque; and quinacridone red, illustrating a transparent red. The effects of a nonabsorbing, highly scattering white over black and of a nonscattering, highly absorbing black over white will also be illustrated.

Each pigment was applied in four different circumstances:

(a) in masstone, that is at complete hiding without any other pigment mixed with it,

(b) as a glaze over white, to illustrate its use as a transparent glaze,

(c) as a glaze over black, to illustrate its translucency or scattering, and

(d) in mixture with white.

The reflectance curves of the panels of chrome yellow prepared in these four ways are illustrated in Figure 2. Note that the color of this pigment applied as a glaze over the black substrate is green, the result of its scattering characteristics. Because the reflectance over black is low, it was measured and appears in Figure 2 at five times (5x) the actual reflectance. In the other figures, the reflectances over black were all measured on the 5x scale.

The spectral reflectance curves of the second yellow, flavanthrone, are given in Figure 3. This is a more transparent pigment than the chrome yellow; i.e., it has a higher ratio of absorption to scattering. Its masstone color is unattractive, the result of its low scattering properties. The fact that only a faint green color is visible when it is glazed over black further illustrates the effect of its low scattering ability. In contrast, the color as a glaze over white is brilliant. In mixture with white, the color is also high in chroma (saturation).

In Figure 4, the absorption (K) and scattering (S) curves of the two yellows are illustrated. As can be seen, the flavanthrone has higher absorption (and hence greater tinting strength) and lower scattering than the chrome yellow. The explanation for their appearance as greens of different intensity and purity when glazed over black can be seen in the character of their scattering curves. In the absorption and scattering curves shown in Figures 4 and 7, portions of the curves are shown by dotted lines because, owing to the experimental error in the measurement, they are known only in an approximate way in these regions of the spectrum.

Panels treating the two reds were prepared in the same way. The reflectance curves of these are shown in Figures 5 and 6. Figure 7 gives the curves of their scattering and absorption coefficients. The higher absorption of the quinacridone in the green region and lower scattering in the red region compared to the cadmium red can be seen. Because of its scattering ability, the color of the cadmium red is still visible when a glaze is applied over the black. In contrast, the color of the quinacridone red applied over black is almost invisible, clearly illustrating its higher absorption-to-scattering ratio.

The chromaticity of each of these panels is plotted on the CIE chromaticity diagram in Figure 8. It can be seen that the glaze of the low-scattering flavanthrone over black is a green of lower purity than that of the high-scattering chrome yellow. Similarly, the low-scattering quinacridone glaze over black is lower in purity than that of the higher-scattering cadmium red over black. It is also apparent that, in every case, the glaze over white is higher in purity (saturation) than the mixture with white.

The black and white pigments were treated in an analogous way. The white pigment used was rutile titanium dioxide and the black, carbon black. Figure 9 illustrates the spectral reflectance curves of three panels: white over black, black over white, and black and white mixed. A distinct difference is noticed between the color of the glaze of black over white, which is yellowish, and the mixture of black and white, which is bluish. The yellowish color can be

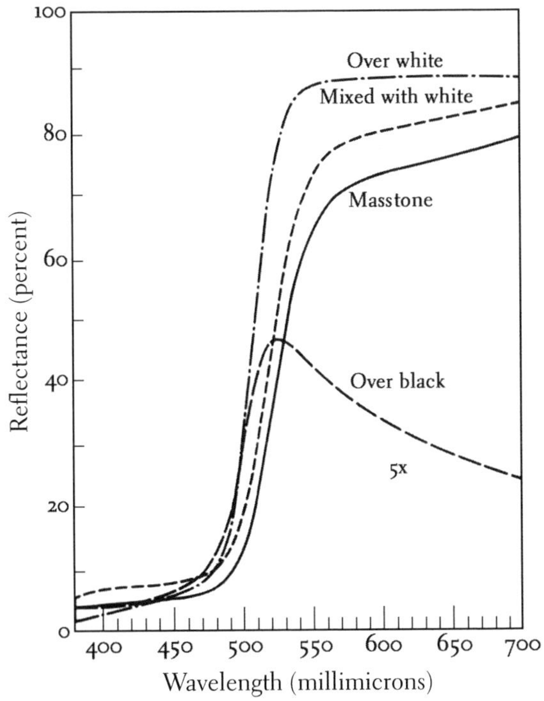

FIGURE 2
Regal Chrome Yellow LT; Imperial X-2777.

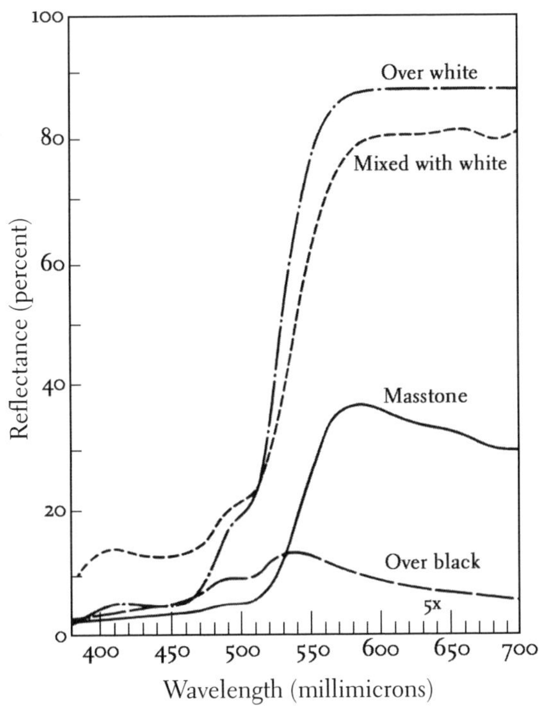

FIGURE 3
Flavanthrone Yellow; Harmon Y-5713.

predicted from the absorption curve of the black. This is shown in Figure 10, where the absorption and scattering curves of the pigments are given. It also can be clearly seen that the absorption coefficient of the black and the scattering coefficient of the white are much higher than the related coefficients of the colored pigments. In other words, a black pigment is an especially high absorber and a white, a high scatterer of light. As can be seen in Figure 10, the titanium dioxide has a higher-scattering coefficient in the blue region relative to the red, long wavelength region. Rutile titanium dioxide is designed to do this; its particle size is controlled to exhibit maximum scattering in the blue region to offset its yellow color It is this greater scattering of the titanium white pigment in the blue wavelengths that is responsible for the bluish cast of the mixture of black and white. A scattering curve for black is not obtainable.

The characteristic effects of the glazes and intermixtures are well known. When working with different pigments, most people, in fact, learn their general characteristics very quickly. Nevertheless, appreciation of the role of the absorption and scattering characteristics of pigments is essential if one hopes to understand and to predict the results that may be expected when pigments are handled in different ways.

Description of the Pigments Used

The chrome yellow selected was Imperial Chemical Company's Regal Chrome Yellow Light, X-2777. The flavanthrone yellow was Harmon Colors' Indofast Yellow Lake, Y-5713, an organic colorant. Because of its transparency, its high tinctorial strength, and its high chroma in dilute clear application, it is a beautiful yellow to use as a glaze over a light-reflecting background or in mixture with white. It is not used commercially in masstone.

The reds selected were cadmium red medium and quinacridone red. The latter is an organic colorant, transparent, tinctorially strong, and of good light stability. DuPont Monastral Red, RT-796-D, was the quinacridone used. Cadmium red is a somewhat opaque pigment, commercially used primarily in masstone and not in mixture with white. Harshaw's No. 100 was used, a cadmium lithopone. The white was duPont's rutile titanium dioxide, R-610; the black, Cabot's carbon black, Carbolac #2.

Chrome yellow is usually considered not to be of the highest lightfastness suitable for artists' use. With the rapid advances being made in pigment and dyestuff technology today, however, classifying the light stability of a colorant on the basis of chemical type is not sufficient. Regal Chrome Yellow, X-2777, is a case in point. It is one of the most lightfast pigments on the market. At a concentration of 20% X-2777 by weight with 80% titanium dioxide in good acrylic vehicle, the average color change of many panels exposed for one year at various times in Florida has been found to be only 2.2 Δe (MacAdam)[14] or about 1 ΔE (NBS). This is an incredibly small color change. Other light chrome

14. MacAdam, D. L., "Experimental No. 6 spacing"; see Simon, F. T., and Goodwin, W. J., "Rapid Graphical Computation of Small Color Differences," Circulation Section-Advertising Department, Union Carbide Plastics Company, 270 Park Avenue, New York. N.Y., 10017.

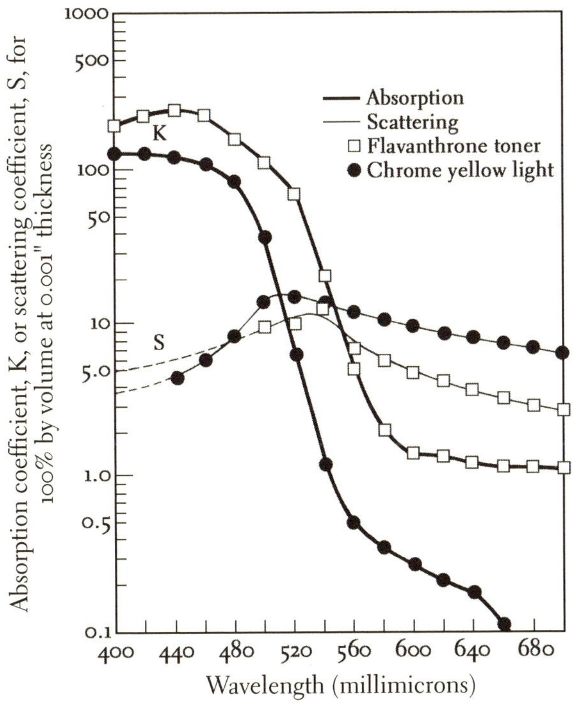

Figure 4
Absorption and scattering coefficients of yellow pigments.

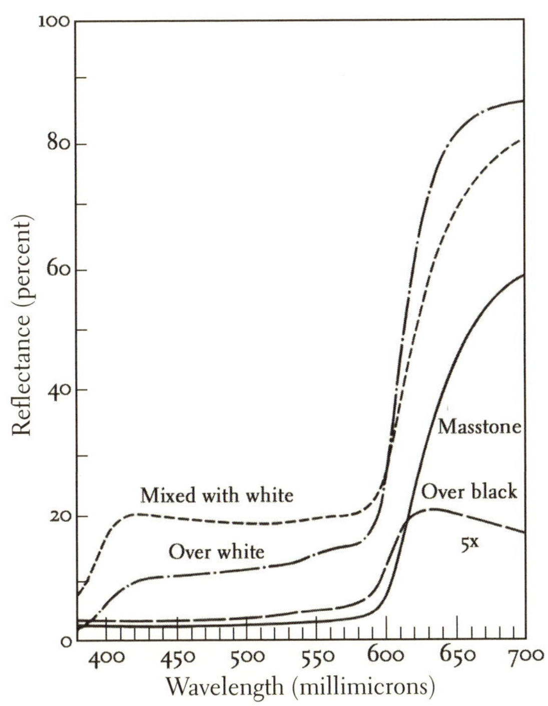

Figure 5
Cadmium Red Medium; Harshaw No. 100.

yellows often show ten times this amount of change for a comparable exposure. Obviously, one cannot discuss the light stability of pigment types per se. It is necessary to be specific and to describe the exact proprietary identification of the pigment concerned.

Flavanthrone yellow has not yet found extensive use in the artists' palette. Its light stability is also good, even when mixed with white. After one year of Florida exposure, a mixture of 20% of the lake (equivalent to 6.1% of toner) with 80% titanium dioxide in an acrylic lacquer exhibited a net color change of only 7.6 Δe (MacAdam), for example. This degree of lightfastness is far better than most Hansa yellows.

The color change observed for the quinacridone pigment, duPont's Monastral Red, RT-796-D, after one year's exposure in Florida, was only 7 Δe (MacAdam) at a concentration of 20% in mixture with 80% titanium dioxide. This pigment, possessing high tinctorial strength, high transparency, and excellent stability to light, is currently being offered to artists by a number of manufacturers of artists' materials. In contrast, the cadmium red pigment, considered to be a reasonably "permanent" artists' pigment,[15] changed 20 Δe (MacAdam) after one year's Florida exposure in an automotive acrylic lacquer at 20% concentration in mixture with 80% titanium white. To provide a further comparison regarding the stability of these pigments, it may be recalled that Vesce has reported that madder lake can change about 17 to 28 ΔE (NBS) (about 35 to 65 Δe, MacAdam) in only three months' exposure in Florida when prepared at a pigment/titanium white ratio of 25/75.[16]

15. "Artists' Oil Paints," *Commercial Standard CS 98-62*, U.S. Dept. of Commerce (1962).

16. Vesce, V. C., "Exposure Studies of Organic Pigments in Paint Systems," *Off. Dig. Federation of Soc. Paint Technol.*, 31, No. 419, Part 2 (1959), p. 3-143.

Chalking

Many other applications of the Kubelka-Munk theory could be illustrated, such as fading, vehicle change, pigment analysis, etc.[17] One phenomenon which has not received much attention in studies of the deterioration of artists' paints is that of chalking. The importance of this type of failure in the study of fading has been pointed out.[18] In general, the phenomenon of chalking may be described as an increase in scattering at the top surface of the film. This is usually caused by the deterioration of the vehicle with age. When the deterioration is advanced, a whitish powder of pigment and brittle vehicle particles can usually be rubbed from the surface, hence the term "chalking." Jacobsen has described details concerning the mechanisms of chalking caused by titanium white pigments.[19]

The effect of chalking on color can be predicted qualitatively by studying the optics of surface reflectances. A smooth glossy upper surface does not scatter the incident light, but reflects a portion specularly, that is by mirror-like or Fresnel reflection. In fresh paint, for example, there is usually a concentration of vehicle at the surface and no particles protrude to scatter the light. The amount reflected from such a surface depends on the refractive index of the

17. Johnston, R. M., and Feller, R. L., "The Use of Differential Spectral Curve Analysis in the Study of Museum Objects," *Dyestuffs*, 44 (1963), p. 277-286.

18. Feller, R. L., and Matous, J. J., "Critical Pigment Volume Concentration and Chalking in Paints, " *Bulletin Amer. Group-II C*, 5, No. 1 (1964), p. 25-26.

19. Jacobsen, A. E., "Chalking of Titanium Dioxide Pigments," *Paint Varnish Prod.*, 55, No. 2 (1965), p. 29-30, 32, 35 and 38.

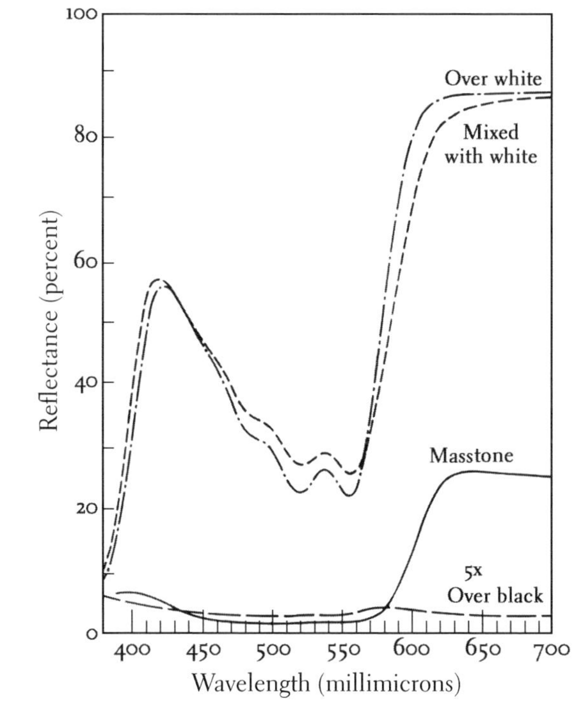

Figure 6
Quinacridone Red; Du Pont RT-796.

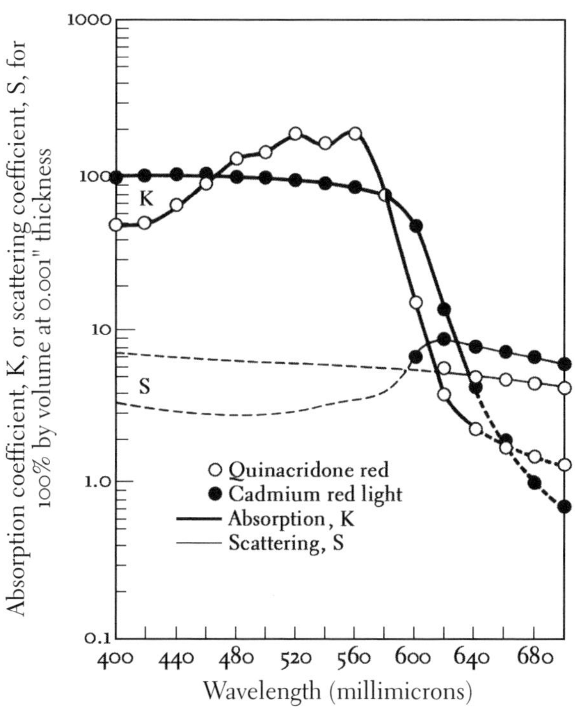

Figure 7
Absorption and scattering coefficients of red pigments.

vehicle, the angle of incidence, and, of course, on the nature of the surface. For a vehicle of refractive index 1.5, the amount lost by specular reflection at near-normal incidence is about 4% of the incident light. By reflection, therefore, some of the light incident at the surface is lost and does not enter the film.

When the light which has entered the film is reflected back and reaches the underside of the surface boundary, some is reflected back into the film again by mirror reflection. The light from within is usually scattered, not a parallel beam. It may arrive at the surface at a low angle of incidence, and, therefore, the amount reflected back into the film can be very large, as much as 60%, for example. This phenomenon helps to explain why glossy paints have higher chroma and appear darker than paints with a rough or "matte" surface.

The equation describing the effect of surface reflectance is as follows:

$$R' = k_1 + \frac{(1 - k_1)(1 - k_2) R}{1 - k_2 R} \qquad (4)$$

where R' is the total reflectance measured, R is the internal reflectance, k_1 is the reflectance loss of the incident beam at the first air-paint boundary, and k_2 is the reflectance loss of the internal light at the paint-air boundary.[5-7]

When the smooth coating becomes rough, scattering occurs at the surface, and the gloss is decreased. The light scattered at the surface is additively mixed with the light scattered from the pigment particles inside the film, with the result that the color is lightened and the chroma or "purity" is decreased. The effect is similar to that which occurs when the surface of a shiny colored object is badly scratched or abraded. The effect on the spectrophotometric curve is that of adding a certain amount of reflectance to all wavelengths, i.e., raising the reflectance curve by a constant amount throughout the spectrum.

It may be recalled that the Kubelka-Munk equation is a special case of subtractive color mixture which may be called "complex subtractive mixture," where absorption and scattering both have to be taken into account. In fact, when scattering approaches zero, the Kubelka-Munk equation reduces to the equation normally used to describe simple subtractive mixture, the Beer-Lambert equation, $T_i = e^{-CKX}$. In *additive* color mixture, the reflectances are additive; so that the resultant reflectance, R, is given by:

$$R = C_1 R_1 + C_2 R_2 + C_3 R_3 + \dots, \qquad (5)$$

where the C's are the concentration in terms of the relative areas (or intensities, in the case of lights), and the subscripts refer to the specific different reflecting surfaces (or lights). The purpose in reviewing the different kinds of color mixture briefly is to point out that, when scattering occurs at the surface, the scattered light is generally mixed additively with the color. Chalking presents an example of the effect of additive mixture. On the other hand, when scattering occurs

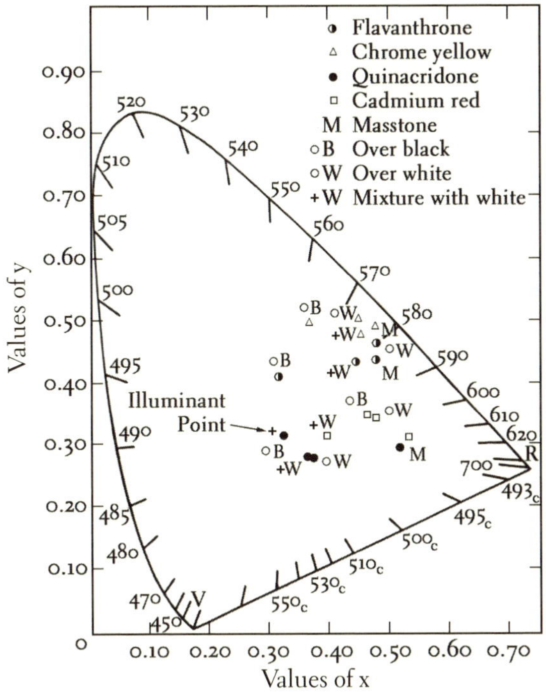

Figure 8
Chromaticity diagram for Illuminant C.

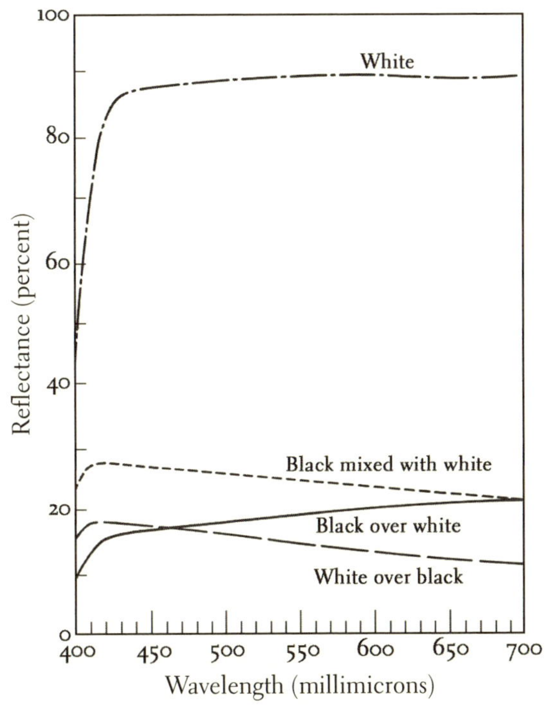

Figure 9
Black: Cabot Carbolac No. 2; White: Du Pont Titanium White R-610.

inside the film, the scattered light is mixed subtractively, according to the Kubelka-Munk prediction. Thus, both glazes and pigment intermixture are examples of subtractive mixture.

The phenomenon of chalking has been described above as an increase in scattering at the film surface, with the resultant effect that the color is lightened and the chroma decreased. As a consequence, this type of film deterioration is frequently confused with fading. Actually, the pigment may not have changed at all; only the optical effects at the surface may have changed.

Whether the loss in color is due to chalking or fading, the net visual result may, of course, be just as objectionable. But there are important differences to be considered when evaluating the light stability of pigment. By scattering the light, a chalked surface may effectively protect the pigment underneath from further ravages of light, so that when the chalk layer is removed, the appearance of the paint is once again close to that of the original.

An increase in the scattering of a paint film may come about through a variety of means: powdering of the vehicle, exposure of the pigment by degradation of the vehicle, separation between vehicle layers if they exist, etc. The color of a paint that has chalked may thus change in a variety of ways, all of which complicate attempts to study the chalking effect in a quantitative manner. When it is remembered that, at the same time, the pigment may be fading and the vehicle may be yellowing, it can readily be seen that the measurement and analysis of each separate effect is indeed difficult.

The equations and concepts given above nevertheless permit some of the general effects of chalking on the measured spectral curves to be predicted semi-quantitatively. When gloss is decreased by surface scattering, equation 4 may be used to predict the increase in reflectance and decrease in chroma. When a layer of scattering material of low absorbance is added as a glaze over the surface, equation 3 will describe the resultant effect. As previously pointed out, equation 2 can be used to determine the magnitude of vehicle yellowing and the amount of pigment fading.[17]

To illustrate the result to be expected from increased scattering at the surface, the two curves in Figure 11 are given. The lower is the spectral reflectance curve of a mixture of titanium white and cadmium red, measured on the General Electric recording spectrophotometer with the specular component of the reflectance excluded. This is done by placing a black surface in the removable ports, shown in the schematic diagram in Figure 12. The lower curve in Figure 11 essentially describes the reflectance of the pigments inside the film. The upper curve is the same paint measured with the specular component included. The reflectance in this case is higher than the lower curve by almost 4% because this amount of light, lost by specular reflection from the top surface of the paint, has been included in the measurement of "total reflectance." If the roughness of the surface of this paint were increased, one would find that the difference between the two curves, that is the difference

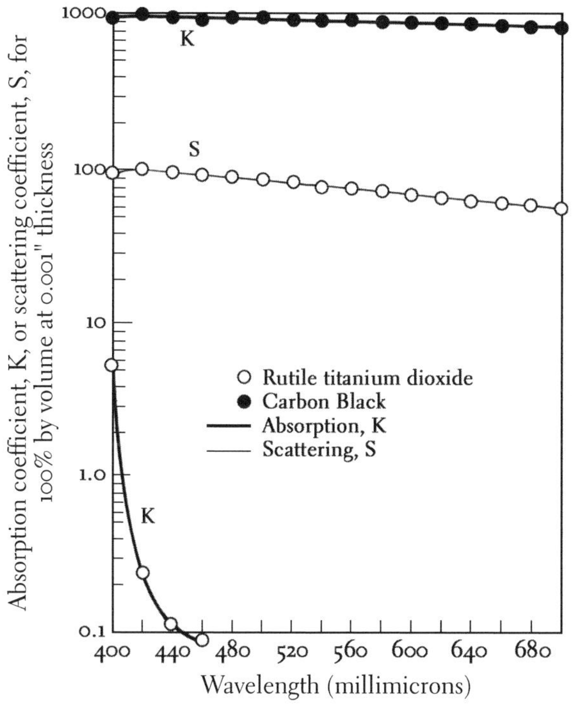

Figure 10
Absorption and scattering coefficients of black and white pigments.

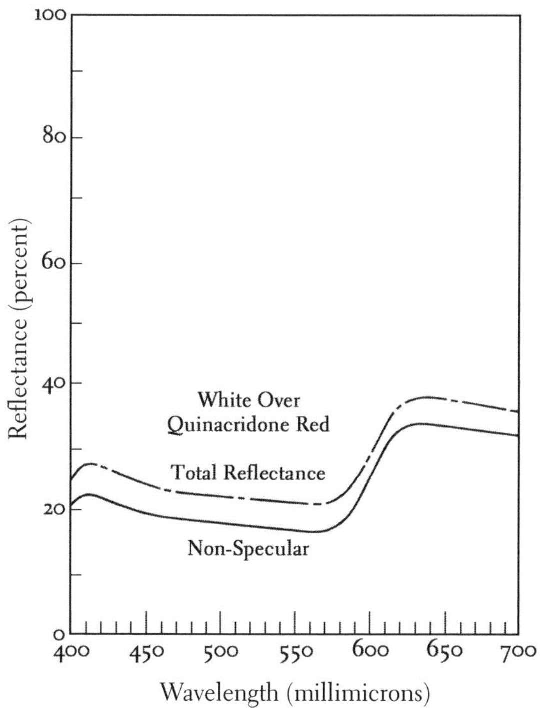

Figure 11
Effect of surface scattering.

between the "total reflectance" and the "nonspecular reflectance" curves, would become less and less as the glossiness decreased.

This example illustrates the change in optical properties of a film, in this case an increase in reflectance, that may be expected if a paint that possesses gloss (represented by the lower curve) later develops extensive scattering on the surface and comes to resemble the "total reflectance" curve above. An important point to note is that the percentage increase in reflectance is about the same regardless of wavelength. Such an increase is an example of what may be expected from chalking that is caused only by the development of increased roughness of the top surface of a paint film. Chalking may also result in selective scattering, depending probably on the scattering character of the exposed particles. Thus, there are at least two different ways that the phenomenon of chalking may affect color. In one case, the percentage increase in reflectance is much the same at all wavelengths; in the other, the increase is greater at certain wavelengths depending upon the variation of the scattering and absorption coefficients of the pigment exposed.

It is obvious that chalking, if not recognized, can interfere with the quantitative study of fading, particularly under accelerated aging conditions.[18]

EXPERIMENTAL

The reflectance measurements were made with a General Electric Recording Spectrophotometer equipped with a digital integrator. The reflectance curves illustrated were measured with the specular reflectance excluded, except for one curve in Figure 11. Pressed barium sulphate was used as the white standard.

The Kubelka-Munk calculations for scattering and absorption were based on the absolute internal reflectance. To obtain this, the total reflectance measurement was first converted to absolute reflectance and then corrected for surface reflectance.

The panels were prepared by draw-down technique using a wire-wound draw bar over steel panels. Film thicknesses were measured with a Mikrotest thickness gage, Electro-Physical Instruments, Ogdensburg, New York.

The concentration of chrome yellow in the glazes was 2.16% of the film by weight or 0.778% by volume. The thickness of the films was 1.4 mils. The ratio of chrome yellow to white in the panel illustrating mixture was 66% to 34% by weight. The volume percentage in the film was 13.7% chrome yellow and 9.36% titanium dioxide.

The flavanthrone glazes were made at 1.78% by weight or 1.21% by volume of the lake, which contains 40% of the pure toner and is 30% of the pure toner in strength. The film thickness was 1.5 mils. The white mixture was made at a weight ratio of 45 parts yellow to 55 parts titanium dioxide. The volume percentage in the films was 12.3% flavanthrone and 6.3% titanium dioxide.

The quinacridone red glazes were made at 0.319% by weight or 0.241% by volume in the film. Film thickness was 1.4 mils. The white mixture was made with a weight ratio of 4.12% quinacridone to 95.88% titanium dioxide. On a volume basis, the percentage of the film was 1.74% quinacridone and 12.83% titanium dioxide.

The cadmium red glazes contained 2.4% by weight or 0.662% by volume. They were applied at a film thickness of 1.25 mils. The mixture with white contained 43 parts by weight of red to 57 parts by weight of titanium dioxide. The volume fractions of the film were 7.52% cadmium red and 10.37% titanium dioxide.

The glaze of white over black was made at 2.0% titanium dioxide by weight or 0.566% by volume in the film. The film thickness was 0.7 mils. The glaze of black over white was made at 0.2% carbon black by weight or 0.125% by volume. Film thickness was 1.1 mils. The white and black intermixture was made at 2.92 parts black to 97.08 parts white on a weight ratio basis, or 0.82% black and 12.7% titanium dioxide by volume in the final film.

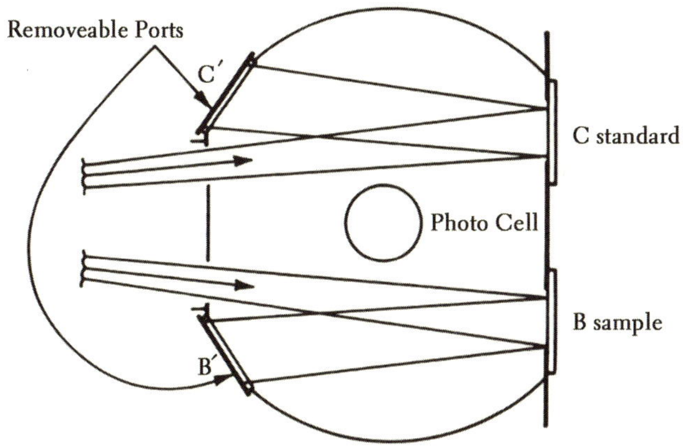

FIGURE 12
Design of Integrating Sphere GE Spectrophotometer.

SUMMARY

Certain optical effects in paint films have been explained by use of the Kubelka-Munk two-constant theory and by use of an equation to describe boundary reflections. The effects of glazes were illustrated by pairs of yellow and of red pigments that exhibit different degrees of transparency. Chrome yellow light, flavanthrone yellow, cadmium red, and quinacridone red were the specific pigments prepared as glazes over white and over black and were compared with their color in mixture with white.

The optics of the phenomenon of chalking were also discussed. This is an important problem because, when evaluating the lightfastness of pigments, chalking may be confused with fading.

Acknowledgment

The National Gallery of Art Research Project is most grateful to the Pittsburgh Plate Glass Company for its kind cooperation in the current studies of color and color change in artists' pigments. Their interest and assistance in these problems have done much to advance our knowledge of pigment identification, fading, and related matters in the care of works of art. The authors wish to thank Mr. Hubert Golden of the Pittsburgh Plate Glass Company for preparing the many panels used in the study of glazes. Miss Johnston particularly wishes to thank Dr. H. L. Gerhart, Director of Research and Development of the Coatings and Resins Division of the Pittsburgh Plate Glass Company, for his interest and encouragement in these investigations, and to thank the Pittsburgh Plate Glass Company for the privilege of publishing these remarks.

The Effect of Pigment Volume Concentration on the Lightness or Darkness of Porous Paints

ROBERT L. FELLER AND NOEL KUNZ

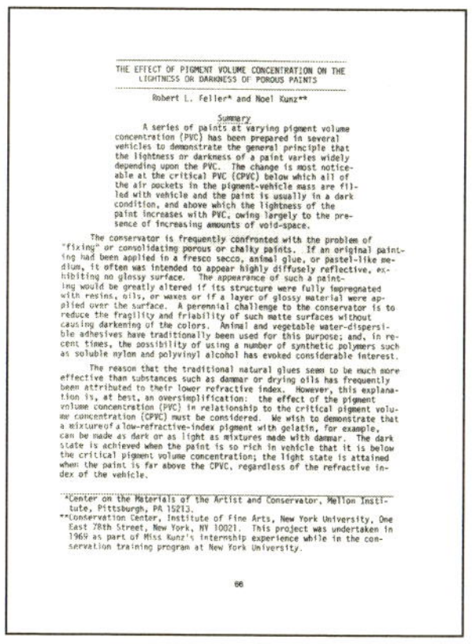

THE EFFECT OF PIGMENT VOLUME CONCENTRATION ON THE LIGHTNESS OR DARKNESS OF POROUS PAINTS

Robert L. Feller* and Noel Kunz**

Summary

A series of paints at varying pigment volume concentration (PVC) has been prepared in several vehicles to demonstrate the general principle that the lightness or darkness of a paint varies widely depending upon the PVC. The change is most noticeable at the critical PVC (CPVC) below which all of the air pockets in the pigment-vehicle mass are filled with vehicle and the paint is usually in a dark condition, and above which the lightness of the paint increases with PVC, owing largely to the presence of increasing amounts of void-space.

The conservator is frequently confronted with the problem of "fixing" or consolidating porous or chalky paints. If an original painting had been applied in a fresco secco, animal glue, or pastel-like medium, it often was intended to appear highly diffusely reflective, exhibiting no glossy surface. The appearance of such a painting would be greatly altered if its structure were fully impregnated with resins, oils, or waxes or if a layer of glossy material were applied over the surface. A perennial challenge to the conservator is to reduce the fragility and friability of such matte surfaces without causing darkening of the colors. Animal and vegetable water-dispersible adhesives have traditionally been used for this purpose; and, in recent times, the possibility of using a number of synthetic polymers such as soluble nylon and polyvinyl alcohol has evoked considerable interest.

The reason that the traditional natural glues seem to be much more effective than substances such as dammar or drying oils has frequently been attributed to their lower refractive index. However, this explanation is, at best, an oversimplification: the effect of the pigment volume concentration (PVC) in relationship to the critical pigment volume concentration (CPVC) must be considered. We wish to demonstrate that a mixture of a low-refractive-index pigment with gelatin, for example, can be made as dark or as light as mixtures made with dammar. The dark state is achieved when the paint is so rich in vehicle that it is below the critical pigment volume concentration; the light state is attained when the paint is far above the CPVC, regardless of the refractive index of the vehicle.

*Center on the Materials of the Artist and Conservator, Mellon Institute, Pittsburgh, PA 15213.
**Conservation Center, Institute of Fine Arts, New York University, One East 78th Street, New York, NY 10021. This project was undertaken in 1969 as part of Miss Kunz's internship experience while in the conservation training program at New York University.

66

Originally printed in:
Preprints of Papers Presented at the Ninth Annual Meeting, Philadelphia, PA
American Institute for Coservation, Washington, DC (1981), Pages 66-74.

The conservator is frequently confronted with the problem of "fixing" or consolidating porous or chalky paints. If an original painting had been applied in a fresco secco, animal glue, or pastel-like medium, it often was intended to appear highly diffusely reflective, exhibiting no glossy surface. The appearance of such a painting would be greatly altered if its structure were fully impregnated with resins, oils, or waxes or if a layer of glossy material were applied over the surface. A perennial challenge to the conservator is to reduce the fragility and friability of such matte surfaces without causing darkening of the colors. Animal and vegetable water-dispersible adhesives have traditionally been used for this purpose; and, in recent times, the possibility of using a number of synthetic polymers such as soluble nylon and polyvinyl alcohol has evoked considerable interest.

The reason that the traditional natural glues seem to be much more effective than substances such as dammar or drying oils has frequently been attributed to their lower refractive index. However, this explanation is, at best, an oversimplification: the effect of the pigment volume concentration (PVC) in relationship to the critical pigment volume concentration (CPVC) must be considered. We wish to demonstrate that a mixture of a low-refractive-index pigment with gelatin, for example, can be made as dark or as light as mixtures made with dammar. The dark state is achieved when the paint is so rich in vehicle that it is below the critical pigment volume concentration; the light state is attained when the paint is far above the CPVC, regardless of the refractive index of the vehicle.

THE CONCEPT OF CRITICAL PIGMENT VOLUME CONCENTRATION (CPVC)

If pigment particles are considered as spheres and are stacked or packed as closely as possible, a minimum volume of air spaces will remain between them. In paint formulation, the "critical pigment volume concentration" is reached when all of the air in the spaces has been replaced with vehicle (resin or oil).[1-3] In the case of rhombohedral packing of spheres, the calculated value for the air or vehicle volume is 26%; the theoretical CPVC is thus 74%. Because of non-ideal "real" conditions, however, CPVC's must be determined experimentally for each pigment-vehicle system, and generally range between 30 and 65%. As noted previously, many physical properties of paint films change markedly at the critical pigment volume concentration: porosity increases and gloss drops quickly as the CPVC is surpassed.[4] Of interest in the present context is the observation that, at pigment concentrations above the CPVC, the perceived color is lighter than that observed when the pigment concentration is below the CPVC.

1. Asbeck, W. K., and Van Loo, M., "Critical Pigment Volume Relationships", *Ind. Eng. Chem.*, 41 (1949), p. 1470-1475.

2. Steig, F. B., "Color and CPVC", *Official Digest, Fed. Paint Varn. Prod. Clubs*, 28 (1956), p. 695-706.

3. Steig, F. B., "Pigment/Binder Geometry," in *Pigment Handbook*, Vol. 3, ed. Patton, T. C., J. Wiley & Sons, New York (1973), p. 203-217.

4. Feller, R. L., and Matous, J. J., "Critical Pigment Volume Concentration and Chalking in Paints," *Bulletin of the American Group-IIC*, 5, No. 1 (1964), p. 25-26.

EXPERIMENTAL

Ultramarine blue pigment was chosen for the demonstration because it is sufficiently close in refractive index (about 1.51) to that of many vehicles that changes in lightness promised to be marked as the vehicle was varied. Two different ultramarines were selected: Holland-Suco's UB6917, now no longer available, and Reckitts' RS16.

Two studies were conducted, one involving preparation of paint films of known composition; the other, simply slurries of pigment in liquids of known refractive index. The paint films were made in the following way: solutions of a known percentage of resin were mixed with known weights of the pigments. The composition of the resulting dried films was calculated and expressed in terms of pigment *volume* concentration, using the densities of the resins and pigments reported in the manufacturer's literature. Vehicles used include Singapore dammar, Peter Cooper Glue No. 4 (Gowanda, N.Y.), Soluble Nylon Calaton CB from Imperial Chemical Industries, and polyvinylacetates, B-7 and B-15 from Air Reduction (viscosity grades 60 and 80, respectively),[5] and Palmer Cement A70 (viscosity grade 9) from Palmer Products, Inc. (Worcester, Pa.). All paints were coated on 2 1/2" x 6" commercially prepared acrylic-gesso panels, one half of which had been painted black, the other white. The prepared paints were allowed to air-dry for one week before the reflectances were measured.

5. Feller, R. L., Stolow, N., and Jones, E. H., *On Picture Varnishes and their Solvents*, The Press of Case-Western Reserve University, Cleveland (1971).

Color measurements were made either with a small-sphere Color-Eye® abridged spectrophotometer (once produced by the Instrument Development Laboratories, Kollmorgen, Inc., Attleboro, Mass.) or with the Leres Trilac® recording spectrophotometer (manufactured by Kollmorgen/Leres).

Reflectances were measured relative to a barium sulfate white standard. The measurements with the Color-Eye® were made with the specular (mirror-like) reflectance from the surface excluded (SCE); those on the Trilac® were made with the specular surface reflectance included (SCI) and, in certain instances, excluded (SCE). The SCI configuration yields a measure of the *total* reflectance; the SCE involves only the *diffusely reflected light*, excluding the brilliant image of the light source reflected from a glossy surface. The difference between these two measurements, $R_{SCI} - R_{SCE}$, then, is a measure of the specular, mirror-like reflectance which an observer rejects from his observation of color and subjectively associates with the smoothness and glossiness of the surface. As this glossiness decreases, the specular reflectance decreases, until the two measurements, SCI and SCE, become identical for a diffusely reflecting surface.

Lightness-Refractive Index Relationship

If one pours water into a glass containing crushed ice, the whitish mass of ice will practically disappear, becoming nearly transparent. The amount of light reflected by the individual ice particles would decrease sharply because of the similarity in the refractive indexes of the ice and water as compared to the dissimilarity in the refractive indexes of ice and air. If, instead of ice, crushed colored glass had been used to fill the glass one would see a darkening of the color in addition to an increase in transparency. If these experiments were repeated using a frosted glass container instead of a clear one, the increase in transparency upon addition of the water would be less than in the case of the clear glass because part of the light would be scattered by the frosted surface of the glass. In the case of the colored crushed glass, the color would not be darkened as much. The light scattered from a matte surface is *additively* mixed (visually) with the light reflected from within, and the perceived color is lightened significantly by a nearly constant amount at all wavelengths.

When pigments are mixed with a vehicle, much the same changes occur. The amount of light reflected from the surface of a particle of crushed ice, glass, or pigment depends on the difference between its refractive index and that of the surrounding medium; the greater the difference, the more light that is reflected and the greater the opportunity to "see" the particle. A demonstration of this fact is familiar to all who have looked at pigments under the microscope, where, when the refractive index of the particle and the mounting medium match, the particle is practically invisible.

Paint films made below the critical pigment volume concentration generally exhibit smooth, glossy surfaces, largely comprised of vehicle. When the critical pigment volume concentration is reached and then exceeded, there is a rapid decrease in the gloss of the dried paint film because some of the pigment particles must then protrude from the film. The surface becomes matte owing to the light scattered by these particles; the color appears lighter by the visual addition of the light scattered by the surface to the light reflected by particles within the film. As the concentration of vehicle is further reduced, scattering increases owing to the increased reflectance from the many pigment-air interfaces within the paint film.

To demonstrate, first of all, the effect of vehicles of different refractive indexes on the observed reflectance below the critical pigment volume concentration, without surface reflectance effects, ultramarine blue was simply mixed with liquids of different refractive index and the reflectance measured. The pigment was placed in a 1 x 6 x 6 cm quartz cell such as used in absorption spectrophotometry and the air spaces in the mass of pigment powder filled with liquids of various refractive indexes. The flat surface of the cell was then placed at the port of the Color-Eye® and the intensity of the light reflected from the pigment/vehicle mass was measured from 400 to 700 nm at 20 nm intervals. (The specular reflectance from the glass surface was excluded.) From an examination of the curves in Figure 1, one can see that the intensity of the reflected blue light decreased as the refractive index of the various liquids (water, acetone, benzene-acetone mixtures) increased towards the refractive index of the pigment itself (1.51); the perceived color of the pigment-solvent mixtures became increasingly darker. The lightest color occurred with the dry pigment; the medium in this case was air, refractive index 1.00.

Figure 2 summarizes the measured reflectances at 440 nm when the pigment is entirely surrounded with the liquids of various refractive indexes. As can be seen, the lowest reflectance is obtained when the refractive index of the pigment closely matches that of the surrounding liquid, and increases on each side as

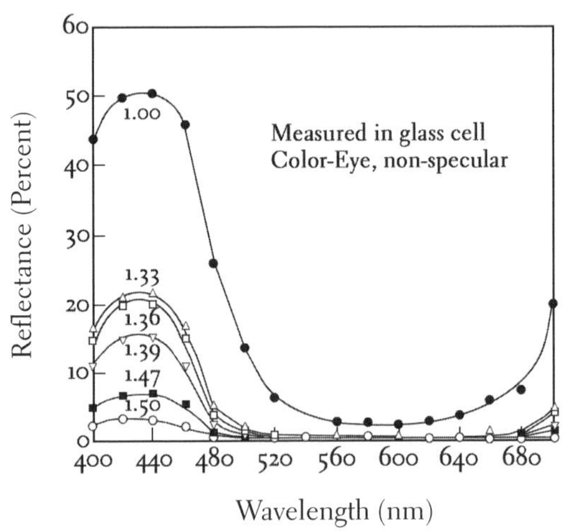

Figure 1
Effect of refractive index of surrounding medium on reflectance of Reckitts RS16 Ultramarine Blue.

the differences in refractive indexes increase, whether lower or higher than the pigment's index. This behavior is as predicted on the basis of the principles regarding the reflectance of light as a function of the difference in the refractive indexes of two phases in contact. The fact that the minimum does not fall precisely at 1.51 may be due to various experimental factors. However, the basic principles are well demonstrated.

One of the objectives in this study was to determine the range of lightness of a dispersion containing ultramarine blue which would be affected solely by the refractive index of the vehicle. The refractive indexes of the various vehicles to be studied in the paint films are noted above the curves in Figure 2. From these we see that, when completely surrounded by typical vehicles, the pigment mass reflects from about 4 to 11% in the case of UB6917. This range, no more than 7%, can be attributed to differences in refractive index alone.

LIGHTNESS-PVC RELATIONSHIP

Dammar: How much will the reflectance of a paint film vary as a function of pigment volume concentration (PVC)? Figure 3 is a graph of the percent reflectance at 440 nm plotted versus the PVC of ultramarine pigment UB6917 in dammar. One can see that there is a sharp increase in the reflectance of the paints at about 41% PVC. Much below this concentration all interstices in the pigment-vehicle mass are filled with vehicle and the surface is smooth and glossy. The difference between the measurements made with the specular reflectance included, SCI, and excluded, SCE, is the specular reflectance, about 5%. (See Figure 4). Above 41% PVC, the gloss rapidly decreases until

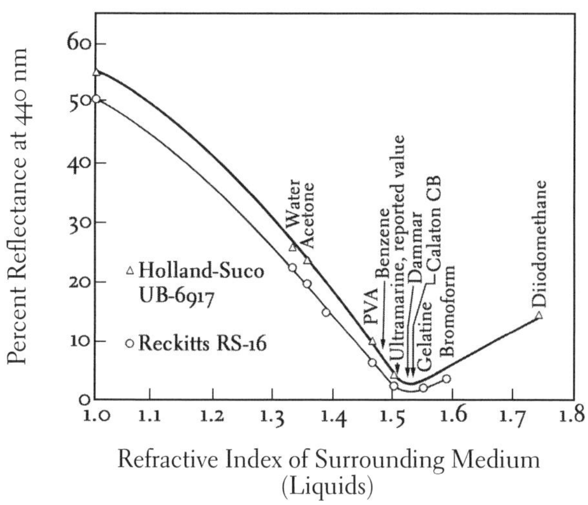

FIGURE 2
Percent reflectance (440nm) of Ultramarine Blue vs. refractive index of surrounding liquid. (Color-Eye vs. 7B, Non-specular insert, percentage reflectance of pigment in glass cell adjusted to $BaSO_4$ from "IDL Average Vitrolite")

the SCI and SCE measurements are the same, and the surface is matte in appearance. With increasing PVC, the color continues to become lighter because more and more of the spaces between the pigment particles are filled with air instead of vehicle. (Measurements were made over the black and over the white underpainted areas to show that the paints were nearly at "complete hiding", that the reflectance values were not seriously influenced by the lightness of the underpaint.)

Figure 3 shows that at low PVC's the diffuse reflectance at 440 nm of ultramarine blue paint made with dammar is about 2%, in reasonable agreement with the values which would be expected from Figure 2. Above the CPVC, the reflectance increases to as much as 60%. The refractive index of dammar, about 1.53, primarily controls the color, or diffuse reflectance, when the ultramarine paint is below the CPVC; above this, the PVC primarily controls the lightness of the color.

Polyvinylacetate: Figure 5 shows that polyvinylacetate, a synthetic resin which has one of the lowest refractive indexes among the polymers employed by conservators (1.467), behaves in much the same manner. The average molecular weights of the three polyvinylacetates increase in the order A-70, B-7, and B-15, and the molecular weight of the polymers may have some influence on the CPVC. The curves of dammar and lowest molecular weight polyvinylacetate, A-70, are very similar, however. The refractive index of the vehicle definitely has an influence below the CPVC, where the pigment particles should be entirely surrounded by the vehicle (the diffuse reflectance at 440 nm is about 10%, in good agreement with prediction from Figure 2), whereas in the dammar, it is about 2%. Because the surfaces of these panels were not as smooth as in the case of dammar, the specular reflectance was low (Figure 4). Subsequent

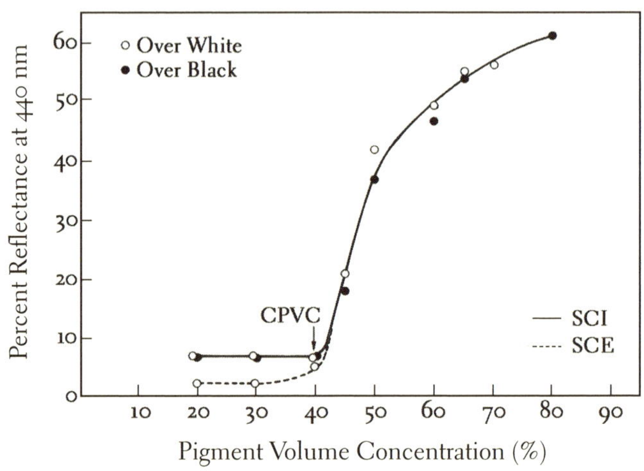

Figure 3
Percent reflectance vs. pigment volume concentration, ultramarine UB-6917 in dammar (taken on Leres, ref: $BaSO_4$).

work with ultramarine blue in the B-7 polyvinylacetate, using drawdowns on glass panels, gave the theoretical specular reflectance, an average of 3.57%.

Gelatin: A traditional consolidant used when minimum darkening of porous paints is desired is animal glue or gelatin. The same basic behavior is observed in Figure 6. At low PVC's, a gelatin-based paint can be very dark. (No diffuse, SCE, measurements were made, but the total reflectance, SCI, is only about 6% at 20% PVC.) Allowing for some specular reflectance, the diffuse reflectance must be low, approaching the color of the pigment in dammar, having about the same refractive index (the reported value for gelatin is 1.53). The CPVC is lower than for dammar, in the range of 35%, with a less steep slope of PVC versus reflectance. The ultimate color at high PVC's is in the same range as with dammar, about 60%.

Soluble Nylon: In the last demonstration, we prepared a series of paints with a type of soluble nylon that has become popular in certain laboratories as a consolidant-fixative that has little tendency to cause darkening of porous paints. We see the same general behavior: around 10% total reflectance at 440 nm when the paint is below its CPVC; high reflectances at high PVC's (Figure 6).

DISCUSSION

If a porous paint is "scarcely darkened" when a fixative is applied, it must mean that the interstices between the pigment particles are not filled with vehicle but are still filled with air; the PVC must be very high, perhaps 70%. Why is it then, when porous paint films are "fixed" with solutions of glue or soluble nylon, that the color often does not darken as much as when dammar or organic-soluble resins are used? One probable answer is that the molecular weights—

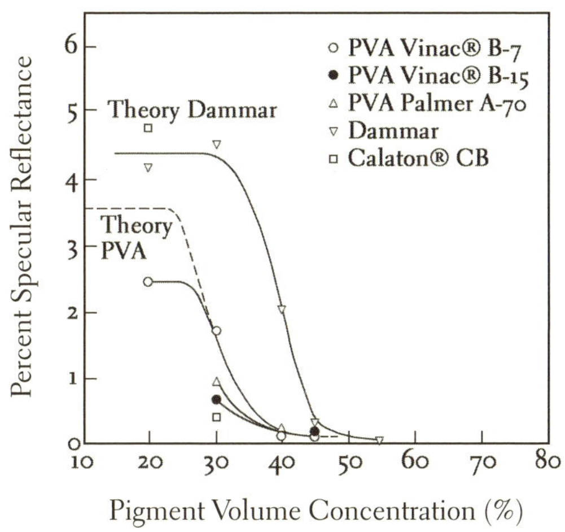

FIGURE 4
Specular reflectance, S, of films formulated near or below the CPVC ($S = R_{SCI} - R_{SCE}$ or $Y_{SCI} - Y_{SCE}$).

and hence the intrinsic viscosities—of glue and soluble nylon are so high that only very dilute solutions can be prepared, perhaps 2-5%, with the result that only a small amount of vehicle is usually available to fill the air interstices, leaving the PVC of the dried film very high. In contrast, more concentrated solutions of dammar (as much as 35%) or of A-70 (~15%) have viscosities in a similar range, allowing as high as seven times as much vehicle to be applied with the same technique, filling more of the air spaces with vehicle and

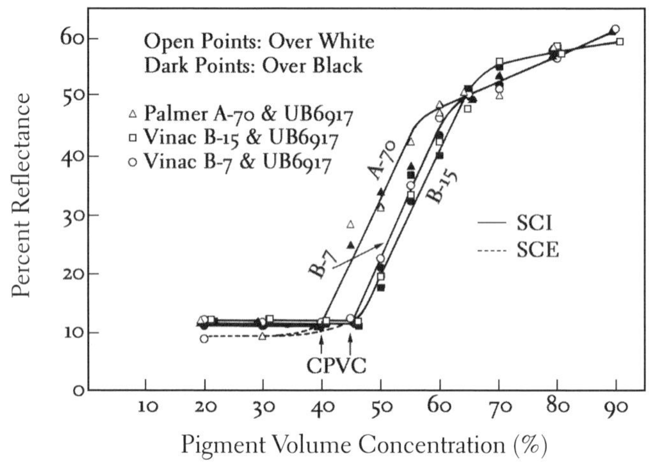

FIGURE 5
Percent reflectance vs. pigment volume concentration, ultramarine UB6917 in poly(vinyl acetate) (taken on Leres, ref: $BaSO_4$).

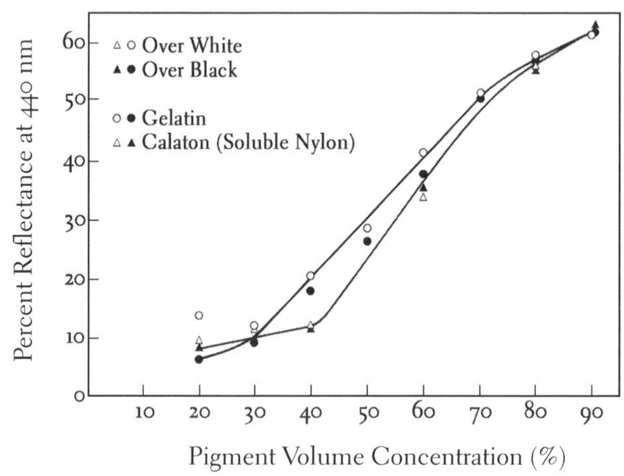

FIGURE 6
Percent reflectance vs. pigment volume concentration, ultramarine UB6917 in gelatin and soluble nylon (taken on Leres, ref: $BaSO_4$, specular included).

significantly reducing the PVC, darkening results. A second possible reason why fixatives based on high-molecular-weight vehicles may have less tendency to darken a porous paint is that, because of the high viscosity of their solutions, penetration into the porous mass and wetting of the pigment particles may be hindered.

Extensive interpretations of results of this series of paints should not be made without further experiments. The role of the polymer's molecular weight on the CPVC, which seems to be displayed in Figure 5, should not be taken to represent a general conclusion; there are many factors that control the CPVC. The much less rapid rise in reflectance with increasing PVC displayed in Figure 6 needs further study before a reasonable explanation can be presented.

Conclusion

In the study of fixative-consolidants, we recommend that the relative lightness or darkness of the resulting consolidations be considered on the basis of the PVC of the "fixed" paint and in the light of the maximum range of lightness or darkness that can be achieved (information such as presented in Figures 3, 5 and 6). In most cases in which very little darkening is achieved, we believe that an analysis of the situation will show that the resultant consolidated paint is at a high PVC.

Acknowledgment

The authors thank Ruth Johnston-Feller for helpful discussions and criticisms regarding the manuscript. This project was undertaken in 1969 as part of Miss Kunz's internship experience while in the conservation training program at New York University.

Section 5: The Color Wheel

Along with our intensive laboratory research program in color and fading phenomena, I developed an interest in the history of color description and color order systems. In 1963, Faber Birren, well-known as a prolific writer of popular books on the subject of color, brought out a handsome facsimile edition of perhaps the rarest known book in the literature of color, Moses Harris's *The Natural System of Color*, dated 1766. The 10 1/4" x 14" privately printed limited edition was distributed by the Whitney Library of Design. Consisting of just ten pages of original text along with six pages of Birren's introduction, the facsimile excited considerable interest because it was claimed that two large diagrams therein constituted the "first known presentation of the color circle in full hue," the familiar red, yellow, and blue system of subtractive color mixture. An unexpected opportunity over the next few years to examine two editions of the original work led to a book review that was given the hard-to-resist title, "Rediscovery of the Wheel."

At a conference on color science in Stockholm in 1969, I noticed a poster that depicted a circular diagram attributed to Sigrid Forsius, 1611. Surprisingly, no lecture or exhibit devoted to Forsius's work had been scheduled for the meeting, but colleagues informed us that the manuscript itself could be seen at the National Library located just a few blocks away. Arrangements were immediately made to have the manuscript brought from the rare book vault the next day for my inspection. In view of Birren's claim of Harris's "invention," it seemed important to call attention to Forsius's extraordinary early depiction of a spherical "color space." The distinguished Swedish color scientist, Åke Stenius, kindly agreed to assist in this effort.

The allusion to the adage, "Someone is always rediscovering the wheel" proved prescient. A few years after these incidents, a hand-colored circle was discovered in a 1708 edition of a small book bearing the title, *Traite de la Peinture en Mignature*. At about that time Charles Parkhurst, a renowned scholar of early color systems, became Chief Curator of the National Gallery of Art. Eventually we were able to collaborate on an extensive review of color circles and color space systems, some extending as far back as 1230-35.

R. L. F.

Rediscovery of the Wheel

ROBERT L. FELLER

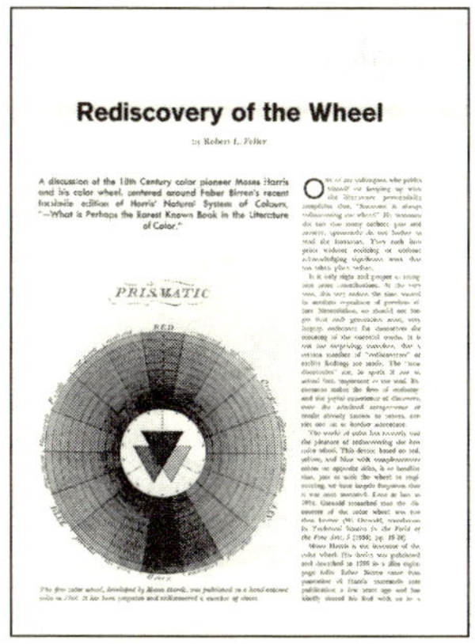
Rediscovery of the Wheel

by Robert L. Feller

A discussion of the 18th Century color pioneer Moses Harris and his color wheel, centered around Faber Birren's recent facsimile edition of Harris' Natural System of Colours, "—What is Perhaps the Rarest Known Book in the Literature of Color."

PRISMATIC

RED

Originally printed in:
COLOR ENGINEERING
Volume 4, Number 6 (1966), Pages 20-23.

One of my colleagues, who prides himself on keeping up with the literature, perennially complains that, "Someone is always rediscovering the wheel!" He bemoans the fact that many authors, past and present, apparently do not bother to read the literature. They rush into print without realizing or without acknowledging significant work that has taken place before.

It is only right and proper to recognize prior contributions. At the very least, this may reduce the time wasted in needless repetition of previous effort. Nevertheless, we should not forget that each generation must, very largely, rediscover for themselves the meaning of the essential truths. It is not too surprising, therefore, that a certain number of "rediscoveries" of earlier findings are made. The "new discoveries" are, in spirit if not in actual fact, important to the soul. Excitement stokes the fires of curiosity, and the joyful experience of discovery, even the admitted re-experience of truths already known to others, carries one on to further adventure.

The world of color has recently had the pleasure of rediscovering the first color wheel. This device, based on red, yellow, and blue with complementary colors on opposite sides, is so familiar that, just as with the wheel in engineering, we have largely forgotten that it was once invented. Even as late as 1934, Ostwald remarked that the discoverer of the color wheel was not then known.[1]

1. Ostwald, W., translation in *Technical Studies in the Field of the Fine Arts*, 3 (1934), p. 19-28.

Moses Harris is the inventor of the color wheel. His device was published and described in 1766 in a slim eight-page folio. Faber Birren came into possession of Harris' extremely rare publication a few years ago and has kindly

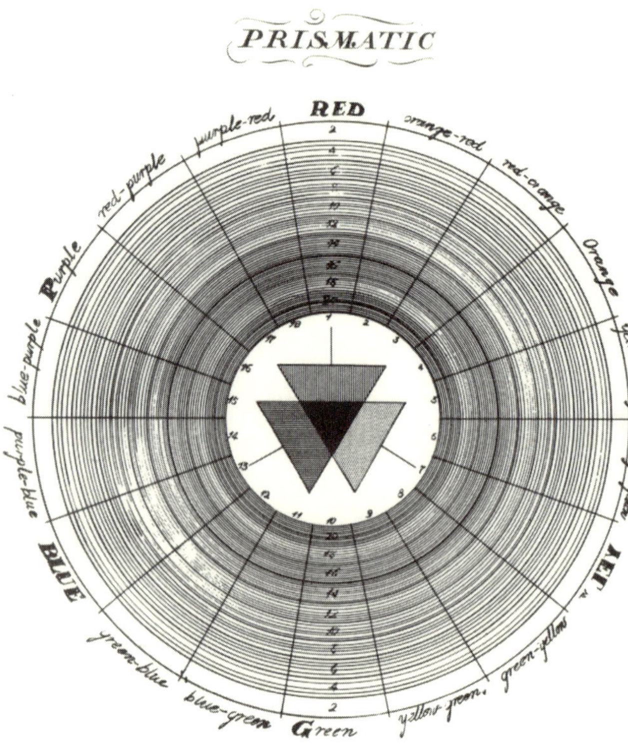

FIGURE 1
The first color wheel, developed by Moses Harris, was published in a hand-colored folio in 1766. It has been forgotten and rediscovered a number of times.

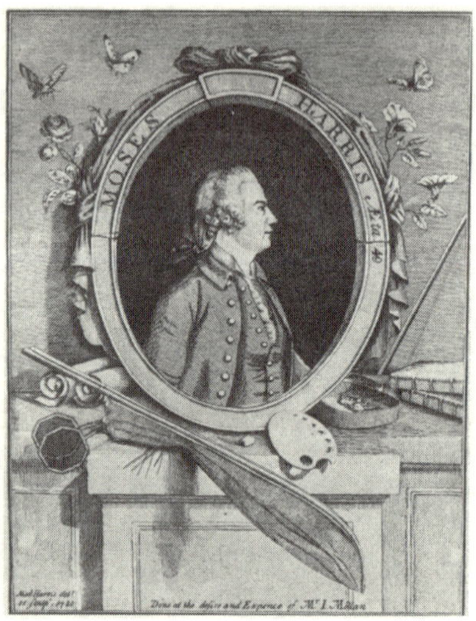

FIGURE 2
A self-portrait of Moses Harris, done about 1780. A noted entomologist as well as a talented painter, his detailed paintings of 18th-century English insect life are comparable to Audubon's work with bird life. Note the insect nets next to his palette.

shared his find with us in a handsome facsimile edition.[2] In the introduction, he shares the adventure of his discovery and traces some of the history of this largely forgotten work. In his recent book on *The History of Color in Painting,* [3] the background of the Harris story is further extended.

2. Harris, M., *The Natural System of Colours* (1766), distributed by the Whitney Library of Design and privately printed in a limited edition, New York (1963).

3. Birren, F., *The History of Color in Painting*, Reinhold Publishing Corporation, New York (1965).

FEW COPIES REMAIN

Only three copies of the Moses Harris book are on record: Mr. Birren's first edition (1766); another first edition owned by Dr. F. Schmid, a scholar of art history living in Zurich; and an 1811 edition owned by the British Museum Library. (*However,* recently a colleague at the Courtauld Institute of Art in London told me that he too had a copy of the second edition, and then in November 1966, Mr. Birren informed me that a fine copy of the first edition is at the Royal Academy in London.) Apparently from the very beginning, copies of Harris' book have always been extremely rare. No doubt its folio size and the brevity of the text are partially responsible for this state of affairs.

Much of the fascinating story behind Harris' color wheel was published about 15 years ago by Dr. Schmid, as Mr. Birren acknowledges in the introduction to the facsimile edition.[4] Dr. Schmid, a well-known authority on the history of the teaching of painting and drawing, illustrated his book with black-and-white

4. Schmid, F. , *The Practice of Painting*, Faber and Faber, London (1948); *The Art Bulletin*, 30 (1948), p. 227-230.

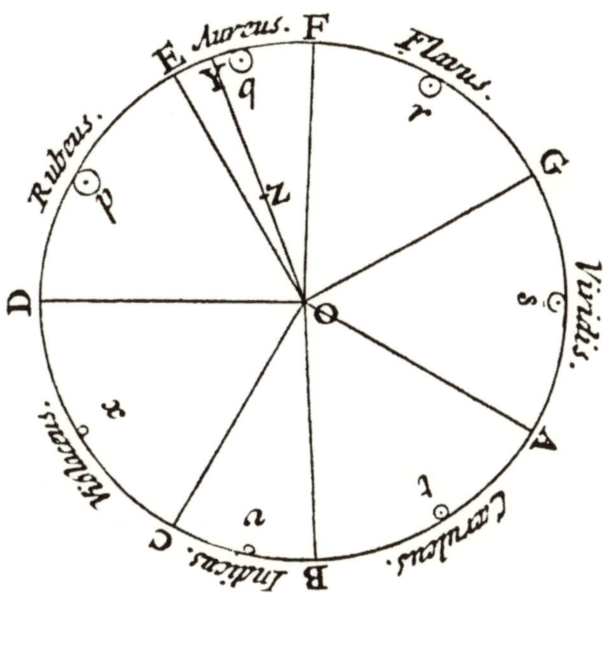

FIGURE 3
Sir Isaac Newton's Color Circle. Based on additive mixture, this was probably the first systematic arrangement of colors in a circle, and the forerunner of Harris' color wheel, which was based on subtractive color mixture.

photographs of the two color wheels. He was able to point out that Harris' contribution was well appreciated by a number of authors in the early nineteenth century although for the last 100 years it had been essentially forgotten.

Schmid argues the case well for the development of the artists' color wheel as we know it today through Sir Isaac Newton, Jacques-Christopher LeBlon, and finally to Moses Harris. Newton seems to have been the first to arrange colors in a circle. LeBlon did not arrange them in a circle but is said to have been the first to emphasize the fact that mixtures of red and yellow pigments made orange, red and blue made purple, and blue and yellow made green. Through the device of placing these three primary colors 120° apart in a circle, Harris was the first to show the important relationships of subtractive color mixture. Newton's color wheel, on the other hand, was plainly based on additive mixture. LeBlon, himself, pointed this out in his *Coloritto*, saying "I am only speaking of material colors, or those used by painters, for a mixture of all the primitive impalpable colors, that cannot be felt, will not produce black, but the very contrary, white; as the great Sir Isaac Newton has demonstrated in his optics."

My previous experience with hand-colored diagrams in Mérimée, in Field's *Chromatography*, and in other books, led me to believe that the transparency and saturation of the colors in the color wheels in the facsimile edition were "all wrong". Yet I found the color wheels in the 1811 edition at the British Museum to be painted in by hand with transparent washes of high saturation. Recently, when Dr. Schmid kindly brought his copy along to lunch for me to see, I found that the hand-painted colors in his edition were equally striking.

In Dr. Schmid's copy and in the British Museum's, the gradation from light at the edge to dark at the center apparently was accomplished by different thicknesses of the wash, rather than by additions of a black or a white pigment. Severe darkening of the washes in the center has occurred in Mr. Birren's copy, as he mentions and shows in a photograph that accompanies the facsimile edition. In the British Museum copy, this darkening also occurs and appears, at least in part, to be the result of the picking up of dirt by the thick, vehicle-rich washes in the center. This was particularly true of the "prismatic" red and orange. The thick paint in the center tends to be glossy and leads one to wonder what vehicle was used. Could the glossy vehicle be the usual one in watercolor, gum arabic? Many of the pigments must have been lakes based upon natural coloring matter, for the usual mineral pigments would be expected to exhibit less transparency, a fact that Harris himself comments upon. The transparency of the washes, as well as their continued high saturation after more than 100 years, is remarkable. Their essential character seems well captured in the facsimile.

Anyone who has ever tried to paint a color wheel or a CIE diagram knows how difficult a problem it is to blend each area in a smooth transition from a lighter to a darker color, as well as from a warmer to a cooler color. If we may judge from the photograph of Mr. Birren's copy and my inspection of the other two, Harris and the artist who colored the engraved plates in 1811 did not attempt to make a continuous blend but, instead, painted several of the pie-shaped segments of the wheel with different pigment mixtures, each graded from dark in the center to light on the border. For this reason, the transitions between hues around the wheel are not smooth. Indeed, it appears that quite a few individual colorants were used in going about the circle of hues. The facsimile also reproduces this lack of smooth transition in hue between many of the radial segments.

There is a curious point about the color wheels in the facsimile edition, however, that apparently represents an error in the attempt to reproduce their "original" appearance. The problem of faithful reproduction is a difficult one, of course, for the colors in the original copies undoubtedly have darkened and discolored somewhat. Mr. Birren and his associates had to, make a compromise in their re-creation and seem to have been successful in doing so, for the essence of the original hand-painted colors in the wheel has been well captured. Yet, curiously, the shading in the facsimile, from light at the outer edge to dark in the center, is graded only by the density of the concentric engraved black lines, not by a change in the intensity of the printed color. While the concentric engraved lines are also present in the original engravings, I have the distinct impression that the gradation from light to dark in the original was also accomplished by increasing the thickness of the color wash. Dr. Schmid confirms the fact that this is the case in his copy. The lack of tapering of the color in the facsimile from dark in the center to light on the outside was previously noted by Kenneth Kelly in his review.[5] (Dr. Schmid has kindly called my attention to the fact that Hundertpfund[6] had the same criticism to make regarding the hand-colored reproduction of Harris' color wheel in Mérimée's *The Art of Painting in Oil*, translated by W. B. S. Taylor, London, 1839. I am grateful for this and the many other suggestions which Dr. Schmid kindly made to me in his continued scholarly interest in Moses Harris.)

5. Kelly, K., *J. Optical Soc. Amer.*, 54 (1964), p. 133-134.

6. Hundertpfund, *The Art of Painting*, London (1849).

Missing Plate

Moses Harris' text describes a third hand-colored plate, which is not present in the facsimile edition but which is present in the British Museum copy and in Dr. Schmid's copy as well. (In Dr. Schmid's copy, the parallel lines opposite each color at the bottom of page 4 are also painted in with the appropriate color.) In the 1811 edition, this hand-colored page is based on two right-hand pages printed with the same engraved plate. This bears the label "Examples". In Dr. Schmid's copy, the printed word "Examples" is pasted on the first page and the numbers put in by hand. The pages consist of nine squares, each about 1 1/8" on a side. As Dr. Schmid described in his book, the squares were cut out of the first page so that the openings permit one to see through this "overlay" to the colored squares on the following page. The coloring on the second page consists of a series of washes, hand-painted across each square generally first in one direction then in another, the second color washed over at right angles to the first. The process illustrates subtractive mixtures of color, not by physical mixture of pigment, but largely by the artist's technique of the application of a translucent "glaze" or "scumble" over another color. Harris used it to illustrate the low saturation that would be obtained by mixing pigments that were nearly complementary on opposite sides of the color wheel. Since these two pages are missing from the facsimile edition, in spite of the reference to them in the text, we must assume that they are missing from Mr. Birren's copy and that Dr. Schmid's is the only known complete copy of the first edition in existence.

1811 EDITION

The 1811 edition in the library of the British Museum is sufficiently different that it deserves mention. Hundertpfund referred to its presence in the library in 1849 so it, too, has been there for a long time awaiting rediscovery. Thanks to the kindness of Dr. E. A. Werner, Keeper of the British Museum Research Laboratory, I had the pleasure of inspecting this copy about two years ago. The edition was edited by Thomas Martyn and printed by L. Harrison and J. C. Leigh, 373 Strand. Dr. Schmid kindly pointed out to me that, like Harris, Martyn was also an entomologist. He also called my attention to the fact that this edition is slightly smaller and that the text contains minor changes, omissions, and additions. Perhaps one of the most interesting changes is Martyn's substitution of the word "hue" for "color." The dedication is to Benjamin West, Esq. President of the Royal Academy, successor to Sir Joshua Reynolds. The plates of the color wheels are apparently newly engraved, for the lettering is different from Harris'; and they no longer carry at the bottom "Moses Harris invt and sculpt".

SOMETHING OF HARRIS

The two-page foreword in this edition tells something of the little known author and bears repeating. It states that Harris was nearly 30 years deceased in 1811, although Schmid has pointed out that this may be slightly in error: the dates of his life are probably 1731 to 1785. The foreword goes on to say, in the language of the day, that Harris was ". . . an eminent painter in watercolours of every class of insects, and author of several admired works in this branch of natural history, including THE AURELIAN, in small folio, and AN EXPOSITION OF ENGLISH INSECTS, in quarto; exhibiting nearly eight hundred different species of various tribes of the *dipterae* or two-winged flies, and others of the *lepidoptera* family. In these interesting works, accuracy of drawing, correct classification, and inimitable colouring of the object are equally conspicuous.

Indeed, to his unwearied assiduity, ingenuity, and perseverance for upwards to 20 years in these pursuits, were owing the assemblage of the multifarious non-descript insects of the several *classes, orders,* and *genera,* which had either been bred, or collected by himself, within twenty miles of the metropolis, and which formed his extensive cabinet for their destined purpose of publication: so that altogether, Mr. Harris' pretensions and qualifications, whether as a zealous collector, scientific naturalist, or exquisite painter of this minute department of animated nature have never been equaled! . . . To these rare merits the present work will be a further and lasting memorial to the author's comprehensive mind and superior genius in this profession; . . ."

RARE EXPERIENCE

It is indeed a great pleasure to rediscover this forgotten publication. How many of us have the opportunity to discover a book so rare that only three copies of the first edition and two of the second are known? How many take the opportunity to read the language and enjoy the typography of the eighteenth century (the language also appealed to reviewer Kelly).

Rediscovery of the wheel? Mr. Harris' contribution has been lost and rediscovered any number of times. Until Schmid reconstructed much of its history in 1948, it has been lost for nearly 100 years. Now, thanks to the current interest and to the facsimile edition of 1,500 copies, Harris' work is not likely to be readily lost again.

Mr. Martyn proudly said in 1811 that the work was only then ". . . rescued from oblivion after more than three years owing to the acute discernment of one of the few of our principal English miniature artists, who, had formerly seen a copy in the possession of the Editor, and appreciating its just value, liberally encouraged the reproduction of it." Our compliments to Mr. Birren, and to the men in 1811 as well, who recognized and appreciated the importance of Mr. Harris' modest publication and who liberally encouraged its reproduction.

The illustrations in this article are through the courtesy of Faber Birren.

On the Color Space of Sigfrid Forsius 1611

ROBERT L. FELLER AND ÅKE S:SON STENIUS

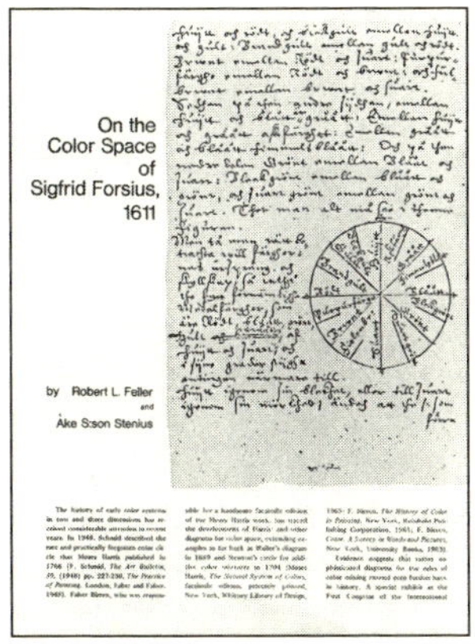
On the Color Space of Sigfrid Forsius, 1611

by Robert L. Feller and Åke S:son Stenius

Originally printed in:
COLOR ENGINEERING
Volume 8, *Number* 3 *(1970),*
Pages 48-51.

The history of early color systems in two and three dimensions has received considerable attention in recent years. In 1948, Schmid described the rare and practically forgotten color circle that Moses Harris published in 1766.[1] Faber Birren, who was responsible for a handsome facsimile edition of the Moses Harris work, has traced the development of Harris' and other diagrams for color space, extending examples as far back as Waller's diagram in 1689 and Newton's circle for additive color mixtures in 1704.[2]

Evidence suggests that rather sophisticated diagrams for the rules of color mixing extend even further back in history. A special exhibit at the First Congress of the International Color Association, held in Stockholm, June 1969, for example, reminded the delegates of a color circle and sphere that appear in a 1611 manuscript on physics by Sigfrid Forsius, kept in the Royal Library, Stockholm (Figure 1). Our hosts at the Conference urged us to visit the library to see the original manuscript. We did this one morning and the librarian kindly arranged to have the four-inch thick manuscript brought to the rare-book and manuscript room where we spent a memorable hour examining it.

A transcription of the Forsius manuscript has been made by Johan Nordostrom in a thesis submitted to Uppsala (Sweden) in 1954. A translation by one of us of the appropriate passages in the chapter "on the vision"[3] follows. Apology is made that the full flavor of the 17th-century Swedish cannot be conveyed:

1. Schmid, F., *The Art Bulletin*, 30 (1948), p. 227-230; *The Practice of Painting*, Faber and Faber, London (1948).

2. Harris, M., *The Natural System of Colors*, facsimile edition, privately printed, Whitney Library of Design, New York (1963); Birren, F., *The History of Color in Painting*, Reinhold Publishing Corporation, New York (1965); Birren, F., *Color, A Survey in Words and Pictures*, University Books, New York (1963).

3. Forsius, p. 419-420, Nordostrom, p. 316-317.

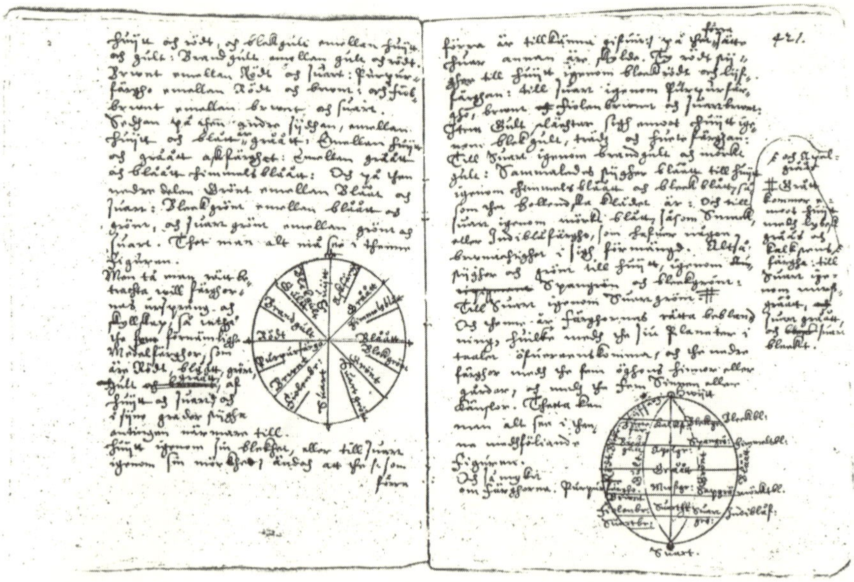

FIGURE 1
From a 1611 manuscript on physics by Sigfrid Forsius, kept in the Royal Library, Stockholm.

The things one can see are either Colors, or Light and Brilliant objects, or the things that can be seen alone at night. The color is outermost or uppermost part of an object that falls under the sight.

And there are three things needed for a color to be noticed: (1) The outermost part of an object, (2) Translucency and clarity, since in darkness nothing can be seen, (3) A moderate distance, because objects that are far away cannot be judged by their color.

Among the Colors there are two prime colors, white and black, from which all others have their origin. The white color belongs to the cold, the weather and the dryness. The black to the cold, the dampness and the putrefaction.

In the middle between these colors, red since olden-times has been placed on one side and blue on the other one: Yellow between white and red, and pale yellow between white and yellow: Orange between yellow and red. Brown between red and black: Purple color between red and brown: and violet brown between brown and black.

Then on the other side, between white and blue is gray. Between white and gray is ash color: Between gray and blue is sky blue. And on the lower part green between blue and black. Pale green between blue and green and black green between green and black. This is all seen in this Figure (Figure 2).

But if you want right to consider the origin and relations of the colors, you should start front the five principle middle colors which are red, blue, green, yellow, and gray of white and black (Figure 3). And their gradings, they rise either closer to white by their paleness or to black by their darkness; albeit they are (as above has been made known) related to one another as previously shown. Because red rises to white through pale red (pink) and skin color: to black through purple, brown, violet brown and black brown. Similarly yellow relates towards white through pale yellow, wooden and wheat color: To black through orange and dark yellow. Equally blue rises to white through sky blue and pale

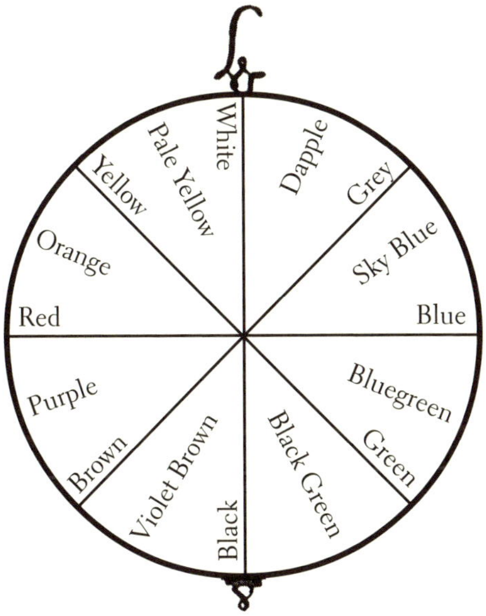

Figure 2
Forsius' color circle in two dimensions.

blue, like Dutch cloth: And to black through dark blue like indigo color that has some brownish in it. So rises also green towards white through Spanish green and pale green: To black through blackgreen. Gray approaches white by the color of Lübeck gray and limestone and dapple gray color: To black through mouse gray black gray and pale black.

And this is the correct relationship of colors, that in their number agree with that of the planets as do the lower colors with the five membranes of the eye, and with the five senses. All this can be seen from the accompanying figure (Figure 3).

And so much about the colors."

Of the two diagrams represented by Forsius, the first is a color circle in two dimensions (Figure 2), an arrangement of colors which we may assume has been employed by others before him, for he says, "In the middle between these colors red *since olden times* has been placed on one side and blue on the other: Yellow between white and red, and pale yellow between white and yellow: Orange between yellow and red. . . ." This diagram is not very close to the arrangement of complements in modern color circles nor is it a proper cross section of the other diagram.

Forsius' second diagram (Figure 3) represents color space as a *sphere*, remarkably close to the 1810 diagrams of Runge and one that we might construct today. White is at the top of his sphere, black at the bottom. Around the circumference, the colors are arranged essentially according to color-perception space arising from the ideas of black-white, red-green, and yellow-blue.

Little seems to have been written about Forsius; in the two biographies found, one makes no mention at all of his *Physica*. The other reads as follows:

The Finnish astronomer and clergyman, Sigfrid Aronus Forsius, was born in Helsingfors, Finland, shortly after the foundation of the town in 1550. He studied, mainly astronomy, at German universities, but also in Uppsala, Sweden, and was ordained to the priesthood in 1597. In 1601-02, he performed mapping in Lapland, the most northern part of Finland. He was then for some time parson in Kimito, Finland. After varying fortunes, he was in

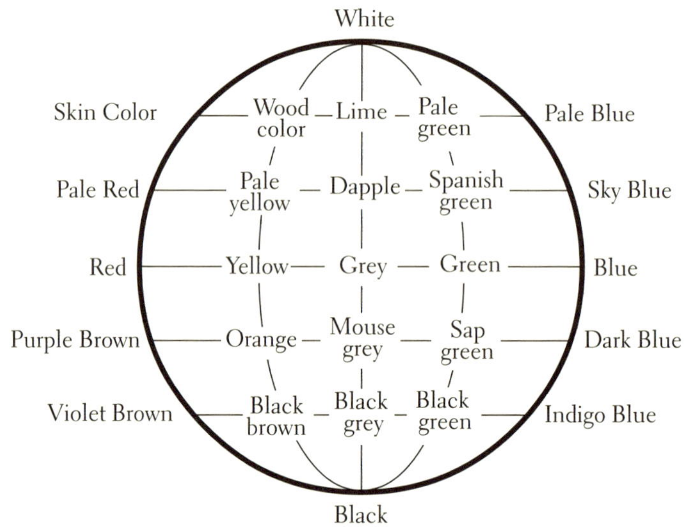

FIGURE 3
Forsius' second diagram representing color space as a sphere.

1608-10 professor in astronomy in Uppsala. In 1613, he became parson in Stockholm, but was suspended in 1615. From 1621 to his death in 1624, he was parson in Ekenäs, Finland.

Sigfrid Forsius is mostly known as an author in prognostics, in astrological predictions and above all for his almanacs. He was the first one who worked out almanacs valid for districts in Sweden (Stockholm) and Finland (Abo); previous almanacs had been translations from German with indications of time (for instance, sunrise and sunset) that were not valid for Sweden and Finland. Forsius was one of the most remarkable representatives in Sweden of the Neo-Platonist mystics inspired by the Swiss philosopher and alchemist Paracelsus. He is said to have been quarrelsome and adventurous of nature.

Sigfrid Forsius wrote his Physica in 1611 and published a collection of "Religious psalms and songs" in 1614.

The sphere in Forsius' manuscript is the earliest diagram that we are aware of that corresponds to present-day arrangements of color space. Perhaps Forsius himself is the inventor for he introduces it, immediately following his description of the first diagram, saying: "But if you want *right* to consider the origin and relations of the colors . . ."

Original Editor's Note: It is of interest, especially for those who attended the Color 69 meeting in Stockholm in June 1969, to note that the exhibit referred to in the second paragraph of this article was the work of the Swedish Color Center, whose director, Anders Hard, also referred to the Forsius manuscript in his paper "Quality Attributes of Colour Perception," presented at that meeting.

WHO INVENTED THE COLOR WHEEL?

CHARLES PARKHURST AND ROBERT L. FELLER

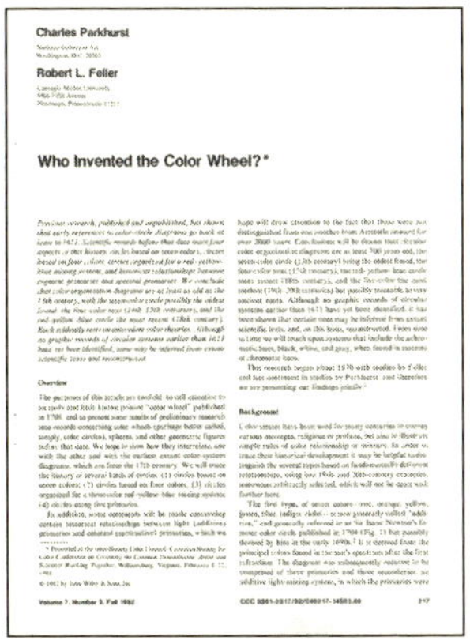
Charles Parkhurst

Robert L. Feller

Who Invented the Color Wheel?*

Volume 7, Number 3, Fall 1982

Originally printed in:
COLOR RESEARCH AND APPLICATION
Volume 7, Number 3 (1982), Pages 217-230.

OVERVIEW

The purposes of this article are twofold: to call attention to an early and little known printed "color wheel" published in 1708, and to present some results of preliminary research into records concerning color wheels (perhaps better called, simply, color circles), spheres, and other geometric figures before that date. We hope to show how they interrelate, one with the other and with the earliest extant color-system diagrams, which are from the 17th century. We will trace the history of several kinds of circles: (1) circles based on seven colors; (2) circles based on four colors; (3) circles organized for a three-color red-yellow-blue mixing system; (4) circles using five primaries.

In addition, some comments will be made concerning certain historical relationships between light (additive) primaries and colorant (subtractive) primaries, which we hope will draw attention to the fact that these were not distinguished from one another from Aristotle onward for over 2000 years. Conclusions will be drawn that circular color-organization diagrams are at least 700 years old, the seven-color circle (13th century) being the oldest found, the four-color next (15th century), the red-yellow-blue circle more recent (18th century), and the five-color the most modern (19th-20th centuries) but possibly traceable to very ancient roots. Although no graphic records of circular systems earlier than 1611 have yet been identified, it has been shown that certain ones may be inferred from extant scientific texts, and, on this basis, reconstructed. From time to time we will touch upon systems that include the achromatic hues, black, white, and gray, when found in systems of chromatic hues.

EDITOR'S NOTE: In this paper the references are found as Footnotes beginning on p. 313.

This research began about 1970 with studies by Feller and has continued in studies by Parkhurst, and therefore we are presenting our findings jointly.[1]

BACKGROUND

Color circles have been used for many centuries to convey various messages, religious or profane, but also to illustrate simple rules of color relationship or mixture. In order to trace their historical development it may bc helpful to distinguish the several types based on fundamentally different relationships, using late 19th- and 20th-century examples, somewhat arbitrarily selected, which will not be dealt with further here.

The first type, of seven colors-red, orange, yellow, green, blue, indigo, violet-is now generally called "additive," and generally referred to as Sir Isaac Newton's famous color circle published in 1704 (Figure 1) but possibly devised by him in

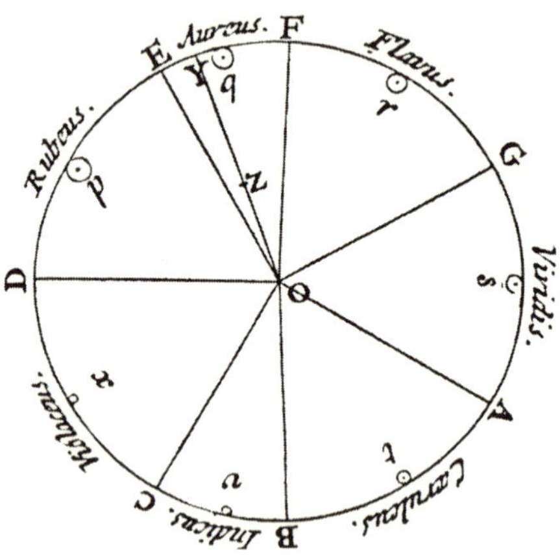

FIGURE 1
Sir Isaac Newton (1704), seven-color circle.

the early 1690s.[2] It is derived from the principal colors found in the sun's spectrum after the first refraction. The diagram was subsequently reduced to be comprised of three primaries and three secondaries, an additive light-mixing system, in which the primaries were variously named, consisting usually of a red, a green, and a blue-violet.

A second circle arranges colors according to principles of color perception, in four primaries—red, green, yellow, and blue—the characters of which have direct phenomenal meaning, providing strong emotional contact and making conceptual abstraction easier.[3] Indeed, these four color categories have been shown through linguistic studies to be the first to enter a language in the abstract form of color names (after white and black) in the order red, green or yellow, and blue.[4] This is the well-known circle used by Ostwald (Figure 2), and before

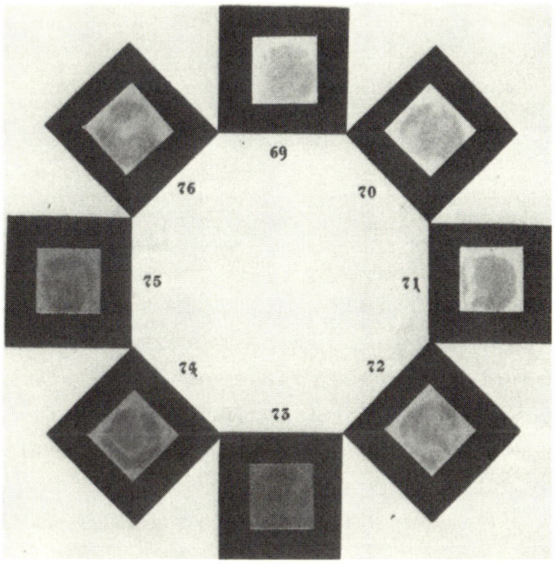

FIGURE 2
Wilhelm Ostwald (1916), four-color circle with intermediaries.

him by the psychologist A. Höfler and also by E. Hering.[5] We shall see later that there are a number of four-color theories lying historically behind this one.

A third type is a five-color circular diagram similar to the Munsell color notation system (Figure 3), in which red, yellow, green, blue, and purple—all of the same value between light and dark—are spaced equally around the circle.[6]

A fourth type, perhaps most familiar to artists, illustrates the relationships of "subtractive" mixture of colorants, by which term we mean here the results obtained from the physical mixing of pigments or dyestuffs. Red, yellow, and blue are placed at the points of an inscribed equilateral triangle. Probably the

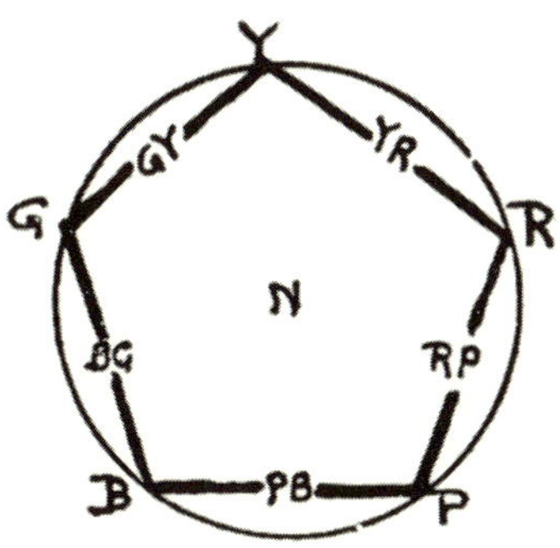

FIGURE 3
A. H. Munsell (1905), five-color equal-value circle (reproduced by permission of Munsell Color Co., Baltimore).

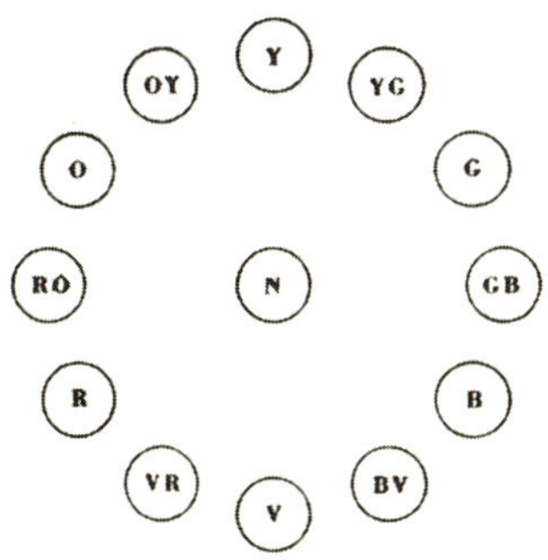

FIGURE 4
Arthur Pope (1929), red-yellow-blue painter's circle, with binary-mixed colorants between.

name most painters in this country would associate with this circle would be Arthur Pope (Figure 4) or, earlier, Hardesty Gillmore Maratta (about 1909).[7]

THE LATE 18TH CENTURY

Intending now to trace the historical development of these four types of color diagrams, let us look first at examples of color circles from about 1769 to 1810 that have been identified to date, as follows in reverse chronological order. We are concerned here only with color circles, although other geometric shapes have been used and we will refer to such diagrams if it is relevant to do so.

Between 1790 and 1823 Johann Wolfgang von Goethe wrote over 2000 pages on color and color theory, much of it published in his *Zur Farbenlehre* of 1808-1810. In 1793 he diagrammed a red-yellow-blue circle with three intermediate binary mixtures producing orange, green, and violet, all of which he derived from a study of prismatic colors as well as paint, but he wished to universalize his theory and present it "as the beginning and end of all color theories." He relied exclusively on the physiologically conditioned colors (complementaries) to explain his color circle.[8] Figure 5 shows sketches of circles he drew, probably

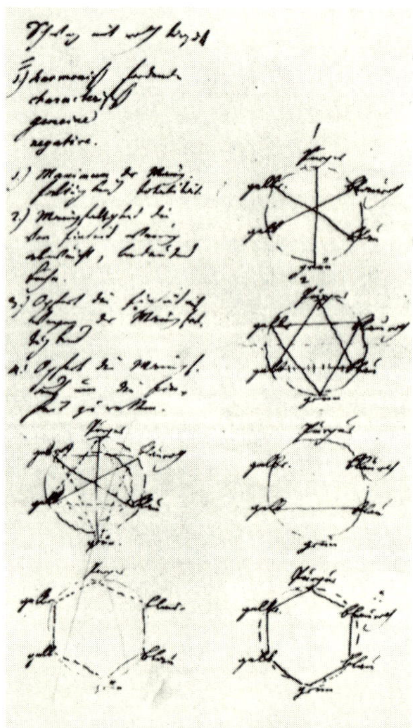

FIGURE 5
Johann Wolfgang von Goethe (1798?), sketches of red-yellow-blue primary circles, with orange-green-violet secondaries between. From *Goethe's Color Theory*, edited by R. Matthaei, translated by H. Aach, © 1971 by Van Nostrand Reinhold, reprinted by permission of the publisher.

in 1798, for his friend J. C. F. Schiller, whose name amusingly relates to the word *schillern*, "to change color."

Goethe is followed closely in his circle by P. O. Runge between 1803 and 1806, for pigment mixing. He had developed a color solid (sphere) by 1807, and published it in 1810.[9]

In 1772 the Jesuit Ignaz Schiffermüller published *Versuch eines Farbensystems*, with a hand-colored plate depicting a color circle based on the four psychological primaries, our Ostwald type, in 12 divisions: 4 reds, 4 blues, 3 greens, and 2 yellows (Figure 6). His is a scholarly work with many bibliographical references to earlier literature, both classical and later sources on color, including earlier three and four-color theories.[10] On his color plate Schiffermüller refers all the prime (blühenden) colors to the rainbow in vignettes at the four corners.

Despite his circle of four colors, Schiffermüller's exposition of his ideas demonstrates the overlap of systems because we find imbedded in it a red-yellow-blue triangle with its points on the circle, very much like Pope's later circle (Figure 4). He subscribes to what he calls the "famous mathematicians" theory of three primaries for paint mixing, red, yellow, and blue, which through mixture provide nine other "lebhaften und glanzenden Farben," "such as we often marvel at in the rainbow, or more clearly separated on a wall from a

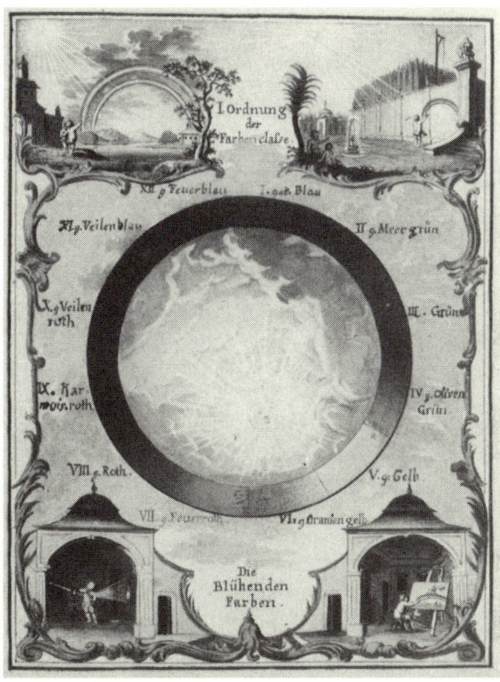

Figure 6
Ignaz Schiffermüller (1772), four-color circle with mixtures between red-blue-green-yellow.

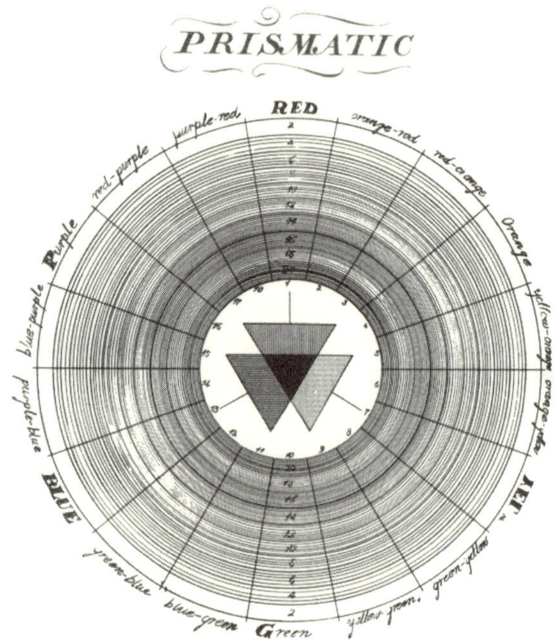

FIGURE 7
Moses Harris (c. 1769-1776), "Prismatic" wheel of red-yellow-blue (after Birren, 1963 facsimile with permission).

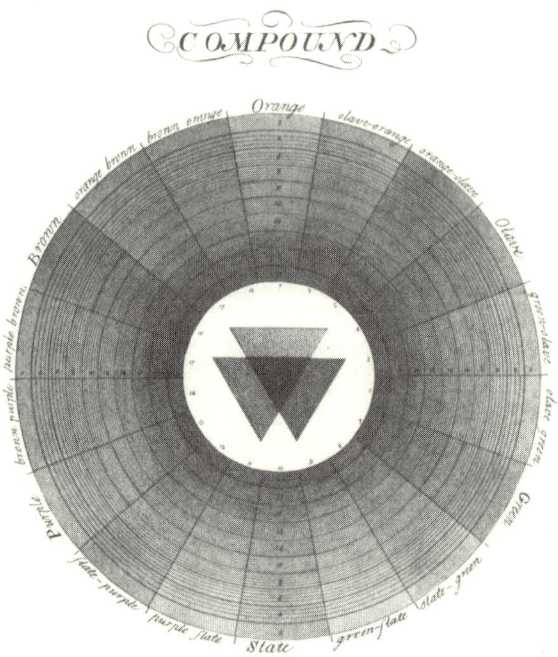

FIGURE 8
Moses Harris (c. 1769-1776), "Compound" mixture of orange-green-violet (after Birren, 1963 facsimile with permission).

sunbeam projected through a prism." He footnotes many earlier works such as those by Newton, Castel, and others. Schiffermüller seems not to make any clear distinction between prismatic and pigment colors, but he suggests that further mixing beyond nine "neben farben" does not yield lively colors.[11] He does not show on his circle two of the three unqualified secondaries, orange and violet. For this reason it should be construed as a four-color circle.

Sometime between about 1769 and 1776 Moses Harris' *Natural System of Colours* was printed with two plates to be hand colored.[12] Harris states that the primary colors of his first circle are those of light, "because it admits of no other colors but such as are shown by the prism," and calling them "prismatic" colors (Figure 7); the other colors are "compound" from them, and shown on the second plate (Figure 8). Nonetheless, his diagrams relate to subtractive color mixture because as his "prismatic" colors he names red, yellow, blue; 15 other colors result from binary mixtures of these 3. His second chart demonstrates the consequences of further binary mixture, this time among the secondaries, creating a third set of binaries—slate, brown, and olive, which may be among the earliest identifications of tertiary hues by names. But, he says, "if the colors were divided into a greater number of tints, or divisions...,it would not render it more useful but rather tend to create confusion between the tints," adding rather sourly, "and would be one step toward reducing it to its primitive chaos, from which the author has taken so much pains to extricate it." Mixing of all his "prismatic" colors will yield a deep black, he says in a brief text, as will mixing of contraries (opposites on the chart). However ambiguous his terminology, Harris' system is, of course, based on the painter's red-yellow-blue subtractive circle.

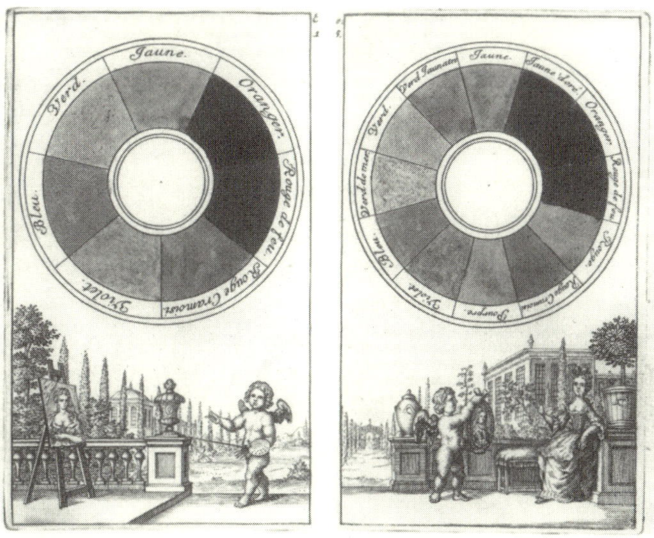

Figure 9
Anonymous (André Félibien?) (1708), color "Encyclopedia" based on red-yellow-blue primaries.

THE "FÉLIBIEN" CIRCLES, 1708-1766

Next let us refer to an even earlier 18th-century printed color circle. In 1708, a new edition of an established book on miniature painting was published by Louis and Henry van Dole, printers in The Hague. It bore the title, *Traité de la Peinture en Mignature*[13] and in its opening paragraph one reads that it is a "new edition with many very considerable additions." Although of uncertain authorship, the dedicatory letter (to a Mademoiselle Foucquet) in the previous edition of this treatise published by van Dole in 1688 is signed with the initials "C.B.," which have been taken most recently, rightly or wrongly, to be those of one Claude Boutet, but sometimes of others. There were at least three later versions of this publication, the latest in 1766. Without going into the full history of this treatise, we should point out that it is based on a first edition of 1673, from Paris, with a slightly different title, *École de la Mignature*.[14] By 1700, 14 editions with one or the other title had appeared, and by 1800 at least 33, such was its popularity. As its fame spread, later editions were published in English, Dutch, Italian, or German.[15]

What is unusual about the 1708, or the second, van Dole edition from The Hague is the addition of new features, as alluded to in the title and in the preface, including a short treatise on pastel which had not been included in the first van Dole edition of 1688. Chapter VIII of the book, which is this treatise on pastel painting, incorporates two printed color "wheels" each with wedges of color between its spokes, on facing pages (Figure 9) with colors applied by hand,[16] of which the author says: "Here are two figures by which one will be able to see how the primitive colors, yellow, red, crimson and blue generate the other colors, and which one might call *The Encyclopedia of Colors*. The first figure includes the four primitive colors and three composed from them, and the second includes those same colors with five others which are produced as much from the primitive as of their composites."

The author has made the three primaries, red, yellow, and blue, into four by adducing two reds, one vermilion (fiery or yellowish red), the other lake or crimson (a little toward blue) which together with three binary mixtures producing orange, green, and violet provides a seven-color circle. Is it more than coincidental that Newton had only recently published the famous seven-color circle in his *Opticks* of 1704, and again in a Latin edition of 1706?

The author of the text of the 1708 van Dole edition states that "all the colors are vivid, but had one mixed them otherwise, for example, orange with violet, fiery red with blue, violet with green, green with orange, or with fiery red, the mixture would have produced only colors dirty and disagreeable." In short, he does not allow binary mixing of secondary colors to produce tertiary colors.

Who could have written this little treatise on pastel, and was he the deviser of its two circles of a type so familiar to our present day artists and which has endured nearly 300 years? At one time we thought it would be a simple research chore to answer these questions, but no longer. Since the treatise on pastel and the circles are not included in van Dole's first (1688) edition, one might presume they were produced in the 20-year period between 1688 and 1708; but by whom? A passage in the foreword[17] to the 1708 edition provides material for speculation, for it reads: "As pastel painting has its partisans as well as miniature painting, there has been added here in their favor, a little treatise which will instruct them in the method of making all the pastels or crayons necessary to practice this sort of painting. Something rather curious is found here with respect to primary colors, and the generation of composite colors; several precepts for actual practice are also inferred. This should be more pleasing for amateurs of this science in that, up to the present one has seen nothing on this subject except a small article which is found in Monsieur Félibien." The phrase "this subject" clearly refers to color mixing from red-yellow-blue primaries. André Félibien died in 1695. There is further reference to Félibien in connection with an alphabetical glossary of terms added at the end of this 1708 edition which is a modified and improved version of a dictionary of terms done earlier by Félibien.

We have found discussion of a three-color system in two of Félibien's earlier essays, the primary one being from his six volumes of essays on the lives and works of the most excellent painters, ancient and modern, first published between 1666 and 1685, and probably widely accessible.[18] In this work Félibien, "speaking as a painter," writes: "There are three primary colors which cannot be perfectly composed of any other, but out of which all the others may be composed, to be known as yellow, red, and blue."[19] The same is put forward in his book on principles of architecture, of sculpture, and of painting, in a significant passage in Chapter 10 dealing with the history of painting in enamels. There the author describes at some length the use of black, red, yellow, and blue upon a white base. In a key paragraph, he says, "The colors which I have just named are the base or often the material from which all others which are used in enamel painting are composed; for it is only necessary to mix them together in order to produce the various shades. This is the same as painters do on their palettes [:] when blue and yellow are mixed, green results; blue and red produce violet, and the others as well."[20]

We do not adduce this passage to prove that Félibien invented or discovered the red, yellow, blue primaries. Far from it; he got these from earlier sources which lead back well over a hundred years, perhaps even to the early 1500s in Italy.[21] But one could adduce it to show that Félibien, or someone close to him in thinking on color, could be author of the 1708 treatise on pastels. On the other hand, since the foreword of 1708 referred to Félibien, why didn't it say more openly that he wrote the pastel treatise, if indeed he did? We can not provide another name at this time.

THE 17TH CENTURY

That the red-yellow-blue mixing system was regarded in some circles around 1708 as a rather novel concept, despite its long prior history, is suggested by the author's statement in the treatise on pastels that it is "something rather curious." We must remember, however, that about this time J. C. Le Blon was probably experimenting with three-color (red-yellow-blue) printing and was to publish what he called this "mighty secret" of coloring in the 1720s.[22] Dr. Alan E. Shapiro has informed us in a recent letter that "Newton invokes the three primaries red, yellow, and blue in his *Optical Lectures* (1670-72), having learned of them from Boyle and a paper on dyeing in Sprat's *History of the Royal Society* (1667). He did not, however, mention them in his *Opticks* (1704)."

Nor, of course, is the idea of arranging colors in a circle new in 1708. We have already noted that it was published by Newton in 1704 in the prismatic color circle. Tracing further the historical appearance of the color circle, one finds in a manuscript on physics at the Royal Library, Stockholm, two color circles drawn by its author, Sigfrid Forsius, Swedish astronomer, clergyman, and neo-Platonist (Figure 10). These color circles, the earliest known, are dateable in 1611 and have been published by Feller and Stenius after the Forsius manuscript came to their attention in a special exhibition for the First Congress of the International Colour Association in 1969. Within 20 years of Forsius' work the first printed color circle appears, in a medical text by Robert Fludd (Figure 11), with blue, green, red, and two yellows. It contains vestiges of Aristotle in its references to the chromatic colors in terms of the contraries, white and black.[23]

The Forsius circles are found in Chapter VII, "On Vision." After a brief review of classical notions regarding the five senses, including vision and how one sees colors, the author discusses his diagrams: "Among the colors there are two prime colors, white and black, from which all others have their origin.... In the middle between these colors, red since classical antiquity has been placed on one side and blue on the other; yellow between white and red, pale yellow between white and yellow, orange between yellow and red. . ." and so on he goes all around the circle. It is worth comment that this scheme is indeed, as he says, based on classical sources, for it contains vestiges of Aristotle's color theories, as we shall presently see.

For us, however, the second color circle of Forsius is perhaps even more interesting for it lays out in curved lines, suggesting a sphere, four principal colors, red, yellow, green, and blue. On this circle Forsius comments as follows: "But if you want right to consider the origin and relation of the colors, you should start from the five principal middle colors which are red, blue, green, yellow and gray of white and black, and their gradings, as they rise either closer to white by their paleness or to black by their darkness. . ." He then describes fully this diagram.

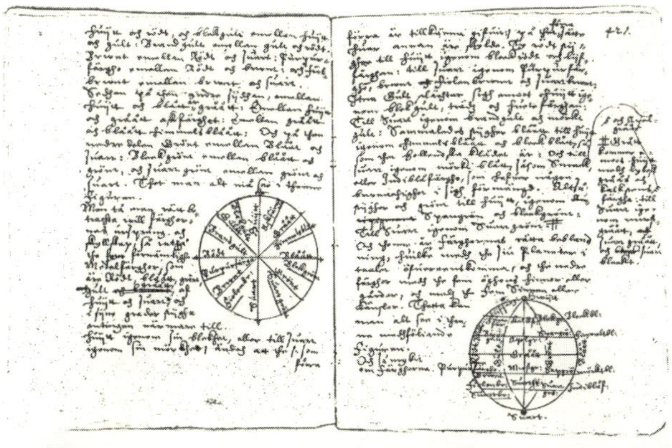

Figure 10
Sigfrid Forsius (1611), manuscript with four-color and twelve-color circular diagrams.

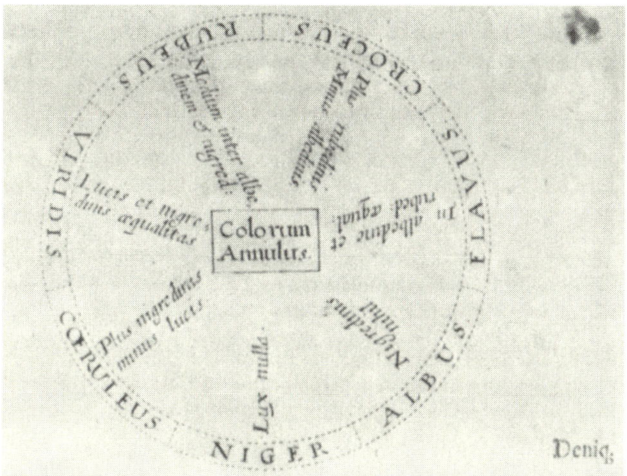

Figure 11
Robert Fludd (1629-1631), the earliest known printed color circle, designating four colors (blue, green, red, and two yellows) with reference to their values in terms of white and black.

In short, he has set down four principal colors, each in scale of light to dark, with gray scale down the center.

Forsius' is the earliest drawing of a geometrical color scheme we have found. One may read it as a sphere without hesitation. We say "without hesitation" because behind this display of color geometry lie several centuries of color mathematics, to which we should now turn. To do so, let us reverse our direction and, instead of continuing our pursuit backward through time, leap directly to color theories of classical antiquity and work forward, returning by this route through nearly 1900 years once again to Forsius.

ANCIENT, MEDIEVAL, AND RENAISSANCE COLOR SYSTEMS

The works upon which medieval and Renaissance science was founded, as all know, were preponderantly those of Aristotle, and this Aristotelian foundation is strikingly demonstrated in the science of color, which has always been of consuming interest to optical branches of science.[24] Aristotle's color notions, and especially his one-dimensional color scale, were a consistent part of his photometric system of optics, including color, light, and vision as found throughout his scientific corpus, but particularly found in connection with his general color theory in *De sensu et sensibili*, and with his spectral or rainbow color theory in *Meteorologica*. Setting the latter aside for the moment, and referring to his system as presented in *De sensu*, one finds (Figure 12) the contraries, white and black, placed at the extremes of a line or "scale" of colors, each representing a fixed position between light and dark (or its "value," as the term is used today by Munsell et al.) signified by the name of a hue which is therefore held to be fixed and invariable as to its value or its position between light and dark.

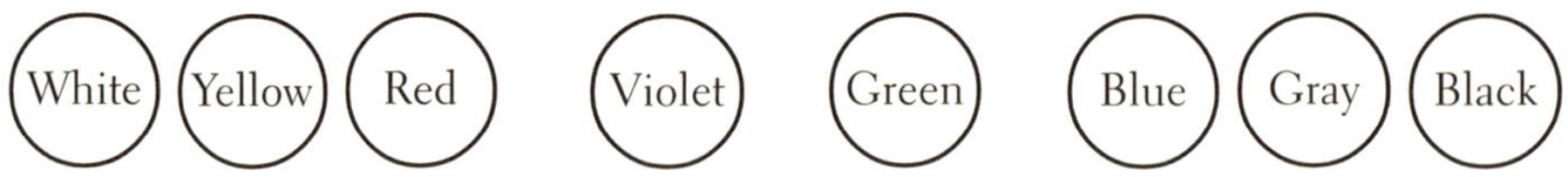

FIGURE 12
Aristotle's color scale reconstructed as a photometric scale of fixed hues between black and white.

Aristotle is certain only of the number of steps in his scale, for he says there are seven, including the extremes of white and black. He is less firm about the hue names to be used, providing eight, with yellow and gray as alternatives. Aristotle's hue names, nonetheless, were each symbolic of a fixed value between lightness or darkness, providing artists with a series of "value points" on his color scale, and these indeed were used by artists over the centuries that followed, blue signifying darkness, for example, and red lightness.

By the early 13th century, if not earlier, marked changes had taken place in Aristotle's theory, as is confirmed by a little treatise of 50 lines, *De colore*, written about 1230-1235 at Oxford, one of four books on optics contributed by the first chancellor of Oxford University, Robert Grosseteste (c. 1168-1253). Grosseteste offers a general theory of color with the following features.

Seven colors, which he does not identify by name (their number, however, is reminiscent of Aristotle) recede from *lux clara* or whiteness, by "remission," descending and changing. By the same reasoning, there is the ascent from *lux obscura* or blackness by "intention" toward the same seven colors. But, Grosseteste continues, they run together in the middle between the extremes of lightness and darkness, so yielding nine colors all together including white and black. Furthermore, he adds that the number of grades of these colors through "intension" and "remission" between whiteness and blackness will be infinite. Thus is the wide range in values of the individual colors (hues) propagated. Although the author provides no diagram for his text, one may reconstruct from his description the geometric model given here (Figure 13). It would appear that Grosseteste may have been the first to set aside the achromatic hues and build a system based on equal-value chromatic hues.

There are several other startling color notions presented here, not the least of which is the 90° rotation of the value-change axis from Aristotle's photometric, linear, one-dimensional color scale. Here, like Aristotle, Grosseteste proposes seven median colors, but unlike Aristotle his seven are of equal value, all midway between light and dark.

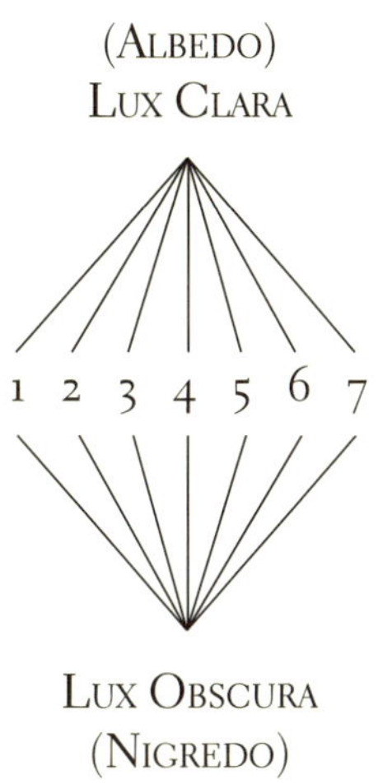

FIGURE 13
Robert Grosseteste (c. 1230-1235), seven-color scheme, reconstructed.

Since, according to Grosseteste, this propagation of species followed geometrical laws, it was understandable through, and demonstrable by, the use of geometrical means. In Grosseteste's own words, "The usefulness of considering lines, angles and figures is the greatest because it is impossible to understand natural philosophy without these." Grosseteste's reference to geometrical "figures" has been shown to mean three-dimensional figures.[25] Thus, while we may visualize light emanating from a point bearing all seven colors with it as a two-dimensional diagram, Grosseteste refers to three-dimensional pyramids of light when writing on optical phenomena like reflection and refraction. Therefore we consider it acceptable to visualize his model with the seven colors descending and seven ascending in cones with a common circular base. Thus from Grosseteste's treatise of the 13th century we have inferred a color solid, with an equal-value color circle between two conical figures (Figure 14). If this is acceptable it anticipates in several aspects color circles and solids that appear later. How Grosseteste's ideas on color were transmitted from the 13th through later centuries is another story; but one can say that his Oxford color system appears to have been popularized and disseminated by the most famous Oxford scientific encyclopedia; Bartholomew's *On the Properties of Things*, probably written around 1250 while Grosseteste was still active.

Thus Aristotle's linear color scale became a color circle, and became incorporated into a color solid. Notably pertinent among the features of Grosseteste's color geometry are: (1) the seven equal-value chromatic colors; (2) his separation of white and black from the chromatic series and correlation of them with lightness and darkness; (3) the tacit implication that white and black each contain all colors; and (4) the grading of chromatic hues into an infinite number of lighter tints and darker tones.

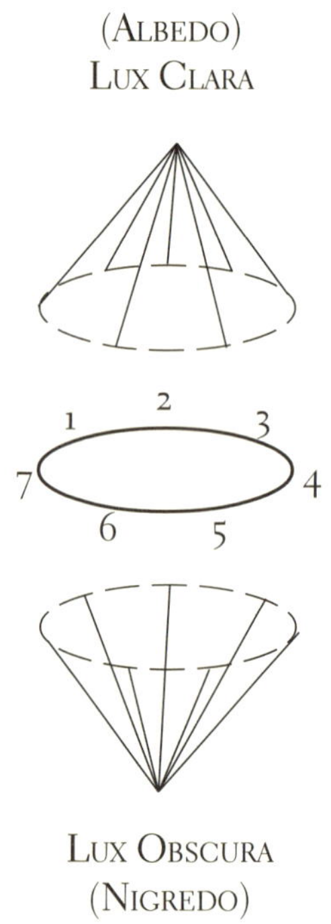

FIGURE 14
Robert Grosseteste (c. 1230-1235), seven-color solid and color circle, conjectural reconstruction.

THE RAINBOW COLORS

Before turning to Renaissance color mathematics one should go back again to Aristotle and examine his other color theory, by which he sought to explain the generation of colors in the rainbow, because it will have a bearing on what will be found later. In his *Meteorologica*, Book III, Aristotle discusses rainbows and says each is made of but three colors, which he names basically as red, green, and blue. Although yellow sometimes appears between red and green, he says it is not primary and its appearance there is owing to contrast of the red and the green (he digresses at this point in his text on simultaneous contrast of colors juxtaposed in weaving, anticipating Chevreul). He emphasizes these three rainbow colors in several other places, as, for example, toward the end of Chapter 4, where he writes: "This, then is why the rainbow is three-colored and why the rainbow is made up of these three colors only."

Moreover, in this same book he reasserts the count of the three colors, this time in connection with paint colors for, he says, "These [three] colors are almost the only ones that painters cannot manufacture; for they produce some colors by a mixture of others, but red, green and blue cannot be produced in this way . . ." Thus Aristotle identified three rainbow (additive) primaries and aligned three painters' (subtractive) primaries with them. The clear separation of the two will not occur until over 2,100 years later.

As in the case of Aristotle's seven-color photometric theory, his three-color theory of the rainbow in due course was also modified, becoming a four-color rainbow theory, although the latter did not replace the former, but ran parallel alongside the older, Aristotelian version. This began to take place as early as the 6th or 7th century, and there is also evidence of changes taking place in Arabic color science from about the year 1000, where four colors are recognized in the rainbow, sometimes corresponding to the four qualities (hot, cold, moist, and dry), the four elements (earth, air, water, and fire), the four humours (blood, phlegm, black bile, and yellow bile), and the four seasons. The four colors were, not surprisingly, red, yellow, blue, and green.

In this way, without tracing in any further detail here the modification of Aristotle's three-color rainbow theories, we discover a shift to a rainbow with four colors. This may have occurred as early as the year 1000, all in the experimental vein of medieval optical science, but still equated with painters' primaries.

By the time we reach the Dominican scientist Theodoric of Freiberg, who wrote a treatise on the rainbow about 1310, we find a report of the author's experimental studies and conclusions, supported by marvelous and famous geometrical diagrams of rainbow color generation. We read in Chapter I, consequent upon several experiments, all of which are described, that ". . . it can be most certainly asserted that the said colors which appear in the rainbow are truly four," which he names as red, yellow, green, and blue. In Chapter III, containing a more general theory of color, Theodoric calls these four "primary middle colors," and says they can be mixed "each with each," i.e., by binary mixing.[26]

Theodoric also deals specifically with the phenomenon of loss of color in shadow, and states that all colors other than the four rainbow colors are not simple, or primary, but the consequence of mixture. He likewise aligns his four colors with the four elements, and he continues to regard the spectral (additive) or rainbow primaries and the pigment (subtractive) primaries of painters as one and the same. Once established by Aristotle, this identification of the two persisted for 16 centuries to Theodoric of Freiberg, and as we have seen will persist for at least another six. Put another way, until the mid-19th century no distinction was made between the origins of reflected object colors and refracted rainbow colors.

Renaissance color theorists received such scientific knowledge based on optical experiments and made it the basis for using four primary hues in a general color and paint-mixing system. Probably the first to do so was Leon Battista Alberti, who wrote the first known statement on color doctrines of the Italian Renaissance. Concentrated in a few lines of his book on painting of 1435, his ideas are somewhat obscure as written. It has been shown recently[27] that one may reconstruct mathematical models of his color system (Figure 15) with four colors. The colors are identified with four colors in the rainbow and the four elements—fire, air, water, earth—as red, blue, green, and yellow, in that order, which as paints may be mixed in pairs. The four colors are arranged to close on themselves in a circle (or possibly a square) and are, as in Grosseteste's scheme, chromatic "middle colors" of equal value with respect to achromatic white and black. Alberti's scheme may also, of course, be therefore reconstructed as a double cone or as a double pyramid (octahedron) (Figure 16), but might as well have been spherical (Figure 17), with complementary pairs, red-green and blue-yellow, opposite each other in whichever figure. In any of these figures a gray scale might be inferred and would be axial. This, of course, could be the ancestor of the color sphere of Forsius (1611) (Figure 10).[28]

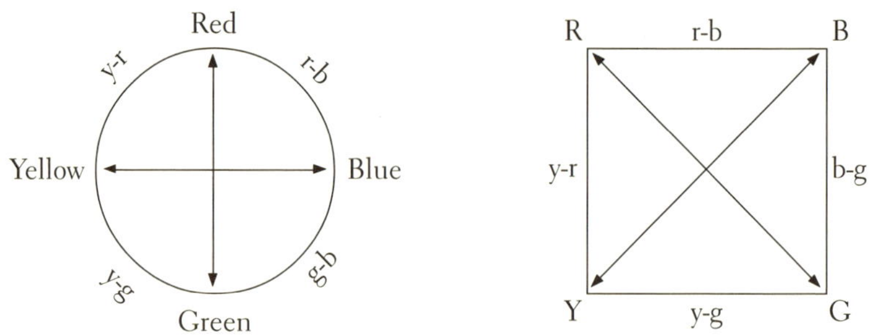

FIGURE 15
Leon Battista Alberti (1435), four-color circle or square, reconstructed.

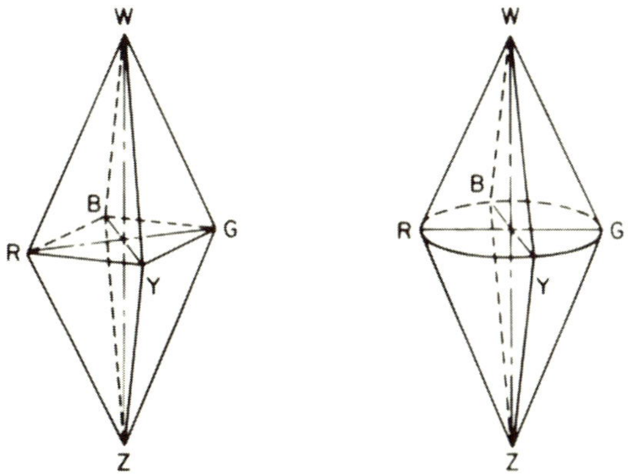

FIGURE 16
Leon Battista Alberti (1435), four-color solid, reconstructed as octahedron or double cone.

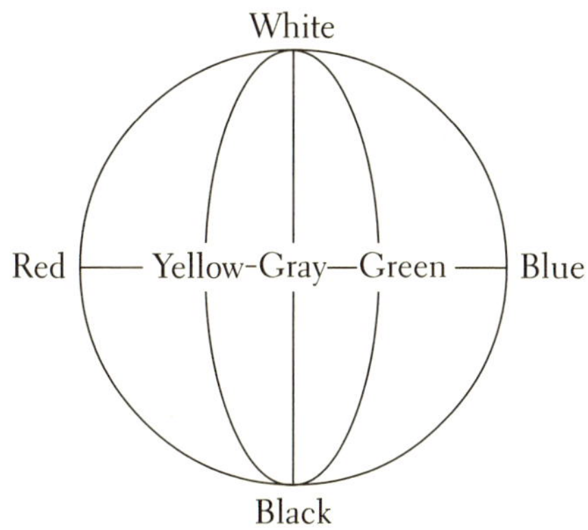

FIGURE 17
Leon Battista Alberti (1435), four-color solid, reconstructed as sphere.

Alberti's scheme, which was a practical one intended for use by artists, went on strongly for two or three generations, and evidence of its influence is found in the notebooks of Leonardo da Vinci (c. 1505-1515). However, there were some undercurrents of misgiving on his part. Leonardo was disturbed by inclusion of the color green, which, as any painter knows is not a primary subtractive color but is instead a secondary, and may be obtained by mixing yellow and blue pigments. Thus, although green may be among the principal colors in additive color mixture and unique in psychological terms, it cannot be among the subtractive primary colors of empirical paint mixing.

Thus, the four-color circle of Ostwald also has a long and mixed ancestry, going back to Alberti in 1435 as to paint, and farther back to Aristotle's rainbow theory as spectral color. The Aristotelian identification of spectral with painters' colors was retained as late as Sir David Brewster, 1834, who thought he could reunite them in a red-yellow-blue color system; but the fallacies of his arguments for this were promptly exploded by other scientists.[29]

From Leonardo onward for almost a century—about four generations—few written documents of color systems have been found, and they are chiefly classical, and based directly upon Aristotle, whose writings became a widely disseminated "Bible of Science," owing largely to the development of printing in the late 15th century. (George Sarton has estimated conservatively that over six million printed books existed by the year 1500.) The more modern four-color system does not appear to have found its way into printed works, perhaps because it was already being superseded in the studio practices of painters. What seems to have happened is that, starting possibly in Leonardo's work, there are discoveries that painters are best served by an empirical three-color mixing system of red, yellow, and blue. Although we do not know that any such system was recorded in a published color circle until 1708 (van Dole), a three-color scheme can be shown to underlie 16th-century Italian painters' color choices after about 1516. Two are recorded in 1609 and in 1613, and as a semicircle we find the earliest known red-yellow-blue diagram (Figure 18).[30]

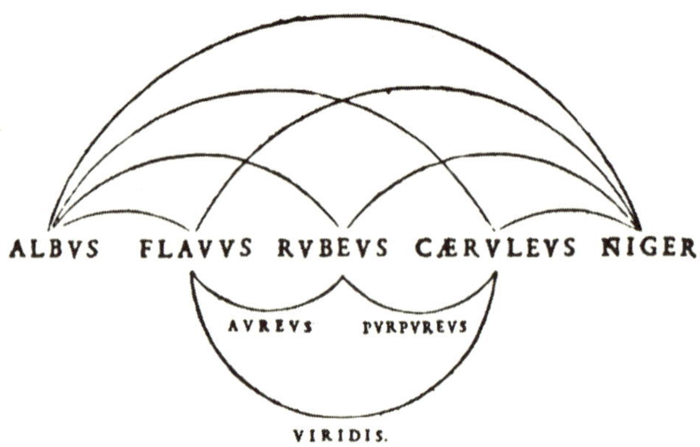

FIGURE 18
Franciscus Aguilonius (1613), red-yellow-blue color diagram with secondaries, orange-green-purple, all between black and white extremes.

CLOSING THE CIRCLE

What then can one say further on this subject, pending more research and perhaps a lucky discovery of some heretofore-unknown color circles? The history of color wheels seems to have no beginning and no end. We hope to have demonstrated that seven-color theories go back to Aristotle, and became circular by 1235 in Oxford; and insofar as chromatic hues only were concerned, Aristotle's was a five-hue system (though he included the achromatic white and black in his seven-color presentation) very similar in its color names to the five of Munsell; four-color theories are more complicated, being psychologically important in primitive and later societies, but this quartet became synonymous with four-color rainbow theories in the middle ages and this relationship between "additive" and "subtractive" colors lingered into the 19th century. Red-yellow-blue as painters' primaries emerged in the early 16th century, but mathematical diagrams for this exist only from the 17th century.

It would have been more to our liking to have achieved what was our first goal, to find a precise moment and place for the "invention" of the color wheel, and to name its inventor. But this goal has eluded us, if indeed there was such an instant in history. Even the authorship of the first printed color wheel of 1708 remains uncertain. We have shown, nonetheless, that the idea of a color circle was not new in the 18th century, nor does it appear to have originated in the 17th century, but much earlier. We can only say we hope further research will both confirm and augment our findings of a very long prior history for such a common and logical device.

Footnotes

1. We are indebted to several persons who over the years have contributed suggestions, corrected errors, provided new bibliographical leads, or assisted in various other ways, among them Ruth Johnston-Feller, Ulrich Middeldorf, William Heckscher, Carl Nordenfalk, Kay Silberfeld, Paul D. Schweizer, Barbara J. Balsiger, Anna M. Link, J. Melvin Edelstein, Carolyn Backlund, Thomas McGill, Paul Arnold, and Fred Cain, and to Jean Stensland for patiently retyping many drafts. The most recent and useful studies in the science of color are Heinz Matile, *Die Farbenlehre Philipp Otto Runges, 2nd ed.*, Mäender Kunstverieg Munich-Mittenwald (1979), and Thomas Lersch, "Farbenlehre," in *Reallexikon zur deutschen Kunstgeschichte*, VII (1974), cols. 157-274. We shall use the modern terminology "additive" and "subtractive" throughout to indicate the principles of mixing colored lights or colored pigments, respectively, rather than historical usages, although some of the older terms are perhaps more descriptive: "colors . . . which our mind can attain free from matter. . . . [as opposed to] those which we see interwoven in matter" [F. Aguilonius (1613)[30]] or similarly the ingenious terms employed by James Christopher Le Blon, as "impalpable colors" for our additive, and "material colors" for our subtractive colors [as reported by Cromwell Mortimer (1731)[22]]. Sometimes the phrase "color circle" is spoken of as a "color wheel" and in at least four instances as a "color spiral."[14]

2. Newton's published circle has seven colors, although he held that sunlight contained innumerable colors. On the evolution of Newton's theory of color see the excellent recent study by Alan E. Shapiro, The evolving structure of Newton's theory of white light and color, *Isis* 7, p. 211-233 (1980). Newton (1642-1727) for 18 months after graduation from Trinity College, Cambridge, returned home to Woolsthorpe to think and figure, a period spoken of as the "most fruitful eighteen months in all the history of creative imagination." By January 1666 he had his theory of colors. In February 1672 he sent a letter to the Royal Society on this theory of light and colors (*Philos. Trans.*, No. 80, February 19, 1672). See I. Bernard Cohen, "Isaac Newton," *Sci. Am.* 193, 6 (1955), p. 73-136; George Sarton, "Discovery of the dispersion of light and of the nature of color," *Isis* 14 (1930), p. 326; and Michael Roberts and Ebenezer R. Thomas, *Newton and the Origin of Colours*, G. Bell and Sons, London (1934). Since the opposite pairs of colored lights in Newton's diagram are complementary and should produce a perfect, or sunlight, white, we believe this was the basic reason for his using a circular diagram, although he wrote, "I could never yet by mixing only two primary colors produce a perfect white" (Bk. I, pt. 2, prop. 6, prob. 2), and he observed not a circle, but an oblong spectral image. However, Newton completed the

colors of his spectrum "with those purple hues which, although not present in the spectrum, were familiar to painters and dyers, and, in this way, closed up the color circle into a band returning on itself" [Wilhelm Ostwald, *Colour Science*, tr. J. Scott Taylor, Winsor and Newton, London (1931), I, p. 6]. The little circles on his remarkable diagram near the circumference apparently represent weighting factors, proportional to the intensity of each color, and are suggestive of a spiral concept.[14] For discussions sec K. T. A. Halbertsma, *A History of the Theory of Colour*, Amsterdam (1949), p. 45-47, and Richard S. Westfall, "The development of Newton's theory of color," *Isis* 53 (1962), p. 339-358.

3. Sometimes referred to as "psychological primaries." See Benjamin J. Kouwer, *Colors and Their Character, a Psychological Study*, M. Nijhoff, The Hague (1949), p. 30 ff., and p. 160. On the circular image, as opposed to the oblong image for a projected prismatic spectrum, see Westfall (1962), p. 351 ff.

4. Berlin, Brent and Kay, Paul, *Basic Color Terms*, University of California Press, Berkeley and Los Angeles (1969).

5. Ostwald, Wilhelm, *Die Farbfibel*, Unesma, Leipzig and Grossbothen (1916). Alois Höfler, *Psychologie*, Leipzig, Holder-Pichler-Tempsky (1897) (2nd ed., augmented, Leipzig (1930), p. 307 ff.). Höfler's four colors were presented not in circular but in square format, creating an octahedral color solid. E. Hering, *Zur Lehre vom Lichtsinne* (1905-1911) (but based on earlier work), translated by Leo M. Hurvich and Dorothea Jameson in *Outlines of a Theory of the Light Sense*, Harvard University Press, Cambridge, MA (1964), especially p. 42-48.

6. Munsell, A. H., *A Color Notation*, Munsell Color Co., Boston (1905). This system is today in wide use for notation and standards, and in industrial and technical applications (12th printing, Munsell Color Co., Baltimore (1971)).

7. Pope, Arthur, *An Introduction to the Language of Drawing and Painting*, Harvard University Press, Cambridge, MA (1929). Pope's three-dimensional contruction provides a double (cool-warm) spiral of light-to-dark hues. See also the exhibition catalogue on Pope's work, *Color in Art*, Fogg Museum, Cambridge (1974), *passim*. On Maratta see William I. Homer, *Robert Henri and His Circle*, Cornell University Press, lthaca and London (1969), p. 184 ff. We owe this reference to Paul D. Schweizer.

8. *Goethe's Color Theory*, arranged and edited by Rupprecht Matthaei [American edition translated and edited by Herb Aach, Van Nostrand Reinhold, New York (1971), p. 37-56, and illustrations 42, 50, 78, and 79]. Speaking of additive and subtractive colors, Goethe, writing on red, said, "As a pigment it cannot be produced by inter-mixture or union (as in physicists' prismatic experiments), but only by arresting the hue in substances chemically acted on, at the high culminating point. Hence the painter is justified in assuming that there are *three* primary colors from which he combines all others." Matthaei-Aach (1971) p. 159, par. 705, and p. 174 (footnote).

9. Runge, Philipp Otto, *Die Farbenkugel*, F. Perthes, Hamburg (1810). See Matile, Heinz, *Die Farbenlehre Philipp Otto Runges* (1979), p. 154-64.

10. Schiffermüller, Ignaz, *Versuch eines Farbensystems*, A. Bernardi, Wien (1772). Other, earlier, color theorists referred to include, for example, Anselm de Boodt, 1609 (a three-color, red-yellow-blue theory), but Schiffermüller cites also a four-color prototype, Leo Battista Alberti's *De pictura* (1435) (p. 20, fn.). Professor William S. Heckscher called our attention to Schiffermüller in 1954.

11. Schiffermüller (1772), p. 8-18, and footnotes.

12. Harris, Moses, *The Natural System of Colours*, Laidler, London, n.d., new edition with additions (edited by Thomas Martyn) London (1811). In copies we have examined the colors of the hand-painted plates have deteriorated badly. A facsimile edition in which the plates have been approximated in modern printing inks has been published with notes and commentary by Birren, F., privately printed, New York (1963). On the date of this work see J. Gage, *Colour in Turner*, Frederick A. Praeger, New York and Washington, DC (1969), p. 222, note 13.

13. van Dole, L. and H., *Traité de la peinture en mignature . . . auquel on a ajouté un petit traité de la peinture au pastel, avec la méthode de composer les pastels*, The Hague (1708). A copy purchased by R. L. Feller in 1968 was presented by him to the library of the National Gallery of Art, Washington, DC.

14. The three later 18th-century editions we have found are (1) *Verhandeling van de schilderkunst in miniatuur* . . . Utrecht (Lobedanius) (1744); (2) *Verhandeling van de schilderkunst in miniatuur . . . Verbetert en vermeerdert . . . Tweede druk* . . . Amsterdam (Gerrit de Groot) (1759) (translated into Dutch by P. J. Verly?); (3) *Élémens de peinture pratique, par M. de Piles* . . . Amsterdam and Leipzig (Arkstee and Merkus) and Paris (C.-A. Jombert) (1766). This last conflation with de Piles' text has led to much confusion. See, for example, Ellen Heuck, *Die Farbe in der französischen Kunsttheorie des 17 Jahrhunderts*, J. H. Heitz, Strasbourg (1929), who mistakenly assumed that the Jombert edition of 1766 was identical with an original de Piles edition of *Les premieres Élémens de la peinture pratique* of 1684, which although it contains a chapter on pastels, does not treat color mixing or contain the color wheels. See also Bernard Teyssèdre, *Roger de Piles et les débats sur le coloris* . . . , La Bibliothèque des Arts, Paris (1957), p. 648. For other information on the problems posed by editions of Boutet see also M. Paillot de Montabert, *Traité complet de la peinture*, Paris (1829-1851), Vol. 1, p. 298-299, where it is noted that Roger de Piles (in *Élémens de la peinture*) attributed the work on miniature painting to Mademoiselle Foucquet. See also Stev. F. Constant-Viguier, *Manuel de miniature et de gouache*, Paris (c. 1829), who notes [3rd ed., Paris (1836), p. 29] several editions and authors, as follows: "attributed in turn to Mlle. Foucquet, after de Piles, to [Claude?] Bailly and to Claude Boutet, according to a handwritten note in the copy of the library of the King; and to La Voye-Mignot, according to the author of the *Dictionnaire des pseudonymes* (Barbier). For the first time this book has been cited in this *Manuel* as the work of Christophe Ballard, printer of music in Paris." The suggestion here that La Voye-Mignot is our author is intriguing because of the traditional connection between color and musical sound first developed fully by Aristotle. La Voye-Mignot wrote *Traité de Musique*, Paris (1656), and 2nd ed., Paris (1666), and this work was published by Robert Ballard, "seul Imprimeur du Roy pour la musique," the same house that published *École* in 1673 and later. However, we can find no evidence to support this authorship. Little seems to be known about La Voye-Mignot except that he was a geometer who published comments on Euclid in 1649. For more recent comment, see R. D. Harley, *Artists' Pigments c. 1600-1835*, American Elsevier, New York (1970), p. 21 (there given to Boutet). Moreover, color circles adapted from the 1708 pastel treatise appeared in 1740, in a work by the Jesuit, Louis Bertrand Castel, deviser of the "clavecin de couleurs." In *L'optique des couleurs* he describes two color circles (p. 167-173) in relationship to musical tones, but does not diagram either. However, the first may be reconstructed as a seven-color circle of the red-yellow-blue type, the intermediate hues being, of course, the secondaries orange, green, and violet, the seventh color being a second red, crimson, just as in the pastel treatise of 1708; he also describes a twelve-color circle, and here also appears to be following the 1708 wheel, albeit with a somewhat different set of hue names. It seems reasonable to suggest here that Castel knew and adapted the 1708 scheme to his musical analogy. Equally interesting is his description of his color circle as a spiral, even to diagraming a small "figure de limaçon" within his text (p. 165), a geometric notion that will be picked up only much later and three dimensionally by Goethe, A. Kirschmann (1895) and A. Pope (1939).[2, 7] On Kirschmann see H. Matile, *Die Farbenlehre Philipp Otto Runges* (1979), p. 211 ff., and on Castel p. 67 ff. Matile appears not to be aware of the color circles in the 1708 treatise on pastels we are examining. As noted above,[8] Goethe also devised a six-color circle in 1793, repeated in 1798, and a color spiral. See H. Matile (1979), p. 232 and references. Goethe was at least aware of Castel's work. See Matthaei-Aach (1971), p. 169 and ills. 49, 50.

15. By another count, we have found 14 editions from 1673 to 1817 with the title beginning "School of"; 17 editions from 1674 to 1802 with the title beginning "Treatise on"; and in addition four editions from 1708-1766 of works containing the added section on pastels. Our research has been greatly aided especially by Professor Ulrich Middeldorf of the German art history institute in Florence, Italy, by Professor Paul Arnold and the staff of the Oberlin College art library, and by Barbara Butts, graduate student in art history, Harvard University, when she was an intern at the National Gallery of Art.

16. There seems little doubt they were intended to be hand colored, probably by their owners, and each of the four copies we have examined (Darmstadt; E. Berlin; Harvard; National Gallery of Art, Washington) seems to vary from the others, or be painted with different pigments. Such

owner participation was not a new idea. For example in W. Goeree, *Verlichterie-kunde . . .*, Middlelburgh (1670), opp. p. 2, has a plate of empty rectangles numbered from 1 to 37, which correspond to the list of 37 paints given opposite. In the copy, which may be found in the library of the Rijksbureau voor Kunsthistorisches Documentatie, The Hague, 13 of these rectangles have been filled in, rather sloppily, by some previous owner.

17. "Avertissement" fol. 3, par. IV, and par. VI.

18. A. Félibien, *Entretiens sur les vies et les ouvrages des plus excellens peintres anciens et modernes*, vols. I-VI, Paris (1666-1685). It appeared in a second, two-volume edition, Paris (1685-1688); a third, four-volume edition, London (1705); and a fourth in Amsterdam in 1706. See B. Teyssèdre, *Les débats . . .*, Paris (1957), p. 590-591. The dictionary of terms is from Félibien's *Principes*.[20]

19. *Entretiens . . .* edition of 1685-1688, vol. 1, p. 624-625.

20. A. Félibien, *Des Principes de l'architecture, de la sculpture, de la peinture, et des autres arts qui en dependent . . .*, Paris (1676). Other editions in 1690, 1699. See B. Teyssèdre, *Les débats . . .*, (1957), p. 591. The quotation given is from the 1699 ed., Bk. III, p. 311.

21. On earlier red-yellow-blue systems for color mixing see C. Parkhurst, "Aguilonius' optics and Rubens' color," *Nederlands Kunsthistorisch Jaarboek*, 12 (1961), p. 35-49; "A color theory from Prague: Anselm de Boodt, 1609," *Allen Mem. Art Mus. Bull.*, 29, 1 (1971), p. 3-10; "Red-Yellow-Blue: A color triad in seventeenth-century painting," *Baltimore Mus. Art Ann. (Studies in Honor of Gertrude Rosenthal)*, 4 (1972), p. 33-39 ; "Camillo Leonardi and the Green-Blue Shift in Sixteenth-century painting," in *Intuition und Kunstwissenschaft: Festschrift für Hans Swarzenski*, Berlin (1973), p. 419-425; and "Louis Savot's 'Nova-Antiqua' Color Theory, 1609," in *Album amicorum J. G. van Gelder*, M. Nijhoff, The Hague (1973), p. 242-247. Later in the 17th century the red-yellow-blue system appears to be accepted and published by several others in addition to Félibien,[18, 19] among them A. Kircher, *Ars magna lucis et umbrae . . .*, Amsterdam (1646); it is reflected in lists of dyes and paints in John Tradescant, the Younger, *Museum Tradescantianum*, London (1656) (we owe this reference to Barbara J. Balsiger in 1954); Johannes Scheffer, *Graphica, id est de arte pingendi*, Nurnberg (1669); Elias Brenner, *Nomenclatura et species colorum . . .*, Stockholm (1680); Edme Mariotte, *Essai de la nature des couleurs*, Paris (1681); Filippo Buonanni, *Recreatio mentes, et oculi . . .*, Rome (1684); Johannes Zahn, *Oculae artificialis telediopticus sive telecopium* III, Würzburg (1685); R. Waller, "A Catalogue of Simple and Mixt Colours," *Philos. Trans. R. Soc. London*, 16, No. 179 (1686-1687), p. 24-32. On several of the above see Matile, H., *Die Farbenlehre Philipp Otto Runges* (1979), p. 54-56.

22. Le Blon, J. C., *Coloritto, or the Harmony of Colouring . . .*, London (c. 1726?). Le Blon may have invented and developed the three-color printing process. On the dating of Le Blon's color printing, see H. Singer, "Le Blon's three-color prints," *The Studio*, 28 (1903), p. 261; see also C. Mortimer, *An Account of . . . Le Blons Principles of Printing . . .*, London (1731).

23. Sigfridus Aronus Forsius, *Physica*, Cod. Holm. D. 6, Royal Library, Stockholm. Transcription by J. Nordström, in *Acta Universitatis Upsaliensis*, 10 (1952). The color circles were first discussed by Robert L. Feller and Åke S:son Stenius, "On the color space of Sigfrid Forsius 1611," *Color Eng.*, 8, No. 3 (1970), p. 48-51. We are indebted to Mr. Stenius and to Britt Reeves for translations of key passages, and to Professor Carl Nordenfalk for clarification of one very difficult passage. For Fludd's circle see Robert Fludd, *Medicina Catholica . . .*, 2 vols., C. Rötelli, Frankfurt (1629-1631). The circle is printed in vol. II, p. 154. We owe this reference to Dr. Alan E. Shapiro, Associate Professor of the History of Science and Technology, University of Minnesota (letter to Parkhurst, November 30, 1981).

24. The history of these earlier color theories, which is put forward here in abbreviated form, is based upon C. Parkhurst, "Alberti's Color Scheme and Some Possible Roots," a paper read at *A Conference on Color and Technique in Renaissance Painting, Italy and the North*, Temple University, September 22 and 23, 1980 (to be published).

25. A. C. Crombie, *Robert Grosseteste and the Origins of Experimental Science 1100-1700*, Clarendon Press, Oxford (1953), especially p. 112, fn. 1 and p. 120, fn. 5.

26. J. Würschmidt, *Dietrich von Freiberg über den Regenbogen . . . , (Beiträge zur Geschichte der Philosophie des Mittelalters*, XII, 5-6) Münster (1914), p. 60-67; and William A. Wallace, *The Scientific Methodology of Theodoric of Freiberg*, The University Press, Fribourg, Switzerland (1959), *passim*.

27. See Parkhurst on Alberti's color scheme.[24]

28. We have found later reflections of the Albertian scheme in W. Ostwald (1916), A. Höfler (1897), E. Hering (1905-1911),[10] but it is worth pointing out that Schiffermüller in 1772 acknowledges Leon Battista Alberti correctly as his forerunner in employing red, green, blue, yellow as his primaries, if somewhat obliquely [Schiffermüller (1772), footnote p. 20].

29. See, for example, Ogden N. Rood, *Students' Textbook of Color . . .* (The International Scientific Series), D. Appleton and Co., New York (1899), p. 108-109. For Brewster's analysis see D. Brewster, "On a new analysis of solar light, indicating three primary colors, forming coincident spectra of equal length," *Trans. R. Soc. Edinburgh* 12 (1834), p. 123-126. Likewise Paul D. Schweizer has found that the painter John Constable was torn between red-yellow-blue rainbow primaries and Newton's heptapartite spectrum when he introduced a rainbow into his painting of *Salisbury Cathedral from the Meadows* (1831) (Collection Lord Ashton of Hyde). Schweizer writes: "For Constable the question as to whether the rainbow was composed of seven homogeneous colors as Newton claimed, or whether this heptapartite spectrum could be further reduced to a more fundamental triad of red, yellow and blue, appears to have been confounded by the optical theories of Sir David Brewster . . ." (from a manuscript to be published by Mr. Schweizer, *Three Colors or Seven? Constable's Rainbow Diagrams and English Color Theory* (1979-1980), generously made available to us).

30. On a red-yellow-blue system see Anselm de Boodt, *Gemmarum et Lapidum historia . . .* , Hanau (1609), and Louis Savot, *Nova, seu verius nova antiqua de causis colorum Sententia . . .* , Paris (1609). For the earliest known red-yellow-blue diagram see Franciscus Aguilonius, *Opticorum libri sex . . .* , Plantin, Antwerp (1613).[21]

Photographs of Authors

Over the years the files have accumulated a number of photos of members of laboratory staff, colleagues who, through their personal interest, enthusiasm and pride in their work, contributed in a major way to the scope and success of the Fellowship's and Center's efforts. It is a pleasure to acknowledge their significant role as co-workers and co-authors of so many of the studies reported herein.

Plate 1
Robert Feller
(© 1959 Sun-Telegraph)

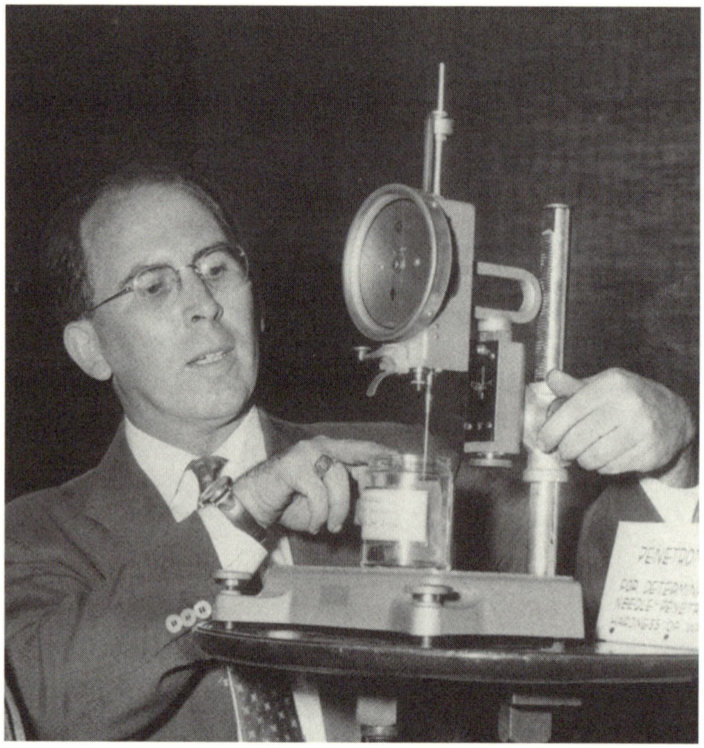

Plate 2
Robert Feller

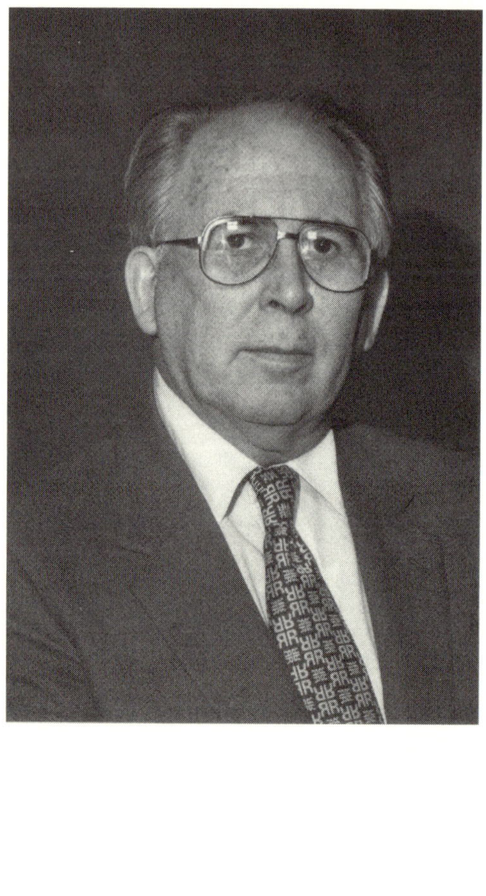

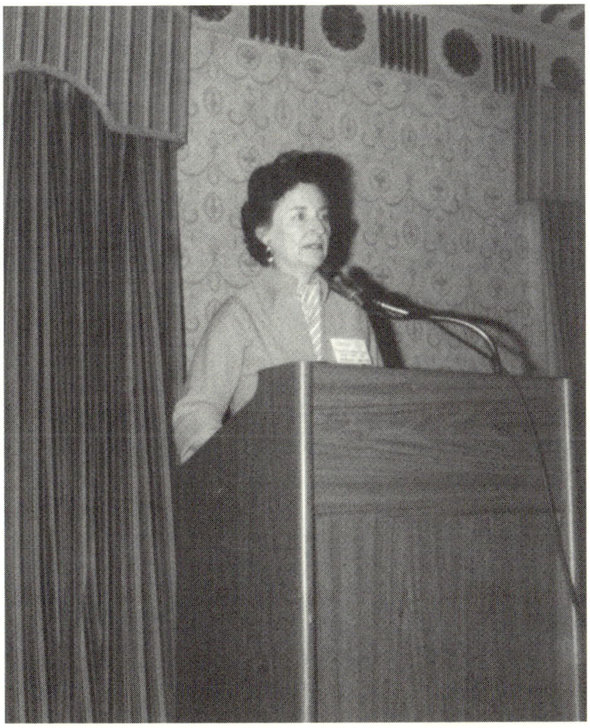

PLATE 3
Ruth Johnston-Feller, senior scientist at the Center, 1975-1989

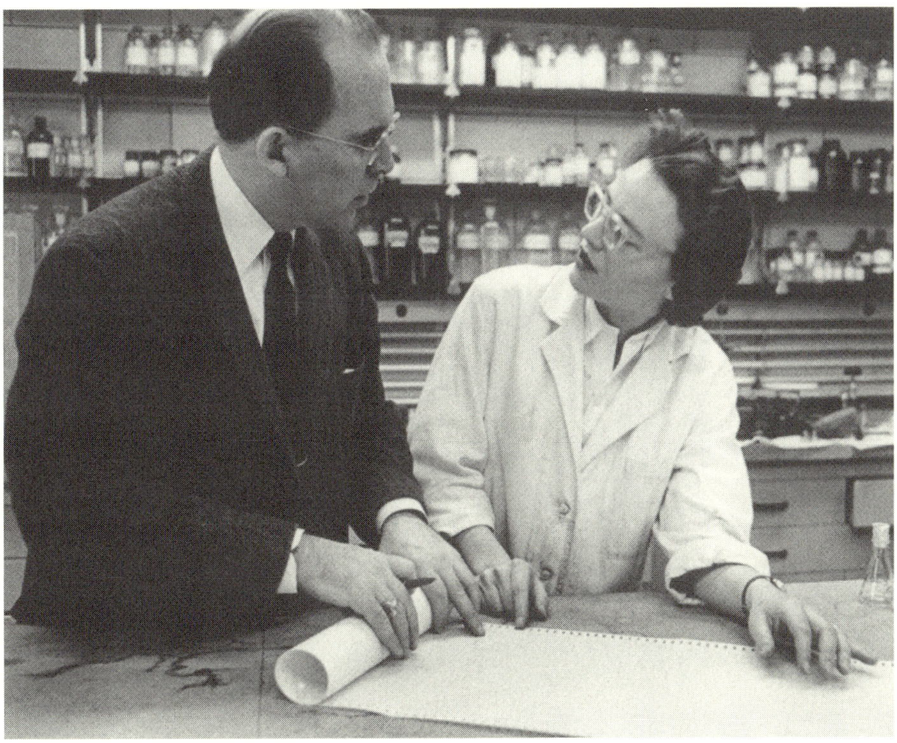

PLATE 4
Robert Feller and Catherine Bailie, staff scientist at the Center, 1958-1961, 1977-1999 (© Clyde Hare)

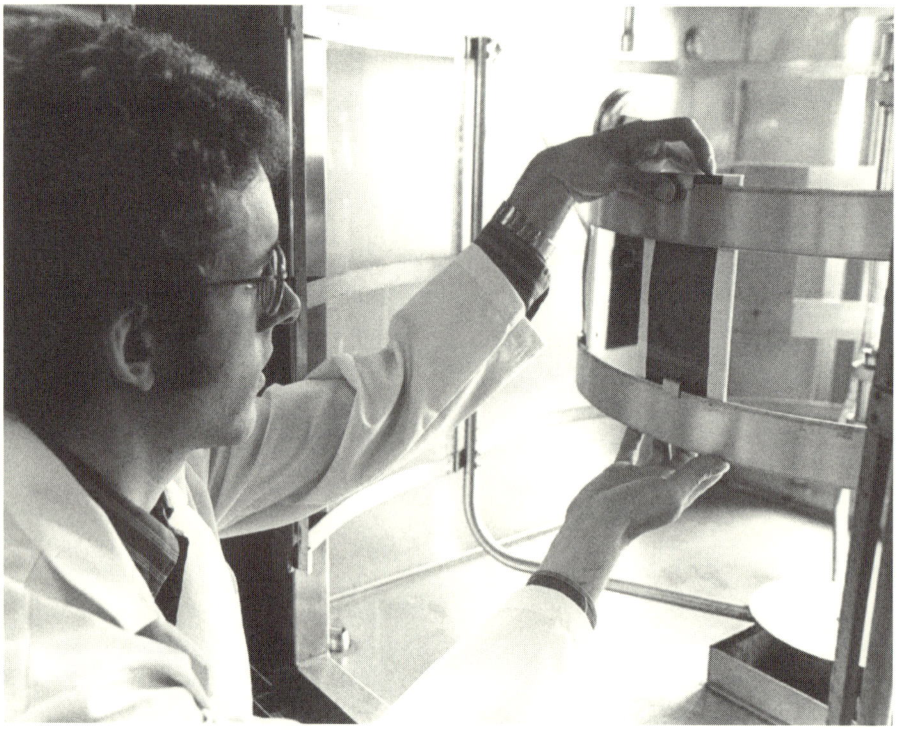

Plate 5
John Bogaard, staff scientist at the Center, 1978-present

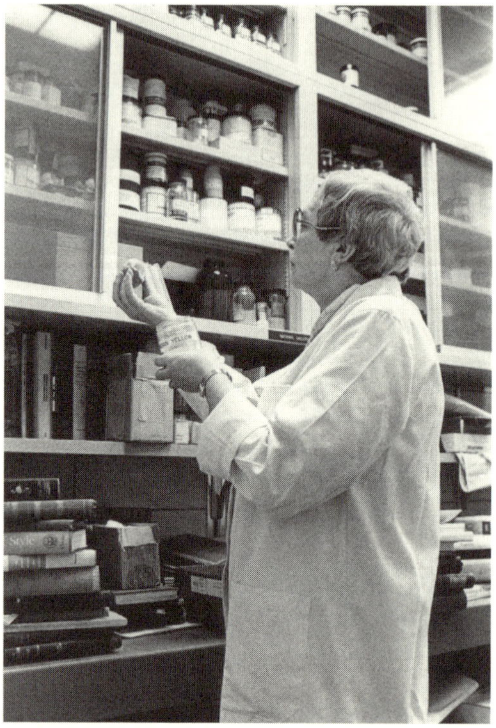

Plate 6
Mary Curran, staff scientist at the Center, 1967-1988 (© Fisher Scientific Company)

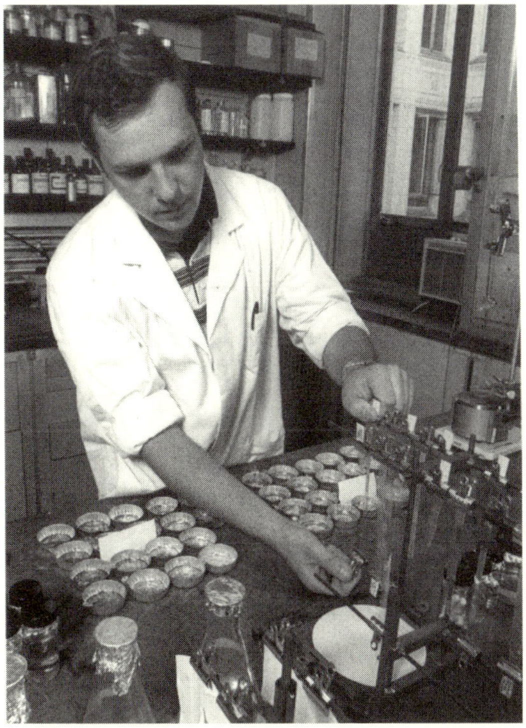

PLATE 7
David Encke, research technologist at the Center, 1978-1980

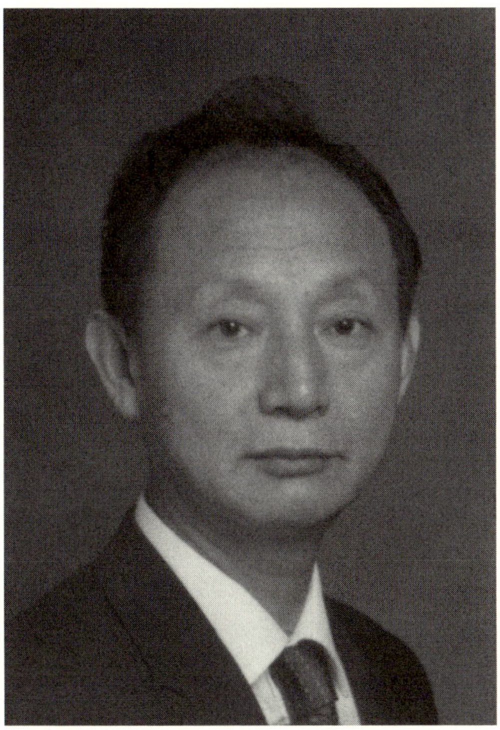

PLATE 8
Dr. Sang B. Lee, staff scientist at the Center, 1978-1988 (photo courtesy of Dr. Lee)

SECTION 6: POLYMERS

In the 1950s, conservators became interested in the use of two acrylic polymers that could be conveniently formulated in petroleum solvents: polymers and copolymers of n-butyl and iso-butyl methacrylate. In contrast to poly(methyl methacrylate), the polymer in Plexiglas well known for its long-term stability, it was found that these two polymers could undergo chemical changes rather readily when exposed to the illumination of ordinary fluorescent lamps or to diffuse daylight passing through window glass, gradually becoming insoluble. Our laboratory became aware of this possibility during accelerated aging tests in 1955, but it required some time to become thoroughly convinced that this "crosslinking" behavior could occur under conditions that might be encountered in museums. Nevertheless, by 1957 it was possible to prepare a brief report on the types of acrylic polymers that exhibited the greatest tendency to crosslink, methods for the detection of crosslinking, and a potential method of inhibition. That same year, the Seminar on Surface Coatings was held at Oberlin College, and by the time the principal papers were published two years thereafter in the book *On Picture Varnishes and Their Solvents*, it was possible to describe in some detail the characteristics of swelling, gel formation, and removal of coatings that undergo crosslinking. At the 1961 triennial conference of IIC, I reported the effects of wavelength of light on crosslinking, confirmed the importance of swelling in the ease of removal by solvents, and drew attention to the excellent stability of the Rohm and Haas copolymers having the composition essentially that of Acryloid® B-72.

While crosslinking may be of great concern because of the consequent loss of solubility, it need not proceed exclusively; chain breaking may occur at the same time and delay or prevent solubility loss, depending on the relative rates of the two processes. A publication in 1966 described how polymer scientists are able to determine the ratio of these two reactions when they occur simultaneously. From this point of view, we studied a series of copolymers having various contents of ethylene and vinyl acetate, demonstrating that there would be little tendency for these particular polymers to become insoluble, provided the proportion of vinyl acetate in the polymer was higher than about 40 percent.

These extensive studies led the Center in 1977 to underscore the fact that the various processes involved in photochemically induced degradation tended to occur in stages. Common rubber cement adhesive and transparent mending tapes provided particularly clear examples of several distinct stages. One of the most significant stages proved to be the build-up and subsequent decline in the concentration of peroxide components, events that often occurred before yellowing and brittleness became apparent. In 1980 I was invited to sum up the knowledge that had been gained regarding the crosslinking behavior of methacrylate polymers at a meeting of the American Chemical Society.

R. L. F.

Crosslinking of Methacrylate Polymers by Ultraviolet Radiation

R. L. FELLER

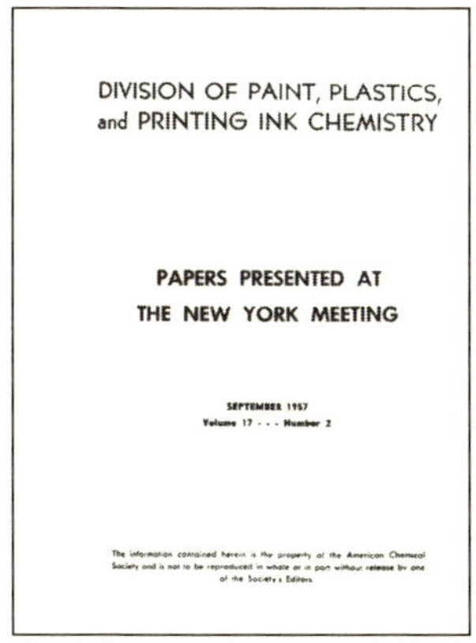
DIVISION OF PAINT, PLASTICS, and PRINTING INK CHEMISTRY

PAPERS PRESENTED AT THE NEW YORK MEETING

SEPTEMBER 1957
Volume 17 - - - Number 2

The information contained herein is the property of the American Chemical Society and is not to be reproduced in whole or in part without release by one of the Society's Editors.

Originally printed in:
AMERICAN CHEMICAL SOCIETY DIVISION OF PAINT, PLASTICS AND PRINTING INK CHEMISTRY: PAPERS PRESENTED AT NEW YORK MEETING
Volume 17, Number 2 (1957), Pages 465-470.

The durability of coatings is generally judged on the basis of discoloration or mechanical failure. In seeking thermoplastic polymers which, in addition to being highly resistant to discoloration and mechanical failure, also retain their solubility in petroleum hydrocarbons over long periods of time, our laboratory found that certain members of the methacrylate family crosslink under the influence of light and heat in advance of visible failure of the coatings.

The ability of "butyl" polymethacrylate to form a tri-dimensional polymer under the influence of heat has been reported by Drinberg and Yakovlev.[1] In 1955 Raynolds,[2] investigating the depolymerization of poly(isoamyl methacrylate)[3] at 160 to 200°C in this laboratory, observed that within 10 to 20 hours the film had become insoluble in hot benzene and toluene. At that time, a report of a paper by Shultz and Bovey[4] stating that poly(t-butyl acrylate) did not crosslink under the conditions of their experiments and one by Wall[5] stating that poly(methyl methacrylate) did not crosslink during degradation, prompted the writer to question the behavior of the other members of the methacrylate family and to investigate the effects of ultraviolet radiation.

1. Drinberg, Y., and Yakovlev, A. D., *Zhur. Priklad. Khim.*, 26 (1953), p. 53; *Chem. Abstr.*, 47 (1953), Abstr. no. 10318.

2. Present Address, Chemistry Dept., University of Pittsburgh, Pittsburgh, PA (as of 1957—Ed.)

3. Poly(isoamyl methacrylate) in this paper refers to a polymer made with monomer based on a 50/50 mixture of 2-methylbutyl and 3-methyl butyl methacrylates.

4. Anon., *Chem. and Eng. News*, 33 (1955), p. 1655; see Schultz, A., and Bovey, Y., *J. Polymer Sci.*, 22 (1956), p. 485.

5. Anon., *Chem. and Eng. News*, 33 (1955), p. 3390.

Results of Exposure in Fadeometer

At first, simple tests were made of the ability to remove the films with solvents. This was done by rubbing with cotton swabs, twisted on the end of a probe and dipped in solvent. "Resistance to solution" was taken when the time required to dissolve, or otherwise wear away, the 1 to 3 mil-thick film through to its support required 75 to 105 seconds. Results of exposure in a National Accelerated Fading Unit Type XV, to radiation from a National Sunshine Carbon arc with Corex D filter, are summarized in Table 1. Samples attained temperatures of about 60°C during exposure.

TABLE 1
Hours of Exposure* of Methacrylate Thermoplastics Before Developing Resistance to Solution in Various Solvents

Alcohol Radical in Polymer	Viscosity Grade** of Polymer in Centipoises	Solvent		
		Cyclohexane	Toluene	Acetone
25/75 p-Methylcyclohexyl/Isoamyl Copolymer	18.5	50	130	
3-Methylbutyl	22	84	160	
Isobutyl, also 2-Ethylbutyl	50, 13	125	200	
Isoamyl	23	133	250	~700
2-Methylbutyl	22	148	300	
n-Butyl	48	425	560	1100
n-Propyl	16		1050	
(Polyvinylacetate for comparison)	80		>1500	

* Actual hours multiplied by 1.9 to approximate effect of reflection from aluminum foil.
** Viscosity of solution in toluene of 20% solids by weight at 70°F.

The shortest wavelengths of ultraviolet radiation are most responsible for these reactions. In tests upon the p-methyl-cyclohexyl/isoamyl copolymer, isobutyl, and isoamyl polymers, it was demonstrated that a filter of ordinary window glass increased the lifetime of these polymers in the Fadeometer by a factor of 1.9.

Films were cast on aluminum foil from solution in toluene, allowed to stand for several days and baked at 70°C in a Fisher "Iso-temp" oven for two or three days before exposure. The isobutyl and n-butyl polymers were du Pont "Lucite" 44 and 45. Other polymers were prepared by Raynolds by polymerization in solution using benzoyl peroxide initiator.[6] The method of measurement is related to the technique of removing coatings from objects of art and is sufficient to demonstrate that certain of the polymers, originally removable in cycloparaffin hydrocarbons, are removed more slowly and with increasing difficulty during the course of exposure to ultraviolet light.

6. Raynolds, S., Master of Science Thesis, University of Pittsburgh (1954).

Apparently, methacrylate polymers having hydrogen atoms located on tri-substituted carbon atoms in the alkyl radical of the alcoholic group are sensitive to crosslinking under the action of ultraviolet radiation. The preparation and testing of methacrylate polymers which do not have this structure has been undertaken.

ESTIMATION OF DURABILITY INDOORS

What do these results mean in terms of exposure to ordinary levels of daylight illumination indoors? The Harrison Report, published by the Metropolitan Museum of Art[7] mentions eight standards of light-fastness, each representing a standard of durability double that of the one below. The report also stated that classification VIII as equivalent to between 930 to 1920 hours of sunlight and that 24 hours in the carbon-arc illumination is supposed to be equivalent to 60 hours of full sunlight. From these data, the following table may be established.

7. Harrison, L. S., *Report on the Deterioration Effects of Modern Light Sources*, The Metropolitan Museum of Art, N. Y. (1954).

TABLE 2
Exposure in Fadeometer Equivalent to Standards of Light-Fastness
Values Derived from L. S. Harrison[7]

Fastness	Standard	Hours in Fadeometer to Cause Noticeable Fading	Average Hours
Slight	I	3-6	4.5
	II	6-12	9
Limited	III	12-24	18
	IV	23-49	36
Satisfactory	V	46-99	72
Good	VI	93-197	144
Very Good	VII	186-384	285
Excellent	VIII	372-768	570

Durability of 250 hours, before resistance to solution in toluene occurs, would place a material in Classification VII. Equating 6 years in a "gallery without sun-louvers" to 72 ± 27 of arc illumination for a class V dye,[7] one may estimate 15 to 33 years before resistance to toluene would be developed by the isobutyl or isoamyl polymers in a gallery without sun louvers. (Attention is drawn to the lack of precision in such an estimate in that these figures would be 45 to 99 years, if one based the estimate on values for a "day-lighted gallery with sun-louvers properly operated," and a minimum of 200 years, if based on values for artificially-lighted galleries at 60 fc. level of illumination.)

In outdoor exposure under window glass on the roof of Mellon Institute, films of poly(isobutyl methacrylate) of 50 centipoise viscosity grade and poly(isoamyl methacrylate) of 23 centipoise viscosity grade, cast on glass from petroleum hydrocarbons of boiling range 160 to 190°C, did not discolor or crack in one year, and were removable in toluene within two minutes of rubbing, but with considerable resistance. Comparing the exposure index for wall exhibits in day-lighted galleries without sun-louvers in Harrison's Table VI to an average (annual) exposure index for outdoor exposure in New York of 51.7 and allowing for the factor of damage per footcandle to be reduced to about 1/2 by a filter of window glass, one may estimate that one year on the roof is equivalent to about 24 years in a day-lighted gallery without sun-louvers.

Crude as these estimates are, figures of about 24 years suggest that, if isobutyl or isoamyl polymers were to be used in the care of museum objects, the problem deserves consideration. Methacrylate polymers have not been of interest to museums for more than about 18 years. Of these, the n-butyl polymer has been of more universal interest than the isobutyl. In a sense, an investigation now is in anticipation of a problem which might someday, but apparently has not yet, become important in removing protective coatings based on methacrylate polymers.

DETECTION OF CHANGE BY PAPER CHROMATOGRAPHY

How can one detect if a change has begun to take place in the polymers? Paper chromatography might indicate if the polymer had become of higher average-molecular-weight owing to crosslinking. Tests of polymers of different viscosity-grade suggested this technique would be useful.[8] Such a technique has been used to distinguish polyethylene oxides.[9]

Strips of 1 1/2"-wide Whatman No. 1 filter paper were used in a descending development. The solvent was a mixture of equal volumes of xylene and non-aromatic, naphthenic, petroleum of boiling range 170 to 190°C, density at 70°F, 0.81. Samples were removed from the test panels with a cotton swab and the swab applied to the marked origin. In five hours development the solvent front migrated to the extent of about 30 to 40 centimeters. The polymers generally developed in a streak, frequently extending to and including the original spot. Since the paper is "varnished" with the resin, spraying with an aqueous solution will reveal areas where the paper is not wet. A dye may be added to the water for permanent record.

Figure 1 shows a curve of Rf values relative to unexposed polymer, based upon the position of the most advanced tip of the polymer. Emperically, the data relates relative Rf to time of exposure. This technique differs from that of Thomson[10] in that he uses a good solvent for the polymer and seeks to sweep all but the insoluble (crosslinked) portion away. His use of stain in the developing liquid no doubt provides excellent reproducibility in revealing the developed pattern. The two techniques could be combined on one paper. After development of the "streak" in poor solvent, the origin could be examined by Thomson's technique.

Many inhibitors were tested. Few were satisfactorily soluble in hydrocarbons. Ultraviolet-absorbing compounds 2,4-dihydroxybenzophenone and 2,2'-dihydroxy-4, 4'-dimethoxybenzophenone were found to increase times in Table 1 by about 150 hours, in the case of the isoamyl, isobutyl, and n-butyl polymers in the presence of 0.5% inhibitor based on the weight of polymer. In Figure 1, the effectiveness of these two compounds and the lack of effect by 2-hydroxy-4-methoxybenzophenone in protecting poly(isoamyl methacrylate) is indicated

8. For absorption chromatography of poly(vinyl acetate) and poly(methyl, methacrylate) see Claesson, S., *Discussions Faraday Soc.*, No. 7 (1949), p. 321.

9. Kume, S., Yamamoto, T., Otozai, K., and Fukushima, S., *Bull. Chem. Soc., Japan*, 26 (1953), p. 93.

10. Thomson, G., *Nature*, 178 (1956), p. 807.

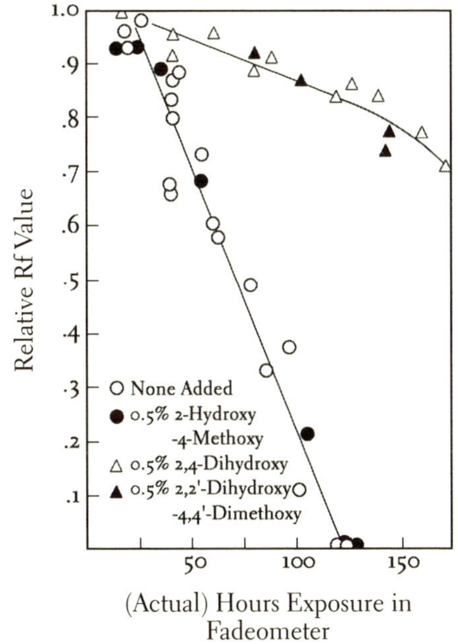

FIGURE 1
Effect of adding substituted benzophenones to poly(isoamyl methacrylate).

by the results of chromatographic analysis. The mechanism of inhibition by such compounds is discussed by Burgess.[11]

11. "Polymer Degradation Mechanisms," *National Bureau of Standards Circular* 525 (1953), p. 156.

The chromatographic technique should indicate if a significant change in a polymer is in progress. Values taken to demonstrate the technique and to determine if change had occurred in various samples exposed indoors in well-lighted rooms are given in Table 3. The ten-year samples were taken from test panels at the Fogg Museum of Art, Harvard University, through the kind cooperation of Miss Elizabeth Jones of the department of conservation.

The lowest relative Rf values found for the isobutyl or isoamyl polymers exposed under normal conditions indoors were about 0.91 for five-year-old panels. If one assumes that the decrease in relative Rf will be linear, and that at

TABLE 3

Methacrylate Polymer	Aged in years	Relative Rf
0.75 to 3.0 mg. of sample		1.00
8 cp. Isoamyl vs. 23 cp. visc. grade		1.06
18 cp. Isobutyl vs. 55 cp. visc. grade		1.08
n-Butyl	10	0.84 ± 0.05
Isobutyl/n-Butyl copolymer	10	0.79 ± 0.01
Isoamyl	4	0.95 ± 0.03
Isobutyl	5	0.93 ± 0.02
Isoamyl in condition of "resistance to removal in cyclohexane"		0.50 ± 0.10

a value of 0.5 the resistance to cyclohexane would become apparent when using solvents on a swab, the results imply that this state will arise in about 28 years of exposure similar to that which has already taken place. According to data in Table 1, resistance to toluene by these polymers develops in about twice this time. This compares favorably with estimates based on exposures in the Fadeometer.

SUMMARY

It has been shown that linear polymers of the methacrylate class, in which the alkyl radical in the alcoholic group of the methacrylic ester contains hydrogen atoms located on tri-substituted carbon atoms, are sensitive to crosslinking under the influence of ultraviolet radiation. A filter of ordinary window glass reduces the rate of crosslinking of certain methacrylate polymers by a factor of about 1/1.9. The compounds 2,4-dihydroxybenzophenone and 2,2'-dihydroxy-4,4'-dimethoxybenzophenone reduce the tendency of the isoamyl, isobutyl and n-butyl polymers to lose solubility in hydrocarbons.

A paper chromatographic technique has been described which permits measurement of the action by ultraviolet light on linear polymers during the early stages, before apparent failure of the film.

New Solvent-type Varnishes

ROBERT L. FELLER

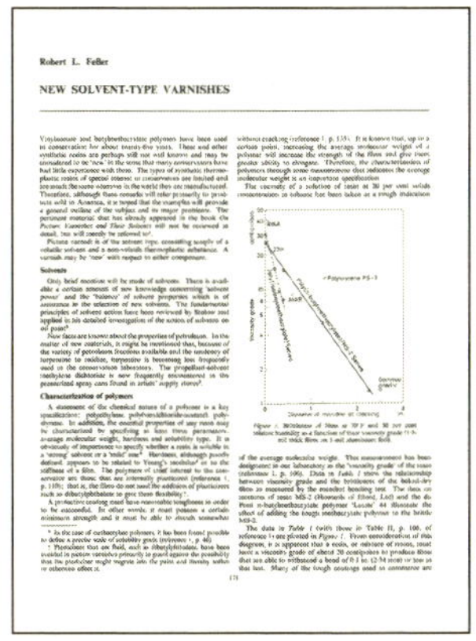
Robert L. Feller
NEW SOLVENT-TYPE VARNISHES

Originally printed in:
RECENT ADVANCES IN CONSERVATION
ed. Thomson, G., Butterworths, London (1963), Pages 171-175.

Vinylacetate and butylmethacrylate polymers have been used in conservation for about twenty-five years. These and other synthetic resins are perhaps still not well known and may be considered to be "new" in the sense that many conservators have had little experience with them. The types of synthetic thermoplastic resins of special interest to conservators are limited and are much the same wherever in the world they are manufactured. Therefore, although these remarks will refer primarily to products sold in America, it is hoped that the examples will provide a general outline of the subject and its major problems. The pertinent material that has already appeared in the book *On Picture Varnishes and Their Solvents* will not be reviewed in detail, but will merely be referred to.[1]

Picture varnish is of the solvent-type, consisting simply of a volatile solvent and a non-volatile thermoplastic substance. A varnish may be "new" with respect to either component.

1. Feller, R. L., Stolow, N., and Jones, E. H., *On Picture Varnishes and Their Solvents*, Intermuseum Conservation Association, Oberlin, Ohio (1959).

Solvents

Only brief mention will be made of solvents. There is available a certain amount of new knowledge concerning "solvent power" and the "balance" of solvent properties which is of assistance in the selection of new solvents. The fundamental principles of solvent action have been reviewed by Stolow and applied in his detailed investigation of the action of solvents on oil paint.[2]

2. Stolow, N., *J. Oil and Colour Chemists' Assoc.*, 40 (1957), p. 337, 448; Reference 1, p. 60.

New facts are known about the properties of petroleum. In the matter of new materials, it might be mentioned that, because of the variety of petroleum fractions available and the tendency of turpentine to oxidize, turpentine is becoming less frequently used in the conservation laboratory. The propellant-solvent methylene dichloride is now frequently encountered in the pressurized spray cans found in artists' supply stores.[3]

3. Brown, J. A., *Paint Varn. Prod.*, 50 (1960) p. 86; Sciarra, J. J., *ibid.*, 51 (1961), p. 3.

CHARACTERIZATION OF POLYMERS

A statement of the chemical nature of a polymer is a key specification: polyethylene, poly(vinylchloride-acetate), polystyrene. In addition, the essential properties of any resin may be characterized by specifying at least three parameters: average molecular weight, hardness and solubility type. It is obviously of importance to specify whether a resin is soluble in a "strong" solvent or a "mild" one.[4] Hardness, although poorly defined, appears to be related to Young's modulus[5] or to the stiffness of a film. The polymers of chief interest to the conservator are those that are internally plasticized (Ref. 1, p. 110); that is, the films do not need the addition of plasticizers such as dibutylphthalate to give them flexibility.[6]

4. In the case of methacrylate polymers, it has been found possible to define a precise scale of solubility grade (Ref. 1, p. 46).

5. Inoue, Y. and Ikeda, K., *Chem. High Polymers (Japan)*, 11 (1954), p. 409; *Chem. Abstr.*, 50 (1956), Abstr. no. 6086e.

6. Plasticizers that are fluid, such as dibutylphthalate, have been avoided in picture varnishes primarily to guard against the possibility that the plasticizer might migrate into the paint and thereby soften or otherwise affect it.

A protective coating must have reasonable toughness in order to be successful. In other words, it must possess a certain minimum strength and it must be able to stretch somewhat without cracking (Ref. 1, p. 135). It is known that, up to a certain point, increasing the average molecular weight of a polymer will increase the strength of the films and give them greater ability to elongate. Therefore, the characterization of polymers through some measurement that indicates the average molecular weight is an important specification.

The viscosity of a solution of resin at 20 percent solids concentration in toluene has been taken as a rough indication of the average molecular weight. This measurement has been designated in our laboratory as the "viscosity grade" of the resin (Ref. 1, p. 106). Data in Table 1 show the relationship between viscosity grade and the brittleness of the baked-dry films as measured by the mandrel bending test. The data on mixtures of resin MS-2 (Howards of Ilford, Ltd.) and the du Pont n-butylmethacrylate polymer Lucite 44 illustrate the effect of adding the tough methacrylate polymer to the brittle MS-2.

The data in Table 1 (with those in Table 2, p. 108, of Ref. 1) are plotted in Figure 1. From consideration of this diagram, it is apparent that a resin, or mixture of resins, must have a viscosity grade of about 20 centipoises to produce films that are able to withstand a bend of 0.1 in. (2.54 mm) or less in this test. Many of the tough coatings used in commerce are based on resins of considerably higher viscosity grade than this (Ref. 1, p. 111).

The selection of suitable viscosity grades for the resin in solvent-type varnishes remains open to question and discussion. From an analysis of the data in

Table 1
Viscosity Grade vs. Cracking of Baked-dry Films
Test Conditions: 70°F, 50% Relative Humidity

Resin formulation in film	Viscosity grade (centipoises)	Sward hardness*	Diameter of mandrel at cracking (in.)
Lucite 44	48	32	< 0.1
Lucite 44/MS-2 80/20	26	53	< 0.1
Lucite 44/MS-2 60/40	14.8	60	0.6
Lucite 44/MS-2 20/80	5.2	77	1.3
Lucite 44/MS-2 10/90	1.3	–	2.3
Lucite 44/AW-2 10/90	1.3	–	1.6
AW-2	1.2	91	4.0
Dammar, Mastic	1.5	81	2.5

* Hardness of window glass base equals 100.

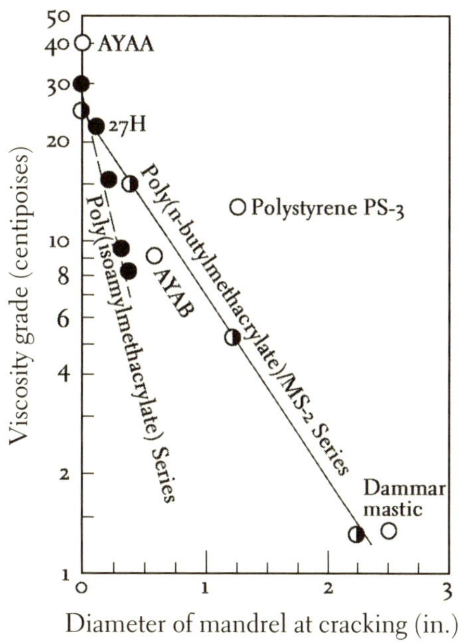

Figure 1
Brittleness of films at 70°F and 50 percent relative humidity as a function of their viscosity grade (1.5 mil thick films on 1 mil aluminum foil.)

Figure 1, however, it does not seem necessary to select grades much higher than 25 centipoises for most protective coatings. In our research, we have been especially interested in types between 9 and 25 centipoises.

SYNTHETIC POLYMERS IN SOLVENT-TYPE VARNISHES

A brief history of the introduction of synthetic thermoplastics in conservation has been presented elsewhere (Ref. 1, p. 104). Polyvinyl acetate was described in the early 1930s by Stout and Gettens. In view of the importance of this polymer, the National Gallery of Art Research Project at Mellon Institute undertook the preparation of a list of the different molecular weight grades available in various countries (Ref. 1, p. 218). Lower molecular weight varieties, particularly resin AYAB, have been introduced in recent years in a number of applications where higher fluidity and gloss were desired.[7]

Because they are available as granular powders and because of their solubility in "mild" solvents, the butylmethacrylate polymers from the du Pont company have attracted much attention: the n-butyl polymer, Lucite 44,[8] the isobutyl, Lucite 45, and the 50-50 copolymer of these, Lucite 46. Most conservators have found it convenient to use mixtures of Lucite 44 and 45, varied according to the need. Lucite 44 is one of the toughest thermoplastic resins available: literature from the du Pont company states that test samples will stretch three times their original length before breaking. Poly(n-butylmethacrylate) is readily identified by its hardness and solubility characteristics.[9]

Resin AW-2, a resin in the viscosity-grade class of dammar and mastic, has received considerable attention abroad and is now available in the solid form in America. The data in Table 1 show that AW-2 is considerably brittle when used alone. A number of proprietary varnishes are said to be prepared with resin AW-2 and additions of stand oil. Howards of Ilford (England) markets a variety of AW-2, designated as MS-2. They have also produced a fully hydrogenated type, MS-2A, which has been evaluated by G. Thomson[10] at the National Gallery, London. Resins AW-2 and MS-2A are of high quality, but the author has had little experience with them in comparison to polymers of higher molecular weight.

Artists' supply stores in America offer a variety of paints and varnishes put up in spray cans. Dammar resin is available in this form, but it is believed that most of the coatings are based on cellulose acetate-butyrate and the Rohm and Haas acrylatemethacrylate copolymer Acryloid B-72 (Paraloid in Europe). These proprietary formulations are seldom used on fine paintings. Tests show, however, that many of them are based on excellent materials and have their use in the conservation laboratory. As mentioned above, the development of pressurized containers has introduced methylene dichloride and related solvents.

An experimental varnish known as 27H was developed at Mellon Institute in 1953.[11] This polymer was unique in that it was specially prepared with selected film strength, high resistance to yellowing and solubility in a petroleum solvent having no aromatic compounds (Ref. 1, p. 200). Extensive studies of this and related polymers led to the discovery and investigation of the tendency of thermoplastic polymers to crosslink under the effect of light, a problem of

7. Polymer AYAB exhibits cold flow in bulk, but in thin films it seems to be satisfactory in tests thus far. If used as the final coating, it may have a slight tendency to become dirty, for the reasons noted (Ref. 1, p. 134) in the case of poly(n-butylmethacrylate).

8. Bedacryl 112 X (Imperial Chemical Industries) is similar in viscosity grade (Ref. 1, p. 217).

9. We have identified in proprietary varnishes a second variety of the n-butyl polymer, having a viscosity grade of 27 centipoises. This is supplied by an American manufacturer other than du Pont and is not available in the solid form.

10. Thomson, G., in *Recent Advances in Conservation*, Butterworths, London (1963), p. 176-184.

11. Varnish 27H was made only in limited quantities in the laboratory at the time and samples are no longer available.

TABLE 2
Selected Resins for Solvent-type Varnishes

In order of decreasing fluidity, brittleness and penetration in porous areas:

Resin	Viscosity grade	Sward hardness
AW-2	1.3	86
Dammar	1.4	81
Polyvinyl acetate AYAB	9.0	63
27H (experimental varnish)	23.0	49
Poly(n-butylmethacrylate)	27.0	25
Acryloid B-72	29.0	65
Lucite 44 poly(n-butylmethacrylate)	48.0	30
Lucite 46 (n-butyl/isobutylmethacrylate copolymer)	50.0	61
Polyvinyl acetate AYAF	80.0	63

In order of increasing strength of solvent needed to formulate the varnish:

Resin	Solubility grade*
27H (experimental varnish)	8
AW-2	10 (?)
Lucite 44, 45 and 46	25
Dammar	—**
Acryloid B-72	80
Polyvinyl acetate AYAF, AYAB	89

* Percentage toluene by weight in n-dodecane necessary to make a clear solution at 25°C.
**Soluble in "mild" petroleum fractions when fresh, requires toluene or "stronger" solvent when aged.

fundamental interest that will be discussed below.

Table 2 summarizes the properties of the resins of principal interest to conservators in America.

APPLICATION OF VARNISH

A number of problems is associated with the introduction of new and unfamiliar materials. In the application of a varnish, perhaps the central problem has been the selection of a solvent sufficiently "mild" to be used with little danger of attacking the paint during application, and, at a later time, during the removal. The risk of attack, of course, depends on the character of the paint or other material directly beneath the varnish. In certain cases, the ability of the substrate to withstand solvents of the "strength" of toluene and cellosolve ($C_2H_5O(CH_2)_2OH$) has made the application of polyvinyl acetate possible. The Rohm and Haas Acryloid B-72 is also soluble in toluene and forms a slightly cloudy solution in xylene. The butylmethacrylate polymers are soluble in turpentine and in petroleum fractions having about 25 to 35 percent aromatics. The experimental polymer 27H was designed to be soluble in paraffin and cycloparaffin hydrocarbons. Resin AW-2 is soluble in similar solvents. Table 2 lists these polymers in order of their solubility grade (Ref. 1, p. 45).

In our research, polymers that may be applied and removed in xylene have been sought as a standard of the maximum solubility parameter or "strength" of solvent to be used. To establish such a specification, the investigator must bear in mind that stronger solvents are generally required to remove aged resins than to dissolve them originally. More information is needed on the removal of aged coatings before this question can be properly considered. Miss Elizabeth Jones' re-forming process, for example, represents one of the many facets of the problem that should be further explored. Certain new facts regarding the aging and removal of polymers are discussed below.

The polymers listed in Table 2 require solvents no "stronger" than toluene to dissolve them. Yet, having thus limited the number, one still has the problem of learning to apply these materials, particularly if their molecular weight is significantly higher than mastic, dammar and AW-2 resins (Ref. 1, p. 120). High viscosity grade is, of course, associated with the desirable properties of increased strength and reduced brittleness. In order to take full advantage of these properties, American conservators have learned to apply the coatings by spray application. A discussion of the relative merits of brushing and spraying will not be undertaken here. Suffice it to say that many conservators have found spraying to be highly satisfactory once they have learned the technique. The use of spraying equipment requires skill and experience, however, just as does successful brushing.

To furnish additional information concerning the application of these materials, the National Gallery of Art Research Project undertook a study of the relationship of viscosity grade to film-forming properties. This led to data of the type shown in Figure 1. Commercial resins of lower viscosity grade than those which were in general use in 1950 were also sought out. For example, the use of polyvinyl acetate AYAB, AYAA and several other polymers in the 20-30 centipoise range of viscosity grade has been suggested.

While the spraying of coatings has introduced problems in the selection of solvents, the desire to brush solutions of high polymers in certain applications has also required special consideration. A sound review of petroleum chemistry was required in order to select solvents of a given solubility parameter and with varying evaporation rate (Ref. 1, p. 17). This subject was briefly reviewed in the *Application of Science in the Examination of Works of Art*, Museum of Fine Arts, Boston, Mass., 1959, pp. 70-72. Attention has been drawn to the wide variety of proprietary solvents described in the *Petroleum Thinner Index* of the National Paint, Varnish, and Lacquer Association, Washington 5, D.C.

Problems in removal of coatings

A fundamental problem associated with materials used in conservation involves their ease of removal. The use of thermoplastic polymers in the viscosity-grade range of 9 to 80 centipoises has raised a problem of basic interest: among the ways that such a substance may become altered in time, one manner is through the formation of crosslinks. Although this phenomenon is not of major importance with materials of the average molecular weight of dammar, mastic and AW-2 resins, it is one the chemist must face if he wishes to use thermoplastic polymers of much higher molecular weight. As has been pointed out, it does not take very many chemical crosslinks to cause such a polymer to become insoluble (Ref. 1, p. 203).

The tendency for thermoplastic polymers to crosslink upon aging has been found to be more the general rule, and the lack of this tendency the exception, than one might at first expect. Activation of certain chemical bonds by light and heat imparts the tendency to form new chemical bonds. Bond formation, of course, may take place with neighboring molecular chains in the polymer. If this proceeds, the chains may become so linked together that they eventually fail to be dispersed in solvents. Thus, our investigation of polymers of the higher alkylmethacrylate esters traced the tendency to lose solubility particularly to the presence of tertiary hydrogen atoms in the side-chain (Ref. 1, p. 152). The tendency to crosslink has also been observed in poly(vinylstearate-acetate) copolymers, poly(vinylbutyral) and cellulose acetate-butyrate.

Table 3
Average Rate of Crosslinking of Poly(n-butylmethacrylate) and Poly(isoamylmethacrylate) on Aluminum Foil, Under Various Filters in the Fadeometer at 62°C

Filter	Filter, wavelength at 50% transmission	Relative rate for films to reach 50% insolubility	Relative rate calculated by the equation
Corex D glass	305	1.0	0.97
Window glass	330	0.44	0.49
Uvinul M-40 filter	358	0.19	0.21
Plexiglas UF-1	396	0.05	0.05

The first phase of an extensive investigation of crosslinking was to determine the structures in the methacrylate side-chain that possessed the greatest tendency to become insoluble. This led to the conclusion just stated and to the selection of materials with a reduced tendency. The second phase was to determine the nature of the radiation that caused crosslinking. Data in Table 3 demonstrate that ultraviolet radiation is primarily responsible. The influence of wavelength upon the rate is almost precisely that predicted by the probable damage factor for a given wavelength (D_λ), described in the Harrison report of the Metropolitan Museum of Art.[12] The rate at which the films reached 50 percent insolubility was found to follow the equation

12. The results of these experiments were reported to the British Group, IIC, at their meeting of September 29, 1960.

$$\text{Rate} = 1/T_{50} = 0.16D_\lambda - 0.05 \qquad (1)$$

where T_{50} is the time required to attain 50 percent insolubility. Investigation also showed that an increase in the temperature of the samples increased the rate of crosslinking.

How important is this phenomenon under museum conditions? Most of the accelerated-aging tests first used to detect and study the problem involved both elevated temperatures and ultraviolet radiation not normally encountered in a picture gallery. To minimize the effect of temperature, our laboratory conducted tests under "daylight" fluorescent lamps in an air-conditioned room at 80°F and 50 percent relative humidity. Under these conditions, the butylmethacrylate polymers from du Pont were found to develop 50 to 80 percent insolubility in 8 to 10 million footcandle hours of exposure. (After this exposure, the polymers could still be removed easily with toluene.) Such a degree of insolubility might be expected to develop in a matter of 50 to 100 years if the objects were to be illuminated at a level of about 50 footcandles during the day. From this first estimate, we may conclude that the phenomenon is not one to cause immediate alarm. Nevertheless, crosslinking can take place during exposure in museums

TABLE 4
Ease of Removal of Pairs of Films in Relationship to Their Ability to be Swollen by Toluene

Aged Film	Percentage insoluble in toluene	Fadeometer exposure (hours)	% gain in weight (absorbed toluene after 3 days)	Time (sec) required to remove film with swabs dipped in:	
				Toluene	Acetone
Isobutyl MA* polymer	89.1	111	598	105	65
n-Butyl MA* polymer	100	228	77.2	> 180	220
1/1 AW-2/n-propyl MA* polymer	96.2	416	91.4	75	6
1/1 MS-2/n-propyl MA* polymer	95.0	416	43.3	> 180	25

* MA = methacrylate

and the possibility warrants particular attention wherever ordinary fluorescent lamps are to be used without ultraviolet filters or wherever daylight enters through ordinary window glass.[13]

Since practically all films become more difficult to remove in time, the laboratory is prompted to return again to fundamental problems regarding the removal of films and the rate at which films, once soluble in solvents such as toluene, come to require "stronger" solvents for removal. This led to studies, which suggested that the ability of a film to swell under the action of solvents was important in facilitating its removal (Ref. 1, p. 200). Recent data, given in Table 4, demonstrate that the amount of insoluble material present in a film is not necessarily the major indication of its ability to be removed. Even though crosslinking may proceed to the extent that 50 to 95 percent of the film becomes insoluble, this does not mean that the film cannot be removed with considerable ease as a soft swollen jelly. These preliminary findings concerning the role that swelling plays in removal draw special attention to the techniques used by Stolow in his studies of solvent action.

Polymers of low molecular weight might be expected to become insoluble at a lower rate. We have thus far found little benefit, however, in lowering the average molecular weight of the polymers from a viscosity grade of 27 centipoises to about 10 centipoises. The possibility of inhibiting crosslinking has also been mentioned (Ref. 1, p. 159 and 161).

To conclude these remarks, it should be said first that the problem of crosslinking is a general one in the aging of thermoplastic polymers of reasonably high molecular weight. Research has shown, however, that certain chemical structures have greater tendency than others and that ultraviolet radiations are primarily responsible for the phenomenon. Moreover, the ability of a film to swell, rather than the actual insoluble matter present, controls in large measure the ease of removal.

The investigations continue and new information is being gained about crosslinking. From what has been learned so far, however, the author believes that, with proper supervision and keeping of records, coatings such as poly(butylmethacrylate) may still be used in the conservation of many objects.

13. There are ways of detecting the onset of crosslinking. Thomson's test is simple and highly sensitive (*Nature*, 178 (1956), p. 807; footnote 1, p. 205). Measurements of the degree of soluble matter in samples picked up on a cotton swab have been used in our laboratory to follow the course of crosslinking in a quantitative way. A chromatographic leaching method has also been used (reference 1, p. 115). The conservator can detect the presence of crosslinked material by the tendency of a film to swell considerably under the action of solvents, and by the presence of a jelly-like mass. Such behaviour is considerably different from that observed when a fresh film of a polymer is dissolved with solvents.

Polymers with little tendency to crosslink

The search for high polymers that have little or no tendency to crosslink continues. Polyvinyl acetate has an extremely low tendency to crosslink in the Fadeometer and in sunlight, if it does so at all. This fact was first pointed out by Thomson and has been confirmed in our experiments. Polyvinyl acetate, therefore, retains its important place in the conservation laboratory.

Our search for structures in the methacrylate family that have little tendency to crosslink led us to the prediction that a copolymer of ethylmethacrylate and methylacrylate should possess this property. Tests in the Fadeometer revealed that after 2000 hours on aluminum foil at 62°C such a polymer was still soluble in acetone to the extent of 95 percent or better. This exposure is equivalent to over 40 million footcandle hours. Although originally soluble in toluene, the polymer developed a certain degree of insolubility in this solvent as time went on. This change is apparently due to partial oxidation of the polymer and not to crosslinking, since otherwise the high solubility in acetone would not have occurred.

Following these tests of durability, we have suggested the consideration of this type of copolymer for certain applications. Commercially available polymers that are similar are the Rohm and Haas resin Acryloid B-72 and their emulsion of a related polymer, Rhoplex AC-33. The results of the exposure of these proprietary polymers in the Fadeometer are given in Table 5. The conclusions from these data are the same as those made concerning the behavior of the related polymers prepared in the laboratory. Acryloid B-72 may be applied in xylene; evaporation may be retarded by the addition of solvents such as diethylbenzene or all-aromatic petroleum fractions. Varnish formulated with these solvents is slightly cloudy, but the films are satisfactory in appearance.[14]

14. The use of Acryloid B-72 in the conservation of metal objects dates back a number of years: See *Rohm and Haas Reporter*, 8, No. 3 (1950), p. 14.

TABLE 5
Percentage Insolubility of Acryloid B-82 and Rhoplex AC-33 upon Exposure on Aluminum Foil in Fadeometer at 62°C

Insolubility in: Exposure time:	Toluene 394 h	Toluene 1045 h	Toluene 2027 h	Acetone 2027 h
B-82*	4.2	22.4	62.0	4.4
AC-33	10.6	43.8	77.4	4.8

* Rohm and Haas polymer similar to B-72.

CONCLUSIONS

We have learned much about the properties of solvent-type coatings in recent years, yet much remains to be investigated. Thus far, research has discovered two thermoplastic polymers of reasonably high molecular weight which are soluble in solvents as "mild" as toluene and xylene, and which form films that remain strong, colorless and soluble for extremely long periods of time. It is hoped that ever-increasing experience with these polymers in applications where they are suitable will strengthen the conservator's confidence and satisfaction in their use. In the range of viscosity grade similar to dammar and mastic resins, resins AW-2 and MS-2A also appear to be sound additions to the conservator's stock of materials.

Studies of the Effect of Light on Protective Coatings Using Aluminum Foil as a Support: Determination of Ratio of Chain Breaking to Crosslinking

ROBERT L. FELLER AND CATHERINE W. BAILIE

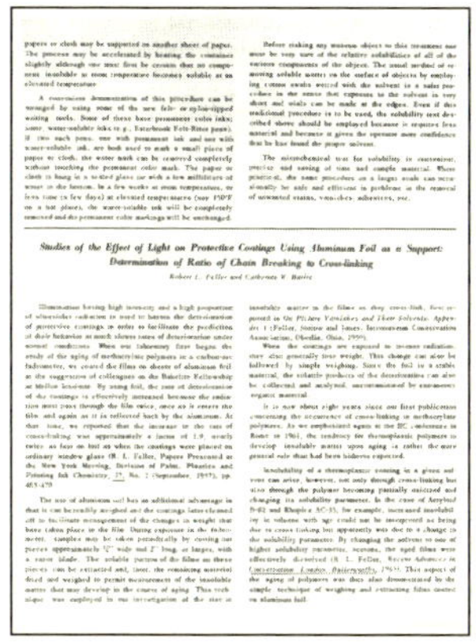

Originally printed in:
BULLETIN OF THE AMERICAN GROUP-IIC
Volume 6, Number 1 (1966), 2 Pages.

Illumination having high intensity and a high proportion of ultraviolet radiation is used to hasten the deterioration of protective coatings in order to facilitate the prediction of their behavior at much slower rates of deterioration under normal conditions. When our laboratory first began the study of the aging of methacrylate polymers in a carbon-arc Fadeometer, we coated the films on sheets of aluminum foil at the suggestion of colleagues on the Bakelite Fellowship at Mellon Institute. By using foil, the rate of deterioration of the coatings is effectively increased because the radiation must pass through the film twice, once as it enters the film and again as it is reflected back by the aluminum. At that time, we reported that the increase in the rate of crosslinking was approximately a factor of 1.9, nearly twice as fast on foil as when the coatings were placed on ordinary window glass.[1]

The use of aluminum foil has an additional advantage in that it can be readily weighed and the coatings later cleaned off to facilitate measurement of the changes in weight that have taken place in the film. During exposure in the Fadeometer, samples may be taken periodically by cutting out pieces approximately 1/4" wide and 2" long, or larger, with a razor blade. The soluble portion of the films on these pieces can be extracted and, later, the remaining material dried and weighed to permit measurement of the insoluble matter that may develop in the course of aging. This technique was employed in our investigation of the rise in insoluble matter in the films as they crosslink, first reported in *On Picture Varnishes and Their Solvents*, Appendix 1.[2]

1. Feller, R. L., Papers Presented at the New York Meeting, Division of Paint, Plastics and Printing Ink Chemistry, 17, No. 2 (September, 1957), p. 465-470.

2. Feller, R. L., Stolow, N., and Jones, E. H., Intermuseum Conservation Association, Oberlin, Ohio (1959).

When the coatings are exposed to intense radiation, they also generally lose weight. This change can also be followed by simple weighing. Since the foil is a stable material, the volatile products of the deterioration can also be collected and analyzed, uncontaminated by extraneous organic material.

It is now about eight years since our first publication concerning the occurrence of crosslinking in methacrylate polymers. As we emphasized again at the IIC conference in Rome in 1961, the tendency for thermoplastic polymers to develop insoluble matter upon aging is rather the more general rule than had been hitherto expected.

Insolubility of a thermoplastic coating in a given solvent can arise, however, not only through crosslinking but also through the polymer becoming partially oxidized and changing its solubility parameter. In the case of Acryloid B-82 and Rhoplex AC-33, for example, increased insolubility in toluene with age could not be interpreted as being due to crosslinking but apparently was due to a change in the solubility parameter. By changing the solvent to one of higher solubility parameter, acetone, the aged films were effectively dissolved.[3] This aspect of the aging of polymers was thus also demonstrated by the simple technique of weighing and extracting films coated on aluminum foil.

3. Feller, R. L., *Recent Advances in Conservation*, Butterworths, London (1963).

At the meeting in Rome in 1961, it was pointed out that light could have a number of effects on polymers. In addition to a weight loss and the change in solubility characteristics just mentioned, a major effect involves the breaking of polymer chains as well as their joining up (crosslinking). Indeed, if a significant amount of chain breaking is also taking place as well as crosslinking, the detection of crosslinked matter within a thermoplastic coating would not necessarily mean that the film would eventually become completely insoluble. Although the subject of simultaneous chain breaking and crosslinking was discussed in detail at the Rome meeting, representing our latest results from the laboratory, the original paper was not changed to include this new interpretation of the results. We wish to take the opportunity now to record how this further significant factor in film deterioration, chain breaking, can be demonstrated by simple means.

The rise in insoluble matter in films of three methacrylate polymers is shown in Figure 1. The method for determining this was the same as that used when such data were first reported by us in Appendix 1 of *On Picture Varnishes and Their Solvents*: at various times during the exposure in the Fadeometer, strips of the coating on aluminum foil are cut with a razor, weighed, placed in fine-mesh wire baskets and extracted in solvent. Following extraction, the films were dried and weighed again to determine the insoluble material that remained. The weight of the foil was finally determined by mechanically removing the solvent-swollen insoluble film from the foil.

Polymer 37 is poly(isobutylmethacrylate); polymer 59 is a copolymer of p-methylcyclohexyl and isoamyl methacrylate previously reported by this laboratory to have a marked tendency to crosslink. Terpolymer A is a commercial acrylic resin.

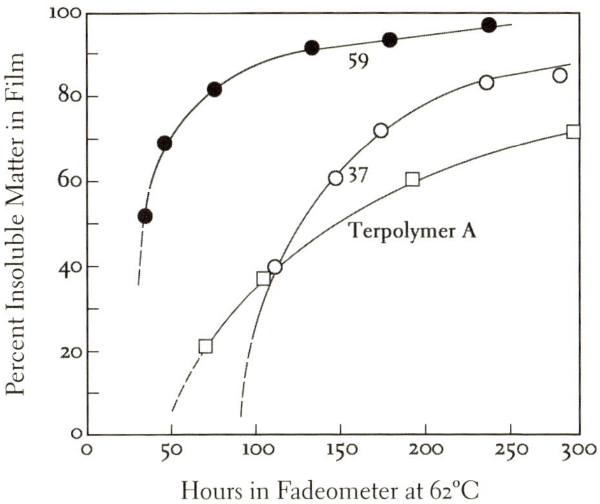

FIGURE 1

In such studies, it was soon noticed that some of the data, such as the curve for the Terpolymer A in Figure 1, did not show a rapid rise to practically complete insolubility. An explanation and interpretation of this slower rate of change is to be found in analysis of the problem given by A. Charlesby.[4] Charlesby has developed equations to predict the rise in insoluble matter in a polymer that is undergoing simultaneous chain scission and crosslinking. He expresses the change in the soluble fraction of such films as a function of the time or "radiation dose" necessary to cause the initial appearance of insoluble gel, R*. A family of curves has been calculated to show the decrease in the soluble fraction as a function of the relative time, R/R*, since the first initial formation of gel.[5] These change in slope as the ratio of chain scission to chain linking varies, resulting in a figure similar to the accompanying Figure 2. The three curves shown were among those presented at the meeting in Rome in 1961.

4. Charlesby, A., *Atomic Radiation and Polymers*, Pergamon Press, London (1960).

5. Charlesby, *ibid.*, p. 211.

In Figure 1, it can be seen that polymer 59 began to form an insoluble fraction at the earliest time; the formation of insoluble matter also rises faster than in the other polymers. Polymer 37 required a longer time of exposure before the initial formation of insoluble matter. According to Charlesby's analysis, the increase in insoluble matter for polymer 37 in Figure 1 should occur at a slower rate than that for polymer 59, as indeed it does. When the data from Figure 1 are plotted in Figure 2 with respect to R/R* according to Charlesby's recommendations, one can see that the data for methacrylate polymers 59 and 37 are practically identical.

In Figure 2 the decrease in the soluble fraction in Terpolymer A does not proceed at the same rate as in the other two polymers. As mentioned, Charlesby gives a family of curves to show that the data may not fall as rapidly as experienced in polymers 59 and 37, but will fall at different rates according to

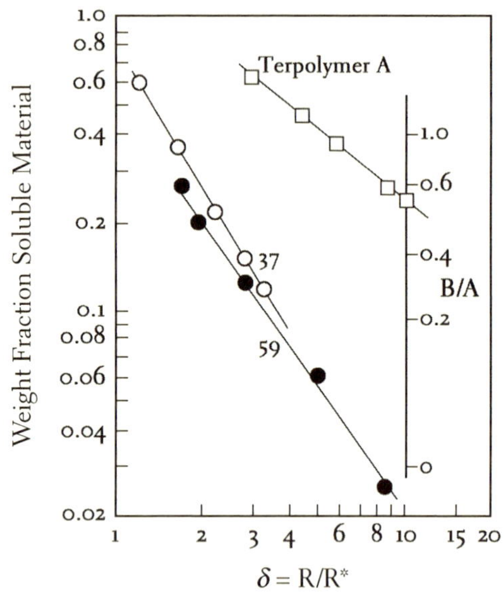

FIGURE 2

the amount of chain breaking, B, and crosslinking, A, taking place simultaneously. The family of curves is not drawn here but only the position that they would reach after a period of ten times the initial gel dose, R/R* = 10. The amount of soluble material that would remain in the film at this point can thus be taken as an indication of the ratio of the number of chain scissions that occur for every chain that is crosslinked, B/A. Terpolymer A thus is seen to exhibit a ratio of the number of polymer chains broken to the number of crosslinks formed of about 0.55. On a basis of this analysis, it can also be seen that polymers 59 and 37 crosslink almost exclusively with no significant amount of chain breaking taking place. Poly(n-butyl methacrylate) behaves in the same way.

The ultimate purpose of these investigations is to predict the behavior at room conditions. In this regard, it should be pointed out that the ratio of chain breaking to chain linking can vary with temperature and, in fact, increased in the case of polymers 37 and 59 when they were tested at a lower temperature (32°C).

A more general and customary way of plotting these data is shown in Figure 3. Here a plot of $S + \sqrt{S}$ is made against $1/\delta$, where S is the soluble fraction that remains in the film and δ is the ratio of the time to cause initial gelation to the time of exposure, that is the inverse of the unit used in x- axis in Figure 2. This approach is used in another analysis published by Charlesby[6] and has the advantage that the slope of the curves should generally be 2. It is difficult to determine precisely the time for initial formation of gel based on the extrapolation of the curves in Figure 1 to 0% insoluble matter. In the data used for the three polymers, apparently errors were made in the selection of the

6. Charlesby, *ibid.*, p. 173.

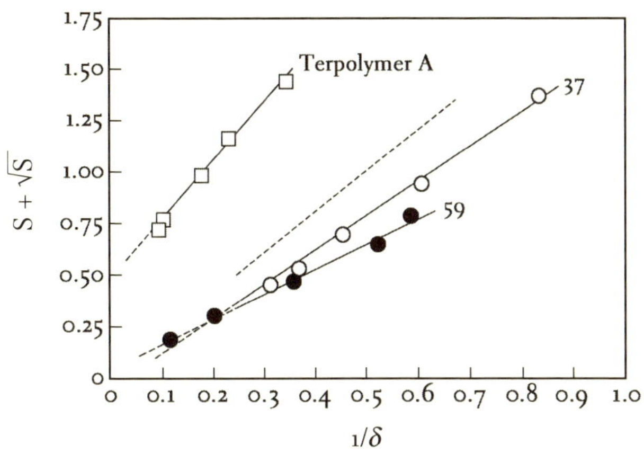

FIGURE 3
S=Soluble fraction, δ = Relative time since initial gel formation

initial gel time, for the plots of the data in Figure 3 do not have a slope of 2 (shown by the dotted line). Correcting these data to bring the slope to a value of 2 can be of help in choosing the most reasonable value for the initial gel time. In this plot, the intercept at $1/\delta = 0$, gives a ratio of B/A of about 0.5 for Terpolymer A.

Consideration of the ratio of B/A in investigations of the effects of ultraviolet light on films has appeared in the literature in recent years.[7] One of the most elaborate analyses of the problem has been made by L. D. Maxim and C. H. Kuist,[8] in which the deterioration of several methacrylate and acrylate polymers was reported. Recently,[9] they have extended their detailed analysis to show how one can detect which bonds in copolymers are the ones most readily broken.

Certainly, the analysis of the loss of weight that takes place when a film ages and the relative importance of chain breaking and chain linking processes will soon become a widespread and standard practice in the study of the deterioration of thermoplastic films.

7. Stephanson, G. B., and Wilcox, W. S., *J. Polymer Sci.*, 1A (1963), p. 2741-2752.

8. Maxim, L. D., and Kuist, C. H., *Offic. Dig. Federation Soc. Paint Technol.*, 36 (1964), p. 723-744.

9. Maxim, L. D., and Kuist, C. H., *Polymer*, 6 (1965), p. 523-529.

SUMMARY

Exposure of protective coatings on aluminum foil to the action of ultraviolet and visible radiation can accelerate the rate of change within the films by a factor of almost 2. The use of aluminum foil as a support also facilitates the weighing of the films. By this method, the loss in weight owing to volatilization can be determined and also, by extraction with solvents, the change in the solubility of the film. An analysis of the rate of change of the fraction of a thermoplastic polymer that remains soluble, considered as a function of the time of exposure, can reveal the ratio of chain breaking to chain linking (chain scission to crosslinking) taking place in the film by means of the mathematical analysis developed by Charlesby.

This report is part of continuing studies on the properties of protective coatings and the effect of light on museum materials supported by the National Gallery of Art, Washington, D. C.

Solubility and Crosslinking Characteristics of Ethylene/Vinylacetate Copolymers

R. L. FELLER AND M. CURRAN

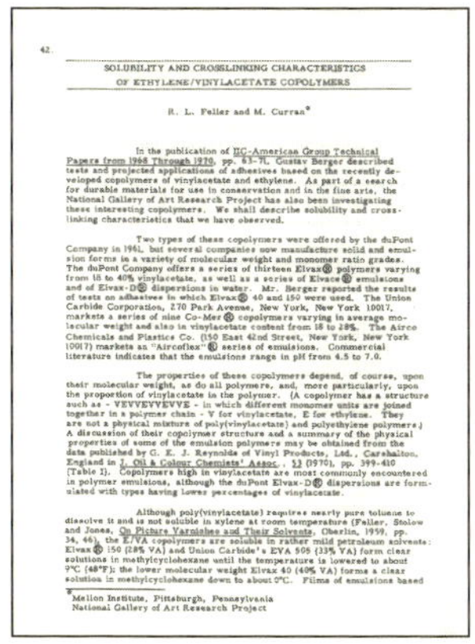
42.

SOLUBILITY AND CROSSLINKING CHARACTERISTICS
OF ETHYLENE/VINYLACETATE COPOLYMERS

R. L. Feller and M. Curran*

In the publication of IIC-American Group Technical Papers from 1968 Through 1970, pp. 63-71, Gustav Berger described tests and projected applications of adhesives based on the recently developed copolymers of vinylacetate and ethylene. As part of a search for durable materials for use in conservation and in the fine arts, the National Gallery of Art Research Project has also been investigating these interesting copolymers. We shall describe solubility and crosslinking characteristics that we have observed.

Two types of these copolymers were offered by the duPont Company in 1961, but several companies now manufacture solid and emulsion forms in a variety of molecular weight and monomer ratio grades. The duPont Company offers a series of thirteen Elvax® polymers varying from 18 to 40% vinylacetate, as well as a series of Elvace® emulsions and of Elvax-D® dispersions in water. Mr. Berger reported the results of tests on adhesives in which Elvax® 40 and 150 were used. The Union Carbide Corporation, 270 Park Avenue, New York, New York 10017, markets a series of nine Co-Mer® copolymers varying in average molecular weight and also in vinylacetate content from 18 to 28%. The Airco Chemicals and Plastics Co. (150 East 42nd Street, New York, New York 10017) markets an "Aircoflex"® series of emulsions. Commercial literature indicates that the emulsions range in pH from 4.5 to 7.0.

The properties of these copolymers depend, of course, upon their molecular weight, as do all polymers, and, more particularly, upon the proportion of vinylacetate in the polymer. (A copolymer has a structure such as - VEVVEVVEVVE - in which different monomer units are joined together in a polymer chain - V for vinylacetate, E for ethylene. They are not a physical mixture of poly(vinylacetate) and polyethylene polymers.) A discussion of their copolymer structure and a summary of the physical properties of some of the emulsion polymers may be obtained from the data published by G. E. J. Reynolds of Vinyl Products, Ltd., Carshalton, England in J. Oil & Colour Chemists' Assoc., 53 (1970), pp. 399-410 (Table I). Copolymers high in vinylacetate are most commonly encountered in polymer emulsions, although the duPont Elvax-D® dispersions are formulated with types having lower percentages of vinylacetate.

Although poly(vinylacetate) requires nearly pure toluene to dissolve it and is not soluble in xylene at room temperature (Feller, Stolow and Jones, On Picture Varnishes and Their Solvents, Oberlin, 1959, pp. 34, 46), the E/VA copolymers are soluble in rather mild petroleum solvents: Elvax® 150 (28% VA) and Union Carbide's EVA 505 (33% VA) form clear solutions in methylcyclohexane until the temperature is lowered to about 9°C (48°F); the lower molecular weight Elvax 40 (40% VA) forms a clear solution in methylcyclohexane down to about 0°C. Films of emulsions based

*Mellon Institute, Pittsburgh, Pennsylvania
National Gallery of Art Research Project

Originally printed in:
Bulletin of the American Group-IIC
Volume 11, Number 1 (1970), Pages 42-45.

In the publication of *IIC-American Group Technical Papers*,[1] Gustav Berger described tests and projected applications of adhesives based on the recently developed copolymers of vinylacetate and ethylene. As part of a search for durable materials for use in conservation and in the fine arts, the National Gallery of Art Research Project has also been investigating these interesting copolymers. We shall describe solubility and crosslinking characteristics that we have observed.

Two types of these copolymers were offered by the duPont Company in 1961, but several companies now manufacture solid and emulsion forms in a variety of molecular weight and monomer ratio grades. The duPont Company offers a series of thirteen Elvax® polymers varying from 18 to 40% vinylacetate, as well as a series of Elvace® emulsions and of Elvax-D® dispersions in water. Mr. Berger reported the results of tests on adhesives in which Elvax® 40 and 150 were used. The Union Carbide Corporation, 270 Park Avenue, New York, New York 10017, markets a series of nine Co-Mer® copolymers varying in average molecular weight and also in vinylacetate content from 18 to 28%. The Airco Chemicals and Plastics Co. (150 East 42nd Street, New York, New York 10017) markets an "Aircoflex"® series of emulsions. Commercial literature indicates that the emulsions range in pH from 4.5 to 7.0.

The properties of these copolymers depend, of course, upon their molecular weight, as do all polymers, and, more particularly, upon the proportion of vinylacetate in the polymer. [A copolymer has a structure such as—VEVVEVVEVVE—in which different monomer units are joined together in

1. Berger, G., *IIC-American Group Technical Papers from 1968 Through 1970*, p. 63-71.

a polymer chain—V for vinylacetate, E for ethylene. They are not a physical mixture of poly(vinylacetate) and polyethylene polymers.] A discussion of their copolymer structure and a summary of the physical properties of some of the emulsion polymers may be obtained from the data published by G. E. J. Reynolds of Vinyl Products, Ltd., Carshalton, England[2] (Table 1). Copolymers high in vinylacetate are most commonly encountered in polymer emulsions, although the duPont Elvax-D® dispersions are formulated with types having lower percentages of vinylacetate.

2. Reynolds, G. E. J., *J. Oil and Colour Chemists' Assoc.*, 53 (1970), p. 399-410.

Although poly(vinylacetate) requires nearly pure toluene to dissolve it and is not soluble in xylene at room temperature,[3] the E/VA copolymers are soluble in rather mild petroleum solvents: Elvax® 150 (28% VA) and Union Carbide's EVA 505 (33% VA) form clear solutions in methylcyclohexane until the temperature is lowered to about 9°C (48°F); the lower molecular weight Elvax® 40 (40% VA) forms a clear solution in methylcyclohexane down to about 0°C. Films of emulsions based on copolymers containing from 10 to 17% vinylacetate, kindly supplied by Dr. Reynolds of Vinyl Products, Ltd., were easily removed in a 50/50 mixture (by volume) of methylcyclohexane/toluene.

3. Feller, R. L., Stolow, N., Jones, E. H., *On Picture Varnishes and Their Solvents*, Intermuseum Conservation Association, Oberlin, Ohio (1959), p. 34, 46.

TABLE 1
Physical Properties of E/VA Copolymer Emulsions[2]

Comonomer	Ethylene					2-ethylhexyl-acrylate	
% Comonomer wt%	4*	10	12.5	17.5	20	15	25
mole%	11.4	25.4	30.5	39.4	43.4	7.6	13.5
Min. Film Form. Temp. °C	9	2	0	-	-	10	5
Tensile Strength Kg/cm²	80.7	40.7	36.7	5.6	4.9	59.7	28.2
Tensile Strength lb/in²	1148	579	522	77	70	849	401
Elongation at Break	210	340	415	615	1220	200	300
Rigidity Modulus, Kg/cm²	500	130	70	55	20	1140	90
Sward Rocker Hardness	27	16	10	4	2	24	12
Brittle Point °C	10	0	-5	-10	-15	10	5
Heat Seal Temp. °C	120	105	95	75	60	120	110
Heat Seal Temp. °F	248	221	203	167	140	248	230

* 96% of the copolymer by weight is vinylacetate.

Near-ultraviolet radiation does not cause poly(vinylacetate) to crosslink but will cause polyethylene to crosslink (unless inhibited[4]). The copolymers thus pose an interesting question and prompted our investigation of their behavior. We have found that a number of the copolymers undergo both crosslinking and chain breaking; the latter process increasing as the vinylacetate content increases. The particular copolymers tested are shown in Table 2 and the results summarized in Figure 1. After 800 hours of exposure on aluminum foil in an

4. Many of the Elvax® resins contain 500-1000 ppm of butylated hydroxytoluene (BHT) antioxidant.

Atlas xenon-arc Fadeometer (using an ordinary glass rather than a quartz filter) the films were removed with ease as a gel in methylcyclohexane even though they had from 14 to 62% insoluble matter present at that point. In contrast, Elvacite® 2046 (a copolymer of isobutyl and n-butyl methacrylate) becomes 30% insoluble in about 80 hours, about 80% after 500 hours exposure. The copolymers tested in Table 2 are 8 to 10 times more resistant to loss of solubility under near-ultraviolet radiation than the easily crosslinked butyl methacrylate polymers.

TABLE 2
Soluble Matter in Ethylene/Vinylacetate Copolymers After 800 Hrs. Exposure on Aluminum Foil in Xenon-Arc Fadeometer (Ordinary Glass Filter)

Polymer	Weight Percent Vinylacetate in Copolymer	Percent Soluble Matter Remaining After 800 Hours Exposure
Co-Mer® EVA 303 (UC)	18	38
Co-Mer® EVA 508 (UC)	28	59
Co-Mer® EVA 505* (UC)	28	92
Elvax® 150 (duP)	33	86

* lower molecular weight than EVA 508; (UC) Union Carbide, (duP) duPont

TABLE 3
Influence of Wax on Tendency to Cross-link
Hours of Exposure on Aluminum Foil in Carbon-Arc Fadeometer (Corex D Filter) Before Developing High Resistance to Removal in Various Solvents (Rubbing with Cotton Swabs Dipped in Solvent)

Film	Methylcyclohexane	Toluene	Acetone
Poly(isoamylmethacrylate) (PIAM)	68	105	-
PIAM + 0.5% Beeswax	115	150	240
PIAM + 1% Beeswax	120	168	240
PIAM + 3% Beeswax	~140	260	480
PIAM + 6% Beeswax	250	340	-

Mr. Berger mentions that the addition of small quantities of wax will reduce the tendency of many polymers to form insoluble matter owing to crosslinking. In our investigations on protective coatings this possibility was tested in 1957 following a question raised by Mr. Mario Modestini. The results, summarized in Table 3, inferred that, at sufficiently low concentrations to yield reasonably transparent varnish (1 to 3%), beeswax would not offer more than a two-fold improvement in the resistance-to-crosslinking, insufficient to warrant a broad endorsement of its use for this specific purpose alone. At higher concentrations

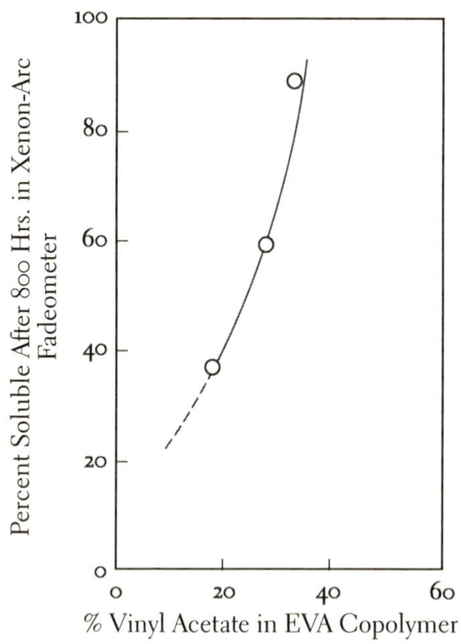

FIGURE 1

of wax, however, the data indicate that the wax will indeed interfere significantly with the tendency to crosslink.[5]

Because the copolymers having a percentage of vinylacetate higher than 25% exhibit a negligible tendency to crosslink under near-ultraviolet radiation, and are soluble in petroleum hydrocarbons of low solubility parameter, we consider that they merit consideration for use in conservation and have suggested that practical applications be explored. In the meantime, tests of the durability of the copolymers, both in solution and emulsion grades, continues; we are currently investigating their chain-breaking characteristics.

5. The National Gallery of Art Research Project's current research program on the inhibition of oxidation has also shown that an antioxidant system based on Geigy's ultraviolet absorber Tinuvin 328 and the synergist, dilaurylthiodipropionate, at concentrations of about 1% each will markedly protect polymers and copolymers of n-butyl and isobutyl methacrylate from crosslinking by near-ultraviolet radiation, but we wish to make further tests before making specific recommendations regarding the application of such a system.

Stages in the Deterioration of Organic Materials

R. L. FELLER

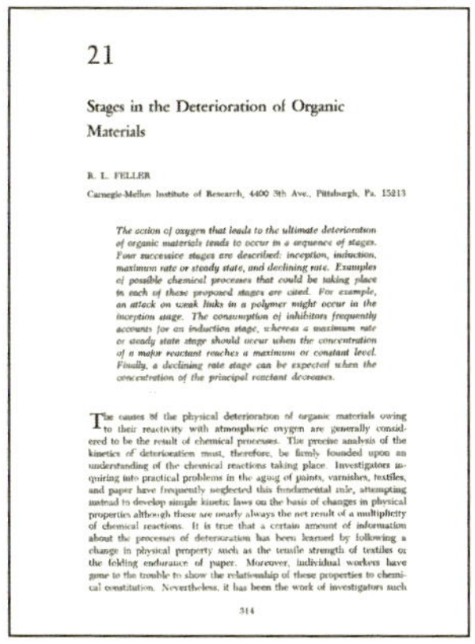
21

Stages in the Deterioration of Organic Materials

R. L. FELLER

Carnegie-Mellon Institute of Research, 4400 5th Ave., Pittsburgh, Pa. 15213

The action of oxygen that leads to the ultimate deterioration of organic materials tends to occur in a sequence of stages. Four successive stages are described: inception, induction, maximum rate or steady state, and declining rate. Examples of possible chemical processes that could be taking place in each of these proposed stages are cited. For example, an attack on weak links in a polymer might occur in the inception stage. The consumption of inhibitors frequently accounts for an induction stage, whereas a maximum rate or steady state stage should occur when the concentration of a major reactant reaches a maximum or constant level. Finally, a declining rate stage can be expected when the concentration of the principal reactant decreases.

The causes of the physical deterioration of organic materials owing to their reactivity with atmospheric oxygen are generally considered to be the result of chemical processes. The precise analysis of the kinetics of deterioration must, therefore, be firmly founded upon an understanding of the chemical reactions taking place. Investigators inquiring into practical problems in the aging of paints, varnishes, textiles, and paper have frequently neglected this fundamental rule, attempting instead to develop simple kinetic laws on the basis of changes in physical properties although these are nearly always the net result of a multiplicity of chemical reactions. It is true that a certain amount of information about the processes of deterioration has been learned by following a change in physical property such as the tensile strength of textiles or the folding endurance of paper. Moreover, individual workers have gone to the trouble to show the relationship of these properties to chemical constitution. Nevertheless, it has been the work of investigators such

314

Originally printed in:
PRESERVATION OF PAPER AND TEXTILES OF HISTORIC AND ARTISTIC VALUE
Advances in Chemistry Series, No. 164, ed. Williams, J. C., American Chemical Society, Washington, DC (1977), Pages 314-335.

The causes of the physical deterioration of organic materials owing to their reactivity with atmospheric oxygen are generally considered to be the result of chemical processes. The precise analysis of the kinetics of deterioration must, therefore, be firmly founded upon an understanding of the chemical reactions taking place. Investigators inquiring into practical problems in the aging of paints, varnishes, textiles, and paper have frequently neglected this fundamental rule, attempting instead to develop simple kinetic laws on the basis of changes in physical properties although these are nearly always the net result of a multiplicity of chemical reactions. It is true that a certain amount of information about the processes of deterioration has been learned by following a change in physical property such as the tensile strength of textiles or the folding endurance of paper. Moreover, individual workers have gone to the trouble to show the relationship of these properties to chemical constitution. Nevertheless, it has been the work of investigators such as Bateman, Gee, and Bolland,[1] Tobolsky,[2] Mayo,[3] and Guillet[4] that has kept the study of oxidative deterioration focused on fundamentals. This has been achieved through their insistence that the proper study of the kinetics of deterioration is the study of the chemical reactions taking place as a function of time.

The pioneering investigations of autoxidation by Bolland provide a prime example of the complexity of reactions that must be considered in determining the mechanism of deterioration. Bolland, Gee, and their associates proposed the involvement of six to eight chemical reaction steps and were not satisfied until techniques had been developed for measuring the rates and energies of

1. Uri, N., "Physico-Chemical Aspects of Autoxidation," Mesrobian, R. B., and Tobolsky, A. V., "Autoxidation of Hydrocarbons Accelerated by Metals, Light and Other Agencies," and Uri, N., "Mechanism of Autoxidation," in *Autoxidation and Autoxidants*, Vol. 1, ed. Lundberg, W. D., Interscience, New York (1961), p. 55-168.

2. Tobolsky, A. V., Metz, D. J., and Mesrobian, R. B., "Low Temperature Autoxidation of Hydrocarbons: The Phenomenon of Maximum Rates," *J. Am. Chem. Soc.*, 72 (1950), p. 1942-1952.

3. Mayo, F. R., and Irwin, K. C., "Some Physical Organic Chemistry of Aging Processes," *Polym. Eng. Sci.*, 9 (1969), p. 282-285.

4. Somersall, A. C., and Guillet, J. E., "Photochemistry of Ketone Polymers VIII," *Macromolecules*, 5 (1972), p. 410-415.

activation for each individual reaction.[1] This landmark series of investigations has served for more than a decade as a model example of a search for an answer to the question, "What is the cause of the physical deterioration of an organic substance in terms of the chemical reactions that are ultimately responsible for the changes observed?"

The conservation scientist responsible for the preservation of the world's cultural patrimony—its art objects, artifacts, archival materials, historic buildings and monuments—is concerned with developing methods and materials that can be used to impede the progress of deterioration in a wide range of organic, inorganic, and metallic materials. Broad as the problems confronting the conservation scientist may be, unless he, too, intends to come to grips with the basic chemical reactions involved, efforts to preserve materials will be spent largely in the superficial treatment of symptoms, not in the discovery and alleviation of causes.

Admittedly, it is not a simple matter to isolate the individual chemical reactions responsible for the observed changes in physical properties. However, it is the author's firm conviction that the observed changes in the rate of oxidative aging of organic materials tend to take place in a well-ordered sequence and that, moreover, the general types of chemical reactions responsible for this behavior can be reasonably accounted for. In this paper, four different stages in the overall rate of deterioration are described. In addition, some of the chemical processes that may be responsible for each stage are discussed, using examples both from the literature and from the author's own work. It is hoped that this synthesis of much of the information that we already know about oxidatively induced changes in organic materials will facilitate a more systematic analysis of the basic problems of deterioration and preservation that confront museums and archives.

Four Stages in the Overall Process of Deterioration

To speak of a "rate of deterioration" is a convenient generalization. However, if one thinks for a moment about the subject, one soon realizes that changes in properties rarely occur at a fixed rate over the entire history of the physical usefulness of materials, from fabrication to the point of complete failure. The ordinary materials that we encounter—textiles, leather, paper—are usually extensive combinations of substances, and the physical properties of these composites, if we may call them that, tend to exhibit different rates of change at various stages in their life histories. Surely this is the consequence of the considerable number of individual chemical processes that take place concurrently each at a different rate relative to the others.

5. Kelleher, P. G., "Thermal Oxidation of Thermoplastics," *J. Appl. Polym. Sci.*, 10 (1966), p. 843-857.

In 1966 Kelleher proposed a set of four curves to represent the rate of oxygen uptake observed in the thermal oxidation of thermoplastics. These are illustrated in Figure 1.[5] In the curve labeled "a," the amount of oxygen uptake is linear with time and the rate of oxygen gain is constant; "b" illustrates the autocatalytic condition in which both the amount and the rate of oxygen uptake increase with time. Curve "c" refers to the circumstance in which oxygen consumption proceeds at a decreasing rate, designated by Kelleher as the autoretardant case. A combination curve, "d," displays an S-shape with the portion "t_i" labeled as the induction time. These four are typical of the types of curves encountered in deterioration studies not only when chemical changes are followed but often when physical changes are observed as well.

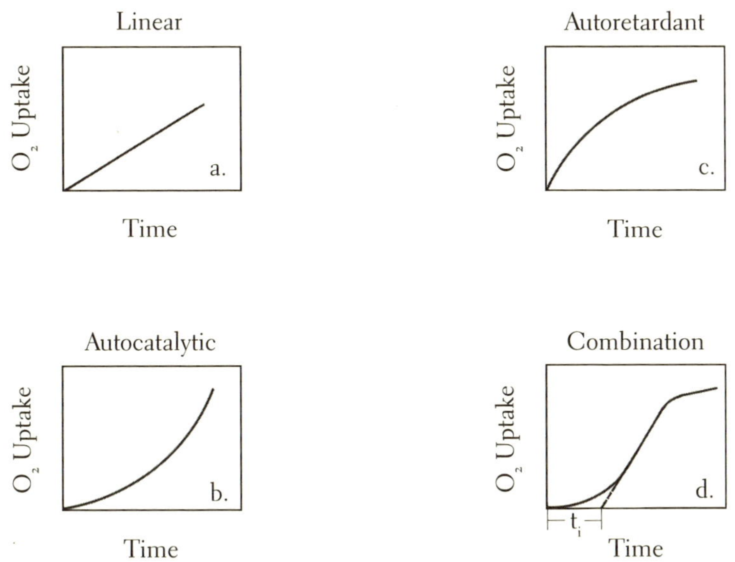

FIGURE 1
Plots of oxygen uptake vs. time, after Kelleher (Ref. 5).

Investigations of the deterioration and degradation of many kinds of organic materials in our laboratory have reinforced the author's conviction that the progress of deterioration resulting from reactions with atmospheric oxygen tends to occur in stages, very much as seen in Kelleher's combination curve, "d." The proposed stages, shown in Figure 2, are designated as follows:

I. Inception stage
II. Induction stage for III
III. Rise and achievement of maximum rate and possible steady state stage
IV. Declining-rate stage

In any particular experiment, one or more of these stages may not be apparent from the measured evidence. For example, the events of inception, Stage I, may occur so rapidly that they escape detection and attention. On the other hand, the experiment may be terminated before Stage IV is reached. Nevertheless, as a conceptual guide for the study of deterioration processes, the possibility of observing all four stages should be kept in mind. A search for the reasons for failing to observe one or more of the stages will frequently prove enlightening. One can postulate, for example, that curves such as Kelleher's "a," "b," and "c" more than likely represent segments of the combined curve, "d," or of the proposed four-stage curve in Figure 2. In the following section, each stage is defined, the kinds of chemical reactions that may be responsible for each are discussed, and a number of specific examples are cited.

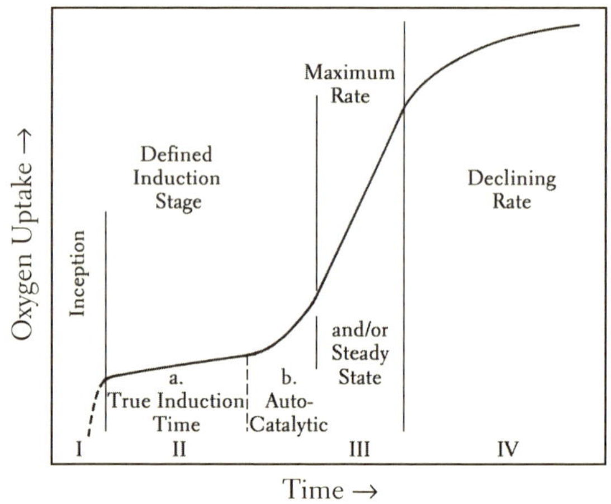

FIGURE 2
Possible stages of oxidative history (after Grassie).[32]

DEFINITION, DISCUSSION, AND EXAMPLES OF THE FOUR STAGES OF DETERIORATION

Stage I: Inception

DEFINITION. The limited, generally rapid change that frequently occurs when samples are first placed under test, is designated—for want of a better term at the moment—as the inception stage. The experimenter often is tempted to ignore the first data points when plotting rate of reaction curves, attributing the deviation to the effects of small amounts of impurities (transition metal ions, residual catalyst, traces of monomer), lack of temperature equilibrium, and other experimental problems. The extent of any such initial event should be carefully observed and the possible reasons for it considered. Chemical reactions in this stage may prove to be the key to the ultimate stability of a particular material.

DISCUSSION. While a search for possible causes for the events in this initial stage should be profitable, there are a number of sound reasons for the initial rate to be rapid without its representing an experimental error or problem. The following are two of the better understood explanations for rapid initial reactivity that should not be overlooked. An initially high rate of oxygen consumption was predicted by Tobolsky, Metz, and Mesrobian in their analysis of the kinetics of autoxidation for the case in which the concentration of hydroperoxide at the beginning, $[ROOH]_0$, is much higher than the concentration under the steady state conditions, that is, the case in which $[ROOH]_\infty / [ROOH]_0 < 1.0$.[2] Such a situation may occur when a substance has accumulated a high concentration of peroxides during storage in the dark and is then suddenly exposed to heat or light.

Another instance of predictably rapid initial rate is the presence of "weak links" in the polymer chain, bonds that degrade more rapidly than the intended primary bonds in the polymer backbone. An example of such links would be the incorporation of traces of oxygen in the main sequence of polyethylene. Although much discussion, pro and con, has taken place in the past concerning whether weak links actually exist, there is little question at present that sites in polymers do occur that tend to be more susceptible to chain breaking than others. Cameron and Grazzie's study of the degradation of polystyrene provides a particularly convincing case.[6]

EXAMPLES. Our investigation of the photochemical stability of polyvinylacetate (PVA) and of the Rohm and Haas methylmethacrylate-ethylacrylate copolymer, Acryloid B-72, revealed that these highly stable resins tend more to chain break rather than to crosslink upon exposure to visible and near-ultraviolet radiation.[7] This is indicated by the decrease in intrinsic viscosity shown in Figures 3 and 4. To guard against unexpected effects that might be caused by slight differences in spectral distribution, this conclusion was verified by using two different sources of radiation, a General Electric RS-type sunlamp and an ordinary "daylight" fluorescent lamp. (Compare Figures 3 and 4.)

Bryce and Greenwood as well as others before them have pointed out that the presence of weak links cannot be demonstrated simply by considering the fall in the average degree of polymerization or in the intrinsic viscosity as shown in Figures 3 and 4.[8] The degradation rate constant in the initial stages is not proportional to the average degree of polymerization ($\overline{DP}$) but to ($1/\overline{DP}$), and it is a plot of ($1/\overline{DP}$) vs. time that should be linear, at least in the initial stages. Thus, degradation may be conveniently followed by use of Equation 1:

$$[1/\overline{DP}_t - 1/\overline{DP}_o] = [\eta]_t^{-1/\alpha} - [\eta]_o^{-1/\alpha} = kt \qquad (1)$$

where DP = average degree of polymerization; $[\eta]$ = intrinsic viscosity; t = time of exposure; o = time zero; α = the exponent in the equation, $[\eta] = K\overline{DP}^{\alpha}$; and k = kinetic rate constant divided by K from $[\eta] = K\overline{DP}^{\alpha}$. When data are plotted according to this equation, the presence of a fraction of rapidly degraded

6. Cameron, G. C., and Grassie, N., "The Thermal Depolymerization of Polystyrene IV: Depolymerization in Naphthalene and Tetralin Solutions," *Polymer*, 2 (1961), p. 367-373.

7. Feller, R. L., "Studies on the Photochemical Stability of Thermoplastic Resins," *Preprints of the Fourth Triennial Meeting, Venice*, ICOM Committee for Conservation (1975), Paper 75/22/4.

8. Bryce, W. A. J., and Greenwood, C. T., "The Degradation of High Polymers," *J. Polym. Sci.*, 25 (1957), p. 480-483.

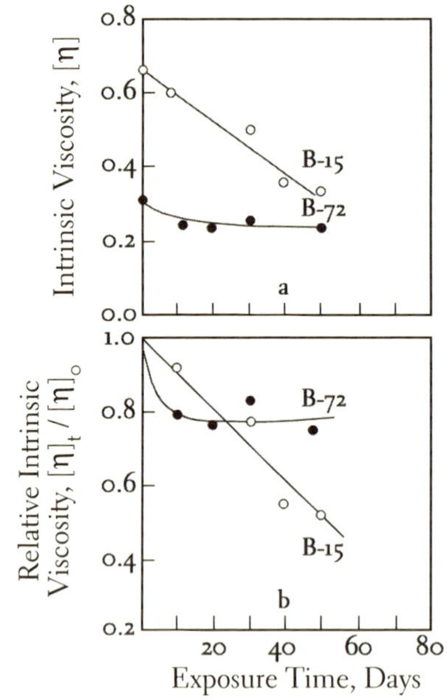

FIGURE 3
Fall of intrinsic viscosity of Acryloid B-72 and Vinac B-15 (PVA) upon exposure to RS-type sunlamp.

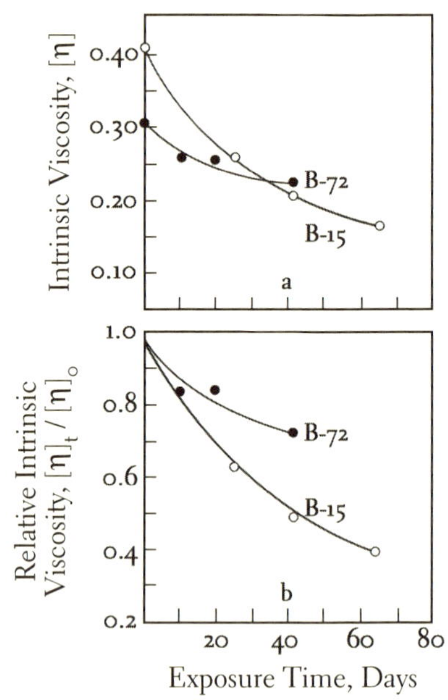

FIGURE 4
Fall of intrinsic viscosity of Acryloid B-72 and Vinac B-15 (PVA) upon exposure to fluorescent lamps.

weak linkages is shown by an intercept on the ordinate, the curve not passing through the origin.

Our principal objective in the above experiment was to demonstrate that these two polymers tended more to chain break than crosslink upon exposure to light. Having confirmed this behavior, we further decided to plot the data of Figures 3b and 4b according to Equation 1, using Moore and Murphy's value of 0.5 for α, both for poly(vinyl acetate), and for convenience, Acryloid B-72.[9] As seen in Figure 5, the data for poly(vinyl acetate) indicate no weak links. An apparent intercept at 10-15 days even suggests that there may be an induction time before chain breaking sets in. The data for Acryloid B-72 may be interpreted in several ways, but it seems fair to say that they indicate the presence of a limited number of links that rapidly break after which the rate of degradation becomes negligible. This initial effort to analyze the behavior of these polymers in terms of Equation 1 has thus suggested new facets concerning their degradation that will stimulate further investigation.

9. Moore, W. R., and Murphy, M., "Viscosities of Dilute Solutions of Polyvinyl Acetate," *J. Polym. Sci.*, 56 (1962), p. 519-532. Intrinsic viscosities were determined in toluene at 30°C except for the data for PVA under the RS-type sunlamp, an early experiment in which ethyl acetate was used. For Figures 3b and 4b, practically the same curves are obtained and the same conclusions reached if the values of α is taken as 1 instead of 0.5.

The intrinsic viscosity fell greatest in the case of the polyvinylacetate to about 45% of the initial value. An estimate of the number of scissions per molecule can serve to indicate whether this represents severe degradation or not. If we use Equation 2 to calculate the average number of scissions at the end of the

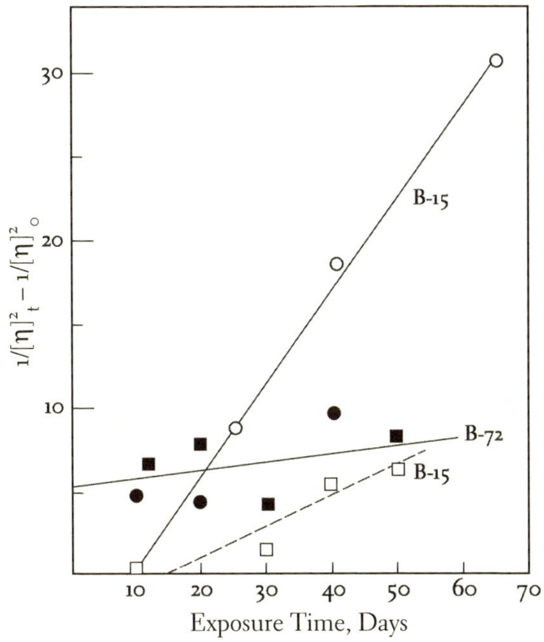

FIGURE 5
Plot of data on relative intrinsic viscosity vs. time from Figures 3b and 4b according to Equation 1. ○ Vinac B-15 and ● Acryloid B-72: exposure under CW fluorescent lamp; □ Vinac B-15 and ■ Acryloid B-72: exposure under RS-type sunlamp.

experiment, as recommended by Shultz and Leahy,[10] we find that the reaction had proceeded to the extent of about 4 scissions in the initial PVA molecule.

$$\text{scissions per chain} = ([\eta]_o/[\eta]_t)^{1/\alpha} - 1 \cong ([\eta]_o/[\eta]_t)^2 - 1 \qquad (2)$$

Members of the present conference are primarily interested in the stability of cellulose. When the data of Demeny on the photochemical deterioration of cotton yarns are plotted according to Equation 1, (essentially as he has done in plotting the number of scissions per 10,000 glucose units), there is an indication of a rapid initial reaction followed by a slower rate.[11] The data of Daruwalla, D'Silva, and Mehta on the exposure of cotton in a carbon-arc Fadeometer also indicate a more rapid initial reaction.[12]

Investigators have been cautioned that the weak links can be broken at the time the cellulose is put into solution in preparation for the measurement of intrinsic viscosity rather than at the time of its exposure to ultraviolet radiation.[8, 13] Nevertheless, regardless of the immediate cause of rupture, the existence of links weaker than normal bond strength can be detected through the use of Equation 1.

10. Shultz, A. R., and Leahy, S. M., "Random Chain Scission of Polyethylene Terephthalate by Light, Determination of Active Wavelength," *J. Appl. Polym. Sci.*, 5 (1961), p. 64-66.

11. Demeny, L., "Degradation of Cotton Yarns by Light from Fluorescent Lamps," *London Conference on Museum Climatology*, International Institute for Conservation, London (1968) p. 53-64.

12. Daruwalla, E. N., D'Silva, A. P., and Mehta, A. C., "Photochemistry of Cotton and Chemically Modified Cotton, Part I: Behaviour During Exposure to Carbon-Arc and Solar Radiations," *Text. Res. J.*, 37, No. 3 (1967), p. 147-172.

13. Sharples, A., "The Hydrolysis of Cellulose, Part II. Acid Sensitive Linkages in Egyptian Cotton," *J. Polym. Sci.*, 14 (1954), p. 95-104.

Stage II: Induction

DEFINITION. The concept of an induction stage has proved useful in many studies of reaction kinetics. There are a number of ways that the induction period may be defined, but the following conforms to the proposed scheme in Figure 2: it is the stage, following any inception stage that may be detected, that occurs before the maximum reaction rate is achieved.

This definition is intended to apply to the study of chemical reaction kinetics. The concept of induction time has also been used, and will continue to be used, to describe the period before a marked and relatively rapid change in a physical property occurs, but the author has suggested that the designation "apparent induction time" should be used for all situations other than those that fit the precise definition presented above.[14] Thus, when the chemical processes that are occurring are not being specifically monitored, the adjectival modifier "apparent" should be used to indicate the lack of precise measure of a specific chemical reaction.

14. Feller, R. L., "Fundamentals of Conservation Science: Induction Time and the Autoxidation of Organic Compounds," *Bull. Am. Inst. Conserv.*, 14, No. 2 (1974), p. 142-151.

In reality, the induction period is ended when the autocatalytic stage "II-b" begins (Figure 2), but the moment of this event is not easily determined. For the sake of a convenient operational definition, the induction stage of a particular reaction is defined as occurring before its maximum rate is attained.

DISCUSSION. The apparent induction period frequently appears to be a time when almost nothing is happening. For example, if one is measuring the change in gloss or loss of weight of paints exposed to sunlight, there frequently will be a period of time during which the properties remain practically unchanged (the apparent induction time) following which a marked and generally abrupt change is observed. A bit of thought concerning this behavior leads to the conclusion that some of the chemical reactions responsible for the rapid change that ultimately occurs must have been in progress during this induction stage, albeit at a very slow rate, and that the reactions taking place in Stages I and II are the very ones that we should be concerned with in deterioration studies.[14]

To preserve paper, textiles, and coatings, it would be necessary to treat them *before* the apparent induction stage is over to extend as long as possible the chemical processes that predominate during this period (essentially these are inhibition reactions). After the apparent induction period is passed—after paint has lost its gloss or after leather and paper have become brittle—it is likely to be too late to preserve the material effectively.

EXAMPLES. There are many reasons why a real or an apparent induction stage will occur in a sequence of chemical reactions. Tobolsky, Metz, and Mesrobian developed equations to show that the induction time in the autoxidation of hydrocarbons is related to the ratio of the steady state concentration of hydroperoxide to the initial concentration, $[ROOH]_\infty/[ROOH]_0$ or expressed another way, to the chain length of the propagation steps Equations (3) and (4) below.[2]

$$R\bullet + O_2 \rightarrow ROO\bullet \qquad (3)$$

$$ROO\bullet + RH \rightarrow R\bullet + ROOH \qquad (4)$$

During the induction stage there is a fast reaction that competes with the particular reaction that the experimenter is attempting to monitor. For example, an inhibitor that is a free-radical scavenger tends to delay the dominance of Reactions (3) and (4) in the overall consumption of oxygen until the inhibitor is essentially used up (see Equations (5) and (6) below.[15]

One must speak of an induction time with respect to a particular reaction or effect because it is to be expected that the rate of change in one particular component will show an induction time while that of another will not be apparent on the same time scale. Thus, peroxides in a polymer such as polyethylene may form almost immediately, whereas the formation of carbonyl groups and carbon dioxide may be delayed; the fall in molecular weight of a polymeric coating may begin almost at once, whereas cracking, chalking or brittleness—changes in physical properties—may be delayed. In the general sequence of reactions in which compound A yields B, which in turn yields C, one is much more likely to observe an induction time as a result of attaining a particular concentration of C than to observe an appreciable induction time for the disappearance of A.

If there is an appreciable initial rate of reaction and if the investigator doesn't carry the experiment far enough, he may miss the second stage, not recognizing the first, relatively modest rate of reaction as an induction stage. The inhibitor itself or its free-radical product frequently may react with oxygen sufficiently to give rise to a measurable rate of oxygen uptake even during the period before the maximum rate sets in [Reactions (5) and (6), where HA is the inhibitor].

$$R\bullet + HA \rightarrow RH + A\bullet \qquad (5)$$

$$A\bullet + O_2 \rightarrow AO_2\bullet \qquad (6)$$

The data of Wilson and Forshee on the change in intrinsic fluidity of plasticized cellulose acetate provide a particularly good example of an appreciable rate during an induction stage (Figure 6).[16,17] Other examples are the subtle change in rate of oxygen uptake in the studies of the inhibited rate of oxidation of polyisoprene by Shelton and Vincent[18] and the initial rate of increase in peroxide values in cottonseed and linseed oils reported by Lea.[19] All of these investigations involved an appreciable reaction at an early stage which could have caused an advanced stage such as III to have gone undetected if the studies had not been carried sufficiently far.

In contrast, induction times may be so short or the early data points so erratic

15. Vink, P., "Changes in the Concentration of Some Stabilizers during the Photooxidation of Polypropylene Films," *J. Polym. Sci., Symp. No. 40* (1973), p. 169-173.

16. Wilson, W. K., and Forshee, B. W., "Degradation of Cellulose Acetate Films," *Soc. Plast. Eng. J.*, 15, No. 2 (1959), p. 146-156.

17. Wilson, W. K., "Deterioration and Preservation of Library Materials," ed. Winger, H. W., and Smith, R. D., University of Chicago Press, Chicago, IL (1970), p. 174, Figure 1.

18. Shelton, J. R., and Vincent, D. N., "Retarded Autoxidation and the Chain-Stopping Action of Inhibitors," *J. Am. Chem. Soc.*, 85 (1963), p. 2433-2439.

19. Lea, C. H., "Antioxidants in Dry Fat Systems: Influence of the Fatty Acid Composition of the Substrate," *J. Sci. Food Agric.*, 11 (1960), p. 143-150.

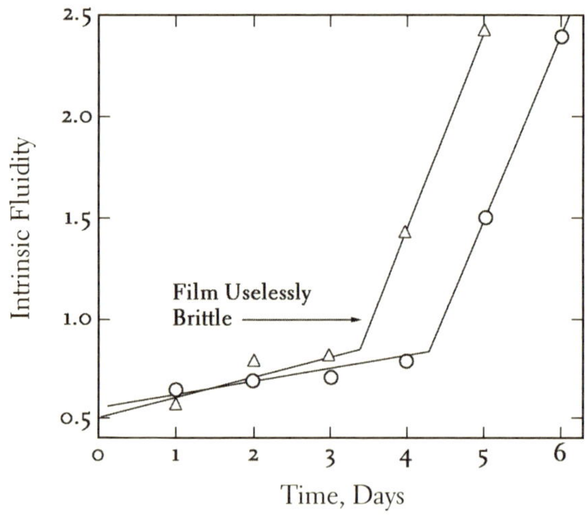

FIGURE 6
Degradation of plasticized cellulose acetate at 110°C (Ref. 16).

that investigators will simply draw their curves through the origin, overlooking an induction stage that may be present. An example of this can be seen in the data of Harris and Jessup on the deterioration of silk, previously cited.[20, 21] Even though the authors chose to draw the curve through the origin, an induction time seems apparent from their data for the samples treated with 0.1N NaOH.

An example of the manner in which the scheme in Figure 2 can stimulate investigation may be illustrated by our recent search for evidence of an induction stage which we believed should be present in the oxygen uptake of paper. Shortly before preparing this manuscript, we had initiated experiments intended both to measure the steady state rate of oxidation of paper and to obtain evidence of an induction stage. We did not succeed initially in achieving the latter objective, but immediately before presenting these remarks, our attention was drawn to a publication by Hernandi on the oxygen uptake of cellulose.[22] Using an instrument that permits the rate of oxygen consumption to be observed within minutes of starting the experiment, Hernandi reported several curves that can be divided into three zones clearly corresponding to the proposed stages I, II, and III. The data display an induction time of about 1-1/2 hr for the uptake of oxygen by cellulose at 190°C and one of about 2 hr at 160°C. At 130°C an induction time of about 9 hr is noted before the evolution of CO_2 becomes appreciable.[23] These times are shorter than we had originally been seeking. Particularly intriguing is Hernandi's observation of a zone in the curves for oxygen uptake that corresponds to Stage 1. The evidence thus seems to be that cellulosic materials tend to oxidize very much as proposed in the scheme in Figure 2.

20. Feller, R. L., "The Deterioration of Organic Substances and the Analysis of Paints and Varnishes," in *Preservation and Conservation: Principles and Practices*, ed. Timmons, S., Preservation Press, Washington, DC (1976), p. 287-299.

21. Harris, M., and Jessup, D. A., "The Effect of pH on the Photochemical Decomposition of Silk," *J. Res. Nat. Bur. Stand.*, 7, No. 6 (1931), p. 1179-1184.

22. Hernadi, S., "Thermal Ageing in Oxygen of Paper Made from Cellulose at Different Degrees of Beating," *Svensk Papperstidning*, 79, No. 13 (1976), p. 418-423.

23. Waller, R. C., Bass, K. C., and Roseveare, W. E., "Degradation of Rayon Tire Yarns at Elevated Temperatures," *Ind. Eng. Chem.*, 40 (1948), p. 138-143; report on induction time of about 4 hr. for the loss of strength of the rayon yarn in air at 135°C in the presence of 0.12 g water/g cellulose.

Stage III: Maximum Rate and Steady State

DEFINITION. To express the overall rate of a sequence of reactions, a special mathematical treatment is often used, known as the steady state treatment. This is based on the assumption that the concentration of certain intermediate compounds or complexes is never large, that their concentration rises at the beginning of the reaction and soon reaches a constant (or steady) value, and that, at this point, the rate of change in the concentration, dc/dt, can be assumed to be zero. If the overall rate of reaction depends on the concentration of this intermediate, then the rate will have reached its maximum at this time.

DISCUSSION. In the autoxidation of organic compounds that takes place through the mechanism of hydroperoxide formation, one finds that, following the induction stage, steady-state mathematics can be applied to account for the observed dependence of the rate of oxidation upon the concentration of the principal compound undergoing oxidation and upon temperature.[2] If the concentration of the principal component is sufficiently great that it remains practically constant, one may expect to observe an appreciable steady state stage. This is a commonly encountered situation. If, on the other hand, the concentration of the principal reactant is low, the system will tend to pass into Stage IV rapidly, and the investigator will be confronted with an S-shaped curve, that is, with an *inflection point of maximum rate* rather than an appreciable period of time during which the rate is constant.

EXAMPLES. In the accepted theory of autoxidation, through the formation of organic hydroperoxides, the rate-controlling step in the maximum-rate stage is Equation (4). For this case, the observed rate of oxidation in Stage III is that of the key compound in question, RH. Thus, Tobolsky, Metz, and Mesrobian expressed the equation for maximum rate of oxygen uptake as Equation (7), where K stands for their $k_3^2 k_6^{-1}$.[2]

$$-\left[\frac{d[O_2]}{dt}\right]_{max} = K[RH]^2 \qquad (7)$$

If Equation (7) is indeed the generally expected explanation for the rate of oxidation in the steady state stage, then it is very much in keeping with the objective of this volume to point out that the rates in Stages I or II do not represent the "characteristic rate of oxidation" of the key substance under consideration (RH in Equation (7)).

In practice one seldom encounters protective coatings, adhesives, and plastics in a condition in which Stage III is very far advanced.[24] The reason for this is simply that, by the time the maximum-rate stage is reached, most materials will have lost so much strength and tenacity that they are no longer in a useful condition. A good example can be found in the work of Shelton and Winn on

24. Hawkins, W. L., Bell Laboratories, Murray Hill, NJ, private communication.

the oxidation of GRS-type rubber.[25] When the reactivity of this material is considered over a range of temperatures, their data show all four stages. At one point the authors state that, before the final upswing in oxygen uptake is reached, that is, before the maximum rate is attained, the physical properties of the stock are degraded "to such an extent as to be of little practical value."

25. Shelton, J. R., and Winn, H., "Oxidation of GR-S Vulcanizates," *Ind. Eng. Chem.*, 38 (1946), p. 71-76.

The findings of Hutson and Scott on embrittlement of inhibited polyethylene shown in Figure 7 also illustrate this point; their presentation is unusual in that the point of embrittlement is indicated on the curve.[26] As can be seen, this occurred before Stage III was well advanced. A similar situation can be found in the data of Wilson and Forshee; the intrinsic fluidity of polymer that had degraded to a highly brittle state is indicated in Figure 6.[16]

26. Hutson, G. V., and Scott, G., "Mechanisms of Phosphite Stabilizers in Polyolefins Subjected to Ultraviolet Irradiation," in *Mechanism of Inhibition Processing Polymers: Oxidation and Photodegradation*, ed. Sedacek, B., *J. Polym. Sci., Polym. Symp.*, No. 40, Wiley, New York (1973), p. 67-71.

Stage IV: Declining Rate

DEFINITION. The declining rate stage is defined as a period of decreasing rate of reaction which tends to approach a limit in the amount of reactant consumed, or product produced (zero rate), under the prevailing conditions of temperature, irradiance, partial pressure of oxygen, concentration of reactants, etc.

DISCUSSION. There are a number of obvious reasons why a limit in the extent of reaction should be reached. The most compelling reason is simply that under given experimental conditions the most readily attacked sites in the molecules become used up. In many cases, the experiment cannot be carried to this advanced stage, and one may fail to observe the period of declining rate. Nonetheless, such a stage should be anticipated, especially under mild conditions.

While it may be true that many materials will become severely embrittled or discolored before Stage III is far advanced, some special situations exist in which a polymeric system can be regarded as being in Stage IV. For example, in the life history of polymers that tend to crosslink when exposed to near-ultraviolet radiation, the conservator may have to remove them or at least have a sound understanding of their swelling properties in solvents under conditions in which the films are in Stage IV with respect to the formation of insoluble matter.[27] In a similar sense, a film of highly crosslinked linseed-oil paint can be considered to be in Stage IV with respect to the formation of the intended paint film properties.

27. Feller, R. L., Stolow, N., and Jones, E. H., *On Picture Varnishes and Their Solvents*, Press of Case-Western Reserve, Cleveland (1971), p. 202-210.

EXAMPLES. Data from an early experiment in which the gain in weight of dammar resin was observed, illustrated in Figure 8, indicate that a limit had been reached after one month.[20] One may assume that the tertiary alcohol groups or the double bond in the vinyl groups, known to be present in this natural resin, were readily oxidized and that thereafter, one would need to increase the temperature or the energy of the photons in order to cause further reaction with oxygen.[28] If we assume that the average molecular weight of many of the triterpenes present is about 442, then the addition of two oxygen

28. Mills, J., and Werner, A. E. A., "The Chemistry of Dammar Resin," *J. Chem. Soc.*, No. 9 (1955), p. 3132-3140.

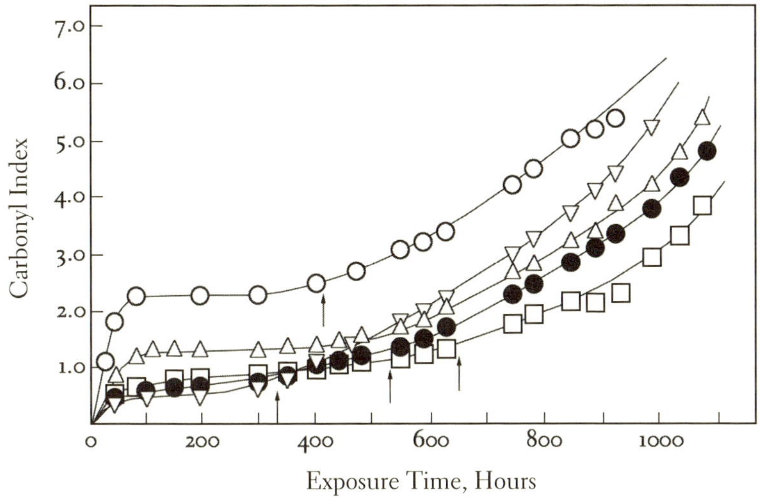

FIGURE 7
Effect on carbonyl index during UV exposure of HDPE with 1% inhibitor processed at various temperatures.[26] ▽, control – no additive; ○, 100°C, 5 min; △, 200°C, 30 min; □, 200°C, 60 min; ●, 200°C, 90 min; ↑, embrittlement time.

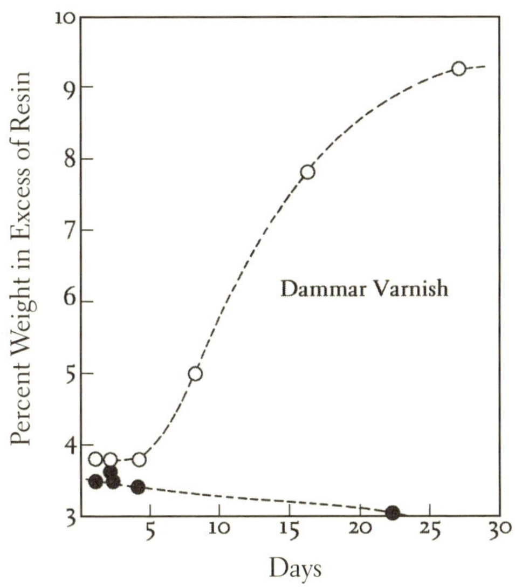

FIGURE 8
Oxidative gain in weight of film of dammar varnish prepared from xylene solution. ○, under "daylight" fluorescent lamps at ~ 240 fc intensity and 90°F, 20% RH; ●, partially protected from the light in dust-free box.

atoms and a loss of four hydrogen atoms would represent an increase in weight of 6.3%. This is close to the gain of 5.5% attained in the as yet incompleted process illustrated in Figure 8, suggesting that a stoichiometric gain of only one or two atoms of oxygen can readily account for the limited uptake of oxygen.

Another reason for an apparent, but not a true, limit to the amount of a particular product formed would be the existence of competing reactions such as chain breaking and crosslinking. Charlesby has developed equations to permit calculation of the ratio of chain breaking to crosslinking in thermoplastic polymers exposed to high-energy radiation.[29] These have been applied to analyze the effect of exposure of Acrylan, Mylar, nylon and polyethylene to 2537 Å ultraviolet radiation.[30]

We have used Charlesby's principles in investigating the exposure of certain acrylic resins to near-ultraviolet radiation to account for the limited build-up of insoluble matter in cases where the chain-breaking reaction is significant.[31] Thus, when exposed in the carbon-arc Fadeometer (Corex D filter) at 32.2°C, the poly(isobutyl methacrylate) polymer in Figure 9 seems to be tending towards a limit in the total amount of insoluble matter; in contrast, at 62°C, the polymer rapidly approaches a condition of near complete insolubility. (The isobutyl methacrylate polymer was prepared in refluxing toluene solution with benzoyl peroxide as a catalyst.) Charlesby, in plots of the log of the soluble matter against the log of elapsed time/gel time (or, what is similar, r/r_{gel}, the total irradiance relative to the radiation dose necessary to produce the first detectable insoluble matter), calculated a series of curves for the rate of decrease in the soluble matter as a function of the ratio of chain breaks to crosslinks taking place, β/α. When the data of Figure 9 are superimposed on Charlesby's family of curves as shown in Figure 10, it can be seen that at the lower temperature the isobutylmethacrylate polymer was experiencing a ratio of about seven chain breaks for every 10 crosslinks formed ($\beta/\alpha = 0.7$), whereas at the higher temperature, the polymer tended to crosslink almost exclusively ($\beta/\alpha = 0.0$).

Analysis of the data in Figure 9 permits the investigator to reach a number of conclusions: (a) because the formation of insoluble matter begins at about the same time in Figure 9 despite the difference in temperature, the initiation step must largely be photochemical in character. (b) In contrast, because the ratio

29. Charlesby, A., *Atomic Radiation and Polymers*, Pergamon Press, London (1960).

30. Stephenson, C. V., Moses, B. C., Burks, B. E., Jr., Coburn, W. C., and Wilcox, W. S., "Ultraviolet Irradiation of Plastics II, Cross-linking and Scission," *J. Polym. Sci.*, 55 (1961), p. 465-475.

31. Feller, R. L., and Bailie, C. W., "Studies of the Effect of Light on Protective Coatings Using Aluminum Foil as a Support: Determination of Ratio of Chain Breaking to Crosslinking," *Bull. Am. Group-IIC*, 6, No. 1 (1966), p. 10-12.

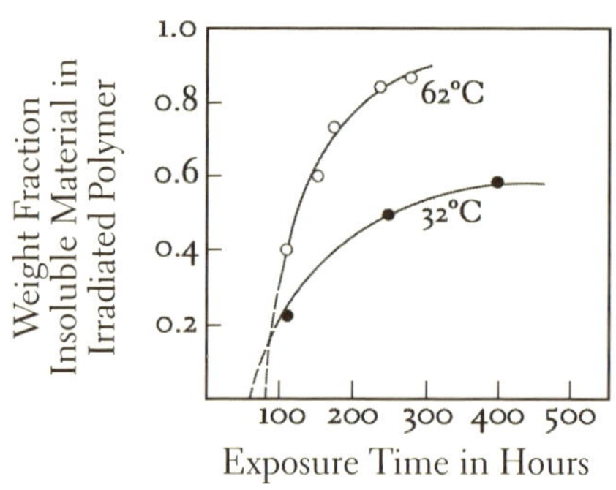

FIGURE 9
Loss of solubility of poly(i-butyl methacrylate) exposed on aluminum foil in single carbon-arc Fadeometer, Corex D filter.

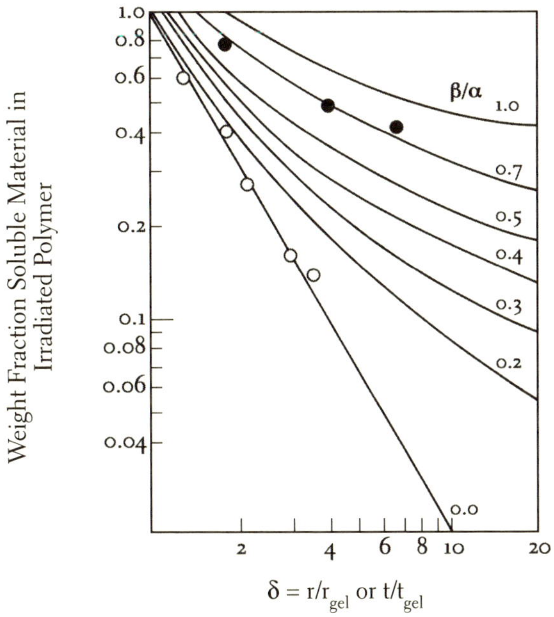

FIGURE 10
Data for loss of solubility of poly(i-butyl methacrylate) exposed in carbon-arc Fadeometer, Corex D filter superimposed on Charlesby's calculated curves. ○, data at 62°C; ●, data at 32.2°C.

of chain breaks to crosslinks is markedly influenced by temperature, the propagation and termination steps are evidently thermally activated. (c) At 62°C the polymer undergoes crosslinking almost exclusively. (d) At the lower temperature of 32.2°C the relative rates of the propagation and termination steps are sufficiently altered so that there is an opportunity for appreciable chain breaking to occur.

The latter observation has important implications in accelerated aging tests for the experiment has clearly demonstrated that raising the temperature can alter the relative importance of competing reactions and thereby alter the practical consequences of exposure. Thus, in Figure 10 one sees that at the temperature usually attained in the single-carbon-arc Fadeometer, about 62°C, crosslinking was observed almost exclusively, whereas at the lower temperature considerable chain breaking occurred. The implication is that exposure of poly(isobutyl methacrylate) to near-ultraviolet radiation at room temperature will result in a greater ratio of chain breaking to crosslinking than would be expected from the results of the usual accelerated-aging exposure in the older Fadeometers without temperature control or even from our special test in which the sample was cooled to 32.2°C.

In investigations carried out in our laboratory some years ago, but as yet unpublished, neither poly(isoamyl methacrylate) or the more commonly encountered poly(n-butyl methacrylate) were found to exhibit a similar tendency to chain break at lower temperatures. Analysis of the data by Charlesby's method indicated that these two polymers still tended to undergo crosslinking almost exclusively even when exposed to visible and near-ultraviolet radiation at temperatures as low as 25°- 35°C.

Kelleher called curve "b" in Figure 1 the autoretardant case.[5] Indeed, a third possible cause for a declining rate stage would be the generation of chemical entities that could interfere and retard the reaction of principal interest.

CONCLUSION

I had three purposes in presenting these remarks. The first was to call attention to the fact that in the course of thermally and photochemically activated deterioration, organic materials such as paper, paints and varnishes, textiles, and adhesives tend to react with oxygen in a reasonably well defined series of stages. This concept provides a guide for analyzing and investigating deterioration problems further.[32]

32. Since the preparation of this manuscript, the author has found that the scheme illustrated in Figure 2 was described more than 20 years ago by N. Grassie (*Chemistry of High Polymer Degradation Processes*, Interscience, New York (1956), p. 201, Fig. 61). It is curious that Grassie's concept of four stages should have attracted so little interest among subsequent authors.

The second point stressed was that the explanation of the rate of deterioration at any given stage must be based on an understanding of the *chemical* reactions taking place. If deterioration studies continue to rely heavily upon measurements of physical properties rather than upon attempts to elucidate the chemistry behind these effects, our understanding of the problems must by necessity remain limited.

As a corollary to this point, the potential advantages in applying kinetic analysis to chemical reaction rates can be emphasized. The investigator is often confronted with a rate of change that resembles Figures 1b, 3, 4, and 9. When these empirical results are analyzed according to the principles of chemical kinetics, as in Figures 5 and 10, order can frequently be achieved and valuable insights can be gained concerning the possible transition from Stages I and II (Figure 5) or from III and IV (Figure 10). Hope for elucidating the complex overall process of deterioration comes from just such kinetic analyses of the rate processes that occur in individual stages and in the transitions between them.

The third purpose was to discuss some of the types of chemical reactions that may occur and in a sense may be responsible for the rate of oxidation in each of the four proposed stages. More than a dozen general and specific examples have been cited.

In general, problems are most amenable to solution if they can be analyzed into their components. I hope that the concept of stages in oxidative deterioration will increase the effectiveness with which problems in preserving organic materials can be attacked and with which improved methods and materials for conservation can be developed. At the moment, many more questions are raised than answered. Nonetheless, the ideas presented here promise to stimulate and give direction to our research for some years to come.

Acknowledgment

The experiments represented in Figure 9 were conducted a number of years ago by Catherine Westervelt Bailie. The data in Figures 3 and 4 were obtained by Mary Curran, to whom I also gratefully acknowledge editorial assistance in preparing the manuscript.

Stages in Deterioration: The Examples of Rubber Cement and Transparent Mending Tape

ROBERT L. FELLER AND DAVID B. ENCKE

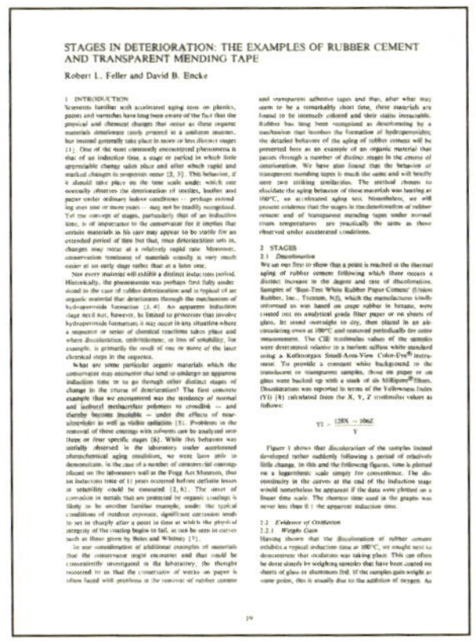
STAGES IN DETERIORATION: THE EXAMPLES OF RUBBER CEMENT AND TRANSPARENT MENDING TAPE
Robert L. Feller and David B. Encke

Originally printed in:
SCIENCE AND TECHNOLOGY IN THE SERVICE OF CONSERVATION
ed. Brommelle, N. S., and Thomson, G., International Institute for Conservation, London (1982), Pages 19-23.

Introduction

Scientists familiar with accelerated aging tests on plastics, paints and varnishes have long been aware of the fact that the physical and chemical changes that occur as these organic materials deteriorate rarely proceed in a uniform manner, but instead generally take place in more or less distinct stages.[1] One of the most commonly encountered phenomena is that of an induction time, a stage or period in which little appreciable change takes place and after which rapid and marked changes in properties occur.[2, 3] This behavior, if it should take place on the time scale under which one normally observes the deterioration of textiles, leather and paper under ordinary indoor conditions —perhaps extending over one or more years—may not be readily recognized. Yet the concept of stages, particularly that of an induction time, is of importance to the conservator for it implies that certain materials in his care may appear to be stable for an extended period of time but that, once deterioration sets in, changes may occur at a relatively rapid rate. Moreover, conservation treatment of materials usually is very much easier at an early stage rather than at a later one.

Not every material will exhibit a distinct induction period. Historically, the phenomenon was perhaps first fully understood in the case of rubber deterioration and is typical of an organic material that deteriorates through the mechanism of hydroperoxide formation.[3, 4] An apparent induction stage need not, however, be limited to processes that involve hydroperoxide formation; it may occur in any situation where a sequence or series of chemical reactions takes place and where discoloration, embrittlement, or loss of solubility, for

1. Feller, R. L., "Stages in the Deterioration of Organic Materials," *Preservation of Paper and Textiles of Historic and Artistic Value*, Advances in Chemistry Series No. 164, ed. Williams, J. C., American Chemical Society, Washington, DC (1977), p. 314-335.

2. Feller, R. L., "The Deterioration of Organic Substances and the Analysis of Paints and Varnishes," *Preservation and Conservation: Principles and Practices*, ed. Timmons, S., Preservation Press, Washington, DC (1976), p. 287-299.

3. Feller, R. L., "Fundamentals of Conservation Science: Induction Time and the Autoxidation of Organic Compounds," *Bulletin of the American Institute for Conservation*, 14, No. 2 (1974), p. 142-151.

example, is primarily the result of one or more of the later chemical steps in the sequence.

What are some particular organic materials which the conservator may encounter that tend to undergo an apparent induction time or to go through other distinct stages of change in the course of deterioration? The first concrete example that we encountered was the tendency of normal and isobutyl methacrylate polymers to crosslink—and thereby become insoluble—under the effects of near-ultraviolet as well as visible radiation.[5] Problems in the removal of these coatings with solvents can be analyzed into three or four specific stages.[6] While this behavior was initially observed in the laboratory under accelerated photochemical aging conditions, we were later able to demonstrate, in the case of a number of commercial coatings placed on the laboratory wall at the Fogg Art Museum, that an induction time of 11 years occurred before definite losses in solubility could be measured.[2,6] The onset of corrosion in metals that are protected by organic coatings is likely to be another familiar example; under the typical conditions of outdoor exposure, significant corrosion tends to set in sharply after a point in time at which the physical integrity of the coating begins to fail, as can be seen in curves such as those given by Bates and Whitney.[7]

In our consideration of additional examples of materials that the conservator might encounter and that could be conveniently investigated in the laboratory, the thought occurred to us that the conservator of works on paper is often faced with problems in the removal of rubber cement and transparent adhesive tapes and that, after what may seem to be a remarkably short time, these materials are found to be intensely colored and their stains intractable. Rubber has long been recognized as deteriorating by a mechanism that involves the formation of hydroperoxides; the detailed behavior of the aging of rubber cement will be presented here as an example of an organic material that passes through a number of distinct stages in the course of deterioration. We have also found that the behavior of transparent mending tapes is much the same and will briefly note two striking similarities. The method chosen to elucidate the aging behavior of these materials was heating at 100°C, an accelerated aging test. Nonetheless, we will present evidence that the stages in the deterioration of rubber cement and of transparent mending tapes under normal room temperatures are practically the same as those observed under accelerated conditions.

STAGES

Discoloration

We set out first to show that a point is reached in the thermal aging of rubber cement following which there occurs a distinct increase in the degree and rate

4. Scott. G., *Atmospheric Oxidation and Autoxidants*, Elsevier Publishing Company, New York (1965).

5. Feller. R. L., Curran, M., and Bailie, C., "Photochemical Studies of Methacrylate Coatings for the Conservation of Museum Objects," in *Photodegradation and Photostabilization of Coatings*, Advances in Chemistry Series No. 151, ed. Pappas, S. P., and Winslow, F. H., American Chemical Society, Washington, DC (1981), p. 183-196.

6. Feller. R. L., Stolow. N., and Jones, E.H., *On Picture Varnishes and Their Solvents*, revised edn., Press of Case Western Reserve University, Cleveland (1971).

7. Bates, R. P., and Whitney, R. L., "Basic Lead Silica Chromate; An Anti-corrosive Pigment," *Official Digest, Federation of Societies for Paint Technology*, 33, No. 432 (1961), p. 27-39. See also Denison, G. H., "Oxidation of Lubricating Oils; Effect of Natural Sulfur Compounds and Peroxides," *Industrial and Engineering Chemistry*, 36 (1944), p. 477-482.

of discoloration. Samples of "Best-Test White Rubber Paper Cement" (Union Rubber, Inc., Trenton, NJ), which the manufacturer kindly informed us was based on crepe rubber in hexane, were coated out on analytical grade filter paper or on sheets of glass, let stand overnight to dry, then placed in an air-circulating oven at 100°C and removed periodically for color measurement. The CIE tristimulus values of the samples were determined relative to a barium sulfate white standard using a Kollmorgan Small-Area-View Color-Eye® instrument. To provide a constant white background to the translucent or transparent samples, those on paper or on glass were backed up with a stack of six Millipore® filters. Discoloration was reported in terms of the Yellowness Index (YI)[8] calculated from the X, Y, Z tristimulus values as follows:

$$YI = \frac{128X - 106Z}{Y} \qquad (1)$$

8. Billmeyer, F. W., Jr., "Yellowness Measurement of Plastics for Lighting Use," *Materials Research and Standards*, 6 (1966), p. 295-301.

Figure 1 shows that discoloration of the samples indeed developed rather suddenly following a period of relatively little change. In this and the following figures, time is plotted on a logarithmic scale simply for convenience. The discontinuity in the curves at the end of the induction stage would nonetheless be apparent if the data were plotted on a linear time scale. The shortest time used in the graphs was never less than 0.1 the apparent induction time.

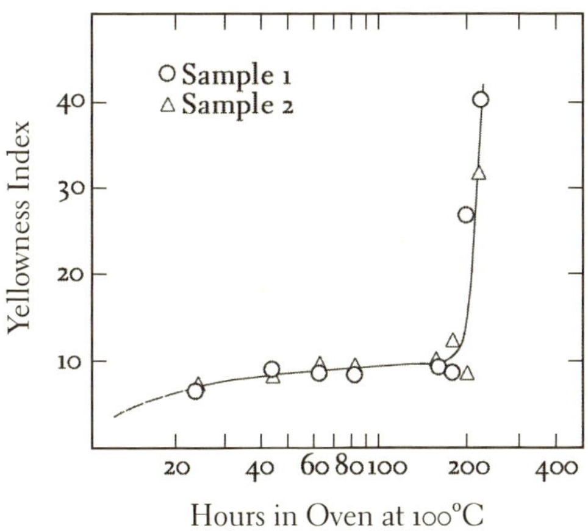

FIGURE 1
Increase of yellowness index of rubber cement aged at 100°C (draw-bar coated on glass).

Evidence of Oxidation

Weight Gain

Having shown that the discoloration of rubber cement exhibits a typical induction time at 100°C, we sought next to demonstrate that oxidation was taking place. This can often be done simply by weighing samples that have been coated on sheets of glass or aluminum foil. If the samples gain weight at some point, this is usually due to the addition of oxygen. As we see in Figure 2, after losing between 1 and 2% weight, owing to traces of retained solvent being driven off, a moment came at which a sharp and rapid gain of about 6 to 9% took place; we take this to be clear evidence of oxidation. The exact time at which this occurred varied, probably owing to different thicknesses of the film, for thickness would control the rate of diffusion of oxygen into the adhesive.

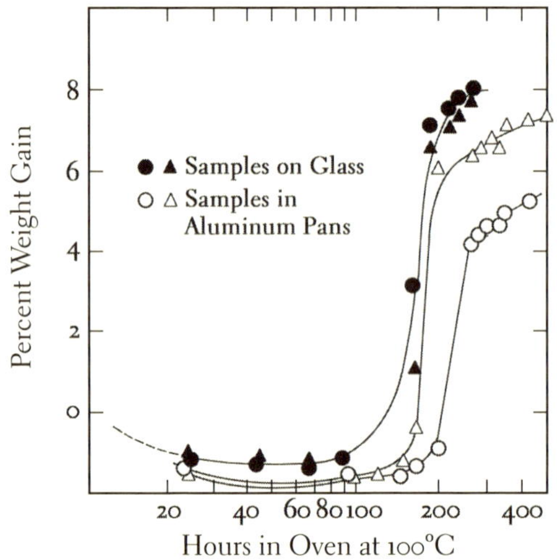

FIGURE 2
Weight gain of rubber cement aged at 100°C (samples in aluminum pans or draw-bar coated on glass).

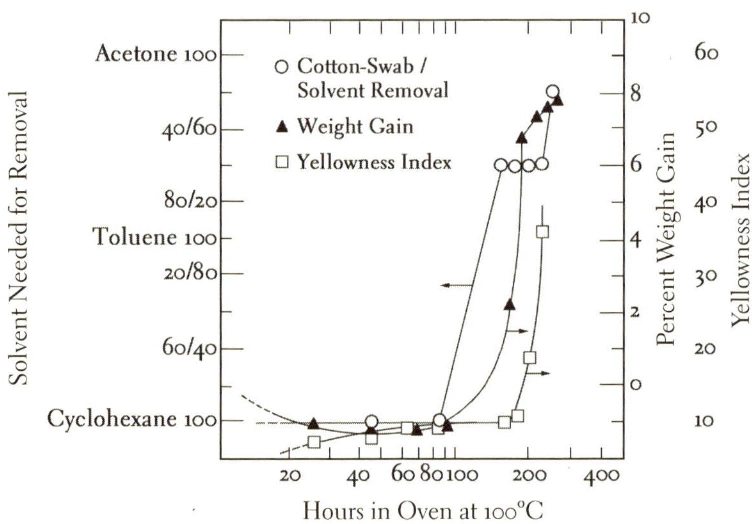

FIGURE 3
Interrelationship of changes in three properties of rubber cement during aging at 100°C.

Increased Polarity or Solubility Parameter of Adhesive

Another potential measure of oxidation would be the finding that solvents of increasing polarity or "strength" were required to remove an aged varnish or adhesive.[9] Accordingly, as the rubber cement was aged in the oven, the samples were withdrawn from time to time and tests made of their ease of removal with cotton swabs moistened with solvent. A series of cyclohexane/toluene and toluene/acetone mixtures of increasing solvent power was used, as described earlier.[10] Again, a point in time came when a sharp increase in polarity was needed to facilitate the removal of the rubber cement. When the data from these tests are plotted along with those shown in Figures 1 and 2, we see in Figure 3 that there is a distinct sequence of events or stages in the course of the accelerated aging of this material.

Tests on two other brands of rubber cement have confirmed that an early stage in this aging sequence is the need for solvents of increasing polarity in order to remove the coating; perhaps even before this, a gain in weight may be detected. It is not until a comparatively late stage is reached that discoloration is observed in the aging of this and many other initially colorless materials.

Hydroperoxide Formation

These findings are not new; the deterioration of rubber has been thoroughly studied for years.[11-13] Its behavior has long served as the basic model for the mechanism of oxidative aging that involves the formation and subsequent breakdown of hydroperoxides (chemical structure ROOH).[3, 14] In such a process, it is usually assumed that, after a period of induction, for which naturally or unintentionally present inhibitors may be responsible, hydroperoxides begin to form in an organic substance. This is usually detectable as a gain in weight. Depending on the particular conditions of temperature, humidity and radiant energy, one may observe a reasonably continuous gain in weight; as seen in Figure 2, or, very commonly, a subsequent loss in weight is observed at the point at which the hydroperoxides or more highly oxidized substances begin to degrade, breaking down into such volatile "oxidation products" as carbon dioxide, water, formaldehyde, formic acid and other small-molecular fragments of the original chemical substances. (The thin film on aluminum foil in Figure 4, for example, exhibited a gain and subsequent loss in weight.)

Hydroperoxides were analyzed by the usual technique of adding potassium iodine and titrating the iodine liberated by the action of the peroxidic oxygen, using essentially the method of Wagner, Smith and Peters.[15] To provide sufficient material for analysis, somewhat larger samples than before were placed in glass petri dishes. The results, shown in Figure 5, yielded a typical curve of build-up of peroxides and their subsequent breakdown. Because the samples had been sacrificed each time in order to carry out the peroxide analyses, we had to

9. Feller, R. L., and Curran, M., "Changes in Solubility and Removability of Varnishes with Age," *Bulletin of the American Institute for Conservation*, 15, No. 2 (1975), p. 17-48.

10. Feller, R. L., "The Relative Solvent Power Needed to Remove Various Aged Solvent-type Coatings," *Conservation and Restoration of Pictorial Art*, ed. Brommelle, N. S., and Smith, P., Butterworths, London (1976), p. 158-161.

11. Peachey, S. J., "The Action of Gaseous Oxygen on Indiarubber," *Journal of the Society of Chemical Industry*, 32 (1913), p. 179.

12. Kohman, G. T., "The Absorption of Oxygen by Rubber," *Journal of Physical Chemistry*, 33 (Jan-June 1929), p. 226-243.

13. Kemp, A. R., Bishop, W. S., and Lasselle, P. A., "Oxidation Studies of Rubber, Gutta-Percha, and Balata Hydrocarbons," *Industrial and Engineering Chemistry*, 23, No. 12 (1931), p. 1444-1449.

14. Tobolsky, A. V., Metz, D. J., and Mesrobian, R. B., "Low Temperature Autoxidation of Hydrocarbons: The Phenomenon of Maximum Rates," *Journal of the American Chemical Society*, 72 (1950), p. 1942-1952.

15. Wagner, C. D., Smith, R. H., and Peters, E. D., "Determination of Organic Peroxides, Evaluation of a Modified Iodometric Method," *Analytical Chemistry*, 19 (1947), p. 976-979.

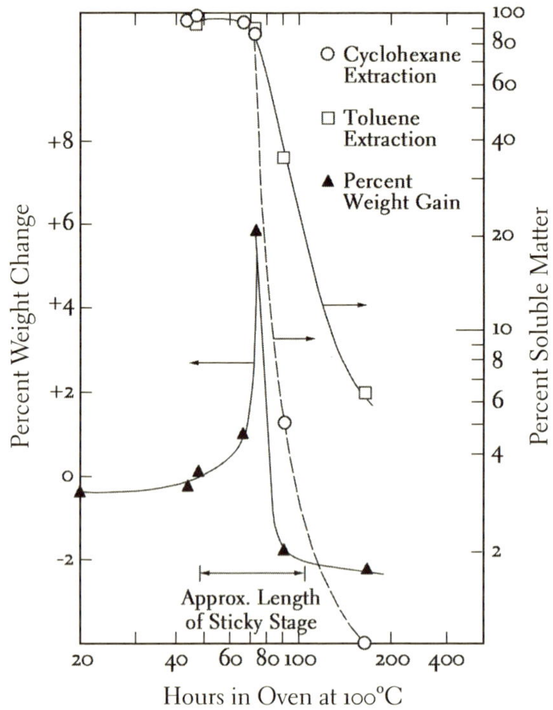

FIGURE 4
Comparison of solubility and weight change for rubber cement samples coated on aluminum foil and aged at 100°C.

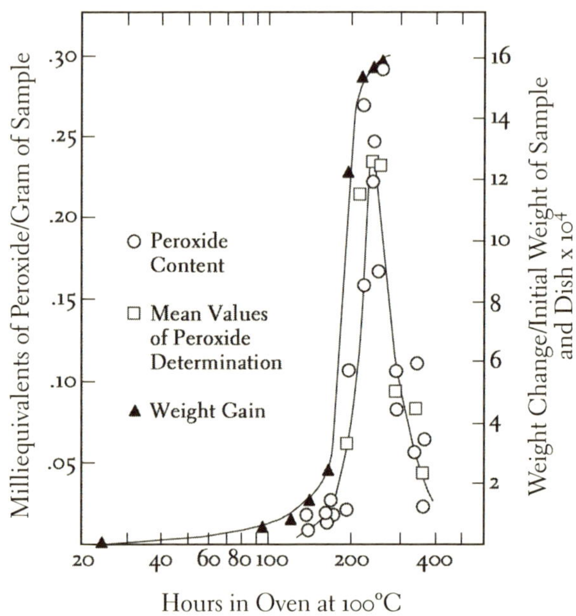

FIGURE 5
Comparison of peroxide formation and weight gain in separate sets of rubber cement samples aged at 100°C.

follow the weight gain of samples placed in separate dishes. As a consequence, the correspondence of marked weight gain and peroxide formation is only qualitatively demonstrated here but others have shown that the two events generally coincide.[16, 17]

16. Farmer, E. H., and Sundralingam, A., "The Course of Autoxidation Reactions in Polyisoprenes and Allied Compounds, Part VI. The Peroxidation of Rubber," *Journal of the Chemical Society* (1943), p. 125-133.

17. Iving, M., Laszlo-Hédvig, S., Kelen, T., Tüdos, F., Füzes, L., Samay, G., and Boder, G., "Study of Thermal Oxidation of Polyolefins VI," *Journal of Polymer Science: Symposium*, No. 57 (1976), p. 55-63.

Final, Highly Oxidized State

As Figure 4 shows, following the initial gain in weight, solubility rapidly falls off, first in a relatively non-polar solvent, cyclohexane, then in a more polar solvent, toluene. In the late, highly discolored and oxidized stage (after 250 hours at 100°C), all of the colored material cannot be removed from a porous substrate such as filter paper by solvent extraction. For example, hot methylene dichloride left behind 41.6% of the aged adhesive; hot methanol, 37.5%. Some of the material may be crosslinked, but other components are known to be acidic.[13, 16] Indeed, dilute ammonium hydroxide reduced much but not all of the color that remained after the methylene dichloride extraction.

The Sticky Stage

One does not have to be involved with these aging studies very long before becoming aware that, at what may be the earliest obvious stage in the process, the samples pass through a particularly sticky state. Apparently this begins at the time of the very first gain in weight and ends about when solubility in cyclohexane is lost, as indicated in Figure 4.

More than 50 years ago, Kohman[12] described this aging sequence—tacky, weak-and-cheesy, and finally, hard-and-brittle—stating that tackiness begins after the rubber has absorbed about 0.5 to 1.2% oxygen; that this begins to disappear after about 12% oxygen has been absorbed and that the hard, brittle stage is reached after an absorption of about 23.5% oxygen. The stickiness may be due in part to the plasticizing or fluidizing result of the breaking up of the rubber polymer. Farmer and Sundralingam show a curve of the rapid fall in molecular weight that occurs during the absorption of the first few percent by weight of oxygen.[16] Rubber cement and many mending tapes lose much of their adhesiveness while in this state; the adherends often peel away from one another. If a conservator finds the adhesive in this state, he should remove it promptly, for solvents of increased polarity are usually required to remove the material in the time period immediately thereafter.

Effect of Light

Light contributes to the initiation step in free-radical-involved processes. In most instances, therefore, exposure to light tends to shorten the induction time but not to speed up the basic rate at which the predominant deterioration reactions in polymers and resins occur.[3,13] Figure 6 shows the results of exposing rubber cement under a bank of "daylight" fluorescent lamps at about 28°C and 50% RH. A second set of the samples was placed in a covered cardboard box next to the exposed samples, to demonstrate the effect of essentially the same temperature without illumination. Whereas an induction time may still be present in the exposed samples, it is obviously much shorter than for those protected from the light.

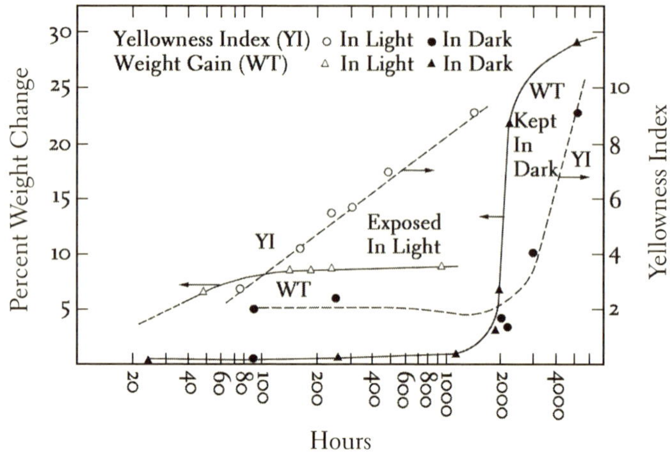

FIGURE 6
Increase in yellowness index and weight of rubber cement at 28°C and 50% RH with and without exposure to "daylight" fluorescent lamplight.

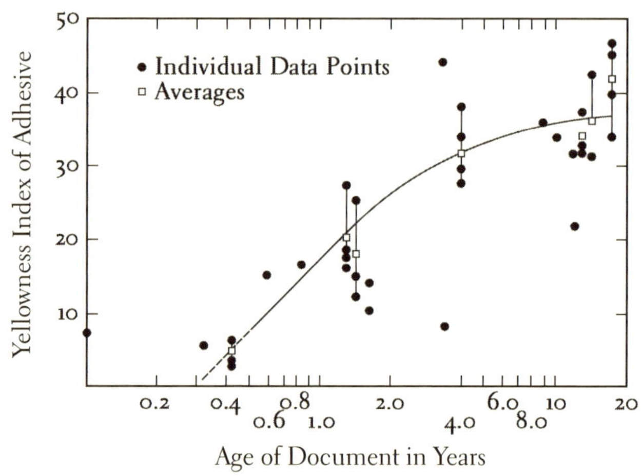

FIGURE 7
Increase of yellowness index of separated typescript pages that had been glued with rubber cement at various dates and stored in office files over a 20-year period.

INDUCTION TIME UNDER NORMAL CIRCUMSTANCES

The experiment shown in Figure 6 was carried out under conditions very much closer to room temperature than the accelerated tests described earlier. We considered further evidence that we might provide, which would confirm that the behavior we have been describing is indeed what one can expect to observe under normal conditions. Our office files contain numerous typescript copies of formal reports that had been prepared at various dates over many years; many graphs, charts and corrected passages had been pasted into these with rubber cement that has since become severely discolored. The thought occurred to us to take the glued sections of these reports, separate the sheets and measure the color of various areas in the usual manner. Because the adhesive had been applied unevenly and would not be separated in a uniform manner when the pages were pulled apart, a wide scatter of the data points was expected. Nonetheless, as we can see in the results shown in Figure 7, there appears to be an induction time of about 0.3 to 0.4 of a year before discoloration sets in. The maximum values of yellowness index found are close to the highest values recorded in the prior experiment (see, for example, Figure 1); hence, the curve in Figure 7 essentially represents the full course of deterioration from initial to highly deteriorated state. We were somewhat disappointed that the induction time was not longer or the increase in discoloration more rapid. The reasonably short induction time might arise from the fact that the papers would normally have been exposed to light for a number of weeks while they were being processed. Newly initiated experiments indicate that the normal induction time for samples well protected from the light may be on the order of 1-1/2 to 2 years. The lack of a sharp transition may be due to the fact that oxygen can attack a sheet of paper from both sides, making the build-up of oxidized material more gradual than in the case of samples on an impervious support such as glass or aluminum foil. A gradual slope of increasing discoloration has indeed been observed in accelerated tests with samples coated on filter paper.

TRANSPARENT MENDING TAPES

The adhesive systems in poorer grades of transparent mending tapes frequently behave in much the same manner as rubber cement. Many are not formulated with rubber-based adhesive systems and, moreover, many now apparently contain inhibitors that delay the onset of discoloration. Nonetheless, in the poorer grade tapes, an induction time with respect to discoloration and loss of solubility can be observed under accelerated aging conditions. A sticky stage is often apparent. One finds also that the small molecules which result from the chemical breakdown of the adhesive often migrate into the paper or other substrate upon which the tape is affixed, resulting in deeply penetrating stains. It occurred to us that the latter could be objectively measured by determining the "contrast ratio", the reflectance of the translucent paper/tape combination measured first over a black background and then over a white background (luminous reflectance of the backgrounds 0.5% and 98%, respectively).[18] As the pores of the filter paper become filled with the migrating sticky adhesive, the translucency increases. The results, in Figure 8, indicate that the translucency of the taped area indeed increased (the contrast ratio decreased) well ahead of the marked discoloration. Stickiness and the first indication of migration are warning signs; the removability of the adhesive material in mild solvents is likely to be still relatively easy at this stage, whereas considerable difficulty can be expected when the adhesive has reached a highly discolored condition.

18. "Opacity of Paper and Paper Products," ASTM Method D-589-44.

Recently we had the occasion to observe some glass-fiber-backed tape that had held a packing slip on a package stored in a laboratory cupboard since May 1962. Where the tape had fallen away after nearly 20 years of normal storage conditions the adhesive was yellow and brittle but, in other spots, it was extremely sticky and the tape easily pulled away. Analysis of the sticky material revealed the presence of 0.44 milliequivalents of peroxide per gram of adhesive in one sample and 0.71 milliequivalents in another, high concentrations indeed when one compares these values with those measured under accelerated aging conditions (Figure 5). Here is a clear example of the "sticky stage" occurring under normal conditions.

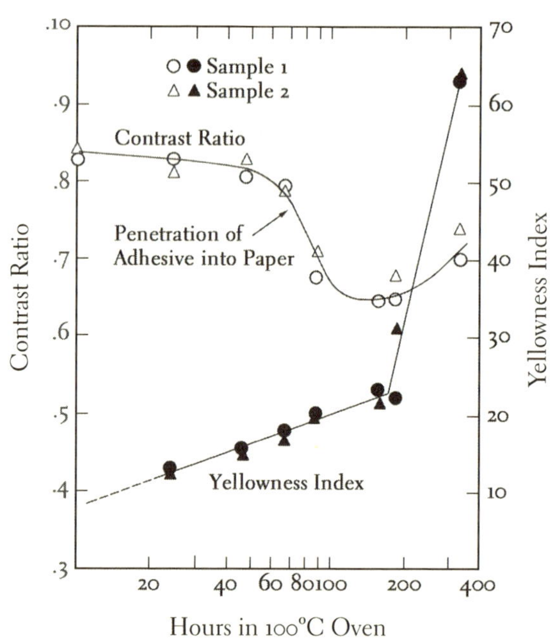

FIGURE 8
Relationship of contrast ratio and yellowness index of translucent adhesive-tape/filter-paper combination aged at 100°C.

SUMMARY

This investigation of the behavior of rubber cement and transparent mending tape under both accelerated and natural aging conditions has demonstrated that these commonly encountered materials tend to deteriorate in a manner typical of processes that involve the formation of hydroperoxide intermediates—in particular, that they display an induction-time phenomenon. Rubber cement provides a clear example of an organic material that passes through a sequence of events or stages in the course of aging. The first is a period of relatively little change (the induction time), the end of which is usually signaled by a soft, sticky condition. Beyond this point, as the rubber becomes further oxidized, one finds that solvents of increasing polarity are needed to remove it. If care is taken, a gain in weight, which may or may not be followed by a decline, can be measured. Finally, the material becomes hard, brittle, discolored and resistant to solvents.

In closing, we must again emphasize that not all organic materials that the conservator may encounter tend to deteriorate by a mechanism that will clearly exhibit an induction time. Moreover, the development of color does not always represent a late stage in deterioration. Nonetheless, if a conservator is called upon to treat organic materials that display the induction-time phenomenon, hopefully he may be guided by some of the concepts and principles described. It is important to realize that there may be a critical point in the history of such materials after which marked deterioration may set in at an increased rate. So far our laboratory has investigated three familiar materials that tend to display induction-time behavior: the crosslinking of higher alkyl methacrylate polymers and the deterioration of rubber cement and of poorer grades of transparent mending tape. It remains the challenging task of the conservation scientist to point out other practical examples of induction-time behavior and of distinct stages in the deterioration of organic materials with which the conservator must contend.

ACKNOWLEDGMENT

In 1978-79, the Center on the Materials of the Artist and Conservator at Mellon Institute received a research grant from the National Endowment for the Arts to investigate the formation of hydroperoxide intermediates as an aspect of the deterioration of organic materials. The investigations just described are based in part on the findings of that research.

Photochemical Studies of Methacrylate Coatings for the Conservation of Museum Objects

ROBERT L. FELLER, MARY CURRAN AND CATHERINE BAILIE

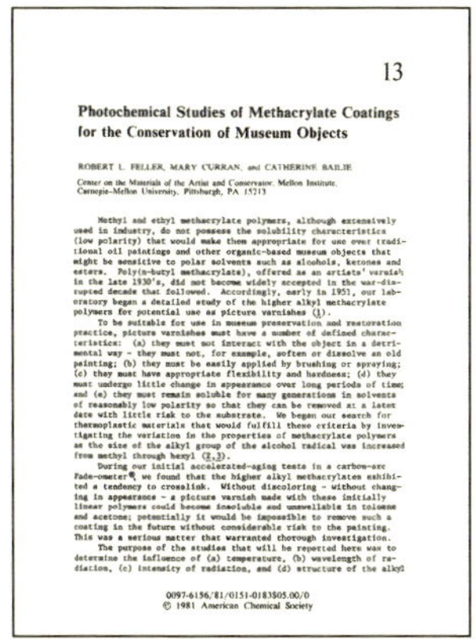

13

Photochemical Studies of Methacrylate Coatings for the Conservation of Museum Objects

ROBERT L. FELLER, MARY CURRAN, and CATHERINE BAILIE

Center on the Materials of the Artist and Conservator, Mellon Institute, Carnegie-Mellon University, Pittsburgh, PA 15213

Methyl and ethyl methacrylate polymers, although extensively used in industry, do not possess the solubility characteristics (low polarity) that would make them appropriate for use over traditional oil paintings and other organic-based museum objects that might be sensitive to polar solvents such as alcohols, ketones and esters. Poly(n-butyl methacrylate), offered as an artists' varnish in the late 1930's, did not become widely accepted in the war-disrupted decade that followed. Accordingly, early in 1951, our laboratory began a detailed study of the higher alkyl methacrylate polymers for potential use as picture varnishes (1).

To be suitable for use in museum preservation and restoration practice, picture varnishes must have a number of defined characteristics: (a) they must not interact with the object in a detrimental way - they must not, for example, soften or dissolve an old painting; (b) they must be easily applied by brushing or spraying; (c) they must have appropriate flexibility and hardness; (d) they must undergo little change in appearance over long periods of time; and (e) they must remain soluble for many generations in solvents of reasonably low polarity so that they can be removed at a later date with little risk to the substrate. We began our search for thermoplastic materials that would fulfill these criteria by investigating the variation in the properties of methacrylate polymers as the size of the alkyl group of the alcohol radical was increased from methyl through hexyl (2,3).

During our initial accelerated-aging tests in a carbon-arc Fade-ometer®, we found that the higher alkyl methacrylates exhibited a tendency to crosslink. Without discoloring - without changing in appearance - a picture varnish made with these initially linear polymers could become insoluble and unswellable in toluene and acetone; potentially it would be impossible to remove such a coating in the future without considerable risk to the painting. This was a serious matter that warranted thorough investigation.

The purpose of the studies that will be reported here was to determine the influence of (a) temperature, (b) wavelength of radiation, (c) intensity of radiation, and (d) structure of the alkyl

0097-6156/81/0151-0183$05.00/0

Originally printed in:
PHOTODEGRADATION AND PHOTOSTABILIZATION OF COATINGS
ACS Symposium Series Number 151, ed. Pappas, S. P., and Winslow, F. H., American Chemical Society, Washington, DC (1980), Pages 183-196.

Methyl and ethyl methacrylate polymers, although extensively used in industry, do not possess the solubility characteristics (low polarity) that would make them appropriate for use over traditional oil paintings and other organic-based museum objects that might be sensitive to polar solvents such as alcohols, ketones and esters. Poly(n-butyl methacrylate), offered as an artists' varnish in the late 1930s, did not become widely accepted in the war-disrupted decade that followed. Accordingly, early in 1951, our laboratory began a detailed study of the higher alkyl methacrylate polymers for potential use as picture varnishes.[1]

To be suitable for use in museum preservation and restoration practice, picture varnishes must have a number of defined characteristics: (a) they must not interact with the object in a detrimental way—they must not, for example, soften or dissolve an old painting; (b) they must be easily applied by brushing or spraying; (c) they must have appropriate flexibility and hardness; (d) they must undergo little change in appearance over long periods of time; and (e) they must remain soluble for many generations in solvents of reasonably low polarity so that they can be removed at a later date with little risk to the substrate. We began our search for thermoplastic materials that would fulfill these criteria by investigating the variation in the properties of methacrylate polymers as the size of the alkyl group of the alcohol radical was increased from methyl through hexyl.[2,3]

During our initial accelerated-aging tests in a carbon-arc Fadeometer, we found that the higher alkyl methacrylates exhibited a tendency to crosslink. Without discoloring—without changing in appearance—a picture varnish

1. Feller, R. L., Stolow, N., and Jones, E. H., *On Picture Varnishes and Their Solvents*, Press of the Times, Oberlin, Ohio (1959); revised edition of Press of Case-Western Reserve, Cleveland (1971).

2. Raynolds, S., *The Dependence of Physical Properties on the Constitution of Alkyl Polymethacrylate-ester Resins*, Thesis, Master of Science, University of Pittsburgh (1954).

3. Feller, R. L., "Identification and Analysis of Resins and Spirit Varnishes," in *Application of Science in Examination of Works of Art*, Museum of Fine Arts, Boston (1959), p. 51-76.

made with these initially linear polymers could become insoluble and unswellable in toluene and acetone; potentially it would be impossible to remove such a coating in the future without considerable risk to the painting. This was a serious matter that warranted thorough investigation.

The purpose of the studies that will be reported here was to determine the influence of (a) temperature, (b) wavelength of radiation, (c) intensity of radiation, and (d) structure of the alkyl group upon the tendency of a number of the polymers to crosslink, as evidenced by the rate of formation of insoluble matter.

PREPARATION OF POLYMERS

Both commercial and laboratory-synthesized polymers were used. Those made in the laboratory were generally prepared by solution polymerization, refluxing commercially available monomers in toluene using benzoyl peroxide as the catalyst. Other preparations were made in which azo-bis-isobutyronitrile (AIBN) was used as initiator, ethanol was employed as the refluxing medium, and monomers were especially synthesized in the laboratory. These variations in preparative procedure did not significantly affect the ranking of the polymers with respect to their tendency to crosslink, as reported in Table 1.

Films of the polymers were generally cast with a drawdown bar from 20% solutions in reagent-grade toluene. All the data reported in the accompanying tables and graphs are based on films thus coated on aluminum foil. The use of the foil speeded up the formation of insoluble matter 1.9 times over that of coatings on glass and also permitted convenient measurements of changes in weight upon exposure and extraction.[4] Ordinary 0.001"-thick household

4. Feller, R. L., and Bailie, C. W., "Studies of the Effect of Light on Protective Coatings Using Aluminum Foil as a Support: Determination of Ratio of Chain Breaking to Crosslinking," *Bulletin of the American Group-IIC*, 6, No. 1 (1966), p. 10-12.

TABLE 1
Tendency of Methacrylate Polymers to Become Insoluble
Samples on Aluminum Foil, Single-Carbon-Arc Fadeometer (Corex D Filter), 62°C

Alkyl Radical in Poly(R-Methacrylate)	Reactive Hydrogen on Alkyl Carbon Atom No.	Intrinsic Viscosity MEK, 23°C	Hours to Attain 50% Insolubility
3-methylbutyl	3t (a)	.27	4
25/75 p-methylcyclohexyl/isoamyl*	3, 4t	~.22	35
2-methylbutyl	2t	.27	49
n-butyl	3, 2s	~.25	91
60/40 neopentyl/n-amyl	4, 3, 2s	~.41	110
isobutyl (Elvacite® 2045)	2t	.59	126
n-propyl	2s	.21	660
ethyl	1s	1.6	>> 1800 (b)
~50/50 methylacrylate/ ethylmethacrylate (Acryloid® B-72)	1s	.20	>> 2023 (b)

* Polymer 59, Figure 2.
(a) 3t = tertiary hydrogen on third carbon from alcoholic oxygen; s = secondary hydrogen.
(b) no more than 5% insoluble matter detected.

aluminum foil was rubbed flat on the surface of glass plates, using a few drops of acetone to aid in the adherence of the foil to the glass and to wet the cotton swabs used to clean the foils and rub them flat. After standing for 24 hours, the cast films were baked 48 hours at 70°C. The retained solvent after baking was usually no more than a few tenths of a percent by weight of resin. The small amount of toluene that remained was considered insufficient to alter the conclusions. Tests have shown, however, that the retention of a chemically active solvent such as turpentine in the film can shorten the induction time considerably.[5]

Major differences in molecular weight can be expected to influence the radiation dose necessary to give rise to insoluble matter; because of this, intrinsic viscosity values are given in Table 1 to indicate that the laboratory-prepared polymers were generally similar in this respect.

CROSSLINKING

a. Influence of Wavelength of Radiation

Our initial indications of crosslinking were observed during exposures in an Atlas Electric Devices Co. carbon-arc Fadeometer equipped with a Corex D filter. Later, the same behavior was found when exposures were made in an Atlas 600WRC xenon-arc Fadeometer having Pyrex-glass filters. When crosslinking was first encountered, a number of colleagues advised us that the phenomenon was not likely to be of importance in a museum; there, the natural or fluorescent illumination would consist only of those wavelengths that would pass through ordinary glass. We soon found, however, that, although the reaction would be very slow under museum conditions, it was indeed activated by the visible as well as by the near-ultraviolet radiation. For example, when an ultraviolet filter was introduced that removed most of the radiation below 400 nm, crosslinking still took place, but the rate of formation of insoluble material in the carbon-arc Fadeometer was reduced to about one-half. The rate of crosslinking was eventually shown to vary almost logarithmically with the shortest wavelength in the illumination, as illustrated in Figure 1.[6] It is interesting to note in Figure 1 that a number of photochemical degradations follow much the same relationship: loss of weight from alkyd paint films,[7] degradation of low-grade paper,[8] evolution of hydrogen from rubber[9] and development of carbonyl groups in poly(vinyl chloride).[10] The reason for this particular sensitivity to wavelength is perhaps the absorption of radiation by functional groups present at very low concentrations.

Later investigations have fully confirmed the conclusion that the crosslinking of the higher alkyl methacrylate polymers will take place on a gallery wall. We have been able to demonstrate that an induction time of about 11 years occurs before insoluble matter begins to form in commercial normal and isobutyl polymers on a well-illuminated museum wall.[1, 11] Protection against such loss

5. Feller, R. L., "Problems in the Investigation of Picture Varnishes," in *Conservation of Paintings and the Graphic Arts*, ed. Brommelle, N., and Smith, P., Butterworths, London (1976), p. 137-144.

6. Feller, R. L., "New Solvent-type Varnishes," in *Recent Advances in Conservation*, ed. Thomson, G., Butterworths, London (1963), p. 171-175.

7. Miller, C. D., "Kinetics and Mechanism of Alkyd Photooxidation," *Ind. Eng. Chem.*, 50 (1958), p. 125-128.

8. Harrison, L. S., "An Investigation of the Damage Hazard in Spectral Energy," *Illumination Eng. (NY)*, 49 (1954), p. 253-257.

9. Bateman, L., "Photolysis of Rubber," *J. Polymer Sci.*, 2 (1947), p. 1-9.

10. Martin, K. G., and Tilley, R. I., "Influence of Radiation Wavelength on Photooxidation of Unstabilized PVC," *British Polymer J.*, 3 (1971), p. 36-40.

11. Feller, R. L., "The Deterioration of Organic Substances and the Analysis of Paints and Varnishes," in *Preservation and Conservation: Principles and Practices*, ed. Timmons, S., Preservation Press, Washington, DC (1976), p. 287-299.

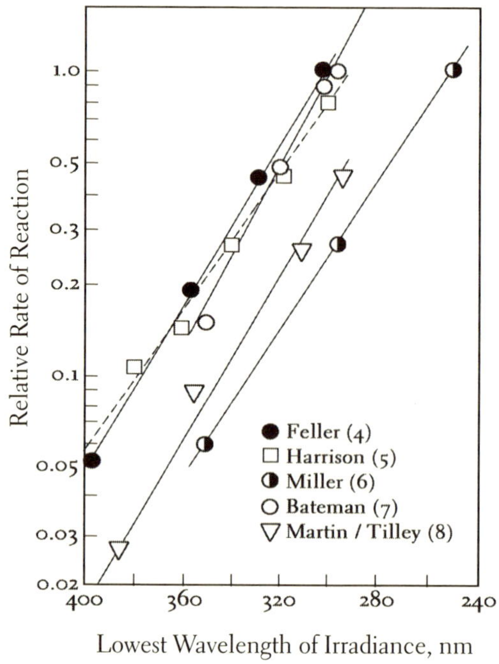

FIGURE 1
Data of various investigators on the influence of the lowest wavelength of irradiance on the rate of a variety of chemical reactions.

of solubility is one of a number of reasons for recommending the use of ultraviolet filters over windows and over fluorescent-lamp sources in museums.[12]

The relationship illustrated in Figure 1 should be extended below 313 nm only with caution. The shorter wavelengths may induce photolytic decomposition, resulting in much more chain scission. Thus, when samples of poly(n-butyl methacrylate) were exposed to a high-pressure mercury lamp (emitting intense radiation at 254 nm), we found that, although gel formation took place in the beginning, considerable chain breaking and volatilization eventually occurred.[13, 14] Under this lamp, the films developed bubbles and blisters at an early stage, perhaps owing to the formation of monomer by unzipping reactions. We have warned colleagues not to attempt to use lamps that emit 254 nm radiation in "accelerated-aging" tests of museum materials, because the photo-activated mechanism of deterioration may be distinctly different from that caused by the near-ultraviolet and visible radiation.

For extensive studies of the effects of 254 nm radiation on acrylates and methacrylates, the reader is referred to the work of Morimoto and Suzuki[14] and of Grassie.[15, 16] In 1964, Oster reported on the crosslinking of these and other polymers by the near-ultraviolet, sensitized by the presence of 2-methylanthraquinone.[17]

12. Feller, R. L., "Control of Deteriorating Effects of Light on Museum Objects," *Museum*, 17 (1964), p. 57-98.

13. Feller, R. L., "Speeding Up Photochemical Deterioration," *Bulletin de l'Institut Royal du Patrimoine Artistique*, 15 (1975), p. 135-150.

14. Morimoto, K., and Suzuki, S., "Ultraviolet Irradiation of Poly(alkyl Acrylates) and Poly(alkyl Methacrylates)," *J. Applied Polymer Science*, 16 (1972), p. 2947-2961.

b. Influence of Alkyl Group

Colleagues and manufacturers further advised us in 1952 that acrylics would not undergo crosslinking because the polymers do not obviously absorb in the near-ultraviolet. (However, the field now realizes that trace components at very low levels may absorb radiation in this range.) Our advisors may be forgiven, for their greatest familiarity at the time was with polymers of methyl and ethyl methacrylate. As the data in Table 1 show, we found the tendency-to-crosslink to reside primarily in the butyl and amyl esters that contain tertiary hydrogens, and secondarily, in the n-amyl, n-butyl and n-propyl esters which have secondary hydrogens removed from the backbone of the main chain by the carboxylic carbon and oxygen and by two alkyl carbons.[1, 18] The normal alkyl polymers that we prepared may have contained traces of an isopropyl, isobutyl or isoamyl impurity; nonetheless, the presence of tertiary hydrogens is not necessary to account for the tendency to crosslink. A six-membered-ring intermediate structure, such as proposed by Grassie[15] and others, may accelerate the loss of hydrogen atoms from the alkyl groups. If so, then the normal butyl, amyl and higher esters should tend to crosslink more readily than polymers made with the methyl and ethyl esters. Thus, Barton,[19] who induced crosslinking in n-butyl and n-nonyl methacrylate polymers through the addition of dicumyl peroxide, clearly demonstrated that the nonyl ester had a higher crosslinking efficiency than the polymer with the shorter side chain.

If the loss of solubility of these initially linear polymers takes place through a free-radical-chain process in which the crosslinking reaction represents the termination step, one may be hesitant at first to explain the fact that a free radical, generated at one point on a relatively sluggish polymer chain, can find a radical on a neighboring chain with which to terminate. However, perhaps reactions of the type

$$R\bullet + R'H \rightarrow RH + R'\bullet \qquad (1)$$

or

$$ROO\bullet + R'H \rightarrow ROOH + R'\bullet \qquad (2)$$

can take place rapidly between pendant alkyl groups along the chain until a radical is encountered on a neighboring chain; this would explain in part the enhancement of crosslinking caused by the tertiary hydrogens being located increasingly further from the main chain backbone (Table 1; Figure 2). Grassie, Semenov,[20] Chien[21] and others have proposed such intramolecular reactions by pendant groups along the main chain.

c. Influence of Intensity of Illumination

There is a strong tendency for those who employ intense sources of illumination in accelerated photochemical aging tests to make quick calculations on the basis of the reciprocity principle—that is, to assume that, as the intensity is

15. Grassie, N., and MacCallum, J. R., "Thermal and Photochemical Degradation of Poly(n-butyl Methacrylate)," *J. Polymer Science, Part 2A*, 2 (1964), p. 983-1000.

16. Grassie, N., "Photodegradation of Methacrylate/Acrylate Copolymers," *Pure and Applied Chemistry*, 34 (1973), p. 247-257.

17. Oster, G., "Photochemical Crosslinking of Non-aqueous Polymers by Near-Ultraviolet Light," *J. Polymer Science, Part B, Polymer Letters*, 2 (1964), p. 1181-1182.

18. Feller, R. L., "Crosslinking of Methacrylate Polymers by Ultraviolet Radiation," *American Chemical Society Dvision of Paint, Plastics, and Printing Ink Chemistry, Papers Presented at New York Meeting*, 17, No. 2 (1957), p. 465-470.

19. Barton, J., "Peroxide Crosslinking of Poly(n-alkyl Methacrylates)," *J. Polymer Science, Part 1A*, 6 (1968), p. 1315-1323.

20. Gordon, G. Ya., "Stabilization of Synthetic High Polymers," *Israel Program for Scientific Translations* (1964), p. 45.

21. Chien, J. C. W., and Boss, C. R., "Polymer Reactions. V. Kinetics of Autoxidation of Polypropylene," *J. Polymer Science, Part 1A*, 5 (1967), p. 3091-3101.

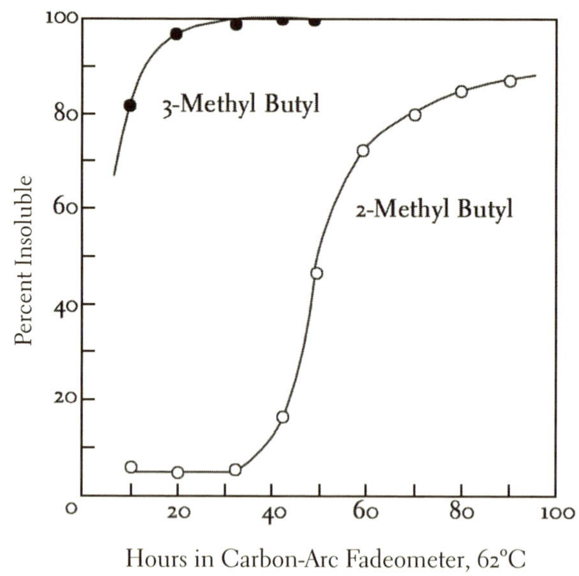

FIGURE 2
Comparison of the rate of formation of insoluble matter in polymers based on 3-methyl and 2-methyl-butyl methacrylate.

lowered, the time for equivalent damage will be correspondingly lengthened. For a number of reasons, the reciprocity principle need not hold true, and it is advisable to check this point. We did so by exposing a readily crosslinkable polymer (duPont's Elvacite® 2046, a copolymer of normal and isobutyl methacrylate) under a series of intensity-reducing screens. The experiments were carried out in the xenon-arc Fadeometer with Pyrex-glass filters, in which the circulating air was maintained at 32°C and about 25% relative humidity. Aluminum wire screens were employed to decrease the intensity without seriously altering the spectral distribution of the illumination. The results, given in Table 2, indicate that the reciprocity law held true over a 30-fold range of intensity.

TABLE 2
Induction Time for Crosslinking of Elvacite® 2046
Under Various Intensities of Daylight Fluorescent Lamps

Intensity, I	Induction Time, T_i	IT_i
1.00	15	15.0
.71	22	15.6
.49	29	14.2
.35	40	14.0
.135	79	10.7
.033	530	17.5
		14.5 ± 2.2

Further experiments were conducted under the relatively mild conditions of exposure to a bank of "daylight" fluorescent lamps in a room maintained at 70°F (21°C) and 50% relative humidity. Under these conditions, the intensity of the near-ultraviolet radiation amounted to 2.5% of the energy in the visible range and the temperature of the samples reached no more than 26°C. Such experiments clearly demonstrated that commercial normal and isobutyl methacrylate polymers, as well as those that are prepared in the laboratory by solution polymerization, develop 50 to 80% insoluble matter after about 9 million footcandle hours of exposure to "daylight" fluorescent lamps. By extrapolation, one may estimate that these coatings would develop the same degree of insolubility in about 50 years of exposure in a daylighted museum under an average level of illumination of about 50 footcandles.[6] We concluded in the 1960s that there was no cause for immediate alarm concerning the past uses of the normal and isobutyl methacrylate varnishes; the polymers had not been introduced to the museum world until about 1939. The oldest coatings in 1960 would not have been more than about 20 years old. Nonetheless, in order to develop a sound understanding of the properties of acrylic resins that the conservator of museum objects might wish to employ with confidence in the future, a thorough investigation of the problem was required.

d. Effect of Film Thickness, Oxygen

In the range of 25- to 50-microns film thickness, the percentage of insoluble matter formed during initial stages of crosslinking decreased with film thickness. We have further shown that the crosslinking induced in these polymers by near-ultraviolet radiation is retarded when the films are exposed in an inert atmosphere and is enhanced when the polymer contains oxygen that may have been inadvertently introduced into the chain during polymerization.[5]

Loss in Weight

During the exposure tests, we found, following an induction period in which little change was observed, that the films tended to lose weight at a rate that was linear with time. This behavior was not extensively investigated, but we believe that the induction time after which the marked rise in insoluble matter occurred corresponded closely with the time after which the distinct loss in weight began. The observation of weight loss may be of practical interest in the selection of photochemically stable acrylic coatings. A number of years ago, Berg, Jarosz and Salanthe[22] suggested that coatings that lost the minimum of weight during exposure tended to be the most durable. In a qualitative way, this was borne out in our results in the carbon-arc Fadeometer (Corex D filter) in which the rate of weight loss, following the induction period, was observed to be as follows: commercial normal and isobutyl methacrylate polymers, 1.0 to 2.0% loss in weight per 100 hours; poly(n-propyl methacrylate), 0.6%; Acryloid® B-72 (copolymer of ethyl methacrylate and methyl acrylate), 0.2%. The ranking is the same as the tendency-to-crosslink reported in Table 1.

22. Berg, C. J., Jarosz, and Salanthe, G. F., "Performance of Polymers in Pigmented Coatings," *J. Paint Technology*, 39 (1967), p. 436-453.

RATIO OF CHAIN BREAKING TO CROSSLINKING

a. Measurement of Ratio.

Early in our investigation we encountered a commercial polymer that, although it eventually formed about 50% insoluble matter in the Fadeometer, did not rapidly rise to a state of more than 90% insolubility as did most of the other polymers then under investigation. The significance of this result was not fully appreciated at first; we realized later that we were observing a case of simultaneous chain breaking and crosslinking. Since then, we,[4, 11, 23] as others before us,[24] have found that the equations relating loss of solubility with exposure time that were developed by Charlesby and Pinner[25, 26] for analyzing the effects of high-energy radiation also apply in the case of the photo-activated changes under study here. We have recommended to colleagues that the decrease in soluble matter during accelerated photochemical aging tests be evaluated on a Charlesby log-log plot as seen in Figures 3 and 4. For convenience, we frequently place a scale on these graphs at ten times the gel dose to indicate the approximate ratio of chain breaks to crosslinks (Figure 3). The location of the scale divisions at ten times the gel dose are as follows: the ratio of breaks to links, b/a, is 1.0 at a soluble fraction of 0.495; it is 0.7 at 0.328, 0.6 at 0.275, 0.5 at 0.225, 0.4 at 0.178, 0.2 at 0.094, and 0.1 at 0.059.

Charlesby's method of plotting $s + \sqrt{s}$ against $1/\delta$ perhaps provides a more objective analysis of the data:

$$s + \sqrt{s} = \beta/\alpha + 2/\delta \qquad (3)$$

where s = soluble fraction, β/α the ratio of degradation to crosslinking for the case of an initially random distribution of molecular weights, and δ is R/R^*, in which R is the accumulated exposure time and R^* is the time required for the initial formation of insoluble gel, essentially the "gel dose."[25] We have used graphs of this equation as an alternate method of plotting the data and obtaining β/α.[1, 4] The locations of the reference-scale divisions at ten times the gel dose were calculated from this equation, using $1/\delta = 0.1$ and the theoretical slope of 2.0.

b. Influence of Temperature

Panels in the carbon-arc Fadeometer reach 62°C; the black-panel temperature in our present 600WCR xenon-arc Fadeometer is 59°C. To learn how seriously temperature affects crosslinking, we built a water-cooled surface upon which to mount our test panels of coated aluminum foil during exposure in the carbon-arc equipment. The results of studies at various temperatures are shown in Figure 3. Here we see that, at the usual high temperature of panels in the carbon-arc Fadeometer, 62°C, copolymer No. 59 (based on 1:3 combination of p-methylcyclohexyl and isoamyl methacrylate) and the isobutyl polymer both

23. Feller, R. L., "Stages in the Deterioration of Organic Materials," in *Preservation of Paper and Textiles of Historic and Artistic Value*, ed. Williams, J. C., American Chemical Society, Washington, DC (1977), p. 314-335.

24. Maxim, L. D., and Kuist, C. H., "The Light Stability of Vinyl Polymers and the Effect of Pigmentation," *Official Digest, Fed. of Soc. for Paint Technology*, 36 (1964), p. 723-744.

25. Charlesby, A., and Pinner, S. H., "Analysis of the Solubility of Irradiated Polyethylene and Other Polymers," *Proc. Roy. Soc.*, A249 (1959), p. 367-386.

26. Charlesby, A., *Atomic Radiation and Polymers*, Pergamon Press, London (1960), p. 143, 173.

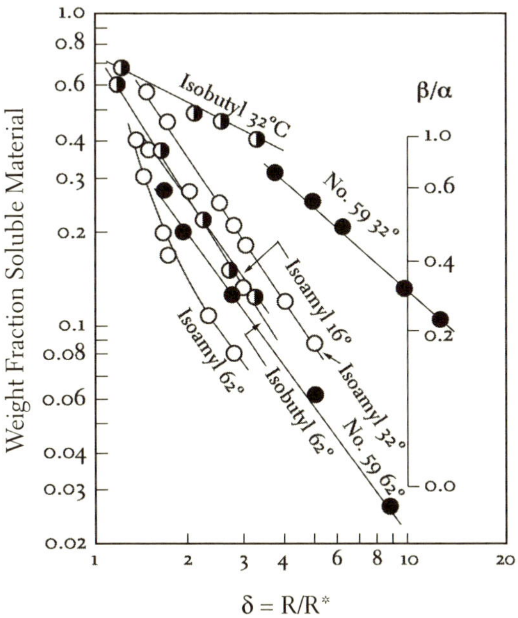

Figure 3
Influence of temperature on the rate of crosslinking of three methacrylate polymers plotted in terms of time relative to the gelation dose, R*, insolubility at R/R* = 10 used as a measure of the ratio of chain breaking to crosslinking.

tend to crosslink almost exclusively; that is, the ratio of chain breaking to crosslinking, β/α, ≈ 0. At a lower temperature, 32°C, considerable chain breaking occurs; the ratio of links broken to crosslinks formed becomes about 0.3 in one case, 0.55 in another. Perhaps much of this behavior can be explained by considering whether the polymers are above or below their second-order transition temperature during irradiation, but we have not pursued this line of investigation. As seen in Figure 3, an isoamyl methacrylate polymer based on a mixture of 2-methyl and 3-methyl esters, with its highly labile tertiary hydrogens, crosslinked almost exclusively even at 16°C, even though the polymer should have been below its T_g (about 26°C) at this temperature.

These findings clearly illustrate how the results of so-called "accelerated-aging" tests can be affected by the high temperatures in ordinary xenon- and carbon-arc equipment and can, therefore, lead to erroneous conclusions regarding the photochemical behavior of materials at near-normal temperatures.[23]

Grassie's studies on the decomposition of acrylate and methacrylate polymers at high temperatures may be of interest in suggesting mechanisms by which some of the alkyl groups may thermally decompose,[27] although his investigations were conducted primarily under vacuum or under inert gases.

27. Grassie, N., "Recent Work on the Thermal Degradation of Acrylate and Methacrylate Homopolymers and Copolymers," *Pure and Applied Chem.*, 30 (1972), p. 119-134.

PETROLEUM-SOLUBLE POLYMERS RESISTANT TO CROSSLINKING UNDER NEAR-ULTRAVIOLET

In view of the above findings, what may we choose for use as a picture varnish that will have little or no tendency to crosslink? Polymers of the perfluoroacrylic esters exhibit no such tendency but require solvents of questionable appropriateness.[5] Moreover, if the coatings are used indiscriminately in making repairs to the structure of a painting, future problems regarding adhesion and wetting may be introduced. Poly(n-propyl methacrylate) is also highly resistant to crosslinking (Table 1) and might be considered, although we do not believe that a suitable commercial polymer is available.

The most stable resin for many of our purposes has proven to be a copolymer of ethyl methacrylate and methyl acrylate. This comes as little surprise; the Rohm and Haas Company has for years sold a durable resin based on these two monomers, Acryloid® B-72.[6, 28] We have also prepared polymers of similar physical properties based on methyl methacrylate and ethyl acrylate and have found that their behavior is practically the same—the methyl and ethyl groups apparently do not become seriously involved in crosslinking. As reported elsewhere,[23] rather than crosslink, Acryloid® B-72 tends to chain break under visible and near-ultraviolet radiation, although at a very slow rate. Polyvinylacetate is another polymer used in the care of museum objects that tends more to chain break than crosslink under these conditions,[23] but it is not our purpose to discuss its properties at this time.

28. DeWitte, E., Goessens-Landrie, M., Goethals, E. J., and Simons, R., "The Structure of 'Old' and 'New' Paraloid B-72," *Preprints of the 5th Triennial Meeting, Zagreb*, ICOM Committee for Conservation (1978), Paper 78/16/3.

One of the recognized challenges in creating a durable polymer is to create one in which there is a proper balance between crosslinking and chain breaking tendencies. This objective was described, pursued and demonstrated in the work of Maxim and Kuist.[24, 29] These authors showed that a distinct maximum in durability was frequently obtained at an intermediate composition when monomers were copolymerized. Much the same behavior was reported by Graham, Crowne and MacAlpine in the evaluation of copolymers and terpolymers of thermosetting acrylics.[30] Maxim and Kuist's studies, which involved acrylate rather than methacrylate polymers, confirm our findings that crosslinking tends to increase with the length of the alkyl side chain.

29. Kuist, C. H., and Maxim, L. D., "The Ultraviolet Degradation of Scissioning Copolymers," *Polymer*, 6 (1965), p. 523-530.

30. Graham, N. B., Crowne, F. R., and MacAlpine, D. E., "A Maximum Durability Prediction Scheme for Thermosetting Acrylics," *Official Digest, Fed. of Soc. for Paint Technology*, 37 (1965), p. 1228-1250.

INHIBITION OF CROSSLINKING

Our initial objectives in undertaking these investigations were to seek an understanding of the causes of the photochemically induced crosslinking and to pinpoint at least a few thermoplastic polymers—originally soluble in hydrocarbons no more polar than toluene—that had little tendency to undergo loss of solubility under ordinary museum conditions. Having achieved these objectives, we returned our attention to the matter of inhibiting crosslinking, a practical possibility that had been accomplished through the addition of 2,4-dihydroxybenzophenone, as described in our 1957 publication.[18] We have since found that adding LTDP (dilaurylthiodipropionate) and Ciba-Geigy's Tinuvin® 328 ultraviolet absorber—each at 1% concentration relative to the weight of the resin—increases the induction time 10- to 15-fold for a 1:1 copolymer of normal and isobutyl methacrylate, Elvacite® 2046, during xenon-arc exposure (Pyrex glass filter).[5] Ciba-Geigy's Tinuvin® 770 alone at 1% concentration in this same resin prolongs the induction time in the xenon-arc Fadeometer about 5-fold. When plotted as we see in Figure 4, data on loss of solubility will readily reveal which combination of ultraviolet absorber and inhibitor affects primarily the induction time, which affects the ratio of chain breaking to crosslinking, and which may affect both.[5]

SUMMARY OF MAJOR FINDINGS

The tendency to crosslink under near-ultraviolet and visible radiation has been demonstrated in polymers prepared from the n-propyl, n-butyl and isobutyl, n-amyl and isoamyl methacrylate esters, particularly in those with alkyl groups that contain tertiary hydrogens. The ratio of chain breaking to crosslinking has been shown to vary with temperature, the higher temperature favoring an increase in the tendency to crosslink. A number of toluene-soluble polymers have been found that exhibit little tendency to crosslink under the near-ultraviolet. Moreover, for those methacrylate polymers that do tend to crosslink, inhibitor systems can be used to delay the onset of insolubility as much as 15-fold. We have determined that the reciprocity law (intensity of illumination times induction time) essentially holds true over a 30-fold range of intensity. Further, the rate of formation of insoluble matter has been shown to increase logarithmically with the decrease in the lowest wavelength of irradiance in the

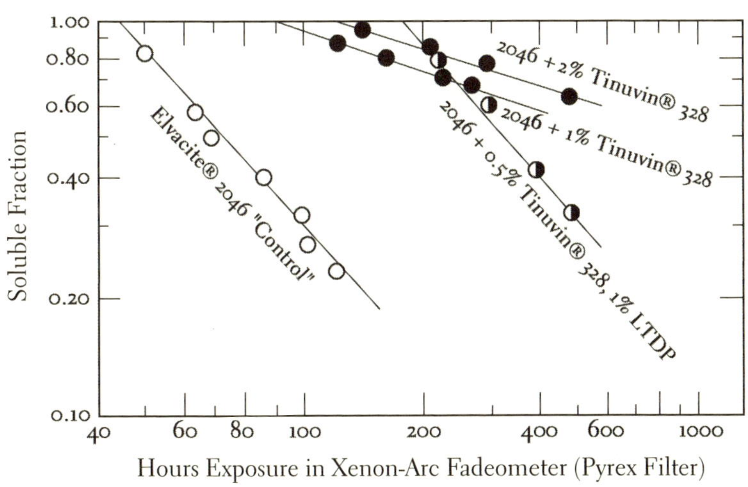

FIGURE 4
Loss of solubility of Elvacite® 2046 with and without u.v. absorber and antioxidant additives.

range of 400 to 300 nm. Crosslinking of these polymers can take place under the normal conditions of daylight or fluorescent-lamp illumination encountered in a museum, although at an extremely low rate.

IMPORTANCE WITH RESPECT TO LONG-TERM USAGE

These studies, sponsored for over two decades by the National Gallery of Art, have established some general principles that can guide conservators both in the selection of durable protective coatings and in the application, maintenance, and repair of such coatings. Based upon tests made under the relatively mild conditions of exposure to "daylight" fluorescent lamps and also to natural illumination on a gallery wall, we can now reasonably predict that—on an art gallery wall experiencing about 110,000 footcandle hours of diffuse daylight through window glass annually—polymers such as Acryloid® B-72 and polyvinylacetate should remain colorless, and soluble in the solvents in which they were originally soluble, for more than 200 years.[31] Truly, these are first-class materials to have been placed at the artists' and conservators' command.

31. Feller, R. L., and Curran, M., "Changes in Solubility and Removability of Varnish Resins with Age," *Bull. of the American Institute for Conservation*, 15, No. 2 (1975), p. 17-26.

ACKNOWLEDGMENTS

We are grateful to the organizers of this symposium for the opportunity to enlarge upon our reports of these investigations that have appeared in a wide variety of publications in the past, publications that have been directed more towards the museum conservator and conservation scientist than to the polymer chemist.

Many of the methacrylate polymers used at the beginning of these investigations were prepared by Stuart Raynolds. Richard A. Tauson carried out the initial studies on the effect of temperature. The authors wish particularly to thank Dr. John Walker, Director Emeritus of the National Gallery of Art, for his great personal interest, support, and encouragement in this research. The work of the Center is principally made possible through the generosity of the Andrew W. Mellon Foundation.

Section 7: Paper Degradation

The laboratory had not engaged in research on cellulose and paper until about 1978. At that time Dr. S. B. Lee joined the Center to devote full attention to this topic, assisted by John Bogaard. As an introduction to the subject, reviews of the literature were shared with colleagues in the form of an annotated bibliographic supplement to *Art and Archaeology Technical Abstracts*. Paper embrittlement and weakening are known to derive from the decrease in degree of polymerization of the constituent cellulose molecules in the paper fibers. Paradoxically, insight regarding the rate at which bonds in the cellulose chains are broken can be gained, not by following changes in the degree of polymerization, but by considering the changes in the reciprocal of that quantity. Concurrent to the cellulose breakdown, oxidation can also take place, resulting in an increase in material that is soluble in hot alkali. Pursuing this point of view, the Center came to support the idea that thermally-induced degradation of paper apparently takes place in two or more stages. Exposure to light is also regarded as taking place in distinct stages. Initially, there is an immediate effect that involves oxidation and the breaking of bonds. Thereafter, an enhanced rate of thermal degradation occurs. Certain conflicting results in thermal aging tests that have been reported in the literature were later shown to be attributable to differences arising from oven aging when it is carried out under dry conditions as opposed to moist.

R. L. F.

Three Fundamental Aspects of Cellulose Deterioration

ROBERT L. FELLER, SANG B. LEE AND MARY CURRAN

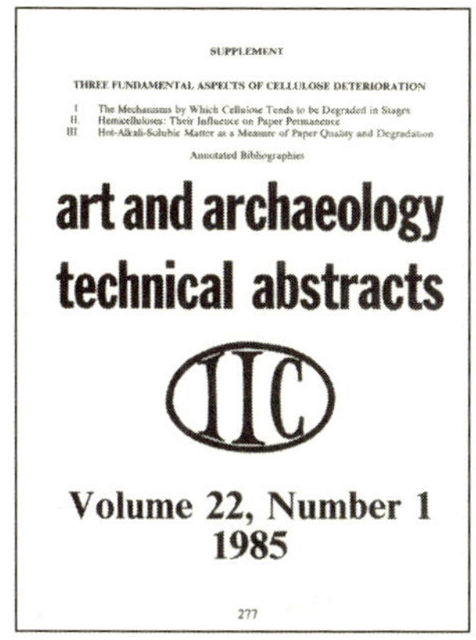

Originally printed in:
ART AND ARCHAEOLOGY TECHNICAL ABSTRACTS
Supplement, Volume 22, Number 1 (1985), Pages 277-358.

Introduction

The following annotated bibliographies review major literature citations on three fundamental aspects of cellulose deterioration. Valuable insights are to be found in journals from such widely diverse fields as wood, carbohydrate, textile, polymer, and paper chemistry. The literature related to any particular aspect of paper deterioration is so extensive that one cannot readily undertake an exhaustive search. Nonetheless, it is important to identify the particularly pertinent references and to relate this information to the general advancement of our understanding regarding the long-term stability of paper and of cellulose-based textiles and ethnographic materials.

Why have these three particular topics been chosen? Concerning the first, it is generally known that acid causes cellulose to degrade by hydrolytic action: the higher the acidity (hydrogen ion activity), the faster the degradation. Yet there are other factors besides acidity that affect the speed at which acid acts upon cellulose, factors inherent in the chemical and physical character of cellulose-based fibers. The net result of such influences is that cellulose-containing fibers can be considered to degrade in stages, or in sequences of decreasing rate, as one or another factor becomes dominant over the entire course of degradation from initial fiber to the final sugar-based fragments that represent the ultimate constitution of cellulose and hemicellulose. Bibliographic topic I reviews a number of intrinsic (as opposed to extrinsic) factors that govern the rate at which acid causes cellulose fibers to break down, principally the factors arising from differences in crystallinity and from differences in the chemical structure of hemicelluloses and oxidized cellulose.

It is perhaps traditional to blame most of the instability of low-grade papers on lignin. However, hemicelluloses also contribute to discoloration and degradation. Bibliographic topic II reviews this subject.

The extent of oxidation of pulp and cotton fibers is usually monitored by measurement of their reducing power, frequently expressed as copper number. The solubility of pulps in boiling dilute sodium hydroxide provides another convenient assessment of quality. If the latter procedure is to be used as a measure of the initial quality of pulp, or of the extent of oxidative attack on the cellulose chain, it is necessary to realize that low molecular weight degradation products and portions of the hemicellulose fraction will also be removed by hot alkali. Bibliographic topic III reviews the chemistry of hot-alkali solubility. This subject cannot be considered without some reference to the solubility of cellulose in more concentrated solutions of alkali at low temperature.

After having located the key references, reviewed and abstracted their contents, it is most desirable to have the information critically reviewed and the relevance of particular research findings to the practice of conservation delineated. Much of the material under topic I has been reviewed from time to time by a number of authors. The latest, scheduled to appear in the American Chemical Society's Advances in Chemistry Series No. 212, *Conservation and Characterization of Historical Textile and Paper Materials*, is entitled "Concerning the Kinetics of Cellulose Deterioration," by R. L. Feller, S. B. Lee, and J. Bogaard. Unfortunately, to gather the publications from so many disparate sources, read, and interpret the contents has been such a formidable task that there has not as yet been sufficient time to prepare fully considered critical reviews on topics II and III. To make up for this in part, brief overviews have been prepared as an introduction to each of the three bibliographies. These do not refer to every article included in the bibliographies but indicate why the major articles were selected, and suggest the threads of thought that extend throughout several sequences of publications. In a few cases, an abstract title appears in more than one of the three bibliographies if the publication is considered particularly pertinent; this cross-reference is found at the end of each section.

Precisely what an individual publication or collection of publications may offer which is of immediate and direct assistance to the conservator or conservation scientist is not elaborated at this moment; that must await further consideration. The information here assembled should, nonetheless, enlarge the conservator's general fund of knowledge and understanding, laying ground work for future advances in our ability to care for cellulose-based materials.

The preparation of these annotated bibliographies was made possible in part by a grant from the National Endowment for the Humanities.

I. THE MECHANISMS BY WHICH CELLULOSE TENDS TO BE DEGRADED IN STAGES

INTRODUCTION

The annotated bibliography that follows is not intended to be complete or exhaustive, but to serve as an introduction to the subject. Cited are the important research contributions that have led to the concept that cellulose tends to degrade under heterogeneous conditions in at least three reasonably distinct stages. The first, an initially rapid stage, is one in which an apparently limited number of 1C-O-4'C glucosidic bonds are present that are more sensitive to rupture than the normal bonds linking the anhydroglucose units. A second period may be considered in which the observed rate of deterioration represents the breakage of the "normal" 1C-O-4'C betaglucosidic bonds located in the amorphous regions of the cellulose-based fibers. A final principal stage is that in which a very much slower rate of scissioning is observed owing to the fact that the bonds being broken at this stage are those on the surface of crystalline or "crystallite" bundles, the remnants of the original crystalline fraction of the fibers.

The vast majority of the articles cited serves to elucidate the mechanism of acid-catalyzed hydrolytic degradation of cellulose. Nonetheless, photolytic, photochemical, and enzymatic degradations of cellulosic fibers also seem to exhibit a number of distinct stages under heterogeneous conditions of degradation. The explanation of why this occurs has not been as extensively put forth, although it may possibly be that the mechanism in all cases is essentially hydrolytic in character.

This brief narrative review is not intended to mention every reference in the accompanying bibliography but only to provide an overview of the subject matter in the abstracts, indicating the principal reasons for their inclusion in the bibliography. The development of salient ideas is presented roughly in chronological sequence but no effort is made to designate the authors who were precisely the first to undertake a particular aspect of the work.

RELATIONSHIP OF TENSILE STRENGTH TO THE FLUIDITY OF CELLULOSE SOLUTIONS

As the early concepts were being formed that came to regard cellulose and proteins as very large molecules, investigations were initiated that sought to demonstrate the relationship between the fluidity (the inverse of the viscosity) of cellulose solutions and the strength of the fibers from which the solutions were derived. Ridge of the Shirley Institute was one of the earliest to investigate this question (Clibbens and Ridge, 1925). The father of polymer science, Herman Staudinger, also investigated the fundamental relationship between tensile strength (TS) and the number of links broken in acid-degraded cellulose

(Staudinger and Sorkin, 1937). Shortly thereafter, Davidson studied the relation of TS to the fluidity values of cellulose which had been intentionally degraded by periodate (1940) and by chromic acid (1941) oxidation. In an extensive investigation that took advantage of nearly a decade of prior research, Conrad, Tripp, and Mares (1951) carried out studies on the number of links that needed to be broken during the thermally-induced oxidation of cotton and rayon automobile-tire cords before all strength was lost. They reached the conclusion that this amounted to considerably less than 1% of the glucosidic bonds.

LEVELING-OFF DEGREE OF POLYMERIZATION (LODP)

At the same time that investigations regarding the relationship of intrinsic viscosity and the tensile strength of cotton and rayon were going on, there was much interest devoted to the phenomenon of the leveling-off degree of polymerization (LODP). Very early it had been observed that the degree of polymerization (DP) of cellulose rapidly fell to a certain value during acid hydrolysis, beyond which further decrease took place at a considerably slower rate. The value of the DP at that point was designated as the LODP. Based on such observations, it was postulated that natural cellulose fibers must be comprised of amorphous and crystalline regions. During acid-catalyzed hydrolysis, the initial period of degradation is considered to represent an attack on the 1C-O-4'C glucosidic bonds in the amorphous phase. Thereafter, when all the glucosidic bonds in these regions have been hydrolytically severed, there remain primarily the crystalline zones (microcrystalline cellulose); the bonds within the latter structures are much more difficult to attack. The average length of the cellulose chains in these crystalline areas essentially represents the LODP (Nickerson and Harble, 1947; Nelson and Trip, 1953). If this explanation were true, then it was suggested that measurement of (a) the LODP, (b) the rate of weight loss, or (c) the ratio of the rate of breaking glucosidic bonds in the initial stage relative to that in the slower subsequent stage would provide a measure of the degree of crystallinity of the particular fiber (Philipp, Nelson, and Ziifle, 1947). By 1953, McBurney was able to give an authoritative overview of acid-induced hydrolysis of cellulose, essentially presenting our present-day views of the process.

THE KINETICS OF ACID-CATALYZED HYDROLYSIS

An analysis of the rate of hydrolysis of cellulose from the point of view of chemical kinetics—essentially the measurement or determination of the rate at which the glucosidic bonds are broken—began very early with the work of Freudenberg in 1921 to 1935 [see Freudenberg, F., and Blomqvist, G., *Berichte der Deutschen Chemischen Gesellschaft*, 68B (1935), p. 2070] and of af Ekenstam (1936). It was in 1954, however, that Sharples began his extensive studies of a particularly rapid initial stage of hydrolysis, using measurements of the change in the inverse of the degree of polymerization as an indicator of the rate at which bonds were being broken, rather than the less informative measurement of the weight of cellulose solubilized. Sharples introduced the concept of "accessibility," the factor by which the rate of hydrolysis of the "normal" 1C-O-4'C glucosidic bonds in a fiber or in regenerated cellulose needed to be multiplied in order to yield the measured rate under heterogeneous conditions. The hydrolysis of cellopentaose, a water-soluble polymer, was taken to represent the rate of hydrolysis of the "normal" bonds (Sharples, 1954).

So-Called Weak Links

Besides the somewhat slower-than-normal rate of hydrolysis under heterogeneous conditions (governed by the accessibility), Sharples focused considerable attention on a brief but particularly rapid initial stage which was considered to represent an attack on a limited number of so-called weak links. These rapidly broken 1C-O-4'C bonds were found to vary in concentration from about 3 per 10,000 glucosidic linkages to as high as 30 or 40 under special conditions in which cellulose had been regenerated (dissolved and then reprecipitated) before being subjected to hydrolysis. Under homogeneous conditions, the so-called weak links practically disappeared. In 1961, Michie, Sharples, and Walter summed up their views on the cause of the especially-sensitive links, concluding that they were the result of physical strains, perhaps at the point where the cellulose chains looped back upon themselves in the crystalline regions.

Stages

By 1961 the concept was well established that there were at least three stages of hydrolytic deterioration under heterogeneous conditions, representing distinctly different rates at which acid could attack the glucosidic bonds in the cellulose fibers owing to: the existence of a limited number of physically strained bonds (Sharples' "weak links"), the "normal" bonds situated in the amorphous regions of the fibers, and finally, the bonds in the crystalline zones, the often-called "microcrystalline" cellulose that remains after the traditional LODP has been reached. By this time, van Nederveen and van Royen (1955) had proposed distinct stages in the rate of thermally-induced discoloration. Paulson (1958) used the concept of chemical kinetics to demonstrate that during hypochlorite bleaching the rate of breaking of bonds in the cellulose chains was a relatively minor factor so long as lignin was present. After the lignin had thoroughly reacted, Paulson showed that the rate of breaking of the cellulose chains would then become a serious factor. This explanation was confirmed the following year by Herbst and Krässig (1959). Shortly before these applications of 1/DP values, apparently it had been necessary for Bryce and Greenwood (1957) to remind colleagues that, rather than simply plotting the change in DP or intrinsic viscosity with time, it was the plot of the change in the inverse of the number-average DP ($1/DP_n$), or of fluidity ($1/[\eta]$), with time that was of fundamental significance in evaluating the rate of deterioration of linear-chain molecules.

Inductive Effects

Interest in measuring accessibility (or degree of crystallinity) by measurement of the differences in the rate of solubilization of cellulose fibers—based upon the measurement of the weight of cellulose remaining after any given time—declined shortly after 1960 (Nelson). Marchessault and Rånby (1959) had begun to examine much more subtle influences on the rate of degradation, again using the determination of the rate of change of $1/DP_n$ (the inverse of the number-average degree of polymerization). They demonstrated that substitutions on the anhydroglucose ring, such as might occur as the result of oxidation of the alcohol groups on carbon atoms 2, 3, and 6 to carbonyl groups, influence the rate of hydrolysis of the nearby 1C-O-4'C glucosidic bonds owing to electronic "inductive" effects within the pyranose ring. These subtle differences in rate were demonstrated during homogeneous hydrolysis. Rånby (1961) clearly demonstrated that substituents on the glycosyl ring in the beta glucosides had the greatest such influence, whereas changes in the aglycon ring had far less effect.

Meller (1963) made some refined studies to demonstrate that cellulose deteriorated by acid-catalyzed kinetics, showing that the rate of hydrolysis in strong acids was proportional to the acidity (specifically to the Hammett acidity function, H_o), correcting his measurements by taking into account the fact that the end groups in a chain tended to hydrolyze more rapidly than groups within the chain. He based the latter assumption on the different rates of hydrolysis of cellobiose, cellotriose, and cellotetrose that Freudenberg had demonstrated more than thirty years before. An overview of the early work that led to ideas concerning the mechanism of catalysis in the hydrolysis of cellulose is found in Nelson's work (1960). She cites studies on model compounds, methyl and phenyl glucosides, carried out in 1955.

Stages in Photochemical Deterioration

An initially rapid stage began also to be reported in studies concerning the photolytic and photochemical degradation of cellulose under heterogeneous circumstances. Bayley was one of the first to consider the photolytic deterioration of cotton fabrics from this point of view (Bayley, et al., 1963). Daruwalla and co-workers soon pointed out that there were at least two major changes in the rate at which bonds in cotton cellulose were broken during enzymatic and photolytic, as well as hydrolytic, degradation under heterogeneous circumstances (Daruwalla and Narsian, 1966). Extensive studies of these stages during degradation in the carbon-arc Fadeometer and under sunlight were later published (Daruwalla, D'Silva, and Mehta, 1967). Desai (1969) also expressed the rate of degradation of cellulose under the very short wavelength ultraviolet radiation (254 nm) in terms of the change in $1/DP_n$. Fung (1969) has regarded degradation at very high temperatures (200°-280°C) in terms of stages.

PRESENT CONCEPTS OF DETERIORATION

The relationship of tensile strength and the number of bonds broken, expressed in terms of 1/DP, is now well appreciated. In 1973, Elema proposed a general scheme for the loss of strength of cotton textile fibers in terms of the change in 1/DP, making the key suggestion that all practical tensile strength would be lost when the LODP was reached. Thus, tensile strength was not considered simply as a function of $1/DP_n$ but was more closely related to $(1/DP_{500} - 1/DP_n)$, or, more generally $(1/DP_{LODP} - 1/DP_n)$. Krässig (1976) related physical properties to the basic morphological units within the fiber, expressing his views in terms of a "morphological unit," something very similar to the LODP and $1/DP_{LODP}$. Philipp and Stöhr (1977) used this view to follow rates of hydrolytic degradation.

Santucci (1963) had early appreciated the contributions of Sharples and of van Nederveen and van Royen, and had begun to express the loss of folding endurance, F, in terms of the equation, $\log F_o/F = a(1/DP - 1/DP_o)$. Marx-Figini (1981) discussed the possible use of many of these concepts in studying the properties of very old cellulosic textiles, as did Stoll and Fengel in that same year. A review of the use of 1/DP to express the kinetic laws of cellulose degradation has been prepared by Feller (1985), who also drew attention to the fact that sodium borohydride (SBH) reduction apparently reduced the initially-rapid stage of heterolytic deterioration; Rånby (1961) had earlier pointed out that SBH reduction would reduce the initially-rapid stage under homogeneous conditions of hydrolysis. We have thus come to the point where these concepts are beginning to be applied to conservation problems.

ABSTRACTS

1932

1. Ridge, B. P. and Bowden, H.
"VII - The tensile strength and fluidity in cuprammonium hydroxide solution of chemically modified rayon and cotton yarns. Part I - Oxidation by hypochlorite solutions."
Shirley Institute Memoirs, 11 (1932), p. 119-166. 26 tables, 13 references.
This is another of the key early publications which demonstrated that there was an essentially linear relationship between cuprammonium fluidity and the strength of cotton and viscose

yarns. Clibbens and Ridge (1928) had demonstrated the essential findings for cotton; this work stressed the results with rayon fibers. A lack of linearity appeared in the relationship of the initial changes in fluidity and a loss of strength up to about 87 to 90% retention of initial strength in the case of cotton yam. This was shown most likely to be due to the influence of short fibers on the strength of yarns, for the non-linearity was not present when continuous filament rayon yams were involved. Comments are also included on the influence of hot-alkali solubility and the action of boiling water on fluidity and loss of strength.

1936

2. af Ekenstam, A.
"Über das Verhalten der Cellulose in Mineralsäurelösungen, Il Mitteil.: Kinetisches Studium des Abbaus der Cellulose in Säure-lösungen."
Concerning the behavior of cellulose in mineral acid solutions, Part II: Kinetic investigation of the degradation of cellulose in acid solutions.
Berichte der Deutschen Chemischen Gesellschaft, 69, Part 1 (1936), p. 553-559. In German. 7 references.
The author may be the first to have proposed following the rate of acid-catalyzed hydrolysis of cellulose kinetically. Hydrolysis was carried out at 20.00 and 0. 12°C in phosphoric acid solution, as described in part I of a pair of articles (p. 549-552). Here, in part II, the kinetic equations are derived and shown to be similar to equations proposed earlier by others, based on empirical results. The equation introduced here is:

$$K = 1/t \ln \frac{1 - m/M}{1 - m/M_t} \qquad (1)$$

where m is the average molecular weight (mw) at the end of the degradation, M the average mw at the beginning, and M_t the average mw at time, t.

1937

3. Staudinger, H. G., and Sorkin, M.
"Über hochpolymere verbindungen, 162 Mitteil: Über Hydro-cellulosen."
On high polymeric compounds, Communication 162: On hydrocelluloses.
Berichte der Deutschen Chemischen Gesellschaft, 70, Part 2 (1937), p. 1565-1577. In German. 41 references.
A series of hydrocelluloses of different average DP, prepared by acid hydrolysis, was used to show that physical properties, such as percent elongation and breaking strength, were directly related to the number of links broken in the cellulose chains (Lückenzahl) having values of DP less than 1,650.

1940

4. Davidson, G. F.
"7 - The properties of the oxycelluloses formed in the early stages of the oxidation of cotton cellulose by periodic acid and metaperiodate."
The Journal of the Textile Institute, 31 (1940), p. T81-T96. 12 references.
The relationship of fluidity, tensile strength, and hot-alkali solubility is reported for oxycelluloses prepared by oxidation with periodic acid. The extreme alkali sensitivity can be seen in the relationship of the cellulose-nitrate fluidities of the modified celluloses and their fluidity as measured in cuprammonium solution, for the latter induces a considerable lowering of the DP owing to the sensitivity of the oxycelluloses to the alkaline copper-ammonia solution. A nearly linear relationship is observed between tensile strength and the fluidity of the nitrated cellulose samples and the cuprammonium fluidity after treatment with boiling dilute NaOH, particularly for the case of hydrocelluloses.

1941

5. Davidson, G. F.
"10 - The progressive oxidation of cotton cellulose by chromic acid over a wide range of oxygen consumption."

The Journal of the Textile Institute, 32 (1941), p. T132-TI48. 12 references.
The effects of oxidation by chromic acid were investigated in order to contrast them with the prior results of periodic acid oxidation (Davidson, 1940). The author notes that the x-ray diffraction pattern is not much affected at advanced stages of chromate oxidation. [Information obtained since these early studies indicates that chromic acid mainly attacks the OH group on carbon 6, apparently first largely converting it to an aldehyde, then slowly to an acid group.]

1943

6. Davidson, G. F.
"9 - The rate of change in the properties of cotton cellulose under the prolonged action of acids."
The Journal of the Textile Institute, 34 (1943), p. T87-T96. 12 references.
This important early paper considers the difference between homogeneous and heterogeneous acid hydrolysis of cellulose. The author's Figure 1 shows the changes in cuprammonium fluidity (0.5% solution) of cotton hydrolyzed by 6 to 10N hydrochloric or sulfuric acid at 20°C, pointing out that the rate of decrease in fluidity (inverse of the viscosity) is very rapid at first. If the author's fluidity data are plotted over the first 100 to 150 hours of hydrolysis, one can visualize that there may be two or three "stages" through which the rate of chain breaking passes before the tensile strength of the cotton is completely lost (an event that occurs at a fluidity of about 44). Curves of loss of solubility in boiling 1% sodium hydroxide solution indicate that the point of complete loss of tensile strength occurs when the solubility in hot alkali is about 35 to 40%. The author points out that the hot-alkali solubility of mildly modified celluloses is roughly equivalent to the copper number.
After considerable treatment to the acids, the rate of attack on the cellulose slowed down considerably. This occurred at a DP of about 190 for normal cotton; 95 for mercerized. The mercerized cotton was attacked by acid more rapidly in the initial stages than cotton cellulose. This was explained by differences in accessibility, the mercerized cotton being more amorphous.

1947

7. Nickerson, R. F., and Harble J. A.
"Cellulose intercrystalline structure; study by hydrolytic methods."
Industrial and Engineering Chemistry, 39, No. 11 (1947), p. 1507-1512. 10 references.
In the hydrolysis of cellulosic fibers with aqueous hydrochloric and sulfuric acids at boiling temperatures, the disordered intercrystalline chain network appears to be attacked first. This inference is drawn from sharp reductions in cuprammonium viscosity and moisture regain coincident with the rapid accumulation of glucose. Simultaneously the fibers are reduced to powdery hydrocellulose. With further hydrolysis glucose continues to form and moisture regain slowly increases, but viscosity remains practically constant. Attack in this stage may be chiefly on lateral crystallite surfaces. Lengths of crystallites calculated from viscosity data range from 280 glucose units for cotton to 110 for high tenacity viscose rayon. Sulfuric acid is much less active than hydrochloric in hydrolyzing cellulose and, after neutralization, does not interfere with the direct volumetric determination of glucose.

8. Philipp, H. J., Nelson, M. L., and Ziifle, H. M.
"Crystallinity of cellulose fibers as determined by acid hydrolysis."
Textile Research Journal, 17, No. 11 (1947), p. 585-596. 22 references.
A study of the kinetics of the heterogeneous hydrolysis of cellulose indicates that there are two distinct rates which have been attributed to the rapid hydrolysis of the loose amorphous regions and to the slow hydrolysis of the dense crystalline portions of the fiber. Based on this interpretation, a simple method for the determination of the relative proportions of crystalline and amorphous cellulose has been developed. The method was applied to a series of natural and regenerated cellulose fibers and the resulting "degree of crystallinity," defined as the amount of crystalline cellulose in a sample expressed as a decimal fraction of the total cellulose, was tabulated. Results obtained on four samples of rayon are very well correlated with crystallinity data derived from x-ray diffraction patterns.
The authors express the rate of heterogeneous hydrolysis of cotton and rayon according to the following equations:

$$r = c + a = c_o e^{-k_c t} + a_o e^{-k_a t} \qquad (2)$$

in which, when a becomes negligible, then

$$\ln r = \ln c_o - k_c t \qquad (3)$$

In these equations r is the total unhydrolyzed cellulose at time t; c and a are the concentrations of unhydrolyzed crystalline and amorphous cellulose at that time and c_o and a_o, are the initial concentrations of these two components; k_c and k_a are the respective specific rate constants. If the linear portion of a plot according to equation (3) is extrapolated back to time zero (t_o), the intercept should be approximately equal to the logarithm of the initial percentage of crystalline cellulose in the unhydrolyzed starting material. The percentage of amorphous material initially present can be estimated by subtracting the percentage of crystalline material thus obtained from 100.

On this basis the rate constants for the hydrolysis of the cellulose in the amorphous areas of cotton and Fortisan rayon were found to be about two times faster than that for the rate of hydrolysis of the crystalline cellulose under the conditions of 100°C and 2.5 to 6N hydrochloric acid.

Considerable differences were found in the rates of hydrolysis of the crystalline fractions of materials such as rayon versus cotton linters but no explanation for this was given by the authors.

1948

9. Nelson, M. L., and Conrad, C. M.

"Effect of grinding on the crystallinity of cellulose fibers, as indicated by the acid-hydrolysis and other techniques."

Textile Research Journal, 18 (1948), p. 155-164. 16 references.

Using samples of purified linters and cut cotton, ground under conditions of increasing severity, the authors demonstrate that even mild grinding causes a detectable reduction in indicated degree of crystallinity, and that as the severity of grinding increases the degree of crystallinity drops correspondingly. When the ground samples were moistened with water and allowed to dry, a portion of the lost crystallinity was regained.

The acid-hydrolysis results are supported by accessibility, heat of wetting, and moisture adsorption measurements, which, with few exceptions, indicate increasingly larger amounts of amorphous material in the samples receiving the more severe grinding. The progressively greater rates of hydrolysis with increased grinding suggest a decreasing crystalline particle size.

Since the x-ray diffraction patterns did not reveal any effect from the most severe grinding, it is assumed that the changes were not sufficient to permit their detection by this means.

Some implications of the results on the interpretation of cellulose structure are discussed.

The results of the present study are interpreted in terms of a continuous network of more or less amorphous cellulose chains connecting crystalline cellulose areas. The observation that, only 15 minutes after the beginning of hydrolysis in 2.5N hydrochloric acid, cotton linters attain almost their maximum fluidity, which scarcely changes for the following 6 3/4 hours, whereas, in 15 minutes, only 2 percent of the residue has been hydrolyzed away, is taken as evidence that the crystalline particles are liberated very quickly, through breaking and partial hydrolysis of the amorphous network, and that hydrolysis thereafter takes place principally by surface attack on the crystalline zones and largely from the lateral surfaces. The estimated degree of polymerization of about 195 glucose units after this short period suggests that this is probably the order of magnitude of the crystallite length. It is postulated that, by grinding, forces are communicated through the amorphous chains to the ends of the crystalline areas, causing splitting and the development of new loose amorphous chain structures.

1951

10. Algar, W. H., Giertz, H. W., and Gustafsson, A.-M.

"Fiber structure and pulp properties. 1. A working hypothesis."

Svensk Papperstidning, 54, No. 10 (1951), p. 335-344. In Swedish [translation available from the Institute of Paper Chemistry]. 13 references.

At the time of this publication. the theory of the physical structure of cellulose into crystalline and non-crystalline sugars had not been completely formulated. The authors used Nickerson's [see Nickerson, R. F., and Harble, J. A., *Industrial and Engineering Chemistry*, 39, p. 1507-1512 (1947)] ideas of easily accessible and difficultly accessible cellulose—based on the loss of weight upon acid hydrolysis—to relate the easily accessible component of pulps to gamma cellulose and to moisture regain.

11. Conrad, C. M., Tripp, V. W., and Mares, T.
"Thermal degradation in tire cords. Part 1: Effects on strength, elongation, and degree of polymerization."
Textile Research Journal, 21 (1951), p. 726-739. 19 references.
Changes in tensile strength and elongation-at-break for cotton and rayon tire cords were measured during thermal degradation at temperatures between 110 to 150°C and were related to the number of links broken in the cellulose chain. The latter was calculated from the equation $L_B = 100N = 1/\overline{DP}_t, - 1/\overline{DP}_o$. Relative humidity was kept high by saturating the air with water vapor at 77°C before passing the air through the aging oven. A curve showing the loss of strength versus the number of links broken demonstrated that 100% loss of strength occurs when 0.3% of the links in cotton cellulose are broken and 0.6% in rayon. These figures apparently are based on viscosity average DP's and would be about twice as high if based on number average values. The results indicated that all strength is lost at a viscosity-average degree of polymerization (DP_v) of 300 in the case of cotton, 125 in the case of rayon (DP_n ca. 150 and 62.5, respectively). Curves are also given that compare the percentage retained strength to the fluidity for both hydrolytically degraded and thermally degraded cotton and rayon. For the series of hydrocelluloses, the relationship between fluidity (related to $1/\overline{DP}_t$) and retained strength is linear.

1953

12. Nelson, M. L., and Tripp, V. W.
"Determination of the leveling-off degree of polymerization of cotton and rayon."
Journal of Polymer Science, 10, No. 6 (1953), p. 577-586. 21 references.
The effects of variations in hydrolytic conditions upon change in estimated DP of cotton linters and viscose rayon have been explored by varying the acid concentration downward over the range of 2.5N to 0.01N and by carrying out the hydrolysis at 100° and at 80°C. Much as expected, it was observed that temperature and acid concentration do not affect appreciably the value obtained for the leveling-off DP. They do affect the rate of hydrolysis.
An alternative way of determining the leveling-off DP has been devised. The leveling-off DP is that value on the curve of DP versus residue percentage, under conditions of continuous hydrolysis, at which the slope of the tangent to the curve first attains a minimum absolute value.
The moisture regain data on the hydrocelluloses indicate that cotton linters recrystallize to a slight extent while rayon does so in significantly greater amount. This crystallization may take place either in the aqueous hydrolytic medium or during the drying of the hydrocellulose, the relative amount occurring at either stage being unknown.

1954

13. McBurney, L. F. C.
"Chemical nature of cellulose and its derivatives: subsection C1. Kinetics of degradation reactions, (b) hydrolysis." (Chapter III)
In book. *High Polymers*, Volume 5, *Cellulose and Cellulose Derivatives, Part I*, 2nd Edn., ed. Ott, E., Spurlin, H. M., and Grafflin, M. W., Interscience Publishers, New York (1954), p. 101-124. 43 refs.
Acid-catalyzed homogeneous hydrolysis of cellulose is reviewed. The convincing evidence for the lack of "weak links" being revealed in cellulose during homogeneous acid degradation is critically discussed and a straightforward derivation is given of the first-order kinetic equation used to calculate the number of bonds broken per unit of time. (The change in fluidity is shown to be proportional to the time of exposure to acids.) [The edition is too early to be able to present the work of Sharples and others on heterogeneous hydrolysis.]

14. Sharples, A.
"The hydrolysis of cellulose. Part I. The fine structure of Egyptian cotton."
Journal of Polymer Science, 13 (1954), p. 393-401. 15 references.
In this opening paper of an important series of publications on the heterogeneous acid hydrolysis of cellulose, Sharples separated the rate of hydrolysis into three stages, attributing the first very rapid stage to abnormally sensitive links in the chain, observable in these experiments in cellulose of an average DP of greater than 1,870. The second stage, during which the number of bonds broken in a period of time clearly proceeded at a steady pace, the author attributed to the hydrolytic rate of attack in the accessible regions of the fibers. After a number average DP of

about 900, the rate of hydrolysis tended to slow down, a stage during which the author concludes that recrystallization of the cellulose chains is occurring, reducing accessibility.

Expressing the rate of breaking bonds during acid hydrolysis as

$$\frac{1}{\overline{DP}_{\varepsilon}} - \frac{1}{\overline{DP}_{o}} = \alpha kt \qquad (4)$$

the author describes the determination of k—the rate of hydrolysis of completely accessible bonds in the model soluble cellulose, cellopentaose—and of α, which he calls the accessibility; t is the time of hydrolysis, carried out in 0.1N H_2SO_4 at 50°C.

15. Sharples, A.
"The hydrolysis of cellulose. Part II. Acid sensitive linkages in Egyptian cotton."
Journal of Polymer Science, 14 (1954), p. 95-103. 15 references.

In this second article (see also Part I, Sharples, 1954), the author pursues his evidence for extranormal ("weak") links in cellulose as revealed by the rate of heterogeneous acid hydrolysis. Plots were made of equations of the type $1/(\overline{DP}_n)_t - 1/(\overline{DP}_n)_o = k\alpha t + zJ_o$ where k was the rate of change, t the hydrolysis time and α the "accessibility." The intercept, I, can be used to estimate the fraction of extranormal (fast hydrolyzing) links, J_o, from the relationship $I - (1/\overline{DP}_n)_o = zJ_o$ where the factor z is a function of accessibility. Sharples concluded that the number of weak links in cellulose regenerated after solution in oxygen-free alkali was no more than about 1 per 2900 anhydroglucose units (AGU) or about 0.03 per 100 AGU. The rates of hydrolysis in the first ("weak link") stage and the second stage were measured, the latter generally occurring between DP_n of 500 and 150 as the regenerated cellulose degraded. The ratio of the rate constants, k′/k, in the two stages was estimated to be greater than 10,000. The author points out that the estimate of heterogeneity, $U = \overline{DP}_w / \overline{DP}_n - 1$, of Schulz and Husemann [*Zeitschrift für Physikalische Chemie*, B-52, No. 1 (1942), p. 23-49, 50-60] has sufficient experimental error that their method of proving the existence of weak links by the changes in heterogeneity was not reliable.

1955

16. van Nederveen, G., and van Royen, A. H. H.
"Quelques recherches sur la durabilité des pâtes cellulosiques." Investigations concerning the durability of cellulosic pulps.
Bulletin Association Technique de l'Industrie Papetiere, No. 4-5 (1955), p. 160-166. In French. 4 references.

These authors undertook studies to compare the number of bonds broken and the content of hemicelluloses to discoloration and the change in folding endurance of paper made from various types of pulp during thermal aging at 100°C and 65% RH. The loss of folding endurance was found to be related to the number of bonds broken. The number of bonds broken during a given time, however, did not seem to be related to the content of hemicellulose. The loss of folding endurance and the yellowing of pulps were shown to be related to hemicellulose content although not entirely. The loss of fold was estimated in relation to the function $1 - \log V_t/V_o$, where V_o and V_t were, respectively, the folding endurance before and after aging.

Yellowing is considered to occur in two stages, an initially very rapid stage and a later stage that proceeds at a slower, linear rate. The first stage is considered to involve the molecules most apt to react and is controlled by the diffusion of oxygen into the fibers; the second stage, the authors suggest, is that in which molecular entities with sufficient energy to react with oxygen are formed. Discoloration was measured by Δb, from the colorimetric $L^*a^*b^*$ system; and the two stages were separately analyzed by extrapolating the so-called linear (later) stage of the change in Δb to time zero, making the assumption that this value, Δb_m, represented the limit to which the first, very rapid change was directed. The authors made a rough calculation of the diffusion-controlled rate in the first process and showed that the rate was not influenced in a major way by temperature, which they considered to be a good indication of a process controlled by diffusion. The second stage was sufficiently responsive to temperature that a typical Arrhenius plot was obtained.

1957

17. Bryce, W. A. J., and Greenwood, C. T.
"The degradation of high polymers."
Journal of Polymer Science, 25, No. 111 (1957), p. 480-483. 8 references.
The authors, citing the work of af Ekenstam (1936) and Schulz and Husemann [*Zeitschrift für Naturforschung*, 1 (1946), p. 268-280] point out the value in plotting the degradation of polymers as a function of 1/DP versus time. Equations for both zero order and first order chain breaking are derived.
The authors replotted some of Schulz and Husemann's data for the oxidative degradation of cellulose, and showed that at high concentrations there was apparently evidence for the existence of weak links. Similar data for the homogeneous hydrolysis of methylated celluloses gave no evidence of weak links.

1958

18. Paulson, J. C.
"Aspects of hypochlorite bleaching."
Tappi, 41, No. 11 (1958), p. 655-657. 4 references.
Hypochlorite is considered to be a rather strong, non-specific oxidant capable of reacting with both the lignin and the carbohydrate fractions of pulp. At first, the lignin loses methoxyl groups and takes up chlorine; later the lignin is oxidized sufficiently that it no longer shows the typical 280 nm ultraviolet absorption peak. In the present study, the author measured both the rate of consumption of hypochlorite and the change in pulp fluidity (the inverse of the intrinsic viscosity) as a function of the hypochlorite consumed. The results clearly show an initial period of rapid consumption of hypochlorite during which the fluidity of the cellulose fraction did not change; the conclusion was reached that the bleach first attacked the lignin without causing a lowering of the DP of the cellulose. For sulfate pulps that contained about 2 to 4% lignin, the attack on the cellulose in the later stages of the process took place in two stages, the first, relatively slow stage, probably due to the slowing action exerted by the presence of lignin. The extent of attack on the cellulose component per unit of hypochlorite consumed increased with increasing temperature in the range 30 to 60°C.
The period during which hypochlorite is consumed at a slower rate is accompanied by a measurable attack on the carbohydrate fraction as indicated by a progressive increase in pulp fluidity. For a series of pulps of different lignin content, at constant temperature and initial hypochlorite level, the change in fluidity as a function of hypochlorite consumed is similar over the same fluidity range throughout the slow part of the reaction. The consumption curves differ only in the extent to which they are displaced owing to the initial rapid consumption by the attack on the lignin.
The significance of this work in the present context is, first of all, the clear demonstration that, during bleaching with alkaline hypochlorite, the deterioration of the cellulose component takes place at two distinct rates, the initial period being very slow, because it is the lignin that rapidly consumes the chlorine at this stage. Secondly, the author finds that a plot of fluidity versus hypochlorite consumed is linear, clearly illustrating the fact that the number of bonds broken is directly related to the amount of bleach consumed.

1959

19. Herbst, J. H. E., and Krässig, H.
"The distribution of oxidant consumption in bleaching."
Tappi, 42, No. 8 (1959), p. 660-664. 8 references.
Following the example of Paulson (1958), the authors concentrated on the intercept and the slope of the second stage of the bleaching process in which the number of cellulose bonds broken is compared to the amount of bleach consumed. As in Paulson's research, the intercept of the two-stage curve of the number of bonds broken versus bleach consumed was shown to be a function of the lignin in the pulp; thus, the amount of bleach consumed without bond breaking increases with increasing amounts of residual lignin in the pulp. The slope of the later, linear stage of the curve of cellulose bonds broken versus bleach consumed decreases with increased lignin or hemicellulose content. The brightness of the pulp usually increases during the linear part of the slope, further suggesting that lignin is present during this stage.

20. Marchessault, R. H., and Rånby, B. G.
"Weak links as inductive effects in the hydrolysis of cellulose."
Proceedings of the 2nd Cellulose Conference, Syracuse (1959), p. 34-52. Table, 26 references.

A number of workers had, prior to this date, detected that wood celluloses hydrolyzed in homogeneous solutions in phosphoric acid at 25-30°C at a more rapid rate than cotton cellulose did. A table is given that shows the increased rate range from 1.2 to 2.3 times greater. The increased rate can be explained both on the basis of the greater amorphous character of wood hemicelluloses and branched cellulose structures and on the basis of "inductive" effects owing to the substituent groups in the anhydropyranose rings. Examples of the effect of substituent groups in simple glucosides and disaccharides are given. The fact is pointed out that xylans from hardwoods, containing one glucuronic acid side group in 10-15 xylose units, are known to be more resistant to homogeneous hydrolysis by acid than xylans from softwoods in which there is one such side group in about five xylan units. Also cited is the evidence that oxycelluloses formed by oxidation by nitrogen dioxide and periodate-chlorous acid are more susceptible to heterogeneous acid hydrolysis.

A plot of $1/DP_t - 1/DP_o$ versus time of hydrolysis should be curved if a certain fraction of the bonds is more easily hydrolyzed than others. The authors show the slightly curved effect that should be observable if 2% of the bonds were 25 and 40 times more susceptible to hydrolysis than the normal beta-glucosidic bonds.

21. Marchessault, R. H., and Rånby, B. G.
"Hydrolysis of cellulose in phosphoric acid solution - inductive effects."
Svensk Papperstidning, 62, No. 7 (1959), p. 230-240. 37 references.

The authors amplify their previously published article (Syracuse Conference) on the possibility of inductive effects (chemical modification of the basic anhydroglucose unit in the cellulose chain) in causing modified cellulose to hydrolyze more rapidly in acids. The homogeneous acid hydrolysis of two wood celluloses, when compared to the hydrolysis of highly purified cellulose, is shown to be 1.9 ± 0.3 times faster and to possess a higher energy of activation. Moreover, when one of the wood celluloses is reduced with sodium borohydride, there was a marked decrease in the rate of hydrolysis (to 1.5 times that of the original sample). Evidence is also supplied from the work of others to support the thesis that chemical modification of a glucose unit in the chain either speeds up or retards the rate of homogeneous hydrolysis.

1960

22. Nelson, M. L.
"Apparent activation energy of hydrolysis of some cellulosic materials."
Journal of Polymer Science, 43 (1960), p. 351-371. 20 references.

This paper from the Southern Regional Laboratory of the U.S. Department of Agriculture presents data and conclusions of fundamental interest to the subject of the heterogeneous acid hydrolysis of cellulose, including explanations of the rate of weight loss during attack in the amorphous and crystalline regions, the degree of order in the crystallite region, the possible mechanism, the relationship of the rate to the acidity, and the energy and entropy of activation.

Measuring the rate of hydrolysis by plotting the logarithm of the weight of insoluble material remaining after any given period of heterolytic acid catalyzed hydrolysis, against the time, Mary Nelson demonstrated that degradation takes place in two principal stages: the first, rapid stage represents the attack on the amorphous regions of the fibers; the subsequent slower stage represents the attack on the crystallites. Extrapolation of the straight line first-order plots back to zero time reveals the apparent proportion of the amorphous and crystalline fractions in the pulp.

The overall rates of acid attack increased as the percentage of amorphous material increased. The higher rates of attack in the crystallite fractions in viscose rayon and recrystallized cellulose were attributed to the shorter length of the chains in the crystallites (or, in other terms, the "leveling-off degree of polymerization, LODP); the proposed explanation is that attack takes place at the ends of the crystallites. In rayons and recrystallized cotton, the division of the plots into two clearly defined first-order stages was difficult, which the author explains by the fact that there are widely different degrees of order in the crystallite regions of such materials and, hence, a series of declining rates rather than simply two clearly differing rates of attack. Data on the distribution of order within a given weight of fiber are provided to substantiate this hypothesis.

Evidence is put forth that does not preclude the concept that the rate-determining step is bimolecular, involving a step such as:

$$glucose \cdot H^+ + H_2O = products \qquad (5)$$

At one point in the paper the author concludes that the rate equation $2/(DP_w)_t - 2/(DP_w)_o = k\alpha t$ is only valid for the very early stage of heterogeneous hydrolysis. A considerable discussion of the energy of activation leads to the conclusion that it is dependent on the temperature. The energy of activation based on the breaking of bonds (1/DP) is lower than that based on the loss of weight and probably is more nearly equivalent to the activation energy for homogeneous hydrolysis.

1961

23. Krässig, H., and Kitchen, W.
"Factors influencing tensile properties of cellulose fibers."
Journal of Polymer Science, 51 (1961), p. 123-172. 38 references, bibliography.

The authors present a major exposition of the factors that govern the tensile strength of cellulose fibers, citing that it was Staudinger, Sorkin, and Franz [*Melliand Textilberichte*, 18 (1937), p. 681] who first demonstrated a linear relationship between 1/DP and tensile properties. Electron microscopical studies have shown the morphological units of native cellulose fibers to be about 300 to 400 glucose units in length. A model has been developed that suggests the tensile properties be directly related to (a) the number of interlinks between these morphological units, (b) their alignment with respect to the fiber axis (degree of orientation), and (c) the perfection of their spatial arrangement (degree of crystallinity). This is expressed as follows:

$$\text{Tensile Strength} = f\,(1/DP_L - 1/DP);\ f\,(\text{orientation});\ f\,(\text{perfection of order}) \qquad (6)$$

where f stands for "function of" and DP_L is the length of the morphological units in the fiber, expressed in glucose units, and DP is the number-average degree of polymerization determined on the particular fiber sample. Objective measures of each of these functions are proposed.

Thirty-seven figures—in which various measurements of crystallinity, orientation, and DP were plotted against such physical properties as tensile strength and stretch—confirm the general validity of the proposed model.

Data are also presented to show that the crystallinity increases as samples are hydrolyzed, more so in the case of rayon-type fibers than for cotton. In fact, the increase in the crystallinity (recrystallization) in the course of breaking 0.1 to 0.2% of the bonds initially present occurred in the inverse order in relation to the initial degree of crystallinity. Orientation was also shown to increase during the hydrolysis of rayons.

The initial rate of acid hydrolytic breakdown was shown to be many times higher in the more disordered rayon types than in the case of cotton, and the decline in rate of hydrolysis with time was more pronounced in cotton than in the case of the rayons.

Plots of the rate of loss of tensile strength (TS) versus 1/DP indicated that the rate of loss, $\Delta TS/\Delta 1000/DP$, was a function of the lateral order of the crystallites in the fibers, falling off rapidly in the case of cotton and least in the case of textile and tire rayon. Cotton which had degraded below DP 300 lost practically all its tensile strength. The extrapolated data of TS versus 1/DP for cotton predicted negligible strength at a DP_L of 515 (authors' Figure 25).

24. Michie, R. I. C., Sharples, A., and Walter, A. A.
"The nature of acid-sensitive linkages in cellulose."
Journal of Polymer Science, 51 (1961), p. 85-98. 24 references.

This work, extending the studies of Sharples, made in 1954 and 1957, reports the calculated number of "weak links" revealed by the heterogeneous acid hydrolysis of cellulose samples prepared and regenerated in many different ways. These calculated values vary between 0.03 and 0.13 per 100 anhydroglucose units in the chains of cellulose samples that varied in initial DP from 900 to 7,800. Through a systematic series of experiments, illustrated in eight figures, the authors conclude that the "weak links" represent an enhanced rate of hydrolysis caused, not by a chemical change in the cellulose, but by a physical or physicochemical process associated with the precipitation of the cellulose from solution, perhaps strains on the glucosidic bonds. The rate of attack on these bonds is approximately 10,000x that of the normal bonds in cellulose.

The curves of the change in $1/DP_n$ or $2/DP_w$ versus time of hydrolysis rise rapidly, then settle down to an apparently linear slope or rate. The extrapolation of the straight portion of the curve to the y axis yields the apparent number of weak links (Intercept - $2/DP_w$). The ratio of the slope of the linear portion of the curve to that for the rate of hydrolysis of normal 1,4-beta-glucosidic bonds, as represented by cellopentaose, is reported as the "accessibility" of the particular cellulose.

25. Rånby, B. G.
"Weak links in polysaccharide chains as related to modified groups."
Journal of Polymer Science, 53 (1961), p. 131-140. 30 references.

The author points out that in the past much of the experimental evidence for "weak links" in the structure of cellulose chains has been based on the heterogeneous (two-phased system) acid hydrolysis of cellulose. Thus, the fast cleavage of certain bonds may simply mean that they are situated in more readily accessible locations in the fiber structure. In the overall mechanism of hydrolysis, Bunton, C. A., et al. [*Journal of Chemical Society (London)*, Part 4 (1955), p. 4419-4423] showed that the glucosidic acetal oxygen (oxygen on the number 4 carbon) is in the nonreducing end group. The author sets out in this research by citing a number of studies on the effect of substituents in the glycosyl (G) and the aglycon (A) portions of the structure on the rate of hydrolysis. It is well known, for example,

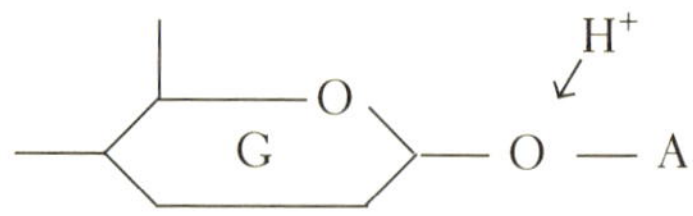

that the more electronegative the aglycon (A), the more rapidly both alpha and beta D-glycopyranosides hydrolyze [Heidt, L. J., and Purves, C. B., *Journal of the American Chemical Society*, 60 (1938), p. 1206; *ibid.* (1944), p. 1385]. Fructosides (furanosides) hydrolyze much faster, by a factor of 10^4 to 10^5, perhaps by a different reaction path.

To confirm this behavior in the case of cellulose itself, cellulose from cotton and from a soft and a hardwood was hydrolyzed in a homogeneous solution in 81.25% phosphoric acid at 20, 30, and 40°C. The wood samples hydrolyzed at rates about 1.5 to 2.0 times faster than cotton cellulose and were shown to have a higher content of acidic and carbonyl groups than the cellulose from cotton linters. However, when one of the wood celluloses was reduced with sodium borohydride (SBH), its rate of hydrolysis was reduced nearly to that of the cotton cellulose. On the other hand, when the carbonyl groups were oxidized to acid groups (COOH) with sodium chlorite, the rate of hydrolysis was practically unchanged. The percentages of carbonyl and carboxyl groups were analyzed in the various samples used in the investigation.

The author concludes "the initial rate of hydrolysis is a function of the amount of modified groups while the final rate is practically the same for all samples. . . . When the main fraction of the fast-cleaving bonds are [*sic*] broken in the initial reaction, the rate of the subsequent reaction is expected to gradually approach that for the normal bonds."

1963

26. Bayley, C. H., Rose, G. R. F., Clifford. J. B., and Cooney, J. D.
"The effect of noncellulosic material on the actinic degradation of cotton fabric."
Textile Research Journal, 33, No. 10 (1963), p. 849-854. 24 references.

The authors demonstrate that small amounts of copper compounds (about 0. 1%) generally enhance the degradation of cotton duck fabric exposed in a Weatherometer. In one illustration they obtained a linear plot of fluidity versus time of exposure up to 20 Rhes for bleached nainsook. A similar curve for bleached and unbleached cotton duck seemed to show an initial fast stage (linear) followed by a slower rate of degradation. Copper oxinate was the principal compound in the series tested that retarded photochemical breakdown.

27. Meller, A.
"Some considerations on the kinetics of the acid hydrolysis of poly- and oligo-saccharides."
Journal of Polymer Science, Part C, No. 2 (1963), p. 97-107. 26 references.

Cellobiose and cellotriose hydrolyze at slightly different rates because the bonds between the anhydroglucose units are slightly affected by differences in the neighboring groups.

The rate data of Freudenberg and Blomqvist [*Berichte der Deutschen Chemischen Gesellschaft*, 68 (1935), p. 2070-2082] on the acid-catalyzed hydrolysis of cellotriose have been used to test the validity of the rate equation of Kuhn by plotting $\ln [(1 - \alpha)/e^{-k_2 t} - k_1/(k_2 - 2k_1)]$ against t (where α is the degree of hydrolysis, t is the time of hydrolysis, and k_1 and k_2 are the hydrolysis rate constants for cellotriose and cellobiose, respectively). A departure from linearity of the plots was found. It is possible that the hydrolysis is accompanied by other reactions (destruction of the monomer formed, condensation of the monomer to a dimer and of the dimer to a tetramer), or that under the experimental conditions (i.e., strong sulfuric acid) the measured rate is not entirely that of the hydrolysis of cellotriose. In support of the latter hypothesis, rate constants for the hydrolysis of cellobiose with strong sulfuric acid solutions at low temperatures and with dilute hydrochloric and sulfuric acid solutions at high temperatures have been collected from the literature. On plotting log k against 1/T (where k is the hydrolysis rate constant of cellobiose and T is the absolute temperature) for 1N HCl at high temperatures extrapolation to 40°C gives a value for log k very close to that obtained by extrapolation of the plot of log k against $-H_0$ values (Hammett's acidity functions) at 40°C.

Weak sulfuric acid solutions at high temperatures seem to be similar to hydrochloric acid in their action on cellobiose.

In conclusion, it is believed that a clarification of the problems involved in the hydrolysis of cellobiose with strong cold sulfuric acid solutions may help to formulate the theoretical basis for the interpretation of the kinetics of hydrolysis of cellotriose, cellotetraose and, most likely, of cellulose with these hydrolyzing solutions.

1966

28. Daruwalla, E. H., and Narsian, M. G.
"Detection and identification of acid-sensitive linkages in cellulose fiber substances."
Tappi, 49, No. 3 (1966), p. 106-111. 26 references.

The course of heterogeneous hydrolysis of cotton, cotton swollen with different swelling agents, and regenerated cellulose substances under mild conditions of acid action can be best represented by two rates of reaction. Both obey first-order kinetics, but the initial apparent rate of acid action is 1,600-2,300 times higher than that for the subsequent stage of hydrolysis. The activation energy required for the initial stage of hydrolysis is 20-22 kcal/mole as compared with 27-28 kcal/mole required for the later stages of acid action. On the other hand, homogeneous hydrolysis in 85% phosphoric acid can be best represented by a single rate of reaction obeying first-order kinetics, and requiring an activation energy of 27-28 kcal/mole. The course of enzymatic hydrolysis and photochemical degradation of a cellulose substance has been compared with that of heterogeneous acid hydrolysis and it is shown that viscosity-average DP values corresponding to the points of transition from the initial high rate to the subsequent lower one are very close to one another, ranging from about 400 to 470. Several possibilities that are likely to give rise to two different rates of acid hydrolysis have been considered, and experimental evidence has been put forth to discard most of these possibilities. The particularly acid-sensitive linkages, numbering about 0.05 to 0.7 per 100 anhydroglucose units in cotton and alkali-swollen cotton, increase in number to as many as 27 to 33 per 100 anhydroglucose units in cellulose regenerated from the dissolved condition.

29. Santucci, L.
"Resistenza e stabilità della carta. VI. Degradazione per trattamento con ossidanti. Effetto della laminazione con acetilcellulosa."

Resistance and stability of paper. VI. Degradation upon treatment with oxidants. Effect of lamination with cellulose acetate.

Bollettino dell'Istituto di Patologia del Libro, 25 (1966), p. 51-72. In Italian. 10 references.

The investigation on bleaching agents has further emphasized the importance of washing with water paper containing harmful impurities; when washing followed 20-hour immersion in 0.5% available chlorine Chloramine-T, paper was practically unharmed, whereas dispensing with washing resulted in serious damage, even much worse damage than was brought about by a 10-mmute soaking in 1.5%-available-chlorine sodium hypochlorite, though the latter is a much stronger oxidant than chloramine. But, apart from the shorter exposure, hypochlorite-treated samples were all subsequently washed in running water, and thus damage was slight; however, the greater danger for weaker papers should be realized.

The second investigation has proven the good stability of paper laminated with one brand of cellulose acetate, which, however, decreases the folding endurance of sheets. A possible way of removing such inconvenience is being studied.

During this investigation, the author used the relationship $\log F_0/F = a(1/DP_t - 1/DP_0)$ as a measure of the relationship of folding endurance (F) versus the degree of degradation. He reported the use of this relationship at the Moscow-Leningrad meeting of the ICOM Committee on Laboratories (Paper 63/24) and earlier in the *Bollettino dell'Istituto di Patologia del Libro*, 20 (1961), p. 158 and *ibid.*, 22 (1963), p. 67.

30. Vink, H.
"Degradation of cellulose and cellulose derivatives by acid hydrolysis."
Die Makromolekulare Chemie, 94 (1966), p. 1- 14. 20 references.

The acid hydrolysis of cellulose and the cellulose derivatives hydroxyethyl cellulose, methyl cellulose and carboxymethyl cellulose has been investigated. The degradation was followed viscometrically and by end group analysis. It was found that the viscosity-molecular weight relationship deviated from the modified Staudinger equation, $[\eta] = Km_n^a$, the exponent a is increasing with decreasing molecular weight. A slight drift in the rate was shown to be caused by the change in the value of the exponent a in the Staudinger equation. The influence of acid concentration, expressed as the Hammett acidity function, and of temperature on the degradation rate was studied. It was found that the different cellulose derivatives exhibit quite similar degradation characteristics. An early rapid stage is detected, a stage usually attributed to the presence of "weak links." Determination of the molecular weight by end group analysis enabled the author to demonstrate that the rate of degradation remained constant over the course of degradation. A straightforward review of the kinetic calculations accompanies the experimental data.

1967

31. BeMiller, J. N.
"Acid-catalyzed hydrolysis of glycosides."
In *Advances in Carbohydrate Chemistry*, Volume 22, ed. Wolfrom, M. L., Academic Press, New York (1967), p. 25-108. 293 references.

An extensive review discusses the reaction mechanism of acid-catalyzed hydrolysis of glucosides in dilute solution at temperatures of 100°C or below. The review covers the mechanism of hydrolysis of pyranosides and furanosides, kinetics of hydrolysis, and steric and electric charge effects of the glycose residue and of the aglycon. Much more briefly discussed is the hydrolysis of polysaccharides.

32. Daruwalla, E. H., D'Silva, A. P., and Mehta, A. C.
"Photochemistry of cotton and chemically modified cotton. Part 1. Behavior during exposure to carbon-arc and solar radiations."
Textile Research Journal, 37, No. 3 (1967), p. 147-160. 36 references.

Both chemical and physical changes were monitored during the exposure of cotton, mercerized cotton, acetylated cellulose, and cyano-ethylated cellulose to the carbon arc and to sunlight. Eight tables present detailed changes in fluidity, peroxide content, alpha-cellulose and carbonyl, carboxyl, and uronic acid content. The degradation in degree of polymerization passed through a minimum at about 5 to 7.5% moisture regain, at the point where the cellulose was completely saturated with chemically-bound water. Clear evidence was presented to show that the principal photochemical action of near-ultraviolet and visible radiation was the oxidation of primary alcohol groups in the cellulose to carboxyl and carbonyl groups. Moreover, at 0% RH, the principal attack appeared on the C6 hydroxyl groups, whereas at 52 and 90% RH, attack seemed primarily on the C2 and, to a lesser extent, on the C3 hydroxyls, converting them to carbonyl groups. At 0% RH the primary water-soluble degradation products were glucose, arabinose, xylose, and cellobiose; at 90% RH they were glucose, arabinose, and 3-beta-D-glucosido-D-arabinose.

Chain breaking was shown to take place by random attack and no post-irradiation effects were noted in samples irradiated at 0% RH in the Fadeometer and stored at either 0 or 100% RH. A series of filters was used to demonstrate that the greatest amount of degradation was caused by the near-ultraviolet. The behavior was also demonstrated to indicate that degradation tended to take place in two stages during each of which the rate at which bonds were broken followed first-order kinetics.

The partially acetylated and cyanoethylated samples serve to demonstrate the effects of substituting other groups on the primary alcohol group at C2 and C3. Immediately following this article, part II of these studies deals with the photosensitization of these same samples by the presence of vat dyes.

1969

33. Desai, R. L., and Shields, J. A.
"Photochemical degradation of cellulose material."
Die Makromolekulare Chemie, 22 (1969), p. 134-144. 13 references.
Mercury-vapor lamps, which emit short-wavelength ultraviolet radiation, tend to decompose cellulose severely. The authors demonstrated that a plot of 1/DP and the calculated number of bonds broken versus time of exposure was linear over a considerable period of time. A graph of the percentage weight loss versus the number of chain scissions indicated that considerable chain breaking occurred before an appreciable loss occurred; this suggests that much of the volatile matter is the result of secondary degradation reactions. A weight loss of up to 6% was incurred over a period of 16 hours of exposure five inches away from a 550-watt lamp.

34. Fung, D. P. C.
"Kinetics and mechanism of the thermal degradation of cellulose."
Tappi, 52, No. 2 (1969), p. 319-321. 24 references.
The pyrolytic (thermal) degradation of cellulose in vacuum at 200, 225, 250, and 280°C was shown to obey first-order kinetics with random chain scission, exhibiting an activation energy of 35.4 kcal/mole. Viscosity measurements were used to follow the reaction kinetics according to the equation $1/DP_t - 1/DP_o = kt/2$, which the author traces back to af Ekenstam, 1936. An initially rapid stage, which one might attribute to the presence of "weak links," was noted.

35. Philipp, B., Baudisch, J., and Ruscher, Ch.
"Influence of the supramolecular structure on structural and chemical changes in the thermal degradation of cellulose fibers."
Tappi, 52, No. 4 (1969), p. 693-698. 13 references.
In this investigation a distinction was made between the rate at which bonds are broken in nitrogen (thermal degradation) and in moist nitrogen (thermohydrolytic degradation). The effect of water vapor was very great in the case of rayons at temperatures of 140, 170, and 200°C, but far less in cotton; the authors suggest that this is the result of the greater accessibility of water vapor in the less-crystalline rayons in contrast to cotton cellulose. The average DP of the initial samples of rayon were about 300 to 500, and some evidence of recrystallization of the rayons could be detected during thermal and thermohydrolytic degradation. Zero-order reaction rate constants for the rate of bond scission are reported.

Various cellulose samples (cotton linters, textile rayon, tire cord rayon) were isothermally degraded in the range of 140 to 300°C, using dry nitrogen, moist nitrogen, and glycerol as degradation media. Degradation residues were characterized by measurement of DP, water vapor sorption, weight loss, and leveling-off DP after ethanolysis, and by x-ray diffractograms. With regard to changes in DP, thermohydrolytic degradation in moist N_2 proceeds much faster than pure thermal degradation in dry N_2. Accessibility methods for structure determination are applicable only in the range up to 200°C where chemical changes in the cellulose chain are still rather small. Results of water vapor sorption and ethanolysis reveal two processes in the regions of low and medium order, counteracting each other, i.e., a loosening of structure due to thermal damage, and a recrystallization of chains under moist N_2, the latter being most intense with tire cord rayon. Degree of order as measured by x-ray diffraction is maintained even in the range of pyrolysis at 200 to 250°C. Lattice transformation to cellulose IV on thermal treatment was never observed with cotton, and barely with textile rayon, but to a considerable extent with tire cord rayon already containing a certain amount of cellulose IV in the original sample. Formation of cellulose IV was enhanced by raising the temperature of treatment and by changing the medium of degradation in the order dry N_2, moist N_2, and glycerol. The question is still open whether an increase in cellulose IV content occurs at the expense of x-ray amorphous material or by a real lattice transformation from cellulose II to cellulose IV.

1971

36. Sharples, A.
"Degradation of cellulose and its derivatives. A. Acid hydrolysis and alcoholysis." (Chapter XVIII)
In *High Polymers*, Volume 5, *Cellulose and Cellulose Derivatives, Part V*, 2nd Edn., ed. Bikales, N. M., and Segal, L., Wiley-Interscience, New York (1971), p. 991-1006.
In this chapter Sharples reviews his own and related work on the subject, compiling a total of 89 references. The subject of hydrolytic degradation is described on p. 130-140 of the same text.

The author points out that the hydrolysis of cellulose proceeds in a manner similar to that of the disaccharides: (1) it is acid-catalyzed involving the rapid formation of an intermediate complex between the glucosidic oxygen and a proton, followed by a rate-determining scission of the glucosidic bond adjacent to C1, (2) in the presence of excess water the reaction is essentially first order, but because many studies only follow the early stages of hydrolysis, the data may be fitted equally well as a zero-order plot. The most sensitive way to follow degradation is by measurement of the DP and essentially plotting 1/DP versus time, but workers have also measured the percentage of soluble matter formed. The latter is not so precise because not all of the fragments formed are soluble. Study of the rate of hydrolysis has been of interest, first of all, because it can indicate whether the (1-4) bond in various glucosidic linkages differs in its sensitivity from that in ordinary alpha-D-glucose. These properties can be investigated if the cellulose is completely dissolved (homogeneous hydrolysis).

1973

37. Chang, M., Pound, T. C., and St. Manley, R.
"Gel-permeation chromatographic studies of cellulose degradation. I. Treatment with hydrochloric acid."
Journal of Polymer Science, 11 (1973), p. 399-411. 21 references.
In the course of acid hydrolysis of cotton linters, gel-permeation chromatography (GPC) clearly showed the decrease in the average molecular weight, and the change in the distribution of molecular weights down to a DP of about 100, that is, to a point in which primarily the crystallites remained. In an investigation of the next stage of degradation in which the acid attack on the crystallites occurs GPC studies provided excellent evidence that, while the hydrolytic attack occurred both at the ends and on the lateral faces, the primary loss of weight at this stage was due to the attack on the lateral faces. Although a completely satisfactory explanation could not be offered at this time, the GPC results clearly showed that fragments of about DP 8 were among the chief products of hydrolytic attack on the crystallites. The GPC curves provide valuable data on the changes in the distribution of molecular sizes that occur throughout the course of acid hydrolysis.

38. Elema, R. J.
"A model for the degradation of cotton cellulose."
Journal of Polymer Science, Symposium No. 42 (1973), p. 1545-1562. 19 references.
The author has been associated with the investigations of Dr. Leene at the Laboratory for Textile Technology, Delft University of Technology, The Netherlands. A model for the rate of degradation of cotton cellulose and the loss of tensile strength is proposed in which both properties are related to $1/DP_n$. Correlations were established experimentally on degraded cotton cellulose between degree of polymerization (DP) on the one hand and the polymolecularity (quotient of weight-average DP and number-average DP), tenacity, and rate of degradation, on the other. It was found that these correlations may be accounted for by the assumption of weak and reactive spots in the cellulose macromolecules at regular intervals of about 500 base units. The same characteristic length of 500 base units follows also from recent concepts of cotton fiber structure.
According to this model all ends of the composing macromolecules in a newly matured yet undegraded cotton fiber cause areas of disorder in the crystalline elementary fibrils at intervals of 500 base units. These disorders bring about stresses that activate the chemical bonds in the rigid cellulose chains. The model is in agreement with current ideas and is supported by various other experimental results published in the literature.

1973-1974

39. Santucci, L.
"Degradazione della cellulosa in presenza di composti inorganici. I. Influenza dell'umidità sul comportamento di cellulosa contenente carbonati di calcio e magnesio."
Degradation of cellulose in the presence of inorganic compounds. I. Influence of humidity on the behavior of cellulose containing magnesium and calcium carbonates.
Bollettino dell'Istituto Centrale per la Patologia del Libro, 32 (1973-1974), p. 57-72. In Italian. 35 references.
In the neighborhood of 100°C, the effect of magnesium carbonate on the heterogeneous degradation of cellulose is strongly dependent on environmental moisture: the rate of change

in the inverse of the intrinsic viscosity and the rate of formation of carbonyl groups in magnesium-containing samples, unlike those that are untreated, appear to decrease with increasing humidity. The net result is that in a dry environment samples previously soaked in 0.012N $Mg(HCO_3)_2$ degrade faster than untreated or even de-ashed samples, the opposite being true at about 70% RH. The rate of degradation of samples treated similarly with calcium bicarbonate is less affected by moisture, and is fairly close to that of untreated samples during dry heating. Correlation of increments in carbonyl groups with the inverse of the viscosity does not show any appreciable influence of the addition of magnesium carbonate on the nature of the process.

1974

40. Chang, M. M. Y.
"Crystallite structure of cellulose."
Journal of Polymer Science: Polymer Chemistry Edition, 12 (1974), p. 1349-1374. 31 references.
The crystallite structure of cellulose has been elucidated through analyses of the degradations of model compounds by gel-permeation chromatography. These compounds include single crystals of cellulose triacetate, regenerated cellulose from single crystals, precipitated celluloses from solution, and commercial regenerated cellulose. The corresponding distribution profiles were shown to follow the Keller transformation, which is a characteristic of folded arrays. It is also found that the leveling-off DP of cellulose is a first-order approximation of the corresponding fold length. A detailed folding chain model for the regenerated cellulose was constructed, and the various aspects of the structure are discussed. The kinetic data and the degradation products of native celluloses were analyzed and found to obey the rule of the chain fold model but not the conventional fringe-micellar model. In addition, an iodine-staining experiment pinpoints a minimum of 1.5% of true amorphous material; this is in quantitative agreement with the conformation analysis for the six-fold helical beta-loop bonds. It is therefore suggested that the same chain-fold conformation is applicable to the native polymer. The hydrolysis of cellulose is found to be composed of a triple mode of degradation, i.e., a first order random scission at the folds, a zero-order peeling reaction at the ends of the crystallites, and a first-order random scission on the lateral surfaces of the cellulose.

1975

41. Krässig, H.
"Structure investigations on cellulose fibers using infrared spectroscopy and x-ray diffraction."
In *Applied Polymer Symposia*, No. 28, Proceedings of the Eighth Cellulose Conference, held at Syracuse, NY, May 19-23, 1975, John Wiley & Sons, New York (1976), p. 777-790. 17 references.
Following up on earlier work [Krässig, H., and Kitchen, W. J., *Journal of Polymer Science*, 51 (1961), p. 123-172], the tensile strength of cellulose fibers is considered to be a function of the number of intact chains going through regions of lower order, interlinking the morphological units to the network structure of the fiber. As a measure of this number, the parameter $1/DP_{mu} - 1/DP$ has been proposed where DP_{mu} refers to the length of the morphological units. Plots of tenacity of fibers versus 1000/DP are used to extrapolate the linear part of the curve to zero tenacity in order to obtain the value of $1/DP_{mu}$.
From the work reported in this paper the following conclusions can be drawn:
The studies on the loss of tensile strength with progressing hydrolytic degradation, using osmotic pressure determinations for DP_n, indicate that the basic building unit of the fiber network seems to be the elementary crystal. The values obtained for the length of the morphological units agree with the values of the DP_L observed in hydrolytic degradations, the values of the long periods detected by small-angle x-ray scattering, and the crystallite lengths determined by electron microscopy on degraded fiber samples. [Abstractor's note: $DP_L \approx$ LODP.]
The analysis of the wide-angle x-ray diffraction of cellulose fibers by mathematical band separation yields additional information from which dimensions of well-ordered and less-ordered regions, number of accessibility sites, magnitude of internal surfaces, etc. can be derived. From this information, properties, such as water adsorption, deuterium exchange, or dyeing characteristics can be explained.

42. Lewin, M., Guttman, H., and Saar, N.
"New aspects of the accessibility of cellulose."
In *Applied Polymer Symposia*, No. 8, Proceedings of the Eighth Cellulose Conference, held at Syracuse, NY, May 19-23, 1975, John Wiley & Sons, New York (1976), p. 791-808. 47 references.

Existing approaches for assessing order/disorder and accessibility of cellulose are based on physical methods (x-ray, infrared, density), on chemical methods (hydrolysis, deuteration, oxidation, formylation) or on physicochemical sorption (moisture regain, iodine sorption). The changes occurring in the structure of cellulose during the accessibility determinations are, in most cases, not taken into account. In the present paper, a new system for the determination of accessibility is developed. It is based on the equilibrium sorption of bromine from bromine water on cellulose. Equilibrium sorption is obtained in about one hour, the sorption accurately described by a partition isotherm. The extent of the oxidation of the cellulose during the sorption under the experimental conditions was shown to be negligible.

Accessibility was determined by this system for seven varieties of cotton fibers known to differ in their crystallinity ratios: ramie, rayons, cellulose triacetate, and hydrolyzed ramie and cotton. The accessibility values obtained with the bromine system were in the range of 6 to 70%. Straight-line relationships were obtained between the bromine accessibilities and the IR crystallinity indices as well as with the x-ray crystallinity index. The bromine molecule apparently is absorbed on the glycosidic oxygen in the less-ordered regions.

As a check on the accuracy and significance of the new accessibility system, a study was undertaken on the effect of bromine sorption on the fine structure of the cellulose. It was found that the crystallinity of the cellulose increased and the accessibility decreased with the increase in the concentration of bromine and the time of the treatment. The moisture regain, the diffusion coefficient, and the maximum uptake of direct dyestuffs were found to decrease simultaneously with the decrease in accessibility. In order to minimize the error in the accessibility caused by this recrystallization effect, the bromine concentrations used in the accessibility determination should be low and preferably less than 0.02 M of bromine, at a cellulose concentration of 1:100 and sorption times of 1-2 hours. Under these conditions, the error is 1-2%.

1977

43. Philipp, B., and Stöhr, W.
"Untersuchungen zu Abbau und Vernetzung bei thermischer Behandlung von Cellulose."
Investigation on degradation and crosslinking in thermal treatment of cellulose.
Faserforschung and Textil-technik, 28, No. 7, Zeitschrift für Polymerforschung (1977), p. 355-361. In German. 11 references.

Measurements were made of the change in DP of cotton linters degraded at 100 and 200°C in various atmospheres (N_2, N_2/O_2, N_2/H_2O, $N_2/H_2O/HCl$). At 140° and 170°C, degradation was clearly greater in oxygen than in nitrogen. In the presence of water vapor and nitrogen, the initial loss of DP was as fast or faster than in air. Evidence was provided that crosslinking could occur, particularly in celluloses of low DP (about 160) when heated in N_2 at these temperatures, possibly owing to the formation of acetal-like linkages between chains. Constants for the rate of chain breaking were calculated on the basis of the following equation:

$$1/(DP_t - LODP) - 1/(DP_o - LODP) = kt \qquad (6)$$

where LODP is the so-called "leveling-off" degree of polymerization that is observed after extensive hydrolytic breakdown.

1980

44. Saad, S. M., and El-Kholy, A. E.
"Kinetic investigation of thermal aging of unbleached Egyptian kraft bagasse paper-pulp."
Holzforschung, 34 (1980), p. 147-149. 14 references.

A brief continuance of studies reported in this same journal in 1979 (Vol. 33, p. 125-128) reports the changes in hot-alkali solubility, solubility in water, crystallinity, breaking and tear upon aging at 105, 135, and 145°C. The authors plot the logarithm of the properties versus time, considering that these plots are essentially linear, at least in the first 24 hours of aging, and that they are obeying the first-order kinetic law. Combining the results of several of the physical tests into an arbitrarily-defined "quality number," the authors found a sharp break in the rate of decline in the properties after 24 hours of aging. A similar break was not observed in the changes observed in the chemical properties.

1981

45. Marx-Figini, M., and Coun-Matus, M.
"On the kinetics of hydrolytic degradation of native cellulose."
Makromolecular Chemistry, 182 (1981), p. 3603-3616. 37 references.
A thoughtfully designed experiment to elucidate the kinetics of acid-catalyzed hydrolysis of cellulose was based on carefully isolated untreated cellulose from unopened cotton balls, having initial values of viscometrically determined DP of between 7,600 and 11,000. The authors concluded that hydrolysis takes place in two simultaneously occurring first-order reactions, the first taking place at a rate about 10^5 times faster than the second. Calculations suggested that there were about 3.5 of the weaker (more susceptible) links in a molecule of initial DP 13,000. This means that the number of standard links in the chain between the weaker links numbers about 3,600. The explanation given is that a limited number of units in the chain, as produced in nature, is not based on glucose but some modification such as 2-deoxyglucoside. What this structure actually is must await further research.

46. Stoll, M., and Fengel, D.
"Untersuchung über die Alterung von Cellulose anhand von altägyptischer Leinenproben."
Studies of cellulose aging using linen samples from ancient Egypt. Part 2. Determination of degree of polymerization and carboxyl content.
Das Papier, 35, No. 5 (1981), p. 178-185. In German. 29 references.
The authors' first report [*Das Papier*, 35 (1981), p. 141-149] concerned the microscopic study of 20 Egyptian and 2 Peruvian linen samples that were 1,000 to 4,000 years old. The present article reports values of degree of polymerization (DP) and carboxyl group content. Most values of DP fell in the range 200 to 800 with the highest values being 1,330 to 1,500. An important conclusion was that the lowest values rarely have been found below 200, strongly suggesting that a "leveling-off" DP occurs during natural aging just as is found in many accelerated-aging studies. The carboxyl groups per 100 glucose units were found to range from 0.1 to 4.4. The authors plot the analytical data in several ways relative to the age of the samples and their DP. They consider that the data relating the DP with age may follow a rate law of the type $\ln(DP_t - DP_e/DP_o - DP_e) = kt$, where DP_o is the original DP of the cellulose, DP_e the "leveling off" DP, and DP_t the DP at the present time. The authors further suggest that the samples thus far studied tend to fall in several groups based on different values of k, and further conclude that this is good evidence of the fact that a number of different degradation processes contribute to the overall deterioration of the cellulose. There was also rather poor correlation between acid-group content (COOH) and age, leading the authors to conclude this to be additional evidence of various mechanisms contributing to deterioration. Before more precise relationships between chemical constituents and age can be developed, a greater number of samples needs to be investigated. Twenty-nine references given include three reports of others who have had the opportunity to study very old samples.

1986

47. Feller, R. L., Lee, S. B., and Bogaard, J.
"The kinetics of cellulose deterioration."
In *Conservation and Characterization of Historical Textiles and Paper Materials*, Advances in Chemistry Series No. 212, ed. Needles, H. L., and Zeronian, S. H., American Chemical Society, Washington, DC (1986), p. 329-346. 32 references.
The degradative scission of cellulose chains can be regarded as taking place in two or three principal stages, frequently appearing to follow zero-order chemical kinetics, the situation in which an equal number of bonds are broken in a given length of time. Although these characteristics have been reported in the past—whether the degradation is essentially thermal, hydrolytic, photolytic, or photochemical in character—this concept of deterioration has not been widely stressed in the literature on conservation. In addition to reviewing the major publications on the subject, the authors provide examples from the literature and from recent laboratory findings concerning ways in which kinetic analysis may provide new insights into the process of cellulose deterioration under the influence of heat and light. Authors' abstract.

II. THE INFLUENCE OF HEMICELLULOSES ON THE DETERIORATION OF PAPER

Hemicelluloses as Wood Components

In addition to the cellulose, lignin, and pectin components, hemicelluloses are important constituents of the cell walls of plants. Polysaccharidic in character, most of the hemicelluloses are water-insoluble, alkali-soluble substances and are more readily hydrolyzed by acid than cellulose. In contrast to cellulose, they are generally branched and have lower molecular weights.

Softwood hemicellulose consists of both pentosans—xylan (in which the chains are chiefly composed of xylose units) and araban (in which the chains are chiefly composed of arabinose units)—and hexosans (the chief types might be designated as mannan and galactan) (see Figures 1 and 2, illustrating entry no. 90). Hardwood hemicellulose consists mainly of pentosans. When produced at the same yield as softwoods, hardwood pulps will retain more hemicellulose. The pentosan content in softwood pulp, for example, is about 7 to 10%, while in hardwood pulp it is about 19 to 25% (Tappi Standard T223, 1971). The softwood xylan does not contain acetyl groups and is more highly branched and more acidic than the corresponding hardwood xylan.

The pentosan content of pulp, a convenient although imprecise indicator of hemicellulose content, reflects the net result of retention and loss of the hemicellulose components during pulping and bleaching processes. A rapid and simple method of determining total pentosans, without reference to the relative amounts of individual sugars, is based on the formation of furfural (2-furaldehyde), generated when pentosans or pentose sugars are distilled with strong hydrochloric acid.

The cooking of wood to reduce it to pulp causes major changes in the amounts, locations, and structures of the various hemicelluloses. The end result is dependent on the wood species and cooking process. Present kraft and sulfite cooking conditions degrade hemicelluloses; thus, their percentage is lower in the pulp than in the original wood.

Hemicelluloses have an important influence on pulp quality with respect to ease of beating, strength developed on beating, swelling capacity, etc.—a fact long recognized. However, pulps with too much hemicellulose hydrate too quickly and do not develop adequate strength, producing a translucent, rattly paper, low in strength. Hemicelluloses also can contribute to the deterioration of cellulose, as will be considered in the paragraphs below.

Changes in Chemical Composition During Acid and Alkaline Pulping

Iversen and Lindgren (1983) describe the changes in the chemical structure of hemicellulose that occur during pulping of wood. In the sulfite process, the hemicelluloses are hydrolyzed to simple sugars or to polysaccharides of lower molecular weight. These then tend to undergo reactions with sulfite, forming aldonic acids or sulfonic acids. Hemicelluloses are also broken down in the kraft process, mainly by the peeling reaction, in which the alkaline degradation starts from an aldehyde end group and proceeds stepwise. The chemical composition of pulps varies, not only with the nature of the wood stock, but also with the pulping process [Annergren, Rydholm and Vardheim, *Svensk Papperstidning*, 66 (1963), p. 196-210]. The kraft process, for example, results in a higher glucuronoxylan, lower glucomannan, and lower alpha-cellulose content than the sulfite process. The variation in chemical composition of pulps, which arises from the use of different woods and from different pulping processes, influences the long-term stability of papers made therefrom.

The presence of hemicellulose in pulp introduces carboxylic acid groups, largely in the form of aldonic and uronic acids. The average acidity of xylan in various pulps was investigated by Sjöström and Enström (1967). Wood xylan is far from homogeneous, but is composed of fractions of greatly varying uronic acid content. In bleached and unbleached pulps, sulfonic and carboxylic acids from the lignin may also be present, together with the native aldonic and uronic acids. In addition to these strong and moderately weak acids, phenolic groups may be found in the unbleached pulps. A quantitative analysis of some of these groups for a number of pulps has been reported by Sjöström, Haglund and Enström [*Journal of Polymer Science: Part C*, 11 (1965), p. 221-241].

The hemicellulosic components which remain associated with wood cellulose produced from western hemlock by the sulfite process have been shown to be a mixture of two families of polysaccharides, namely, glucomannans and xylan polyuronides. The predominant polysaccharides were found to be the glucomannans in which the ratio of D-glucose to D-mannose was one to three. Oxidation studies carried out on these indicated that both of the above hemicellulose types were linked predominantly through 1-4-beta-glycosidic bonds; the glucomannans were predominantly straight chain and the xylan polyuronides, slightly branched (Hamilton and Thompson, 1957).

The DP of native glucomannan has been found to be about 80-100, and of glucuronoxylan around 130-200. This DP is largely preserved in kraft pulps but at a fairly early stage is reduced to less than half that value in sulfite cooking (Rydholm, 1965).

Extraction of Hemicellulose Components

Sodium and lithium hydroxide are much more effective than potassium hydroxide in their relative extractive power for the mannose-containing polysaccharides, whereas these three alkalis have nearly equal solvent power over a wide range of concentration for xylose-, arabinose-, galactose-, and uronic acid-containing polysaccharides (Hamilton and Quimby, 1957). The solubility of sulfite pulp in cold 18% NaOH gives a fair indication of hemicellulose content; errors are more pronounced in kraft pulps, especially those from hardwood.

Rydholm (1965) recommended hot-alkali purification for sulfite pulps, because very little swelling occurs and yet useful chemical reactions occur. The dominating chemical reaction is the "peeling reaction" which starts at the aldehydic end groups in the carbohydrate chains. Among the products found in the alkali solution after this treatment were pyruvic acid, $CH_3COCOOH$, and formic acid, lactic acid, and beta, gamma-dihydroxybutyric acid. Aldehydes such as methyl glyoxal, CH_3COCHO, were also found. On the other hand, purification in cold concentrated alkali mainly involves physical changes (swelling); little alkali is consumed and yet the hemicelluloses, almost exclusively intermicellar materials, are removed.

Beélik and coworkers (1967) developed a technique for the isolation of three main hemicelluloses from softwood holocelluloses, based on the principle of selective extraction with barium and sodium hydroxide.

The hemicellulose isolated from white birchwood contained an average of 197 1,4-linked beta-D-xylopyranose residues with, on the average, every eleventh unit carrying a single side chain of 4-O-methyl-D-glucuronic acid, attached by an alpha-glycosidic bond to C2 of the xylose. The number-average DP of hemicelluloses from four hardwoods ranged from 197 to 178. Softwood xylans contain fairly large amounts of terminal L-arabinofuranose residues and contain more acid groups than hardwood hemicellulose [Timell, Glaudemans, and Gillham, *Pulp and Paper Magazine of Canada*, 59, No. 10 (1958), p. 242-251]. In the case of coniferous woods, preliminary delignification is generally essential for effective isolation of hemicelluloses. Extraction has been done in 1N NaOH for 48 hours at room temperature. This prevents the extreme swelling caused by 16 or 24% alkali. The glucuronoxylans are readily soluble.

Characterization of the general chemical composition of hemicelluloses involves an analysis of the sugars, uronic acid groups, methoxyl, and acetyl groups (Browning, 1967). Further discussions of the chemistry of alkali extraction can be found in the following bibliography, topic III.

Hemicelluloses Susceptible to Hydrolysis and Oxidation

In contrast to cellulose, the hemicelluloses are almost entirely amorphous, situated in amorphous regions of the native woods and plants. Most chemical agents, therefore, reach the hemicelluloses much more easily than the components in or near crystalline regions of the cellulose (Rydholm, 1965). Studies of the heterogeneous acid hydrolysis in pulps have clearly demonstrated that the hemicellulosic material is hydrolyzed at a much greater rate than cellulose (Springer, 1966).

The hydrolysis process of both cellulose and hemicelluloses can be catalyzed by the organic acids that arise through the cleavage of acetyl groups and possibly through the oxidized products derived from the methoxyl groups found in the hemicelluloses. The oxidation and subsequent degradation reactions will affect the hemicelluloses more rapidly than cellulose (Rydholm, 1965).

Koura and Krause (1977) investigated a number of tests that can be used to indicate the extent of deterioration. Cotton linters, a sulfate pulp, and two sulfite pulps having a range of matter soluble in cold 10% NaOH, ranging from 0.5 to 15.0%, were used. Their investigation also led to the conclusion that the hemicelluloses were more easily oxidized than cellulose. During thermal aging, the amount of matter soluble in 10% alkali increased greatest in those samples that were originally highest in 10% cold-alkali-soluble matter (highest in hemicellulose content).

ROLE OF HEMICELLULOSE ON PAPER STRENGTH

Our primary purpose is to review the contribution of hemicelluloses to degradation. The role of hemicelluloses in affecting the utilitarian physical properties of paper is an extensively investigated subject, which will not be treated at length here. Nonetheless, a few words must be said about the beneficial aspects of hemicelluloses on the initial properties of papers. Their contribution to the development of initial strength is well recognized. It is related to fibrillation of the fibers. Hemicelluloses function as binders through the mutual bonding of their hydroxyl groups. In wood pulp papers, a large fraction of the bonding material may consist of hemicelluloses; hence, their chemistry is especially germane to the behavior of low-grade papers, particularly with respect to the stability of interfiber bonding [Wilson and Park, *Restaurator*, 3 (1979), p. 40]. The maximum strength of papermaking pulps increases with an increase in the amount of alkali-soluble hemicellulosic material (in a sense, the net hemicellulose) in the pulp rather than solely the pentosan content. Because alkalinity in the soda process removes the hemicelluloses, soda pulps do not "wet" as readily as sulfite pulps. Softwood hemicellulose is much more effective in fiber bonding than hardwood hemicellulose due to the higher content of the hexose sugar, mannose, or the mannose and other hexose content [Klingstedt, *Svensk Papperstidning*, 40 (1937), p. 412-416]. Gamma-cellulose content is the most important factor, because it contains the hemicellulosic material involved in the fiber bonding. Even though the gamma fraction is partially removed during alkaline pulping, a sufficient amount remains in sulfite and sulfate pulps to affect the paper strength [Casey, *Pulp and Paper*, Volume II, 2nd Ed., Wiley-Interscience, New York (1960), p. 690]. Studies on aspen holocellulose indicate that the optimum content of alkali-extractable hemicellulose for maximum bursting strength is about 20% (March, 1948).

Experiments Using Model Compounds to Elucidate Color Reversion

Sheets of paper dipped in solutions of pentoses (xylose and arabinose), aldohexoses (mannose and glucose), a ketohexose (fructose), and two uronic acids (galacturonic and gluconic) resulted in appreciable discoloration during thermal aging at 105° and 120°C at add-on weights from 0.1 to 1.0% of these materials on the paper (Roudier and Saulquin-Bisson, 1959). It was found that the greatest discoloration was attained with the uronic acids, followed by xylose and arabinose. Roudier and Sauret (1960) also added xylose, glucose, and methylated products of these two sugars to pulps to study their effect during thermal aging. In these experiments, xylose and glucose caused the greatest discoloration and loss of DP, with xylose causing the greater degree of change. The conclusion was reached that depolymerization, accelerated by the presence of aldoses, precedes discoloration. The acidity that develops in the course of aging accelerates the hydrolytic decrease in DP.

Rapson and Corbi [*Pulp and Paper Magazine of Canada*, 64 (1964), p. T459-T466] selected about 34 acidic, hydroxy, ketonic, and aldehydic compounds having six to two carbon atoms, representing substances that might be formed as decomposition products in cellulose. Filter paper was impregnated with these, after which the test sheets were thermally aged to learn which compounds tended to yield the greatest discoloration. Arabinouronic acid led to intense discoloration; 4-O-methylglucuronic acid, which occurs in the hydrolysis products of hemicelluloses, also resulted in considerable yellowing. The authors concluded that hemicelluloses only cause yellowing if they are oxidized and hydrolyzed (a statement that could just as well be made about cellulose). The authors note that the hemicelluloses are more sensitive to oxidation and are more readily hydrolyzed.

Beélik (1967) also conducted impregnation experiments with model compounds. Upon thermal aging of pulps impregnated with free aldoses (24 hours at 105°C), substantial increases in yellowing, the formation of water-soluble acids, and the formation of water-soluble compounds absorbing at or near 275 nm in the ultraviolet region were observed. All aldoses contributed to the color and it was observed that acidity developed in heat-aged test sheets.

Thermally Induced Discoloration as a Function of Hemicellulose Content

In Spinner's extensive review (1962), the author cites the findings of a number of investigators and reviews what little was then known of the role that hemicellulose plays in the reversion of different types of pulps. Extraction of accelerated aged pulp with water and low-percentage alkali solutions indicated that the color-forming materials in thermally aged pulps were probably degradation products of the hemicellulose components. The work of Rollinson

(1955) and of Giertz and McPherson (1956) is cited to show that much of the water-soluble colored material was found in the gamma-cellulose fraction. Spinner also quotes Centola [*Association Technique de l'Industrie Papetiere*, 5 (1951), p. 111-115] as reporting that reversion could be correlated with pentosan content. Marum [*Tappi*, 39, No. 6 (1956), p. 390-394] is cited as finding that the best correlation with reversion was obtained against the combined beta- and gamma-cellulose content in pulps.

Jappe and Kaustinen (1959) oxidized isolated hemicellulose fractions with periodate to obtain material having various copper numbers, and compacted plaques of the material were oven aged for one hour at 105°C. With highly oxidized materials having copper numbers between 5 and 50, brightness reversion did not increase with copper number. These data raise questions concerning possible relationships between the character of the carbonyl groups and their effect on reversion.

Chang and Kondo (1971) used a fully bleached kraft and a sulfite pulp (KP and SP) made from Japanese red pine, white birch, and beech, in addition to two related commercial pulps. By measuring the degree of discoloration before and after ethanol/benzene extraction, and before and after extraction in 10% NaOH at 20°C, the authors were able to calculate the percentage of loss of reflectance that was due to the extractives, to the hemicelluloses, and to cellulose. The results, expressed as graphs, clearly indicate that the removal of the considerable percentage of hemicellulose in pulps contributes most to the reduction in color reversion upon dry and humid aging. Per gram of material extracted, however, the extractives contributed most to thermally induced discoloration. Thermal aging at high humidity increased the proportion of the net discoloration due to the cellulose and hemicellulose components; the greatest increase under humid conditions was attributed to pulps having high xylose content.

Cold- or hot-alkali extraction of sulfite or kraft pulp prior to bleaching increases brightness stability, as does ethanol extraction to remove resins (Rydholm, 1965). Removal of hemicelluloses during sulfite pulping, or by means of subsequent alkali extraction, may increase the brightness stability of bleached pulps.

Croon, Dillén and Olsson (1965) focused their study on the cellulosic and extractive compounds which cause brightness reversion in birch sulfate pulp. The colored aqueous or alkali extracts of aged pulps were shown to contain oligo- and monosaccharides and acidic products. The increase in brightness stability on alkaline treatment of pulp is due to the dissolution of the most strongly oxidized part of the cellulose as well as to the removal of the hemicellulosic materials from alkali treatment. For a number of pulps a relation has been found between brightness stability and uronic-acid content.

Von Roffael (1979) found that the color reversion of two spruce sulfite pulps during thermal aging correlated well with their hydrolyzibility to simple sugars and their degradation products (e.g., furfural), but not with their pentosan content as such. He concluded that not only the hemicellulose content but, to a greater extent, the content of easily-hydrolyzable pentosans is a relevant factor for yellowing.

Sjöström and Eriksson (1968) tested four different types of bleached pulps (pine and spruce sulfite pulps, and pine and birch kraft) on their brightness stability. Pulps that had been both reduced and esterified are highly stable. Whatever brightness reversion is observed after such treatment is evidently due to remaining carbonyl and carboxyl groups and to new groups that may be formed during aging. The "acidic xylan" carboxyl groups, attached to a xylan backbone as methylglucuronic acid residues, appear to be the most sensitive polysaccharide component with respect to thermally induced discoloration.

COLORED MATERIALS FORMED UPON AGING

The colored aqueous or alkali extracts of reverted pulps have been shown to contain oligo and monosaccharides and acidic products, the greater part of which is derived from hemicellulose. Rolfinson (1955) studied the colored water-soluble substances that were formed during the accelerated aging of bleached kraft pulp, both by light-absorbance methods and by chemical analysis. It was shown that the rate of formation of colored water-soluble material paralleled the total discoloration of the pulp and that the water-soluble material accounted for 50 to 60% of the total discoloration formed during accelerated thermal aging. The colored material was acidic, and about 23% of the water-soluble substances was found to be oligosaccharides composed of sugars typical of hemicellulose.

Giertz and McPherson (1956) stated that almost all the color formed by accelerated aging of sulfate pulp can be extracted with 10.9% lithium hydroxide. The colored material was found to be concentrated in the "gamma" fraction. They concluded, on the basis of chromatographic examination of the hydrolysates of these extracts, that the color-forming materials were probably degradation products of hemicellulose, perhaps largely owing to carbonyl groups introduced in this fraction during bleaching. This agrees with the 1945 finding of Giertz that hemicellulose-rich pulps have a greater tendency to discolor during thermal aging than pulps with lower hemicellulose content.

Giertz and McPherson (1956) also found a high carbonyl-group content in the "gamma" fraction of caustic extracts of an aged pulp. The reduction of the pulp with sodium borohydride before aging decreased the color of the alkali extract. Thus, borohydride treatment exerts a beneficial effect on color reversion.

It is well established that an alkaline treatment which removes most of the hemicellulose increases the brightness stability of pulps [Tongren, *Paper Trade Journal*, 107, No. 8 (1938), p. 34-42; Scribner, *Journal Research National Bureau of Standards*, 23 (1939), p. 405-413]. Croon, Dillén and Olsson (1965) found the colored aqueous or alkali extracts of reverted pulps contain oligo- and monosaccharides and acidic products, the greater part of which was derived from hemicellulose. Thus, there is no doubt that color reversion is related to the hemicellulose content of bleached pulp.

Light-induced Degradation

Lewis and Fronmuller [*Paper Trade Journal*, 121, No. 14 (1945), p. 25-27] showed that the content of pentosan in groundwood sheets was decreased upon exposure to the light of the carbon-arc fadeometer for 10 hours on each side of the sheet.

Yasuda, Nagaoka, and Hanzawa [*Research Bulletins of the College Experiment Forests Hokkaido University*, 13, No. 11 (1974), p. 1-7] used three kinds of pulp (untreated, extractive-free, and bleached groundwood pulps) for irradiation with a mercury lamp (2500-4000 Å radiation). The colored materials extracted with 10.9% LiOH from the bleached pulps after irradiation were acidic, including uronic acid components. Monosaccharides—such as glucose, mannose, galactose, xylose, and arabinose—were hardly discolored. It was concluded that the monosaccharides derived from the hemicelluloses had little effect as intermediates in the formation of colored materials by ultraviolet radiation. To demonstrate further whether the hemicelluloses would contribute to discoloration under photochemical circumstances, Lee and Feller (1985) isolated the hemicellulose fractions from four different pulps and added-on these to filter paper sheets in order to evaluate the tendency of such components to cause discoloration under fluorescent "daylight" and "blacklight" lamps and to compare their influence under the conditions of thermal aging at 90°C and 50% RH. Only negligible changes in color occurred under the various light sources, whereas thermal darkening was intense. Similar results were observed with a series of handsheets containing reduced levels of gamma-cellulose.

One may conclude that the tendency of hemicelluloses to lead to discoloration is principally a thermally induced effect.

Abstracts

1908

48. Armstrong, H. E., and Glover, W. H.
"Studies on enzyme action. XI. Hydrolysis of raffinose by acids and enzymes."
Proceedings of the Royal Society, Series B, 80 (1908), p. 312-321. 2 references.

The authors demonstrate that the alpha and beta methylgalactosides are hydrolyzed about four to five times faster than the alpha and beta methylglucosides. Raffinose is hydrolyzed at about one-fifth the rate of cane sugar.

1929

49. Rasch, R. H.
"Quality of purified wood fibers."
Paper Trade Journal, Technical Association Section (1929), p. T125-T140.

In tests on 152 paper stocks, the influence of pentosans on thermal stability was one of the factors investigated. The rag half-stocks here studied have extremely low, almost negligible amounts of furfural-yielding substances (largely pentosans) present, whereas all the wood fibers contain from 2% to more than 3%; the ordinary sulfite wood pulp from over 3% to nearly 4%; while the one sample of soda pulp tested had pentosans present to the amount of 12%. The fact that soda pulps can contain such large amounts of pentosans and still have relatively good permanence qualities as compared to ordinary grades of sulfite pulps or poorer grades of half-stock, suggests that this substance does not contribute to the deterioration of cellulose to anywhere near the extent that the modified celluloses do. The pentosan content of the purified wood fibers may also account for the fact that these fibers, approximately 4% lower in alpha-cellulose content than high-grade rag half-stocks, are as stable toward the accelerated aging test. That is, a great part of the difference in the alpha-cellulose contents of the two materials represents this content of relatively inert pentosans.

It was noted that soda pulps had lower copper numbers than the higher alpha-cellulose containing sulfite pulps. This is undoubtedly due to the fact that the pentosans, which are known to have no aldehyde groupings and consequently no reducing power toward Fehling's solution, are present in the former to a much greater extent.

1945

50. Meller, A.
"The hot alkaline purification of cellulose I."
Paper Trade Journal, 121, *Tappi Section* (1945), p. T119- T131. 79 references.

This survey of the literature on the hot-alkaline purification of cellulose aims to bring together the results and discussions on the subject published in the past twenty years. It is not claimed that the review is complete but is confined as much as possible to published work which contributed to the advance of knowledge about the action of dilute alkali solutions at higher temperatures on cellulosic materials for the production of purified cellulose. Patent literature is not included.

The action of strong cold-alkaline liquors on cellulosic fibers has been surveyed, and comprehensive reviews are included in various textbooks and handbooks on cellulose chemistry or on pulping. However, these textbooks or reviews do not contain separate chapters dealing with the action of hot dilute-alkaline liquors on cellulose, or on cellulosic materials. Thus, the present survey served to fill this gap.

The survey treats the subject under the following headings: (A) introduction, (B) the hot and cold purification processes and their applicability, (C) a comparison of the behavior of sulfite and sulfate pulps in the hot alkaline refining, and (D) the action of alkaline solutions at higher temperatures on cellulose, pulps, and wood. Among the topics discussed is the strength of the linkage between cellulose and pentosans, the action on bleached pulps and on wood.

The purification or refining processes for cellulosic materials are designed for the removal of all constituents of the fibers except cellulose, and by doing so obtain products similar to cotton.

In order to obtain very pure fibers from cellulosic materials, such as woods, special purification processes must be applied since the common fiber liberation or cooking processes and the bleaching processes do not result in fibers of high purity as are required for certain purposes. The commonly processed wood pulps, i.e., cooked and bleached products, contain considerable quantities of hemicelluloses, such as degraded cellulose, pentosans, polyuronides, and some hexosans of noncellulose nature. Thus, the purification processes are primarily concerned with the removal of these substances. Secondly, it is required that this removal not be accompanied by a serious degradation of the high polymeric, long-chained cellulose. Hence, the principal aim for an efficient purification process is to apply a chemical which under certain conditions acts selectively insofar as it should remove only hemicellulosic materials and not degrade below certain limits the longer-chained cellulose of the pulp. This chemical must necessarily be obtainable easily and inexpensively, and its consumption and recovery will furthermore influence the economy of the purification process.

For these two reasons—efficiency and economy—it seems not accidental that caustic soda is perhaps the most commonly used chemical in the wood pulp purification industry. Caustic soda applied to wood pulps (and to other cellulosic fibers) exerts a selective action, insofar as it removes hemicelluloses extensively under certain conditions without degrading considerably the cellulose; therefore, it may be regarded as an efficient purifying chemical. Its consumption in the process does not influence materially the economy of the purification process.

1948

51. March, R. E.
"The effect on pulp quality of the stepwise removal and replacement of the hemicelluloses from aspen holocellulose."
Paper Trade Journal, 127, No. 17, *Tappi Section* (1948), p. 431-437. 28 references.

That hemicelluloses have an important influence on pulp quality with respect to rate of beating, strength developed on beating, swelling capacity, etc., is a fact that has long been recognized. Many investigators have commented on the rapid hydration and greater strength of pulps high in hemicellulose when compared with those of a low hemicellulose content. However, much of this work has not involved a thorough investigation of the relationship between hemicellulose content and pulp quality.

Aspen holocellulose was extracted with potassium hydroxide to effect stepwise removal of the hemicelluloses; the fibrous residues were converted into handsheets for testing. Within a hemicellulose range, these sheets showed certain maxima and minima in physical properties. Thus the maximum bursting strength, minimum opacity, and minimum tearing strength were found in sheets prepared from holocellulose containing about 20% alkali-extractable hemicelluloses. The rate of beating was most rapid when the hemicellulose content was about 26.5%. The extracted holocellulose pulps were compared with a bleached aspen kraft pulp (three-stage bleaching operation). In general, the former possessed a far greater bursting strength and the latter a much higher tearing resistance.

Difficulties were encountered when attempts were made to reintroduce the extracted hemicelluloses into the fibrous holocellulose residues. The addition of small quantities of hemicelluloses increased the bursting strength slightly; very large amounts did not improve pulp strength.

1950

52. Meyer, K. H. F.
"Animal and vegetable polysaccharides and lignin. I. Hemicelluloses: pentosans, hexosans, and polyuronic acids."
In *Natural and Synthetic High Polymers*, 2nd edn., Interscience Publishers, a division of John Wiley & Sons, New York (1950), p. 407-418. 38 references.

In a brief review of the subject, it is pointed out with structural diagrams that hydrolysis of hemicelluloses usually finds D-glucose associated with D-glucuronic acid and L-xylose, while D-galactose is usually associated with D-galacturonic acid and L-arabinose. For rough fractionation the crude hemicelluloses extracted in alkali are first precipitated with acetic acid producing a dextro-rotatory A-fraction consisting chiefly of glucosans and mannans. The B-fraction, obtained by precipitation with alcohol, is levorotatory and contains polyuronic acids and pentosans. The latter increases with the age of wood. Much can be learned about the structure of hemicelluloses through methylation of the free hydroxyl groups and subsequent hydrolysis. Beta- 1-4 xylan can be isolated from almost all lignified cell walls through extraction with 5% alkali. Xylan from Esparto grass contains small amounts of L-arabinose in the furanose form. Wood also contains low-molecular-weight glucosans which are found in the beta-cellulose fraction along with hemicelluloses. Most woods contain traces of galactans, usually arabogalactans; larch contains 8% of this material with a 1:6 ratio of arabinose to galactose.

Mannans, difficultly separable from the cellulose, are found in coniferous woods but are scarcely present in deciduous woods. Straw contains polyglucuronic acid. Galacturonic acid is found among the hydrolysis products of pine and beech wood hemicelluloses.

1952

53. Purves, C. B.
"Chemically modified cellulose."
In *Wood Chemistry*, Volume 1, American Chemical Society Monograph Series No. 97, 2nd edn., ed. Wise, L. E., and Jahn, E. C., Reinhold Publishing Corporation, New York (1952), p. 162-183. 160 references.

A detailed review of the ways of producing and confirming the chemical structure of variously oxidized celluloses is given, such as the preparation of gluconic-acid types by oxidation of the C6 hydroxyl groups, or periodate celluloses, where aldehyde groups are to be found on C2 and C3. The chapter is useful in reviewing what specific chemical changes are caused by various frequently employed oxidizing agents: nitrogen dioxide, chlorous acid, acid chromate with oxalic acid added. The author closes by reviewing methods for determining carbonyl, aldehyde, and carboxyl groups.

54. Rånby, B. G.
"The physical characteristics of alpha-, beta- and gamma-cellulose."
Svensk Papperstidning, 55, No. 4 (1952), p. 115-124. 22 references.

Electron microscope evidence of dispersed holocellulose from spruce indicates that it contains two different phases: micelle strings (= cellulose) and a dispersed phase (= hemicellulose). Further investigations of alpha-, beta- and gamma-celluloses isolated from bleached technical softwood pulps have shown that the alpha-fraction is built up of micelle strings (~80 Å wide), while the gamma-fraction contains no strings. The beta-fraction is a mixture of short-string fragments and small particles. An x-ray investigation has proven that alpha- and beta-celluloses have the same lattice, that of mercerized cellulose or Cellulose II. The gamma-celluloses contain lattices almost identical when isolated from sulfite and sulfate pulps but quite different from Cellulose II. From a comparison of the x-ray diagrams of two gammafractions, it is assumed that one fraction might be a pure phase while the other fraction is a mixture of this and at least one additional phase.

For a discussion of the chemical composition and properties of the alpha-, beta- and gammacelluloses, the reader is referred to the reviews by Hägglund, in *Chemistry of Wood*, New York (1951), p. 129-131.

If this interpretation is accepted, the beta-cellulose of soft-wood pulps is nothing but disordered and broken micelle strings, containing only depolymerized cellulose. This assumption is supported by the well-known fact that the beta-fraction increases upon different types of degradation of technical wood pulps. Holocellulose and unbleached sulfate pulps (kraft pulps) have been reported to contain very small amounts of beta-cellulose.

The physicochemical data, described in this paper, strongly support the assumption that the cellulose strings in a pulp fiber(spruce and pine) form a pure phase with a definite boundary to the other carbohydrates. This assumption is in agreement with the hypothesis by Algar et al. [Algar, W. H., Giertz, H. W., and Gustafsson, A.-M., *Svensk Papperstidning*, 54 (1951), p. 335] that the gamma-fraction is identical with the wood polyoses and with the amorphous interfibrillar substance of a pulp. Furthermore, the assumption of a pure cellulose phase is supported by the fact that the amount of true cellulose, which can be isolated by different technical processes, is fairly constant.

1953

55. Meller, A.
"The alkaline extraction of wood pulps."
The Bleaching of Pulp, Tappi Monograph Series No. 10 (1953), p. 31-77. 45 references.

Weak alkali tends largely to penetrate only the less-ordered regions present in the wood pulp. Hot alkali will dissolve some of the chlorinated lignin in bleached pulps, a subject that is reviewed in a five-page section. It is considered that the chlorination, sulfonation, and oxidation of the lignin all contribute to its solubility in hot-alkali solution. The efficiency of extraction is lower in calcium hydroxide than for the hydroxides of alkali metals, possibly because of relative insolubility of the salt-like compounds formed between the treated lignin and calcium.

In a second section, the effects of temperature, alkalinity, and time of extraction on lignin removal are reviewed. The results support the conclusion that the removal of lignin is largely diffusion controlled. At the same time hemicelluloses are also removed. The effectiveness of hemicellulose removal in the so-called alkaline purification step is also covered. The industrial use of alkali in the range of 0.5 to 2.0% is governed largely by economic considerations. With increasing concentrations of alkali, the extraction of hemicelluloses rises to a maximum. Many of the curves of data refer only to improvements no more than a few percent in pentosan or alpha-cellulose content. Pentosan levels varied from 2 to 6% in many of the refined products. Data of Swedish investigators are given, showing a linear relationship between a decrease in copper number and alkali solubility up to 28% weight loss (the author's Figure 28). An appendix discusses cold extraction.

1954

56. Vinogradova, L. M., Konkin, A. A., and Rogovin, Z. A.
"Investigation of the hydrolysis process of cellulose and galactan."
Nauchno-Issledovatel'skie Trudy Moskovskii Tekstil'nyi Institut, 13 (1954), p. 87-98. In Russian.
The hydrolysis of cellulose (I) and galactan (II) and of model compounds, such as betaisopropyl glucoside (III) and galactoside (IV), was studied in homogeneous and heterogeneous media. The rate of hydrolysis of II in a homogeneous medium was 2.6 times greater than the hydrolysis of I and that of IV was 4.3 times greater than that of III. In a heterogeneous medium the rate of hydrolysis of II was 300 times that of I (from Chemical Abstracts 1956, 14221d).

57. Wise, L. E., and El-Taraboulsi, M. A.
"The rind and pith of papyrus—an analytical study."
Tappi, 37, No. 1 (1954), p. 25-28. 35 references.
Quantitative chemical data of the rind and pith of Egyptian papyrus, *Cyperus papyrus L.*, are presented. The hemicellulose and alpha-cellulose fractions were isolated from holocellulose. The chlorite process used resulted in white, powdery hemicellulose products. Ash determinations were done by charring and ignition in the presence of concentrated H_2SO_4. The "sulfated ash" was extracted with hot water, and the water-soluble ash was assumed to be present as K_2SO_4, and, furthermore, that the K originally comes from the -CO•OK in the uronide portion of the hemicellulose fraction. The hemicellulose content ranged from 21 to 25%. The authors' Table X presents a detailed analysis of the components in the rind and the pith: ash, extractives, acetyl, alpha-cellulose, lignin (16.1 to 19.2%) and hemicellulose (21 to 25%).
Chromatographic data revealed appreciable amounts of xylan, whereas only small amounts of arabans are found.

1955

58. Huffman, G. W., Rebers, P. A., Spriestersbach, D. R., and Smith, F.
"Nature of a hemicellulose extracted from cellulose with water."
Nature, 175 (1955), p. 990-991. 7 references.
Although filter papers are generally considered to represent very pure cellulose, this letter to the editor describes the finding of a total of 0.086% water-soluble polysaccharidic material extractable from Whatman No. 1 filter paper. Apparently hemicellulosic in character, the material was found to contain 53.6% xylose, 15.8% arabinose, 11.1% glucose, 1.9% galactose, and 17.71% uronic acids. This behavior may also be related to the "brown line" effect. There was also non-carbohydrate material extractable with methanol, possibly a beater additive.

59. Richter, G. A.
"Production of high alpha-cellulose wood pulps and their properties."
Tappi, 38, No. 3 (1955), p. 129-150.
Kraft pulps are richer in pentosans than sulfite pulps, dependent in part on the wood used in the process. For example, pentosans might run as high as 20% in krafts made from birch; 6% in that from pine. Pentosans can be found in the alpha- and beta-cellulose fractions but are chiefly found in the gamma.
The author reports that alpha cellulose is comprised of a resistant and a hot-alkali-soluble fraction. Evidence leads to the conclusion that some of the original hemicellulose resists the action of hot sodium hydroxide solution and, in the process of extraction, some of the alpha-cellulose is progressively degraded. Thus, the author points out that there may be considerable deficiencies in the character of the alpha-cellulose fraction obtained from various pulps.
Extraction with 5 to 6% sodium hydroxide at 20°C removes chiefly the gamma-cellulose.
Analytical values were given for the gamma-cellulose and pentosan contents of six unbleached sulfite pulps which revealed 8.3 to 10.3% gamma and from 2.4 to 9.8% pentosans. Curves are also given to show the change in the alpha-cellulose, yield, pentosan and hot-alkali-soluble matter with variations in the conditions of sulfite pulping. Depending on the type of pulp, the pentosan content ran from about 1 to 2% up to as high as 7.1 %. Curves are also given to show the change in pentosan content with increased concentration of cold NaOH used in the refining.
Although the definition of alpha-cellulose is based on an empirical analytical procedure, the values so obtained have had practical importance in the wood cellulose industry for many years. Work reported here suggests that a more complete understanding of its significance would result if such figures were regularly corrected for other alkali-resistant components and if the

alpha-cellulose were further characterized by its resistance to a boiling 7.14% sodium hydroxide solution. Methods that can be employed to prepare high alpha-cellulose wood pulps are reviewed. Refinement of raw stocks by the well-known hot-alkali digestion procedure is reviewed and much is reported on the lesser used cold-alkali extraction process. Multiple-stage pulping as a means of obtaining alpha-cellulose levels of 95% and higher is discussed in some detail. Reference is made to refining procedures with weak alkaline buffer solutions at temperatures as high as 200°C. Although at this time of academic interest only, methods are described for the elimination of hemicellulose by weak cuprammonium solutions and also by a freezing process that depends on a localized concentration of sodium hydroxide by separation of ice from a weak alkali solution that has been used in a preliminary stage to saturate the fiber.

60. Rollinson, S. M.
"The colored water-soluble materials of heated bleached kraft pulps."
Tappi, 38, No. 3 (1955), p. 186-192. 9 references.

It has been shown that the rate of formation of colored water-soluble material parallels the total discoloration of the pulp and that this material accounts for from 50 to 60% of the total yellow color formed during accelerated aging. Even though this substance is soluble, it is of considerable importance commercially since it would not normally be removed.

A complete analytical description of this colored material has not been obtained, but all of the evidence obtained support the hypothesis that it is carbohydrate-derived. About 23% of the water-soluble materials are oligosaccharides composed of simple sugars typical of the hemicelluloses. Apparently ligneous materials are absent; this statement is based on the low methoxyl content, dioxane insolubility, and the absence of any of the typical lignin oxidation products when the material is subjected to an alkaline nitrobenzene oxidation. The marked maximum at 280 nm in the ultraviolet absorbance curves for these water extracts is probably attributable to the presence of small amounts of furfural. Although iron is present in these water extractables, the data indicate that it does not noticeably contribute to their color either by its own or by color intensification.

Although no direct evidence has been obtained concerning the water-insoluble colored materials formed during accelerated aging, it is postulated that this color is of the same general nature as the water-soluble color and owes its lack of solubility to either its location in the fiber or to condensation with itself or with the cellulosic constituents of the fiber.

A comparison was made of the color of the water extracts of aged pulp with that formed on heating alkaline hemicellulose solutions. It was found that the color of the water extracts resembles the visible light absorption of pectic acid or arabogalactan solutions while it resembles the ultraviolet light absorption of xylan. It was found that hemicelluloses do not yellow when heated in neutral solution, but they do yellow appreciably when heated in alkaline solution.

About 1/2 the change in $\Delta K/S$ (darkening of color) on thermal aging could be removed by hot-water extraction. The extracted materials contained polysaccharides of average DP about 3.5. The content of sugars in the acid-hydrolyzed extract was glucose, galactose, arabinose, mannose, and xylose in the order 1, 2.5, 4.5, 5, and 10.0 by weight. A total of about 24% sugars was found in the extracted material, but no lignin residues (vanillin). Ash content was 5 to 25%, principally containing Ca, Mg, traces of Al, Mn, Cr, Cu, etc. A major portion of the colored material was precipitated when $Ca(OH)_2$ was added.

With charcoal, 87% of the color in the extract could be removed, representing 16% of the total organic materials in the water extract. At 98% discoloration, 50% of the organic matter was removed. Dialysis removed 71% of the organic material and reduced the ash content to 5.7% of the initial (undialyzed) condition.

1956

61. Giertz, H. W., and McPherson, J.
"Brightness reversion of bleached pulps. II. Alkali extraction of yellowed material."
Svensk Papperstidning, 59, No. 3 (1956), p. 93-97. 5 references.

Sulfate pulp that had been overbleached with hypochlorite at pH 8 was heated 40 hours at 110°C to discolor it, then extracted to yield fractions similar to alpha-, beta-, and gamma-cellulose. The latter amounted to 13.6% of the original spruce wood. The alpha-cellulose material was white and highly resistant to further thermally induced discoloration. The beta-cellulose fraction was also white. The orange-colored gamma-cellulose fraction could be further separated by precipitation in ethanol to yield perhaps a quarter of the total components as a white powder. A lesser portion was isolated as a brown powder and also as a brown syrup; the latter was strongly acidic and had a high content of carbonyl groups.

The authors demonstrated that practically all the colored material could be extracted with 10.9% LiOH. It was concluded that most of the colored substances were formed as degradation products of the hemicellulose fraction, perhaps largely owing to carbonyl groups induced in this fraction during bleaching. This agrees with 1945 findings of Giertz (*Svensk Papperstidning*, 48, No. 13, p. 317-323) that hemicellulose-rich pulps have a greater tendency to discolor than ones with a lower hemicellulose content. Sodium borohydride treatment of the bleached pulp reduced the degree of discoloration.

62. Roudier, A.

"Progrès récents dans la chimie structurale des hémicelluloses. Répercussion possible dans le domaine pratique."

Recent progress in the structural chemistry of hemicelluloses. Possible practical consequences.

Association Technique de l'Industrie Papetiere Bulletin, No. 3 (1956), p. 90-96. In French. 21 references.

An overview of the chemistry of hemicelluloses as of 1956, by a chemist from the Laboratoire des Services Chimiques de l'État, Paris, who has been active in the field. One of the points made cites recent work that implies that the sugar moieties—glucose, glucuronic acid, and xylose—may represent a sequence in which the basic glucose structure has been modified through oxidation and then decarboxylation (loss of CO_2) in the process of synthesizing the components of hemicellulose in nature.

1957

63. Hamilton, J. K., and Quimby, G. R.

"The extractive power of lithium, sodium, and potassium hydroxide solutions for the hemicelluloses associated with wood cellulose and holocellulose from western hemlock."

Tappi, 40, No. 9 (1957), p. 781-786. 27 references.

It has been found that sodium and lithium hydroxide are much more effective than potassium hydroxide in their relative extractive power for the mannose-containing polysaccharides (hemicelluloses) which are associated with wood cellulose or holocellulose produced from western hemlock. Results of these experiments showed that these three alkali hydroxides have nearly equal solvent power over a wide range of concentration for xylose, arabinose, galactose, and uronic acid-containing polysaccharides (hemicelluloses). On the other hand, potassium hydroxide is much less effective and sodium and lithium hydroxides are much more effective in dissolving mannose-containing polysaccharides.

The experiments reported herein conclusively show that nearly all of the resistant mannose-containing polymers (glucomannans) can be removed from wood cellulose or holocellulose by using a suitable extraction procedure involving sodium hydroxide at a concentration of 2.77N (10%) to 5.39N (18%) at a temperature of 25°C. In conjunction with previous mechanical disintegration and mild acid hydrolysis, in order to disrupt the primary wall, the extracted holocellulose residue upon hydrolysis has been shown to contain only 0. 18 to 0.20% mannose based on wood. Similar results have been observed with other softwoods. In the case of certain hardwoods treated in analogous fashion, the mannose content of the extracted residue was found to be as low as 0.03 to 0.05% based on wood.

These results taken together indicate that most of the resistant mannose-containing polymers are not chemically combined with the cellulose. If suitable extraction conditions are followed, they can be removed to a very large extent, if not completely, leaving a residue (28 to 32%) of high viscosity that is essentially pure cellulose.

The least soluble hemicelluloses, the glucomannans, have a short chain length and a ratio of D-glucose to D-mannose of 1 to 3. The xylan polyuronides vary widely in DP and have a ratio of D-xylose and 4-O-methyl-D-glucuronic acid of about 4 or 8 to 1. A table is given showing the effect of the concentration of lithium and potassium hydroxide on the ratio of alpha-, beta-, and gamma-cellulose left behind in an extracted pulp which originally contained 87.8% alpha cellulose.

64. Hamilton, J. K., and Thompson, N. S.

"Graded acid hydrolysis studies of a xylan polyuronide associated with wood cellulose from western hemlock."

Journal of the American Chemical Society, 79 (1957), p. 6464-6469. 36 references.

The hemicelluloses which remain associated with wood cellulose produced from western hemlock by the sulfite process have been shown to be a mixture of two families of polysaccharides, namely, glucomannans and xylan polyuronides. The predominant polysaccharides were found to be the glucomannans in which the ratio of D-glucose to D-mannose was one to three. The xylan polyuronides varied in their ratio of 4-O-methyl-D-glucuronic acid to D-xylose from one to four to one to eight. Oxidation studies carried out on these indicated that both of the above hemicellulose series were linked predominantly by 1-4-beta-glycosidic bonds, also that the glucomannans were predominantly straight chain and the xylan polyuronides were slightly branched. This paper is concerned with the partial degradation of the xylan polyuronide components of this wood cellulose system by graded acid hydrolysis and the isolation, identification, and characterization of certain members of the oligouronides so obtained. The significance of this publication in the present context is that it shows how detailed information can be obtained on the structure of hemicelluloses.

1958

65. Machell, G., and Richards, G. N.
"Stabilization of cellulose towards alkaline degradation."
Tappi, 41, No. 1 (1958), p. 12-16. 20 references.
The authors note that the degradation of cellulose by boiling in dilute hot alkali is a process that begins at the reducing (aldehydic) end group of the cellulose molecule and proceeds by a progressive uncoupling of terminal glucosidic moieties until the process is eventually "stopped" by the creation of a D-glucometasaccharinic acid end unit. Their research demonstrates that, when pre-swollen by 2.5 to 5N NaOH, an initially highly crystalline cellulose I of DP 150 has an increased tendency to form such "stopping" end groups and, consequently, suffers a much lower loss of weight when treated with dilute alkali at 100°C. The presence of calcium ions also increases the tendency to form the "stopping" metasaccharinic acid end group. Forming a methyl glucoside by methylating the OH group on carbon atom number 1 also stabilizes the cellulose. Prior oxidation with hypohalite ions (such as hypochlorite), which yields acidic end groups, or reduction with sodium borohydride, which yields an alcoholic end group, reduces the solubility of both cellulose and hemicelluloses in dilute hot alkali. Apparently hemicelluloses degrade in oxygen-free alkali in much the same mechanism as cellulose itself, for the authors mention that xylan degrades in dilute sodium hydroxide to yield mainly xyloisosaccharinic acid. In concentrated alkali the amount of cellulose lost by degradation with the oxygen-free alkali increases with increased alkali concentration from about 10 to 20%. However, in the range of 0.5 to 5% alkali, the concentration has very little effect on the amount dissolved.

66. Timell, T. E., Glaudemans, C. P. J., and Gillham, J. K.
"Recent studies on the polysaccharides of white birch and other hardwoods."
Pulp and Paper Magazine of Canada, 59 (1958), p. 242-252. 90 references.
As of the date of this paper, the authors state that it has only been in recent years that information has been obtained concerning the structure of the main wood hemicelluloses. The hemicellulose isolated from white birchwood contained an average of 197 1,4-linked beta-D-xylopyranose residues with, on the average, every eleventh unit carrying a single side chain of 4-O-methyl-D-glucuronic acid, attached by an alpha-glycosidic bond to C2 of the xylose. The chain length distribution had one maximum. The number-average DP of hemicelluloses from four hardwoods (white elm, yellow birch, white birch, sugar maple) ranged from 197 to 178. The hemicelluloses studied varied in structure from those having about one single unit of 1,2-linked 4-O-methyl-D-glucuronic acid group for every seven to ten anhydroxylose units in the chain. Softwood xylans contain fairly large amounts of terminal L-arabinofuranose residues and contain more acid side groups than hardwood hemicelluloses.
Evidence obtained thus far does not suggest that there is a direct chemical link between the cellulose and hemicellulose portions of white birchwood.

1959

67. Jappe, N. A., and Kaustinen, D. A.
"The carbonyl group and brightness reversion."
Tappi, 42, No. 3 (1959), p. 206-208. 8 references.
Increasingly drastic periodate oxidations of sulfite and kraft pulps and of hemicellulose isolated from aspen-sulfite pulp material gave the same type of unexpected reversion behavior pattern, in that further oxidation above a low copper number of 4 did not result in further increases in reversion. It was recognized that the periodate oxidation used in these experiments would

selectively cleave the 2-3 bond. When compared at copper numbers of 1, 2, 4 and 8, the oxidized sulfite pulp gave the greatest darkening after dry aging one hour at 105°C; the oxidized kraft and the oxidized isolated hemicellulose fraction, an intermediate degree of darkening. Oxidized cotton linters darkened the least. The authors conclude that the results raise questions regarding the relationship between carbonyl content and color reversion. Their data suggest that a maximum in the degree of darkening under dry heat aging is reached at values of copper number between 12 and 16 for periodate-oxidized sulfite pulp.

68. Rånby, B. G., and Marchessault, R. H.
"Inductive effects in the hydrolysis of cellulose chains."
Journal of Polymer Science, 36, No. 130 (1959), p. 561-564. 8 references.
It has been observed that wood celluloses hydrolyze in concentrated phosphoric acid at about double the rate that cotton cellulose does. Reduction of the sulfite cellulose with $NaBH_4$ decreased the rate of hydrolysis under these homogeneous conditions about half. Since it could be shown that the wood cellulose chains in the present study had, on the average, one carboxyl or carbonyl group for every 50 to 100 glucose units, it is not surprising that the rate of hydrolysis of the C1-O-4´C bonds in the chains was enhanced by inductive effects on this acetal linkage.

69. Roudier, A., and Saulquin-Bisson, A. (Part 1); Roudier, A., and Sauret, G. (Part 2).
"Recherches sur le jaunissement des pâtes cellulosiques et des papiers."
Investigations regarding the yellowing of cellulosic pulps and papers.
Association Technique de l'Industrie Papetiere Bulletin, No. 3, Part 1 (1959), p. 109-122; No. 4, Part 2 (1959), p. 187-198. In French. 18 references cited in the two Parts.
Part 1 discusses the hypothesis that much of the colored matter in paper pulps is derived from low-molecular weight decomposition products of the cellulose and hemicellulose components. Characteristics of the ultraviolet absorption curves of these products are discussed, particularly an absorption band at about 258 nm. The idea is put forth that some of the colored products arise owing to condensation. The coloration developed by certain amino acids and cysteine, known to condense with furfural, was demonstrated in the presence of known quantities of furfural. Finally, sheets of paper were dipped in solutions of two pentoses (xylose and arabinose), two aldohexoses (mannose and glucose), a ketohexose (fructose), and two uronic acids (galacturonic and gluconic) which resulted in appreciable discoloration at 105°C and 120°C at add-on weights from 0. 1 to 1.0% of the pulp. The greatest discoloration was attained with the uronic acids, followed by xylose and arabinose. Part 2 reports investigations on the effect of humidity, nitrogen and oxygen on thermal discoloration, leading to the conclusion that discoloration depends on two successive stages: one is hydrolysis which is enhanced by high humidity, and the second, a process of condensation of the low-molecular-weight fragments, which is not aided by oxygen. Oxygen, in fact, resulted in a slight degree of bleaching.

1960

70. Roudier, A., and Sauret, G.
"Recherches sur le jaunissement des pâtes cellulosiques et des papiers."
Investigations regarding the yellowing of cellulosic pulps and papers.
Association Technique de l'Industrie Papetiere Bulletin, No. 2, Part 3 (1960), p. 53-64. In French. 12 references.
Xylose, glucose, and methylated products were added to pulps to learn the effect of these compounds during thermal aging. Xylose and glucose caused the greatest discoloration and loss of DP, with xylose causing the greater degree of change. Blockage of the terminal aldehyde groups almost completely suppressed this action. Washing with water recovered 50 to 80% of the discoloration after thermal aging at low humidity, but only 30 to 65% after aging at higher moisture levels. The conclusion is also reached that depolymerization, accelerated by the presence of aldoses, precedes discoloration. Increased humidity enhances the loss of DP. The acidity that develops in the course of aging accelerates the hydrolytic decrease in DP. Hence, degradation is autocatalytic, in the sense that the products of hydrolysis lead to agents that will contribute to further degradation.

1962

71. Axelsson, S., Croon, I., and Enström, B.
"Solution of hemicelluloses during sulfate pulping. I. Isolation of hemicelluloses from the cooking liquor at different stages of a birch soda cook."
Svensk Papperstidning, 65 (1962), p. 693-697. 22 references.
The solution and eventual readsorption of the xylan during the cook, and the modification of the xylan in the cooking liquor at different stages of the alkaline cook were studied. Precipitable polysaccharides from the cook liquors at 100° and 150°C amounted to 4.5% and 8% of the wood, respectively. Up to 160°C, the neutral carbohydrate composition was galactose: glucose: mannose: arabinose: xylose = 4:0.4:0.3:0.9:94.4. Up to 160°C, 30-50% of the dissolved xylan material consisted of 4-O-methyl-D-glucuronoxylan, DP 130-180.

72. Spinner, I. H.
"Brightness reversion: a critical review with suggestions for further research."
Tappi, 45, No. 6 (1962), p. 495-514. 155 references.
Purified as well as unpurified wood pulps, of course, contain hemicellulose. It is therefore of interest to inquire whether the color-forming groups in these pulps are in the cellulose or the hemicellulose component. In this extensive review, the author cites the findings of a number of investigators and reviews what little is known of the role that hemicellulose plays in the reversion of different types of pulps. The effect of alkali treatment at various stages of pulp preparation is also noted. Extraction of accelerated-aged pulp with water and a low percentage of alkali solution indicated that the color-forming materials were probably degradation products of the hemicellulose components. The investigations of Rollinson [*Tappi*, 38, No. 10 (1955), p. 625-631] and Giertz and McPherson (1956) are cited to show that many of the water-soluble colored materials are in the gamma-cellulose fraction. He also quotes Centola [*Bulletin Association Technique de l'Industrie Papetiere*, 5 (1951), p. 111-115] as reporting that reversion in his research could be correlated with pentosan content. Marum [*Tappi*, 39, No. 6 (1956), p. 390-394] is also cited as finding that the best correlation with reversion was obtained against the beta- and gamma-cellulose content.

1963

73. Capon, B.
"Intramolecular catalysis in glucoside hydrolysis."
Tetrahedron Letters, No. 14 (1963), p. 911-913. 4 references.
To follow up the suggestions of Rånby and Marchessault (1959) that the hydrolysis of polysaccharides containing glucuronic acid residues could be speeded up by an intramolecular general-acid catalysis, the author shows that the hydrolysis of para-orthocarboxyphenyl-beta-D-glucoside is about 10^4 times faster than would be expected without an intramolecular assist. These results indicate an additional reason why hemicelluloses may hydrolyze faster than cellulose itself.

74. Schuerch, C.
"The hemicelluloses."
In *The Chemistry of Wood*, ed. Browning, B. L., Interscience Publishers, a division of John Wiley & Sons, New York (1963), p. 191-246. 287 references. Addendum by N. S. Thompson. 41 references.
The chemistry of hemicelluloses is reviewed under the following headings: isolation and purification, proof of structure, distribution in plant tissues, detailed structure of hemicelluloses, minor linkages, practical importance of hemicelluloses, their biochemistry. The addendum surveys the general direction that investigations appear to be taking in the time immediately prior to publication.

75. Shafizadeh, F.
"Acidic hydrolysis of glycosidic bonds."
Tappi, 46, No. 6 (1963), p. 381-383. 16 references.
In this paper the basic structural and chemical reactivity of the compounds forming cellulosic materials are reviewed. The pyranose (6-membered ring) form in the chair conformation is in its lowest energy, with all substituents staggered; there is, therefore, no non-bonding interactions and no angle strain. The furanose (5-membered ring) and septanose (7-) are strained rings.

In methyl beta-D-glucopyranose the CH_2OH and OH groups are in the equatorial position, whereas in methyl beta-D-galactopyranoside, an OH group is in the axial position. This gives the latter a higher energy.

The rate of hydrolysis of glycosidic (C-O-C) bonds is related to the structure and conformation of these sugars. Empirical information shows that the more strained molecules hydrolyze faster. Methyl alpha-D-galactofuranoside and -septanoside at .01N HCl acid and at 95°C hydrolyzed 100 times faster than methyl alpha-D-galactopyranoside. It has also been shown that the rate of hydrolysis generally increases with the number of axial groups in the favored chair conformation. This was shown for a pentose, hexose, and heptose series by selecting methyl alpha-D-mannopyranoside's rate of hydrolysis as the standard for comparison.

An important consideration in the rate of hydrolysis, of course, is the type of substituents in the ring. Deoxy sugars, for example, are hydrolyzed much faster than normal sugars. When the hydroxyl group is replaced with an amino group the hydrolysis effect is reversed, the aminosubstituted methyl glycoside becomes highly resistant to acid hydrolysis. The presence of a carboxyl group on carbon atom 6 also has a small effect.

76. Wayman, M.
"Alkaline extraction."
In *The Bleaching of Pulp*, Tappi Monograph Series No. 27, ed. Rapson, W. H. (1963) p. 67-103. 61 references.

The various sections of this exposition include a discussion, first of all, of alkali extraction as a first stage; curves are shown of the pulp yield, pentosans, soda solubles and alpha-cellulose content versus the concentration of cold alkali in the case of unbleached sulfite and sulfate pulps. Data also illustrate the different results between hot and cold extractions. There follows a section on neutralization after chlorination. Among the topics mentioned here is the fact that "stopping reactions" seem to halt the removal of alkaline degradation products at a level of about 96% alpha cellulose, the so-called "alpha ceiling." There is also reference to alkali-resistant pentosans that can show up in the alpha, beta and gamma fractions of cellulose according to the work of Richter [*Tappi*, 38, No. 3 (1955), p. 120-150]. Much of the discussion, of course, focuses on the effects of extraction conditions on yield.

Cold extraction with alkali is required to attain alpha-cellulose above the 96% "ceiling." Cold solutions at 4 to 20% concentration are said to have a more selective action on the nonalpha-cellulose components than hot dilute alkali; swelling of cellulose is greatest. Alkali extraction also helps eliminate resins. The main disadvantage of the much-used industrial practice of hot-alkali extraction is the loss of yield. Yields might be improved by facilitating the "stopping reaction," but these procedures correspondingly increase the same tendency in hemicelluloses.

1964

77. Rapson, W. H., and Corbi, J. C.
"Colour reversion caused by degradation products of oxidized cellulose."
Pulp and Paper Magazine of Canada, 65 (1964), p. T459-T466. 12 references.

In this important study, the authors selected about 34 acidic, hydroxy, ketonic, and aldehydic compounds having six to two carbon atoms, representing substances that might be formed as decomposition products in cellulose, and impregnated filter paper with them after which they proceeded to age the test sheets thermally to learn which compounds tended to yield the greatest discoloration. Aging was carried out at 150°C, at 70°C, and 75% RH for 18 hours, and for one year at 25°C and 50% RH. The relative behavior of the different materials was not always the same under the three different conditions of aging. Uronic acids, compounds with a carboxylic acid group at one end and an aldehyde at the other, proved to be particularly prone to color formation. As far as hemicellulose chemistry was investigated, the authors showed that xylan and polygalacturonic acid did not lead to strikingly yellow discoloration. They concluded that hemicelluloses only cause yellowing if they are oxidized and hydrolyzed, a statement that could just as well be made about cellulose. However, they do mention that hemicelluloses are sensitive to oxidation and are readily hydrolyzed. Arabinouronic acid led to intense discoloration. 4-O-Methylglucuronic acid, which occurs in the hydrolysis products of hemicelluloses, also resulted in considerable yellowing.

1965

78. Croon, I., Dillén, S., and Olsson, J.-E.
"Brightness reversion of birch sulfate pulp."
Journal of Polymer Science, Part C, No. 11 (1965), p. 173-195. 51 references.

This study focuses on the cellulosic and extractive compounds which cause brightness reversion or yellowing in birch sulfate pulp. Regarding hemicelluloses, it has been determined that alkaline treatment increases the brightness stability of pulps since it removes most of the hemicellulosic compounds. Among certain pulps there seems to be a relationship, found by others, between brightness reversion and pentosan content. The colored aqueous or alkali extracts of aged pulps were shown to contain oligo- and monosaccharides and acidic products. Carbonyl groups render the cellulose unstable in alkali. The increase in brightness stability on alkaline treatment of pulps is due to the dissolution of the most strongly oxidized part of the cellulose as well as to the removal of the hemicellulosic materials from alkali treatment. For a number of pulps a relation has been found between brightness stability and uronic-acid content. Uronic acid increases the yellowing but only when it is the monomer form. Treatment with chlorine can introduce chlorine into the lignin and extractive components which can yield hydrochloric acid during accelerated-aging tests. In contrast, the use of chlorine dioxide does not result in as much harmful liberation of hydrochloric acid.

The variables controlled in these experiments were time, temperature, RH percent, available oxygen and exposure to light (ultraviolet and visible). The pulps were aged in darkness, in a dry and wet atmosphere (2% and 80% RH). The dry-aged sheet reflected less of the shorter wavelengths and so it appeared more yellow. The wet-aged sheet absorbed longer wavelengths. Thus, the differing aging conditions have resulted in a difference in the colored products.

Alkaline high temperature treatment (AHTT) improved the brightness stability of both dry and wet pulps but there was a 2-4% loss of yield. A decrease was noticed in the following: in uronic acid content, in the copper number of the xylan fraction, and in the total carboxyl content. The material extracted from the pulp was mostly composed of xylan. The xylan removed had a high uronic-acid content. Uronic acids extracted appear to be levulinic acid and xylosaccharinic acids. This work is in agreement with the fact that alkaline treatment does increase the brightness stability.

79. Rydholm, S. A.
"I. Sulfite process (3E). Carbohydrate reactions of the sulfite cook." (Chapter 9)
In *Pulping Processes*, Interscience Publishers, a division of John Wiley & Sons, Inc., New York (1965), p. 503-539. 109 references.

An extensive review of the literature is given which refers to the fate of the hemicelluloses during sulfite pulping. The process is used to produce grease proof and glassine papers because lignin removal is efficient. Along with the ability to control hemicellulose content, however, the properties of the carbohydrates remaining in the pulp are also impaired. The extent of hydrolytic attack depends not only on the chemical structure of the hemicelluloses but also on their location within the wood-fiber structure. For example, whereas galactosidic and xylosidic bonds are able to hydrolyze 5 times and mannosidic bonds 3 times as fast as glucosidic, because of differences in accessibility, glucomannan and xylan dissolve out of the pulp during acid sulfite processing about 15 to 30 times faster than cellulose. Curves are given showing the mannose and xylose composition remaining in the pulps, up to 13% and 5%, respectively, depending on treatment. Data are also given on the distribution of molecular-weight fractions in pulps processed to different average DP's. See also 109.

80. Rydholm, S. A.
"Carbohydrate-removing methods." (Chapter 15)
In *Pulping Processes*, Interscience Publishers, a division of John Wiley & Sons, Inc., New York (1965), p. 992-1023. 29 figures, 63 references.

Chapter 15 provides an excellent overview of the literature and subject of pulp purification by extraction in both hot and cold alkali. Purification in cold concentrated alkali mainly involves physical changes (swelling); little alkali is consumed; the hemicelluloses, almost exclusively intermicellar materials, are removed. A curve is given, Figure 15.3, to show the changes in alkali solubility of rayon pulp (a degraded pulp) with increasing concentration of alkali, and the considerable fraction of beta-cellulose present. In contrast, the solubility of a less degraded pulp (Mitscherlich type) is also shown in which only a slight difference in solubility between 10% and 18% NaOH is seen, representing principally the hemicellulose fraction. Swelling takes place almost instantly.

In the case of hot-alkali purification, very little swelling occurs and chemical reactions are involved. Hot-alkali purification is ordinarily used only for sulfite pulps, the author states. The dominating chemical reaction is the "peeling reaction" which starts at the aldehydic end groups in the carbohydrate chains. An average of 50 glucose units per chain are peeled off in the case of glucose. Glucomannan is less stable than glucan and glucan less stable than xylan. Thus, glucomannan is the main hemicellulose removed. Moreover, hardwood pulps, containing higher xylan content than softwoods, are more difficult to obtain in a highly purified form.

The loss of yield (increased solubility in hot dilute alkali) is related to the copper number (or carbonyl content). Among the products found in solution after this treatment were pyruvic acid, $CH_3COCOOH$, and formic acid, lactic acid, and beta, gamma-dihydroxy-butyric acid. Aldehydes such as methyl glyoxal, CH_3COCHO, are also found. The author presents detailed information of the chemistry of the alkaline cooking processes in Chapter 9 (see also 79).

81. Rydholm, S. A.
"Characterization of pulp properties." (Chapter 20)
In *Pulping Processes*, Interscience Publishers, a division of John Wiley & Sons, Inc., New York (1965), p. 1108-1110, 1116-1118. 37 references.

The first sub-section cited is entitled, "Determination of residual hemicelluloses." The author remarks that alkali solubility methods to characterize the content of hemicelluloses are not reliable and are therefore included under methods of physically characterizing pulps rather than chemically. Paper chromatographic methods are currently used to detect the individual hemicellulose monomers. Acetyl and O-methyl groups can be determined by separate chemical steps, although the first is present only in unbleached sulfite pulps in some quantity; the latter groups are also found in lignin. Pentosans are determined by converting them to furfural and the author mentions possible errors that can occur in the presence of lignin or uronic acids.

The solubility of sulfite pulp in cold 18% NaOH gives a fair indication of hemicellulose content; errors are more pronounced in kraft pulps, especially hardwoods. Use of cold 5% NaOH yields a xylan-rich fraction but the method is now little used. In 10%, much of the degraded cellulose dissolves so that the difference between the solubility in cold 10% and in 18% NaOH is sometimes called the "rest-hemi" fraction. In the case of unbleached pulps, the lignin should be removed by chlorite oxidation first.

1966

82. Croon, I., Dillén, S., and Olsson, J.-E.
"Brightness reversion of birch sulphate pulp."
Svensk Papperstidning, 69 (1966), p. 139-149. 51 references.

The authors begin with a one-page review of the subject of brightness reversion. Memorable statements are the following: "While a high degree of delignification during digestion has been found to decrease the brightness reversion of the pulp after bleaching, it is invariably accompanied by removal of the other wood constituents so that it is not certain that the effect is due entirely to removal of lignin . . . the increase in brightness stability resulting from alkali treatment is due partly to the dissolution of the most strongly oxidized part of the cellulose as well as to the removal of hemicelluloses." A correlation between reversion and uronic acid content has been found for certain pulps; apparently, the hydrolytic liberation of free uronic acids can lead to considerable discoloration. The principal new research described in this publication involves the effect of alkaline high-temperature extraction as a method for reducing thermally induced discoloration of pulps.

83. Springer, E. L.
"Hydrolysis of aspenwood xylan with aqueous solutions of hydrochloric acid."
Tappi, 49, No. 3 (1966), p. 102-106. 20 references.

Aqueous solutions of hydrochloric acid were employed in hydrolysis of finely-divided aspenwood. Rate of removal of the first 60% of the xylan from the wood was determined by means of weight loss measurements. Reaction temperatures from 60 to 120°C and acid concentrations from 0.05 to 6.0M were investigated. The rate of weight loss was fitted to a first-order rate equation. A linear relationship was found to exist between the logarithm of the first-order rate constant and the Hammett acidity function. The activation energy was found to be 28.2 kcal per gram mole. These results suggest that the rate of disappearance of xylan from the solid phase may be directly proportional to the rate of the rate-controlling reaction taking place. On the basis of the evidence, it was not possible to establish the identity or the mechanism of the rate-controlling reaction. The evidence, however, is in agreement with the hypothesis that the rate-controlling reaction is hydrolysis of the beta-D-(1→4) acetal linkages in the xylan hemicellulose polymer.

1967

84. Beélik, A.
"Transformation of free monosaccharides within pulp sheets during heat aging."
Pulp and Paper Magazine of Canada, 68, No. 3 (1967), p. T135-T142. 38 references.
The isolation of crystalline L-arabinose and the paper chromatographic identification of several mono- and oligo-saccharides proved that both cellulose and hemicellulose chains were cleaved in the heat aging of pulp and cotton linters. The isolation of L-ribose indicated epimerization of liberated L-arabinose. Experiments with model compounds showed epimerization occurs generally when free aldoses are exposed to heat-aging conditions in cellulose sheets. All aldoses contributed to the color and acidity which developed in the heat-aged celluloses. Modified aldoses that lacked reducing power and the ability to epimerize underwent no transformation during heat aging and did not contribute to pulp reversion. Modified aldoses that lacked only one of these two functions underwent limited transformation and contributed less to pulp reversion than glucose. Modified aldoses that combined in one molecule reducing and acidic functions as well as two sites for epimerization were completely consumed in heat aging and contributed much more to pulp reversion than glucose. The enediol intermediates of epimerization are also key intermediates in the formation of furfurals, which have long been regarded as color precursors in heat reversion. The sequence by which color is formed from polysaccharides via monosaccharides, enediols, and furfurals has thus been substantiated by the identification of some of the products arising in the earliest phases of this process.
When free sugars were added to test sheets, the greatest yellowing of the pulps was generated by mannose, followed by galactose and xylose, in that order. The color and the acidity of the eluate, on the other hand, followed in the order xylose (greatest), arabinose, galactose, mannose, glucose (least). When kraft pulps were hydrolyzed, arabinose, xylose, xylobiose, and xylotriose were found in the degradation products. Sulfite pulp yielded mannose rather than arabinose but the latter was observed in the products from cotton linters.

85. Beélik, A., Conca, R. J., Hamilton, J. K., and Partlow, E. V.
"Selective extraction of hemicellulose from softwoods."
Tappi, 50, No. 2 (1967), p. 78-81. 15 references.
A new procedure based on the principle of selective extraction was developed for the isolation of the three main hemicelluloses from softwood holocelluloses. Selectivity is assured by impregnating the holocellulose with 1-2% barium hydroxide in the first of three extraction steps. Impregnation is particularly effective when used in conjunction with 10% potassium hydroxide. Xylans readily dissolve in this medium, while the dissolution of mannose-containing polymers is largely suppressed. After removal of the barium hydroxide, the easily soluble galactoglucomannan can be extracted from the holocellulose with 1% sodium hydroxide in the second step of the sequence, and the glucomannan in the last step with 15% sodium hydroxide. All three hemicelluloses obtained in this fashion are reasonably homogeneous, and thus tedious secondary purifications by precipitation can be kept to a minimum.

86. Browning, B. L.
"Isolation and separation of the hemicelluloses." (Chapter 26)
In *Methods of Wood Chemistry*, Volume II, Interscience Publishers, a division of John Wiley & Sons, New York (1967), p. 561-587. 80 references.
Although the major portion of hemicelluloses can be extracted from deciduous woods and straws, the remaining portion is not available to alkali and can be extracted only after the lignin has been removed. In the case of coniferous woods, preliminary delignification is generally essential. Extraction has been done in 1N NaOH (about 4%) for 48 hours at room temperature. This prevents the extreme swelling caused by 16 or 24% alkali. The glucuronoxylans are readily soluble, whereas the glucomannans are sparingly soluble. Characterization of hemicelluloses involves an analysis of the sugars, uronic acid groups, methoxyl, and acetyl groups.

87. Sjöström, E., and Enström, B.
"Characterization of acidic polysaccharides isolated from different pulps."
Tappi, 50, No. 1 (1967), p. 32-36. 31 references.
The content and properties of acidic polysaccharides remaining in softwood (pine) and hardwood (birch) pulps after kraft, two-stage sulfite, and bisulfite cooking procedures have been compared. The hemicellulose components were separated into fractions. The carbohydrate composition of the resulting fractions and the pulp residues were analyzed by paper and gas-liquid chromatography. The xylan-rich fractions were further characterized by potentiometric titration, methoxyl analysis, and viscometry. Other analyses were performed to determine the carboxyl content of the original and completely extracted (hemicellulose-free) pulps. The results indicate a considerable heterogeneity of the xylan fractions, both with respect to 4-O-methylglucuronic acid content and molecular weight. Weight-average DP's of 50 to 125 are reported for various xylans, and the compositions are given.
Some comments are offered regarding the remarkably low uronic acid content of the bisulfite pulps. Considering the acidic hydrolysis of polysaccharides, recently reported results indicate that the hydrolysis rate of uronosidic linkages in relation to glucosidic linkages increases with increasing pH. This could be one reason for the low uronic acid content of the bisulfite pulps, since the cooking takes place under moderately acidic conditions and at a high temperature. However, other factors are probably also involved. For example, only a very slight decrease in the uronic acid content has been reported to occur when wood xylan fractions are heated under conditions similar to those prevailing during a bisulfite cook. This demonstrates that experiments with model substances are not very representative for the heterogeneous system of wood pulping. The conditions inside the wood structure, where the reactions during pulping take place, obviously deviate markedly from those in a homogeneous solution. It must further be taken into consideration that wood xylan itself is far from homogeneous, but is composed of fractions of greatly varying uronic acid contents. In view of this, it is probable that a fractionation takes place during pulping. Thus, such factors as temperature and pH can greatly influence the distribution of acidic fragments between the liquid and solid phases.

1968

88. Sjöström, E., and Eriksson, E.
"The influence of carboxyl and carbonyl groups on the brightness stability of bleached pulps."
Tappi, 51, No. 1 (1968), p. 16-19. 20 references.
Four different types of bleached pulp were tested for their brightness stability: pine and spruce sulfite pulps and pine and birch kraft. The resin was removed with a mixture of 1 ethanol: 1 benzene, then ethanol and finally with water. Reduction of the carbonyl groups in the pulps was done with a $NaBH_4/Na_2CO_3$ solution. The esterification of the carboxyl groups was done with propylene oxide. After each treatment, the pulps were converted to an H^+ ionic form with .1N HCl or to Na^+ with .1M NaCl solution.
Brightness stability after aging at 105°C was increased by extracting the resins. Even birch kraft pulp, which had a low resin content of .07%, showed an increased brightness. The spruce sulfite pulp, with a high resin content (.60%) showed the most pronounced increase in brightness.
$NaBH_4$ treatment improved the brightness stability and the initial brightness of the pulps. However, esterification with propylene oxide stabilized the brightness more than any other treatment.
The effect of esterification was less pronounced after reduction than before, indicating an interaction between the carbonyl and the carboxyl groups and in the fact that the influence of the carbonyl is negligible in the absence of carboxyls.
Pulps that have been both reduced and esterified are highly stable. Whatever brightness reversion is observed after such treatment is evidently due to remaining carbonyl and carboxyl groups and new groups that may be formed during aging. The pulps tended to be more stable in the hydrogen ion form than in the sodium ion form.
Aging the pulps at a higher temperature, 120°C, increased the brightness reversion. In the case of the birch kraft pulp, the esterification alone stabilized the pulp to a high degree.
Aging the pulps in a wet atmosphere at 80°C showed brightness reversion of the same order of magnitude as in the dry atmosphere at 120°C. In a wet atmosphere the individual pulps seemed to show more variation among themselves. Extraction had a lower effect on brightness reversion.
The "acidic xylan," carboxyl groups attached to a xylan backbone as methylglucuronic acid residues, appears to be the most sensitive polysaccharide component with respect to brightness stability.

89. Springer, E. L., and Zoch, L. L.
"Hydrolysis of xylan in different species of hardwoods."
Tappi, 51 (1968), p. 214-218. 12 references.

Wood species appear to have little effect on the rate of hydrolysis of xylan, the most easily hydrolyzed hardwood constituent. Thin cross-sections of quaking aspen, paper birch, American elm, and red maple were hydrolyzed in 0.100M HCl at 120°C and in distilled water at 170°C; initial and final xylan-removal rates were observed. Initial-rate data for each set of conditions were fitted to a first-order rate equation. With 0. 1M HCl at 120°C, only small variations, probably caused by differences in the proportions of 4-O-methyl-glucuronic acid side chains on xylan polymers, were found among the initial removal rates of the species. With the distilled water at 170°C, much larger differences in initial rates were found. Some of this variation can be attributed to differences in the proportions of glucuronic acid side chains on the xylan polymers, but the other causes are unknown. Removal rates for the more resistant portion of the xylan were remarkably similar for all species under both sets of conditions.

1970

90. Haun, J. L.
"Hemicelluloses."
In *Handbook of Pulp and Paper Technology*, 2nd edn., ed. Britt, K. W., Van Nostrand Reinhold, New York (1970), p. 25-31. 35 references.

In this review article Haun discusses hemicelluloses as they exist in wood and in pulps, after the wood has been treated. Hemicelluloses of wood consist of several types of amorphous, polymeric carbohydrates that occur throughout the woody structure of plants.

Wood Hemicelluloses

Hardwoods contain more hemicelluloses than softwoods. Both hard- and softwoods contain essentially the same hemicelluloses. Hardwood hemicelluloses consist mainly of the pentose sugar xylose, while in softwoods they consist mainly of the hexose sugar mannose. Both wood types contain small amounts of miscellaneous hemicelluloses consisting largely of the hexose sugar galactose. The five basic sugars composing wood hemicelluloses are:

three hexoses – glucose, mannose, galactose
two pentoses – xylose and arabinose.

All of these sugars exist in the 6-membered pyranose ring form, except for arabinose which is in the 5-membered furanose ring. Only xylose does not have a free $-CH_2OH$ group on C-5. Two uronic acids are found in wood: galacturonic and glucuronic acid. The $-CH_2OH$ group of the corresponding sugar has been oxidized to a $-COOH$ group. Both occur as side chain units and galacturonic is also present as a polymer, pectin (see Figure 1).

Structure:

Xylans—Hardwoods contain high amounts of xylan and it is the principal source of pentosan content. Xylan is known as 4-O-methyl-D-glucuronoxylan (see Figure 2). In birch the xylan backbone is made up of 150-200 anhydro-beta-D-xylose units in a straight chain with 1-4' links. The side chains of 4-O-methyl-D-glucoronic acid are, on the average, on every 10th xylose unit linked by a 1-2' bond. The O-acetyls are on 7 of the 10 xylose units, on the C-3 more so than on C-2.

Softwoods have a medium amount of 4-O-methyl-D-glucuronoarabinoxylan (in Western Hemlock this constitutes 4-5% of the wood) and it is the major source of pentosan in these woods (see Figure 2). The softwood xylan is similar to the hardwood xylan but differs in that it has an L-arabinofuranose group attached to roughly every 8th xylose unit at C-3, and the 4-O-methyl-D-glucuronic acid attached to about every 6th xylose unit, and no acetyl groups. Among the softwood species, structural variations between the xylans have been observed.

Glucomannans—In hardwoods they are present in small amounts, 2-3%, but in softwoods they can approach 1/6th of the total weight. It is the principal source of non-cellulosic hexosans of most woods. Structurally, it is a straight chain with 1-4' beta bonds. The ratio of D-mannose to D-glucose in hardwoods is 3:4 and in softwoods 1:2. It is partially acetylated with O-acetyls on C-2 and C-3 of about every 4th mannose unit. The DP of glucomannan is low; for example, the DP of the material from Norway Spruce is 60-140 (see Figure 2).

Galactoglucomannans—These have a beta-D-galactose attached to the C-6 of the glucose and mannose units of glucomannans. The galactose content varies from 0-20%.

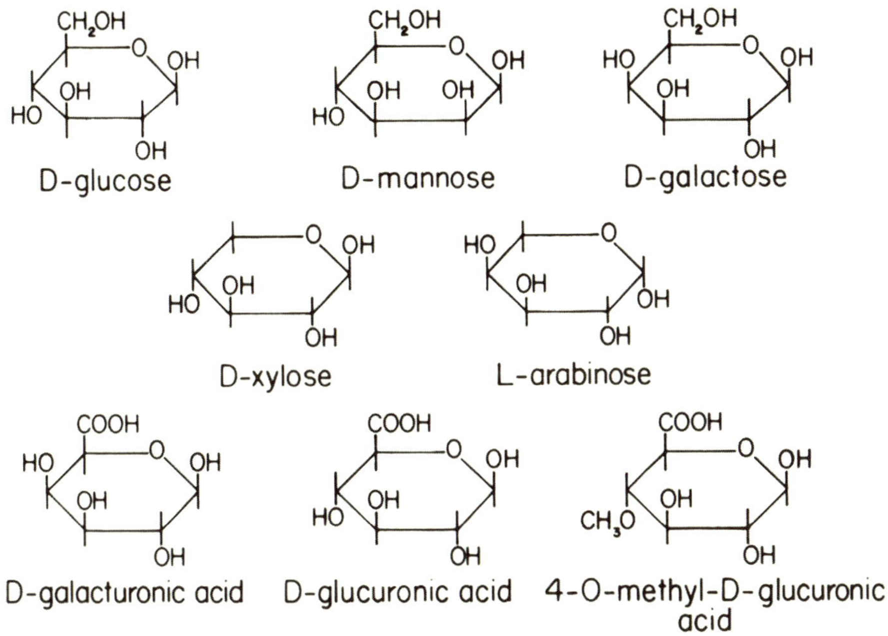

FIGURE 1
Structure of principal sugars that are constituents of wood hemicelluloses.

Arabinogalactans—These are found mostly in softwoods. They are highly branched and water soluble. There is some variation between softwood species but the basic structure is similar. The backbone consists of beta-1,3-linked galactose units with many short chained branches of beta-1,6-linked arabinose and galactose units.

Location:
Two main techniques have been used to determine the distribution of hemicelluloses across the cell wall of wood fibers: microdissection and microphotometric technique.

Isolation:
In isolated state, all the hemicelluloses are water soluble, but when in the wood they are difficult to extract with water. Thus, it is necessary to delignify the wood before extractions are made. So hemicelluloses are extracted from holocellulose preparations.

For extraction, alkaline solvents are used. KOH is preferred over NaOH since KAc (potassium acetate) is more soluble than NaAc in the alcohol that is used to precipitate the hemicelluloses after neutralization of the basic extracts with CH_3COOH. The concentration of the base used ranges up to 25%. At a concentration below 5%, the easily accessible hemicelluloses are removed. Above 5%, all hemicelluloses are removed with optimum removal at concentrations ranging front 10-25% depending on the wood species. A concentration of 18% is most commonly used to get the total hemicellulose content of wood. Two disadvantages in doing a basic extraction: 1) reducing end-groups are attacked by an alkaline chain peeling reaction, resulting in low DP's; and 2) the O-acetyl groups are saponified. To avoid this DMSO has been used.

Reactivity:
Glycosidic bonds of the hemicelluloses are susceptible to catalytic cleavage by H^+. The condition for acid hydrolysis to occur depends on the nature of the glycosidic bond, the accessibility of the hemicellulose in the wood, the concentration of the acid used, the temperature, and the time of reaction. The acid hydrolysis reaction of hemicellulose has an activation energy (E_A) of 28 to 30 kcal/mole, similar to that of cellulose. Delignification reaction E_A is 22 kcal/mole.

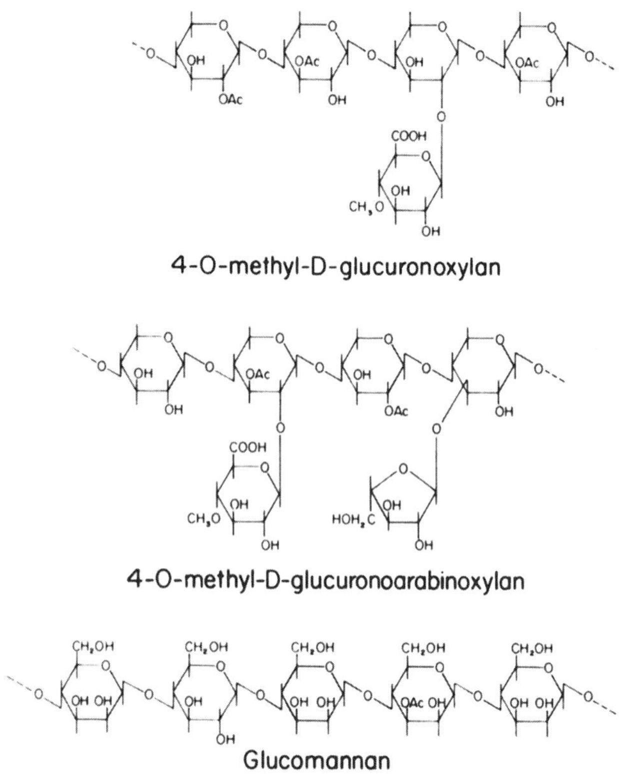

Figure 2
Idealized representation of three major types of hemicelluloses. Ac = acetyl group, CH_3CO-.

The glycosidic linkage between glucuronic acid and the main xylose chain has been shown to be quite resistant to acid hydrolysis. In acid sulfite pulping, acid hydrolysis results in the loss of side chains and in a loss in DP and yield. Alkaline hydrolysis does not occur as readily as acid hydrolysis. It requires high temperatures (150-160°C) and a minimum pH. In alkaline pulping, retention of side chains (arabinose and galactose) and a higher DP are observed when compared to acid sulfite pulping. But alkaline pulping has been found to attack the 4-O-methyl-D-glucuronic acid side chains in hard- and softwood xylans.

Oxidative degradation is another reaction affecting the stability of hemicelluloses. The reducing end-groups are attacked by a hydroxyl ion, "alkaline chain peeling." This reaction is faster than alkaline hydrolysis at most temperatures and alkalinities. Both hydroxyl and hydrogen ions hydrolyze the O-acetyl groups of hemicelluloses.

Pulp Hemicelluloses:

The cooking of wood to pulp causes major changes in the amounts, locations, and structures of the various hemicelluloses. This is also dependent on the wood species and cooking process. Present kraft and sulfite cooking conditions degrade hemicelluloses; thus, their percentage is lower in pulp than in wood.

Hardwoods pulped to the same yield as softwoods will retain more hemicelluloses. Alkaline pretreatment of a sulfite cook (Stora process), and when $NaBH_4$ or polysulfide is injected into the kraft process, will increase the hemicellulose content of the pulp by 50%.

Structure:

The hemicellulose composition found in pulps is a function of the pH at which the wood was cooked. In softwood pulps cooked under acidic conditions, the mannose to xylose ratio is about half of what it is in wood. Glucomannan is still the principal hemicellulose in softwood pulps (Stora and high bisulfite pulps). Arabinose is lost during cooking so glucuronoarabinoxylan changes to glucuronoxylan, an artifact hemicellulose. In softwood pulps cooked under alkaline conditions, the mannose to xylose ratio lowers and arabinose is partly lost. Softwood pulps prepared under very strong alkaline conditions have a very low ratio (about 1) of mannose to xylose. If the alkalinity is too high then artifact hemicelluloses are formed in both soft- and hardwoods.

Hardwood hemicellulose is mostly glucuronoxylan and is affected by pulping also.

The DP of hemicelluloses from the various pulping processes varies considerably. The DP of cooked isolated hemicelluloses is lower than those in the wood, implying that the hemicelluloses are not completely accessible to the liquor. The high DP of hemicelluloses in alkaline pulps is due to the rapid formation of stable metasaccharinic acid end groups.

The alkaline pretreatment of a sulfite process, Stora process, for softwoods is believed to stabilize the softwood hemicelluloses by losing the O-acetyl from the glucomannan and galactoglucomannan.

The uronic acid of both hard- and softwood xylan is fairly resistant to hydrolysis under acidic, neutral and weakly basic cooking conditions.

91. Kass, A., Wangaard, F. F., and Schroeder, H. A.

"Chemical degradation of wood: relationship between strength retention and pentosan content."

Wood Fiber, 2, No. 1 (1970), p. 31-39.

A prior study on the effect of acids and alkalis on wood supported the contention that a low original pentosan content in wood favors resistance to chemical degradation. In the present study, the relationship of pentosan content and its retention to the resistance of wood to chemical attack is quantitatively established. Thirteen wood samples were used: two softwoods, Douglas fir (*Pseudotsuga manziesii*) and Caribbean pine (*Pinus caribaea*); one temperate-zone hardwood, white oak (*Quercus alba*); and ten tropical hardwoods. A low original pentosan content and a high pentosan retention correlate well with resistance to acid degradation as measured by strength retention. The relationship between these two pentosan values and alkaline degradation is weaker, but still significant. Hemicellulose retention is a better indicator of the behavior of wood with respect to alkaline attack. The most useful result is the fact that resistance to chemical degradation can be indicated through chemical analysis for pentosans in a wood.

1971

92. Chang, F.-J., and Kondo, T.

"Studies on brightness reversion of pulps. II. Chemical constituents of bleached pulp involved in brightness reversion."

Journal Japan Wood Research Society, 17, No. 4 (1971), p. 160-167. In Japanese. 10 references.

The authors used a fully bleached kraft and a sulfite pulp (KP and SP) of Japanese red pine, white birch, and beech. In addition, two related commercial pulps were used in the study. By measuring the degree of discoloration before and after ethanol/benzene extraction, and before and after extraction in 10% NaOH at 20°C, the authors were able to calculate the percentage of loss of reflectance that was due to the extractives, hemicellulose, and cellulose. The lignin content of the laboratory-prepared pulps was less than 0.5%. Bar graphs clearly indicate that the removal of hemicellulose contributes greatest to the reduction in color reversion upon dry and humid aging. If one calculates the effect of these constituents on the basis of the percentage of each constituent by weight (0.03 to 0.29% organic extractives and 12.1 to 25.2% hemicelluloses), one finds that the organic (resinous) extractives contribute about 20 to 70 times more to yellowing than the hemicelluloses and the latter 10 to 60 times more than cellulose itself. Thermal aging at high humidity increased the proportion of the net discoloration due to the cellulose and hemicellulose components; the greatest increase under humid conditions was attributed to a high xylose content. [The library of the Institute of Paper Chemistry provides a translation.)

93. Technical Association of the Pulp and Paper Industry.
"Pentosans in wood and pulp."
Tappi Standard T223 os-71, Tappi, Atlanta, GA (1971). 2 references.
Wood contains a certain amount of non-cellulosic carbohydrates called hemicelluloses which consist of both pentosans (xylan and araban) and hexosans (mannan and galactan).
Pentosan content in pulp can be taken as an indication of the retention or loss of hemicellulose during the pulping and bleaching process. Pentosan content in softwoods is about 7 to 10%; in hardwoods 19 to 25%. Since hemicellulose contributes to the strength of paper pulps, a high pentosan content is desirable in certain pulps.
This method is designed for the determination of pentosan content in wood and unbleached or bleached pulp. It involves the boiling of the sample in 3.85N hydrochloric acid to convert the pentosans to furfural, distilling, and determining the furfural in the distillate colorimetrically using orcinol-ferric chloride reagent.

1974

94. Technical Association of the Pulp and Paper Industry.
"Alpha-, beta- and gamma-cellulose in pulp."
Tappi Standard T203 os-74, Tappi, Atlanta, GA (1974). 7 references.
The determination of the three groups as listed in the heading can only be carried out on bleached or delignified pulps. Unbleached and semi-bleached pulps must be delignified before the method can be applied.
The pulp is extracted at 25°C. first with 17.5% followed by a rinse of 9.45% sodium hydroxide. The pulp suspension is separated by means of a coarse glass-frit funnel. After discarding the first 10 to 20 cc of the filtrate, about 100 cc is collected. An aliquot is taken for the alpha determination by oxidation of the soluble components with potassium dichromate (17.5% NaOH and water are used in the blank) and subtracting the percent soluble matter from 100 (see below).
For determination of the beta- and gamma-cellulose a second aliquot of the filtrate is used to precipitate the beta fraction with sulfuric acid at 70-90°C. The precipitate is allowed to settle overnight and 50 cc of the top clear solution analyzed with potassium dichromate.
The first titration determines the beta- plus gamma-cellulose. The alpha-cellulose is calculated by determining the difference between the weight of total pulp specimen and total material found in the dissolved portion. In the second aliquot after precipitation of the beta fraction, the gamma-cellulose is determined in the clear liquid. Beta is then determined by the difference between the first (beta plus gamma) titration analysis and the second (gamma).

1976

95. Czirnich, W., and Patt, R.
"Investigations of hemicellulose stabilization during sulfite pulping with magnesium base."
Holzforschung, 30, No. 1 (1976), p. 19-27. In German. 15 references.
When pulping pine and beech together using magnesium bisulfite, hemicelluloses are exchanged. Xylan is adsorbed on pine pulp in larger quantities. Correspondingly, larger amounts of glucomannan are found in the beech pulp.

1977

96. Aspinall, G. O.
"The selective degradation of carbohydrate polymers."
Pure and Applied Chemistry, 49 (1977), p. 1105-1134. 62 references.
Methods for the selective chemical degradation of polysaccharides and glycoproteins are reviewed. Such techniques are used in elucidating the structure of hemicelluloses, providing information concerning the sugar sequences and linkages. Special attention is given to a variety of base-catalyzed degradations and to the introduction of acid-sensitive units within carbohydrate chains. The information is also helpful in suggesting possible pathways by which oxycellulose and hemicelluloses may degrade at specifically-sensitive sites in the chain.

97. Koura, A., and Krause, Th.
"Aging of paper. (1) Studies of the ageing of lignin-free pulps."
Das Papier, 31, No. 10A (Nov. 1977), (translation, 17 p.). 12 references.

The authors proposed to investigate a number of tests that can be used to indicate the extent of deterioration. Cotton linters, a sulfate pulp and two sulfite pulps having a range of matter soluble in 10% NaOH (indicative of hemicellulose content) ranging from 0.5 to 15.0% were used which led to the conclusion that the hemicelluloses were more easily oxidized than cellulose. The amount of 10%-alkali-soluble matter increased on thermal aging most in those samples that were originally highest in alkali-soluble matter (highest in hemicellulose content). Tests also showed the much greater rate of formation of carbonyl groups and of discolored material at 70% RH than at 0% RH. [A translation (T-1503) is available from the Institute of Paper Chemistry.]

98. Löwendahl, L., and Samuelson, O.
"Carbohydrate stabilization during kraft cooking with addition of anthraquinone."
Svensk Papperstidning, 80, No. 17 (1977), p. 549-551. 18 references.
The increased yield at a given degree of delignification during a kraft cook on addition of anthraquinone is explained by an increased delignification rate and by a stabilization of the carbohydrates due to the formation of aldonic acid end-groups. Gluconic, mannonic, xylonic, lyxonic, and altronic acid end-groups were found to be the most abundant aldonic acid groups. Thus, the alkaline "peeling reaction" and subsequent loss of cellulose and hemicellulose material are suppressed by the oxidation of aldehydic to carboxylic acid end-groups.

1978

99. Alekseeva, T. V., Belen'kaya, N. G., Chochieva, M. M., Antonovskii, S. D., and Terpukova, A. F.
"Effect of arabinogalactan on paper properties."
Khimiya Drevesiny (Riga), No. 5 (1978), p. 104-109. Through *Abstract Bulletin of the Institute of Paper Chemistry*, 50, No. 2 (1979), Abstr. 1227. In Russian. 4 references.
Surface sizing with 15% aqueous arabinogalactan solution improved paper strength, especially paper made from bleached cotton or flax pulp by improving folding endurance. All samples treated with arabinogalactan were stable to heat and light and more resistant to aging.

100. Koura, A., and Krause, Th.
"Aging of paper. (2) Effect of refining on the aging of lignin-free fibers."
Das Papier, 32, No. 5 (May 1978), p. 189-194. In German. 6 references.
The effects of increased beating on the thermal aging of cotton linters and of a bleached sulfite pulp are reported. One conclusion that the authors draw is that it is the presence of hemicelluloses in the pulp that leads to greater decomposition and discoloration as a result of beating than that which occurs in the case of the higher-quality linters. [Translation available from the Institute of Paper Chemistry.]

101. Sears, K. D., Alexander, W. J., Goldschmid, O., and Hamilton, J. K.
"Characterization of hemicelluloses."
Tappi, 61, No. 9 (1978), p. 105-110. 27 references.
Hemicellulose polymers were selectively extracted from black spruce and balsam fir holocelluloses, acid bisulfite (S) paper pulps, conventional kraft (K) paper pulps, and prehydrolyzed kraft (PK) dissolving pulps. In addition to routine analyses, viscometric degree of polymerization (DP) and gel permeation chromatography DP analyses were carried out on hemicellulose nitrates. Results from the two species were very similar; larger quantities of glucomannans were obtained from balsam fir. Total hemicellulose extract yields from holocelluloses, S, K, and PK pulps, respectively, were 21.0-23.1, 9.5-9.8, 12.0-12.8 and 2.9-3.3%. Additional quantities of glucomannans were isolated from extracted residues by further treatment, and several polymers of high purity were obtained. Lowest purity hemicelluloses were recovered from PK pulps. Nitrate DP and GPC results illustrated differences between hemicellulose polymers present in the different pulps; they did not reflect, however, the true polymolecularity of the hemicelluloses. Cuene intrinsic viscosity data and appropriate equations gave DP values ranging from 18 to 157, consistent with the literature. Xylans and mannans from the K pulps and holocelluloses were of similar size; by comparison, the DP values for hemicelluloses from the S pulps were markedly reduced, and the polymers from the PK pulps also appeared to be of low DP. Authors' abstract.

1979

102. von Roffael, E.
"Über mögliche zusammenhänge zwischen der hydrolysierbarkeit von gebleichten zellstoffen und ihrer vergilbungsneigung."
On possible correlations between hydrolyzibility of bleached pulps and their color reversion by thermal aging.
Holzforschung, 33 (1979), p. 117-124. In German. 34 references.
The results show that a correlation seems to exist between the degradation of cellulosic pulps during heterogeneous hydrolysis under mild conditions to simple sugars and their degradation products and the color reversion of such pulps by thermal treatment. Pulps with a high resistance to hydrolysis such as cotton linters tenaciously resist yellowing by thermal aging. Pulps with little resistance to hydrolysis such as straw pulps experience a marked color reversion. Furthermore, it was found that the color reversion of two spruce sulfite pulps during thermal aging correlates well with their hydrolyzibility to simple sugars and their degradation products (e.g., furfural), but not with their pentosan content as such.
The pH value of pulp hydrolysates correlates clearly with color reversion over a wide range; this confirms anew the well-known influences of pH value on thermal color stability.
It is concluded that not only the hemicellulose content but, to a greater extent, the content of easily hydrolyzable pentosans are relevant factors for yellowing. As the result of hydrolysis, simple sugar compounds such as furfural and formaldehyde are built up, which are able to undergo polymerization reactions.

103. Teratani, F., and Yamashita, T.
"Extraction of hemicellulose with morpholine solution."
Shizuoka Daigaku Nogakubu Kenkyu Hokoku, No. 29 (1979), p. 19-24. In Japanese.
The extraction of hemicellulose (HC) from the holocellulose of *Fagus crenata* and *Cryptomeria japonica* by aqueous morpholine was compared with that by aqueous 24% KOH and dimethyl sulfoxide. The yield was increased by decreasing the morpholine concentration and increasing the extraction temperature.

1980

104. Dickey, E. E., and Roth, L.
"Hemicelluloses and Gums 1: Occurrences, properties, and analysis of hemicellulose."
Bibliographic Series, Number 287, The Institute of Paper Chemistry, Appleton, WI (1980).
An extensive bibliography on many aspects of hemicellulose properties.

105. McGinnis, G. D., and Shafizadeh, F.
"Cellulose and hemicellulose."
In *Pulp and Paper Chemistry and Chemical Technology*, 3rd edn., Volume 1, ed. Casey, J. P., John Wiley & Sons, New York (1980), p. 29-33. 173 references.
Hemicelluloses are mixtures of cell-wall polysaccharides of plants, excluding the cellulose, lignin and pectin components. They are generally water-insoluble, alkali-soluble substances and are more readily hydrolyzed by acid than cellulose. They are branched and have lower molecular weights than cellulose. Those found in the stalk or supporting tissue of woody plants are primarily modified xylans, galactoglucomannans, glucomannans, and arabinogalactans. The arabinogalactans are water soluble and are commonly classed as the extractives. The basic sugars that make up these polysaccharides are: D-xylose, D-mannose, D-glucose, D-galactose, L-arabinose and 4-O-methyl-D-glucuronic acid.
Hemicelluloses from gymnosperms (softwoods) and angiosperms (hardwoods) differ. In hardwoods a partially acetylated xylan (4-O-methylglucurono) predominates with smaller amounts of glucomannan. In gymnosperms a partially acetylated galactoglucomannan predominates with smaller amounts of arabino-(4-O-methylglucurono) xylan. Larches are unique; they contain mostly the arabinogalactan.
The xylan skeleton consists of a linear or singly-branched backbone of 1-4-linked anhydrobeta-D-xylopyranose units. Generally, most land plants contain xylans with short side chains of one or two sugars linked to the backbone.
Hardwood xylans consist of xylose and 4-O-methyl-alpha-glyco-pyranosyluronic acid. The structure is a linear or singly branched beta-D-xylopyranose backbone linked together with 1-4 glycosidic bonds. Branch points generally occur at C-2 positions of the xylose unit with

4-O-methyl-D-glucuronic acid; a small proportion of branching may also occur at the C-3 position. DP reported for hardwood xylans varies from 50-300, but a more careful study determined the number-average DP to be in the 150-200 range.

Softwood xylans consist of an arabino-(4-O-methylglucurono) xylan. They make up 7-15% of softwood. The anhydro-xylose backbone has branch points containing alpha-linked 4-O-methyl-D-glucuronic acid groups and alpha-linked L-arabinofuranose units. The arabinose units are acid labile and can be removed under mild acid conditions without affecting the xylose chain. The uronic acid group is linked at the C-2 position and the arabinose units at the C-3 position of the anhydro-xylose backbone. The softwood xylan does not contain acetyl groups and is more branched and more acidic than the corresponding hardwood xylan.

Hardwoods contain 3-5% glucomannans, made up of D-glucose and D-mannose in a 1:2 ratio. The backbone is a 1-4-linked beta-D-glucopyranose and beta-D-mannopyranose units with no galactose.

Galactoglucomannans are the major hemicellulose component of softwoods. The backbone is a 1-4-linked beta-D-mannopyranose and beta-D-glucopyranose units where at the C-6 position of glucose and mannose is the branching point for alpha-D-galactopyranose units. The major galactoglucomannan is water soluble and contains galactose, glucose, and mannose in a 1:1:3 ratio; a minor water-insoluble type contains the above sugars in a 0.1:1:3 ratio.

Arabinogalactans are water-soluble polysaccharides found in the wood of conifers. Those that are found in larches have backbones of 1-3-linked beta-D-galactopyranose units with side chains linked at the C-6 position of the galactose units. Side chain composition is not fully known; most common is the 1-6-linked beta-D-galactopyranose. Another common unit is 3-O-beta-L-arabinopyranosyl-L-arabinofuranose.

106. Robinson, J. V.
"Fiber bonding sub-section. Effect of hemicellulose on fiber bonding." (Chapter 7)
In *Pulp and Paper Chemistry and Chemical Technology*, 3rd edn., Volume 2, ed. Casey, J. P., John Wiley & Sons, New York (1980), p. 940-944. 11 references.

This chapter/section deals with the effect hemicelluloses have on fiber bonding. They favor bonding in two ways: contributing plasticity to the fiber wall by permitting cellulose surfaces to conform more to each other, and their mobility allows them to contribute more to H-bonding per unit area than the cellulose surfaces do. Adsorption studies have shown that hemicelluloses seek cellulose surfaces and form H-bonds.

Hemicellulose is important in the development of strength properties since it is related to fibrillation of the fibers. They function as binders through the mutual bonding of their hydroxyl groups. By increasing the amount of hemicellulose in wood pulps, the strength of the bonds per unit area is increased. This is due to the increase in the number of bonds per area of contact.

Hemicelluloses appear to have a critical minimum degree of polymerization below which they lose their ability to contribute strength to paper.

The entire hemicellulose fraction is not effective in fiber bonding. Softwood hemicellulose is much more effective in fiber bonding than hardwood hemicellulose due to the higher content of mannose and hexose. Gamma-cellulose is the most important factor, even though it is partially removed by alkaline pulping but a certain amount remains in sulfite and sulfate pulps, since it contains the hemicellulosic material involved in the fiber bonding. Beta-cellulose is ineffective in fiber bonding since it consists of mostly degraded cellulose produced from the cooking process. The bursting strength of bleached wood pulps has been shown to be related to the gamma-cellulose content of the pulp. This relationship does not hold for unbleached pulps due to the high amounts of lignin.

The maximum strength of paper-making pulps increases with an increase in the amount of alkali-soluble hemicellulosic material in the pulp rather than the pentosan content. The low-polymerized polyoses and polyuronides present in the pulp are responsible for the easy beating. This explains why soda pulps do not "wet" as readily as sulfite pulps, because the alkali in the soda process removes all the alkali-soluble hemicelluloses.

Pulps with too much hemicellulose hydrate too quickly and do not develop adequate strength, producing a translucent, rattly and low in strength paper. Holocellulose containing all its hemicellulose produced sheets of low strength; removal of some hemicellulose with weak alkali solution increased the strength to an optimum value beyond which further removal caused a decline in strength. The optimum strength was obtained at a yield of 50-52% of the weight of original wood with hemicellulose content of 18.0% and lignin of 1.09%.

The maximum tensile strength obtained is a function of the area fiber bonding which depends on an optimum hemicellulose content. The intrinsic strength of the bonds depends on the chain length of the hemicellulose. If there is too much hemicellulose present, the proportion of short-chain fraction will exceed the optimum for intrinsic bond strength and the overall sheet strength will be affected unfavorably. The type of wood and cooking process determines what the optimum hemicellulose content will be. The sheet properties desired must also be considered and will also determine the optimum hemicellulose content.

1981

107. Dickey, E. E., and Roth, L.
Hemicelluloses and Gums 2: Hemicelluloses in Pulping, Bleaching, and Papermaking, Bibliographic Series, Number 288, The Institute of Paper Chemistry, Appleton, WI (1981).
Hemicelluloses and Gums 3: Hemicellulose Derivatives and By-products, Bibliographic Series, Number 289, The Institute of Paper Chemistry, Appleton, WI (1981).
Hemicelluloses and Gums 4: Gums and Mucilages, Bibliographic Series, Number 290, The Institute of Paper Chemistry, Appleton, WI (1981).
Continuations of the Institute of Paper Chemistry's bibliographic series.

108. Hon, D. H.-S.
"Yellowing of modern papers."
In *Preservation of Paper and Textiles of Historic and Artistic Value II*, Advances in Chemistry Series 193, ed. Williams, J. C., American Chemical Society, Washington, DC (1981), p. 119-141. 179 references.
Unfortunately, modern papers produced from wood pulps contain a high quantity of vulnerable cellulose and hemicellulose that have been oxidized during the pulping processes. They are relatively unstable toward light, heat, moisture, air, and other environmental factors. Modern "high-yield" papers contain oxidized cellulose and hemicellulose; they also incorporate a considerable amount of lignin and extraneous material from wood, which further impairs permanence and durability. Discoloration or yellowing of modern papers as a consequence of the oxidative reactions of cellulose, hemicellulose, and lignin, as well as color stabilization, is reviewed. Suggestions for further research in color stabilization are made.
It is generally considered that photodegradation of hemicellulose is identical to that of cellulose. Several authors have reported that the colored substances formed in pulp yellowing are water-soluble and related to the oxidative degradation of hemicellulose. Yellowing of pulps through the oxidation of hemicellulose also has been verified by many researchers, and it has been recognized that carbonyl groups on the hemicellulose are probably responsible for thermally induced yellowing.

1983

109. Hunger, G. K.
"Hemicelluloses in papermaking."
Papier, 37, No. 12 (Dec. 1983), p. 582-591. In German. 103 references.
Hemicelluloses affect the papermaking properties in a variety of ways. This review discusses many aspects—the pulping processes, beating and refining, formation, sizing, fiber bonding, fillers and additives, calendering of pulp from annual plants and straw, special papers, aging, and comparison with other hydrocolloids (methylcellulose, carboxymethylcellulose, hydrolyzed celluloses).

110. Iversen, T., and Lindgren, B. O.
"Behavior of hemicelluloses during acid and alkaline pulping of wood."
Papier, 37, No. 12 (Dec. 1983), p. 573-581. In German. 53 references.
The chemical structure of hemicelluloses is described briefly, and their fate during pulping discussed. In the sulfite process, the hemicelluloses are hydrolyzed to simple sugars or to polysaccharides of lower molecular weight. These then undergo reactions with sulfite, forming aldonic acids or sulfonic acids. Hemicelluloses are also broken down in the kraft process, mainly by oxidation in the peeling reaction to saccharinic acids.

111. Wilkie, K. C. B.
"Hemicellulose."
ChemTech, 14 (1983), p. 306-319. 54 references.
The author reviews in general but technically precise terms the current state of information regarding hemicelluloses. Particularly, he points out that the structure of branched hemicelluloses (heteroglucans) is not precisely known; analysis is only able to detect certain structural features. Thus, names such as arabinogalactan may imply a known structure, but in truth the exact structure is rarely known. He also points out difficulties inherent in the study of hemicellulose chemistry, such as the rarity of being able to isolate a specific part of a plant, rather than grinding up the entire leaf, stem, or seed, before analysis of the constituent hemicelluloses is undertaken. Moreover, the composition of the hemicellulose fraction changes with the age of the plant.

1984

112. Das, N. N., Das, S. C., and Mukherjee, A. K.
"Ester linkage between lignin and 4-O-methyl-D-glucurono-D-xylan in jute fiber (*Corchorus capsularis*)."
Carbohydrate Research, 127, No. 2 (1984), p. 345-348. 4 references.
There have long been questions whether cellulose or hemicelluloses are chemically linked to lignin. In this note, infrared spectroscopy was used to resolve some ambiguities in the origins of acidity (weak and very weak acid groups) of jute fibers and in the degree of esterification (ca. 34%) by which uronic acid side-chain residues on methyl-glucurono-xylan are partly linked to the lignin component.

1986

113. Lee, S. B., and Feller, R. L.
"Influence of the hemicellulose fraction on thermal and photochemical discoloration."
In *Conservation and Characterization of Historical Textiles and Paper Materials*, Advances in Chemistry Series No. 212, ed. Needles, H. L., and Zeronian, S. H., American Chemical Society, Washington, DC (1986), p. 377-386. 12 references.
Isolated arabinogalactan, acetylxylan, and neutral hemicellulose fractions extracted from four different pulps were added-on to filter paper and subjected to both thermal and photochemical-aging tests. In addition, pulps containing decreased proportions of gamma-cellulose were prepared from an unbleached kraft pulp. Whereas thermal aging led to marked discoloration in all cases, only negligible color changes were caused by exposure to visible or to near and middle (254 nm) ultraviolet radiation. Clearly the tendency of these hemicellulosic materials to discolor is the result of thermal rather than photochemical changes.

See also:
Investigations concerning the durability of cellulosic pulps, 16.
Weak links as inductive effects in the hydrolysis of cellulose, 21.

III. ALKALI-SOLUBLE MATTER AS A MEASURE OF PAPER QUALITY AND DEGRADATION

There are three major types of degradation of cellulose by alkali, as Casey states in two concise pages, [Casey, J. P., *Pulp and Paper Chemistry and Chemical Technology*, 3rd ed., Volume 1, J. Wiley & Sons, New York (1980), p. 21-22.] The first is an oxidative process that occurs when alkaline solutions are in contact with air at high temperatures. We will not discuss this form of interaction with alkali, although the reader if interested can refer to the review by Golova and Nosora (1973) and to the early publication of Entwistle, D., Cole, E. H., and Wooding, N. S., *Textile Research Journal*, 19 (1949), p. 527-546, 609-624.

Alkali can catalyze the non-oxidative breaking of glucosidic bonds, somewhat analogous to the well-known acid-catalyzed degradation of cellulose, but this only occurs at temperatures around 160 to 180°C and therefore is of little concern to the conservator. A second non-oxidative form of degradation by bases is the so-called peeling reaction in which the cellulose chains are gradually shortened by a beta-elimination mechanism (Figure 1), a chemical change that normally proceeds from the reducing end of the cellulose molecule until the reaction is interrupted by the "stopping" reaction (see Figure 2). These processes are of considerable interest in the refining of pulps and in gaining an understanding of certain procedures used in research and testing. It is in regard to this particular aspect of the action of alkali that bibliographic topic III has been prepared.

Cellulose can be solubilized by the action of alkalis at moderate temperatures. The amount of cellulose taken up by concentrated alkalis is primarily a function of its degree of polymerization, the concentration of the alkali, and the temperature. Upon neutralization, all but the lowest molecular weight material can be reprecipitated, essentially unchanged. A few remarks must be made about this more or less non-degradative solvent action of cold alkali before discussing the degradative reactions in hot alkali.

1. SOLUBILITY IN COLD CONCENTRATED ALKALI

The solvent power of alkaline solutions at moderate temperature is related to the ability to swell the fiber. Maximum solubility is generally observed at about the same concentration as maximum swelling. The degree of swelling by alkali metal hydroxides becomes large if the concentration is greater than about 1.0N (4 grams of NaOH per 100 ml of solution or about 4%) and reaches a maximum at a concentration that depends upon the temperature and the nature of the alkali. The solubility and swelling decrease up to about 50°C; at higher temperatures the solubility again increases because of chemical attack by the alkali solution. Solubility generally increases with decreasing temperature owing to the increased swelling which occurs at lower temperatures.

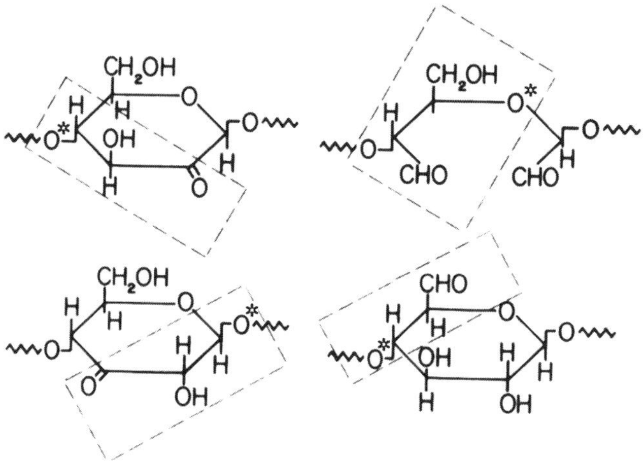

FIGURE 1
Beta-alkoxy groups (*) in carbonyl-oxidized cellulose showing the molecular configurations that lead to chain scission.

Peeling Reaction

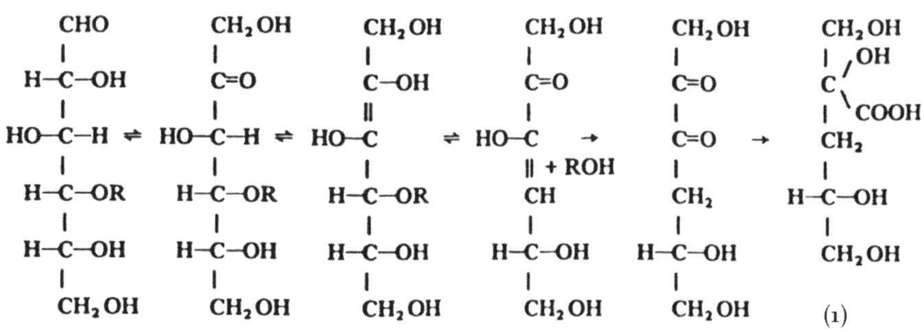

(1)

Stopping Reaction

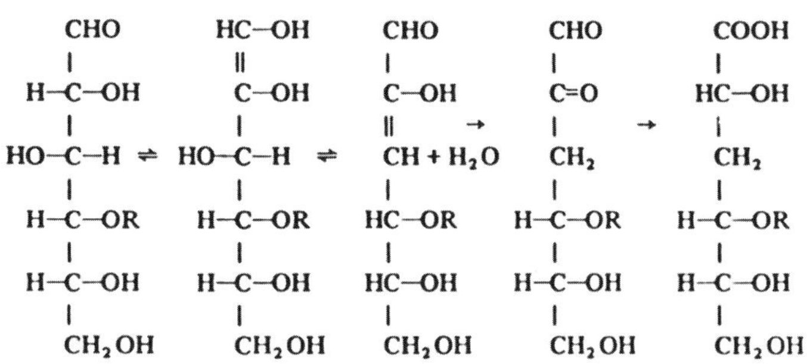

(2)

1 = isosaccharinic acid; 2 = metasaccharinic acid end group; R = cellulose

FIGURE 2
Non-oxidative alkaline degradation reactions of cellulose.

From the very early days of cellulose research, C. F. Cross and E. J. Bevan [*Journal of the Chemical Society*, 53 (1888), p. 889] sought to classify three constituents of cellulosic materials: *alpha*, that which would not dissolve in 17.5% sodium hydroxide at 20°C; *beta*, that which dissolved in this solvent but which was precipitated out of solution upon neutralization with acid; and *gamma*, that which remained in solution. The alpha fraction could be dried and weighed; the beta and gamma were usually estimated by oxidizing the material quantitatively with chromic acid. The alpha content of pulps could also be determined by subtracting the percentage of beta plus gamma, determined by chromate oxidation, from 100 (Launer, 1938). Years of investigation have served to elucidate the fact that the alpha fraction is largely pure cellulose of $(DP)_v$ higher than 200; the beta is similar material but of lower DP; and the gamma fraction is a much lower- molecular-weight material often consisting of degradation products and polysaccharidic material hemicellulosic in character.

Extraction of hemicellulose, degraded cellulose, and cellulose varies depending on the concentration, temperature (hot or cold), and types of alkaline solution. Hot alkali will also dissolve some of the chlorinated lignin in bleached pulps. Chlorination, sulfonation, and oxidation of lignin all contribute to its solubility in hot-alkali solution. With increasing concentrations of alkali, the extraction of hemicelluloses rises to a maximum (Meller, 1953).

With increasing alkali concentration, the solubility of beta-cellulose increases, reaches a maximum at 10% alkali, and then decreases; gamma-cellulose solubility increases up to 10% sodium hydroxide, then is nearly constant up to 18% (Wilson, Ringström and Hedlund, 1952; see Figure 3). The beta-cellulose content of pulp is thus considered to be linearly related to the difference between the 10 and 18% alkali solubilities (Tappi Method T235 os-76). Browning (1967) provides a convenient overview of the analytical aspects of solubility in alkaline solvents in his 13-page Chapter 20.

Staudinger and Jutish [*Kunstseide und Zellwolle*, 21 (1939), p. 6-9] gave values of DP of 10-150 for beta; about 10 for gamma. Using a standard cupriethylenediamine technique of viscosity determination, they found that the $(DP)_v$ of beta-cellulose in rayons was 370 to 240, but that in 18% NaOH, which dissolves little of beta, a fraction of $(DP)_v$ 65 was found. Although there is nothing fixed about the quantities, chemical nature, or DP of beta- and gamma-cellulose, the authors suggested measuring the amount of material that dissolves in 10% alkali (beta and gamma) and that which dissolves in 18% alkali (gamma) to provide useful information. The full curves of solubility and neutralized-precipitated fraction versus alkali solubility, as extensively shown by the authors, are most informative.

Wilson, Ringström and Hedlund (1952) also discussed the determination of beta- and gamma-cellulose in detail, giving curves of alkali solubility versus the percentage of sodium hydroxide. These clearly show that gamma-cellulose

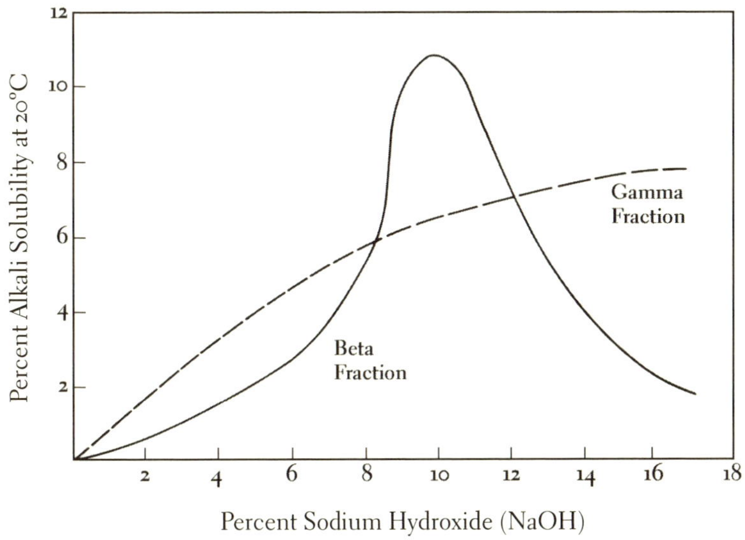

FIGURE 3
Alkali solubility at 20°C of pulp CCA8 at different NaOH concentrations.
(Wilson, Ringström and Hedlund, 1952)

increases in its solubility as the alkali concentration increases up to 18 or 20% NaOH, whereas the beta-cellulose exhibits a fairly sharp maximum at 10% NaOH and negligible solubility at 18% NaOH (Figure 3). A Swedish alpha-cellulose determination involves dissolving pulp in 18% NaOH and washing the precipitate with water. The latter operation can wash out some of the beta owing to the dilution of the concentrated alkali by the wash water. (In contrast, Tappi Method T203 involves washing the pulp in 17.5%, then 9.45%, NaOH.) The organic matter in the filtrate is oxidized with chromic acid. Beta-cellulose is separated out by precipitation with acid and is then oxidized by chromic acid. Alpha is determined by the difference. By this technique, alpha, beta, and gamma are caused to add up to 100%.

Hamilton and Quimby (1957) compared the extractive power of lithium, sodium, and potassium hydroxide solutions for the hemicelluloses associated with wood cellulose and holocellulose from western hemlock. Potassium hydroxide is much less effective and sodium and lithium hydroxides much more effective in dissolving mannose-containing polysaccharides, especially at concentrations in the range of 2.5N and higher.

Wayman (1963) shows data on pulp yield, pentosans, soda solubles, and alpha-cellulose content versus the concentration of cold alkali for the case of unbleached sulfite and sulfate pulps. His data also illustrate the different results between hot and cold extractions. Much of his discussion focuses on the effects of extraction conditions on yield. Cold extraction with alkali is required to attain alpha-cellulose above the 96% "ceiling." Cold solutions at 4 to 20%

concentration are said to have a more selective action on the non-alpha-cellulose components than hot dilute alkali. The main disadvantage of the much-used industrial practice of hot alkali extraction is the loss of yield.

Rydholm (1965) stated that the solubility of a sulfite pulp in 18% NaOH gives a fair indication of the hemicellulose content, though some degraded cellulose may go into solution and some hemicelluloses always remain undissolved. The deviations are more pronounced for kraft pulps, especially from hardwoods. In pulps containing much degraded cellulose, this portion tends to dissolve at about 10% NaOH concentration, under conditions in which the fibers have maximum swelling. Only limited amounts are dissolved in 18% NaOH, and the difference in alkali solubility between 10 and 18% NaOH, sometimes called "rest-hemi," may be taken as a rough measure of the state of degradation.

Kamide and co-workers (1984) established a correlation of the solubility of cellulose in aqueous alkali solution with its so-called "amorphous content." To achieve this end, cellulose was regenerated from its cuprammonium solution under different preparative conditions, and cotton linter was acid hydrolyzed into a fibrous form and the pulp was physically milled to powder. Cellulose having a relatively large molecular weight was found to completely dissolve in aqueous NaOH solution at 4°C. The solubility of cellulose in a 10 weight percent aqueous NaOH at 4°C could generally be correlated to the relative amount of the high magnetic field envelope of the C4 carbon NMR (nuclear magnetic resonance) peak, assigned to the region where intramolecular hydrogen bonds are at least partly broken. The intramolecular hydrogen bond was found to have an important influence on the solubility behavior of cellulose.

2. Solubility in Hot Dilute Alkali

The solubility in hot 1% sodium hydroxide has been used for diagnostic evaluation of wood and pulps. The amount soluble in pulps is usually less than that found with strong cold sodium hydroxide solutions and more nearly approximates the amount of gamma-cellulose as determined by conventional procedures. The solubility of wood pulps can indicate the degree of degradation by fungus, heat, light, and oxidation, while the solubility of high-alpha pulps primarily indicates an extent of cellulose degradation during pulping and bleaching processes. The cellulose is chemically attacked by hot dilute alkali, initiated at sites where aldehydic carbonyl groups are situated and proceeding until a chemical rearrangement of the sugar end-group takes place which prevents further "peeling off" of the sugar entities ("stopping reaction").

Procter and Chow (1973) proposed that a rot index, based on 1% NaOH solubility, be assigned to wood chip samples. The results of tests on the wood chips were indicative of pulp yield.

3. DEGRADATION OF THE CELLULOSE CHAIN IN ALKALI

It is well known among organic chemists that in any situation in which there is a beta-alkoxycarbonyl group present in alkali a rapid elimination of the alkoxyl group (OR) will occur [Isbell, H. S., *Journal Research National Bureau of Standards*, 32 (1944), p. 45]:

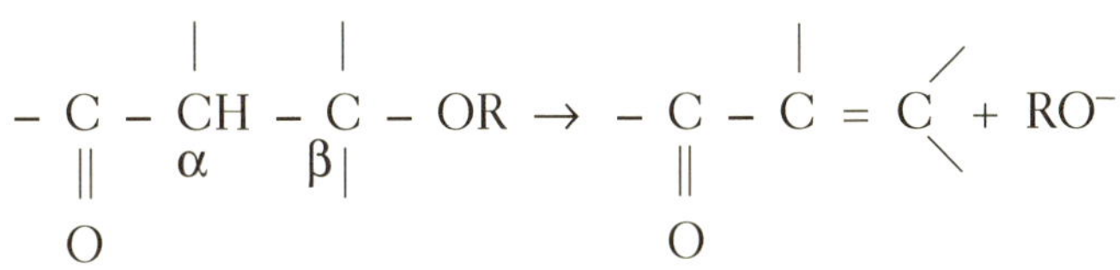

When cellulose is oxidized, with the result that carbonyl groups are introduced into the pyranose rings at various locations, the chains can be readily degraded (broken) by this mechanism (see Figure 1). Such a reaction would degrade a hydrocellulose chain by eliminating one aldehydic glucose "end-group" at a time until some further chemical rearrangement interrupted the process. Thus, the action of hot dilute, oxygen-free, sodium hydroxide on cellulose chiefly involves two competing reactions: degradation via the beta-alkoxycarbonyl group elimination (the peeling reaction which begins at the aldehydic ends of the chains) and a "stopping reaction" (see Figure 2). An excellent review of the subject has been prepared by Richards (1971).

(a) Peeling Reaction

Kenner and Richards extensively investigated model compounds which led to an understanding of the peeling reaction of oxidized cellulose in dilute alkali. In their first article (1953) the importance of the oxidized keto state was demonstrated, for this can form an enol group which then can ionize, liberating a hydrogen ion. In the immediately following paper, Corbett and Kenner (1953) proposed that cellulose oxidized to the stage where there is a carbonyl group on carbon atom 2 can rearrange to yield an alpha and beta isosaccharinic acid. The formation of a related structure, metasaccharinic acid, effectively stops the peeling off of glucosidic groups from the ends of the polymer chains (see *Stabilization*, Figure 2).

A study of the products of alkaline degradation of certain periodate-oxidized monosaccharides suggests that the degradation occurs predominantly by scission of the carbon-oxygen linkage in the beta-position with respect to the aldehyde groups. Subsequent rearrangements of the products probably yield a molecule of glycolic acid from each scission of this type (O'Meara and Richards, 1958).

Hot alkaline treatment of pulps and the accompanying peeling reaction from the aldehyde groups of the carbohydrate molecules preferentially creates and leads to the solution of short-chain material derived from both the hemicelluloses and cellulose. On the other hand, stabilizing reactions, accompanied by adsorption of xylan onto the cellulose microfibrils, cause more hemicelluloses to remain in the pulp than was their initial proportion relative to that of cellulose (Rydholm, 1965).

When cellulose is autoclaved at 170°C in 10% NaOH, Samuelson and Wennerblom (1954) demonstrated that degraded cellulose chains contain one carboxyl (COO-) group per chain (largely the saccharinic acid groups). They calculate that about 50 cellulose units go into solution for each breakage of the chain at the beginning of the cook. Moreover, 1.4 equivalents of weak acid are formed for each glucose unit going into solution; hence, the process must involve chain or sequential reactions.

(b) Stabilization

The alkaline depolymerization (peeling) process just described can be prevented by chemical modification of the reducing end-group or, alternatively, by the extent to which the process is suppressed by adopting conditions that favor the stabilization reaction. In an excellent review of the subject, Meller (1951) pointed out that the action of aqueous solutions of both acids and oxidizing agents on cellulose fibers usually results in considerable changes in their chemical and physical constitution. These changes may significantly affect the resistance (stability) of the modified celluloses toward dilute hot and concentrated cold alkali solutions. The changes in physical structure and chemical composition of cellulose fibers arising from the action of both hydrolyzing and oxidizing solutions usually take place concurrently, particularly during the chemical treatment of pulps.

Meller (*Tappi*, 1953) demonstrated the dominant influence of carbonyl groups present in modified cellulose (oxycellulose) upon its fragmentation and dissolution by hot-alkali solutions, owing to the beta-carboxycarbonyl elimination reaction (see Figure 1). If reduction of oxycelluloses is carried out in alkaline borohydride solution, leading to the formation of alcohol groups in place of the normally alkali-sensitive aldehyde function, hot-alkali solubility is diminished. In contrast, borohydride reduction under *acidic* conditions leads to the production of aldehydic groups from structures containing acid groups; the result is increased sensitivity to hot dilute solutions of alkali.

Hydrolytically degraded cellulose fibers are irreversibly degraded. In contrast, "potential" degradation of cellulose involves oxidation of some of the hydroxyl groups to carbonyls in the hexose or pentose units of the polysaccharide chains. The sensitivity to potential degradation by alkali may be prevented by chlorous-acid oxidation of the fiber (which oxidizes aldehyde to carboxylic acid groups) as well as by reduction with sodium borohydride (which reduces aldehyde groups to alcohols) (Meller, *Tappi*, 1955). The hot-alkali stability of modified cellulose fibers is greatly influenced by the enolizability of chemically modified functional groups present in the anhydroglucose building units. The following equations illustrate the results of the formation of an "enol" intermediate structure; the position of the carbonyl oxygen can be shifted to another, potentially more active, location in the sugar molecule:

$$\begin{array}{cccccccccccc}
 & O & & H & & & OH & & & & OH & & \\
 & \| & & | & & & | & & & & | & & \\
- & C & - & C & - \leftrightarrow - & C & = & C & - \leftrightarrow - & C & - & C & - \\
 & & & | & & & & | & & | & & \| & \\
 & & & OH & & & & OH & & H & & O &
\end{array}$$

Meller (1956) reported that both measured alpha-cellulose content and the cuprammonium viscosity of chemically modified wood pulps increase upon chlorous acid oxidation or sodium borohydride reduction because the loss by the peeling reaction is suppressed. The apparent increase in alpha-cellulose content is low, however, in comparison to the decrease in solubility in hot 1% NaOH.

Variation of sodium hydroxide concentration within the range 0.125 to 1.25N was found to be without significant effect on the extent of degradation of hydrocellulose of DP 150 at 100°C in the absence of oxygen. In the range 2.5 to 5N alkali, under otherwise similar conditions, the extent of degradation increased with increasing alkali concentration. It has been found that the extent of degradation of hydrocellulose in 0.5N NaOH at 100°C in the absence of oxygen can be reduced by applying any one of the following procedures: pre-swelling of the hydrocellulose in cold 5N NaOH; addition of a calcium salt to the sodium hydroxide; pre-methylating the hydrocellulose with small amounts of dimethyl sulfate in cold 0.5N NaOH. Of these procedures, the last proved to be the most effective (Machell and Richards, 1958).

In 1971 Richards was able to review the chemistry of "stopping reactions," dividing the subject into a discussion of the technological importance of alkaline degradation, the mechanism of the erosion of the polymer chains, and stabilization (stopping) reactions, the alkaline scission of cellulose, attempts to minimize alkaline degradation, and the mechanism of alkaline degradation of carbonyl-containing oxidized cellulose.

In recent years, the use of anthraquinone has been adapted to improve the quality and yield of pulps. Löwendahl and Samuelson (1977) stated that the increased yield at a given degree of delignification during a kraft cook owing to the addition of anthraquinone can be explained by an increased delignification rate and by a stabilization of the carbohydrates against the peeling reaction owing to the formation of aldonic-acid end-groups. During NaOH cooking of wood, addition of 0.1% anthraquinone (calculated on dry wood) results in a stabilization of carbohydrates and the formation of a large proportion of aldonic acid end-groups. Experiments with cotton and hydrocellulose show that, with the exclusion of air, the amount of anthraquinone required for an appreciable stabilization was very large (Carlson and Samuelson, 1979).

Ziderman (1980), following up on the work of Menachem Lewin, recognized that the amount of yellow chromogen in alkaline extracts of celluloses and starches provides a quantitative estimate of the extent of chain end-initiated depolymerization ("peeling" reaction) that occurs by beta-elimination, but he pointed out that the results are more difficult to evaluate when concurrent cleavage of non-terminal glucosidic linkages occurs. Keto celluloses are largely resistant to extraction with dilute aqueous alkali, and the extract is not colored. A mechanistic explanation for this stability was offered.

(c) Degradation Products

Oxycellulose, prepared by the action of acid permanganate on linters, had a high copper number and contained uronic acid residues; boiling 1% NaOH dissolved 50% of the oxycellulose. The soluble portion, shown to have undergone decomposition of the type suffered by oligosaccharides in the presence of alkali, was found to contain formic, acetic, lactic, and dihydroxybutyric acids, as well as a mixture of C6-saccharinic acids. The material that remained insoluble in hot 1% alkali had a low copper number and contained no acidic groups (Godman, Haworth and Peat, 1939). Since this work, the major soluble products of the degradation of cotton hydrocellulose by boiling oxygen-free 0.5N sodium hydroxide have been further identified and, in several cases, their relative amounts determined. D-glucoisosaccharinic and formic acid were produced in greatest amount under the conditions used (Richards and Sephton, 1957). Among the products found in solution after this treatment were pyruvic acid, $CH_3COCOOH$, formic acid, lactic acid, and beta, gamma-dihydroxybutyric acid. Aldehydes such as methyl glyoxal, CH_3COCHO, are also found (Rydholm, 1965). The loss in pulp yield (increased solubility in hot dilute alkali) is related to the copper number (or carbonyl content).

Johansson and Samuelson (1975) stated that the determinations of reducing and carboxylic acid end-groups in hydrocellulose of DP ~200 from sulfite pulp showed that the end-wise degradation when treated with hot alkali proceeded to such an extent that some of the molecules were brought completely into solution. The major carboxylic acid end-groups, formed in reactions competing with the end-wise degradation, were 3-deoxyhexonic and 2-C-methyl-glyceric acids; 2-deoxypentonic, 2-C-methyl-pentonic, and minor amounts of aldonic acids were also formed.

Similar reactions occur with starch: the action of 0.5N (2%) sodium hydroxide at 100°C on amylose (from starch) is also of two main types: "degradation" producing mainly D-glucoisosaccharinic, formic, and lactic acid; and a "stopping" reaction affording alkali-stable amylose probably bearing terminal D-glucometasaccharinic acid units (Machell and Richards, 1958).

4. HOT-ALKALI SOLUBILITY AND ITS CORRELATION WITH:

(a) Copper Number

When the chemistry of cellulose first began to be explored, Birtwell, Clibbens and Ridge (1925) were among the early workers to draw attention to the fact that two principal types of oxidized cellulose (oxycelluloses) resulted from the action of chemical oxidizing agents: an acidic and an aldehydic type. The first tended to absorb considerable quantities of methylene blue dye; the second exhibited high copper numbers, an indication of aldehydic content. A considerable portion of the latter dissolved when this type of oxycellulose was treated with hot dilute sodium hydroxide solution. Thus, at an early date, it was realized that the amount of cellulose that would dissolve upon boiling in 1% sodium hydroxide solution for one to four hours provided very much the same indication of the extent of oxidation that the cellulose had undergone as did the customary measure of the "reducing power" of cotton, the well-known "copper number" test. We now know that hot-alkali solubility reflects primarily the aldehydic type of oxidation, rather than the more advanced, acidic, stage.

Both for hydrocelluloses and for cellulose oxidized by various agents, the relationship of hot-alkali solubility versus copper number is practically linear up to a copper number of 2.5 and a loss of solubility of 15%, the solubility being about six times the copper number. Above 2.5, the curves diverge, oxycelluloses suffering a slightly less loss in weight on alkali boiling than hydrocellulose at a given copper number; the data have been extended to 30% loss in weight and to a copper number of 5 (Birtwell, Clibbens and Geake, 1926; Clibbens, Geake and Ridge, 1927) (see Figure 4). Qualitatively speaking, pulps have little folding endurance when their hot 1%-NaOH soluble matter is above about 25% or their copper number above 8.

In the case of hydrocelluloses prepared from relatively pure cellulose, which only have an aldehyde group at the end of each chain, the solubility is completely defined by the copper number. In contrast, among modified celluloses prepared by different oxidizing agents, or with the same oxidizing agent acting under different conditions, there is no general relation between solubility and copper number (Birtwell, Clibbens and Geake, 1928).

Davidson (1943) also pointed out that the hot-alkali solubility of mildly modified cellulose is roughly equivalent to the copper number. The hot-alkali solubility of cellulose which was the result of heterogeneous acid hydrolysis of unmercerized cotton was related to the average molecular weight (or fluidity) of the hydrocellulose.

Data of Bergek, Gustavsson and Lindval [*Svensk Papperstidning,* 50, No. 11B, Hagglund *Festkrift* (1947), p. 22-28, from Meller, 1953] show a linear relationship between a decrease in copper number and alkali solubility up to 28% weight loss.

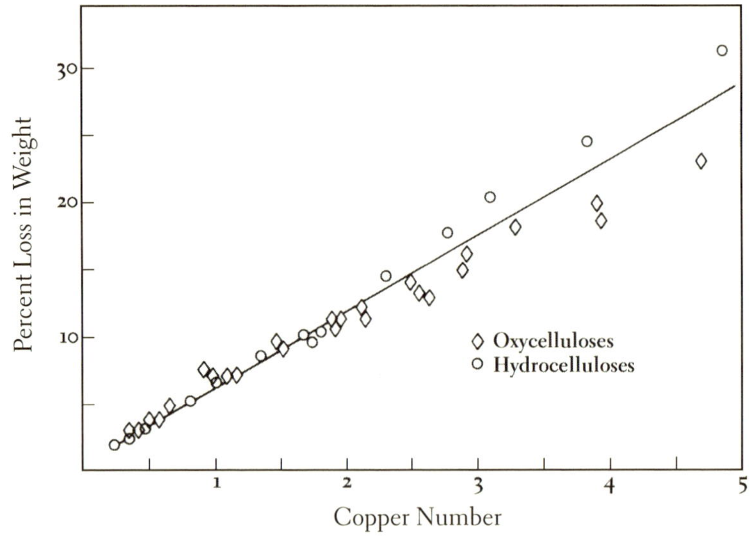

Figure 4
Relationship of loss in weight upon boiling in hot 1% sodium hydroxide solution and copper number of cotton cellulose. (Clibbens, Geake and Ridge, 1927)

Samuelson and Ramsel (1950) present a linear plot of the logarithm of the copper number plotted against hot-alkali solubility over the range of copper numbers from 1 to 10 and hot 7% alkali-soluble matter of approximately 13 to 55%.

(b) Carbonyl Content

Rutherford, Minor, Martin and Harris (1942) found a practically linear relationship between aldehyde content in periodate oxycellulose and solubility in boiling 1% sodium hydroxide solution. (Periodate oxidation tends to put aldehyde groups on carbon atoms 2 and 3, splitting the pyranose ring at that point.)

The content of carbonyl groups found by the borohydride method has been found to be proportional to the copper number by the method of Hagglund [*Cellulosechemie*, 11 (1930), p. 1] but the slope of the curve is not the same if oxidized celluloses are prepared with different oxidants [Lidman-Safwat and Theander, *Svensk Papperstidning*, 61 (1958), p. 42, from Browning (1967), p. 447].

(c) Fluidity and Tensile Strength

Ridge and Bowden (1932) demonstrated that there was a linear relationship between cuprammonium fluidity (in essence, the inverse of the intrinsic viscosity) and the strength of cotton and viscose yarns; Clibbens and Ridge

(1928) had previously demonstrated the essential findings for cotton. They included information on the influence of hot-alkali solubility and the action of boiling water on fluidity and loss of strength. It was concluded that boiling in water must have a degrading and dissolution effect on the modified cellulose.

Davidson (1934) noted that boiling hydrocelluloses in hot dilute alkali causes a slight increase in fluidity in the remaining yarn and in solubility in sodium hydroxide at low temperature. The loss of weight depended on the severity of the boil; this suggested that progressive shortening of the chain-molecules had taken place, the attack being from the end originally bearing the reducing group. The result was that the average DP of the treated material was slightly higher on account of the dissolution of lower-molecular-weight components.

Davidson (1934) also showed that extractions of cellulose with sodium hydroxide solutions of increasing solvent power removed fractions of increasing cuprammonium viscosity. He was the first to suggest that alkali attacked the macromolecules of oxidized cellulose at anhydroglucose units containing carbonyl groups. He demonstrated (1940) that measurements of viscosity in cuprammonium solution reflected a considerable lowering of the DP, apparently owing to the sensitivity of the aldehydic oxycelluloses to the alkalinity of the cuprammonium solution. A nearly linear relationship was observed between tensile strength and fluidity of the nitrated cellulose samples (reflecting the molecular weight of the cellulose before alkaline degradation) and the cuprammonium fluidity of the hydrocelluloses after treatment with boiling dilute NaOH. His later publication (1943) showed that a plot of fluidity versus the loss in weight upon alkali boil confirmed the fact that the two parameters were related regardless of the acid that had been used to cause the hydrolytic degradation, the temperature or time of exposure in the case of unmercerized cotton having undergone heterogeneous hydrolysis.

5. Discoloration Due to Alkali Treatment

(a) Color of Extracts

In Lewin and co-workers' first report (1965), attention is drawn to the parallelism between the color reversion of cotton on aging as measured by the post-color number and the yellowness of the extracts in hot-alkaline solution. The authors' procedure consists of extracting the sample in a 5% sodium bicarbonate solution under reflux (i.e., boiling) for one hour and measuring the intensity of the color of the solutions obtained. The absorbance was found to be proportional to the solubility, linearly related to the amount of cotton used in the extraction and to the extent of hydrolytic or oxidative modification of the cellulose.

In their second report (1965), the authors investigated the factors causing the yellow color of the extracts. The yellowing tendency of modified cotton was shown to depend solely on the aldehyde-group content; the ketone and carboxyl groups had no appreciable influence. Linear plots of yellowing against aldehyde-group content and against 1/DP were found. Reducing sugars showed an absorption spectrum of the yellow solutions similar to that obtained with the modified celluloses. The ultraviolet spectra of bicarbonate extracts of modified celluloses and of glucose were found to be similar to alkaline extracts of aged celluloses. These spectra and the infrared spectra point to the possibility that colored substances are furan derivatives.

In the third report on the subject (1965), Lewin showed that the parallelism between the brightness reversion on aging and the yellow color of the alkaline-extraction solutions does not hold in the case of hydrocelluloses or carboxyl (COOH)-containing celluloses. Possible explanations were offered concerning why the aldehyde (carbonyl, -CHO) end groups in hydrocelluloses contribute very little to color reversion. The author suggests that the tendency of pulps to discolor is closely related to the "peeling-off" process which occurs in the alkaline treatment of pulps and that the colored products formed on aging are closely related to the breakdown products that result from the peeling-off process. Ziderman (1980) made additional suggestions.

(b) Color-Forming Materials

Giertz and McPherson [*Svensk Papperstidning,* 59 (1956), p. 93] investigated the yellow material in heat-aged hypochlorite-bleached sulfite pulp by extracting the material with 10.9% lithium hydroxide solution. The alpha- and beta-celluloses were both obtained in a white condition; the gamma-cellulose solution was deeply colored. Their results clearly indicated that the substances in bleached-sulfite pulps that were responsible for the yellow color are probably degradation products of the hemicelluloses, and that the carbonyl groups produced by oxidation during the bleaching played a large role in the yellowing of the pulp.

Abstracts

1925

114. Birtwell, C., Clibbens, D. A., and Ridge, B. P.
"The chemical analysis of cotton—Oxycellulose, Part 1."
The Journal of the Textile Institute, 16 (1925), p. T13-T52. 9 references.

Some of the most common faults encountered in finished cotton fabrics are traceable to the effects of oxidizing agents on cellulose. Methods by which this detrimental change can be detected in its early stages are most necessary, and therefore the existing tests have been examined critically in order to judge their merits and to learn how to interpret the information they furnish. As a basis of comparison for the tests, standard methods are described for the preparation of oxycelluloses. This paper provides an excellent review of the understanding of this subject at the time.

On investigation, the products are found to fall between two extreme types of oxycellulose, one characterized by great affinity for methylene blue and abnormal retentive power for alkalis (an acidic type of oxycellulose); the other by high reducing power (as expressed by the copper number) and excessive loss of weight in boiling dilute alkali (an aldehydic type of oxycellulose). If cotton has been so affected by oxidizing agents that the second type of oxycellulose has been produced in it, it regains its original chemical properties on scouring with alkali, though, of course, suffering loss in weight. The measurement of the viscosity of a solution of the cotton in "cuprammonium" does, however, reveal changes in the cellulose, no matter what the conditions of oxidation or the subsequent treatment may have been. On the other hand, other changes beside those caused by oxidizing agents may bring about the same reduction in viscosity, but this is not likely to happen as the result of a normal alkali scour.

The copper number is recommended as a useful criterion of the tendency of bleached cotton to yellow, the methylene blue absorption as an indication of level-dyeing properties, and the viscosity as a means of detecting tendering in a normal process of scouring and bleaching. It is concluded that the solubility in boiling 1% NaOH gives practically the same information as copper number; a curve relating copper number and hot-alkali-soluble matter of celluloses oxidized in several different manners is given, covering the range in the former parameter up to 6 and in the latter up to 20%.

Which type of oxycellulose is produced predominantly in hypochlorite bleaching depends on the acidity or alkalinity of the hypochlorite solution. On the alkaline side, the product has a high methylene blue absorption and low copper number, but on the acid side, the absorption is low and the reducing power great.

1926

115. Birtwell, C., Clibbens, D. A., and Geake, A.
"15—The chemical analysis of cotton. xii.—Hydrocellulose."
The Journal of the Textile Institute, 17 (1926), p. T145-T170. 16 references.

This is one of the historically key articles that show the relationship of the logarithm of the viscosity of a 2% solution of cotton celluloses acid-degraded by various means versus breaking load, copper number versus breaking load, and log viscosity versus copper number. At a copper number of 3 to 4, 80 to 90% of the tensile strength of a hydrocellulose is lost. A graph is also given to show the relation of copper number and loss in weight after boiling in hot 1% alkali. The relation is practically linear up to a copper number of 2.5 and a loss of solubility of 15%, both for hydrocelluloses and oxycelluloses, leading to the conclusion that copper number and loss of weight on alkali boiling have the same analytical significance, at least within this limited range. The observation is also made that the copper number of the alkali-boiled material is reduced to negligible values but the viscosity is not significantly reduced. The article closes with data on the rate of attack of sulfuric and hydrochloric acids at various concentrations and temperatures.

1927

116. Clibbens, D. A., Geake, A., and Ridge, B. P.
"24—The chemical analysis of cotton: the action of hot dilute sodium hydroxide solutions on modified cotton cellulose."
The Journal of the Textile Institute, 18 (1927), p. T277-T282. 5 references.

This is one of the key publications in which these workers report the relationship of hot alkali solubility (after four hours' boiling in 1% sodium hydroxide solution) versus copper number. Both for hydrocelluloses and for cellulose oxidized by various agents, the relation is practically linear up to a copper number of 2.5; the solubility being about six times the copper number. Above 2.5, the curves diverge, oxycelluloses suffering a slightly lower loss in weight on alkali boiling than hydrocelluloses at a given copper number; the data are extended up to 30% loss in weight and to a copper number of 5. Data are also given to indicate the loss of viscosity of the samples after alkali boiling and to note the fact that the methylene blue absorption of the boiled product usually is increased as a consequence of the treatment. (Reference only to the authors' previous 5 publications dating back to 1923.)

1928

117. Birtwell, C., Clibbens, D. A., and Geake, A.
"23—The chemical analysis of cotton. The action of sodium hydroxide solutions on modified cotton cellulose at the ordinary temperature."
The Journal of the Textile Institute, 19 (1928), p. T349-T364. 6 references.

The authors point out that it had been known ever since chemically modified celluloses had first been made that they dissolved, often extensively, in caustic soda (alkali) solutions and that the soluble portion could be reprecipitated essentially unchanged by the addition of acids. The authors present a curve that relates the solubility of both hydrocelluloses and variously oxidized celluloses in dilute alkali at room temperature in relation to the fluidity of cuprammonium solutions (their Figure 4). Thus, solubility is related to the size of the cellulose or modified-cellulose molecules (the degree of polymerization, DP) in dilute alkali at 15°C. Solubility does not increase much over 5% until the fluidity increases above 20.

Because there are difficulties in filtering oxidized celluloses in a reproducible manner, the solubility of cellulose in alkali solutions is estimated by oxidizing aliquots of the slurry that are separated from the insoluble pulp by filtration through a glass-frit sieve, using chromic acid and assuming that the consumption of 1 meq of chromate solution is equivalent to 6.75g of cellulose. Using such a procedure, the authors' Figure 1 shows clearly that the maximum solubility of representative modified celluloses in single solution at 15°C occurs at 3N sodium hydroxide (about 12%) and that it falls off sharply at either higher or lower concentrations of alkali. Solubility also falls off rapidly with rising temperature between 15 and 25°C. The conditions of maximum solubility are greatest when the raw cellulosic material is finely divided and is steeped first in 10N sodium hydroxide (NaOH) solution and then diluted to a condition of 2N. For convenience, this is designated as 10N-2N caustic soda.

In the case of hydrocelluloses made by treating bleached cotton with acids under various conditions, the authors show in their Figure 2 that the solubility in 10N-2N NaOH at 15°C is defined by the copper number of the hydrocellulose. When cotton is first mercerized before being degraded by hydrolysis in acids, the solubility in 10N-2N alkali is lowered, a fact also illustrated in the authors' Figure 2.

A very different relationship exists between solubility and copper number in the case of chemically oxidized cellulose, relationships illustrated in the authors' Figure 3. The solubility of oxycelluloses in 10N-2N alkali at 15°C depends on the oxidizing agent used to modify the cellulose. Nonetheless, in any one series of cotton celluloses modified by any one oxidizing agent acting under much the same conditions, the copper number will define the solubility in 10N-2N alkali at 15°C.

Another important aspect of the solubility of modified celluloses in alkali is the "reducing value" of the material that is dissolved, a measurement which is practically the same as a determination of the copper number of the dissolved material. The authors' Figure 5 illustrates the reducing values of alkali-extracted portions of various modified celluloses in relation to the copper number of the modified cellulose. Above copper numbers of about 2, and up to a value of 6, the reducing values of the extracted materials reached practically constant levels: 4 for the material dissolved from hydrocelluloses, 8 for material from oxycelluloses prepared by treatment with neutral hypochlorite and with dichromate, 12 for material from hypochlorous acid oxycelluloses, and 20 for oxycelluloses prepared from dichromate in oxalic acid solution.

The reducing values of extracts from modified celluloses which have been boiled in dilute alkali are low, as are those from oxycelluloses formed from treatment with alkaline hypobromite solution.

This is one of the important early publications that attempts to bring order to the subject of alkali solubility of modified cellulose. The authors state that it is still at this date difficult to explain quantitative relationships in the behavior of the celluloses that have been modified by different oxidative reagents. However, they consider that there are at least two groups of properties that can be correlated with one another. The first is related to reducing properties, the copper number, and the solubility when boiled in dilute sodium hydroxide. The second set of closely related properties concerns measurements of fluidity (related to DP), tensile strength, and loss in weight when treated with concentrated alkali at ordinary temperatures. Measurements of this group of properties are more closely related to one another than to the first group. The authors thus state that high solubility in concentrated alkali at ordinary temperatures is related to high fluidity, but may be associated either with high or low loss of weight when boiled with dilute alkali. Apparently, based on our modern understanding of the "stopping" reaction, hydrocelluloses made from mercerized cotton will fail to peel off many cellulose units before the reaction in alkali is "stopped." In contrast, the attack on the amorphous portion of ordinary celluloses may give liberated, soluble, low-molecular-weight fragments that originate from the breakup of the crystallites.

1932

118. Ridge, B. P., and Bowden, H.
"VII—The tensile strength and fluidity in cuprammonium hydroxide solution of chemically modified rayon and cotton yarns. Part I—Oxidation by hypochlorite solutions."
Shirley Institute Memoirs, 11 (1932), p. 119-166. 26 tables of data, 13 references.
This is another of the early key publications which demonstrated that there was an essentially linear relationship between cuprammonium fluidity and the strength of cotton and viscose yarns. Clibbens and Ridge (1928) had demonstrated the essential findings for cotton; this work stressed the results with rayon fibers. A lack of linearity appeared in the relationship of the initial changes in fluidity and a loss of strength up to about 87 to 90% retention of initial strength in the case of cotton yarn. This was shown most likely to be due to the influence of short fibers on the strength of yarns, for the non-linearity was not present when continuous filament rayon yarns were involved. Comments are also included on the influence of hot-alkali solubility and the action of boiling water on fluidity and loss of strength.

1934

119. Davidson, G. F.
"12—The dissolution of chemically modified cotton cellulose in alkaline solutions. Part I—In solutions of sodium hydroxide, particularly at temperatures below the normal."
The Journal of the Textile Institute, 25 (1934), p. T174-T196. 25 references.
The hydrocelluloses used in many of the experiments reported by the author apparently were very low in molecular weight (or DP); the fluidity values cited were largely above 20, many from 22.4 to 45.4. At this date, little was known regarding the chemical structure of oxycelluloses.
Davidson was able to conclude that the chain-length distribution resulting from the acid hydrolysis of mercerized cotton is different from that obtained from unmercerized. [We now know that an advanced degree of hydrolysis of unmercerized cellulose tends toward the leveling-off DP (LODP) represented by the average length of the chains in the crystallite portion of the cellulose.]
The author further concluded that boiling hydrocelluloses in hot dilute alkali causes a slight increase in fluidity and solubility in sodium hydroxide at low temperatures. "It is possible," he says, "that the action of hot dilute alkali consists in the shortening of each chain-molecule by the removal from the end of a portion containing the reducing group, the very short chain-molecules containing only a few glucose units being decomposed into soluble products. It is difficult, however, to account for the large percentage loss of weight that occurs on alkali boiling, unless the portion removed from the end of each chain-molecule is a considerable fraction of the chain. The loss of weight depends on the severity of the boil, and this suggests that progressive shortening of the chain-molecules takes place, the attack being from the end originally bearing the reducing group. The shortening of each chain-molecule would tend to increase the fluidity, but the removal of the shortest as soluble products would tend to lower it, so that it is possible that the two effects might counter-balance one another in such a way that neither the average chain length, determining the fluidity, nor the proportion of short chain-molecules, determining the low temperature solubility, is greatly affected in spite of a considerable loss in weight."
The effect of alkali boiling on oxycelluloses is to reduce the copper number to a low value and to cause a loss of weight proportional to the original copper number, as with hydrocelluloses but, with certain of the oxycellulose series, the fluidity is considerably increased. These effects can reasonably be explained by the assumption already discussed that oxidation does not necessarily always result directly in the breaking of the chain-molecules, but reduces their chemical stability towards alkalis, so that during alkali boiling, in addition to the processes postulated to occur with hydrocelluloses, further chemical decomposition resulting in scission of the chains takes place. This might also account for the fact that the fluidity-tensile strength and the fluidity-solubility relations of oxy- and hydro-celluloses are substantially the same after the modified materials are alkali-boiled, since it is improbable that dissolution in cuprammonium, or treatment with sodium hydroxide at low temperatures, would produce further scission of chain-molecules after the relatively drastic hot-alkaline treatment.
Today there is ample evidence that Davidson was essentially correct in these explanations.

1937

120. Anderson, O. E.
"Some causes for non-uniformity in sulphite pulp manufacture."
Paper Trade Journal, Tappi Section, 104 (1937), p. 70-72. 2 references.
The one percent caustic soda solubility test affords a reliable, convenient check on wood quality. With respect to its relationship to tear resistance of the pulp made therefrom and yield, both decrease with increasing caustic solubility. Variation in pulp quality as a result of mixing species of wood in the digester is found to vary, the wood of higher lignin content producing pulp of higher lignin content.

1938

121. Launer, H. F.
"Volumetric determination of alpha-, beta-, and gamma-cellulose in pulps and in papers containing sizing, filler, and other materials."
Journal of Research of the National Bureau of Standards, 20, Research Paper RP1068 (1938), p. 87-95. 13 references.
The volumetric method for the determination of alpha-, beta-, and gamma-cellulose, previously described in Bureau Research Paper RP979 (1937), is shown to be applicable to papers containing rosin, glue, starch, oxidizable fillers and lignin, and to pulps containing natural resins in any amount. As the method depends upon the estimation of each cellulose fraction by oxidation with dichromate, and as the sizing materials are all oxidizable, the manner of making corrections for these materials is described in detail. The alpha-cellulose values so obtained are shown to be in good agreement with those of a gravimetric method for a wide range of pulps and papers.

1939

122. Godman, G. L., Haworth, W. N., and Peat, S.
"394. Polysaccharides. Part XXXVII. Oxycellulose. Part 2."
Journal of the Chemical Society (London) (1939), p. 1908-1914.
Oxycellulose, which was prepared by the action of acid permanganate on linters, had a high copper number (14) and contained uronic acid residues. Boiling 1% NaOH dissolved 50% of the oxycellulose and the soluble portion was shown to have undergone decomposition of the type suffered by oligosaccharides in the presence of alkali. The soluble portion was found to contain formic, acetic, lactic, and dihydroxybutyric acids, as well as a mixture of C6-saccharinic acids. The material insoluble in hot 1% alkali had a low copper number (0.27) and contained no acidic groups. It was acetylated and methylated with ease and its molecular size (by end-group assay) corresponded to a chain of 90 glucose units. This insoluble fraction of oxycellulose was treated with 2.5N alkali, in which half of it dissolved. The soluble portion had a chain length of 35 glucose units (by end-group assay) and the insoluble part, a chain length of 60 units.
Cold acid hydrolysis of the oxycellulose led to the isolation of an aldobionic acid which was not further characterized.

123. Rys, L., and Bonisch, A.
"Refining of pulp."
Paper Trade Journal (May 11, 1939), p. 31-40. 4 references.
Experimental evidence is presented on the influence of such factors as temperature, pulp consistency, time and order of treatments, and the presence of additives on the yield and alpha-cellulose content of pulps refined with both hot and cold alkali. Substitution of calcium hydroxide for the sodium resulted in increased yield but lowered alpha-cellulose and intrinsic viscosity.

1942

124. Rutherford, H. A., Minor, F. W., Martin, A. R., and Harris, M.
"Oxidation of cellulose: the reaction of cellulose with periodic acid."
Journal of Research of the National Bureau of Standards, 29, Research Paper RP1491 (August 1942), p. 131-141. 18 references.
The authors report in their Figure 3 the practically linear relationship between aldehyde content in periodate oxycellulose (where aldehyde groups are on carbon atoms 2 and 3) and

solubility in boiling 1% sodium hydroxide solution. The range covered in their data was up to about 0.022 millimoles of aldehyde per gram of cellulose and about 22% solubility in the hot alkali.

1949

125. Samuelson, O., Berrison, I., and Regestad, S. O.
"The influence of the cooking conditions on the hot alkali solubility of unbleached sulphite pulps."
Svensk Kemisk Tidskrift, 61 (1949), p. 234-241. 14 references.

For a series of unbleached sulfite pulps prepared under different conditions in the cook, the solubility in hot 7% NaOH is shown to be related to the pulp viscosity. This is taken to show a close relation between carbonyl content and viscosity of much the same type as found in the case of hydrocelluloses.

1950

126. Samuelson, O., and Ramsel, C.
"Effect of chlorine and chlorine dioxide bleaching on the copper number, hot alkali solubility and carboxyl content of sulphite cellulose."
Svensk Papperstidning, 53, No. 6 (1950), p. 155-163. 24 references.

In a previous paper a report was made of the influence of different conditions during the sulfite cooking process on the hot-alkali (boiling 7% NaOH solution) solubility of the unbleached sulfite pulp. It was shown that a relationship existed between the solubility and viscosity of the pulp, and that this relationship was in only a small degree affected by the cooking conditions. With decreasing viscosity an increase occurs in the hot-alkali solubility and copper number of the pulp. This is in agreement with the assumption that, in the degradation of carbohydrates during sulfite cooking, the reaction mechanism is in principle the same as in acid hydrolysis of pure cellulose. In both cases carbonyl groups are formed.

It is therefore inevitable that a pulp is obtained during the sulfite cooking that has a high solubility in hot alkali and a high content of carbonyl groups. For many fields of use, for instance in the case of pulps for the production of certain cellulose derivatives and paper pulps for various purposes, e.g., photographic paper, it would be a considerable advantage if the sulfite pulp were modified by purification and some suitable form of bleaching so that the hot-alkali solubility and copper number decreased.

The purpose of this work was to study the changes in sulfite cellulose in respect to its hot alkali solubility and copper number during various bleaching processes. Employing a 7% solution of sodium hydroxide (NaOH), the authors found a linear relationship between the logarithm of the copper number and the hot-alkali-soluble matter over the very wide range of copper numbers from 1 to 10 and hot-alkali-soluble matter from about 10 to 55%.

In addition to these studies, it was clearly demonstrated that treatment with chlorine dioxide did not change the copper number (which indicates the carboxyl groups present) or the methylene blue absorption of the cellulose (the latter indicates the carboxyl groups present). It was therefore suggested that excellent pulps might be made by treatment first with sodium hypochlorite and then with chlorine dioxide.

1951

127. Meller, A.
"Studies on modified cellulose. I. The alkali stability of oxidized, hydrolyzed, and methanolized celluloses."
Tappi, 34, No. 4 (1951), p. 171-179. 52 references.

An excellent review of the subject points out that the action of aqueous solutions of both acids and oxidizing agents on cellulose fibers usually results in considerable changes in their chemical and physical constitution. These changes may significantly affect the resistance (stability) of the modified celluloses toward dilute hot and strong-cold alkali solutions. The changes in physical structure and chemical composition of cellulose fibers arising from the action of both hydrolyzing and oxidizing solutions usually take place concurrently. It is, therefore, very often difficult to ascribe the changed stability to chemical or to physical structural effects. Nonetheless, the overwhelming influence of certain chemical modifications of cellulose, brought about by the action of both acid and oxidizing solutions, on their stability toward hot weak alkali solutions has been demonstrated. The hot-alkali stability has been interpreted in terms of rate and extent of loss in weight measured on heating chemically modified celluloses with weak alkali solutions. Close relationships have been established between alkali solubility of modified celluloses and their (1) copper number, (2) solution viscosity in cuprammonia, (3) dialdehyde grouping and other "alkali-sensitive" molecular structures, (4) carboxyl (uronic-acid type), (5) keto-carbonyl, and (6) diketo-groupings. The author cites the publications which refer to each of these topics.

Modified celluloses prepared from cotton fibers by means of periodic acid, potassium dichromate and sulfuric acid, respectively, are found to be sensitive to hot dilute alkali. While the periodate and dichromate oxycelluloses showed comparatively high carbonyl-group content, as estimated by the hydroxylamine method, the carbonyl content of the hydrocelluloses was relatively low. The low "apparent and approximate" carbonyl content of the hydrocelluloses may be attributed in part to a limited accessibility of the aqueous hydroxylamine solution applied under the conditions of analysis and to a hindrance of the intramolecular transfer from hemiacetals to reactive aldehydic groupings of the terminal glucose units of the chain molecules. The limited accessibility would be conditioned by a presumed highly crystalline or dense substructure of the hydrocelluloses.

Oxidation of the periodate and dichromate oxycelluloses and hydrocelluloses by chlorous acid rendered the modified celluloses more stable toward hot weak alkali solutions. Chlorous acid oxidation of the periodate and dichromate oxycelluloses resulted in a marked decrease of carbonyl content paralleled by an increase of carboxyl content. The chlorous acid treated hydrocelluloses are termed "gluconic acid oxycelluloses," presumably containing gluconic acid terminal units in the polymeric molecules and rendering hydrocelluloses stable to hot weak alkali solutions. Prolonged oxidation of modified celluloses by chlorous acid yielded products with increased stability toward hot alkali and decreased carbonyl content. It is inferred from carbonyl-group contents and relative stabilizations arising from chlorous acid oxidation of periodate and dichromate oxycelluloses that their properties would be consistent with the presence of slowly reacting keto-carbonyl groupings and dense substructure in the dichromate oxycelluloses, and fast reacting aldehydic carbonyls and expanded supermolecular structure in the periodate oxycelluloses. Methanolized hydrocellulose showed an increased hot-alkali stability and decreased carbonyl content when compared with the parent material. Chlorous acid treatment of both chlorinated and of hypochlorite bleached softwood sulfite pulps resulted in decreased carbonyl and increased carboxyl group contents.

Oxidation and hydrolytic reactions lead to the formation of carbonyl groups in the cellulose molecules. By enolization of the carbonyls the units may assume an "alkali-sensitive biosidic configuration" and play a dominant role in the chemical mechanism involved in the fragmentation and dissolution of modified celluloses when heated with dilute alkali solutions. The cold-alkali stability of modified cotton celluloses and partially bleached wood pulps, as interpreted in terms of alpha-cellulose content, was lower than that of the parent fibers. Oxidation by chlorous acid of the modified cotton celluloses and wood pulps resulted, in the majority of instances, in a little increase in alpha-cellulose content. It is concluded that cold-alkali stability of modified celluloses is possibly more dominantly governed by factors relevant to supermolecular structure, degree of polymerization, and presence of chemically modified groupings (other than carbonyls, such as methoxyl, carboxyl, etc.) and less influenced by carbonyl groupings. Many of the modified celluloses showed a higher solution viscosity in cuprammonia resulting from oxidation by chlorous acid. This higher solution viscosity in cuprammonia seems to be due to the presence of groupings of greater stability toward alkali and hence toward the cuprammonia solution.

1952

128. Meller, A.

"Studies on modified cellulose. II. The hot alkali stability and acid hydrolytic resistance of modified cellulose."

Tappi, 35, No. 2 (1952), p. 72-75. 11 references.

A relationship appears to exist between moisture regain capacity and acid hydrolytic resistance of modified celluloses. A graph is also given to show a similar relationship between carbonyl-group content of oxycelluloses and their hot-alkali stability. While acid hydrolytic resistance of modified celluloses is overwhelmingly influenced by physical structural characteristics, their hot-alkali stability is predominantly affected by chemical constitution. Extensive oxidation of cotton cellulose by periodic acid resulted in oxycelluloses of increased moisture regain capacity; the crystallinity of the cellulose may have been disrupted. The increase in moisture regain resulting from oxidation of cotton cellulose with an acid solution of potassium dichromate was less pronounced and seemed to be complicated by superimposing effects. Reoxidation of periodate dialdehyde oxycelluloses by chlorous acid had only a comparatively small effect on moisture regain. Chlorous acid oxidation of dichromate oxycelluloses decreased the moisture adsorption significantly, suggesting that uronic acid carboxyls present in oxycelluloses may either possess a lower "affinity" for water or that they may be involved in the formation of cross bonds and contribute to a lower accessibility to water vapor. The moisture regain capacity of hydrocelluloses prepared by means of hot acid solutions showed a decreased moisture regain power, possibly due to an increased degree of crystallinity. Hydrocelluloses prepared by cold acid solution showed a slightly greater moisture adsorption. The moisture adsorbing capacity of gluconic acid oxycelluloses (primarily oxidized on carbon atom 6) was only slightly greater than that of the parent hydrocelluloses. On converting oxycelluloses containing uronic acid carboxyls into calcium salts the moisture regain capacity increased considerably. The increased moisture regain capacity of oxycelluloses in which the carboxyl groups have been converted to calcium salts indicates that in drying of such oxycelluloses the extent of cross-bond formation (involving probably secondary valence type of bonds) is reduced. Author's abstract.

129. Ohlsson, K.-E.
"The alkali solubility of pulps. I. Method."
Svensk Papperstidning, 55, No. 10 (1952), p. 347-357. 33 references.

When determining the alkali solubility of a pulp between 12 and 29°C by oxidation of the dissolved material with dichromate and sulfuric acid, the amount of chromic ions formed may be determined colorimetrically at a wavelength of 600 nm. The method has been shown to be rapid and accurate. Calibration curves have been computed by using glucose and xylose.

The author presents extensive data to demonstrate the dependence of the alkali solubility on the lye to pulp ratio, the temperature, the time and method of treatment, and the nature of the sample. From this information and on the basis of the Swedish standard method CCA 8, an improved method for the determination of the alkali solubility was worked out. The carbohydrates dissolved by the lye are divided into a beta and a gamma fraction by neutralization with phosphoric acid. The rate of dissolution of the beta and gamma fractions was shown to be dependent on the type of pulp and on the lye concentration. Oxygen can increase the beta fraction, but has little effect on the gamma. Lignin and extractives should be removed before the determination.

130. Wilson, K., Ringström, E., and Hedlund, I.
"The alkali solubility of pulp."
Svensk Papperstidning, 55, No. 2 (1952), p. 31-37. 22 references.

The determination of beta- and gamma-cellulose is discussed, giving curves of alkali solubility versus the percentage of sodium hydroxide. These show that gamma-cellulose increases in its solubility up to 18 or 20% NaOH, whereas the beta-cellulose shows a fairly sharp maximum at 10% NaOH and negligible solubility at 18% NaOH. Beta is defined by the authors as that part of the cellulose that goes into solution in concentrated NaOH and is precipitated out by neutralization with acid.

A Swedish alpha-cellulose determination at this date involved dissolving pulp in 18% NaOH and washing the precipitate with water. The latter operation can wash out some of the beta owing to the dilution of the alkali by the wash water. An alternate method, discussed in this paper, is to extract the pulp solely in 18% (or 10%) NaOH. (Tappi T203 involves washing the pulp in 17.5%, then 9.45% NaOH.) The organic matter in the filtrate is oxidized with chromic acid. Beta-cellulose is separated out by precipitation with acid. By this technique, alpha, beta, and gamma are caused to add up to 100%, for they are all done on the same sample and the alpha portion is considered to be the difference, 100 – beta – gamma.

Staudinger and Jurish [*Kunstseide und Zellwolle*, 21 (1939), p. 6-9] gave values of DP of 10-150 for beta; about 10 for gamma. The present authors, using a standard CED technique, found that the DP of beta cellulose in rayons was 500 to 130. In an 18% solution of NaOH, which dissolves little of the beta, a fraction of DP 65 was found.

The authors mention that wood contains no beta-cellulose but a lot of gamma (DP 10-30), which is largely removed during the cook.

Although there is nothing fixed about the quantities, chemical nature, or DP of the beta and gamma fractions of cellulose, the authors suggest measuring the amount of material that dissolves in 10% alkali (beta and gamma) and that which dissolves in 18% alkali (gamma) to provide useful information. The full curve of solubility and neutralized-precipitated fraction versus alkali solubility, as extensively shown by the authors, is most informative.

1953

131. Kenner, J., and Richards, G. N. (Part I); Corbett, W. M., and Kenner, J. (Part II).
"The degradation of carbohydrates by alkali. Part I. Alpha-alkoxy-ketones." "Part II. Lactose."
Journal of the Chemical Society (1953), p. 2240-2244; p. 2245-2247.

Preliminary studies on model compounds are reported which lead to an understanding of the "peeling reaction" of oxidized cellulose in dilute alkalis. In the first article the importance of the oxidized keto state is demonstrated, for this can form an enol-OH group which then can ionize. The second paper proposes that cellulose oxidized to the stage where there is a 2-carbonyl group can rearrange to yield an alpha and beta *iso*saccharinic acid which would effectively stop the peeling off of glucosidic groups from the ends of the polymer chains. It is the enolization that can help explain the formation of two- and three-carbon fragments during the oxidative deterioration of cellulose.

132. Meller, A.
"The hot alkali stability of chemically modified cellulose fibres."
Australian Pulp and Paper Industry, Technical Association Proceedings, 7 (1953), p. 263-315. 90 references.

Chemical constitutional changes of cellulose fibers resulting from oxidation are very often reflected in their behavior towards weak hot alkali solutions. The magnitude of the response of chemically modified cellulose fibers to alkali solutions may serve as an aid to elucidate the nature and extent of the chemical modification. In this paper, the hot-alkali stability of acid hydrolyzed and of oxidized cellulose fibers is extensively reviewed and discussed. Particular reference is made to the chemical mechanisms involved in the fragmentation by hot weak alkali solutions of acid-hydrolyzed and oxidized cellulose fibers. Detailed chemical equations are given to explain the action of alkali on oxidized cellulose and the products that are formed. In this mechanism the role of carbonyl groups is demonstrated. Some reference is also made to the rate of alkaline extraction and to the breakdown products present in alkaline extracts.

133. Meller, A.
"Reduction of chemically modified cellulose by sodium borohydride."
Tappi, 36, No. 8 (1953), p. 366-367. 4 references.

After having shown, in a previous article, that chlorous acid oxidation of aldehyde-group-containing oxycelluloses led to decreased hot-alkali solubility, the author describes in this brief communication the similar effect of sodium borohydride reduction, which tends to convert aldehyde end groups to alcoholic groups.

134. Meller, A.
"Reduction of oxycellulose by sodium borohydride."
Chemistry and Industry, 72 (1953), p. 1204. 3 references.

A short communication on the results that show the dominant influence of the carbonyl groups present in modified cellulose (oxycellulose) on fragmentation and dissolution by hot alkali solutions. Reduction of oxycelluloses was carried out in alkaline borohydride solution, leading to the formation of alcohol groups in place of the normally alkali-sensitive aldehyde groups. In contrast, reduction under acidic conditions leads to the production of aldehydic groups from structures containing acid groups, resulting in increased sensitivity to hot dilute solutions of alkali.

1954

135. Kenner, J., and Richards, G. N.
"The degradation of carbohydrates by alkali. Part III. 3-O-Methyl derivatives of glucose and fructose."
Journal of the Chemical Society (1954), p. 278-282.

In an ongoing study regarding the *peeling reaction*, the results of a study of 3-O-methyl derivatives of glucose and fructose accord with the view that the site of the etherifying group (when this is in the 3 or the 4 position) regulates the result of the attack by dilute alkali.

The general mechanism of the action of dilute alkali on disaccharides has been discussed in Part II and it appears that it is regulated by the point of attachment of the glucoside grouping (when this is in the 3 or the 4 position) to the parent aldose or ketose. Thus, it involves the carbonyl group in the 1 position in 3-glucosidoglucose but only operates from the 2 position in the cases of 4-glucosido- and 4-galactosido-glucoses (maltose and lactose) after their isomerization to the corresponding ketoses. This comparison, as well as the need to explain the behavior of the monosaccharides towards alkali, impelled the authors to study that of the 3-O-methyl derivatives of glucose and fructose.

In a solution of 3-O-methyl fructose in 0.05N calcium hydroxide at 25°C, the conversion to the metasaccharinic acid is much more rapid owing to stimulation of carbonyl reactivity by $[CaOH]^+$.

136. Purves, C. B.
"B. Chain structure." (Chapter III)
In *Cellulose and Cellulose Derivatives, Part I*, 2nd edn., ed. Ott, E., Spurlin, H. M., and Grafflin, M. W., Interscience Publishers, New York (1954), p. 91-98. 57 references.

Section B (d) of this chapter, entitled "Significance of the Alkali-Solubles," provides an excellent overview of the subject. In particular, words of caution are expressed concerning the fact that what is extracted under any particular set of conditions will be dependent on the initial composition of the pulp and on its prior chemical treatment.

137. Samuelson, O., and Wennerblom, A.
"Degradation of cellulose by alkali cooking. Part I. Formation of carboxyl groups."
Svensk Papperstidning, 57, No. 22 (1954), p. 827-830. 8 references.

When cellulose is autoclaved at 170°C in 10% NaOH, the authors demonstrate that the degraded cellulose chains contain one carboxyl ($-COO^-$) group per chain. They further calculate that about 50 cellulose units go into solution for each breakage of the chain at the beginning of the cook. Moreover, 1.4 equivalents of weak acid are formed for each glucose unit going into solution; hence, the process must involve some kind of chain reaction. The mechanism of hot-alkali treatment is much the same and the authors conclude that the carboxyl group is formed at that part of the macromolecule where the chain reaction stops and not at the point where the reaction sets in.

1955

138. Meller, A.
"The actual and potential degradation of cellulose fibres."
Australian Pulp and Paper Industry, Technical Association Proceedings, 9 (1955), p. 192-208. 14 references.

An attempt has been made to clarify the terms "degradation," "over-cooking," and "overbleaching" of cellulose fibers.

Hydrolytically degraded cellulose fibers are "actually" degraded; so far this degradation has not been reversed. In contrast, "potential" degradation involves an oxidation of some of the functional hydroxyl groups to carbonyls in the glucose or pentose units of the polymeric carbohydrate molecules. Such changes do not shorten the molecules but render them vulnerable to attack by alkaline or acid solutions. Examples are quoted showing the importance of these degradations in the processing of cellulose fibers when used in the textile and wood pulp industries. It is also shown that "potential" degradation can be remedied to a large extent. [A short authoritative overview of the subject by one of the leading investigators. Pages 207-208 contain the discussion of various points between the lecturer and members of the audience.]

139. Meller, A.
"Studies on modified cellulose. IV. Oxidation by chlorous acid and reduction by sodium borohydride of chemically modified celluloses."
Tappi, 38, No. 11 (1955), p. 682-687. 16 references.

The hot-alkali sensitivity of hydrolyzed, periodate and dichromate-oxidized cotton cellulose, and of oxidized (bleached) wood pulp decreases considerably upon chlorous acid oxidation or sodium borohydride reduction in alkaline solution. Reduction of "gluconic oxycellulose" by acidic solutions of sodium borohydride decreases its resistance toward hot, dilute alkali. The carbonyl-group content of oxidized cotton cellulose and wood pulps decreases as a result both of chlorous acid oxidation and of alkaline sodium borohydride reduction. The uronic acid carboxyl content of dichromate oxycellulose increases considerably on chlorous acid oxidation, indicating the presence of aldehyde groupings on carbon six of some of the glucose residues in this oxycellulose. The hot-alkali stability of modified cellulose fibers is greatly influenced by the enolizability of chemically modified functional groups present in the glucose anhydride building units. The very high resistance to alkali of methanolized cellulose is consistent with the nonenolizability of glucosidic methoxyl end-groups present in these modified celluloses. Gluconic oxycellulose seems to contain lactonized carboxyl end-groups, which are reducible with acidic sodium borohydride solution. The carboxyl groups in chlorite-periodate and chlorite-dichromate oxycelluloses do not appear to be lactonized, and are not easily reducible with sodium borohydride. "Potential degradation" of cellulose fibers, which manifests itself in subsequent alkaline treatment, can be remedied to a large extent by chlorous acid oxidation or by sodium borohydride reduction in alkaline solution.

1956

140. Meller, A.
"Studies on modified cellulose. V. The cold alkali stability (alpha-cellulose) of chlorous acid oxidized and sodium borohydride reduced wood pulps."
Tappi, 39, No. 10 (1956), p. 722-724. 8 references.
Both alpha-cellulose content and cuprammonium viscosity of chemically modified wood pulps increase upon chlorous acid oxidation or sodium borohydride reduction. The increase in alpha-cellulose content is low, however, in comparison to the decrease in solubility in hot 1% NaOH. The results of these studies indicate that molecular size influences the determination of alpha-cellulose through solubility in cold 17.5% NaOH. On the other hand, chemical modification strongly influences the solubility in hot dilute NaOH. The author further comments that a precipitate occurs when the filtrate from the extraction in concentrated cold alkali is acidified but not in the case of the filtrate from the hot 1% NaOH.
The cuprammonium viscosity test should be treated with due regard to its limitations when applied to bleached pulps for measuring the degradation taking place during bleaching. Bleached pulps not actually degraded but containing alkali-sensitive groupings in the cellulose molecules contribute to low solution viscosity results.

1957

141. Richards, G. N., and Sephton, H. H.
"The alkaline degradation of polysaccharides. Part I. Soluble products of the action of sodium hydroxide on cellulose."
Journal of the Chemical Society (London), Paper No. 905 (1957), p. 4492-4499. 21 references.
The major soluble products of the degradation of cotton hydrocellulose by boiling oxygen-free 0.5N sodium hydroxide have been identified and in several cases their relative amounts have been determined. D-Glucoisosaccharinic and formic acid are produced in greatest amount under the conditions used. Lactic and acetic acids are produced in lower quantities.

142. Machell, G., and Richards, G. N.
"The alkaline degradation of polysaccharides. Part II. The alkali-stable residue from the action of sodium hydroxide on cellulose."
Journal of the Chemical Society (London), Paper No. 906 (1957), p. 4500-4506. 24 references.
The extent of solubility in hot dilute alkali is known to be related to the DP and, hence, the number of aldehydic end groups per given quantity of cellulose. The action of hot dilute, oxygen-free sodium hydroxide on cellulose is shown to take place involving two competing reactions: degradation and a "stopping" reaction. Part I (see 141) discusses the degradation process and Part II the "stopping" reaction, a slower reaction which involves the formation of an alkali-stable structure. The alkali-stable polysaccharide resulting from the treatment of cotton hydrocellulose with hot dilute, oxygen-free sodium hydroxide has been partially hydrolyzed and D-glucometasaccharinic acids isolated and identified. Alpha-D-glucosaccharinic acid was tentatively identified by paper chromatography. It is concluded that these acids originate by alkaline rearrangement of the reducing end-group of the cellulose molecule and so impart stability to alkali. Equations are given postulating the chemical steps involved.

1958

143. Machell, G., and Richards, G. N.
"The alkaline degradation of polysaccharides. Part III. Action of sodium hydroxide on amylose."
Journal of the Chemical Society (London), Paper No. 237 (1958), p. 1199-1204. 25 references.

The action of 0.5N (ca. 2%) sodium hydroxide at 100°C on amylose (from starch) is of two main types: "degradation" producing mainly D-glucoisosaccharinic, formic, and lactic acid; and a "stopping" reaction affording alkali-stable amylose probably bearing terminal D-glucometasaccharinic acid units. The analogy with the similar degradation of cellulose is discussed.

144. O'Meara, D., and Richards, G. N.
"The alkaline degradation of polysaccharides. Part IV. Monosaccharide analogues of periodate oxycellulose."
Journal of the Chemical Society (London), Paper No. 238 (1958), p. 1204-1211. 27 references.
A study of the products of alkaline degradation of certain periodate-oxidized monosaccharides suggested that the degradation occurs predominantly by scission of the carbon-oxygen linkage in the beta-position with respect to the aldehyde groups Subsequent rearrangements of the products probably yield a molecule of glycolic acid from each scission of this type. In certain circumstances an intramolecular Cannizzaro rearrangement can occur in the periodate-oxidation product. The analogy with periodate oxycellulose is discussed.

1963

145. Rapson, W. H., and Spinner, I. H.
"Brightness reversion in bleached pulps."
In *The Bleaching of Pulp*, Tappi Monograph Series No. 27, ed. Rapson, W. H. (1963), p. 299. 50 references.
In their discussion of ways of improving brightness reversion the authors have the following statements under a section entitled "Caustic Extraction." Their remarks are quoted entirely.
"It is well known that hot alkali extraction is the most important treatment for improving the color stability of bleached pulps. Part of this improvement may be attributed to removal of lignin, resin, and hemicellulose, but even in later stages of bleaching where these substances are present to a very small extent, caustic extraction still improves color reversion.
"It is highly probable that carbonyl groups in hemicellulose and cellulose are destroyed by the hot alkali extraction, and that this in turn is responsible for the improvement in color reversion. The higher the alkali to pulp ratio, the higher the temperature and the longer the time, the greater is the improvement in color reversion brought about by hot alkali extraction.
"It has been shown that there is a linear relation between copper number and hot alkali solubility for most pulps, and also between postcolor number and copper number. Therefore there is a relationship between color reversion and hot alkali solubility. Since hot alkali extraction naturally diminishes the hot alkali solubility, it diminishes the copper number and diminishes color reversion. For reasons which will be evident after the discussion of the effects of peroxide, a hot alkali extraction should always precede a final treatment with any bleaching agent, unless the bleaching stage itself is strongly alkaline."

1965

146. Albeck, M., Ben-Bassat, A., Epstein, J. A., and Lewin, M.
"The yellowing of cotton cellulose. Part I: A new method for the determination of yellowing."
Textile Research Journal, 35, No. 9 (1965), p. 836-843. 39 references.
In this first report of three in 1965, attention is drawn to the parallelism between the color reversion of cotton on aging as measured by the PC number and the yellow hue of cotton extracts in hot-alkaline solution. The authors' procedure consists of extracting the sample in a 5% sodium bicarbonate solution under reflux for one hour and measuring the color intensity of the yellow solutions obtained. The solutions obey Beer's Law. The color intensity is proportional to the solubility during extraction and is linearly related to the amount of cotton used in the extraction and to the extent of hydrolytic or oxidative modification of the cellulose.

147. Albeck, M., Ben-Bassat, A., and Lewin, M.
"The yellowing of cotton cellulose. Part II: The influence of functional groups and the nature of the yellowing."
Textile Research Journal, 35, No. 10 (1965), p. 935-942. 20 references.
In this second report, the factors causing the yellow color of the hot-alkaline extracts were investigated. The yellowing tendency of modified cottons is found to depend solely on the aldehyde-group content; the ketone and carboxyl groups have no appreciable influence.

Linear plots of yellowing against aldehyde-group content and against 1/DP were found. Reducing sugars show an absorption spectrum of the yellow solutions similar to modified celluloses; pentoses and alduronic acids yield lower specific absorption coefficients than hexoses, while non-reducing sugars produce no yellowing. The ultraviolet spectra of bicarbonate extracts of modified celluloses and glucose are similar to alkaline extracts of aged celluloses. These spectra and the infrared spectrum of the color substance and its elementary analysis point to the possibility that the colored substances are furan derivatives.

148. Lewin, M.
"The yellowing of cotton cellulose. Part III: On the mechanism of yellowing upon aging and alkaline extraction."
Textile Research Journal, 35, No. 11 (1965), p. 979-986. 59 references.
In this third report on the subject, by Menachem Lewin, it is shown that the parallelism between the brightness reversion on aging and the yellow color of the alkaline-extraction solutions *does not hold* in the case of hydrocelluloses or carboxyl-containing celluloses. The present publication deals with possible explanations of why the aldehyde (carbonyl) end-groups in hydrocelluloses contribute very little to color reversion. This the author attributes to the formation of 1,6-anhydride linkages at the temperatures usually used for accelerated-aging tests, 105-140°C. This would also explain the increased discoloration of pulps when aged under moist conditions. The author suggests that the tendency of pulps to discolor is closely related to the "peeling-off" process which occurs in the alkaline treatment of pulps and that the colored products formed on aging are related to the breakdown products which result from the peeling-off process. On this basis, he gives detailed explanations of the aging behavior of many different oxidized conditions of cellulose.

149. Rydholm, S. A.
"Chemical pulping I. Sulfite process (subsection 3). Reactions of the sulfite process, C. Carbohydrate reactions." "Part IV. Pulp properties and uses, II. Physical properties, 1. Alkali solubility." (Chapter 9)
In *Pulping Processes*, Interscience Publishers, a division of John Wiley & Sons, Inc., New York (1965), p. 596-609; p. 1116-1118.
On p. 596-609 a discussion of the carbohydrate chemistry of alkaline cooks includes the following reactions:
1) Alkaline swelling, causing physical changes in the fiber wall.
2) Alkaline dissolution of carbohydrate matter without chemical degradation.
3) Reprecipitation of dissolved carbohydrates and their adsorption on the fiber structure.
4) Alkaline hydrolysis of acetyl groups.
5) Alkaline "peeling," a reaction sequence involving enolization, hydrolysis of beta-alkoxycarbonyl bonds, and further degradation of the products of hydrolysis, including isomerizations, as well as hydrolysis, to hydroxy acids.
6) "Stopping" reactions, involving dehydration and fragmentation or intramolecular rearrangements to alkali-stable configurations.
7) Alkaline hydrolysis of beta-glycosidic bonds, resulting in depolymerization of the chain molecules to shorter fragments, which are exposed to further degradation by alkaline peeling.
A table summarizes the nature of the polysaccharides that remain in birchwood over the course of five hours of a soda cook. This is accompanied by several pages of detailed discussion of the chemistry of the "peeling" and "stopping" reactions in both cotton and in wood. Also discussed is the alkaline-resistance of xylan and other hemicellulosic components and the tendency of xylan to be reabsorbed on the cellulose. A table (9.25) is also given showing the composition of sugars and the DP of various alkali-extracted components that essentially represent alpha-, beta-, and gamma-celluloses.
As briefly stated on p. 1116-1118 (see also 81) in general, the solubility of a sulfite pulp in 18% NaOH gives a fair indication of the hemicellulose content, though some degraded cellulose may go into solution and some hemicelluloses always remain undissolved. The deviations are more pronounced for kraft pulps, especially from hardwoods. In pulps containing much degraded cellulose, this portion tends to dissolve at about 10% NaOH concentration, where the fibers have maximum swelling. Only limited amounts are dissolved in 18% NaOH, and the difference in alkali solubility between 10 and 18% NaOH, sometimes called "rest-hemi," may be taken as a rough measure of the state of degradation.
Alkali solubility is often determined at concentrations close to 10 and 18% NaOH. This involves the dispersion of the pulp at a consistency of 1.5% in the lye at 20°C, allowing one hour of

swelling, filtering off the residue, and determining the amount of dissolved matter by oxidation with acid dichromate and analyzing the dichromate consumption volumetrically or spectrophotometrically. The classical determination of the alpha-, beta-, and gamma-cellulose content of pulp is similar. The pulp sample is treated with 17.5% NaOH. Later water is added, the pulp suspension filtered, and the filter cake washed with water, acetic acid, and water again before drying and weighing. This insoluble fraction is called the alpha-cellulose. The combined filtrate and washings are acidified with sulfuric acid and the precipitate, or beta-cellulose, allowed to settle, leaving the gamma-cellulose in solution. The former can be determined volumetrically or gravimetrically, the latter always volumetrically by dichromate oxidation as described above.

The alkali solubility of unbleached pulps is sometimes of interest to determine. However, the residual lignin affects the accuracy of the determination and should be removed by chlorite oxidation, which does not affect the carbohydrates and their alkali solubility. See also 79.

1967

150. Browning, B. L.
"Solubility in alkaline solvents." (Chapter 20)
In *Methods of Wood Chemistry*, Volume II, Interscience Publishers, a division of John Wiley & Sons, New York (1967), p. 415-427. 40 references.

After a brief description of the components of cellulose that are dissolved by various concentrations of alkali and at various temperatures, a short history of the development of the measurement of alpha-cellulose content is given (solubility in 17.5% sodium hydroxide). This is followed by descriptions of the procedures for determining alpha-, beta-, and gamma-cellulose. Essentially, the first is filtered off and determined by weighing; the latter two fractions are determined primarily by oxidizing the material that remains in the filtrate with an acidic solution of potassium dichromate. Gamma-cellulose is the material that remains in solution when the highly alkaline filtrate is neutralized; this too can be determined by chromate oxidation. Some discussion is given of potential errors arising from differences in the material that is dissolved; for example, the method is not recommended when significant quantities of lignin are present. Also briefly discussed is the effectiveness of various concentrations of sodium hydroxide from 10 to as high as 73%, as well as a few paragraphs on the nature of the substances dissolved.

151. Perl'shtein, E. Ya.
"Stepwise solubility of paper in caustic soda solutions."
Starenie Bumagi A kademia Nauk SSSR Laboratoriia Konservatsii i Restavratsii Dokumentov (1965), p. 30-45; through *Chemical Abstracts*, 67 (1967), Abstr. 23045r. In Russian. 33 references.

In most cases, the degradation of paper upon aging is evaluated from changes in mechanical properties, which reflect the chemical changes taking place in only an indirect way. Considering that the chemical aspects of paper aging are important for understanding the aging process, an attempt was made to characterize aging by direct determination of changes in the content of alpha-, beta-, and gamma-celluloses. The study was carried out on experimental papers. These papers were made from flax and cotton pulps and from bleached sulfite and kraft wood pulps. Each contained only one type of fiber, no size or fillers, fillers or size only, or both size and fillers, so that the effects of each paper component could be determined. The samples were subjected to accelerated aging at 100°C for 3-30 days. The degradation of the fibrous materials was reflected in a gradual decrease of alpha-cellulose content, a steady increase of beta-cellulose, and alternating increase and reduction of gamma-cellulose. It was assumed that the reduction indicated degradation with formation of gaseous products. The fibers could be arranged according to their decreasing resistance to aging in the order: flax, bleached kraft, cotton, and bleached sulfite pulps. The addition of fillers increased the resistance of the papers, $CaCO_3$ being more effective than kaolin, especially in the case of wood celluloses. On the other hand, the addition of 2% high-rosin size and 2-2.5% alum accelerated the degradation of fibers.

1971

152. Richards, G. N.
"Alkaline degradation. Degradation of cellulose and its derivatives, Part B." (Chapter XVIII)
In *High Polymers*, Volume 5, *Cellulose and Cellulose Derivatives, Part V*, 2nd edn., ed. Bikales, N. M., and Segal, L., Wiley-Interscience, New York (1971), p. 1007-1014. 31 references.

The chapter provides a concise, seven-page, exposition of the subject by one of the investigators who did so much to elucidate the chemistry of the "stopping reaction." The subject is divided into a discussion of the technological importance of alkaline degradation, the mechanism of the erosion, and stabilization (stopping) reactions, the alkaline scission of cellulose, attempts to minimize alkaline degradation, and the mechanism of alkaline degradation of carbonyl-containing oxidized celluloses. The subject of oxidation of cellulose under alkaline conditions is not addressed, but rather the action of oxygen-free alkali.

1973

153. Golova, O. P., and Nosova, N. I.
"Degradation of cellulose by alkaline oxidation."
Russian Chemical Reviews, 42, No. 4 (1973), p. 327-338. 136 references.
An extensive review of the subject, beginning with the process of alkaline degradation in the absence of oxygen, through the topic of oxidation of cellulose by molecular oxygen in an alkaline medium, and finally the effect of the crystalline-non-crystalline regions of cellulose on the degradation process. Extensive equations are given to explain the chemical mechanisms of the degradation reactions. Only about 20 of the references are articles that appear in Russian journals.

154. Procter, A. R., and Chow, W. M.
"A chip quality index for rot."
Pulp and Paper Magazine of Canada, 74, No. 7 (1973), p. T239-T242.
The authors propose that a rot index, based on 1% NaOH solubility, be assigned to wood chip samples. The wood chips were tested by Tappi Standard T4 os59. It is shown that 1% NaOH solubility is indicative of pulp yield.

1975

155. Johansson, M. H., and Samuelson, O.
"End-wise degradation of hydrocellulose during hot alkali treatment."
Journal of Applied Polymer Science, 19 (1975), p. 3007-3013. 22 references.
Determinations of reducing and carboxylic acid end-groups in hydrocellulose of DP ~200 from sulfite pulp showed that the end-wise degradation when treated with hot alkali proceeded to such an extent that some of the molecules were brought completely into solution. The major carboxylic acid end-groups formed in reactions competing with the end-wise degradation were 3-deoxyhexonic and 2-C-methylglyceric acids; but 2-deoxypentonic, 2-C-methylpentonic, and minor amounts of aldonic acids were also formed. The formation of these stable end-groups explains the observed increase in number-average degree of polymerization during the treatment. The ratio between the number of lost glucose moieties and the number of carboxylic acid groups formed in the solid phase was about 170 and was only slightly affected by the duration of the hot-alkali treatment.

1976

156. Technical Association of the Pulp and Paper Industry.
"One per cent sodium hydroxide solubility of wood and pulp."
Tappi Standard Method T212 os-76, Tappi, Atlanta, GA (1976). 3 references.
The method for the determination of one percent sodium hydroxide solubility can be used on wood, unbleached and bleached pulp. Wood or pulp is extracted with hot 1% sodium hydroxide solution for one hour. The method mainly removes degraded cellulose and hemicellulose in pulp and wood. The remaining pulp is filtered, washed principally with hot water, dried and weighed.
The percentage of material extracted indicates the degree of change in the pulp or wood.

157. Technical Association of the Pulp and Paper Industry.
"Alkali solubility of pulp at 25°C."
Tappi Standard Method T235 os-76, Tappi, Atlanta, GA (1976) 5 references.
This room temperature solubility method is carried out on bleached and delignified pulps only. If there is lignin present, it interferes with the extraction and determination of carbohydrates; this method, therefore, is not used on unbleached pulps.
The method extracts low-molecular-weight carbohydrates which are hemicelluloses and degraded cellulose from lignin-free pulp. The 10% sodium hydroxide solution dissolves both hemicellulose and degraded cellulose; the 18% solution dissolves mainly hemicellulose. From

the difference between the results with the 10% and 18% extractions, the content of degraded cellulose can be estimated.

The five references given, the latest from 1961, refer to experiments in which details of the method were worked out.

1979

158. Carlson, U., and Samuelson, O.

"Influence of anthraquinone on the degradation of cellulose in alkaline media."

Svensk Papperstidning, 82, No. 2 (1979), p. 48-52. 14 references.

During NaOH cooking of wood, addition of 0.1% anthraquinone (AQ) (calculated on dry wood) results in a stabilization of the carbohydrates and the formation of a large proportion of aldonic acid end-groups. The results with cotton and hydrocellulose show that in experiments with exclusion of air, the amount of AQ required for an appreciable stabilization was very large. The slight effect in the experiments with hydrocellulose under nitrogen at 107 and 130°C is explained by a rapid oxidation of soluble intermediates formed during the end-wise degradation. AQ is reduced to AHQ in these competing reactions as it dissolves. Only a very small proportion of the additive will therefore be available for the desired oxidation of the glucose end-groups to aldonic acid end-groups.

Model experiments with cellobiose show that at 97°C the oxidation to aldonic acids is a slow reaction compared to other reactions which lead to the reduction of AQ to AHQ. The additive is, however, easily regenerated by oxidation (for instance with air). The results show that, when a modest amount of AQ is used as additive, a reoxidation of reduced AQ by some oxidant is a prerequisite for an effective stabilization towards end-wise degradation.

During NaOH cooking of wood, a small addition of AQ gives a remarkable stabilization in the absence of air. A small amount of oxygen leads to a decreased delignification rate and under certain conditions to a somewhat higher yield of carbohydrates compared with a given lignin content in the pulp. The results obtained in the present work and those obtained in a parallel investigation [Löwendahl, L., and Samuelson, O., *Tappi*, 61 (1978), p. 2, 19] can be explained by the assumption that AQ is regenerated by the reaction of some reduced form of AQ with the lignin.

1980

159. Ziderman, I.

"The resistance of ketocelluloses to chain end-initiated depolymerisation during alkali-catalysed chain-scission."

Carbohydrate Research, 81 (1980), p. 196-199. 22 references.

The amount of yellow chromogen in alkaline extracts of celluloses and starches provides a quantitative estimate of the extent of chain end-initiated depolymerization ("peeling" reaction) that occurs by beta-elimination. However, the results are more difficult to evaluate when concurrent cleavage of non-terminal glucosidic linkages occurs, as ketocelluloses are largely resistant to extraction with dilute aqueous alkali, and the extract is not colored. A mechanistic explanation for this stability is presented.

1982

160. International Organization for Standardization.

"Pulps—determination of alkali solubility."

ISO Standard 692-1982 (E) (1982), 4 p.

A procedure is presented which describes the volumetric determination of the solubility of pulp in cold NaOH solutions of various and fixed concentrations. The method is applied preferably to the control of bleached pulp, but may be used with unbleached pulps.

1983

161. Technical Association of the Pulp and Paper Industry.

"One percent sodium hydroxide solubility of wood and pulp."

Tappi Test Method T212 om-83, Tappi, Atlanta, GA (1983), 2 p. 3 references.

This method for determination of hot-1% NaOH solubility may be applied to wood and to unbleached and bleached pulp.

1984

162. Kamide, K., Okajima, K., Matsui, T., and Kowsaka, K.
"Study on the solubility of cellulose in aqueous alkali solution by deuteration IR and ^{13}C NMR."
Polymer Journal, 16, No. 12 (1984), p. 857-866. 21 references.
A correlation of the solubility of cellulose in aqueous alkali solution was established with its so-called "amorphous" content. To achieve this end, cellulose was regenerated under different preparative conditions from its cuprammonium solution and cotton linter was acid-hydrolyzed into a fibrous form and the pulp physically milled to powder. Cellulose having a relatively large molecular weight was found to dissolve completely in aqueous NaOH solution at 4°C. The solubility of cellulose in a 10 weight-percent aqueous NaOH at 4°C could generally be correlated to the relative amount of the high magnetic field envelope of the C_4 carbon NMR (nuclear magnetic resonance) peak, assigned to the region where intramolecular hydrogen bonds are at least partly broken. The intramolecular hydrogen bond was found to have an important influence on the solubility behavior of cellulose.
Thus, it was demonstrated that cellulose whose intramolecular hydrogen bonds are completely broken or weakened dissolves in aqueous alkali and the solution thus prepared is stable over a long period of time. For a more detailed understanding of the mechanism underlying the dissolution of the cellulose in alkali, additional research must be carried out.

See also:

7 – The properties of the oxycelluloses formed in the early stages of the oxidation of cotton cellulose by periodic acid and metaperiodate, 4.
9 – The rate of change in the properties of cotton cellulose under the prolonged action of acids 6.
Kinetic investigation of thermal aging of unbleached Egyptian kraft bagasse paper-pulp, 44.
The hot alkaline purification of cellulose I, 50.
The alkaline extraction of wood pulps, 55.
The extractive power of lithium, sodium, and potassium hydroxide solutions for the hemicelluloses associated with wood cellulose and holocellulose from western hemlock, 63.
Stabilization of cellulose towards alkaline degradation, 65.
Alkaline extraction, 76.
Carbohydrate-removing methods, 80.
Carbohydrate stabilization during kraft cooking with addition of anthraquinone, 98.

AUTHOR INDEX

Note: Numbers refer to abstracts, not to pages.

Relation of Cellulose Chain Scission to Hot-Alkali-Soluble Content During Thermal and Photochemical Degradation of Paper

SANG B. LEE, ROBERT L. FELLER AND JOHN BOGAARD

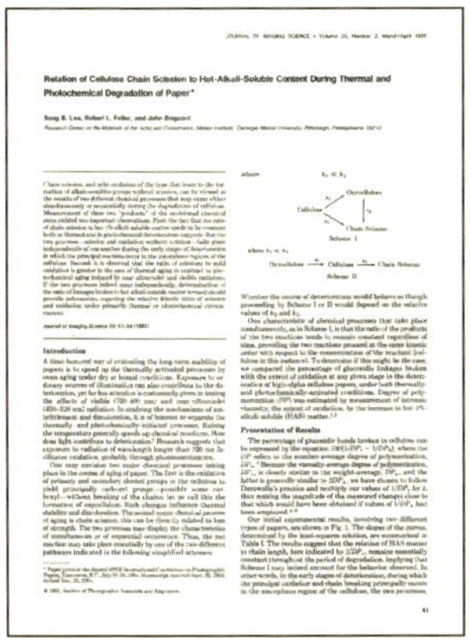

Originally printed in:
Journal of Imaging Science
Volume 29, Number 2 (1985),
Pages 61-64.

Introduction

A time-honored way of evaluating the long-term stability of papers is to speed up the thermally activated processes by oven-aging under dry or humid conditions. Exposure to ordinary sources of illumination can also contribute to the deterioration, yet far less attention is customarily given to testing the effects of visible (700-400 nm) and near-ultraviolet (400-320 nm) radiation. In studying the mechanisms of embrittlement and discoloration, it is of interest to separate the thermally and photochemically initiated processes. Raising the temperature generally speeds up chemical reactions. How does light contribute to deterioration? Research suggests that exposure to radiation of wavelength longer than 320 nm facilitates oxidation, probably through photosensitization.

One may envision two major chemical processes taking place in the course of aging of paper. The first is the oxidation of primary and secondary alcohol groups in the cellulose to yield principally carbonyl groups—possibly some carboxyl—without breaking of the chains; let us call this the formation of oxycellulose. Such changes influence thermal stability and discoloration. The second major chemical process of aging is chain scission; this can be directly related to loss of strength. The two processes may display the characteristics of simultaneous or of sequential occurrence. Thus, the net reaction may take place essentially by one of the two different pathways indicated in the following simplified schemes:

where $k_3 << k_2$

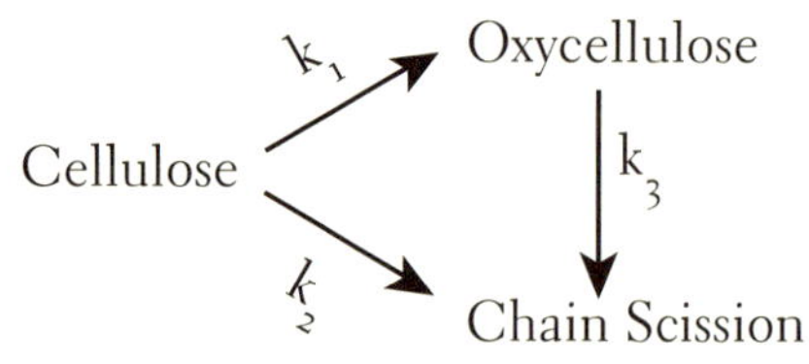

Scheme 1

where $k_2 << k_3$

$$\text{Cellulose} \xrightarrow{k_1} \text{Oxycellulose} \xrightarrow{k_3} \text{Chain Scission}$$

Scheme 2

Whether the course of deterioration would behave as though proceeding by Scheme 1 or 2 would depend on the relative values of k_2 and k_3.

One characteristic of chemical processes that take place simultaneously, as in Scheme 1, is that the ratio of the products of the two reactions tends to remain constant regardless of time, providing the two reactions proceed at the same kinetic order with respect to the concentration of the reactant (cellulose in this instance). To determine if this might be the case, we compared the percentage of glucosidic linkages broken with the extent of oxidation at any given stage in the deterioration of high-alpha-cellulose papers, under both thermally and photochemically activated conditions. Degree of polymerization (DP) was estimated by measurement of intrinsic viscosity; the extent of oxidation, by the increase in hot-1%-alkali-soluble (HAS) matter.[1, 2]

PRESENTATION OF RESULTS

The percentage of glucosidic bonds broken in cellulose can be expressed by the equation $100(1/DP_t - 1/DP_o)$, where the DP refers to the number-average degree of polymerization, DP_n.[3] Because the viscosity-average degree of polymerization, DP_v, is closely similar to the weight-average, DP_w, and the latter is generally similar to $2DP_n$, we have chosen to follow Daruwalla's practice and multiply our values of $1/DP_v$ by 2, thus making the magnitude of the measured changes close to that which would have been obtained if values of $1/DP_n$ had been employed.[4-6]

Our initial experimental results, involving two different types of papers, are shown in Figure 1. The slopes of the curves, determined by the least-squares

1. Tappi Standard T212 os-76.

2. Birtwell, C., Clibbens, D. A., and Geake, A., *J. Text. Inst.*, 19 (1928), T349.

3. Conrad, C. M., Tripp, V. W., and Mares, T., *Text. Res. J.*, 21, No. 10 (1951), p. 726.

4. Daruwalla, E . H., and Narsian, M. G., *Tappi*, 49 (1966), p. 106.

5. Daruwalla, E. H., D'Silva, A. P., and Tandon, G. N., *Text. Res. J.*, 37, No. 3 (1967), p. 160.

6. Daruwalla, E. H., D'Silva, A. P., and Mehta, A. C., *Text. Res. J.*, 37, No. 3 (1967), p. 147.

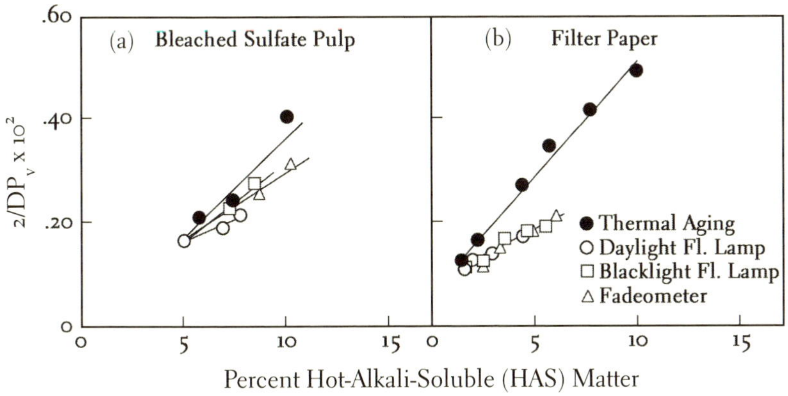

FIGURE 1
Relation of 2/DP_v as a function of hot-1%-alkali-soluble (HAS) matter for test sheets of bleached sulfate pulp and of filter paper following exposure to heat or to light.

TABLE 1
Slopes of Curves of 2/DP Versus Hot-Alkali-Soluble Matter Obtained During the Degradation of Sheets Based on Bleached Pulp and on Filter Paper

Base Paper	Circumstances of degradation		Slope x 10^{4*}	No. of data points
Bleached sulfate pulp	Thermal:	90°C, 50% RH	4.4 ± 2.0	3
	Photochem.:	Fadeometer	2.8 ± 1.2	3
	Photochem.:	Daylight fluorescent lamps	1.6 ± 1.6	3
	Photochem.:	BLB Fl. blacklights	3.2 ± 0.6	3
Filter paper	Thermal:	90°C, 50% RH**	5.0 ± 1.0	5
	Photochem.:	Fadeometer	2.6 ± 1.2	4
	Photochem.:	Daylight fluorescent lamps**	2.2 ± 0.4	5
	Photochem.:	BLB Fl. blacklights**	2.2 ± 0.4	5
Thermally-aged filter paper	Photochem.:	BLB Fl. blacklights**	2.4 ± 0.4	18
Photochemically-aged filter papers	Thermal:	90°C, 50% RH**	4.2 ± 0.4	27
Chemically-bleached filter paper	Thermal:	90°C, 50% RH following various bleaches	4.1 ± 0.2	16
Bleached sulfate pulp	Control:	Hydrolytically produced hydrocelluloses	3.2 ± 0.5	7

* Slope, apparent fraction of linkages broken per 1 % HAS matter developed, calculated on the basis of least squares, with 95% confidence interval.
** Samples reduced with sodium borohydride prior to determination of degree of polymerization.

solution, are summarized in Table 1. The results suggest that the relation of HAS matter to chain length, here indicated by $2/DP_v$, remains essentially constant throughout the period of degradation, implying that Scheme 1 may indeed account for the behavior observed. In other words, in the early stages of deterioration, during which the principal oxidation and chain breaking principally occurs in the amorphous region of the cellulose, the two processes, chain breaking and oxidation, appear to occur independently.

There are significant differences, however. At moderate temperatures, photochemical deterioration induced by near-ultraviolet and visible radiation tends to give rise to oxidation of the cellulose accompanied by little chain breaking. On the other hand, thermal aging results in a much greater degree of bond scission relative to the mild oxidation that leads to the formation of alkali-sensitive groups (largely carbonyl) along the chain.

Additional data to confirm the ratio of chain scission to the formation of HAS matter during thermal aging were provided by a series of filter papers that had been treated with various chemical bleaches and then aged at 90°C and 50% RH for 10 and 20 days. The results, shown in Figure 2, yield a ratio of chain scission to formation of alkali-sensitive oxidation products (represented by HAS matter) of $4.1 \pm 0.2 \times 10^{-4}$, based on 16 data points.

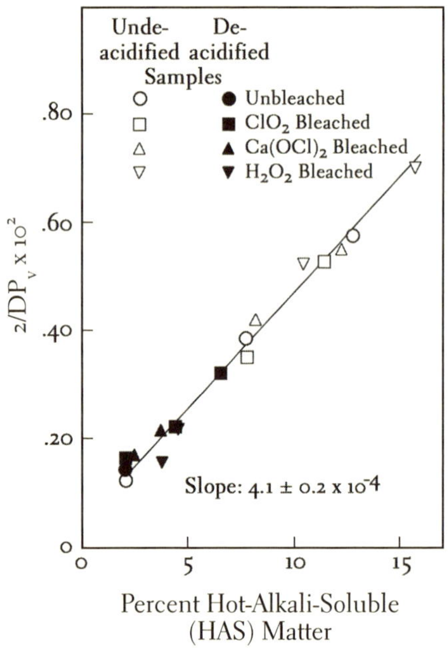

FIGURE 2
Relation of $2/DP_v$ as a function of hot-1%-alkali-soluble (HAS) matter for a series of chemically bleached filter papers upon thermal aging at 90°C, 50% RH for 10 and 20 days.

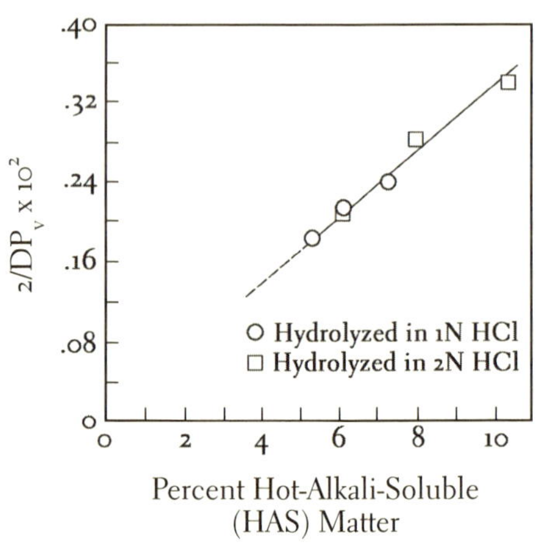

FIGURE 3
Relation of $2/DP_v$ as a function of hot-1%-alkali-soluble (HAS) matter for a series of hydrocelluloses prepared from bleached sulfate pulp by hydrolysis in 1N and 2N HCl.

In considering the relation of 1/DP to carbonyl groups present, it is realized that one aldehydic carbonyl is situated at the end of each chain. Accordingly, there must be a "normal" relationship between 1/DP and hot-alkali-soluble matter for a series of hydrocelluloses of decreasing chain length. To determine this relationship for the case of unaged cellulose, a series of hydrocelluloses were prepared from the bleached sulfate pulp by hydrolysis in 1N and 2N HCl. The data obtained on the hydrocelluloses are shown in Figure 3. The slope is 3.2×10^{-4} based on the least-squares solution, 95% confidence interval ±0.5 (Table 1). A precise analysis of the differences between the thermal and photochemical results should take into consideration this normal relationship between $1/DP_n$ and HAS matter. The slopes of the curves of $\Delta 2/DP_v$ versus ΔHAS are thus apparent rather than true measures of the fraction of chains broken per 1%-HAS-matter developed. Nonetheless, for the purpose of drawing attention to the relative constancy of this ratio throughout the early stages of cellulose degradation and to the differences exhibited between thermal and photochemical aging, we will treat the slopes of the curves simply as the ratio of chain breaks to HAS matter developed, without correction.

The viscosity-average degree of polymerization of the aged samples of bleached sulfate pulp (Figure 1a) was determined using cupriethylenediamine (CED) solution.[7] It has been reported that oxidized cellulose that contains alkali-sensitive groups can be cleaved by the alkalinity of cuprammonium solutions, yielding low values of DP.[5] To check this possibility in the case of CED determinations, a second set of the filter-paper-based samples used in Figure 1b were first reduced by sodium borohydride (SBH) to minimize this potential error. As the data in Table 2 indicate, the difference in DP_v determined in CED solution with and without reduction did not exceed 2-5%. Nonetheless, in subsequent experiments, reduction with SBH was carried out on all samples prior to viscosity determination, as noted in Table 1.

7. ASTM Standard D1795.

To confirm the lower ratio of chain scission relative to the formation of HAS matter during photochemical deterioration, data were taken from a series of filter papers that had been aged at 90°C and 50% RH for various times and subsequently exposed under BLB fluorescent blacklights which emit primarily near-ultraviolet radiation. Based on 18 data points, the least-squares-calculated slope was $2.4 \pm 0.4 \times 10^{-4}$ (Figure 4a).

Finally, four sets of samples of filter paper that had first been exposed to BLB blacklights for periods up to 310 hr were subjected to thermal aging at 90°C and 50% RH; the slope of the resulting thermal-aging data was $4.2 \pm 0.4 \times 10^{-4}$, based on 27 data points (Figure 4b). These values, reported in Table 1, correspond closely to our initial findings based on a fewer number of data points and on samples that had not been reduced with SBH prior to the determination of viscosity.

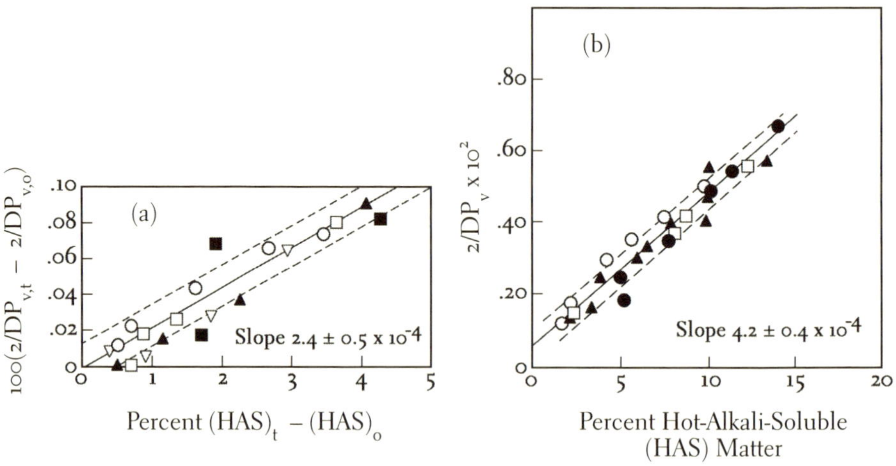

Figure 4
Apparent percentage of linkages broken in relation to hot-1%-alkali-soluble (HAS) matter generated for series of treated filter papers. (a) Results of exposure to BLB "blacklight" lamps following initial thermal aging at 90°C, 50% RH. (b) Results of thermal aging at 90°C, 50% RH following initial exposure to BLB blacklight lamps.

Table 2
Effect of SBH Reduction on the Value of Degree of Polymerization (DP) Determined on a Number of Representative Samples Using Cupriethylenediamine (CED)

Base Paper	Type of degradation and aging circumstances	DP		% Increase following SBH treatment
		No SBH treatment	SBH treatment	
Filter paper (FP)	Control: unaged, unexposed	1600	1666	4.1
	Heated (90°C, 50% RH for 670 hr)	399	412	3.3
	Heated (90°C, 50% RH for 100 hr) + bleaching (ClO_2)	646	680	5.3
	Heated (90°C, 50% RH for 100 hr) + bleaching (ClO_2) + light (D.L. Fl. for 480 hr)	592	619	4.6
Bleached sulfate pulp (BP)	Control: unaged, unexposed	1218	1242	2.0
	Light (daylight fluorescent lamp, 290 hr)	896	925	3.2
	Heated (90°C, 50% RH for 480 hr)	374	386	3.1

Discussion

While measurements of the number of chain scissions or number of carbonyl groups developed per unit of time have been extensively reported during aging studies, few authors have considered the ratio of bonds broken to carbonyl groups formed. Daruwalla[6] used such a plot to demonstrate, in the case of photosensitization of cotton by Vat Orange 9, that equal numbers of carbonyl groups and of carboxyl groups were formed per chain scission. Philipp[8] also plotted the number of scissions versus carbonyl formation at 170°C.

Other than these principal references, experimental results have rarely been expressed from the point of view considered here. Nonetheless, it is possible to construct graphs similar to ours based on published data in which both the DP and the carbonyl-group content have been measured during the course of degradation. Figure 5 summarizes the results of a number of authors. Shown are the data of Reine[9] for the exposure of cellulose to 365 nm ultraviolet radiation in air, along with those of Flynn[10] relating to the exposure of purified cotton cellulose sheets to 254 nm radiation at 40°C in a vacuum. The data of Shafizadeh[11] for thermal degradation at 190°C are also given. The normal relationship of 1/DP to content of terminal aldehyde groups is also illustrated,

8. Phillipp, B., Van Nghi, D., and Baudisch, J., *Faserforsch. Textiltech.*, 21, No. 12 (1970), p. 519.

9. Reine, A. H., and Arthur, J. C., Jr., *Text. Res. J.*, 40, No. 1 (1970), p. 90.

10. Flynn, J. H., Wilson, W. K., and Morrow, W. L., *J. Res. Nat. Bur. Stand.*, 60, No. 3 (1958), p. 229, Table 2.

11. Shafizadeh, F., and Bradbury, A. G. W., *J. Appl. Poly. Sci.*, 23 (1979), p. 1431.

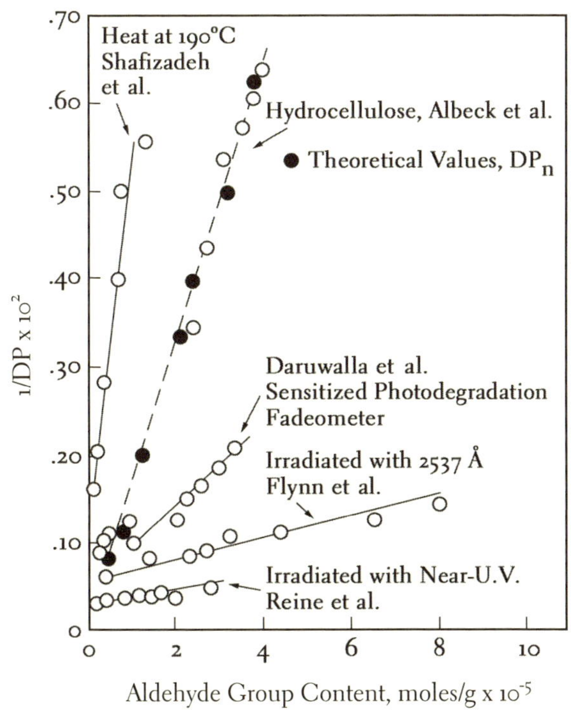

Figure 5
Relation of 1/DP as a function of aldehyde content for the variously degraded samples; data reported in the literature.

12. Albeck, M., Ben-Bassat, A., and Lewin, M., *Text. Res. J.*, 35, No. 10 (1965), p. 935.

based on a series of hydrocelluloses prepared by Albeck[12]; the curve coincides with the theoretical values of the inverse of the number-average molecular weight versus the content of aldehyde end groups.

The published data based on direct measurements of aldehyde-group content support our findings based on hot-1%-alkali-soluble matter. The slopes in Figure 5 will, of course, differ according to whether the reported degree of polymerization was based on weight-average or number-average measurements. The thermal degradation data of Shafizadeh, and Albeck's values for hydrocellulose, apparently involve number-average figures. On the other hand, the photochemical data in Figure 5 were based on viscosity determinations; the values of the degree of polymerization may, therefore, be higher than DP_n by a factor of as much as 2 (the resulting values of 1/DP, low).

Conclusions

Although the use of hot-1%-alkali solubility (HAS) as a measure of the degree of oxidation of the cellulose is less precise than the direct determination of aldehyde, ketone, and carboxyl content, it is convenient and the results corroborate the findings of others who have employed direct chemical measurement of the extent of oxidation. The fact that the ratio of links broken to HAS matter formed remains essentially constant during the initial phases of thermal or photochemical aging of test papers suggests that the chemical processes of chain breaking and chain oxidation proceed independently of one another within the amorphous regions of cellulose. Because the experimental results essentially involve a ratio of two kinetic rate constants, study of the change in this relationship with variations of temperature, humidity, and illumination may prove useful in comparing the relative rates of two principal degradation processes under different circumstances. This potential for providing additional insights into the mechanism of paper deterioration has not been extensively stressed.

Experimental

Papers: Whatman No. 42 filter paper was used as an example of high alpha-cellulose sheets. The sample of bleached pulp (Canadian Standards Freeness of 420 ml) was based on 50% hardwood and 50% softwood kraft pulps having an initial lignin content of 0.24% and hot-1%-alkali-soluble matter of 5.0%.

Chemical Bleaching: The experiment illustrated in Figure 2 was based on sheets that had been bleached by three different treatments: chlorine dioxide, calcium hypochlorite, and hydrogen peroxide, using the procedures described by Hey.[13]

13. Hey, M., *The Conservator*, 2 (1977), p. 10; 4 (1979), p. 74.

Exposure to Light Sources: Three different light sources were used: (a) a bank of six General Electric High-Output (HO) "daylight" fluorescent lamps, (b) a bank of six fluorescent BLB blacklights, and (c) an Atlas 6500-W xenon-arc Fadeometer with Pyrex® filters. Data on the irradiance on samples under the three lamps is given in Table 3, information provided primarily to indicate the percentage of ultraviolet emitted by the various sources.

TABLE 3
Irradiance from Different Light Sources (mW/cm^2)
as Measured by the IL700 Radiometer

Light source	Ultraviolet energy (310-392 nm)		Visible energy (400-750 nm)	Total energy (257-750 nm)	Sample distance from source (in.)
G.E. Fl. blacklight F40 BLB	4.13	(86%)	0.70	4.90	3.5
Fl. "daylight" high-output	0.228	(3.8%)	5.70	5.928	3.5
Xenon-arc Fadeometer[a] pyrex/pyrex filters	13.775[b]	(6.3%)	203.49[c]	217.265[d]	11.6

a. Measurements obtained from Atlas Electric Devices, Inc. catalog multiplied by 1.9 (factor for 11.6 in.-diameter drum).
b. Irradiance from 340-400 nm.
c. Irradiance from 400-750 nm.
d. Irradiance from 340-750 nm.

Both types of fluorescent lamps were mounted about 3 1/2 inches above the samples in a room maintained at 50% relative humidity (RH) and 70°F (21°C). The temperature of the handsheets reached about 24°C. In the 6500-W Fadeometer, the air was maintained at 31°C (± 1.0°) and 27% RH; the black-panel temperature reached 65°C.

Viscosity: Viscosities were determined at 25°C in 0.5M cupriethylenediamine (CED) using a Cannon-Fenske viscometer (size 100).[14] From these data, intrinsic viscosity $[\eta]$ was obtained by reference to Table 3 of ASTM D1795, which gives $[\eta]C$ for the indicated values of η_{rel}, determined by Martin's equation[15]:

$$\log \frac{\eta_{rel} - 1}{C} = \log [\eta] + K [\eta] C \qquad (1)$$

where

$$K = 0.13.$$

Reasonable values for the viscosity-average degree of polymerization are obtained by multiplying $[\eta]$ by 190.

14. Tappi Standard T230 os-76.

15. Martin, A. F., *Tappi*, 34, No. 8 (1951), p. 363.

THE KINETICS OF CELLULOSE DETERIORATION

R. L. FELLER, S. B. LEE AND J. BOGAARD

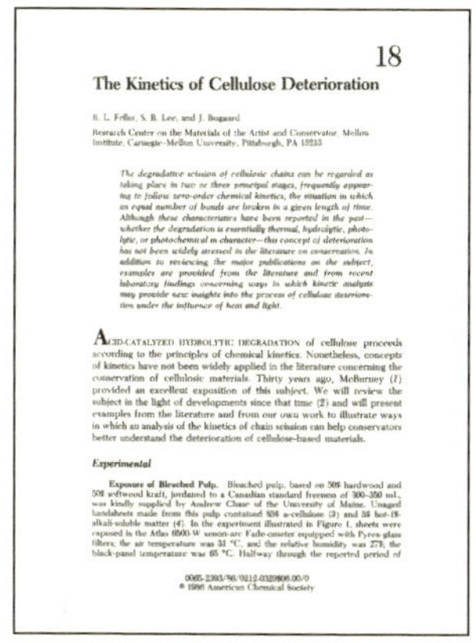
18
The Kinetics of Cellulose Deterioration

Originally printed in:
HISTORIC TEXTILE AND PAPER MATERIALS: CONSERVATION AND CHARACTERIZATION
Advances in Chemistry Series, No. 212, ed. Needles, H. L., and Zeronian, S. H., American Chemical Society, Washington, DC (1986), Pages 329-346.

Acid-catalyzed hydrolytic degradation of cellulose proceeds according to the principles of chemical kinetics. Nonetheless, concepts of kinetics have not been widely applied in the literature concerning the conservation of cellulosic materials. Thirty years ago, McBurney[1] provided an excellent exposition of this subject. We will review the subject in the light of developments since that time[2] and will present examples from the literature and from our own work to illustrate ways in which an analysis of the kinetics of chain scission can help conservators better understand the deterioration of cellulose-based materials.

EXPERIMENTAL

Exposure of Bleached Pulp

Bleached pulp, based on 50% hardwood and 50% softwood kraft, jordaned to a Canadian standard freeness of 300-350 mL, was kindly supplied by Andrew Chase of the University of Maine. Unaged handsheets made from this pulp contained 85% α-cellulose[3] and 5% hot-1%-alkali-soluble matter.[4] In the experiment illustrated in Figure 1, sheets were exposed in the Atlas 6500-W xenon-arc Fadeometer equipped with Pyrex-glass filters; the air temperature was 31°C, and the relative humidity was 27%; the black-panel temperature was 65°C. Halfway through the reported period of exposure, sheets were turned

1. McBurney, L. F., in *Cellulose and Cellulose Derivatives*, Part 1, ed. Ott, E., Spurlin, H. M., and Grafflin, M. W., Interscience, New York (1954), p. 99.

2. Sharples, A., in *Cellulose and Cellulose Derivatives*, Vol. 5, Part 5, 2nd edn., ed. Bikales, N. M., and Segal, L., Wiley, New York (1971), p. 991-1006.

3. *TAPPI Standard T203 os-74*, "Alpha-, Beta-, Gamma-Cellulose in Pulp."

4. *TAPPI Standard T212 os-76*, "One Per Cent Sodium Hydroxide Solubility of Wood and Pulp."

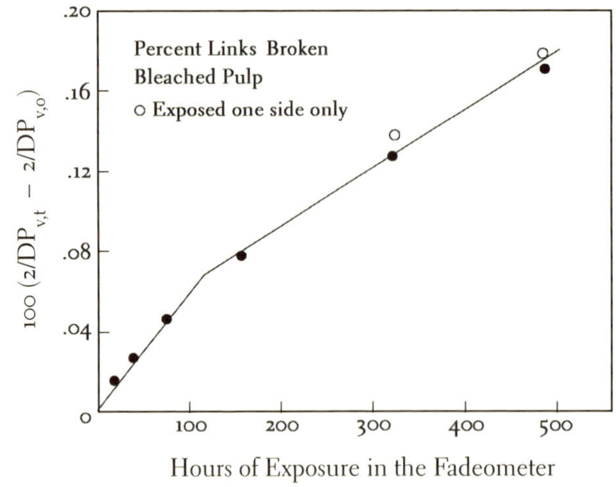

FIGURE 1
Change in $2/DP_v$ with time of exposure of bleached-sulfate-pulp handsheets in xenon-arc Fadeometer.

over to ensure that both sides were equally exposed. This procedure made such little difference in the plots of the simplified rate equation used that handsheets were exposed on one side only in subsequent experiments.

Oxidation of Pulp

In the experiment reported in Figure 2, the bleached pulp stock was chemically modified prior to the formation of handsheets. The stock was first oxidized by dichromate as follows: After a short period of pretreatment in a 0.2 N solution of oxalic acid at a pulp consistency of 3%, the pulp was filtered and potassium dichromate was added to yield a 0.1 N solution. The pulp was allowed to react at room temperature for 30 min and then was filtered and washed. Thereafter the oxidized pulp was soaked overnight in oxalic acid and then was filtered and washed again.

Following chromate oxidation, the pulp was divided. One part was further oxidized by soaking the pulp overnight at 2% consistency in chlorous acid. The other portion was reduced by soaking it overnight in a 0.01 M solution of sodium borohydride (SBH). After these treatments, the samples were washed, and handsheets were prepared.

Carboxyl content was determined by the magnesium-ion-exchange method.[5] For chlorous acid-treated pulp, the carboxyl group concentration was found to be 14.8 meq/100 g; for sheets reduced with SBH, the carboxyl group concentration was 3.4 meq/100 g. Neither had any significant aldehyde group content; the copper number[6] of the chlorous acid sheets measured 0.9 meq/100 g; that of the SBH sheets measured 0.3 meq/100 g.

Handsheets were exposed in the Fadeometer for periods up to 299.0 h.

5. Sjöström, E., and Haglund, P., *Svensk Papperstidning*, 64 (1961), p. 438.

6. *TAPPI Standard* T430 os-75, "Copper Number of Pulp, Paper, and Paperboard."

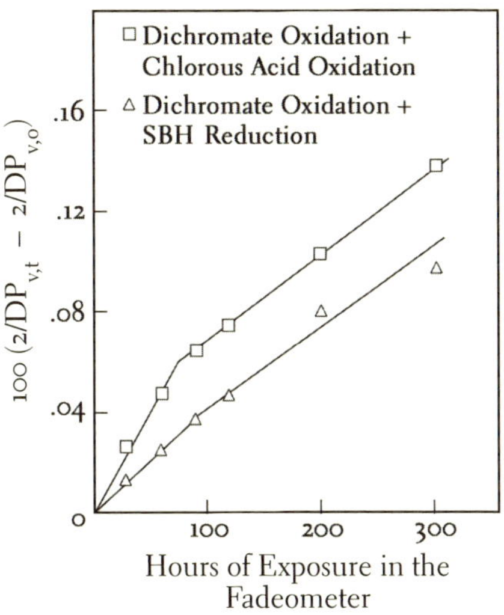

Figure 2
Change in 2/DP$_v$ with time of exposure. Photochemical deterioration of bleached-pulp handsheets, both oxidized and SBH reduced, in Fadeometer.

Thermal Degradation

Whatman No. 42 ashless filter paper has a basis weight of 92 g/cm^2, an α-cellulose content of 98%, and a solubility in hot 1% sodium hydroxide of 1.5%. Test sheets were first exposed under BLB fluorescent black lights. The lamps, which emit primarily near-UV radiation, were mounted 3 1/2 in. above the samples. In the paper-testing room maintained at 50% rh and 21.1°C (70°F), the temperature of sheets under the bank of lamps reached about 24°C.

After exposure to the BLB lamps, the papers were thermally aged at 90°C and 50% rh for 3-30 days.

Viscosity Determination

Viscosities were measured in 0.5 M cupriethylenediamine (CED) at 25°C with a Cannon-Fenske viscometer, size 100.[7] From these measurements, the intrinsic viscosity was determined according to the American Society for Testing and Materials (ASTM) Standard D1795. Values for the viscosity-average degree of polymerization were obtained by multiplying the intrinsic viscosity by 190.

All samples were first reduced with SBH to minimize possible degradation of the aged pulps which might occur during measurement of viscosity because of the presence of alkali-sensitive linkages. The paper was weighed, disintegrated in a blender, and dispersed in a 0.04% solution of SBH. At a pulp consistency of 0.2%, the suspension was left overnight, filtered, and washed. The pulp cake was then ready to be dissolved in the CED solution.

7. *TAPPI Standard T230 os-76*, "Viscosity of Pulp (Capillary Viscometer Method)."

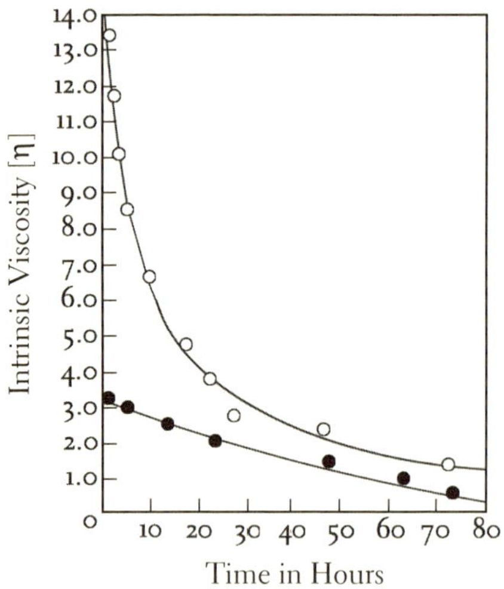

FIGURE 3
Change in intrinsic viscosity of cellulose during homogeneous hydrolysis in H_3PO_4 solutions at 20°C.
Key: ○, surgical cotton; ●, hydro-cellulose. (Reproduced with permission from Ref. 1. Copyright 1954 Wiley.)

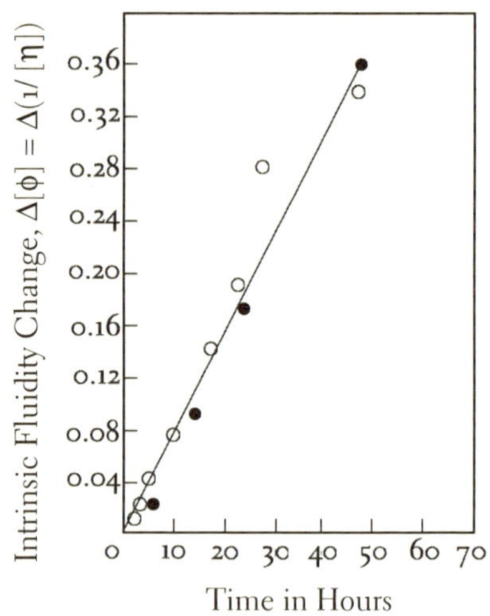

FIGURE 4
Data in Figure 3 plotted in terms of intrinsic fluidity. Key: ○, surgical cotton; ●, hydro-cellulose. (Reproduced with permission from Ref. 1. Copyright 1954 Wiley.)

KINETIC LAWS

For many years it has been the custom to follow the degradation of cellulose simply by measuring the fall in viscosity, as illustrated in Figure 3.[8] A number of researchers have pointed out, however, that it is not the intrinsic viscosity but the inverse, the intrinsic fluidity (more precisely, the inverse of the degree of polymerization), that is the significant unit to use in studying the degradation of linear polymers; this measurement is related to the number of bonds broken.[9-12] Thus, McBurney[1] pointed out that, when the data in Figure 3 are plotted in terms of the change in intrinsic fluidity, the results demonstrate clearly that no decrease occurs in the rate at which the chains are being broken (Figure 4). Moreover, chain breakage is occurring at identical rates both in the case of the high-molecular-weight surgical cotton and the hydrocellulose of lower molecular weight.

If one considers the number of bonds between monomer units in a linear chain polymer that may be broken per unit of time, expressed in terms of the total number available, N, one may propose that random scissioning of bonds obeys a first-order kinetic rate law:

$$dN/dt = k' N \qquad (1)$$

8. Hiller, L. A., Jr., and Pascu, E., *Text. Res. J.*, 16 (1946), p. 564-570.

9. af Ekenstam, A., *Ber. Dtsch. Chem. Ges.*, 69 (1936), p. 553-559.

10. Schulz, G. V., and Löhmann, H. J., *J. Prakt. Chem.*, 157 (1941), p. 238-282.

11. Sharples, A., *J. Polym. Sci.*, 14 (1954), p. 95-104.

12. Bryce, W. A. J., and Greenwood, C. T., *J. Polym. Sci.*, 25, No. 111 (1957), p. 480-483.

where t is the time and k´ is the observed rate constant. This relationship may also be expressed in terms of the fraction of the total number of available bonds, $1 - 1/DP_n$, where DP_n is the number-average molecular weight:

$$d(1 - 1/DP_n)/dt = k'(1 - 1/DP_n) \quad (2)$$

The integrated form of this equation yields the familiar linear first-order rate equation in which ln stands for the logarithm to the base e, and the subscripts o and t refer to the initial value of DP_n and to the value at any time, t, respectively:

$$\ln(1 - 1/DP_{n,o}) - \ln(1 - 1/DP_{n,t}) = k't \quad (3)$$

At very low degrees of degradation, when only a small fraction of the bonds have been broken, Equation 3 can be simplified[1] to yield

$$1/(DP_n)_t - 1/(DP_n)_o = \alpha kt \quad (4)$$

The factor α is a measure of the accessibility of the bonds,[13] and k is the specific rate constant for the rate of breaking of normal 1C-O-4´C bonds in the anhydroglucose chain (k´ in Equations 1-3 equals αk). The expression $100[1/(DP_n)_t - 1/(DP_n)_o]$ gives the percentage of the initial number of bonds that have been broken. Equation 4 is also the equation that would apply if an equal number of bonds were broken in equal periods of time, that is, at a constant rate of chain scissioning (zero-order kinetics). Conformance to Equation 4, therefore, cannot be taken as proof of first-order kinetic behavior.[12]

Many researchers have employed Equation 4 to follow the early stages of degradation of cellulose and other linear polymers and have applied Equation 3 when extending the investigation of an advanced state. [Since the meeting of the International Council of Museums (ICOM) Committee on Conservation in Leningrad and Moscow, 1963 (ICOM Report 63-24), L. Santucci has been using a function of the type $\log(F_o/F) = \alpha(1/DP - 1/DP_o)$ to express changes in DP and folding endurance, F.[14, 15]] Figure 5 shows a typical set of curves, plotted according to Equation 4, taken from the work of Daruwalla and Narsian,[16] concerning the heterogeneous hydrolysis of cotton and of regenerated cellulose in 1 N hydrochloric acid. As these and other researchers have pointed out, the results are typical of the kinetic data related to the degradation of paper and textiles; the course of the heterogeneous degradation of cellulose by acid hydrolysis, and by photochemical and enzymatic attack as well, seems to display two distinct stages: a fast initial stage followed by a significantly slower one.[17, 18]

One may suggest that there are not two but at least three major stages in the overall hydrolytic degradation of cellulosic fibers, extending from the initial

13. Sharples, A., *J. Polym. Sci.*, 13 (1954), p. 393-401.

14. Santucci, L., *Boll. Ist. Patol. Libro*, 22 (1963), p. 67.

15. *Ibid.*, 25 (1966), p. 51.

16. Daruwalla, E. H., and Narsian, M. G., *Tappi*, 49, No. 3 (1966), p. 106-111.

17. Daruwalla, E. H., D'Silva, A. P., and Mehta, A. C., *Text. Res. J.*, 37, No. 3 (1967), p. 147-160.

18. Daruwalla, E. H., D'Silva, A. P., and Tandon, G. N., *Text. Res. J.*, 37, No. 3 (1967), p. 160-171.

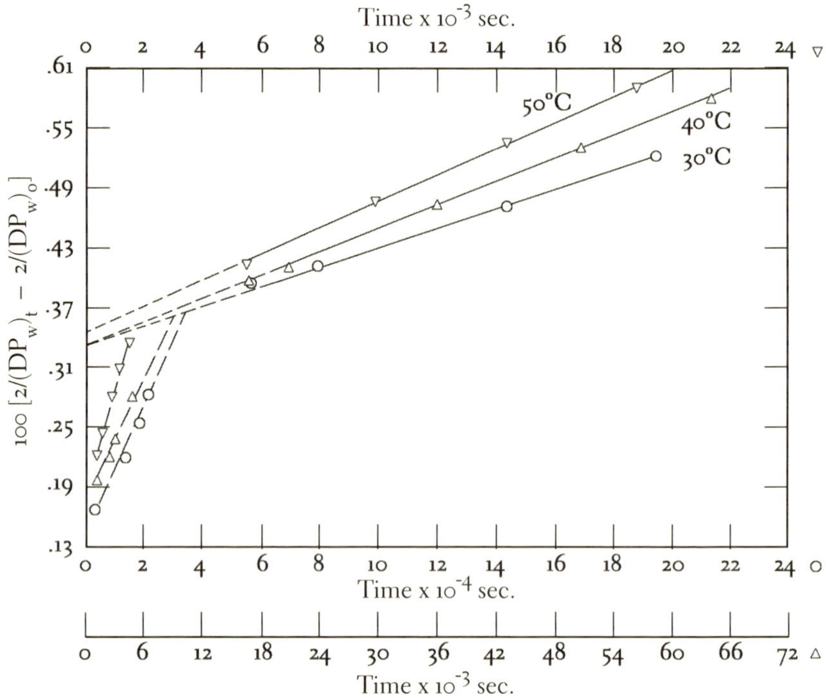

FIGURE 5
Data regarding heterogeneous hydrolysis of regenerated cellulose in 1 N hydrochloric acid at different temperatures. (Reproduced with permission from Ref. 16. Copyright 1966 Tappi.)

attack to the final point when all of the cellulosic material has been converted to soluble fragments. The first, rapid stage is considered to represent an attack on a small fraction of the 1C-O-4´C bonds in the cellulosic chain that are particularly sensitive to scissioning. These have frequently been designated as weak links. The second, slower stage essentially represents the attack on the normal 1C-O-4´C bonds located in the amorphous regions of the fibers. The final major stage represents the much slower rate of attack on the cellulose that is bound into the crystallite bundles or zones that usually make up a significant fraction of the fibers. This last stage of hydrolytic attack begins at a condition close to the leveling off degree of polymerization (LODP).[19, 20] We will review the evidence that supports these concepts.

19. Chang, M. M. Y., *J. Polym. Sci.*, 12 (1974), p. 1349-1374.

20. Millett, M. A., Moore, W. E., and Saeman, J. F., *Ind. Eng. Chem.*, 46 (1954), p. 1493-1497.

THE RAPID INITIAL STAGE

Perhaps it would be convenient to label the initial stage of hydrolytic degradation as the weak-link stage. However, this label tends to focus attention on the possible existence of a specific type of chemical bond and thus draws consideration away from other factors that may influence the phenomenon. Instead, we prefer to call the first stage simply the rapid initial stage, regardless of what may prove eventually to be its cause or causes.

In the 1940s and 1950s, a distinguished body of research grew out of the attempts of researchers[2, 8-11] to demonstrate that certain links in cellulose were particularly sensitive to rupture. Many speculated as to what these links might be. Eventually it was shown that none of the especially weak links were apparent when the hydrolysis was carried out under homogeneous conditions, such as can be accomplished by dissolving the cellulose in concentrated phosphoric acid. Under the conditions of heterogeneous hydrolysis, however, a modest number of highly sensitive linkages could be demonstrated. These are now thought to be the result of physically induced strains in the bonds.[19, 21]

As these concepts were being developed, Marchessault and Rånby[22-24] came to believe that the presence of oxidized groups in the pyranose rings would also be the cause of an increased ease of hydrolysis, attributable to inductive effects of the neighboring 1C-O-4′C bond. Figure 6 illustrates Rånby's[24] demonstration, in the case of oxidized cotton cellulose degraded under homogeneous conditions, of a rapid initial stage of hydrolysis, a phenomenon that could practically be eliminated by SBH reduction. He also demonstrated a similar effect of SBH in the case of wood cellulose (Figure 7).

21. Michie, R. I. C., Sharples, A., and Walter, A. A., *J. Polym. Sci.*, 51 (1961), p. 85-98.

22. Marchessault, R. H., and Rånby, B. G., "Weak Links as Inductive Effects in the Hydrolysis of Cellulose," *Proceedings 2nd Cellulose Conference, Syracuse* (1959), p. 33-52.

23. Marchessault, R. H., and Rånby, B. G., *Svensk Papperstidning*, 62 (1959), p. 230-240.

24. Rånby, B. G., *J. Polym. Sci.*, 53 (1961), p. 131-140.

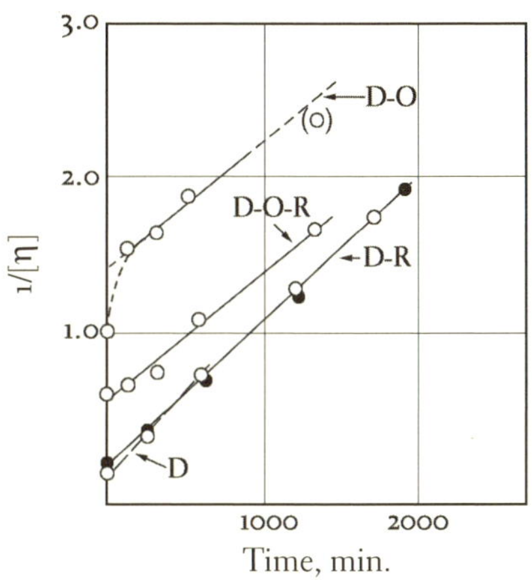

FIGURE 6
Rate of homogeneous hydrolysis of cotton cellulose in 81.25% phosphoric acid (D), chromate-oxidized cellulose (O), and SBH-reduced cellulose (R). (Reproduced with permission from Ref. 24. Copyright 1961 Wiley.)

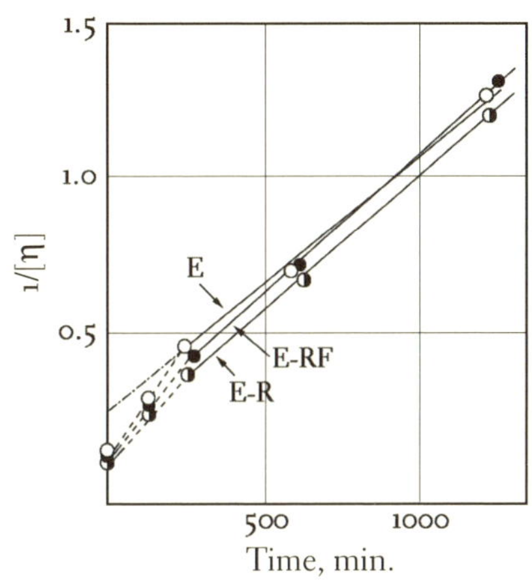

FIGURE 7
Rate of homogeneous hydrolysis in 81.25% phosphoric acid, expressed in terms of intrinsic fluidity, 1/[η]. Key: E, purified wood cellulose; RF, SBH-treated cellulose reduced in fiber form; and R, SBH-treated cellulose reduced as a gel. (Reproduced with permission from Ref. 24. Copyright 1961 Wiley.)

An initial period of rapid hydrolysis is usually most apparent during the heterogeneous hydrolysis of cellulose regenerated from solution. As many as 12-28 particularly sensitive links per 10,000 normal bonds can be introduced, apparently because of physical strains at the folds in the cellulosic bundles.[11, 13, 21, 24] The relative rate of breaking these strained bonds is said to be about 10,000 times the rate of normal 1C-O-4´C bonds. In contrast, Rånby's [24] bonds that were sensitized by the inductive effects of neighboring oxidized units, observed during homogeneous hydrolysis, appear to be only about three to four times more sensitive than the normal bonds.

The Second Stage: Accessibility

In plots of the inverse of DP_n, against time during heterogeneous hydrolysis, a second stage is usually observed in which scissioning also appears to be linear with time, although it proceeds at a noticeably slower rate. Sharples[11] has suggested that this second stage represents the breaking of the 1C-O-4´C bonds between the normal anhydroglucose units located in the amorphous regions of the fibers. Because the amorphous component represents only a fraction of the total fiber, the observed scissioning will appear to take place at a lower overall rate than would be the case if all bonds were equally available. During this second stage the ratio of the observed rate, k´, relative to the rate of breaking of bonds in cellopentaose (a soluble cellulose polymer), k, is considered to represent the accessibility, α, of the particular cellulose fiber under test (see Equation 4).[11] On this basis, the accessibility of cotton is found to be about 9-20%. In contrast, cotton that has been swollen by various treatments prior to hydrolysis can yield accessibility values from 22% to as high as 67%. Such figures agree qualitatively with the percentage of amorphous material in the fibers determined by methods such as moisture regain and X-ray diffraction.

If one considers the rate of hydrolysis of wood cellulose, in which the degree of crystallinity may be lower than in cotton fibers and in which bonds involving

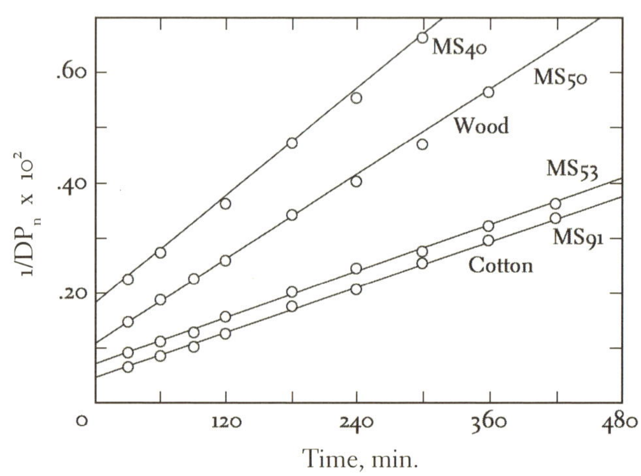

Figure 8
Homogeneous hydrolysis in phosphoric acid, illustrating enhanced rate for wood pulp (MS40, 50) compared to that of cotton samples (MS53, 91). (Reproduced with permission from Ref. 2. Copyright 1971 Wiley.)

units other than glucose, such as xylose and mannose, may be present, the inherent rate of breaking of bonds in the proposed second stage can be expected to be greater than in the case of pure cellulose. A number of workers have pointed out that wood cellulose, under homogeneous conditions, is indeed hydrolyzed at a greater rate, as clearly displayed in Figure 8.[2] These data and Rånby's[24] data indicate that scissioning in wood cellulose occurs as much as two times faster than in cotton.

Whether one can effectively analyze the first or the second stages of scissioning in terms of weak links or of greater accessibility, or whether both factors contribute to each stage, is a matter for further consideration. Nonetheless, the rate at which 1C-O-4′C bonds are broken in the second stage is expected to be influenced in a major way both by accessibility and by the presence of bonds involving pentoses and hexoses other than glucose.

The Final Stage: Attack on Crystallites

In the earliest investigations of hydrolysis, it was readily observed that the DP of the cellulose fell rapidly to a low value, commonly designated as the leveling off degree of polymerization (LODP), beyond which further degradation proceeds at an extremely slow rate.[19, 20] Modern theories of fiber structure envision that the chains of cellulosic molecules are arranged in ordered and in disordered regions and that, furthermore, in the highly ordered regions the chains are aligned in small zones or regions that are called crystallites. These crystalline regions are pictured as involving chain segments that have been folded back and forth upon themselves to form relatively discrete bundles. As a first approximation, the average length of the chain segments in the aligned bundles corresponds to the LODP. Thus, when hydrolytic attack has thoroughly disintegrated the material in the amorphous region, the so-called microcrystalline cellulose consisting largely of the described crystallite bundles is left. Within these, scissioning proceeds at a much slower rate and depends in part on the size of the crystallites. Further attack is on their surfaces; therefore, disintegration is fastest for the residue of fibers such as rayon, which normally have LODP values below 100, and is much slower for the residual material from cotton, which frequently has an LODP of around 200 (number average).[20]

Stages in Deterioration

In recent years, our laboratory has begun to stress that the deterioration of organic materials rarely takes place uniformly in time but tends instead to occur in stages.[25] It is natural to ask if one or more of the stages just outlined is of practical significance to the conservator. We believe they are. The stages described here, however, are not related to those observed in autoxidation; the stages in autoxidation usually involve a different reaction mechanism.

25. Feller, R. L., in *Preservation of Paper and Textiles of Historic and Artistic Value*, Advances in Chemistry Series No. 164, ed. Williams, J. C., American Chemical Society, Washington, DC (1977), p. 314-335.

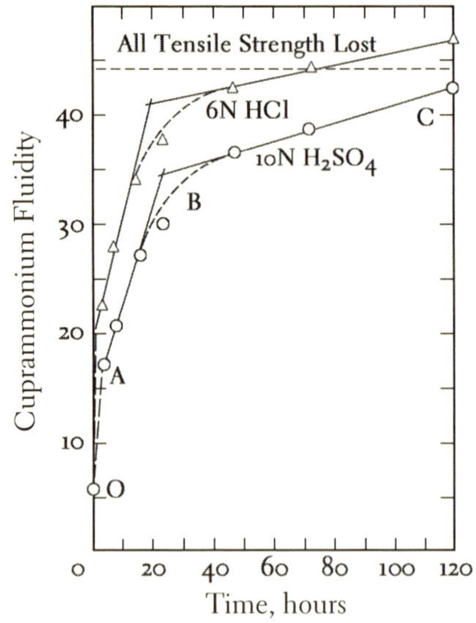

Figure 9
Data showing the change in cuprammonium fluidity of cotton cellulose in acids at 20°C. (Reproduced with permission from Ref. 26. Copyright 1943)

26. Davidson, G. F., *J. Text. Inst.*, 34 (1943), T87-T96.

Figure 9 shows the 1943 data of Davidson[26] on the hydrolysis of cotton cellulose at 20°C. The increase in fluidity with time suggests three stages: the period between points O and A may represent the rapid, initial, weak-link stage; that from A to B may represent the attack on the normal 1C-O-4´C bonds; and that from B to C may represent the slowing down period as disintegration approaches the LODP. Davidson[26] reported that tensile strength became negligible at a fluidity of about 45; Figure 9 shows that the proposed stages occur before all strength is lost.

Daruwalla and Narsian[16] regarded two stages to be of practical importance. They stated,[16] "For all the fiber substances examined, that is, cotton, cotton treated with swelling agents, and cellulosic substances regenerated from different solvents, there exists two rates of heterogeneous hydrolysis. . . . The initial apparent rate of acid action (K_1) is about 1600-2300 times higher than that of the subsequent stage of hydrolysis (K_2)." Their principal findings regarding the heterogeneous attack on regenerated cellulose, illustrated in Figure 10, indicate two-stage behavior for enzymatic and acid-catalyzed hydrolysis, as well as for decomposition by 2540 Å radiation.

The numbers on the vertical axis of Figure 5, and in Figures 1, 2, and 10-12, refer to the percentage of bonds broken. One may evaluate the extent of degradation at any point in these curves by recalling the percentage of 1C-O-4´C bonds that need to be broken in cotton automobile-tire cords before all strength is lost, a subject investigated by Conrad et al.[27] in 1951. Essentially using Equation 4, they came to the conclusion that only 0.3% of the bonds in cotton cords needed to be broken before complete loss of strength occurred. Upon consideration one finds that the calculations that led to this conclusion

27. Conrad, C. M., Tripp, V. W., and Mares, T., *Text. Res. J.*, 21 (1951), p. 726-739.

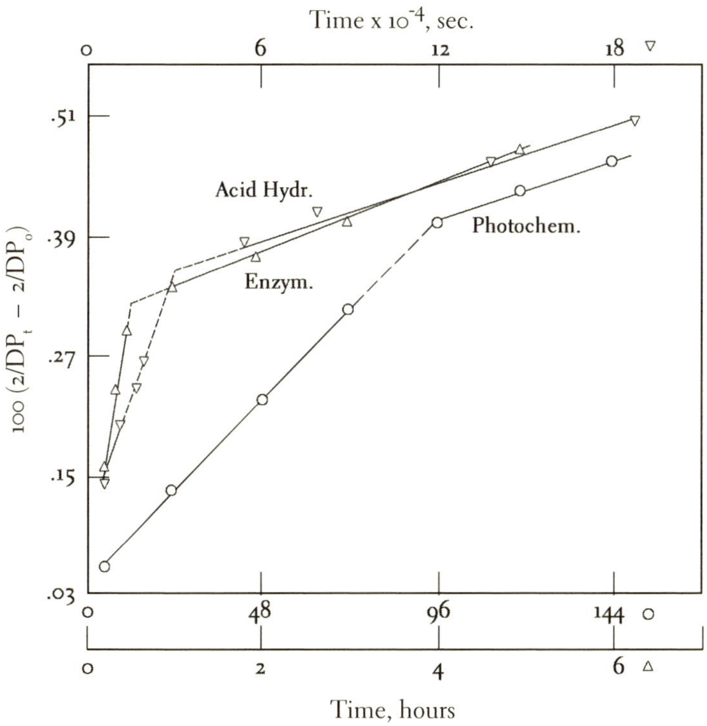

FIGURE 10
Data on the heterogeneous acid hydrolysis, enzymatic hydrolysis, and photochemical decomposition of regenerated cellulose. (Reproduced with permission from Ref. 16. Copyright 1966 Tappi.)

were based on viscosity-average molecular weights, DP_v. Hence, we may estimate, based on number-average molecular weights, DP_n, that the figure should be twice this percentage, or about 0.6%. From Figure 10, we see that the apparent break in the curves occurs when only about 0.33-0.39% of the bonds have broken. Thus, we have additional confirmation that the observed change in rate takes place well before all useful strength is lost.

The two segments in the curves in Figures 5 and 10 may represent the proposed second stage and the beginning of the third, marked-slowdown stage as the LODP is approached; the breaks occurred at DP_v levels between 400 and 470,[16] close to reported values of LODP in cotton (DP_n = 200-235).

Stages also seem to occur during thermal degradation at high temperatures.[28] Figure 11, taken from the work of Shafizadeh and Bradbury,[29] indicates three stages before a DP_n of 200 is reached. Shafizadeh and Bradbury[29] described their results as follows: "An initial small deviation (possibly due to some weak links) is followed by a linear portion, indicating random bond scission, and then a slackening off in the rate of DP_n decrease occurs beginning at $DP_n \approx$ 250, corresponding to a DP_w = 500."

28. Fung, D. P. C., *Tappi*, 52 (1969), p. 319-321.

29. Shafizadeh, F., and Bradbury, A. G. W., *J. Appl. Polym. Sci.*, 23 (1979), p. 1431-1442.

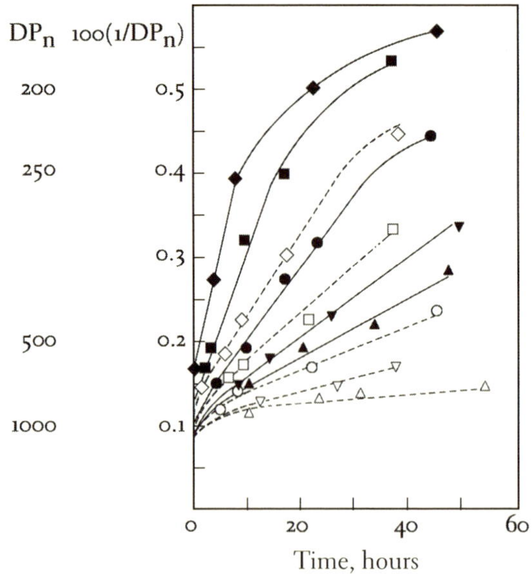

FIGURE 11
Change in $1/DP_n$ with time for cellulose heated in air and in nitrogen at 150-190°C. Key: open symbols represent N_2; filled symbols represent air; △,▲, 150°C; ▽,▼, 160°C; ○,●, 170°C; □,■, 180°C; ◇,◆, 190°C. (Reproduced with permission from Ref. 29. Copyright 1979 Wiley.)

Results

Several years ago we began to express the results of our thermal and photochemical degradation experiments in terms of Equation 4. Figure 1 shows some of the initial data obtained, relating to exposure of test sheets in the Fadeometer. just as Daruwalla and Narsian[16] observed, there seems to be a break in the curve. If the first stage indeed represents weak links, about 3 such bonds per 10,000 in the original cellulosic chain (0.03%) are apparent, a figure similar to the fraction of labile acid-sensitive links determined by Sharples[21] and labile photochemically sensitive links reported by Daruwalla[16] when radiation at 2540 Å was used.

Figure 2 shows the results of a second photochemical study. Handsheets made from bleached sulfate pulp were first oxidized with dichromate solution and then with chlorous acid; following dichromate oxidation, the pulp was reduced with SBH. The data again suggest a rapid initial stage followed by a slower one. Moreover, after SBH reduction, intended to remove carbonyl groups that might be the cause of increased sensitization of neighboring 1C-O-4′C links, we found the initial stage practically eliminated. This finding corresponds to related findings of Rånby for homogeneous hydrolysis (Figures 6 and 7).[24]

Figure 12 shows results of a thermal-aging experiment. Sheets of Whatman No. 42 filter paper were first exposed to near-UV radiation (BLB fluorescent

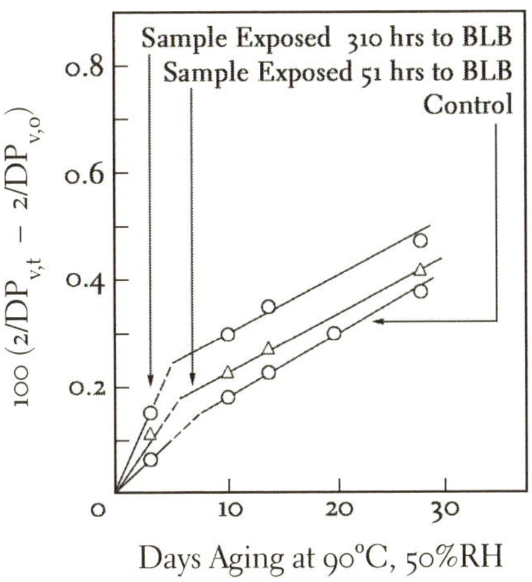

FIGURE 12
Change in $2/DP_v$ during thermal degradation of Whatman No. 42 filter paper previously exposed to near-UV radiation (BLB fluorescent black lamps), compared to control sample.

black lights), after which they were subjected to oven aging. The longer the period of exposure was, the greater the postulated initial stage. If one considers the later data points to represent a linear rate of chain breaking, then the intercepts at time zero imply that a certain fraction of the bonds were particularly sensitive to rupture, perhaps because of a rapid initial stage that occurred before less than 0.2% of the bonds were broken, well in advance of the loss of all useful strength.

Exposure to near-UV radiation in air leads to the formation of carbonyl groups on the cellulosic chain.[30] These groups may weaken neighboring bonds as a result of inductive effects and could thus be the explanation of the increase in the initial stage of thermal deterioration with increased exposure to UV. If the second stage does indeed represent the rate of attack on the normal 1C-O-4′C bonds in the amorphous regions of the cellulose, then it is not surprising that the slopes (the rates of bond breaking) in the later part of the experiment are virtually the same. Kinetic analysis thus seems to provide new insights into the process of deterioration. It is one thing to state simply that exposure to near-UV radiation causes high-α-cellulose pulps to become more sensitive to thermal deterioration; it is quite another thing to suggest that exposure to light primarily increases the initial rapid stage.

30. Reine, A. H., and Arthur, J. C., Jr., *Text. Res. J.*, 40 (1970), p. 90.

DISCUSSION

To express the results in terms of a few distinctly linear stages is a simplification. We have reviewed a number of factors that can influence the ease of chain scissioning. A point that we have not discussed, however, is the fact that several researchers[31, 32] have suggested that high molecular-weight native celluloses may have a significant crystallite or chain-dislocation unit of the order of

31. Elema, R. J., *J. Polym. Sci. Symp.*, 42 (1973) p. 1545-1562.

32. Marx-Figini, M., and Coun-Matus, M., *Makromol. Chem.*, 182 (1981), p. 3603-3616.

DP_n=500. The overall rate must certainly represent transitions between the predominating influence of one or another of the factors noted.

We further realize that the slowest rate shown in many of the curves here may be an artifact resulting from the use of simplified Equation 4; if Equation 3 is applied, a noticeable break between the end of the second stage and the beginning of the final stage (crystallite attack) may be put off to a much more advanced degree of degradation if a discontinuity in the experimental data appears at all.

Whether it is useful to analyze the scissioning of cellulosic chains in terms of two, three, or more stages remains for further investigation. Regardless of what future investigators may conclude, our purpose here has been, first, to remind conservators and conservation scientists that relatively simple means are at our disposal for interpreting the deterioration of cellulose in terms of intrinsic fluidity and the laws of chemical kineties. In particular, the initial stages of chain scissioning appear to follow zero-order kinetics in which a constant number of 1C-O-4′C bonds tend to be broken in a given length of time. Second, when expressed in these terms, the overall course of bond scissioning can be considered to take place in a number of reasonably well-defined stages.

When nothing remains but microcrystalline cellulose, no physical strength is present. Therefore, the rate of attack during the final stage described here will be of little practical concern to the conservator. Nonetheless, if the course of degradation is plotted in accordance with Equation 4, the onset of this last stage, an apparent slowdown as the LODP is approached, may indeed begin before all practical strength is lost.

Although the concepts outlined in this chapter are particularly appropriate for the interpretation of hydrolytic deterioration of cellulose, they show promise as an aid in the interpretation of thermal, photochemical, photolytic, and enzymatic degradation as well. Equations 3 and 4 are generally applicable to the scissioning process in linear polymers.[33, 34]

33. Montrol, E. W., *J. Am. Chem. Soc.*, 63 (1941), p. 1215-1220.

34. Davis, A., and Golden, J. H., *Die Makromol. Chem.*, 78 (1964), p. 16-23.

Acknowledgment

The Research Center gratefully acknowledges the assistance of a grant from the National Endowment for the Humanities to facilitate the preparation of an annotated bibliography concerning this topic and two related topics, which is planned for publication elsewhere.

Damaging Effects of Visible and Near-Ultraviolet Radiation on Paper

S. B. Lee, J. Bogaard and R. L. Feller

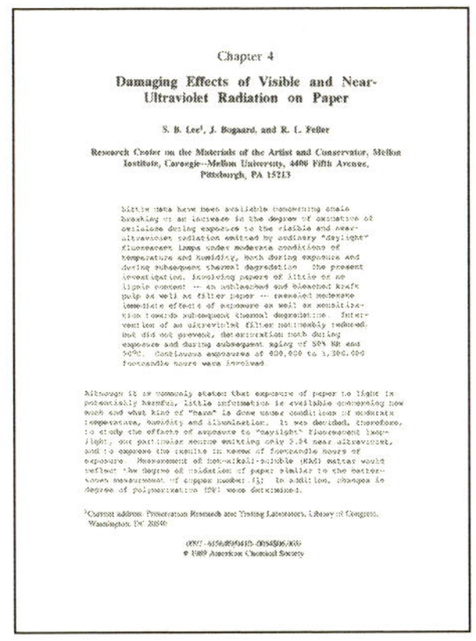
Chapter 4

Damaging Effects of Visible and Near-Ultraviolet Radiation on Paper

S. B. Lee[1], J. Bogaard, and R. L. Feller

Research Center on the Materials of the Artist and Conservator, Mellon Institute, Carnegie–Mellon University, 4400 Fifth Avenue, Pittsburgh, PA 15213

Little data have been available concerning chain breaking or an increase in the degree of oxidation of cellulose during exposure to the visible and near-ultraviolet radiation emitted by ordinary "daylight" fluorescent lamps under moderate conditions of temperature and humidity, both during exposure and during subsequent thermal degradation. The present investigation, involving papers of little or no lignin content -- an unbleached and bleached kraft pulp as well as filter paper -- revealed moderate immediate effects of exposure as well as sensitization towards subsequent thermal degradation. Intervention of an ultraviolet filter noticeably reduced, but did not prevent, deterioration both during exposure and during subsequent aging of 50% RH and 90°C. Continuous exposures of 600,000 to 1,300,000 footcandle hours were involved.

Although it is commonly stated that exposure of paper to light is potentially harmful, little information is available concerning how much and what kind of "harm" is done under conditions of moderate temperature, humidity and illumination. It was decided, therefore, to study the effects of exposure to "daylight" fluorescent lamplight, our particular source emitting only 3.8% near ultraviolet, and to express the results in terms of footcandle hours of exposure. Measurement of hot-alkali-soluble (HAS) matter would reflect the degree of oxidation of paper similar to the better-known measurement of copper number (1). In addition, changes in degree of polymerization (DP) were determined.

[1]Current address: Preservation Research and Testing Laboratory, Library of Congress, Washington, DC 20540

0097-6156/89/0410-0054$06.00/0
© 1989 American Chemical Society

Originally printed in:
Hstoric Textiles and Paper Materials II
ACS Symposium Series Number 410, ed. Zeronian, S. H., and Needles, H. L., American Chemical Society, Washington, DC (1989), Pages 54-62.

Although it is commonly stated that exposure of paper to light is potentially harmful, little information is available concerning how much and what kind of "harm" is done under conditions of moderate temperature, humidity and illumination. It was decided, therefore, to study the effects of exposure to "daylight" fluorescent lamplight, our particular source emitting only 3.8% near-ultraviolet, and to express the results in terms of footcandle hours of exposure. Measurement of hot-alkali-soluble (HAS) matter would reflect the degree of oxidation of paper similar to the better-known measurement of copper number.[1] In addition, changes in degree of polymerization (DP) were determined.

Exposure can result thereafter in increased thermally induced changes. In order to evaluate this aspect of the potential harm that exposure to light represents, the exposed sheets were subjected to thermal aging at 90°C and 50% RH and the changes in HAS matter and DP measured. Thus, not only were the immediate changes that took place during exposure monitored but the effect of exposure upon subsequent thermally induced deterioration were determined as well. This is basically the same technique of investigation used by Launer and Wilson in 1943.[2]

1. Feller, R. L., Lee, S. B., and Curran, M., "Three Fundamental Aspects of Cellulose Deterioration," Supplement to *Art and Archeology Technical Abstracts*, 22, No. 1 (1985), p. 329-354.

2. Launer, H. F., and Wilson, W. K., "Photochemical Stability of Papers," *J. Res. Natl. Bur. Stand.*, 30 (1943), p. 55-74.

EXPERIMENTAL

Characteristics of Stock Pulps

The characteristics of the bleached (BP) and unbleached (UBP) kraft pulps used in these studies have been previously described.[3] Principally the BP pulp contained about 0.24% lignin and an initial solubility in hot 1%-sodium hydroxide of about 4.7% (HAS matter). The UBP pulp contained 4.4% lignin and 2.7% HAS matter. Whatman No. 42 filter paper, having only about 1% HAS matter, was chosen to represent a cellulose of high purity.

3. Lee, S. B., and Feller, R. L., "Influence of the Hemicellulose Fraction on Thermal and Photochemical Discoloration of Paper," in *Historic Textile and Paper Materials*, Adv. in Chem. Ser. 212, ed. Needles, H. L., and Zeronian, S. H., American Chemical Society, Washington, D. C. (1986), p. 377-386.

Measurements of Properties

Water-leaf handsheets at a basis weight of about 75 g/m^2 were prepared from the BP and UBP pulps using a Williams Standard Sheet Mould. Solubility in hot-1%-sodium hydroxide was determined by TAPPI method T212 om 83. Intrinsic viscosity in cuene (cupraethylenediamine solution) and, therefrom, the DP_v were determined according to ASTM Standard D1795.

Exposure to Light Sources

As previously described[3] exposures took place under a bank of six General Electric 48-inch high-output "daylight" fluorescent lamps. The spectral power curve for such a lamp and the tolerance in its correlated color temperature has been published.[4] Spectral distribution data have also been provided by Harrison.[5] Measurements with a calibrated International Light IL700 research radiometer indicated that, of the total milliwatts of visible and near-ultraviolet irradiance, the lamps emitted about 3.8% in the near ultraviolet. A separate sensor head provided readings in terms of footcandles (1 footcandle = 10.67 lux). The lamps were mounted 3 1/2 inches above samples placed on a wire shelf in a room maintained at 50% RH and about 23°C. Owing to an unavoidable heating effect of the lamps, the test papers reached a temperature of about 29°C. Exposures were also carried out under a Plexiglas UF-3 sharp-cut-off filter (Rohm and Haas Company) which transmits less than 5% energy below 400 nm. Under the filter the luminous intensity was reduced about 10%; in the figures this correction has been made for the reduced effective time of exposure. Exposures were only to one side of the test sheets; previous studies indicated that the same net exposure on two sides would have much the same result.[6]

4. *IES Lighting Handbook, Reference Volume*, Illuminating Engineering Society, New York (1981), Figures 8-21 and 8-36.

5. Harrison, L. S., *Report on the Deteriorating Effects of Modern Light Sources*, Metropolitan Museum of Art, New York (1954).

6. Feller, R. L., Lee, S. B., and Bogaard, J., "The Kinetics of Cellulose Deterioration," in *Historic Textile and Paper Materials*, Adv. in Chem. Ser. 212, ed. Needles, H. L., and Zeronian, S. H., American Chemical Society, Washington, D. C. (1986), p. 329-346.

Thermal Aging

Aging at 90°C and 50% RH took place in a humidity-controlled Blue M oven.

Results

Unbleached Pulp

Initially, the development of hot-alkali-soluble (HAS) matter in water-leaf handsheets prepared from unbleached pulp (UBP) was followed both during exposure and during subsequent thermal aging of the exposed sheets at 90°C and 50% RH. The thermally induced generation of HAS matter in the exposed test sheets was compared to that which occurred in an unexposed control. An increase in the degree or rate of generation of HAS matter during thermal aging subsequent to exposure was taken as an indication that potential damage had been done to the paper as the result of the exposure.

As seen in Figure 1, the sheets, which had been exposed to a total of about 800,000 footcandle hours, developed a greater amount of HAS matter upon subsequent thermal aging than did the unexposed control. Exposure to visible radiation alone (exposure under a Plexiglas UF-3 ultraviolet filter) resulted in less generation of HAS matter than when the sheets were exposed to unfiltered radiation. If one may envision thermal deterioration of the papers to occur in at least two principal stages,[6] then the results shown in Figure 1b suggest that exposure to light tends to increase the extent of thermal degradation that occurs subsequently in an initial stage or stages. Thereafter, as suggested by the dashed lines, thermally induced oxidation proceeds at a rate similar to that experienced by the unexposed control.

If this interpretation of the results is valid, one may state that exposure of the UBP sheets to "daylight" fluorescent lamplight tends to cause some immediate photochemical damage, more oxidative in character than directly causing chain scission.[7] This is reflected in a rise in HAS matter during exposure and also during an initial rapid stage of thermal deterioration. Thereafter, the rate of

7. Lee, S. B., Feller, R. L., and Bogaard, J., "Relation of Cellulose Chain Scission to Hot-Alkali-Soluble Content during Thermal and Photochemical Degradation of Paper," *J. Imaging Sci.*, 29, No. 2 (1985), p. 61-64.

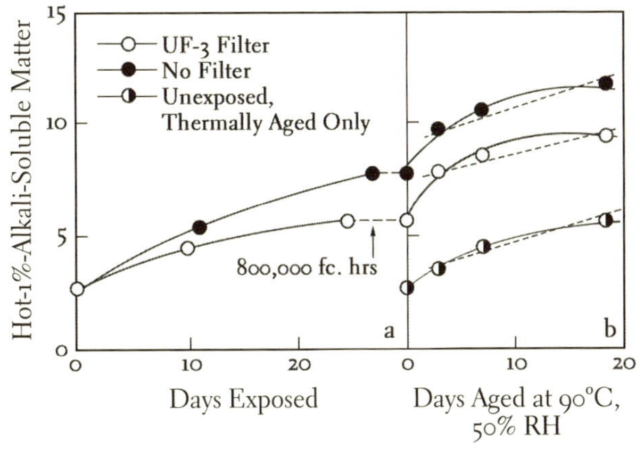

Figure 1
Development of hot-alkali-soluble matter in handsheets of unbleached pulp during exposure to "daylight" fluorescent lamps and subsequent thermal aging.

generation of HAS matter at 90°C and 50% RH resembles somewhat the rate exhibited by unexposed controls. With removal of the ultraviolet portion of the irradiance from the "daylight" fluorescent lamps by use of a Plexiglas UF-3 filter, the extent of subsequent generation of HAS matter is reduced to about one-half the rate without the filter. The immediate damage in this paper, which contained about 0.24% lignin, was reduced, but not eliminated, by removal of ultraviolet radiation.

Bleached Pulp

The initial experiment was followed by similar exposures of handsheets prepared from a bleached pulp (BP). Again the UF-3-filtered situation resulted in a lesser development of HAS matter (Figure 2a). The initial photochemical and subsequent thermally induced changes in HAS matter were less extensive than in the case of the unbleached pulp (2a and b).

Owing to the very low lignin content in this particular pulp, the viscosity in cuene solutions could be determined and from this the fall in DP of the cellulose estimated (Figure 2c and d). In Figure 3, a plot of the change in inverse of the degree of polymerization (1/DP) with time indicates that measurable chain breaking may have occurred even when the UF-3 filter was interposed. The

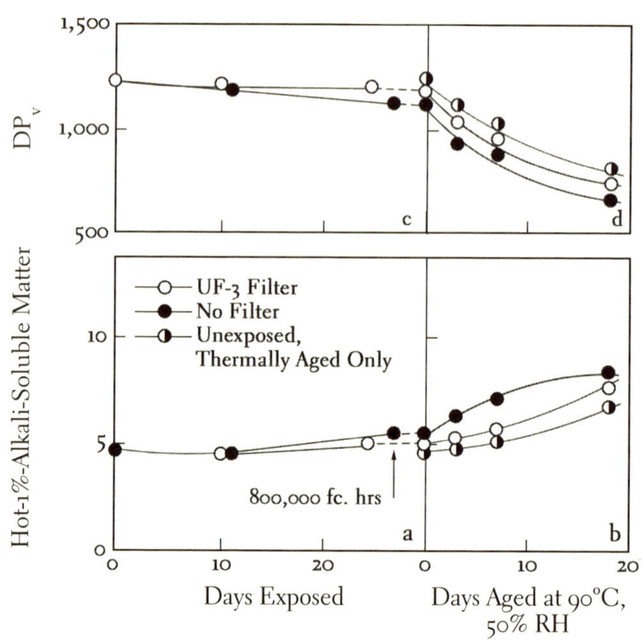

FIGURE 2
Development of hot-alkali-soluble matter and loss of degree of polymerization in bleached pulp during exposure to "daylight" fluorescent lamps and subsequent thermal aging.

lower degree of photochemical damage under the UF-3 filter is also reflected in the subsequent rate of chain breaking when the exposed samples were thermally aged at 90°C and 50% RH (Figure 3d). As with the UBP, the exposed samples of BP seem to exhibit an increase in an initial rapid stage of thermally induced chain breaking followed by a rate similar to that of the unexposed sheets; again, the dotted line is intended to indicate the similarity in rate after the initial rapid rise.

Filter Paper

Lastly, the same test procedures were carried out on Whatman No. 42 filter paper. Folding endurance tests were also performed to indicate in a practical way the immediate effect of exposure and to demonstrate how extensively the pulp had been degraded in the subsequent thermal-aging test (Figure 4f). The results again indicate an increased rate of development of HAS matter and also of chain breaking (Figure 5b) during an initial stage of thermally induced aging followed by a stage in which the rate is similar to that of the unexposed control.

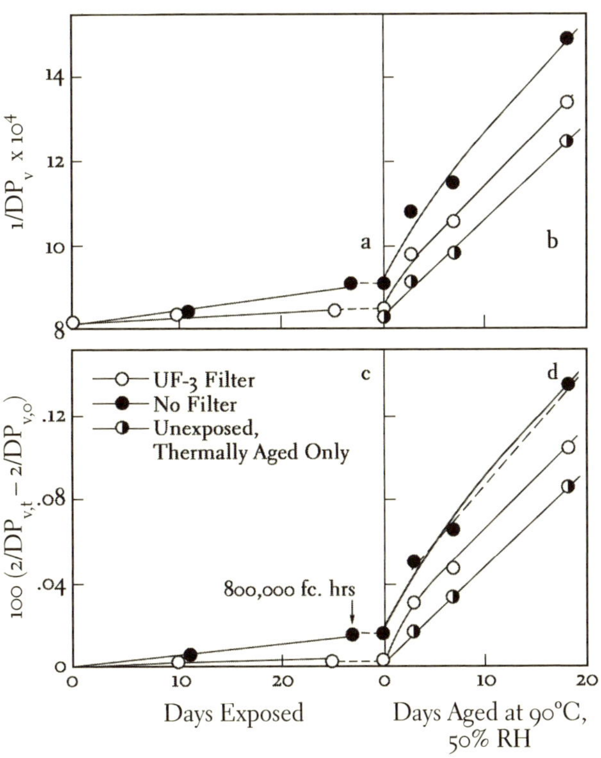

FIGURE 3
Data from Figure 2 expressed in terms of the percentage of links broken, $100\ (2/DP_{v,t} - 2/DP_{v,o})$.

Chain Breaking During Extensive Continuous Exposure

Test sheets based on the bleached pulp (BP) stock and on Whatman No. 42 filter paper have been involved in a number of experiments at the Research Center under these same conditions of exposure. The results of four investigations involving the filter paper and two involving BP are summarized in Figure 6. The approximate percentage of links broken is reflected in the measurement

$$100\,[2/(DP)_{v,t} - 2/(DP)_{v,o}] \qquad (1)$$

We see that, in these two papers, having no more than 0.24% lignin in one case and none in the other, exposure to about 550,000 footcandle hours of illumination from the "daylight" fluorescent lamps resulted in the breaking of between 0.03 and 0.04% of the bonds in the filter paper and between 0.010 to

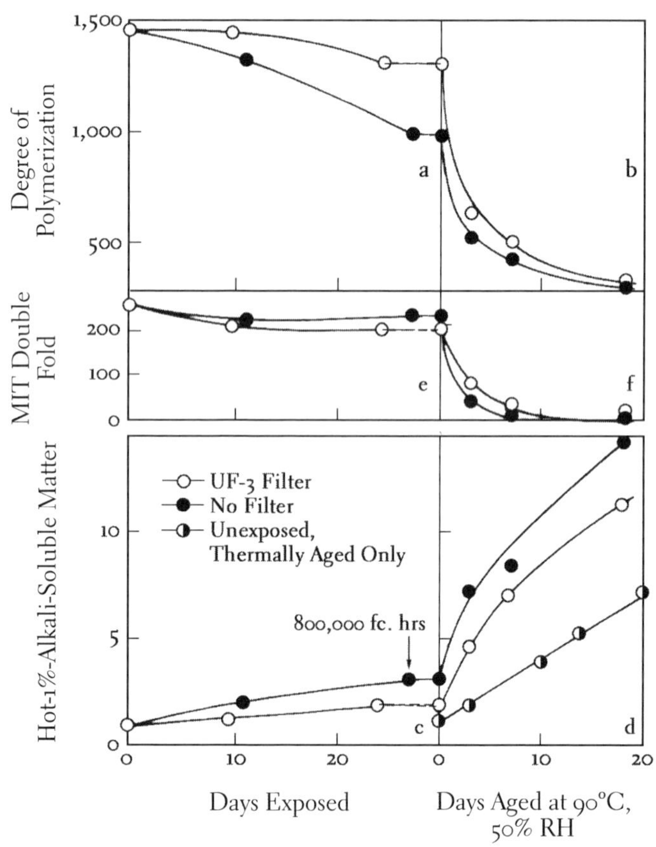

FIGURE 4
Loss of degree of polymerization and folding endurance and rise in hot-alkali-soluble matter in Whatman 42 filter paper during exposure to "daylight" fluorescent lamps and subsequent thermal aging.

0.014% of the bonds in the BP. If it requires the breakage of between 0.3 and 0.6% of the bonds in cellulose to reduce paper to practically zero folding endurance (see Figures 4f and 5b and also reference 6), and if one assumes that a linear rate of chain breaking takes place, one may calculate a minimum "lifetime" before all folding endurance is lost to be about 5.5 million footcandle hours of exposure for the one paper and about 15.4 million footcandle hours for the other. Although these are only rough estimates, the results suggest, in the worst case, that we are talking about something on the order of 50 years'

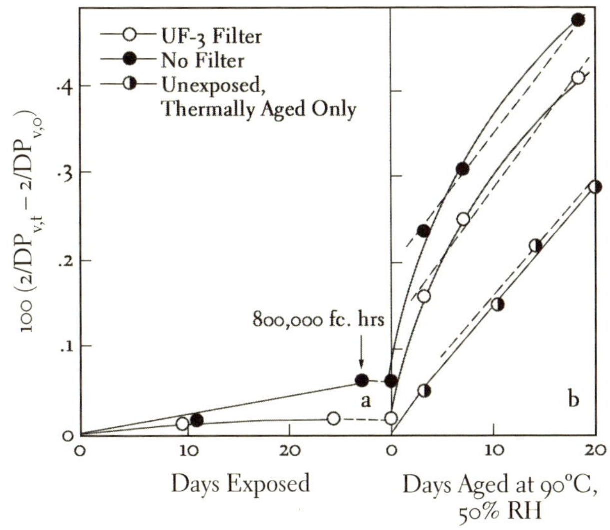

FIGURE 5
Data on loss of DP from Figure 4 expressed in terms of the percentage of links broken, $100\ (2/DP_{v,t} - 2/DP_{v,o})$.

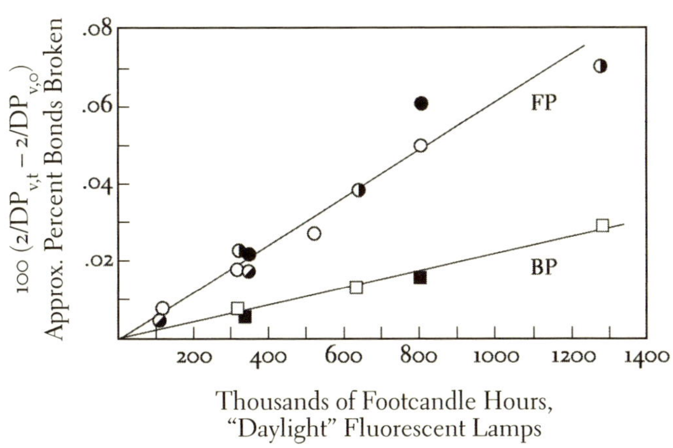

FIGURE 6
Data on the direct exposure of filter paper (FP) and bleached pulp (BP) summarizing the results in terms of the percentage of links broken versus footcandle hours of exposure to "daylight" fluorescent lamps. Each set of symbols signifies a different experimental run; the solid black data points refer to the present results shown in Figures 3 and 5.

exposure on a gallery wall, well-illuminated by diffuse daylight at an average level of about 30 to 38 footcandles.[8] If a moderate level of illumination of about 10 footcandles is employed in the display of works on paper, the estimate becomes well over 100 years for these papers having little or no content of lignin. For papers containing considerable lignin, the potential for damage, of course, is much more significant.[2]

8. Feller, R. L., "Control of Deteriorating Effects of Light Upon Museum Objects," *Museum*, 17 (1964), p. 57-98.

These preliminary studies to evaluate the potentially damaging effects of the direct exposure to test sheets to visible and near-ultraviolet radiation employed exposures tens of times greater than those customarily used in "light bleaching" procedures. Moreover, the test sheets have been exposed directly to the radiation in contrast to the customary exposure of the papers in an aqueous bath to achieve bleaching. Experiments to be reported elsewhere show that the damage to the paper is very much reduced when the sheets are immersed at less than a centimeter depth in water or in aqueous alkaline buffer solutions of about pH 8.[9]

9. Lee, S. B., Bogaard, J., and Feller, R. L., "Bleaching by Light I: Effect of pH on the Bleaching or Darkening of Papers in the Dry and in the Immersed Condition under Visible and Near-Ultraviolet Radiation," in *Conservation of Historic and Artistic Works on Paper: Proceedings of Symposium* 88, ed. Burgess, H., Canadian Conservation Institute, Ottawa, Canada (1994), p. 181-190.

Conclusions

The results provide an indication of the potential rate of deterioration of papers of high or reasonably high quality in terms of footcandle hours of exposure to "daylight" fluorescent lamps under moderate conditions of temperature and humidity. The use of an ultraviolet filter considerably reduces but does not eliminate the harmful effects of the exposure. Papers having little or no lignin content appear to possess considerable resistance to damage by an essentially visible-irradiance source even when several percent near-ultraviolet is present. It is notable that no significant induction time is apparent; photochemical damage began immediately.

A certain amount of chain breaking occurs during exposure, evidenced by a loss in DP and possibly by the increase in HAS matter. This is the "immediately harmful" effect of exposure. Some oxidation of the cellulose chains also seems to take place, suggested by the rise in HAS matter during exposure and by the initially rapid rate of change that occurred upon subsequent thermal degradation of the exposed papers. This represents the "potentially harmful" effect of exposure, leading to loss in degree of polymerization through thermally induced reactions. These concepts are not new but the authors trust that a clear demonstration of the effects has been useful.

Accelerated Thermal Degradation of Pulp Sheets: Effect of Beating and Importance of Humidity

S. B. LEE, J. BOGAARD AND R. L. FELLER

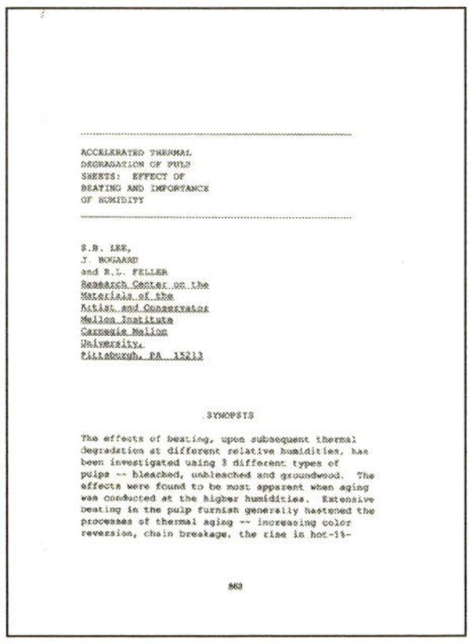
ACCELERATED THERMAL
DEGRADATION OF PULP
SHEETS: EFFECT OF
BEATING AND IMPORTANCE
OF HUMIDITY

S.B. LEE,
J. BOGAARD
and R.L. FELLER
Research Center on the
Materials of the
Artist and Conservator
Mellon Institute
Carnegie Mellon
University,
Pittsburgh, PA 15213

SYNOPSIS

The effects of beating, upon subsequent thermal degradation at different relative humidities, has been investigated using 3 different types of pulps -- bleached, unbleached and groundwood. The effects were found to be most apparent when aging was conducted at the higher humidities. Extensive beating in the pulp furnish generally hastened the processes of thermal aging -- increasing color reversion, chain breakage, the rise in hot-15-

863

Originally printed in: CELLULOSE AND WOOD: CHEMISTRY AND TECHNOLOGY, *ed. Schuerch, C., John Wiley & Sons, New York (1989), Pages 863-884.*

Introduction

Beating is a mechanical process known to cause a number of physical changes in the fibrous structure and colloidal nature of pulp. Well-beaten fibers, for example, can be readily formed into a uniform sheet of paper of high density, whereas unbeaten fibers cannot. This is one of many beneficial effects of beating. As part of this laboratory's general concern for the causes and mechanisms of the deterioration of paper, it was of interest to review the potential detrimental effects of beating, particularly with respect to thermally induced aging.

In the past, there have been conflicting statements made regarding effects of beating upon subsequent thermal aging. In the 1930s, Curran et al.[1] reported scarcely any significant change in the initial 1%-alkali solubility following the beating of sulfite and sulfate pulps for 6 hours. Kress and Bialkowsky[2] also stated that alpha-cellulose content, copper number, and alkali solubility remained practically constant during the course of beating of an unbleached Mitscherlich sulfite pulp, but they did observe a decrease of 16% in cuprammonium viscosity and an increase in the ease of hydrolysis by acid. Richter[3] concluded that highly purified wood pulps (alpha cellulose 94%) showed little effect of beating on the loss of fold at 100°C in spite of a wide range of beating times and freeness. However, after thermal aging, the fold retention of sulfite papers fell to about 50% for small changes in freeness. Such differences in the effects of beating on subsequent thermal deterioration may be due to the ability of certain pulps to withstand the stress of mechanical deformation better than others.[4] Goldhorn investigated and found that the

1. Curran, C. E., Simmonds, F. A., and Chang, H. M., *Ind. Eng. Chem.*, 23, No. 1 (1931), p. 104.

2. Kress, O., and Bialkowsky, H., *Paper Trade J.*, 93, No. 20 (1931), p. 35.

3. Richter, G. A., *Ind. Eng. Chem.*, 23 (1931), p. 371.

4. Leopold, B., and Moulik, S. K. R., *Tappi*, 51, No. 8 (1968), p. 334.

post-color number showed a moderate increase as a result of beating[5] bleached neutral-sulfite straw pulps.

5. Goldhorn, H., *Papierwereld*, 14, No. 2 (1959), p. 367.

In the last decade or so, a number of workers have specifically investigated the effects of beating upon the subsequent thermal degradation of pulps. Luner and Cardwell[6] reported that the rate at which folding endurance was lost during accelerated thermal aging at 80°C and 81% RH increased with an increase in freeness but, upon reconsideration of their data, we do not consider that the evidence overwhelmingly supports this conclusion. A few years later, Hernadi[7] came to the opposite conclusion: that the degree of beating of bleached sulfate pulp did not influence the rate of loss of tensile strength and folding endurance or the rate at which bonds were broken in the cellulose chains during thermal aging at 130 and 160°C. However, beating did increase thermal discoloration in Hernadi's subsequent tests at 190°C.[8]

6. Luner, P., and Cardwell, R. D., 5th Int. Fundamental Res. Symp., Cambridge (Sept., 1973).

7. Hernadi, S., *Svensk Papperstidning*, 79, No. 13 (1976), p. 418.

8. Hernadi, A., *Paperi ja Puu*, 59, No. 9 (1977), p. 566.

The conflicting evidence provided by these two studies may have arisen in part from the difference in behavior of paper when aged under dry thermal conditions (Hernandi) and under moist (Luner and Cardwell). Another possible reason for conflicting results among various workers is the fact that the effect of beating on thermal deterioration may be less apparent on highly purified pulps than in the case of sulfite or kraft pulps, as Richter had pointed out. A further complication in attempting to draw firm conclusions is the fact that beating tends to increase interfiber bonding; paper made from beaten pulps thus may be initially stronger and, owing to the formation of increased fiber bonding (hydrogen bonding) with time, may show an increase in folding endurance and tensile strength during the initial stages of thermal aging, noticeably decreasing only at a later stage.

Since these studies, Koura and Krause have convincingly demonstrated that, under conditions of 90°C and high humidity (70% RH) the stability of bleached sulfite pulp and cotton linters decreased with increased beating.[9] They found that beating led during subsequent thermal aging to a more rapid increase in solubility in 10% alkali. Intrinsic viscosity also fell at a greater rate. Handsheets prepared from the beaten pulps developed acidity and discoloration more rapidly.

9. Koura, A., and Krause, T., *Das Papier*, 32, No. 5 (1978), p. 189.

Freeness, the ease of drainage of water from a slurry of pulp, is customarily used as a measure of the degree of beating; freeness decreases with increased beating. The following report, relating pulp freeness to the subsequent degradation of three different pulp stocks, suggests the importance of humidity in thermal aging tests designed to reveal the effects of beating.

Influence of Pulp Freeness and Humid Aging on Discoloration

Handsheets were prepared from three different stocks: bleached pulp (BP), unbleached pulp (UBP), and groundwood pulp (GP). Initial properties of the pulps are noted in Table 1. The relative rates of color reversion of test sheets

TABLE 1
Initial Properties of Three Pulps Before Aging

Pulp Stock	C.S. freeness (ml)	Lignin content (%)	HAS* (%)	Cold-water extract pH	Copper** number
Bleached Pulp (BP) (50% Hardwood Kraft/ 50% Softwood Kraft)	420 80	0.24 –	4.2 5.3	6.0 6.3	0.5
Unbleached Pulp (UBP) (50% Hardwood Kraft/ 50% Softwood Kraft)	480 80	4.4 –	3.5 4.0	6.3 6.6	0.6
Unbleached Groundwood Pulp (GP)	290 70	29.1 –	12.5 12.5	5.8 6.1	2.9

* Initial solubility in hot 1% sodium hydroxide solution, by Tappi T212.
** Tappi T430.

TABLE 2
Initial Rates* of Discoloration Upon Thermal Aging at Different Relative Humidities

Pulp Stock	C.S. freeness** (ml)	0% RH	50% RH	90% RH
Bleached Pulp (BP)	420 80	20 51	50 119	247 726
Unbleached Pulp (UBP)	480 80	-11 –	34 104	719 2090
Groundwood Pulp (GP)	290 70	52 42	108 106	300 413

* $(K/S)_t - (K/S)_o/t \times 10^7\ min^{-1}$, where t = 72 hours (4320 minutes) and K/S determined at 460 nm.
** C.S. freeness 420, moderately beaten pulps; 80, highly beaten. For groundwood pulp, the corresponding values are 290 and 70.

(between 0 and 72 hours) are reported in Table 2 and are plotted in Figure 1 in terms of the change in post-color number with increased aging time at 90°C under relative humidities of 0, 50, and 90%.[10] At 90% relative humidity, handsheets made with the extensively beaten pulps were clearly more susceptible to thermally induced color reversion. At 50% relative humidity, the greater tendency of the more highly beaten pulps to discolor was less evident and at 0% RH, scarcely demonstrable at all. Initial rates of discoloration, represented by the change in K/S values at 460 nm between the initial and first data points in Figure 1 and given in Table 2, indicate that thermally induced

10. Giertz, H. W., *Svensk Papperstidning*, 48, No. 13 (1945), p. 317.

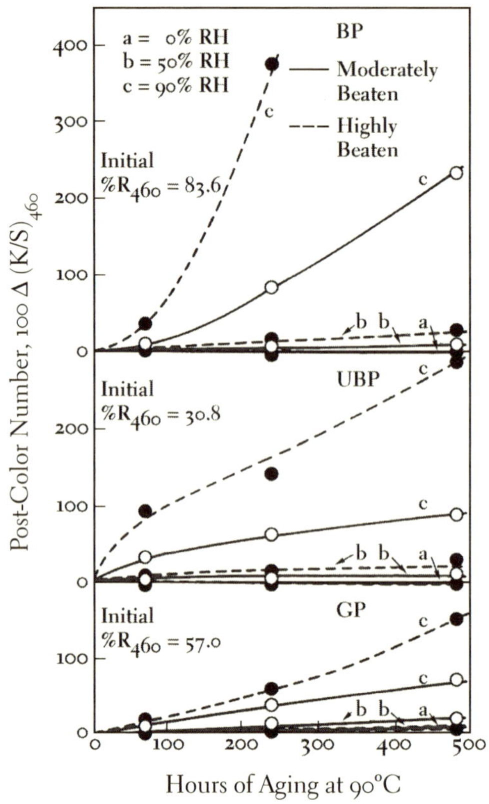

FIGURE 1
Post-color number of test sheets based on moderately and highly beaten groundwood pulp (GP), unbleached pulp (UBP) and bleached pulp (BP) aged at 90°C at 0, 50 and 90% RH.

discoloration is increased in nearly every instance by increased humidity and degree of beating. In the case of the groundwood pulp (GP), an effect of beating is scarcely measurable at 50% RH and may even have been slightly reversed at 0% RH in comparison to the behavior at 90% RH.

Further findings will concentrate primarily on the behavior of the handsheets at 90°C and 50% RH, largely because we, as well as others,[11, 12] consider that moist thermal aging will give a more reliable prediction of the natural aging characteristics of a paper than methods using dry conditions. The consistency of results at 90% RH proved to be unsatisfactory.

11. Wilson, W. K., and Hebert, R. L., *Tappi*, 52, No. 8 (1969), p. 1523.

12. Graminski, E. L., Parks, E. J., and Toth, E. E., NBS Report IR 78-1443 (March 1978).

INFLUENCE OF PULP FREENESS ON THE DEVELOPMENT OF HOT-ALKALI-SOLUBLE (HAS) MATTER

Before aging, in agreement with the findings of Koura and Krause, the initial hot-1%-alkali solubility (HAS) of highly beaten pulps was only slightly higher than that of moderately beaten pulps. Upon subsequent thermal aging, the rate at which HAS matter increased was significantly greater in the case of the more highly beaten pulps, as seen in Figure 2. As suggested by Koura and Krause, these results can be explained by the possibility that, as a result of structural loosening within highly beaten pulps, the cellulose becomes more accessible to aging influences. From a plot of moisture regain versus freeness, Alexander,[13] for example, reported indications that beating produced a reduction in crystallite size.

13. Alexander, S. D., Marton, R., and McGovern, S. D., *Tappi*, 51, No. 6 (1968), p. 277.

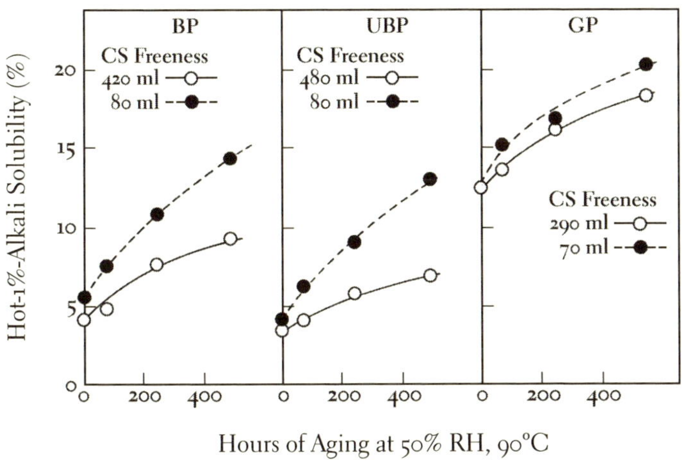

Figure 2
Development of solubility in hot-1%-alkali of test sheets based on bleached pulp (BP), unbleached pulp (UBP) and groundwood pulp (GP), moderately beaten and highly beaten (the latter having Canadian Standard Freeness 70-80 ml), upon aging at 50% RH and 90°C.

If one compares the times in Figure 2 to reach a given level of HAS matter both for the highly beaten (t_2) and for the lightly beaten (t_1) pulps, relative rates for the development of HAS matter can be determined. The results, seen in Figure 3a, clearly indicate a more rapid rate of development in the case of the highly beaten samples. The approximate relative rate increases ($K_{420\ ml}/K_{80\ ml} = t_1/t_2$) are 2.8x for UBP, 2.0x for BP and 2.5x for GP (initial) and 1.0 (later) stages. If a similar comparison is made of the times to reach the same development of acidity (Figure 6), Figure 3b results. The relative increases in acid generation caused by extensive beating is 2.1 for UBP, 1.7 for BP and 1.1 (initial period) to 0.5 in the later stages for GP. If the kinetic order of reaction that governs the rise in HAS matter or acidity is unchanged throughout the time period observed, a linear relationship between t_2 and t_1 should be observed. This is essentially the situation in the case of the changes observed in the case of the BP and UBP test sheets. In the case of GP, however, more complex changes are apparent; the relationship of t_2 and t_1 changed as the extent of thermal aging progressed. Nonetheless, extensive beating clearly increased the rate of development of HAS matter, but had much less effect on the development of acidity.

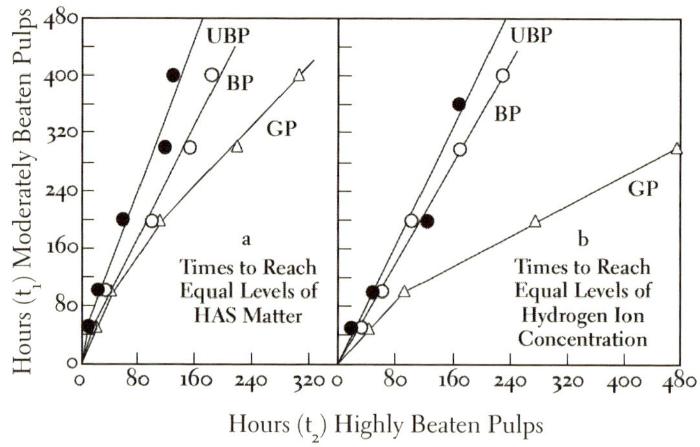

FIGURE 3
Comparison of times to reach equal levels of hot-alkali-soluble (HAS) matter and hydrogen ion concentration for moderately and highly beaten pulps aged at 50% RH and 90°C, based on the data shown in Figures 2 and 6.

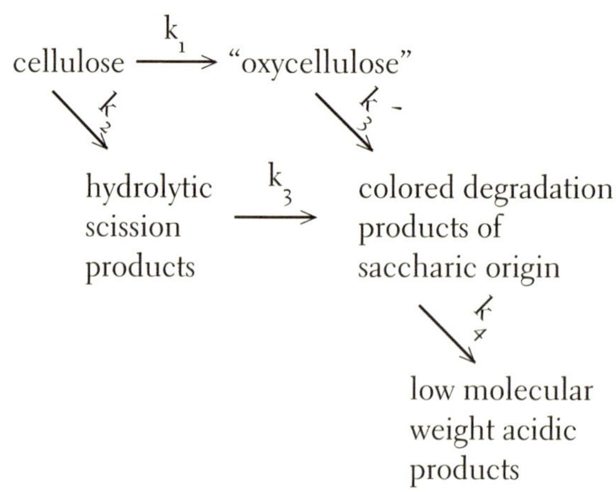

FIGURE 4
Simplified degradation scheme

RELATIONSHIP OF DISCOLORATION TO HAS MATTER FORMED

One may envision potential pathways of thermal deterioration in cellulose as a sequence of chemical reactions that leads to the formation of oxycellulose and hydrolytic degradation products and, eventually, to colored or acidic decomposition products therefrom, as suggested in the above highly simplified scheme. It is logical to consider that the development of color and acidity might occur subsequently to the formation of oxycellulose and hydrolytic degradation products, represented by an increase in matter soluble in hot-1%-sodium hydroxide (hot-alkali-soluble, HAS, matter).

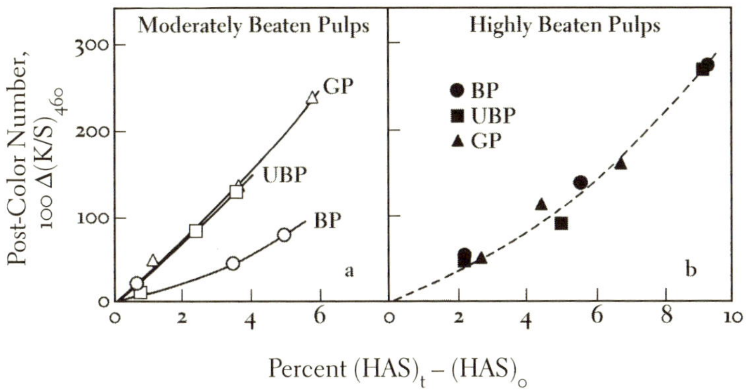

FIGURE 5
Post-color number of pulps compared to the percent increase in hot-alkali-soluble matter (HAS) during aging at 50% RH and 90°C.

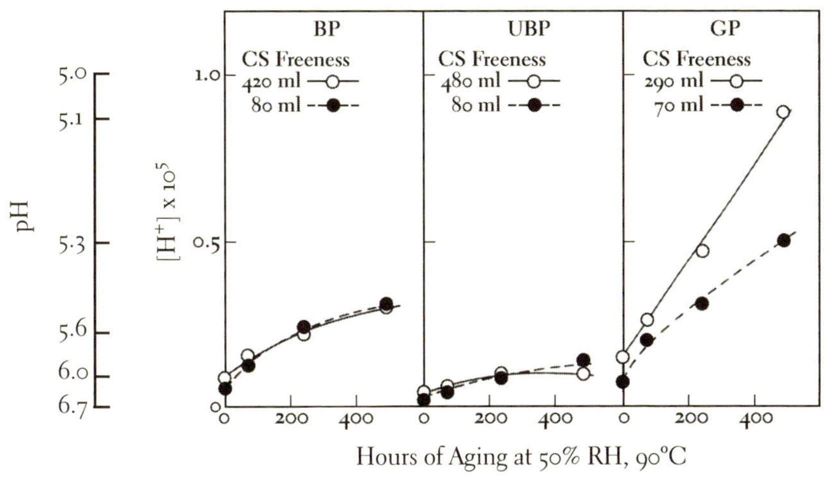

FIGURE 6
Development of acidity in test sheets based on moderately and highly beaten (the latter having Canadian Standard Freeness 70-80 ml) pulps during aging at 50% RH and 90°C.

The increase in HAS matter [$(HAS)_t$ - $(HAS)_o$ or ΔHAS] that occurred during thermal aging of the three paper stocks beaten moderately and aged at 50% RH, 90°C is plotted in Figure 5 against the change in post-color number under the same conditions. The graphs of the data are slightly curved upward, suggesting that the development of color may be accelerated as the amount of hot-alkali-soluble matter increases, especially after the pulps were highly beaten. As the result of extensive beating, there is little change in the relation of discoloration to the increase in hot-alkali-soluble matter in the case of the high quality bleached pulp (BP) (Figure 5b). However, in the case of the less pure pulps UBP and GP, the rate of development of hot-alkali-soluble matter, perhaps reflecting an increased ease of oxidative attack, may have speeded up more than the rate of discoloration.

ACIDITY GENERATED DURING HUMID AGING AND ITS CORRELATION WITH COLOR REVERSION AND HOT-ALKALI SOLUBILITY

In the course of aging at 90°C and 50% RH, samples were extracted with cold water and the pH measured. The acidity, expressed in terms of hydrogen ion concentration rather than pH, developed with time, as shown in Figure 6. In every case, the acid was initially generated at a higher rate in the extensively beaten pulps. This is particularly true in the case of the groundwood pulp (GP); the rise between the initial and first data point is at a nominal rate of 1.38 x 10^{-2} mmol of acid per minute compared to that of the pulp having a higher freeness, 2.29 x 10^{-2} mmol per minute. Although the ultimate overall rate of generation of acidity was less for the more highly beaten GP, only slight differences are found in the change in acidity between the highly and moderately beaten furnishes in the case of BP and UBP sheets.

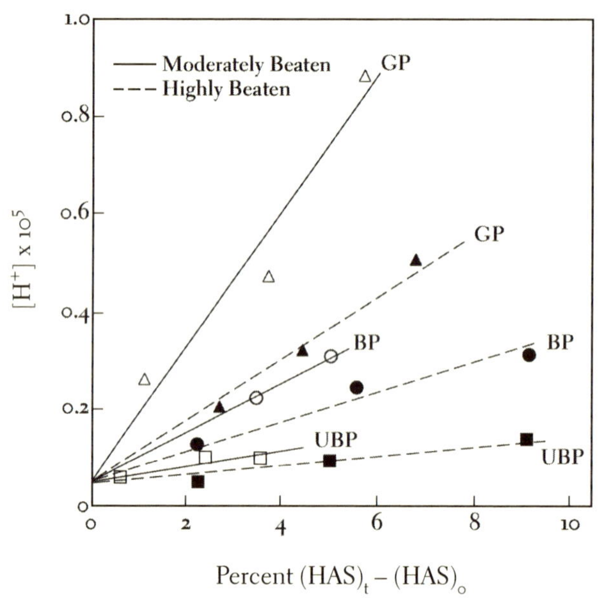

FIGURE 7
Acidity developed in comparison to the increase in hot-alkali-soluble (HAS) matter for moderately and highly beaten pulps during aging at 50% RH and 90°C.

In Figure 7, the increase in HAS matter upon aging, $(HAS)_t - (HAS)_o$, is compared with the acid liberated. The results are expressed here as a linear relationship, not to insist that this is actually the case but simply to indicate the similar effects of beating in all three papers. In every case beating increased the amount of HAS matter that was formed during thermal aging relative to the increase in acidity. If our sequential scheme (Figure 4) is essentially correct, this result may be explained by suggesting that the rate of formation of acid (k_4) may not have been increased as much as k_1 and/or k_2.

Linkages Broken Upon Humid Aging and Its Dependence on Degree of Beating in Pulp Furnish

Because the bleached pulp (BP) was largely free of lignin, cupriethylenediamine (CED) solutions could be readily prepared for viscosity measurements to monitor the effect of beating on the rate of chain breaking of the cellulose. The percentage of chain links broken in degraded cellulose may be calculated by subtracting the reciprocal of the initial number-average degree of polymerization (DP_n) from that of the reciprocal of the DP_n at any given time during degradation and multiplying by 100 according to the following equation: $LB = 100\ (1/DP_{n,t} - 1/DP_{n,o})$.[14] Because DP_n is approximately equal to $2/DP_v$, we have estimated LB from viscosity measurements in CED, multiplying our values of $1/DP_v$ by 2. When the percentage of links broken thus calculated is plotted against time, as in Figure 8, one sees clearly that a higher rate of chain breakage occurs during the aging of the highly beaten pulp and that this is most apparent at high humidities.

As reported earlier,[15] an essentially linear relationship exists between the apparent percentage of chain links broken (during thermal aging) and the increase in HAS matter. As seen in Figure 9, the rise in hot-alkali-soluble (HAS) matter is related to the thermally induced increase in the number of links broken, irrespective of the degree of beating and the rate at which the bonds are broken. This implies that the ratio of k1 to k2 in the possible degradation scheme (Figure 4) is not influenced by beating.

14. Conrad, C. M., Tripp, V. W., and Mares, T., *Tex. Res. J.*, 21, No. 10 (1951), p. 726.

15. Lee, S. B., Feller, R. L., and Bogaard, J., *J. Imaging Science*, 29, No. 2 (1985), p. 61.

Summary and Conclusions

The effects of beating upon subsequent thermal degradation, investigated using examples of bleached, unbleached and groundwood pulps, were found to be most apparent when thermal aging was conducted at high humidities. Extensive beating of the pulp furnish generally hastened the processes of thermal aging—increasing color reversion, chain breakage, the rise in hot-1%-alkali-soluble matter, and the development of acidity.

Experimental

Three types of pulp—bleached pulp (50% hardwood kraft/50% softwood kraft), unbleached pulp (50% hardwood kraft/50% softwood kraft), and unbleached groundwood pulp—were beaten in a ball mill at a 2 percent consistency. Analyses of the initial lignin and hot-alkali-soluble matter are given in Table 1.

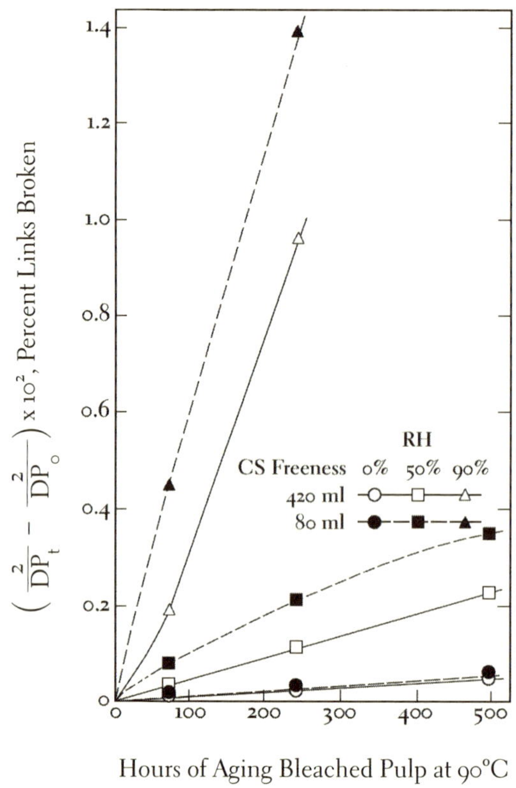

Figure 8
Approximate percent links broken for moderately and highly beaten bleached pulps aged at 90°C and 0, 50 and 90% RH.

Handsheets were made from a moderately and a highly beaten pulp at the basis weight of 75 g/m^2.

Accelerated heat aging was carried out in an oven at 90°C under relative humidities of 0, 50, and 90%.

The amount of color reversion is expressed as post-color number,[10] the change in the Kubelka-Munk K/S number before and after aging, multiplied by 100, where $K/S = (1 - R_\infty)^2 / 2R_\infty$ and R_∞ in this case is the reflectance of a stack of six sheets of the sample papers at 460 nm taken from the spectrophotometric reflectance data using a 10 nm bandpass slit width. The standard Tappi measurement refers to the reflectance determined with a bandpass filter peaked at 457 nm and having a half-band width of about 40 nm. We have found the two methods yield much the same results.

The hot-1%-alkali solubility (HAS) and coldwater extract pH were determined according to Tappi methods.[16] The viscosity of the aged samples was determined in cupriethylenediamine solution at 25°C. The intrinsic viscosity $[\eta]$ was

16. Tappi Standard T212 and T509.

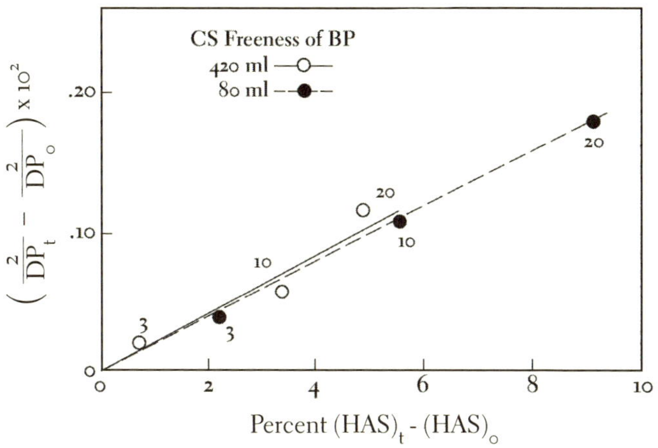

FIGURE 9
Approximate percentage links broken, $(2/DP_t - 2/DP_o) \times 10^2$, compared to the increase in hot-alkali-soluble (HAS) matter for test sheets based on moderately and highly beaten bleached pulps (BP) aged 3, 10 and 20 days at 50% RH and 90°C.

calculated using Table 3 of ASTM Standard D1795 which is prepared from the Martin equation[17]:

$$\log \frac{\eta_{rel} - 1}{C} = \log [\eta] + K [\eta] C \qquad (1)$$

where K = 0.13. The degree of polymerization is obtained by multiplication of [η] by 190.

17. Martin, A. F., *Tappi*, 34, No. 8 (1951), p. 363.

Section 8: Paper Darkening and Bleaching

Many but not all papers noticeably darken over time. Exposure to light often results in darkening, yet light can also induce bleaching of paper. To expand further the chemist's and conservator's understanding of this behavior, the Center investigated the role of visible and ultraviolet radiation, pH, and wet or dry conditions of treatment. Evidence was obtained that typical exposures used to bleach paper do not cause significant scission of the cellulose molecule, and thus light bleaching treatments pose little risk to the strength and flexibility of the paper sheet. The initial study of the literature took note of the changes in absorption in the ultraviolet as well as in the visible wavelengths. Individual sugars are often found among the breakdown products of cellulose degradation. They vary in their tendency to discolor when heated, yet the resulting degradation products were shown to be bleached upon exposure to light. It was observed that these products were bleached even by ultraviolet radiation, whereas representative components of lignin were found to darken under both visible and ultraviolet radiation. The extent of post-irradiation discoloration was shown to be dependent on the concentration of lignin present. In contrast, hemicelluloses, another identifiable component of paper pulp, do not darken when exposed to visible or ultraviolet radiation.

R. L. F.

The Darkening and Bleaching of Paper by Various Wavelengths in the Visible and Ultraviolet

R. L. FELLER, S. B. LEE AND J. BOGAARD

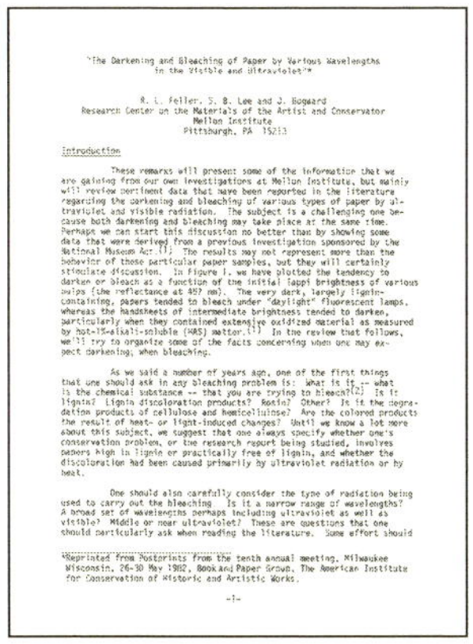

"The Darkening and Bleaching of Paper by Various Wavelengths in the Visible and Ultraviolet"*

R. L. Feller, S. B. Lee and J. Bogaard
Research Center on the Materials of the Artist and Conservator
Mellon Institute
Pittsburgh, PA 15213

Introduction

These remarks will present some of the information that we are gaining from our own investigations at Mellon Institute, but mainly will review pertinent data that have been reported in the literature regarding the darkening and bleaching of various types of paper by ultraviolet and visible radiation. The subject is a challenging one because both darkening and bleaching may take place at the same time. Perhaps we can start this discussion no better than by showing some data that were derived from a previous investigation sponsored by the National Museum Act.(1) The results may not represent more than the behavior of these particular paper samples, but they will certainly stimulate discussion. In Figure 1, we have plotted the tendency to darken or bleach as a function of the initial Tappi brightness of various pulps (the reflectance at 457 nm). The very dark, largely lignin-containing, papers tended to bleach under "daylight" fluorescent lamps, whereas the handsheets of intermediate brightness tended to darken, particularly when they contained extensive oxidized material as measured by hot-1%-alkali-soluble (HAS) matter.(1) In the review that follows, we'll try to organize some of the facts concerning when one may expect darkening, when bleaching.

As we said a number of years ago, one of the first things that one should ask in any bleaching problem is: What is it -- what is the chemical substance -- that you are trying to bleach?(2) Is it lignin? Lignin discoloration products? Rosin? Other? Is it the degradation products of cellulose and hemicellulose? Are the colored products the result of heat- or light-induced changes? Until we know a lot more about this subject, we suggest that one always specify whether one's conservation problem, or the research report being studied, involves papers high in lignin or practically free of lignin, and whether the discoloration had been caused primarily by ultraviolet radiation or by heat.

One should also carefully consider the type of radiation being used to carry out the bleaching. Is it a narrow range of wavelengths? A broad set of wavelengths perhaps including ultraviolet as well as visible? Middle or near ultraviolet? These are questions that one should particularly ask when reading the literature. Some effort should

*Reprinted from Postprints from the tenth annual meeting, Milwaukee Wisconsin, 26-30 May 1982, Book and Paper Group, The American Institute for Conservation of Historic and Artistic Works.

-1-

Originally printed in:
Postprints of the Book and Paper Group at the 10th Annual Meeting, Milwaukee, Wisconsin
American Institute for Conservation, Washington, DC (1982), 17 Pages.

Introduction

These remarks will present some of the information that we are gaining from our own investigations at Mellon Institute, but mainly will review pertinent data that have been reported in the literature regarding the darkening and bleaching of various types of paper by ultraviolet and visible radiation. The subject is a challenging one because both darkening and bleaching may take place at the same time. Perhaps we can start this discussion no better than by showing some data that were derived from a previous investigation sponsored by the National Museum Act.[1] The results may not represent more than the behavior of these particular paper samples, but they will certainly stimulate discussion. In Figure 1, we have plotted the tendency to darken or bleach as a function of the initial Tappi brightness of various pulps (the reflectance at 457 nm). The very dark, largely lignin-containing, papers tended to bleach under "daylight" fluorescent lamps, whereas the handsheets of intermediate brightness tended to darken, particularly when they contained extensive oxidized material as measured by hot-1%-alkali-soluble (HAS) matter.[1] In the review that follows, we'll try to organize some of the facts concerning when one may expect darkening, when bleaching.

1. Lee, S. B., Bogaard, J., and Feller, R. L., "Concerning the Exposure of Paper to Light: Discoloration of Handsheets of Known Initial Lignin and Hot-Alkali-Soluble Content," *Preprints of the 6th Triennial Meeting, Ottawa,* ICOM Committee for Conservation (1981), Paper 81/14/4.

As we said a number of years ago, one of the first things that one should ask in any bleaching problem is: What is it—what is the chemical substance—that you are trying to bleach?[2] Is it lignin? Lignin discoloration products? Rosin? Other? Is it the degradation products of cellulose and hemicellulose? Are the colored products the result of heat- or light-induced changes? Until we know a lot more about this subject, we suggest that one always specify whether one's

2. Feller, R. L., "Notes on the Chemistry of Bleaching," *Bulletin of the American Group - IIC,* 11, No. 2 (1971), p. 39-56.

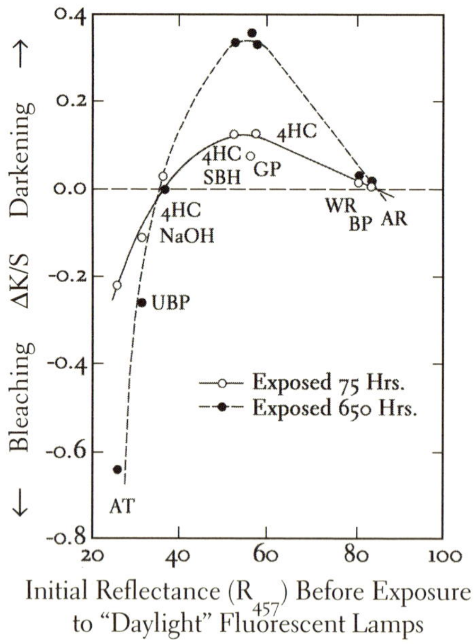

FIGURE 1
The tendency of various pulps to darken or bleach on exposure to "daylight" fluorescent lamps. (Data from Ref. 1)

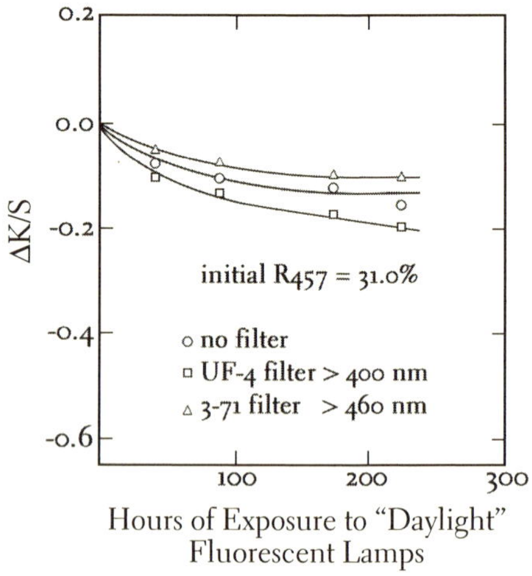

FIGURE 2
Unbleached pulp: 4.4% lignin, 3.7% HAS matter.

conservation problem, or the research report being studied, involves papers high in lignin or practically free of lignin, and whether the discoloration had been caused primarily by ultraviolet radiation or by heat.

One should also carefully consider the type of radiation being used to carry out the bleaching. Is it a narrow range of wavelengths? A broad set of wavelengths perhaps including ultraviolet as well as visible? Middle- or near-ultraviolet? These are questions that one should particularly ask when reading the literature. Some effort should be made also to determine whether the temperature was or will be much above 30°C and whether the humidity is high or low.

IMPORTANCE OF WAVELENGTH IN PAPERS HIGH IN LIGNIN

Let us begin by showing the results of some rather simple bleaching tests of our own involving various wavelengths of radiation and various types of paper. Figures 2 and 3 confirm the statement that we emphasized earlier[1]: under mild conditions lignin tends to bleach. Note in Figures 2 and 3 that considerable bleaching takes place owing to visible radiation alone (under the UF-4 filter). Even under a filter that transmits only visible radiation above 460 nm, bleaching takes place. This is just as predicted by Nolan, Van den Akker and Wink in 1945[3], later confirmed by Leary.[4,5] It is the ultraviolet that may cause darkening, as we see in Figure 4, where the unfiltered "daylight" fluorescent lamps cause

3. Nolan, P., Van den Akker, J. A., and Wink, W. A., "The Fading of Groundwood by Light. II. The Physical Mechanism of Fading," *Paper Trade Journal*, 121, No. 11 (1945), Tappi Section, p. 101-105.

4. Leary, G. L., "The Yellowing of Wood by Light," *Tappi*, 50 (1967), p. 17-19.

5. Leary, G. L., "The Yellowing of Wood by Light: Part II," *Tappi*, 51 (1968), p. 257-260.

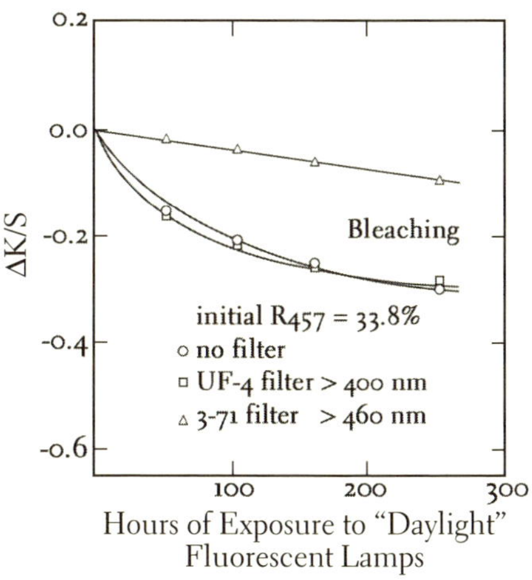

FIGURE 3
Filter paper with lignin added-on: 3% lignin, 5.2% HAS matter.

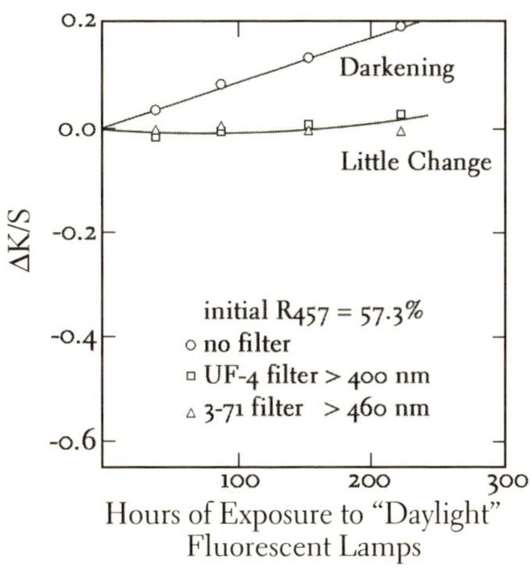

FIGURE 4
Groundwood pulp: 29% lignin, 15% HAS matter.

some darkening in groundwood pulp. We believe that this may be due to the fact that this particular groundwood pulp sample had a high concentration of alkali-sensitive groups, as evidenced by the high hot-1%-alkali (HAS) matter, as well as having a high percentage of lignin. The result that there is little change observed when the sheets made from this pulp were irradiated by visible radiation alone may be due to the fact that both the darkening and bleaching effects were balancing out. The possibility that both processes go on at the same time and that short-wavelength ultraviolet light tends to cause darkening of pulps that contain lignin is demonstrated in Figure 5; here, the same unbleached pulp shown in Figure 2 darkened at first; later bleaching took over. This curve seems rather good evidence that the two processes of darkening and bleaching are going on at the same time. This same pulp, when bleached with chlorine dioxide or calcium hypochlorite, tended also to darken first, then bleach, when exposed to "daylight" fluorescent lamps.

These simple experiments indicate how important it is to consider the influence of wavelengths of radiation involved. Perhaps no better presentation of the basic facts can be cited than the work of Leary,[4] shown in Figure 6 and in Tables 1 and 2. Leary used a monochromator to irradiate newsprint with narrow-band widths of radiation (10 to 20 nm). As Figure 6, a qualitative representation of the information in Tables 1 and 2, indicates, darkening of newsprint, a pulp that undoubtedly contains considerable lignin, takes place

TABLE 1
Effect of Wavelength on Yellowing of Newspaper (from Leary, 1967).

Wavelength of Irradiation, nm	Observed Effect
330	Nil
342	Nil
355	Nil
360	Very slight yellowing
365	Slight yellowing
370	Yellowing
375	Yellowing
380	Yellowing
385	Yellowing
390	Very slight yellowing

TABLE 2
Effect of Wavelength on Bleaching of Photochemically Yellowed Newspaper (from Leary, 1967).

Wavelength of Irradiation, nm	Observed Effect
400	Slight greying
420	Bleaching
448	Bleaching
480	Bleaching
508	Slight bleaching

principally under radiation between 360 and 390 nm in the near-ultraviolet (defined as 320-400 nm) and bleaching occurs at wavelengths perhaps between 420 and 510 nm in the visible. We see in Figure 6 how confusing the results might be if one wished to irradiate such a paper with a source that contained both near-ultraviolet and visible such as the radiation from a carbon-arc Fadeometer. One would have every reason to expect that both darkening and bleaching were going on at the same time under such a source or under a filter that transmitted radiation above 385 nm, as Nolan et al. used.[3] With such a filter, bleaching first occurred, then darkening in their experiments, leading them to conclude that there were at least two reactions going on.

The darkening of pulps that contain lignin was very well investigated in 1945 by Nolan and his co-workers.[3] We see their results in Figure 7 where the absorption curve of various lignin-containing liquors is compared to the sensitivity of lignin to discoloration (called "fading" by these authors, but really referring to darkening). Figure 7 clearly indicates what we mean when we say that lignin tends to darken as increasingly lower wavelengths of ultraviolet are employed.

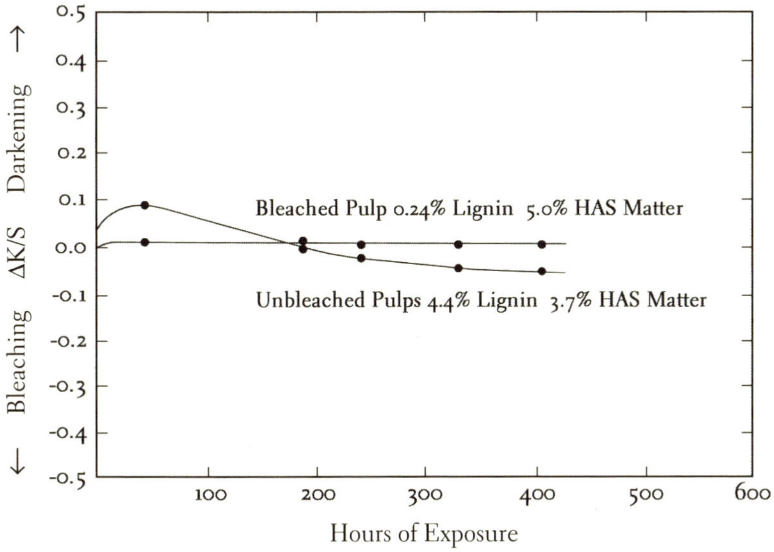

Figure 5
Exposure under fluorescent sunlamp (ultraviolet 290-350 nm).

Figures 6 and 7 scarcely begin to tell the story; we see in Figure 8, taken from the work of Lin and Kringstad,[6] that irradiation by the near-ultraviolet caused the extinction coefficient to decrease in the ultraviolet region; in a sense, the lignin is fading in the ultraviolet. If one looks carefully at the curves in the region of 400 nm, however, one notices that the tail of the absorption curve after exposure is actually slightly higher as it enters the visible region. Apparently we have fading in the ultraviolet and increased absorption (darkening) in the visible.

6. Lin, S. Y., and Kringstad, K. P., "Photosensitive Groups in Lignin and Lignin Model Compounds," *Tappi*, 53 (1970), p. 658-663.

Indeed, the change in the spectral absorption and reflectance curves is complex. If one simply measures the spectral reflectance curves in the visible region as, for example, in the case of wood pulps being extracted in the work of Polcin and co-workers seen in Figure 9,[7] we find that subtle changes in the reflectance curves can hardly be interpreted. If the two sets of curves are subtracted, however, as is done in their Figure 7, we display the spectral absorption curve of the extracted material. Subtraction of curves has frequently proven useful in discoloration work, for by means of subtraction the nature of the colored substances that were developed or removed can be detected.

7. Polcin, J., Wayman, M., Anderson, C. B., and Rapson, W. H., "Sapwood and Heartwood Groundwood of Western Hemlock and Jack Pine. Part 1. Influence of Extractives on Optical Properties," *Pulp and Paper Magazine of Canada*, 69, No. 21 (1969), T410-T417.

In Figures 10 and 11 taken from the work of Claesson et al.,[8] we see in detail the results of discoloration, caused by 365 nm ultraviolet and bleaching brought about by the visible wavelength of 436 nm. These data represent the differences in percent reflectance after irradiation. In these curves, we see bleaching in the ultraviolet and darkening in the visible, with a peak of absorption around 430 nm. The shape of these absorption curves, in a sense, describes the colored matter that was formed upon irradiation by the ultraviolet. We also see the changes brought about during bleaching by 436 nm light.

8. Claesson, S., Olson, E., and Wennerblom, A., "The Yellowing and Bleaching by Light of Lignin-rich Papers and the Re-yellowing in Darkness," *Svensk Papperstidning*, 71, No. 8 (1968), p. 335-340.

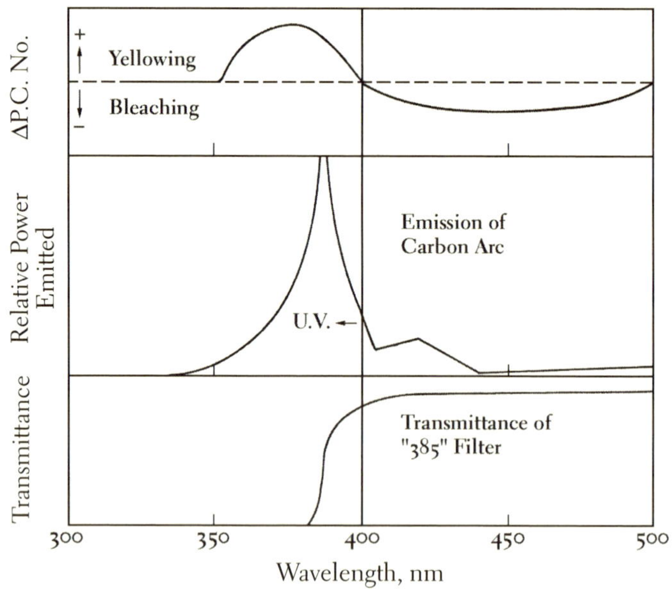

FIGURE 6
Data of Leary, 1967 (Ref. 4).

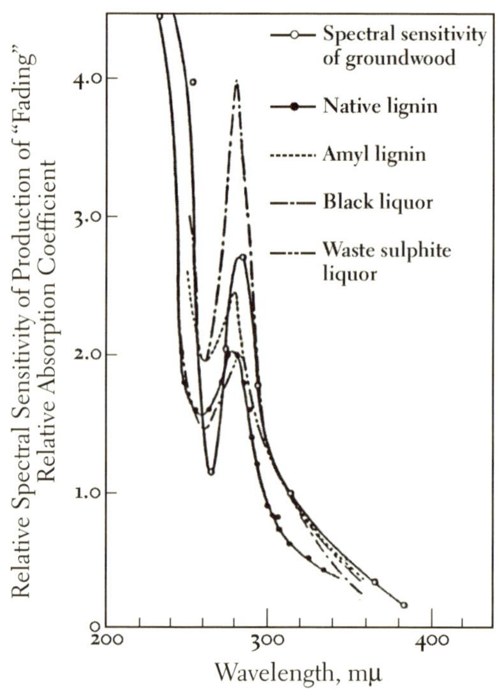

FIGURE 1 (of Ref. 3)
Comparison of Spectral Sensitivity of "Fading" of Groundwood and Spectral Absorption Curves of Lignin Compounds

FIGURE 7
Sensitivity of lignin to darkening (called "fading" here). Data of Nolan, Van den Akker and Wink, 1945 (Ref. 3).

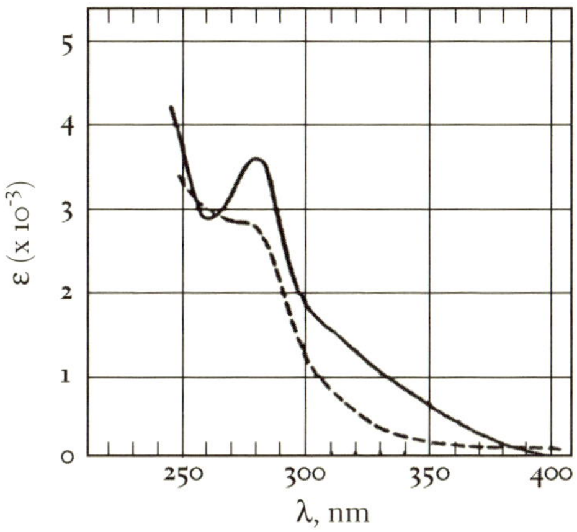

FIGURE 3 (of Ref. 6)
Ultraviolet absorption spectrum of spruce milled-wood lignin in methyl cellosolve before (——) and after (- - -) exposure to near-ultraviolet light for 3 hr.

FIGURE 8
Changes in absorption coefficient, ε, of lignin in ultraviolet region. Data of Lin and Kringstad, 1970 (Ref. 6).

Figures 10 and 11 (the authors' Figures 14 and 15) clearly demonstrate that one cannot learn very much about the discoloration and bleaching of pulps by consideration of the visible region alone (such as shown in Figure 9). It is in the ultraviolet that the absorption peaks, which depend upon the particular substances involved, occur.[9]

9. Pew, J. C., and Connors, W. J., "Color of Coniferous Lignin," *Tappi*, 54, No. 2 (1971), p. 245-251.

CHANGES IN PURE CELLULOSE

Thus far we have been considering the changes in the spectra of pulps high in lignin.[9] The work of Silvy and LeNest,[10] on the other hand, specifically concerned the darkening of pure cellulose. Referring to Figure 12, we see, in Silvy and LeNest's Figure 2, the effects of increased heating of cellulose and note the development of absorption peaks at 260, 310 and 420 nm. (PNCR% is essentially the relative post-color number, ΔK/S/ K/S.) Curves 1, 2 and 3 in this figure are the results of heating the cellulose at 120°C in the dark for 13 hours, 148 hours and 3 months. Curves 4, 5 and 6 show the effects of exposure to simulated "daylight" in a Xenotest apparatus for 40 minutes, 6 hours and 70 hours, respectively. We know from our own work that thermally discolored cellulose tends to bleach, as stated earlier.[1] The authors point out that the peak absorption at 420 nm disappears most readily, the peak at 310 nm less readily and that at 260 nm is scarcely reduced at all by the light. This is a particularly illuminating illustration of the complexity of the darkening and bleaching

10. Silvy, J., and LeNest, J. F., "Mise en évidence par diffusion de la lumiére d'un effet photochrome de la cellulose," *Comptes Rendus Des Seances De L'Academie Des Sciences, Paris, Series C, Sciences Chimiques*, 271 (1970), p. 221-224; see also "A Photochromic Effect Shown During the Ageing of a Cellulose Material by Heat and Light," *Transactions of the Symposium held at Cambridge, September* 1973, ed. Bolam, F., Technical Division, The British Paper and Board Industry Federation, Plough Place, Fetter Lane, London EC4A 1AL (1976).

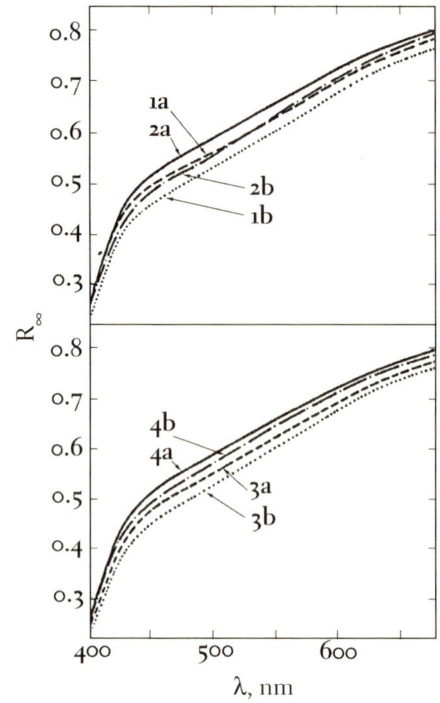

Figure 6 (of Ref. 7)
Reflectance spectra of western hemlock sapwood pulps before (a) and after (b) thermal aging:
1) original pulp, 2) EDTA treated, 3) ethanol-benzene extracted, 4) acetone-water extracted.

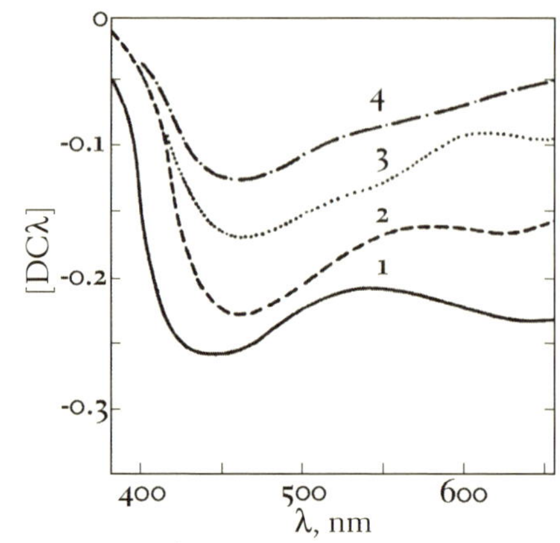

Figure 7 (of Ref. 7)
[DCλ] curves of western hemlock sapwood pulps before and after thermal aging:
1) original pulp, 2) EDTA treated, 3) ethanol-benzene extracted, 4) acetone-water extracted (1:1).

Figure 9
Authors' Figure 7 illustrating changes in reflectance of pulps in the visible region, showing usefulness of subtracting the curves of reflectance seen in authors' Figure 6. Data of Polcin et al., 1969 (Ref. 7). $[DC\lambda] = (K/S_o - K/S_1) / (K/S_o)$

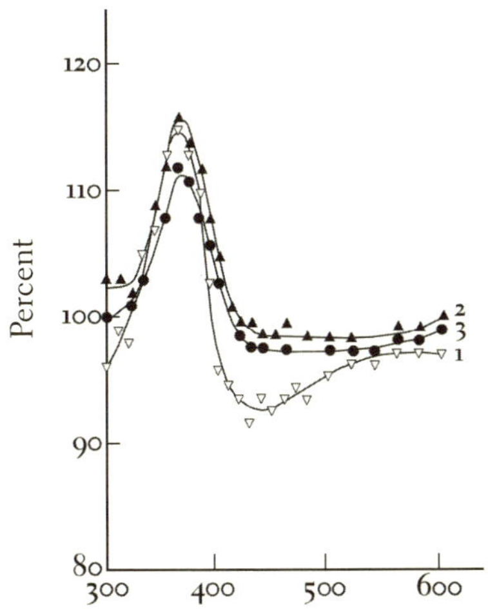

FIGURE 14 (of Ref. 8)

FIGURE 10
Relative reflectivity as a function of wavelength of newsprint showing re-yellowing after irradiation and bleaching. 1. Irradiated at 365 nm for 30 minutes. 2. As 1 but bleached at 436 nm for 60 minutes, measured immediately after bleaching. 3. As 2 but measured 2 weeks later. (from Ref. 8)

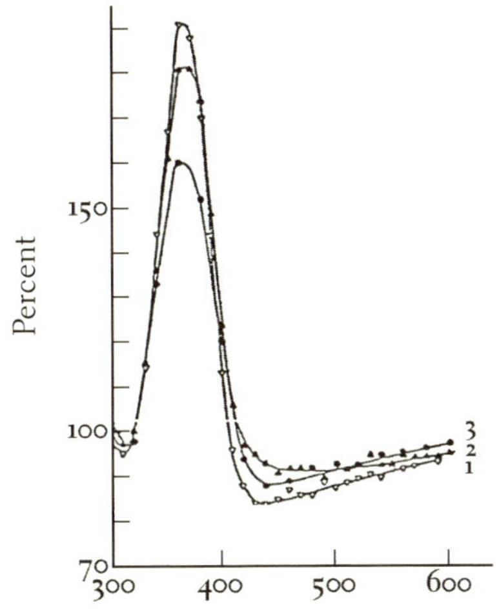

FIGURE 15 (of Ref. 8)

FIGURE 11
Relative reflectivity as a function of wavelength of irradiated wood from spruce. Reference: Unirradiated sample. 1. Irradiated at 365 nm for 8 hours. 2. As 1 but bleached at 436 nm for 2 hours, measured immediately after bleaching. 3. As 2 but measured 3 weeks later. (from Ref. 8)

phenomenon and shows clearly the importance of considering the absorption spectra in the ultraviolet in order to learn what is going on.

As mentioned, the work of Silvy and LeNest concerns the effects on thermally darkened cellulose, not involving lignin. Their Figure 4, curve number 2 (in our Figure 12), shows the result of heating cellulose that had been oxidized by sodium periodate. Such material is known to have a high concentration of aldehyde, and perhaps some ketonic and acidic, groups; the authors wished to show that the absorption peaks, when this chemically oxidized material is discolored by heat (curve 2 in their Figure 4), are much the same as those found in aged papers. Curve number 1 in their Figure 4 is based on the difference in reflectance just after the cellulose had been oxidized by the periodate; the peak at 260 nm, they point out, is likely to be due to carbonyl groups for it is present in this chemically oxidized cellulose even before thermal aging.

RATE OF CHANGE; POST-IRRADIATION EFFECT

Having stressed that the most significant changes in the absorption spectra of pulps must be observed in the ultraviolet region, what else can we say on the basis of what has been published in the literature concerning the bleaching of darkened pulps by visible radiation? We can point out, from the work of

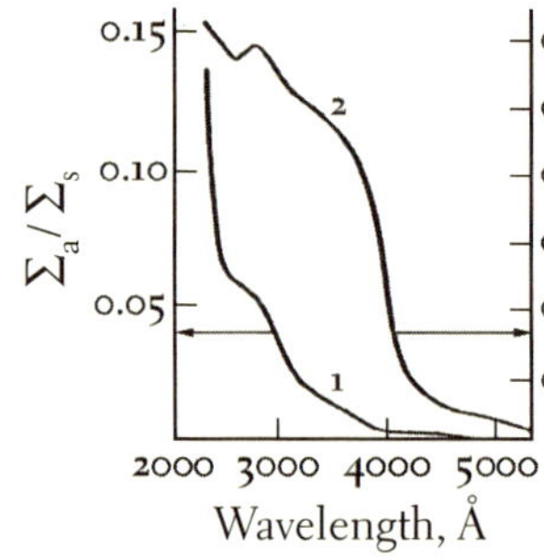

Figure 1 (of Ref. 10)

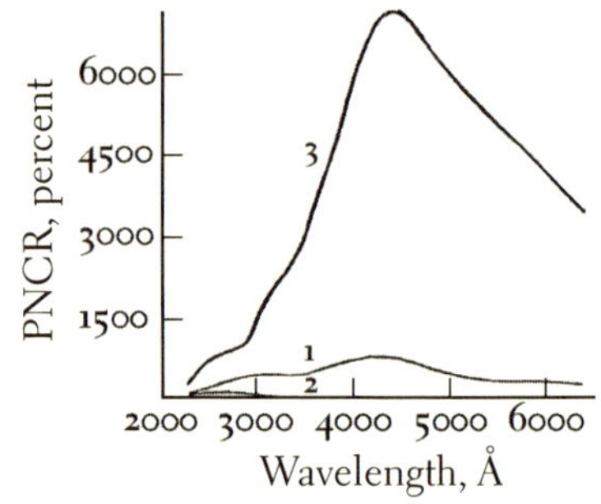

Figure 3 (of Ref. 10)

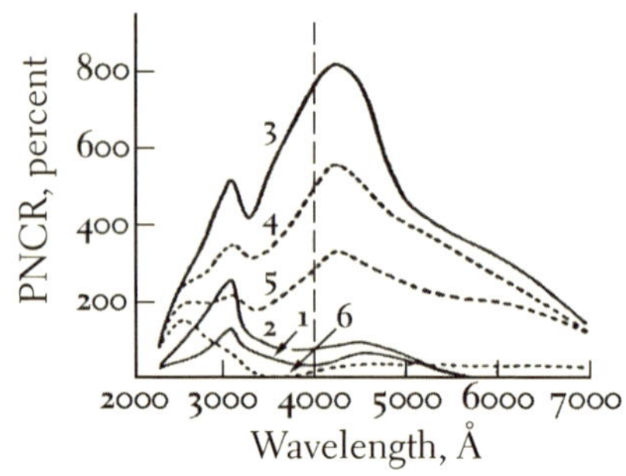

Figure 2 (of Ref. 10)

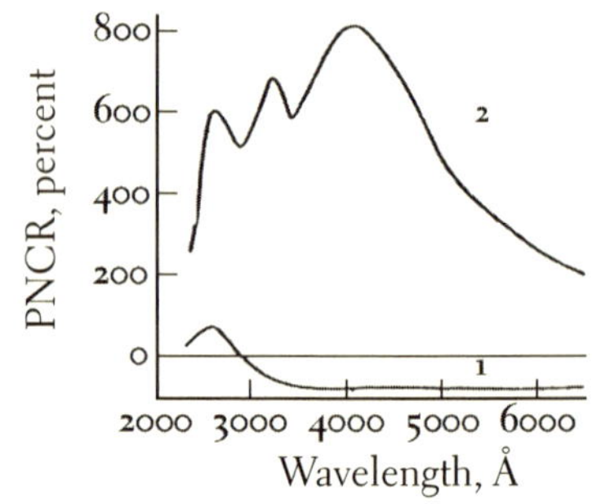

Figure 4 (of Ref. 10)

Figure 12
Authors' Figure 1. Spectra of pure (1) and lignified (2) cellulose. (from Ref. 10)

Authors' Figure 2. Absorption of pure cellulose heated at 120°C (curves 1, 2 and 3) and subsequently bleached in "daylight" (dotted curves 4, 5, 6). (from Ref. 10)

Authors' Figure 3. Pure cellulose thermally darkened (1), bleached in "daylight" at 35°C, (2) and subsequently heated in dark (3). (from Ref. 10)

Authors' Figure 4. Result of periodate oxidation of cellulose (1) and subsequent heating of the oxidized material (2). (from Ref. 10)

Claesson et al.,[8] that the darkening caused by near-ultraviolet takes place very rapidly at first and then tends to slow down, perhaps reaching a constant value (Figure 13; the authors' Figure 4). The same is true for bleaching; very rapid at first and then becoming negligible as the exposure is increased (Figure 14).

We also note that there is likely to be post-irradiation darkening. This is illustrated in Figures 10 through 14 It is a phenomenon that was also pointed out by Silvy and LeNest.[10] Their Figure 3 (our Figure 12), curve 1, shows the absorption curve after heating pure cellulose; curve 2 is the result after bleaching in "daylight" for 70 hours; and curve 3 is the intense absorption after reheating that paper in the absence of light for 25 days. Claesson's results reported in Figures 10, 11, 13 and 14 are particularly convincing. At the present stage of our knowledge regarding the bleaching of papers by visible radiation, we believe that the possibility of a post-irradiation effect—that is, thermally induced

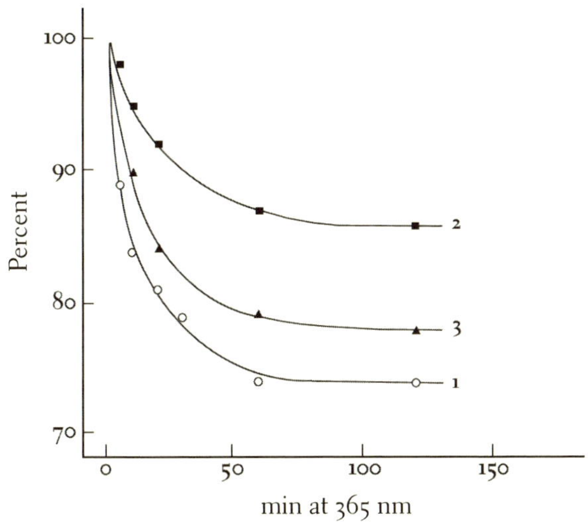

Figure 4 (of Ref. 8)

Figure 13
Relative reflectivity at 470 nm of high yield bisulphite liner as a function of the duration of the irradiation at 365 nm. 1. Irradiated only at 365 nm. 2. As 1 but bleached at 436 nm for 60 minutes, measured immediately after bleaching. 3. As 2 but measured 24 hours later. (From Ref. 8)

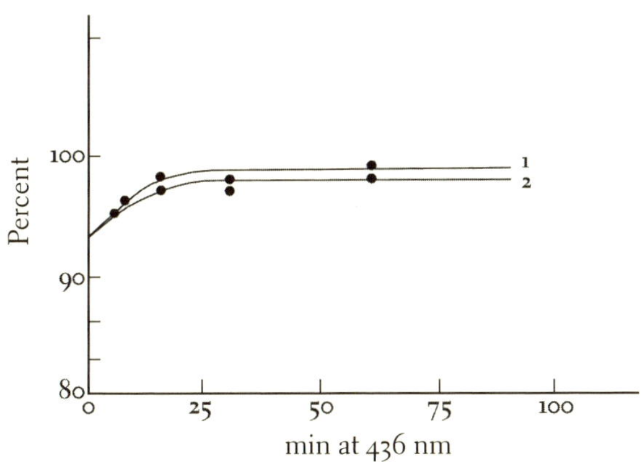

Figure 13 (of Ref. 8)

Figure 14
Relative reflectivity at 430 nm of newsprint, irradiated at 365 nm for 30 minutes, as a function of the duration of the bleaching at 436 nm. 1. Measured immediately after bleaching. 2. Measured 2 weeks later. (From Ref. 8)

darkening that can occur after the papers have been exposed to light—is the aspect of this procedure that must be investigated most carefully in the immediate future.

Summary

What can be said at the end of this brief review of the subject? In general, when moderate temperatures are involved, lignin tends to be darkened by ultraviolet radiation of decreasing wavelength, as shown by the work of Nolan, Van den Akker and Wink in 1945; discolored papers tend to be bleached by visible radiation, perhaps very much as Leary demonstrated for newsprint in 1967 and 1968. If radiation that contains both ultraviolet and visible is involved, both bleaching and darkening processes probably go on at the same time. If either effect predominates, the change appears to be rapid at first, and frequently tends towards a limit in the net change at any given temperature and humidity. The most significant changes in absorption take place in the ultraviolet region; investigation of bleaching or darkening effects by measurement of changes in color and Tappi brightness alone are not likely to lead to major enlightenment. Until much more is learned about the processes, it is recommended that the wavelengths of radiation being employed be specified, the temperature and humidity of the samples noted and consideration given as to whether one is attempting to bleach primarily thermally or photochemically discolored lignin, degradation products of cellulose and hemicellulose, or both. An important aspect of photochemical bleaching of paper for future study is a post-irradiation effect, darkening of the sheets, that may largely be thermally induced.

Bleaching by Light I: Effect of pH on the Bleaching or Darkening of Papers in the Dry and in the Immersed Condition under Visible and Near-Ultraviolet Radiation

S. B. LEE, J. BOGAARD AND R. L. FELLER

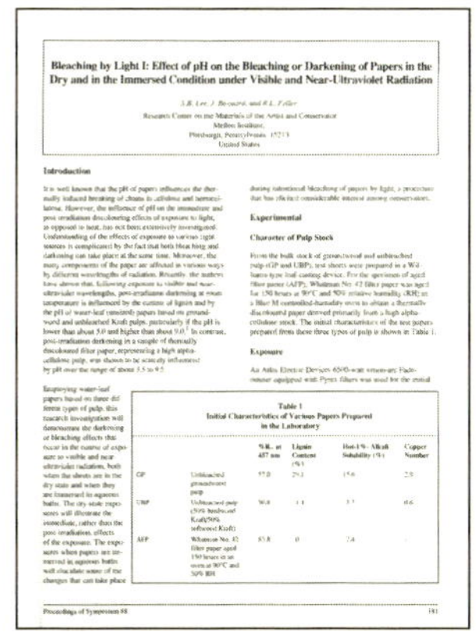
Bleaching by Light I: Effect of pH on the Bleaching or Darkening of Papers in the Dry and in the Immersed Condition under Visible and Near-Ultraviolet Radiation

Table 1
Initial Characteristics of Various Papers Prepared in the Laboratory

Originally printed in:
Conservation of Historic and Artistic Works on Paper: Proceedings of Symposium 88
ed. Burgess, H., Canadian Conservaton Institute, Ottawa, Canada (1994), Pages 181-190.

Introduction

It is well known that the pH of papers influences the thermally induced breaking of chains in cellulose and hemicellulose. However, the influence of pH on the immediate and post-irradiation discoloring effects of exposure to light, as opposed to heat, has not been extensively investigated. Understanding of the effects of exposure to various light sources is complicated by the fact that both bleaching and darkening can take place at the same time. Moreover, the many components of the paper are affected in various ways by different wavelengths of radiation. Recently, the authors have shown that, following exposure to visible and near-ultraviolet wavelengths, post-irradiation darkening at room temperature is influenced by the content of lignin and by the pH of water-leaf (unsized) papers based on groundwood and unbleached Kraft pulps, particularly if the pH is lower than about 5.0 and higher than about 9.0.[1] In contrast, post-irradiation darkening in a sample of thermally discolored filter paper, representing a high alpha-cellulose pulp, was shown to be scarcely influenced by pH over the range of about 3.5 to 9.5.

Employing water-leaf papers based on three different types of pulp, this research investigation will demonstrate the darkening or bleaching effects that occur in the course of exposure to visible and near-ultraviolet radiation, both when the sheets are in the dry state and when they are immersed in aqueous baths. The dry-state exposures will illustrate the immediate, rather than the post-irradiation, effects of the exposure. The exposures when papers are immersed in aqueous baths will elucidate some of the changes that can take place during intentional bleaching of papers by light, a procedure that has elicited considerable interest among conservators.

1. Lee, S. B., Bogaard, J., and Feller, R. L., "Darkening of Paper Following Exposure to Visible and Near-Ultraviolet Radiation," *Journal of the American Institute for Conservation*, 28, no. 1 (1989), p. 1-18.

EXPERIMENTAL

Character of Pulp Stock

From the bulk stock of groundwood and unbleached pulp (GP and UBP), test sheets were prepared in a Williams-type leaf-casting device. For the specimen of aged filter paper (AFP), Whatman No. 42 filter paper was aged for 150 hours at 90°C and 50% relative humidity (RH) in a Blue M controlled-humidity oven to obtain a thermally discolored paper derived primarily from a high alpha-cellulose stock. The initial characteristics of the test papers prepared from these three types of pulp is shown in Table 1.

TABLE 1
Initial Characteristics of Various Papers Prepared in the Laboratory

		$\%R_\infty$ at 457 nm (%)	Lignin Content (%)	Hot-1%-Alkali Solubility (%)	Copper Number
GP	Unbleached Groundwood Pulp	57.0	29.1	15.6	2.9
UBP	Unbleached Pulp 50% hardwood Kraft/ 50% softwood Kraft	30.8	4.4	3.7	0.6
AFP	Whatman No. 42 filter paper aged 150 hours in an oven at 90°C and 50% RH	85.8	0	2.4	–

Exposure

An Atlas Electric Devices 6500-watt xenon-arc Fadeometer equipped with Pyrex filters was used for the initial exposures. The air temperature was approximately 31°C, the relative humidity was 27% and the black-panel temperature was about 65°C.

All exposures to fluorescent lamps took place in a temperature- and humidity-controlled room maintained at 50% RH and 23°C. The 48" (1.22 m) General Electric "daylight" fluorescent lamps and the BLB fluorescent blacklights were arranged on a set of metal-screen shelves so that the samples were about 3 1/2" (8.9 cm) from the lamps. Under these conditions the temperature of the handsheets reached about 28°C to 30°C. The approximate distribution of energy emitted by these lamps is noted in Table 2.

TABLE 2
Irradiance from Different Light Sources (10^{-3} W/cm^2)
as Measured by the IL700 Radiometer

Light source	Ultraviolet energy (310-392 nm)		Visible energy (400-750 nm)	Total energy (257-750 nm)	Measurement distance from source (in.)
GE Fl[a] Blacklight F40 BLB	4.13	(86%)	0.70	4.90	3.5
"Daylight" High-Output Fl[a]	0.228	(3.8%)	5.70	5.928	3.5
Xenon-Arc Fadeometer[b] Pyrex/Pyrex Filters	13.775[c]	(6.3%)	203.49[d]	217.265[e]	11.6

a. Fl stands for "fluorescent".
b. Measurements obtained from Atlas Electric Devices, Inc. catalog multiplied by 1.9 (factor for inner ring specially installed).
c. Irradiance from 340-400 nm.
d. Irradiance from 400-750 nm.
e. Irradiance from 340-750 nm.

When exposures were made in Pyrex glass Petri dishes on these same shelves, the intensity of radiation was measured by placing the glass cover over the detector head of an IL700 research photometer. A similar measurement was made when the Plexiglas UF-3 filter or the Corning 3-71 glass filter was interposed over the covered dishes. The UF-3 filter transmits only a few percent at wavelengths below 410 nm; the 3-71 behaves similarly at wavelengths below 460 nm. The BLB-type fluorescent blacklights emit near-ultraviolet radiation only, in the region from approximately 320 nm to 392 nm, with a maximum intensity at approximately 365 nm.

Determination of Reflectance

In the initial experiments with the papers exposed in the Fadeometer, a Tappi Brightness meter was used in which the band-pass filter has a peak transmission at 457 nm (Tappi Standard T305 su 68). Thereafter, the reflectance of the test sheets was measured on a Kollmorgen ColorEye 2000 abridged spectrophotometer. A barium sulphate white standard was used as reference for these reflectance measurements. The Kubelka-Munk K/S values, where $K/S = (1 - R)^2/2R$, were determined from the spectrophotometrically measured reflectance value at 460 nm.[2] The results of both measurements were practically identical.

2. Feller, R. L., "Comments on the Measurement of Yellowness in Pulp and Paper," *Book and Paper Annual*, 6, ed. Espinosa, R., American Institute for Conservation, Washington, D.C. (1987), p. 40-51.

Preparation of Buffered Test Sheets

The buffers used are listed in Table 3. Test sheets at various values of pH were prepared by dipping the papers in the buffered solutions and then airdrying the sheets overnight. The resultant initially determined pH of the papers was not precisely that of the buffer but was the value shown in the various figures, values determined by cold-water extraction of the impregnated sheets by the method of Tappi Standard T509 su 68. The values of pH thus achieved can be seen in the results plotted in Figures 1 through 8 for the exposure in the "dry" condition, that is, exposures of the test sheets without immersion.

TABLE 3
Common Buffer Solutions Used for Paper Treatment

pH 2.8	Potassium hydrogen phthalate + HCl
pH 3.5	Potassium hydrogen phthalate + HCl
pH 4.01	Potassium hydrogen phthalate buffer
pH 5.3	Potassium hydrogen phthalate + NaOH
pH 6.86	Phosphate buffer
pH 9.18	Borax Buffer
pH 11	Sodium bicarbonate + NaOH
pH 11.7	0.1 % $Ca(OH)_2$ solution

Exposure During Immersion

Exposure of the paper while immersed in variously buffered solutions was carried out by placing 2" x 2" (5 cm x 5 cm) specimens of the text sheets in 4"-diameter (10 cm) Pyrex glass Petri dishes and covering the paper to a depth of about 0.3" (8 mm) with distilled water or with the buffered aqueous solutions. The immersion bath was not changed throughout the exposure period.

To measure the change in color of the immersed papers which were either exposed to light or extracted in the dark, the sheets were removed from the immersion bath, were washed with distilled water until acid- or alkali-free and then were air-dried in the laboratory before measurement.

Thermal Aging

In the experiments in which the papers were thermally aged following exposure, thermal aging was carried out in a 90°C air-circulating oven without control of the relative humidity.

RESULTS AND DISCUSSION

Exposure of Papers Without Immersion

An initial series of exposures of three types of paper was carried out in a xenon-arc Fadeometer equipped with Pyrex filters. The composition of the pulps—groundwood (GP), unbleached Kraft pulp (UBP) and aged filter paper (AFP)—is reported in Table 1. The results are shown in Figure 1; the changes in reflectance (R) at 457 nm are expressed in terms of the Kubelka-Munk K/S function where $K/S = (1 - R)^2/2R$. Note that, regardless of pH, the groundwood pulp darkened and the unbleached pulp sheets bleached. The aged filter paper, although scarcely affected, also bleached.

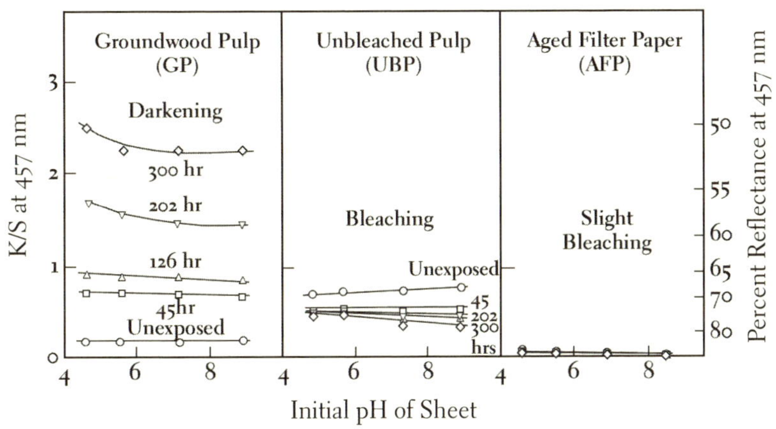

FIGURE 1
Results of exposure in the xenon-arc Fadeometer of test sheets made with three types of pulp vs. pH.

Because exposure in the Fadeometer involves wavelengths both in the ultraviolet and in the visible and also rather high temperatures (the black-panel temperature was about 65°C), further exposure tests were carried out under more moderate conditions using near-ultraviolet-emitting BLB fluorescent blacklights and "daylight" fluorescent lamps (see Table 2). The latter emit primarily visible (96.2%) and a small amount of near-ultraviolet (3.8%) radiation. Exposures took place in a room maintained at 50% RH and 23°C. The test sheets reached a temperature of about 28°C to 30°C under these conditions. As the changes in the aged filter paper sheets were small, it was decided to concentrate on the effects in groundwood and in the unbleached pulp.

The results are summarized in Figures 2 to 8. Again, the changes in reflectance are primarily referred to in terms of the Kubelka-Munk function K/S (where $K/S = (1 - R)^2/2R$) instead of in terms of the change in reflectance. Moreover, instead of a measurement with a narrow-band-pass filter peaked at

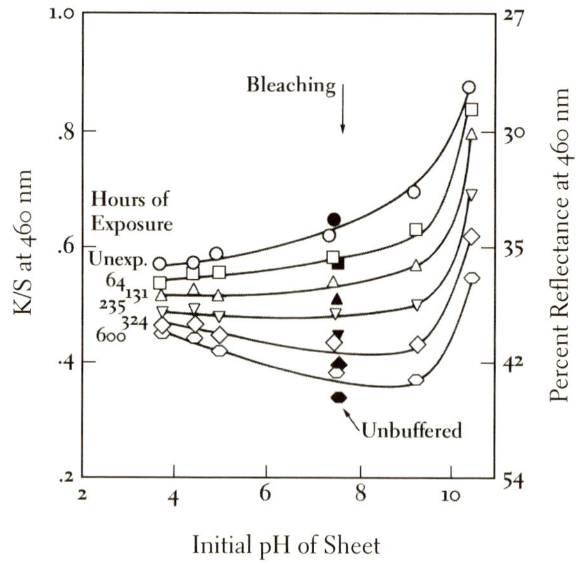

FIGURE 2
Results of exposure to BLB fluorescent blacklights of unbleached pulp (UBP) test sheets having various pH values.

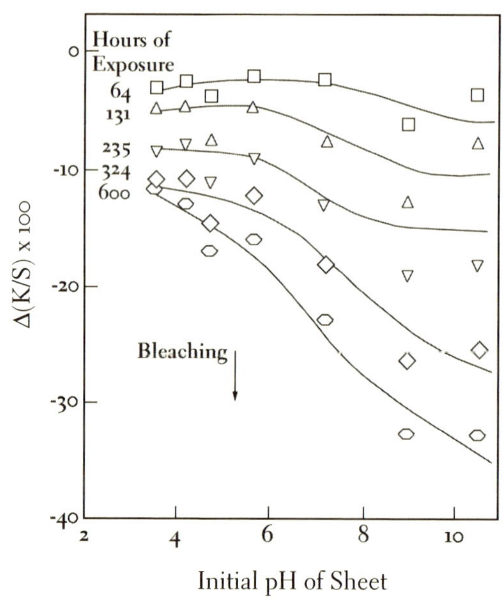

FIGURE 3
Post-color number [100 Δ(K/S) at 460 nm] of test sheets of unbleached pulp (UBP) exposed to BLB fluorescent blacklights at various initial values of pH.

457 nm, as in the usual measurement of Tappi Brightness (Tappi Standard T305 su 68) used in the initial experiments, the reflectance at 460 nm was used instead, taken from spectrophotometric data.[2] The results of both measurements are practically identical. For convenient reference, the percent reflectance at 460 nm is indicated in many of the figures.

As seen in Figure 2, bleaching occurred when the sheets based on the unbleached pulp (UBP) were exposed to the near-ultraviolet radiation emitted by the BLB fluorescent lamps; there is a decided influence of pH. One can also see that, in the initial (unexposed) condition, the darkness of the sheets increased when they were alkaline. This effect has been noted by others; for example, in the data published by Irancic and Rydholm[3] and by Iiyama and co-workers.[4]

The net result of exposure of the sheets can best be seen when, rather than the reflectance or K/S value, the change in K/S [Δ(K/S)] is plotted, as in Figure 3. Then it can be seen that the greatest photochemically induced bleaching of this pulp occurred in the alkaline condition.

The bleaching effects brought about in the UBP sheets by exposure to "daylight" fluorescent lamps were so similar that only the data in terms of the change in K/S are given (Figure 4). The similarity in the results of exposure to both the "daylight" and the BLB (near-ultraviolet-emitting) lamps is noteworthy.

The groundwood pulp containing high levels of lignin was not bleached by near-ultraviolet radiation; it was darkened instead (Figure 5). Again, in the highly alkaline condition, an intensification of the initial color took place.

3. Irancic, A., and Rydholm, S. A., "Technical Color Reactions of Lignin," *Svensk Papperstidning*, 62 (1959), p. 554-566.

4. Iiyama, K., Nakano, J., and Migita, N., *Japan Tappi*, 21, No. 12 (1967), p. 755-759.

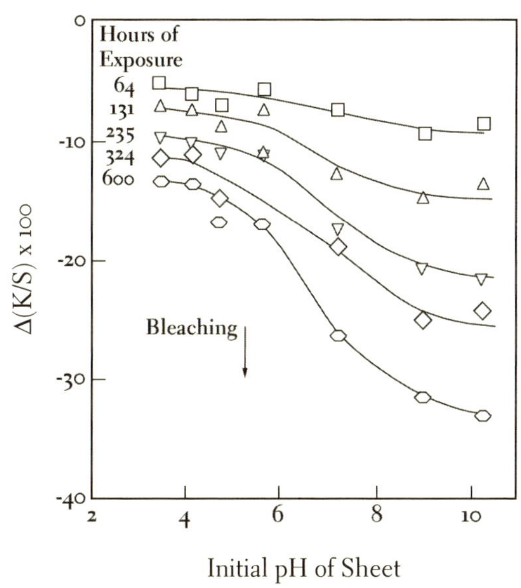

FIGURE 4
Post-color number of test sheets of unbleached pulp (UBP) exposed to "daylight" fluorescent lamps at various initial values of pH.

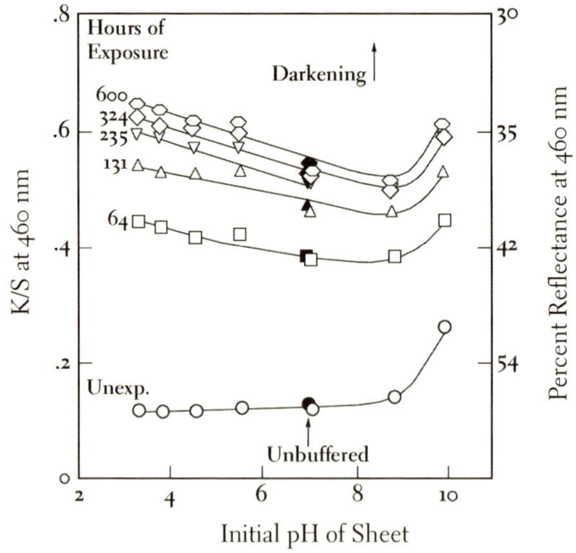

FIGURE 5
Results of exposure to BLB fluorescent blacklights of groundwood pulp (GP) test sheets at various initial values of pH.

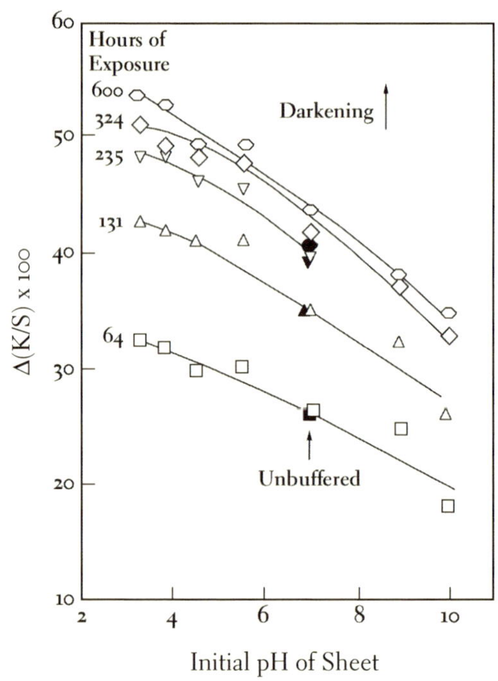

FIGURE 6
Post-color number of test sheets of groundwood pulp (GP) exposed to BLB fluorescent blacklights at various initial values of pH.

However, when the experimental results are again expressed in terms of the change in K/S (Figure 6), an orderly effect of pH is seen. The greatest darkening occurred in the acid papers; the least, in the alkaline papers.

"Daylight" fluorescent lamps emit a low percentage of ultraviolet radiation. Hence, under these lamps, darkening rather than bleaching also occurred in the groundwood pulp (Figure 7). The effect of pH was again much the same as under the BLB lamps, but the net darkening [Δ(K/S)] that took place was less than when ultraviolet lamps alone were used (Figure 8 and Figure 6).

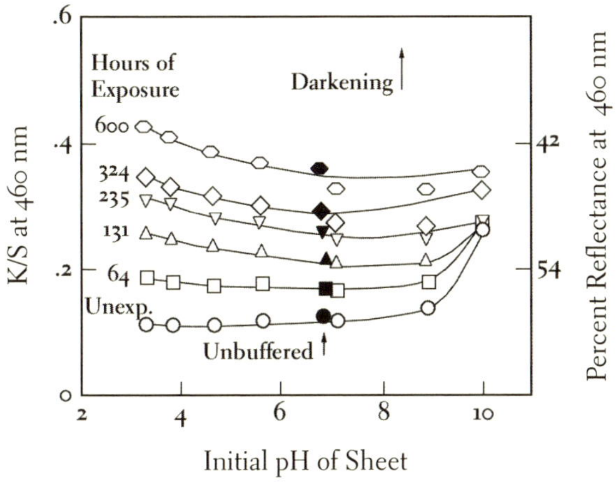

FIGURE 7
Results of exposure to "daylight" fluorescent lamps of groundwood pulp (GP) test sheets at various initial values of pH.

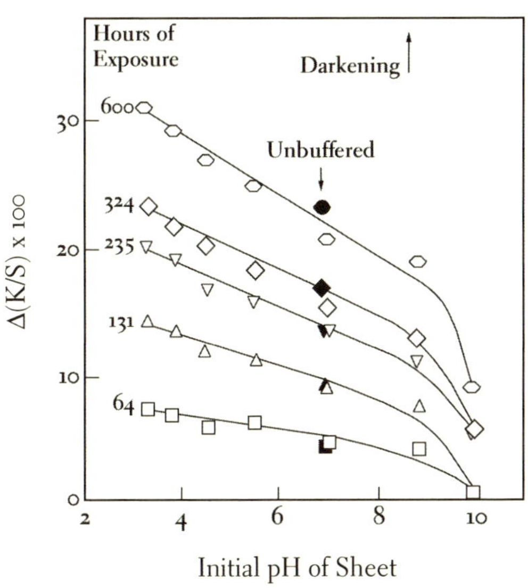

FIGURE 8
Post-color number of test sheets of groundwood pulp (GP) exposed to "daylight" fluorescent lamps at various initial values of pH.

If both darkening and bleaching processes go on at the same time and if, as will be discussed further, alkalinity can speed up the bleaching of certain decomposition products, then the greater tendency to bleach under alkaline conditions can explain the lesser net amount of darkening that took place in the alkaline, lignin-rich sheets (Figures 6 and 8).

Influence of Wavelength

Figures 9, 10 and 11 present the results of exposures of thermally aged filter paper (AFP), unbleached pulp (UBP) and groundwood pulp (GP) in both the dry and immersed conditions using various wavelength regions of irradiance. Sharp cut-off filters were interposed between the specimens and the "daylight" fluorescent lamps. As well, sheets were exposed to the near-ultraviolet radiation alone from BLB lamps. Figure 9 shows that the AFP, in which any thermally discolored material must be derived almost entirely from cellulose and possibly from hemicellulose, was bleached by both visible and near-ultraviolet radiation. When immersed in water, the color decreased somewhat more rapidly with time of exposure. It is our general experience that the discolored material derived from cellulose and hemicellulose tends to be bleached by both visible and

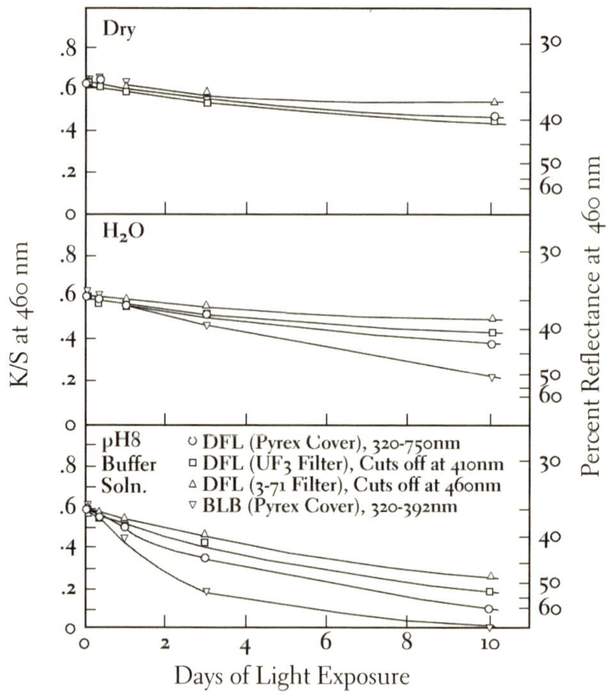

Figure 9
Change in K/S value at 460 nm of aged filter paper (AFP) exposed to visible and ultraviolet radiation directly (dry) and when immersed in water or in an aqueous solution buffered at pH 8. (Note occurrence of bleaching.)

near-ultraviolet radiation. However, the bleaching effectiveness falls off at the longer wavelengths.

The identity is not certain of the principal colored matter in the unbleached pulp (UBP). However, these sheets also bleached similarly as a function of wavelength, although the rate of bleaching apparently was faster in the case of the UBP (Figure 10). The AFP sheets were initially very much darker than the UBP; nonetheless, when the net extent of bleaching is considered, expressed in terms of the change in K/S [Δ(K/S)], nearly as much bleaching occurred in each case when the sheets were immersed in water. When the immersion bath was buffered at pH 8, the extent and rate of bleaching increased in both the AFP and the UBP sheets.

The results with the groundwood pulp (GP) contrast (Figure 11). The near-ultraviolet radiation of the BLB lamps darkened the test sheets, even when they were immersed in water. Slight darkening also took place under "daylight" fluorescent lamps. The slight darkening that occurred can be attributed to the low percentage of ultraviolet radiation present in this source; when the UF-3 filter is interposed, little or no bleaching or darkening occurred. Even when visible radiation above 460 nm was employed, useful bleaching of these high-lignin sheets was not achieved.

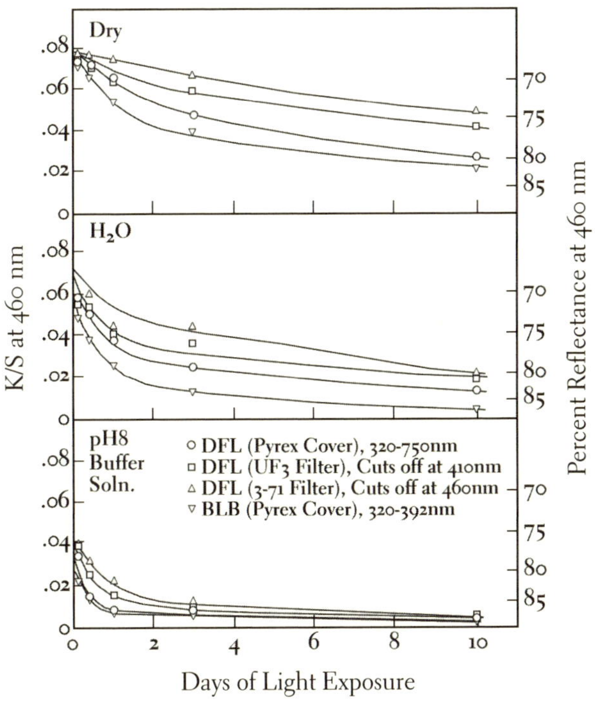

FIGURE 10
Change in K/S value at 460 nm of unbleached pulp (UBP) test sheets exposed to visible and ultraviolet radiation directly (dry) and when immersed in water or in an aqueous solution buffered at pH 8. (Note occurrence of bleaching.)

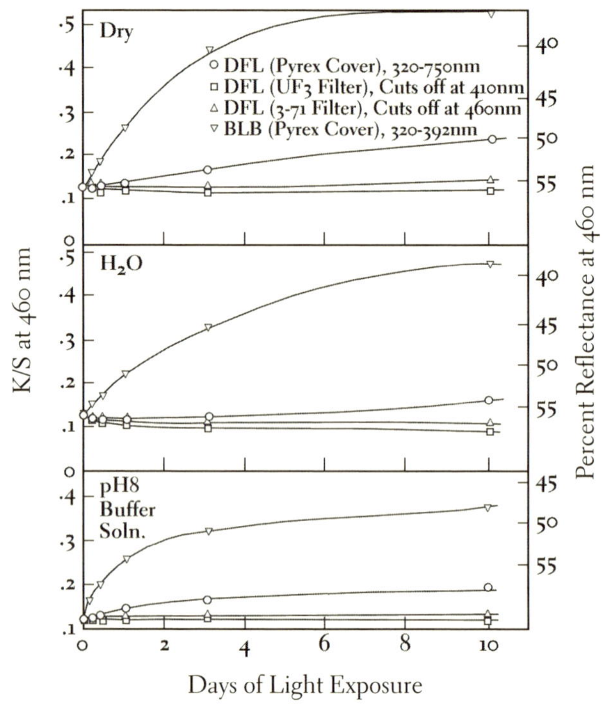

FIGURE 11
Change in K/S value at 460 nm of groundwood pulp (GP) test sheets exposed to visible and ultraviolet radiation directly (dry) and when immersed in water or in an aqueous solution buffered at pH 8. (Note tendency to darken.)

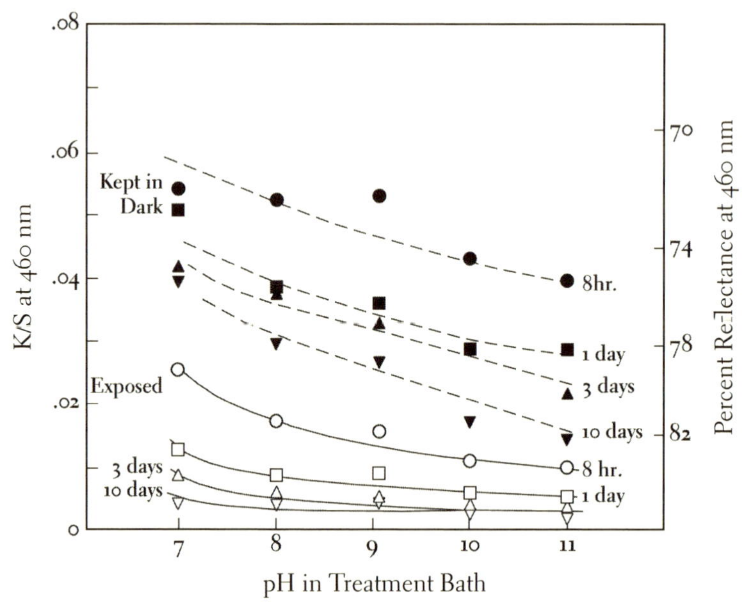

FIGURE 12
Results of exposure of aged filter paper (AFP) when exposed to "daylight" fluorescent lamps while immersed in solutions buffered at various values of pH in comparison to the changes resulting from immersion in the same buffered solutions without exposure to light.

Bleaching of Papers when Immersed in Aqueous Solutions at Various pH Values

As seen in Figures 9, 10 and 11, when bleaching occurred, alkalinity seemed to speed up the process. This is reminiscent of the results observed in the direct exposure of papers at various pH levels.[1] To gain further information on this behavior, papers were immersed in aqueous solutions buffered at various values of pH and were subjected to bleaching with "daylight" fluorescent lamps. It was thought that alkalinity might speed up the bleaching process, both for the apparently inherent chemical processes as seen in the case of exposures in the dry conditions (Figures 4 and 8) and also because some of the colored material may be extracted more effectively in an alkaline bath.

Figure 12 shows the extent of bleaching of aged filter paper (AFP) when immersed in aqueous baths buffered at various levels of pH from 7 to 11. Little effect of pH was seen in the bleaching by light, particularly above pH 8. When the immersed samples were kept in the dark, it can be seen that alkalinity apparently extracted some of the colored material. As noted earlier, the colored matter in this instance must be largely decomposition products derived from cellulose and hemicellulose. As in the earlier experiments summarized in Figures 9, 10 and 11, the K/S values reported are based on the reflectance of the sheets after being removed from the immersion baths, washed free of buffer and air dried.

Similar results using the UBP sheets are shown in Figure 13. Again, alkalinity above pH 8 provided little advantage. The bleaching of colored material in the dark did not at first appear to be extensive. However, the difference in the K/S values at 460 nm between extraction in a bath at pH 7 and at pH 8 or 11 for the papers that had been immersed but kept in the dark was about 0.04. The related change in K/S in the case of the AFP sheets in Figure 12 amounted to 0.023. Thus the amount of color leached out of the much darker UBP, when expressed not as a change in reflectance but as a change in K/S, appeared not much different than the degree of bleaching in the dark that occurred in the AFP sheets.

The evidence seems to be that bleaching of sheets under water is a combination of photochemical and leaching effects. As for employing alkalinity in the bath to facilitate bleaching, there seems little advantage to go above about pH 8. Further tests may show that a lower pH, perhaps 7.6 to 7.8, would be almost as effective.

Consideration of Potential Damage during Light Bleaching of Papers when Immersed

One way to test for potential damage to paper during light bleaching would be to age the sheets thermally after the bleaching procedure to determine if

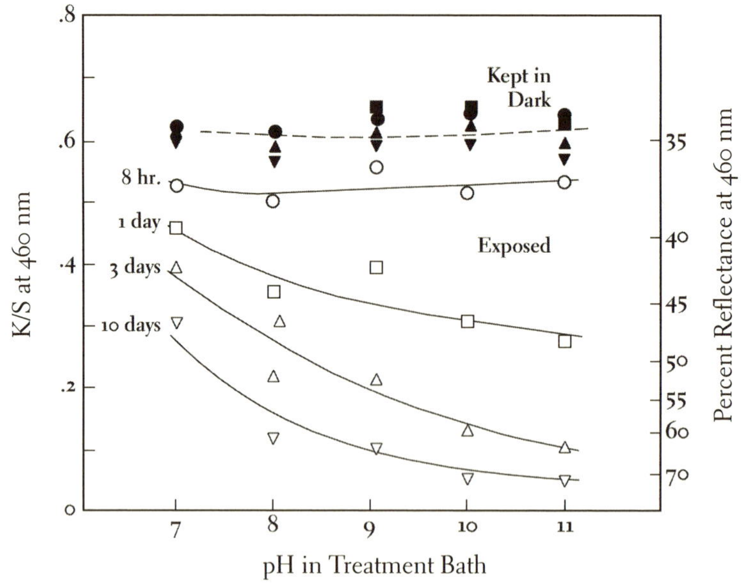

Figure 13
Results of exposure of unbleached pulp (UBP) test sheets when exposed to "daylight" fluorescent lamps while immersed in solutions buffered at various values of pH in comparison to the changes resulting from immersion in the same buffered solutions without exposure to light.

exposure under certain conditions would lead to a markedly different degree of thermally induced discoloration. Figure 14 gives the results for aged filter paper (AFP) exposed without immersion and with immersion in distilled water and in buffered solutions. The sheets were removed after 10 days of exposure, a degree of exposure at least 10 times longer than a conservator would normally employ in practical bleaching. Following 10 days of exposure, the immersed sheets were removed; those immersed in buffers were washed, all sheets were air-dried and then were subjected to thermal aging at 90°C. Under these circumstances, noticeably greater subsequent thermal darkening occurred in the sheets that were exposed without benefit of immersion (dry state). This implies that the AFP sheets were harmed, probably oxidized, by the direct exposure to the "daylight" fluorescent lamplight. In contrast, the extent of thermal darkening induced in the sheets that had been immersed was practically the same regardless of the neutrality or alkalinity of the immersion bath.

As further evaluation of the action of the alkaline bath and any possibly harmful effects, a number of test sheets were not exposed, but were simply immersed in the aqueous solutions for the same number of days. They then were washed, dried and thermally aged at 90°C (Figure 15). Virtually no difference in the degree of thermally induced darkening was indicated regardless of the pH of the immersion bath. The experiment further confirmed that a certain fraction of colored matter was removed from the thermally discolored

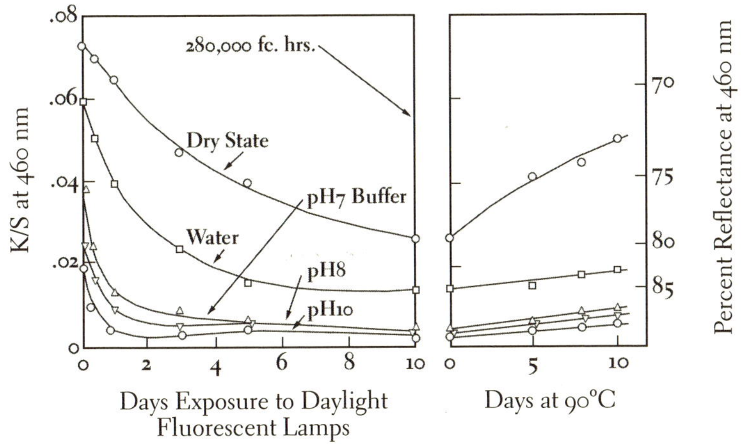

Figure 14
Change in K/S values and percent reflectance at 460 nm of thermally aged filter paper (AFP) exposed to "daylight" fluorescent lamps for 10 days followed by thermal aging of the exposed papers that had, when necessary, been washed free of alkalinity.

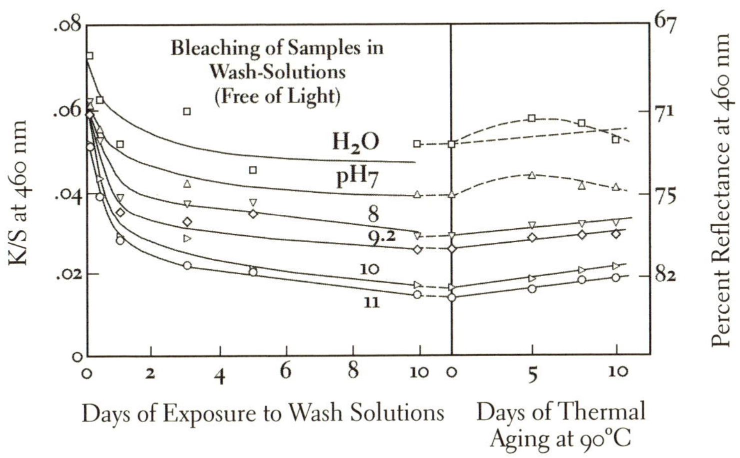

Figure 15
Change in K/S values and percent reflectance at 460 nm of test sheets of AFP after immersion in water or in variously buffered solutions followed by thermal aging of the papers thus immersed. Before thermal aging at 90°C, the papers were washed until they were free of alkalinity.

filter paper (AFP); more was removed in the more alkaline baths. The increase in brightness of the papers that was achieved by simply leaching out the soluble material through immersion in the buffered solutions (Figure 15) was not as great as when the immersed sheets were exposed to light (Figure 14). The results of heating the sheets that had been immersed in water or in a buffer solution at pH 7 are not understood, but may be due to experimental error, for the net change in reflectance was only about 1%. Nonetheless, the increased discoloration after 10 days at 90°C appeared little different from that of the sheets that had been in highly alkaline solutions.

Much the same results were observed when test sheets of the unbleached pulp (UBP) were similarly exposed and were subsequently thermally aged (Figure 16). Again, the consequence of exposure of the sheets without immersion (the "dry" experimental condition) produced greater thermally induced darkening in the exposed sheets. The aged filter paper samples appeared to bleach more completely than this dark pulp. The slower rate of bleaching of the UBP sheets relative to the AFP sheets in several of these experiments may have been due to the fact that some of the coloring matter in the UBP was not as readily removed from the fibers or as oxidatively attacked within the fibers. Moreover, the UBP contained 4% lignin, which could have resulted in a slight darkening process that counteracted some bleaching.

Although experiments reported in Figures 14, 15 and 16 provide useful indications of lack of serious damage owing to exposure when papers are immersed in variously buffered baths, it was decided to measure the effects of exposure directly on the degree of polymerization (DP) of paper treated under similar conditions. Table 4 presents data on the change in DP of Whatman No. 42 filter paper (unaged in this case) when exposed without immersion and when immersed in distilled water or in a solution buffered at pH 8. Apparently, there are considerable experimental errors; the DP of the unexposed paper appears higher than the other measurements. In spite of some unexplained numbers, two conclusions appear justified: (1) significant chain breaking clearly occurs when the sheets are not immersed, and (2) no significant difference can be found between sheets immersed in distilled water and sheets immersed in a buffered solution at pH 8.

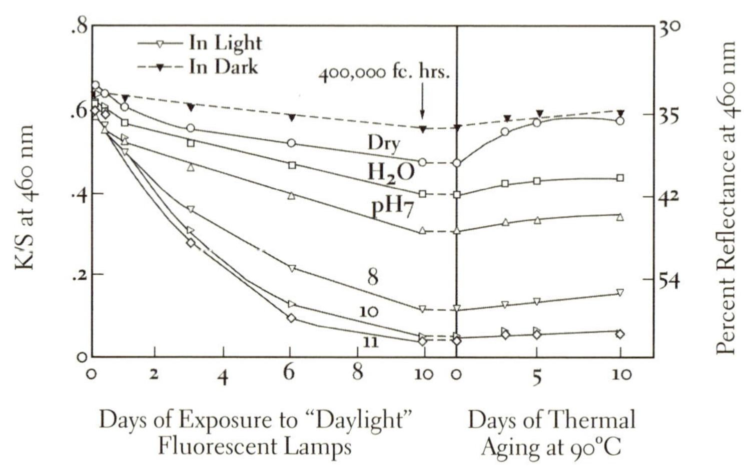

Figure 16
Change in K/S values and percent reflectance at 460 nm of test sheets of unbleached pulp (UBP) exposed to "daylight" fluorescent lamps for 10 days followed by thermal aging of the exposed papers that had, when necessary, been washed free of alkalinity.

Table 4
Chain Scission of Whatman No. 42 Filter Paper during Exposure to Daylight Fluorescent Lamps in Dry State, in Water and in pH 8 Buffer*

		Unexposed	8 hrs	1 Day	3 Days	10 Days
Dry State	DP	1,592	1,465	1,562	1,410	1,313
	2/DP	0.1256	0.1365	0.1280	0.1418	0.1523
	Δ2/DP		0.0109	0.0024	0.0162	0.0267
H_2O	DP**		1,455	1,459	1,524	1,482
	2/DP		0.1375	0.1371	0.1312	0.1350
	Δ2/DP		0.0119	0.0115	0.0056	0.0094
pH 8 Buffer	DP**		1,573	1,467	1,482	1,482
	2/DP		0.1271	0.1363	0.1350	0.1350
	Δ2/DP		0.0015	0.0107	0.0094	0.0094

*Petri dish containing 40 ml of liquid.
**Mean of all 8 values for the immersed exposed papers.
DP = 1491 ± 40 (±2.6%); 2/DP = 0.1343 ± 0.0035 (±2.6%).

Table 5
Light-induced Damage of Whatman No. 42 Filter upon Exposure to Daylight Fluorescent Lamps for 6 Days in Buffer Solutions with Various pH Levels*

Type of Treatment	HAS** Matter (%)	DP
Unexposed	1.05	1,410
Exposed, untreated and dry	1.05	1,341
Exposed, buffer solution of pH 4	1.04	1,410
Exposed, buffer solution of pH 5.3	1.03	1,440
Exposed, buffer solution of pH 6.9	0.97	1,482
Exposed, buffer solution of pH 8	1.04	1,455
Exposed, buffer solution of pH 9	1.03	1,414
Exposed, buffer solution of pH 10	1.04	1,429
Exposed, under water	1.04	1,429

*Petri dish containing 40 ml of liquid.
**HAS refers to "Hot-Alkali Soluble".

In this first set of tests, the apparent decrease in DP of the immersed papers compared to that of the unexposed filter paper may have been an artifact and perhaps not significant. As a further check, a second set of tests were carried out, with the results shown in Table 5 and in Figure 17. Again, when exposed without immersion, a slight but measurable loss in DP occurred but, as Figure 17 indicates, no significant change in DP occurred in the immersed sheets.[5]

5. A similar conclusion has been tentatively reached by Dr. L. Santucci using a number of exposed papers that were selected by Keiko Keyes and that were tested at her request.

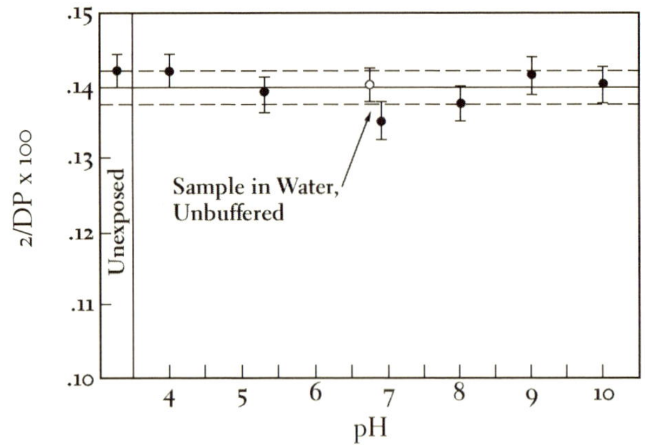

FIGURE 17
Values of 2/DP for AFP immersed in baths at various pH levels after exposure to "daylight" fluorescent lamps for 6 days compared to two control samples. Plot of data from Table 5.

TABLE 6
Viscosity-Average Degrees of Polymerization of Whatman No. 42 Filter Paper Exposed to "Daylight" Fluorescent Lamps or Maintained in the Dark for 3 Days

	Exposed	In Dark
(Two samples), pH 7 and 8	1487 ± 9	1450 ± 19
(Two samples), pH 10 and 11	1453 ± 34	1463 ± 21
(Two controls), original paper: unexposed, not immersed		1467 ± 1
(Avg. of five samples, pH 7, 8, 9, 10 and 11)	1453 ± 33	1455 ± 16

No increase in hot-1%-alkali solubility (HAS Matter) occurred; an increase would have been expected if significant oxidation of the cellulose had occurred as the result of exposure.

A third experiment was carried out with samples of Whatman No. 42 paper immersed in aqueous solutions buffered at four values of pH, exposed to the lamps or kept in the dark for three days. The results, summarized in Table 6, again provide no evidence of significant damage owing to immersion.

Conclusions

It is known that acidity influences the thermal deterioration of paper largely through the enhanced rate of scission of the cellulose chains. One of the initial objectives of this series of experiments was to determine the influence of pH on the immediate effects of photochemically induced darkening or bleaching of paper. Although the range of acidity or alkalinity normally encountered in ordinary papers is limited, a wide range of pH values was employed in these experiments so that confidence could be placed on the direction and type of changes induced. The experimental results show that there is a definite influence of pH on the immediate effects of photochemical exposure, whether bleaching or darkening. The effects are minimal in the case of thermally discolored filter paper, a stock that has negligible content of lignin and hemicellulose.

Of the three types of pulps used, darkening occurred primarily in the groundwood (GP) sheets that contained high levels of lignin, largely induced by the ultraviolet radiation; visible wavelengths had little effect. In sheets of aged filter paper (AFP), in which the discolored material must be largely derived from the decomposition of cellulose and hemicellulose, bleaching occurred, whether near-ultraviolet radiation or visible wavelengths as high as 460 nm were employed. The particular sample of unbleached Kraft pulp (UBP) used in these experiments also bleached under both visible and ultraviolet radiation.

When the test sheets were immersed in water or in aqueous solutions buffered at various pH values, particularly on the alkaline side, a greater speed and degree of bleaching took place in the AFP and UBP samples than when the papers were simply exposed directly to the light without immersion. Good evidence was obtained that the net bleaching effect is the result of a certain degree of leaching out of colored matter as well as of the bleaching effect of light.[6] Evidence also was gained that a greater rate of bleaching took place under alkaline conditions in all three types of pulps, although the effect was relatively minor in the case of the aged filter paper (AFP).

In tests for potential damage to the paper owing to exposure, clear evidence was obtained that damage could occur when the papers were directly exposed to daylight fluorescent lamps for the length of time employed in the experimental tests. However, when the papers were immersed, whether simply in distilled water or in alkaline baths, little if any photochemical damage resulted.[7] The evidence for this was obtained both by measurements of the thermal discoloration of the exposed papers and by measurement of changes in the degree of polymerization. The exposures used in these tests were many times longer than would normally be used in conservation. The intention has been to demonstrate that, even under rather severe conditions of exposure, little damage could be detected.

The work described herein was not designed to make immediate

6. This conclusion was demonstrated in the work of Guy Savard in a report submitted to the Department of Art History, Queen's University, to conform with the requirements of Master of Art Conservation, "An Investigation of Whitening and Colour Reversion Effects on Light-Bleached Artificially Aged Paper," Aug. 18, 1986. The authors are grateful to Professor James Hanlan for sharing this report with us and for calling our attention to related research carried out earlier at Queen's by Michèle Lepage and Johanne Perron. The authors also gratefully acknowledge helpful discussions with Keiko Keyes and with Professor Catherine Baker at various stages of this research.

7. Little evidence of physical changes owing to exposure was also found in the studies by van der Reyden, D., Mecklenburg, M., Baker, M., and Hawill, M., "Update on Current Research into Aqueous Light Bleaching at the Conservation Analytical Laboratory," *The Book and Paper Group Annual*, 7 (1988), p. 73-106.

recommendations for treatment procedures. Much of the information gained was essentially already known to the conservator on the basis of practical experience. Nonetheless, factual data have been scarce. Primarily, these studies have confirmed that (1) little damage to the cellulosic components of paper is likely to occur during the relatively modest periods of exposure normally used in the intentional bleaching of paper, (2) that the bleaching effect is the net result of both leaching and bleaching actions, which are apparently most effective on the degradation products derived from cellulose and hemicellulose, (3) that ultraviolet radiation will induce darkening rather than bleaching in a paper containing high levels of lignin and (4) that both darkening and bleaching frequently go on during photochemical exposure.

It is hoped that this information will contribute to a better understanding of the intentional bleaching of paper by light and to improved establishment of practical procedures. In the future, additional studies should be made concerning the possibility that a slight degree of alkalinity in the immersion bath would be tolerable and advantageous.

Bleaching by Light II: Studies of the Bleaching of Thermally Discolored Sugars and Other "Model" Compounds

S. B. Lee, J. Bogaard and R. L. Feller

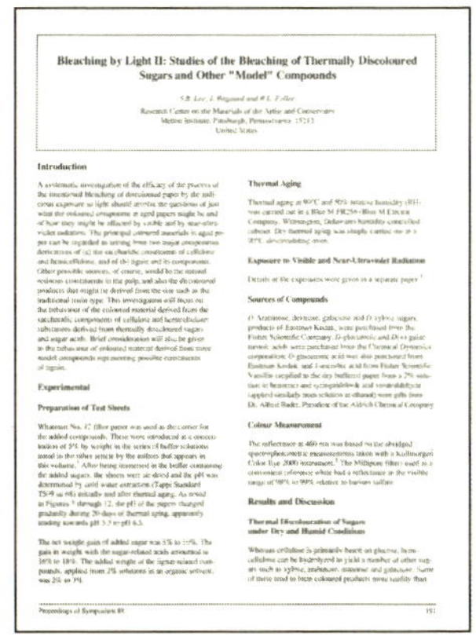
Bleaching by Light II: Studies of the Bleaching of Thermally Discoloured Sugars and Other "Model" Compounds

Originally printed in:
Conservation of Historic and Artistic Works on Paper: Proceedings of Symposium 88
ed. Burgess, H., Canadian Conservaton Institute, Ottawa, Canada (1994), Pages 191-198.

Introduction

A systematic investigation of the efficacy of the process of the intentional bleaching of discolored paper by the judicious exposure to light should involve the questions of just what the colored component in aged papers might be and of how they might be affected by visible and by near-ultraviolet radiation. The principal colored materials in aged paper can be regarded as arising from two major components: derivatives of (a) the saccharidic constituents of cellulose and hemicellulose, and of (b) lignin and its components. Other possible sources, of course, would be the natural resinous constituents in the pulp, and also the discolored products that might be derived from the size such as the traditional rosin type. This investigation will focus on the behavior of the colored materials derived from the saccharidic components of cellulose and hemicellulose: substances derived from thermally discolored sugars and sugar acids. Brief consideration will also be given to the behavior of colored material derived from three model compounds representing possible constituents of lignin.

Experimental

Preparation of Test Sheets

Whatman No. 42 filter paper was used as the carrier for the added compounds. These were introduced at a concentration of 5% by weight in the series of buffer solutions noted in the other article by the authors that appears in this

1. Lee, Sang B., Bogaard, J., and Feller, R. L., "Bleaching by Light I: Effect of pH and the Bleaching or Darkening of Papers in the Dry and in the Immersed Condition under Visible and Near-Ultraviolet Radiation," in *Conservation of Historic and Artistic Works on Paper: Proceedings of Symposium* 88, ed. Burgess, H., Canadian Conservation Institute, Ottawa, Canada (1994), p. 181-190.

volume.[1] After being immersed in the buffer containing the added sugars, the sheets were air-dried and the pH was determined by cold water extraction (Tappi Standard T509 su 68) initially and after thermal aging. As noted in Figures 7 through 12, the pH of the papers changed gradually during 20 days of thermal aging, apparently tending towards pH 5.5 to pH 6.5.

The net weight gain of added sugar was 8% to 10%. The gain in weight with the sugar-related acids amounted to 16% to 18%. The added weight of the lignin-related compounds, applied from 2% solutions in an organic solvent, was 2% to 3%.

Thermal Aging

Thermal aging at 90°C and 50% relative humidity (RH) was carried out in a Blue M FR256 (Blue M Electric Company, Wilmington, Delaware) humidity-controlled cabinet. Dry thermal aging was simply carried out in a 90°C air-circulating oven.

Exposure to Visible and Near-Ultraviolet Radiation

Details of the exposures were given in a separate paper.[1]

Sources of Compounds

D-Arabinose, dextrose, galactose and D-xylose sugars, products of Eastman Kodak, were purchased from the Fisher Scientific Company. D-glucuronic and D(+)-galacturonic acids were purchased from the Chemical Dynamics Corporation; D-glucuronic acid was also purchased from Eastman Kodak, and 1-ascorbic acid from Fisher Scientific. Vanillin (applied to the dry buffered paper from a 2% solution in benzene) and syringaldehyde and veratraldehyde (applied similarly from solution in ethanol) were gifts from Dr. Alfred Bader, President of the Aldrich Chemical Company.

Color Measurement

The reflectance at 460 nm was based on the abridged spectrophotometric measurements taken with a Kollmorgen Color-Eye 2000 instrument.[1] The Millipore filters used as a convenient reference white had a reflectance in the visible range of 98% to 99% relative to barium sulfate.

Results and Discussion

Thermal Discoloration of Sugars under Dry and Humid Conditions

Whereas cellulose is primarily based on glucose, hemicellulose can be hydrolyzed to yield a number of other sugars such as xylose, arabinose, mannose and galactose. Some of these tend to form colored products more readily than others, and some clearly more readily than glucose.[2] The plan, therefore, was to thermally discolor sheets of filter paper that had been impregnated with a number of sugars and to observe the tendency of the discolored materials, thus derived solely from a saccharidic source, to bleach when exposed to light.

2. Béelik, A., "Transformation of Free Monosaccharide within Pulp Sheets during Heat Aging," *Pulp and Paper Magazine of Canada*, 68, No. 3 (1967), p. T135-T142.

Figures 1 and 2 give the results of an initial study showing the effect of heating Whatman No. 42 filter paper that had been impregnated with various sugars to the extent of 8% to 10% pick-up in weight. The reported color changes in terms of the Kubelka-Munk function K/S [where $K/S=(1-R)^2/2R$ and R is the reflectance at 460 nm of the test papers backed by several sheets of Millipore filter as a reference white]. Xylose discolored the most: this was followed by arabinose, then galactose and glucose (dextrose). It was also obvious that aging under dry conditions (90°C in an air-circulating oven) caused less discoloration than under moist conditions (90°C, 50% RH). Mannitol, which contains an alcohol end-group rather than an aldehyde as in the other examples, did not darken during the same period of heating.

Figure 3 confirms the fact that only minor discoloration was generated when these same sheets were exposed either to "daylight" fluorescent lamps or to near-ultraviolet-emitting BLB fluorescent blacklights. Very slight darkening occurred during these exposures, possibly due to a certain degree of thermally induced discoloration owing to the fact that the test sheets were warmed somewhat by the heat of the fluorescent lamps. During exposure, the sheets impregnated with sugars tended to darken less than the control, possibly because of the tendency of some of the impregnated substances to undergo bleaching. However, the change in reflectance during these exposures was no more than 1%.

This series of tests demonstrates that thermal aging, particularly under humid conditions, will generate discoloration, especially in the case of two prominent constituents of hemicellulose: xylose and arabinose. Furthermore, the discoloration of the sugars by heat is far more than that caused by exposure to visible and near-ultraviolet radiation. The authors have previously shown that exposure of hemicellulose to visible and near-ultraviolet radiation results in neglible discoloration compared to that produced by thermal aging.[3]

3. Lee, S. B., Feller, R. L., and Bogaard, J., "Influence of the Hemicellulose Fraction on Thermal and Photochemical Discoloration of Paper," in *Historic Textile and Paper Materials*, Advances in Chemistry Series, No. 212, ed. Needles, H. L., and Zeronian, S. H., American Chemical Society, Washington, DC (1986), p. 377-386.

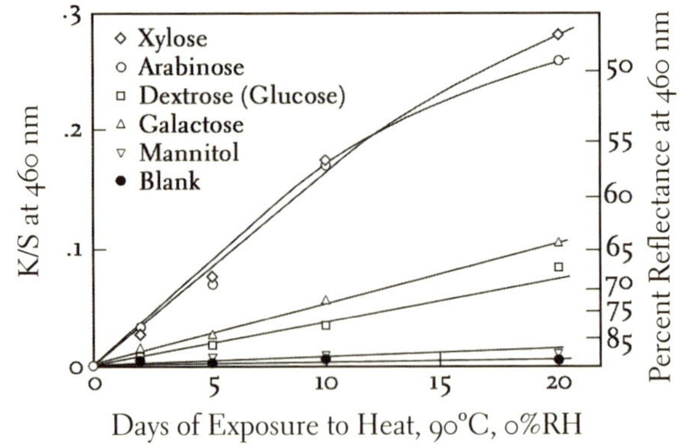

FIGURE 1
Change in K/S and percent reflectance at 460 nm upon thermal aging at 90°C of filter paper impregnated with various sugars.

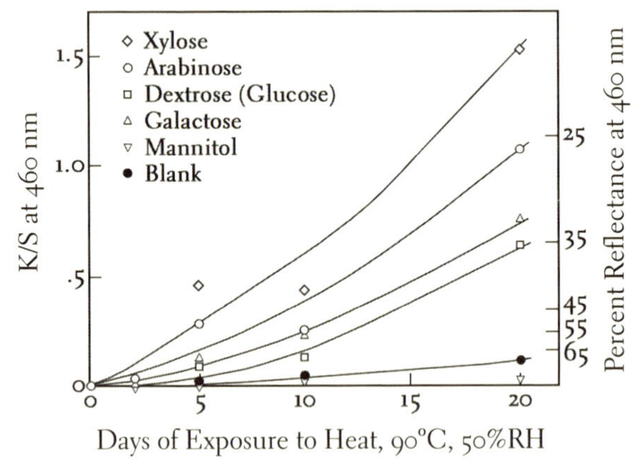

FIGURE 2
Change in K/S and percent reflectance at 460 nm upon thermal aging at 90°C and 50% RH of filter paper impregnated with various sugars.

Effect of Prolonged Exposure of Sugar-Impregnated Filter Papers to "Daylight" Fluorescent Lamps

To gain further insight into the behavior of the sugars, sugar-impregnated sheets prepared in a range of pH values were not purposely discolored through heat, but instead were simply exposed for a total of 795 hours (approximately 33 days) to "daylight" fluorescent lamps in a room maintained at about 23°C and 50% RH. Various initial values of pH were achieved by dipping the sheets of Whatman No. 42 filter paper into buffered solutions containing 5% of the sugar, and air-drying the resultant test papers.

The results of prolonged exposure to the "daylight" fluorescent lamps are

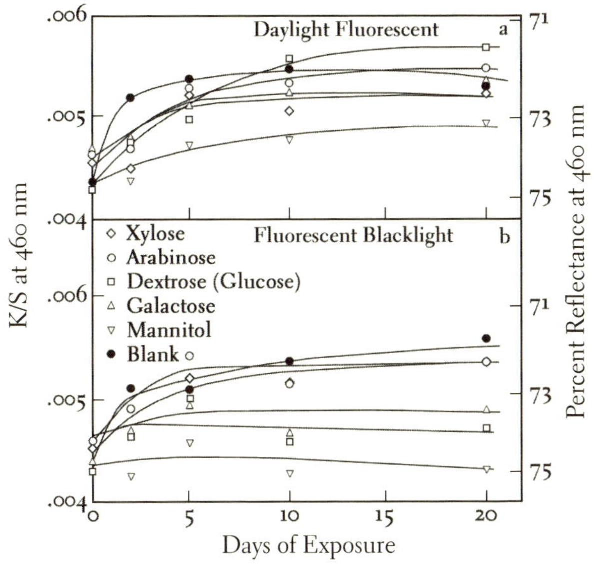

FIGURE 3
Change in K/S and percent reflectance at 460 nm upon exposure to (a) "daylight" fluorescent lamps and (b) BLB fluorescent blacklights of filter paper impregnated with various sugars.

shown in Figures 4, 5 and 6. Papers impregnated with xylose (Figure 4) and with arabinose changed in almost the same manner and degree. Acidity led to a slight net darkening of the test sheets (a change from about 92% to 84% reflectance at 460 nm). One may assume that the darkening that occurred in the sheets at low pH was primarily thermally induced; papers impregnated with these same two sugars, placed under the same test conditions but shielded from the light, darkened to much the same extent within about 1% reflectance. Alkalinity much above pH 9 led to an initially darkened state. Upon light exposure, however, this discoloration rapidly bleached. Similar conclusions regarding the darkening caused by acidity and the significantly reduced darkening in the alkaline condition were reached when papers, without added sugars, at various pH levels were exposed to "daylight" fluorescent lamps and then were allowed to undergo post-irradiation darkening by storage in the dark.[4]

Figure 5 shows that filter paper itself behaved much the same way with respect to pH. However, the net change in color was much smaller; initially the reflectance at 460 nm was 92%, decreasing to about 87% for the most acidic condition. The change in the sheets impregnated with dextrose (glucose) and with mannitol was much the same as when no sugar had been present. In the case of galactose, the degree of change was intermediate (Figure 6).

Acidity apparently can enhance thermally induced darkening of these particular materials. The conclusion is that even under moderate temperatures, darkening in the more acid papers can predominate over the bleaching that might be expected from prolonged exposure to light.

4. Lee, S. B., Bogaard, J., and Feller, R. L., "Darkening of Paper following Exposure to Visible and Near-Ultraviolet Radiation," *Journal of the American Institute for Conservation*, 28, No. 1 (1989), p. 1-18.

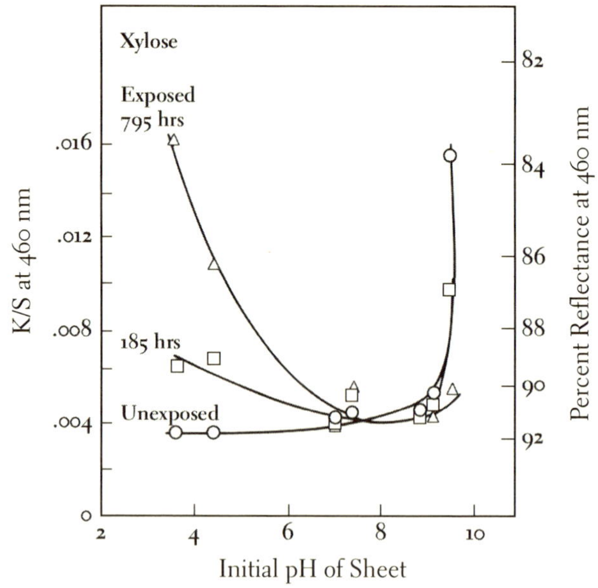

Figure 4
Change in K/S and percent reflectance at 460 nm of filter paper impregnated with xylose at various initial values of pH and exposed to "daylight" fluorescent lamps. Arabinose-impregnated sheets responded similarly.

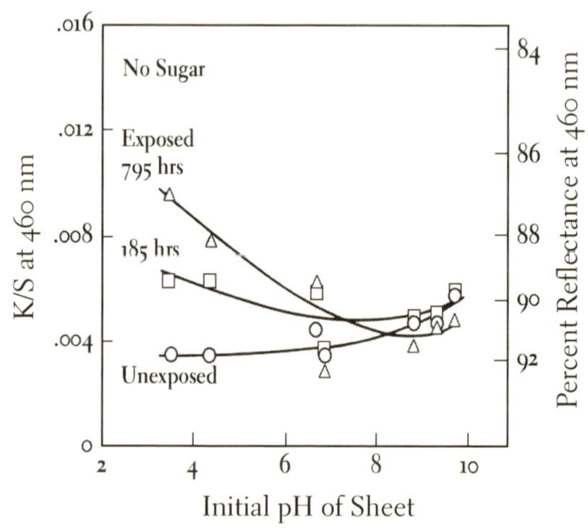

Figure 5
Change in K/S and percent reflectance at 460 nm of unimpregnated filter paper at various initial values of pH and exposed to "daylight" fluorescent lamps. Sheets impregnated with dextrose (glucose) and mannitol responded similarly.

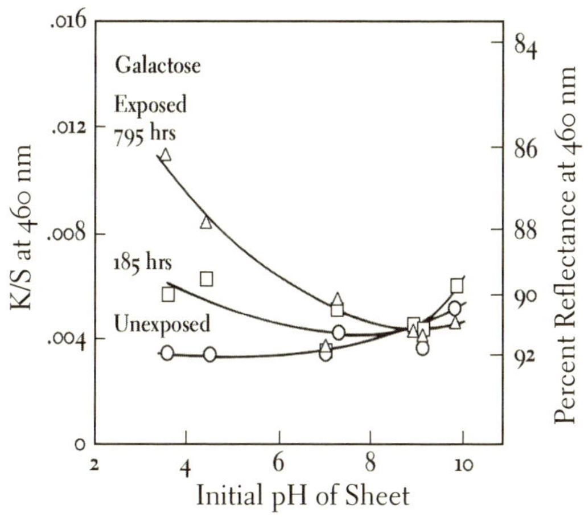

Figure 6
Change in K/S and percent reflectance at 460 nm of filter paper impregnated with galactose at various initial values of pH and exposed to "daylight" fluorescent lamps.

Intentional Thermal Discoloration of Sugars and Subsequent Bleaching

Since it has long been known that the pH of paper can influence thermally induced darkening, the initial experiments, shown in Figures 1 and 2, were repeated, this time with the sugar-impregnated test papers prepared at various pH levels. The results of heating for 20 days at 90°C and 50% RH are shown in Figures 7 to 12, along with the results of the subsequent exposure to light.

Following the thermally induced discoloration, it was found possible to bleach all samples upon exposure to "daylight" fluorescent lamps. To follow the process at some length, tests were carried out for a period of 33 days, amounting to about 800,000 footcandle hours (8.6 Mlx.h) of exposure. The control filter paper that contained no added sugar thermally discolored far less (Figure 12); nonetheless, the colored matter, which must have been derived from saccharidic composition products in the filter paper, also bleached. Again, mannitol, which is not an aldose (aldehyde-group-containing sugar), did not discolor thermally as much as the other sugars which contained an aldehyde group.

The principal conclusion of these tests is that the discolored products that resulted from the thermal aging of sugars are bleachable with the essentially visible radiation from "daylight" fluorescent lamps. It does not appear, however, that the test sheets can be completely bleached during a reasonable period of exposure. The latter conclusion, however, must be verified by further investigation.

Several other interesting observations may be made concerning the experimental results. Although the pH of the papers did not remain constant up to the point of termination of heating, in every case the unbuffered control sheets of filter paper discolored far less than the sheets that had been buffered

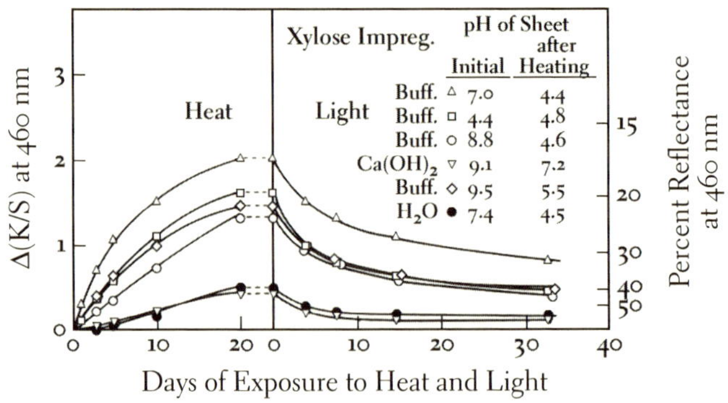

Figure 7
Darkening and bleaching expressed as the change in K/S values at 460 nm: filter paper impregnated with xylose was subjected initially to thermal aging at 90°C and 50% RH followed by exposure to "daylight" fluorescent lamps.

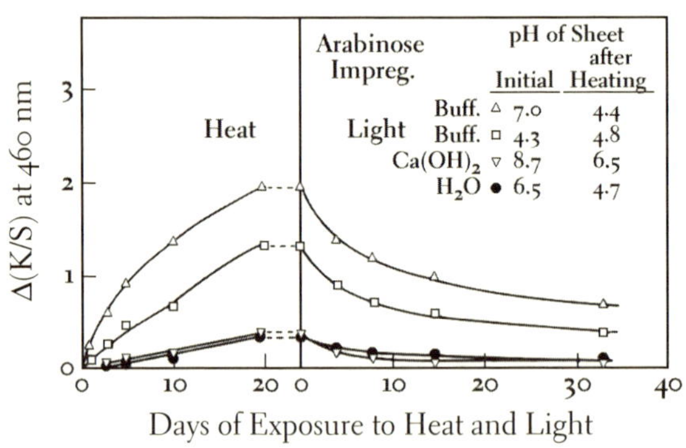

Figure 8
Filter paper impregnated with arabinose. Same circumstances as in Figure 7.

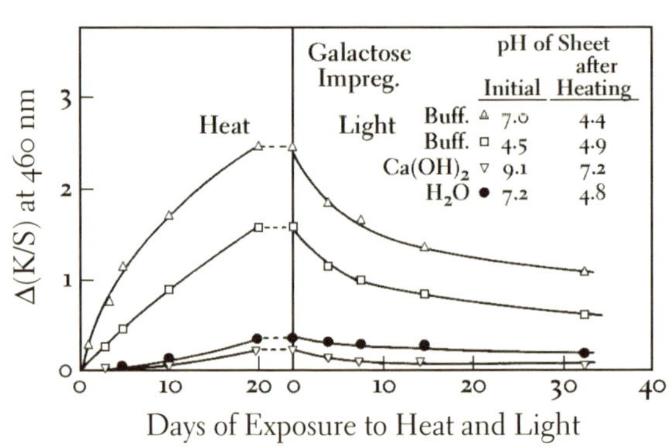

Figure 9
Filter paper impregnated with galactose. Same circumstances as in Figure 7.

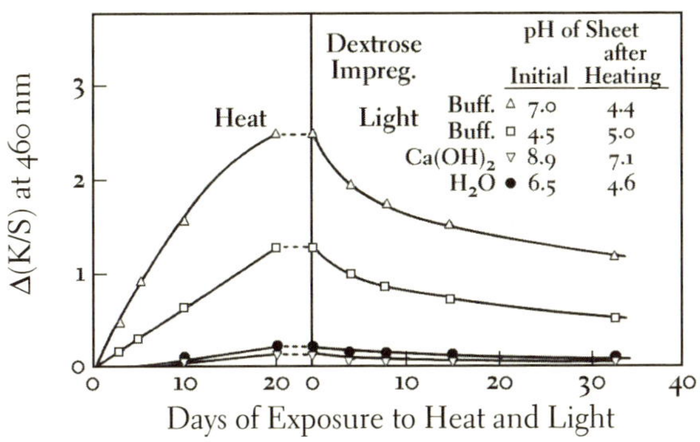

FIGURE 10
Filter paper impregnated with dextrose (glucose). Same circumstances as in Figure 7.

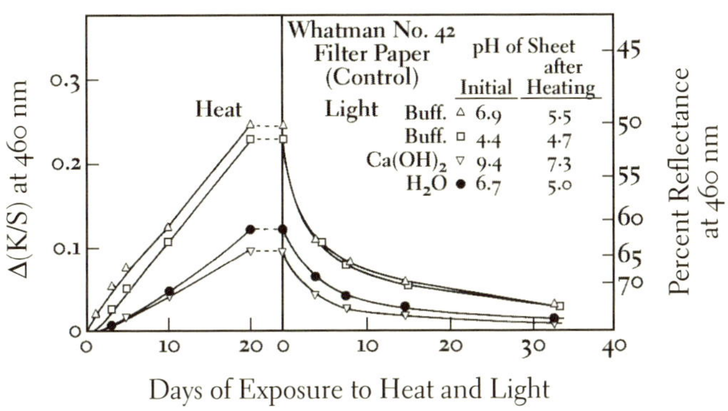

FIGURE 11
Unimpregnated filter paper. Same circumstances as in Figure 7. Note the much lower degree of darkening than in Figures 7 to 10.

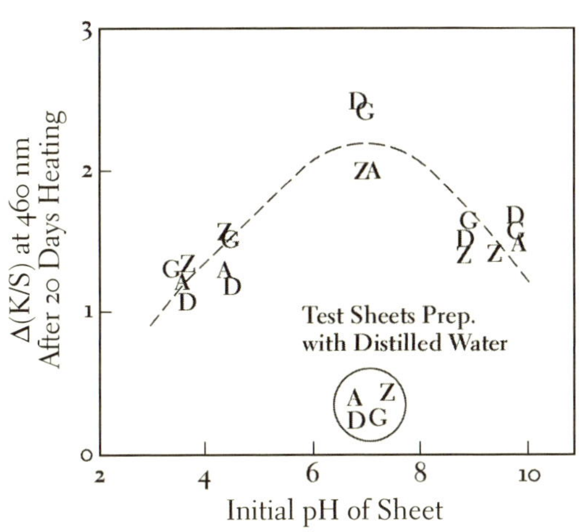

FIGURE 12
Summary of the thermal darkening that occurred in papers impregnated with various sugars as a function of the initial values of pH. The degree of darkening of the same papers immersed in water rather than in buffers is also shown. Symbols: X = xylose; A = arabinose; D = dextrose; G = galactose.

at pH 7. It was not our intention to investigate this matter further, but it is possible that the sodium or potassium ions present in the buffered papers may have been responsible for the greater degree of thermally induced discoloration. The effect of these ions on the thermal aging of paper has been elucidated by the work of Sihtola and his associates.[5]

5. Sihtola, H., and Sumiala, R., "The Influence of Cations on the Yellowing of Monocarboxycellulose," *Paperi ja Puu*, 45, No. 2 (1963), p. 43-48.

Another feature of the thermally induced discoloration is of interest: in the course of heating to discolor them, the test papers prepared at various initial values of pH behaved as summarized in Figure 12. The curve is not the same as that which we have obtained as the result of observations of thermal aging at moderate temperatures[4] or that of the related results seen in Figures 4, 5 and 6: the color development in the acid condition was not as high as might have been expected. The explanation could be that under the conditions of acid-catalyzed aging at high temeprature, some of the colored products may be so extensively oxidized that they in turn produce colorless molecular entities, as has been reported by others.[6] In addition, certain colored substances such as furfural can be volatilized out of the paper.

6. Garrett, E. R., and Dvorchik, B. H., "Kinetics and Mechanisms of the Acid Degradation of Aldopentoses to Furfural," *Journal of Pharmaceutical Science*, 57 (1969), p. 813-820.

The pH did not remain unchanged in the course of heating for 20 days at 90°C and 50% RH, but in most cases tended towards neutrality regardless of whether they were initially very acid or very basic. Nonetheless, when the results are plotted against the initial pH, definite trends in behavior are apparent.

Bleaching of Discolored Products from Ascorbic and Sugar Acids

In 1964, Rapson and Corbi[7] added various model compounds to filter paper to learn which types of chemical structure, representing possible products from the oxidation of cellulose and hemicellulose, tended to thermally darken the most. Using molecules of no more than six carbons in length, they found that a combination of carbonyl, carboxyl and hydroxyl groups in these structures led to severe discoloration. Rather than speculate on the full range of three and four carbon molecules that might be representative of the advanced degradation products of cellulose and hemicellulose, it was decided to degrade thermally just a few simple six-carbon structures that might represent an early stage of oxidation: glucuronic, galacturonic and gluconic acids. As a possible representative of a highly advanced stage of oxidation, ascorbic acid was employed.

7. Rapson, W. H., and Corbi, J. C., "Colour Reversion Caused by Degradation Products of Oxidized Cellulose," *Pulp and Paper Magazine of Canada*, 65 (1964), p. T459-T466.

The greatest thermally induced discoloration was achieved in the sheets impregnated with galacturonic acid and ascorbic acid, followed by those impregnated with glucuronic acid (Figure 13). Gluconic acid, which does not contain an aldehyde group, did not lead to extensive discoloration. As seen in Table 1, these results are similar to those obtained by Rapson and Corbi. Again, the principal conclusion to this investigation is that the discolored products derived from these sugar-related materials tend to bleach under "daylight" fluorescent lamps.

The sheets impregnated with ascorbic acid had a pH of about 3.1. Upon

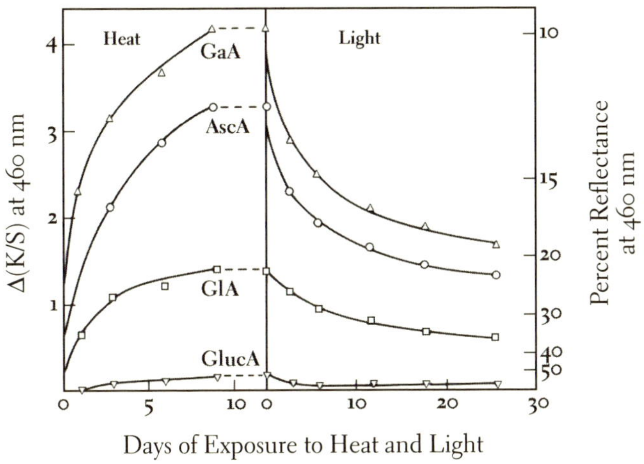

FIGURE 13
Papers impregnated with various sugar acids and with ascorbic acid. Circumstances as in Figure 7. Symbols: GaA = galacturonic acid; AscA = ascorbic acid; GlA = glucuronic acid; GlucA = gluconic acid.

TABLE 1
Comparison of Post-Color Number of Filter Papers Impregnated with Various Acids*

	Darkening	
	Rapson and Corbi's Data*** Aged 105°C 18 hrs.	Our Data** (Figure 13) Aged 90°C, 50% R.H. 9 Days
Glucuronic acid	35.8	127
Galacturonic acid	32.6	417
Ascorbic acid	16.3	328
Gluconic acid	0.75	18

* Post-color number is 100 [$(K/S)_t - (K/S)_o$] where the K/S value was calculated from the reflectance at 457 nm or 460 nm. The o indicates the initial value and the t, the final value after heating.
** Compounds present in the sheet at 16% to 18% by weight.
*** 18.2 millimoles of model compound per 100 g of paper.

consideration, it was thought that the marked discoloration here could have been due not so much to the ascorbic acid but to the acid-catalyzed attack on the filter paper. Indeed, when a glass-fibre filter paper was impregnated with ascorbic acid, the sheets only darkened to the extent of K/S 0.3. The high acidity of the impregnated sheets in this case must have led to the intense discoloration of the filter paper. If this is what occurred, the result provides additional evidence that the colored products resulting from the acid-catalyzed thermal degradation of a high-alpha-cellulose paper, which must have a saccharidic origin, can be bleached by exposure to "daylight" fluorescent lamps.

Effects on Lignin Model Compounds

In 1970, Kringstad and Lin[8] did much to elucidate the nature of the chromophores that were introduced during the exposure of compounds representative of structures that might be found in lignin. They showed that the formation of phenoxy radicals led to the formation of colored substances, often orthoquinones. They proposed that one of the principal sites of photo-excitation was the carbonyl group attached to a benzene ring. For example, veratraldehyde and 3,4,5-trimethoxy-benzaldehyde, which possess carbonyl groups but which have their phenolic groups blocked, did not develop absorption at 410 nm when irradiated in the near-ultraviolet. In contrast, vanillin and syringaldehyde discolored (see structures, Figure 14). Moreover, Kringstad and Lin showed that vanillyl alcohol and syringyl alcohol, which have no carbonyl groups but possess phenolic OH's, could be discolored in the presence of veratraldehyde. Apparently the carbonyl group in the veratraldehyde can be excited by the absorption of ultraviolet radiation. Because of its blocked phenolic OH, however, this compound in itself cannot generate the phenolic radicals that lead to discoloration. Nonetheless, the photo-excitation of veratraldehyde was able to lead to the formation of phenolic radicals in the vanillyl and syringyl alcohols when they were present.

8. Kringstadt, K. P., and Lin, S. Y., "Mechanism in the Yellowing of High-Yield Pulps by Light. Structure and Reactivity of Free Radical Intermediates in the Photodegradation of Lignin," *Tappi*, 53, No. 12 (1970), p. 2296-2301.

Years ago, Nolan, Van den Akker and Wink,[9] and later Leary,[10] showed that newsprint, which undoubtedly contained considerable lignin, was darkened by the ultraviolet radiation lower than 390 nm to 385 nm, and that bleaching occurred in their samples in the wavelength region of about 420 nm to 490 nm. Therefore, in order to induce marked discoloration, sheets of Whatman No. 42 filter paper impregnated with compounds representing possible constituents of lignin—vanillin, syringaldehyde and vertraldehyde—were exposed to near-ultraviolet radiation (emission of BLB fluorescent blacklight lamps). The results are shown in Figure 15: considerable darkening occurred. Because ultraviolet radiation would cause darkening, visible radiation ("daylight" fluorescent lamplight through a Plexiglas UF-3 filter) was used in an attempt to bleach the darkened test sheets. However, the photochemically induced darkening could only be partially bleached by exposure to visible radiation.

9. Nolan, P., Van den Akker, J. A., and Wink, W. A., "The Fading of Groundwood by Light II. The Physical Mechanism of Fading," *Paper Trade Journal, Tappi Section*, 121, No. 11 (1945), p. 101-105.

10. Leary, G. L., "The Yellowing of Wood by Light," *Tappi*, 50 (1967), p. 17-19.

Heat caused very little darkening of papers impregnated with these compounds (Figures 16 and 17). Note that the thermally induced darkening of the sheets impregnated with vanillin and syringaldehyde was probably not due to these substances; the darkening amounted to no more than that of the control filter paper (Figure 16). Moreover, the bleaching was identical in the case of the vanillin-impregnated paper. Bleaching was also identical in the cases of

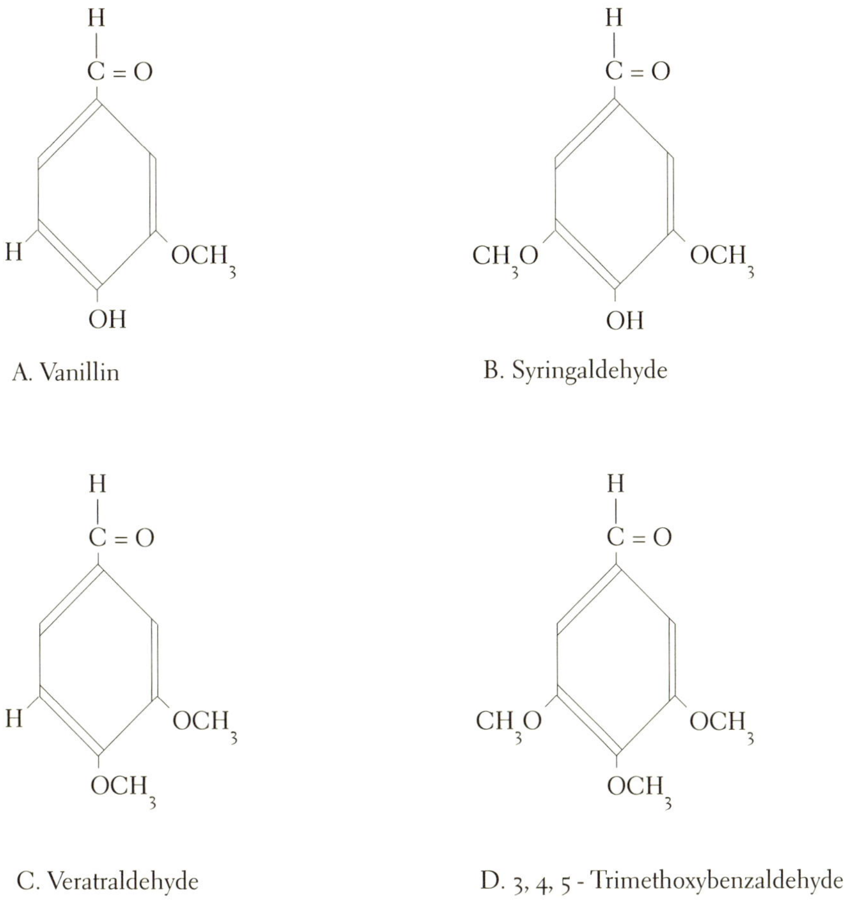

FIGURE 14
Compounds related to the structure of constituents found in lignin.

both vanillin and syringaldehyde impregnation when the 3-71 filter was used (Figure 17); this glass filter transmits visible radiation longer than 460 nm. The implication of the similarity of both darkening and bleaching effects, therefore, is that the slight thermally induced discoloration in these test papers must have been largely due to the degraded constituents of the cellulose in the paper.

The behavior of veratraldehyde was different from the other two compounds. Note in Figure 14 that veratraldehyde does not have a free phenolic OH group whereas the other two compounds do. Perhaps the veratraldehyde also acted as a mild inhibitor in the case of thermally induced darkening; hence, the slightly lower degree of darkening achieved. Possibly syringaldehyde was darkened slightly by the wavelengths between 400 nm and 460 nm (Figure 17 and Figure 16).

The evidence of these brief tests suggests that the bleaching action of visible

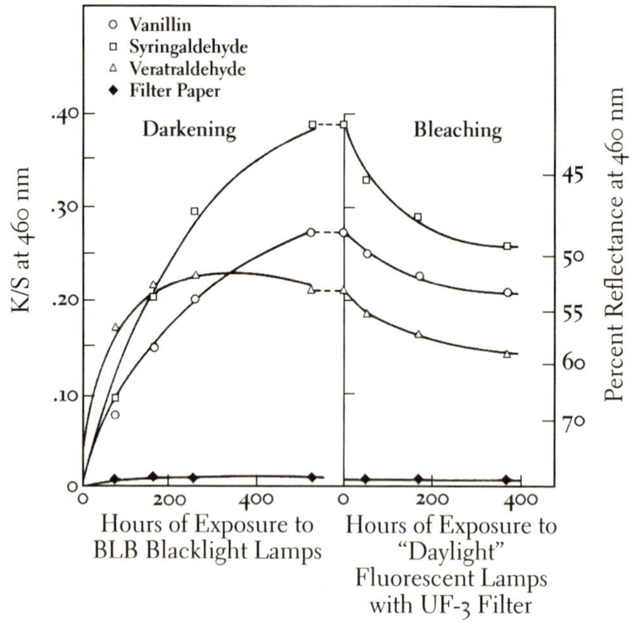

FIGURE 15
Darkening and bleaching of filter paper impregnated with model compounds related to the constituents of lignin. Initially exposed to near-ultraviolet radiation (BLB fluorescent blacklamps), then bleached by exposure to visible radiation (emission of "daylight" fluorescent lamps filtered through a Plexiglas UF-3 filter).

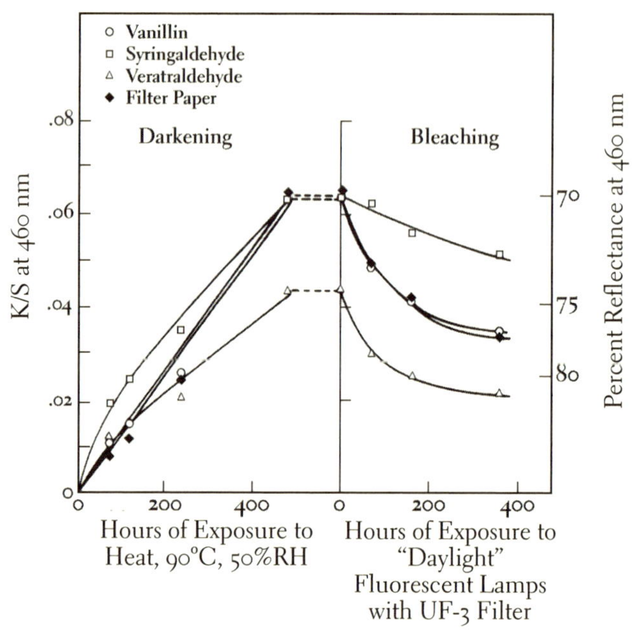

FIGURE 16
Darkening and bleaching of filter paper impregnated with model compounds related to constituents of lignin. Initially subjected to thermal aging at 90°C and 50% RH, then bleached by exposure to visible radiation.

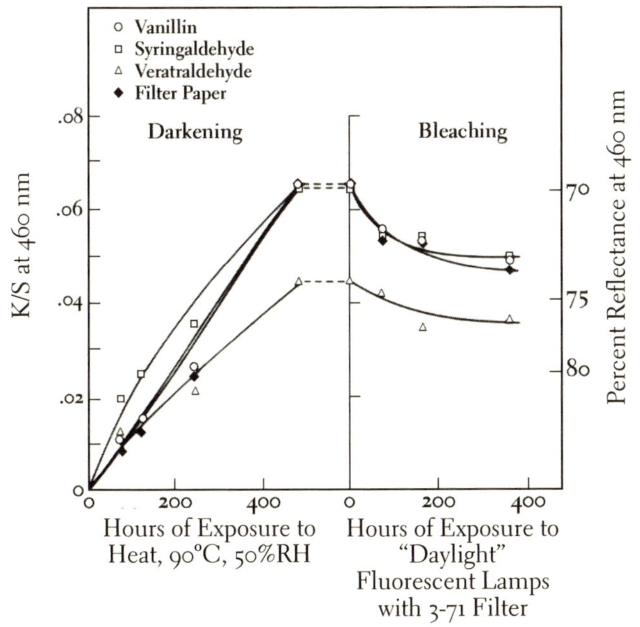

Figure 17
Same circumstances as in Figure 16 except that bleaching is accomplished by exposure to radiation above 460 nm by use of a Corning 3-71 filter.

radiation is only partially effective on the ultraviolet-discolored lignin model compounds (Figure 15). It also appears that bleaching of the slight thermally induced discoloration of these papers is only partially effective, but this point needs further investigation.

Conclusions

Sugars are among the potential products resulting from the deterioration of cellulose and of the hemicellulosic constituents of paper. Four commonly encountered aldoses are extensively discolored by heat: xylose and arabinose discolor more extensively than glucose or galactose. Thermal aging under humid conditions results in far greater discoloration than under dry conditions. In contrast to thermally induced discoloration, papers impregnated with sugars are scarcely discolored by exposure to the radiation from "daylight" fluorescent lamps or from near-ultraviolet-emitting BLB fluorescent blacklights. The slight darkening that occurs under these circumstances can possibly be attributed to a certain degree of thermally induced darkening. Thermal discoloration of these materials under moderate conditions tends to be greater in acid conditions. Alkalinity much above pH 9 tends to cause immediate darkening which is, however, readily bleached upon exposure to "daylight" fluorescent lamplight.

The principal conclusion to these investigations is that the thermally induced discoloration of aldoses and related sugar acids is at least partially bleached by exposure to "daylight" fluorescent lamps. If such materials should be responsible for some of the discoloration of aged paper products, the evidence is that they would be extensively, although perhaps not completely, bleached by exposure to either visible or near-ultraviolet radiation. In contrast, compounds similar in structure to substances that might be found in lignin are not so much discolored by heat as by exposure to ultraviolet radiation.

In neither case did the bleaching by visible radiation of thermally discolored products from sugar nor of photochemically discolored model compounds related to lignin appear to go to completion even after rather extensive exposures. This is an important practical aspect of light bleaching that must be verified by further experiments. Nonetheless, this investigation has demonstrated that the bleaching effect of exposure to "daylight" fluorescent lamps, with 96.2% of the radiant power in the visible wavelengths, is most effective against the thermal decomposition products from sugar or sugar acids and is least effective against photochemically discolored substances derived from model compounds related to the components of lignin.

Influence of the Hemicellulose Fraction on Thermal and Photochemical Discoloration of Paper

S. B. LEE AND R. L. FELLER

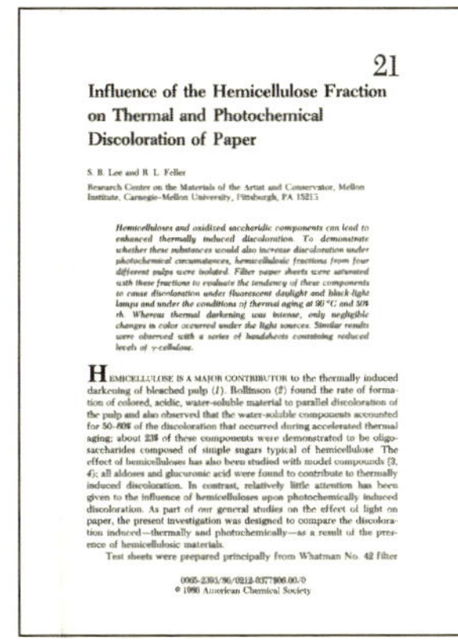
21

Influence of the Hemicellulose Fraction on Thermal and Photochemical Discoloration of Paper

S. B. Lee and R. L. Feller

Research Centre on the Materials of the Artist and Conservator, Mellon Institute, Carnegie-Mellon University, Pittsburgh, PA 15213

Hemicelluloses and oxidized saccharidic components can lead to enhanced thermally induced discoloration. To demonstrate whether these substances would also increase discoloration under photochemical circumstances, hemicellulosic fractions from four different pulps were isolated. Filter paper sheets were saturated with these fractions to evaluate the tendency of these components to cause discoloration under fluorescent daylight and black-light lamps and under the conditions of thermal aging at 90 °C and 50% rh. Whereas thermal darkening was intense, only negligible changes in color occurred under the light sources. Similar results were observed with a series of handsheets containing reduced levels of γ-cellulose.

HEMICELLULOSE IS A MAJOR CONTRIBUTOR to the thermally induced darkening of bleached pulp (*1*). Rollinson (*2*) found the rate of formation of colored, acidic, water-soluble material to parallel discoloration of the pulp and also observed that the water-soluble components accounted for 50-60% of the discoloration that occurred during accelerated thermal aging; about 23% of these components were demonstrated to be oligosaccharides composed of simple sugars typical of hemicellulose. The effect of hemicelluloses has also been studied with model compounds (*3, 4*); all aldoses and glucuronic acid were found to contribute to thermally induced discoloration. In contrast, relatively little attention has been given to the influence of hemicelluloses upon photochemically induced discoloration. As part of our general studies on the effect of light on paper, the present investigation was designed to compare the discoloration induced—thermally and photochemically—as a result of the presence of hemicellulosic materials.

Test sheets were prepared principally from Whatman No. 42 filter

0065-2393/86/0212-0377$06.00/0
© 1986 American Chemical Society

Originally printed in:
HISTORIC TEXTILE AND PAPER MATERIALS
Advances in Chemistry Series No. 212, ed. Needles, H. L., and Zeronian, S. H., American Chemical Society, Washington, DC (1986), p. 377-386.

Hemicellulose is a major contributor to the thermally induced darkening of bleached pulp.[1] Rollinson[2] found the rate of formation of colored, acidic, water-soluble material to parallel discoloration of the pulp and also observed that the water-soluble components accounted for 50-60% of the discoloration that occurred during accelerated thermal aging; about 23% of these components were demonstrated to be oligosaccharides composed of simple sugars typical of hemicellulose. The effect of hemicelluloses has also been studied with model compounds;[3, 4] all aldoses and glucuronic acid were found to contribute to thermally induced discoloration. In contrast, relatively little attention has been given to the influence of hemicelluloses upon photochemically induced discoloration. As part of our general studies on the effect of light on paper, the present investigation was designed to compare the discoloration induced—thermally and photochemically—as a result of the presence of hemicellulose materials.

Test sheets were prepared principally from Whatman No, 42 filter paper saturated with hemicellulosic fractions that had been isolated from various unaged or thermally aged pulps by LiOH extraction.[5] Another set of test sheets was prepared by saturation of filter paper with commercially prepared arabinogalactan and especially isolated acetylxylan. (The term "added-on" will be used to refer to the occlusion of the various fractions to the filter paper as a result of saturation.) Finally, a series of handsheets was made from the pulps of different γ-cellulose contents (10.5-1.5%) that were obtained by the extraction of bleached sulfate pulp with different strengths of caustic solution. The various

1. Chang, F.-J., and Kondo, T., *J. Jpn. Wood Res. Soc.*, 17, No. 4 (1971), p. 160.

2. Rollinson, S. M., *Tappi*, 38, No. 3 (1955), p. 186.

3. Beélik, A., *Pulp Paper Mag. Canada*, 68, No. 3 (1976), T135.

4. Rapson, W. H., and Corbi, J.-C., *Pulp Paper Mag. Canada*, 65, No. 11 (1964), T459.

5. Hamilton, J. K., and Quimby, G. R., *Tappi*, 40, No. 9 (1957), p. 781.

test papers thus prepared with different concentrations of hemicellulosic components were evaluated for their tendency to discolor upon exposure to visible and near-UV radiation (daylight and black-light fluorescent lamps) and to heat.

Experimental

Extraction of Hemicelluloses from Various Pulps

6. Giertz, H. W., and McPherson, J., *Svensk Papperstidning*, 59, No. 3 (1956), p. 93.

The extraction procedure followed was that of Giertz and McPherson.[6] Fifteen grams of unaged pulps [four kinds in all: bleached kraft pulp (BP), unbleached kraft pulp (UBP), unbleached groundwood pulp (GP), and chlorited UBP] was treated with 290 mL of 10.9% lithium hydroxide solution for 1 h at room temperature and was stirred intermittently (see Table 1 for the characteristics of the pulps). After the samples were diluted to about 1 L, the insoluble material was filtered off. The extract (filtrate) was neutralized to pH 6 with 2 M phosphoric acid (300 mL) and was allowed to stand overnight. The precipitated lithium phosphate and β-cellulose were separated by filtration. The filtrate was then concentrated to about 175 mL under reduced pressure at 30-40°C. Salts present in the system were further precipitated with methanol and separated by filtration. The filtrate was again concentrated to about 60 mL. The filter paper was immersed into this concentrate to saturate the test sheets.

This procedure was applied also to BP, UBP, and GP samples that had been thermally aged at 90°C and 50% RH for 20 days prior to extraction. The transmittances of 5% solutions of the neutral extracts at 460 nm, obtained with a Bausch and Lomb Spectronic 20 spectrophotometer, are noted in Table 2. Whatman No. 42 filter paper was used as the substrate for the extracted material. When the sheets were dipped in the 60-mL concentrate, the pickup was found to be about 15%.

Table 1
Characteristics of Stock Pulps Used

Sample	Description	Initial Reflectance at 457 nm (%)	Lignin Content (%)	Initial Hot-1%-Alkali Solubility (%)	Copper Number*
BP	50% hardwood kraft-50% softwood kraft	83.6	0.24	5.0	0.5
UBP	50% hardwood kraft-50% softwood kraft	30.8	4.4	3.7	0.6
GP	–	57.0	29.1	15.6	2.9

* Copper number is defined as the number of grams of metallic copper reduced from the cupric to the cuprous state by 100 g of pulp fibers.

TABLE 2
Light Transmittances of Neutralized Extracts at 5% Solids*

Sample	Extract from Unaged Sample	Extract from Heated Sample**
BP	96	94
GP	2	35
UBP	46	50
Chlorited UBP	91	–

* Values are the percent transmission at 460 nm, 1-cm cell.
** Hemicellulose fractions from pulp stock heated at 90°C and 50% RH for 20 days.

Hemicellulose Samples

Two water-soluble hemicellulosic samples were used: (1) acetyl-4-O-methylglucuronoxylan (molecular weight ≈ 30,000) from white birchwood, isolated by the extraction of chlorine holocellulose with dimethyl sulfoxide and (2) arabinogalactan, Stractan from St. Regis Paper Company (arabinose: galactose = 1:6; molecular weight ≈ 60,000), prepared by the water extraction of heartwood of western larch. Both samples were kindly provided by T. E. Timell (SUNY).

Preparation of Pulp Having Different γ-Cellulose Contents

Unbleached pulp (50% hardwood kraft-50% softwood kraft) was subjected to chloriting treatment to remove lignin. The chemicals and conditions employed were the following: 25 g of sodium chlorite, 12.5 mL of glacial acetic acid, a pulp consistency of 10%, a temperature of 75-80°C, and a reaction period of 30 min. This treatment produced a pulp having 1.06% lignin which was then extracted with different concentrations of alkali to adjust the γ-cellulose content, as indicated in Table 3.[7] The extractions with caustic were carried out at a pulp consistency of 2%.

After caustic extraction, the pulp was separated by filtration (except the 17.5% alkali-treated pulps, which were first diluted and then decanted), 500 mL of 10% acetic acid was added, and the mixture was allowed to soak for 1 min before filtration. This procedure was repeated with a second 500-mL portion of 10% acetic acid. The material was then washed several times until it was acid free. From the pulps thus prepared, handsheets were made at the basis weight of 75 g/m^2.

7. Kaustinen, O. A., and Jappe, N. A., *Paperi ja Puu*, 42, No. 9 (1960), p. 485.

TABLE 3
Characterization of Chlorite-Treated and Caustic-Extracted Pulps

Treatment Conditions for Caustic Extraction	Composition of Resulting Celluloses (%)		
	γ-Cellulose	α-Cellulose	β-Cellulose
Sample unextracted after chloriting	10.5	88.0	1.5
1% NaOH, 80°C, 2h	8.8	88.4	2.8
5% NaOH, 20°C, 1h	5.1	93.2	1.7
17.5% NaOH, 20°C, 1/2h	1.5	96.8	1.7

Exposure to Light Sources

Three different light sources principally were used: (1) a bank of six General Electric high-output daylight fluorescent lamps, (2) a bank of six fluorescent BLB blacklights, and (3) an Atlas 6500-W xenon-arc Fadeometer with Pyrex filters. Measurements of the output of the lamps indicated about 6.3% near-UV radiation (relative to UV-plus-visible radiation) in the Fadeometer, 3.8% under daylight fluorescent lamps, and 86% under the BLB lamps. The brief tests with short-wave UV radiation (involving 254-nm emission) were done with mineral-inspection mercury-vapor lamps that had the dark-blue long-wavelength filters removed. These lamps previously had been used to investigate the depolymerization of methacrylate polymers by short-wave UV radiation.

The fluorescent lamps were mounted 3 1/2 in. above the samples in a room maintained at 50% RH and 21.1°C (70°F). Under these conditions the temperature of the handsheets reached 24°C. In the 6500-W Fadeometer, the air temperature was 31°C, the relative humidity was about 27%, and the black-panel temperature was 65°C.

RESULTS

Filter Paper with Hemicellulosic Fractions Added On

Hemicellulose fractions were isolated from unaged and thermally aged papers by extraction with LiOH, following the procedure described by Giertz and McPherson.[6] The changes in post-color number $[100\,(K/S) - (K/S)_o]$[8] for various test sheets during exposure to visible and UV radiation and to heat are shown in Figures 1a, 1b, and 1c for material extracted from unheated samples and then added on to the filter paper by immersion in a suspension of the neutralized extract. [The value for K_o/S_o is obtained from the reflectance, R, at 460 nm of a stack of six sheets of the sample papers, according to the relation $K/S = (1 - R)^2 / 2R$. $(K/S)_o$ is the initial value, and $(K/S)_t$ is the value following

8. Tongren, J. C., *Paper Trade J.*, Technical Association Section, 107, No. 8 (1938), p. 34.

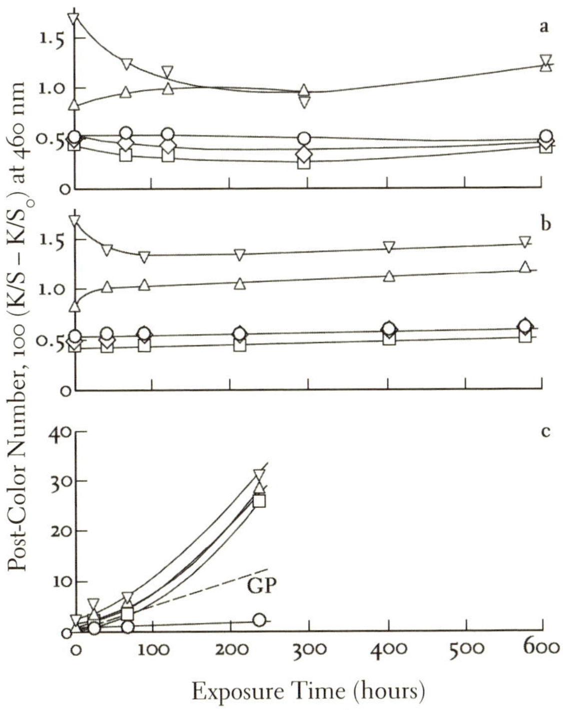

FIGURE 1
Changes in post-color number of filter-paper sheets with 15% add-on of hemicellulose fractions derived from various pulps. For comparison, the thermal discoloration of a test sheet of groundwood pulp, GP, is shown. The initial postcolor number represents the increase in color over that of the untreated filter paper: a, daylight fluorescent lamps; b, black-light fluorescent lamps; c, heat (90°C and 50% RH). Key: ○, blank filter paper; □, BP; △, GP; ▽, UBP; ◇, chlorited UBP.

exposure or other treatment.] The net gain of added-on material was 15% on a dry-weighted basis. Figures 2a, 2b, and 2c show the results for sheets impregnated with hemicellulosic fractions extracted from thermally aged pulps (pulp stocks were aged at 90°C and 50% RH for 20 days before extracts were prepared). Twice as much material was extracted from the thermally aged sheets with LiOH as was obtained from the unaged stock. This result is similar to the finding of Giertz and McPherson[6] that increased material could be extracted from aged sheets. Table 2, which gives the transmittances of the extract solutions at 5% solids, shows that the darkness of the extract solutions decreased in the order GP > UBP > chlorited UBP > BP. Colored alkali-soluble materials formed during thermal aging have been investigated by Lewin et al.[9]

9. Albeck, M., Ben-Bassat, A., Epstein, J. A., and Lewin, M., *Tex. Res. J.*, 35, No. 9 (1965), p. 836.

The initial K/S values in the graphs of Figures 1a, 1b, 2a, and 2b depended on the type of pulp extracted and on the quantity of material added on. (The initial post-color number represents the increase in color over that of the untreated filter paper.) The papers impregnated with the extract from the UBP tended to bleach during the first 100 h of photochemical exposure—the same result that we previously observed with the UBP pulp itself—and then a leveling

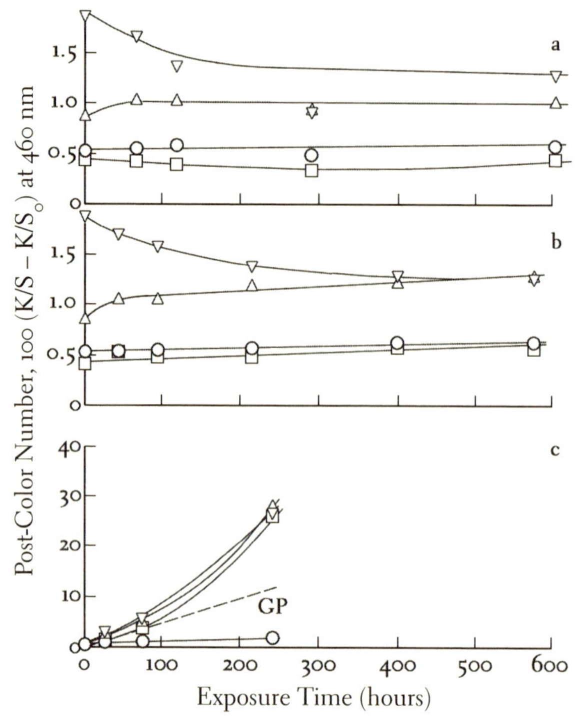

FIGURE 2
Changes in post-color number of filter-paper sheets with 15% add-on of fractions from thermally aged hemicellulose derived from various pulps. For comparison, the thermal discoloration of a test sheet of groundwood pulp, GP, is shown. The initial post-color number represents the increase in color over that of the untreated filter paper: a, daylight fluorescent lamps; b, black-light fluorescent lamps; c, heat (90°C and 50% RH). Key: ○ , blank filter paper; □, heated BP; △ , heated GP; ▽ , heated UBP.

off occurred. The extract material from the GP tended to darken, probably because of the lignin present. In the case of test sheets containing the added-on hemicellulosic material extracted from the BP and from the chlorite-treated UBP, the color scarcely changed, possibly because of the negligible presence of lignin. For these extracted substances, we found little difference in the behavior of the samples under daylight or under BLB black-light fluorescent lamps (Figures 1a, 1b, 2a, and 2b). Little discoloration also was observed when these samples were exposed to mercury-vapor lamps that emitted considerable radiation at 254 nm.

In contrast, marked and very similar darkening occurred upon the thermal aging of added-on test sheets at 90°C and 50% RH, regardless of the particular hemicellulosic extract (Figures 1c and 2c). For comparison, the behavior of handsheets made solely from groundwood pulp (GP), notorious for its tendency to discolor, is shown in each figure.

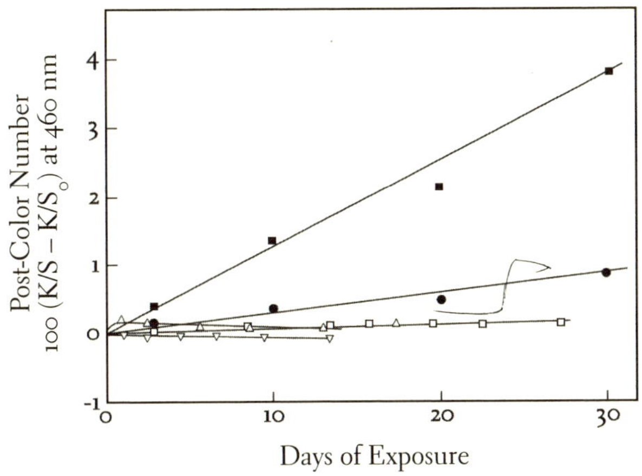

FIGURE 3
Post-color number of filter papers saturated to a level of 3% add-on with arabinogalactan. Key (thermal discoloration):■, 50% RH and 90°C;●, 0% RH and 90 °C; and (photochemical discoloration): □, daylight fluorescent lamps; △, BLB fluorescent lamps; ▽, Fadeometer.

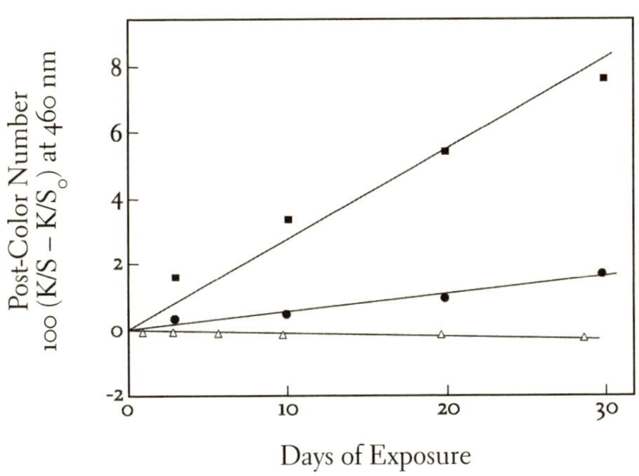

FIGURE 4
Post-color number of filter papers saturated to a level of 3% add-on with acetylxylan. Key (thermal discoloration): ■, 50% RH and 90°C; ●, 0% RH and 90°C; and (photochemical discoloration): △, BLB fluorescent lamps.

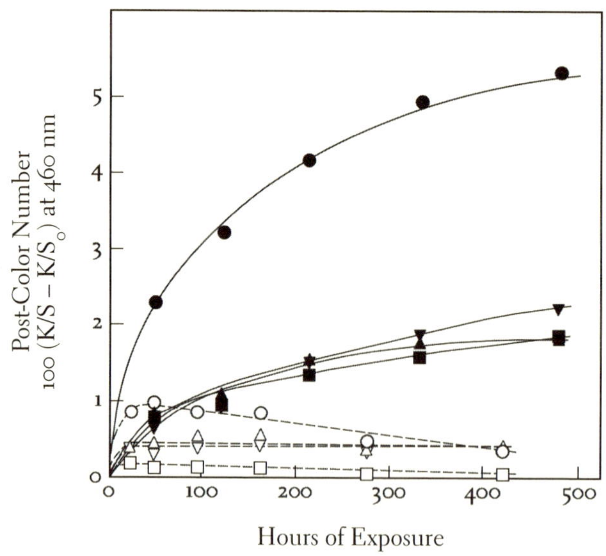

FIGURE 5
Post-color number of test sheets of various contents of γ-cellulose that were prepared from UBP by extraction with alkali. The 10.5% sample is the starting material. Key (percent γ-cellulose): ●,○, 10.5; ■,□, 8.8; ▲,△, 5.1; and ▼,▽, 1.5. Dark symbols represent thermal discoloration; light symbols represent photochemical discoloration.

Arabinogalactan and Acetylxylan Added on Test Sheets

Filter paper that had been dipped into suspensions of arabinogalactan and of acetylxylan to the extent of a 3% weight gain (3% add-on) was subjected both to humid (50% RH) and to dry thermal aging at 90°C and also to exposure to various light sources. As shown in Figures 3 and 4, thermal aging darkened the test sheets considerably under humid conditions, whereas only moderate discoloration occurred under dry conditions. This behavior confirms the effects of humidity on the thermal-aging behavior of model hemicellulosic decomposition products, as reported by Rapson and Corbi,[4] and of kraft pulps, as reported by Chang and Kondo.[1] Exposure to visible and near-UV radiation caused no marked change in color.

Handsheets Made from Pulp Having Different Concentrations of γ-Cellulose

Figure 5 shows that a handsheet made from unextracted UBP darkened considerably upon thermal aging at 90°C and 50% RH, whereas handsheets based on alkali-extracted pulps from this stock discolored only moderately. The results are similar to the findings of Giertz and McPherson[6] that treatment with alkalis, which tends to remove hemicelluloses, increases the brightness stability during thermal aging.[6] Regardless of the γ-cellulose level in these test sheets, exposure to visible and near-UV radiation scarcely affected the color, even though changes in chemical composition could be detected.[10, 11] Little discoloration was observed when these same samples were exposed to mercury-vapor lamps that emitted considerable radiation at 254 nm. Similar results were reported by Yasuda et al. when various monosaccharides were irradiated with a mercury-vapor lamp that emitted radiation between 250 and 400 nm.[12]

10. Lewis, H. F., and Fronmuller, D., *Paper Trade J.*, 121 (1945), p. 25.

11. Lee, S. B., and Feller, R. L., unpublished data.

12. Yasuda, S., Nagaoka, M., and Hanzawa, M., *Res. Bull. Coll. Expt. For., Hokkaido Univ.*, 31, No. 1 (1974), p. 1.

Conclusion

Tests using hemicellulosic extracts from several different pulp stocks indicate that little if any discoloration is caused by exposure of these components to visible and near-UV radiation or even radiation at 254 nm (UV). In the continuing quest to understand the causes of discoloration in paper, these results support the contention that the tendency of hemicelluloses to lead to discoloration is primarily a thermally induced effect.

Darkening of Paper Following Exposure to Visible and Near-Ultraviolet Radiation

S.B. Lee, J. Bogaard and R. L. Feller

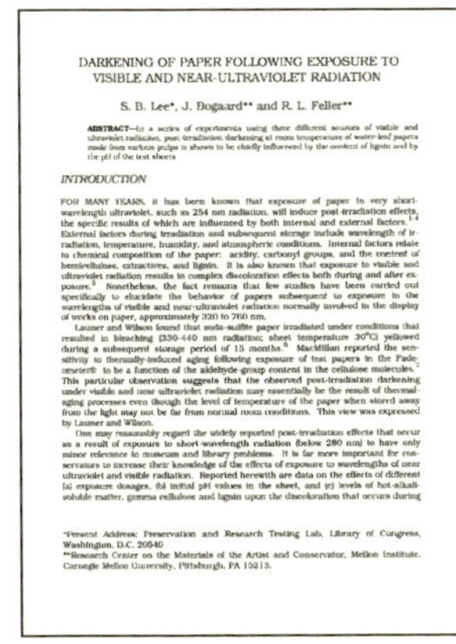
DARKENING OF PAPER FOLLOWING EXPOSURE TO VISIBLE AND NEAR-ULTRAVIOLET RADIATION

S. B. Lee*, J. Bogaard** and R. L. Feller**

ABSTRACT—In a series of experiments using three different sources of visible and ultraviolet radiation, post-irradiation darkening at room temperature of water-leaf papers made from various pulps is shown to be chiefly influenced by the content of lignin and by the pH of the test sheets.

INTRODUCTION

FOR MANY YEARS, it has been known that exposure of paper to very short-wavelength ultraviolet, such as 254 nm radiation, will induce post-irradiation effects, the specific results of which are influenced by both internal and external factors.[1-4] External factors during irradiation and subsequent storage include wavelength of irradiation, temperature, humidity, and atmospheric conditions. Internal factors relate to chemical composition of the paper: acidity, carbonyl groups, and the content of hemicellulose, extractives, and lignin. It is also known that exposure to visible and ultraviolet radiation results in complex discoloration effects both during and after exposure.[5] Nonetheless, the fact remains that few studies have been carried out specifically to elucidate the behavior of papers subsequent to exposure to the wavelengths of visible and near-ultraviolet radiation normally involved in the display of works on paper, approximately 320 to 760 nm.

Launer and Wilson found that soda-sulfite paper irradiated under conditions that resulted in bleaching (330-440 nm radiation; sheet temperature 30°C) yellowed during a subsequent storage period of 15 months.[6] MacMillan reported the sensitivity to thermally-induced aging following exposure of test papers in the Fade-ometer® to be a function of the aldehyde-group content in the cellulose molecules.[7] This particular observation suggests that the observed post-irradiation darkening under visible and near-ultraviolet radiation may essentially be the result of thermal-aging processes even though the level of temperature of the paper when stored away from the light may not be far from normal room conditions. This view was expressed by Launer and Wilson.

One may reasonably regard the widely reported post-irradiation effects that occur as a result of exposure to short-wavelength radiation (below 280 nm) to have only minor relevance to museum and library problems. It is far more important for conservators to increase their knowledge of the effects of exposure to wavelengths of near ultraviolet and visible radiation. Reported herewith are data on the effects of different (a) exposure dosages, (b) initial pH values in the sheet, and (c) levels of hot-alkali-soluble matter, gamma cellulose and lignin upon the discoloration that occurs during

*Present Address: Preservation and Research Testing Lab, Library of Congress, Washington, D.C. 20540

**Research Center on the Materials of the Artist and Conservator, Mellon Institute, Carnegie Mellon University, Pittsburgh, PA 15213.

Originally printed in:
Journal of the American Institute for Conservation Volume 28 (1989), Pages 1-18.

Introduction

For many years, it has been known that exposure of paper to very short-wavelength ultraviolet, such as 254 nm radiation, will induce post-irradiation effects, the specific results of which are influenced by both internal and external factors.[1-4] External factors during irradiation and subsequent storage include wavelength of irradiation, temperature, humidity, and atmospheric conditions. Internal factors relate to chemical composition of the paper: acidity, carbonyl groups, and the content of hemicellulose, extractives, and lignin. It is also known that exposure to visible and ultraviolet radiation results in complex discoloration effects both during and after exposure.[5] Nonetheless, the fact remains that few studies have been carried out specifically to elucidate the behavior of papers subsequent to exposure to the wavelengths of visible and near-ultraviolet radiation normally involved in the display of works on paper, approximately 320 to 760 nm.

Launer and Wilson found that soda-sulfite paper irradiated under conditions that resulted in bleaching (330-440 nm radiation; sheet temperature 30°C) yellowed during a subsequent storage period of 15 months.[6] MacMillan reported the sensitivity to thermally induced aging following exposure of test papers in the Fadeometer to be a function of the aldehyde-group content in the cellulose molecules.[7] This particular observation suggests that the observed post-irradiation darkening under visible and near-ultraviolet radiation may essentially be the result of thermal-aging processes even though the level of temperature of the paper when stored away from the light may not be far from normal room conditions. This view was expressed by Launer and Wilson.

1. Stillings , R. A., and Van Nostrand, R. J., "The Action of Ultraviolet Light upon Cellulose," *J. Am. Chem. Soc.*, 66 (1944), p. 753-760.

2. Launer, H. F., and Wilson, W. K., "The Photochemistry of Cellulose. Effects of Water Vapor and Oxygen in the Far and Near-Ultraviolet Regions," *J. Am. Chem. Soc.*, 71 (1949), p. 958-962.

3. Egerton, G. S., "The Mechanism of the Photochemical Degradation of Textile Materials," *J. Soc. Dyers and Colourists*, 65 (1949), p. 764-780.

4. MacClaren, R. H., Wells, F. L., Rosequist, J. V., and Ingerick, D. F, "Brightness Reversion of Cellulose Exposed to Ultraviolet Light," *Tappi*, 45 (1968), p. 789-793.

5. This subject has been reviewed by Feller, R. L., Lee, S. B., and Bogaard, J., "The Darkening and Bleaching of Paper by Various Wavelengths in the Visible and Ultraviolet," *Postprints*, The Book and Paper Group, American Institute for Conservation of Historic and Artistic Works, the tenth annual meeting, Milwaukee, Wisconsin (1982), p. 26-30.

6. Launer, H. F., and Wilson, W. K., "Photochemical Stability of Papers," *J. Research National Bureau of Standards*, 30 (1943), p. 55-74.

7. MacMillan, W. R., "Color Reversion in Oxidized Cellulose," M. A. Sc. Thesis, Department of Chemical Engineering and Applied Chemistry, University of Toronto, Canada (1958).

One may reasonably regard the widely reported post-irradiation effects that occur as a result of exposure to short-wavelength radiation (below 280 nm) to have only minor relevance to museum and library problems. It is far more important for conservators to increase their knowledge of the effects of exposure to wavelengths of near-ultraviolet and visible radiation. Reported herewith are data on the effects of different (a) exposure dosages, (b) initial pH values in the sheet, and (c) levels of hot-alkali-soluble matter, gamma-cellulose and lignin upon the discoloration that occurs during storage at room temperature subsequent to exposure of test sheets to radiation in the visible and near-ultraviolet region.

The primary objectives of the studies reported here are to determine (1) whether paper exposed to these wavelengths will undergo post-irradiation darkening under rather ordinary conditions of storage away from the light, (2) whether the pH of the paper has an influence on post-irradiation darkening under these circumstances, and (3) whether the content of lignin, rather than hot-alkali-soluble (HAS) matter, has a marked influence on the phenomenon. The answer to all three questions is affirmative. The information is of value with respect to the behavior that may be expected as the result of the display of works on paper. The results are not intended to be informative with respect to the practice of intentionally bleaching by light while discolored papers are immersed in an aqueous bath. That will be the subject of a publication to appear elsewhere.

RESULTS AND DISCUSSION

Effect of Exposure Time

Especially prepared water-leaf (without size) test sheets based on four types of paper stock, designated GP (groundwood pulp), UBP (unbleached pulp), FP (filter paper) and BP (bleached pulp) (see Table 1 regarding symbols and description of the pulps) were exposed during an initial series of tests for 18.8, 37.5, 65.8, and 155 hours in the Atlas Electric Devices xenon-arc Fadeometer equipped with Pyrex® filters. The respective net exposed energy, monitored at 420 nm, was 250, 510, 904, and 2120 kJ/m^2. Reflectance readings at 460 nm, taken from measurements of the spectrophotometric curves, were noted at various times on the samples during exposure in the Fadeometer and during a subsequent period of storage in the dark in a steel cabinet in a constant-temperature and constant-humidity room maintained at 23°C and 50% RH. Darkening or bleaching is expressed in terms of the Kubelka-Munk function for the ratio of absorption coefficient, K, to the scattering coefficient, S, where $K/S = (1 - R)^2 / 2R$, and R is the decimal fraction reflectance at 460 nm of a stack of six sheets of the paper being tested, relative to a standard white of barium sulfate. Table 2 compares K/S values with percent reflectance; increased K/S indicates darkening.

Table 1
Characteristics of Paper Samples Used
(Before Exposure to Light)

Sample Designation	Description	Initial Reflectance at 460 nm (%)	Lignin Content (%)	Initial Hot-1%-Alkali Solubility (%)	Copper Number
BP	Bleached Pulp 50% hardwood kraft/ 50% softwood kraft	83.6	0.24	5.0	0.5
UBP	Unbleached Pulp 50% hardwood kraft/ 50% softwood kraft	30.8	4.4	3.7	0.6
GP	Unbleached Groundwood Pulp	57.0	29.1	15.6	2.9
FP	Filter Paper Whatman No. 42	88.2	0	1.6	0.2

Table 2
Comparison of Reflectance Values with Kubelka-Munk K/S Function*

Percent Reflectance	K/S
30	0.817
40	0.450
50	0.250
60	0.133
70	0.064
80	0.025
85	0.013

* $K/S = (1 - R)^2 / 2R$ where R is the percent reflectance divided by 100.

Changes in the K/S values, Δ(K/S), are plotted against exposure and storage time in Figure 1 for the GP test sheets and in Figure 2 for the UBP, FP, and BP sheets. [The more commonly used expression, post-color number,[8] would be 100 times this, based on measurements with a narrow-band-pass filter peaked at 457 nm. However, with the introduction of modern spectrophotometers, it has proved convenient for this laboratory to use the reflectance at 460 nm. The net change is practically the same whether the reflectance at a wavelength of 457 or 460 nm is used.][9]

8. Giertz, H. W., "The Yellowing of Pulp," *Svensk Papperstidning*, 48, No. 13 (1945), p. 317-323.

9. For further discussion of the relationship of Tappi brightness and the use of K/S values see Feller, R. L., "Comments on the Measurement of 'Yellowness' in Pulp and Paper," *The Book and Paper Group Annual* (AIC), 6 (1987), p. 40-51.

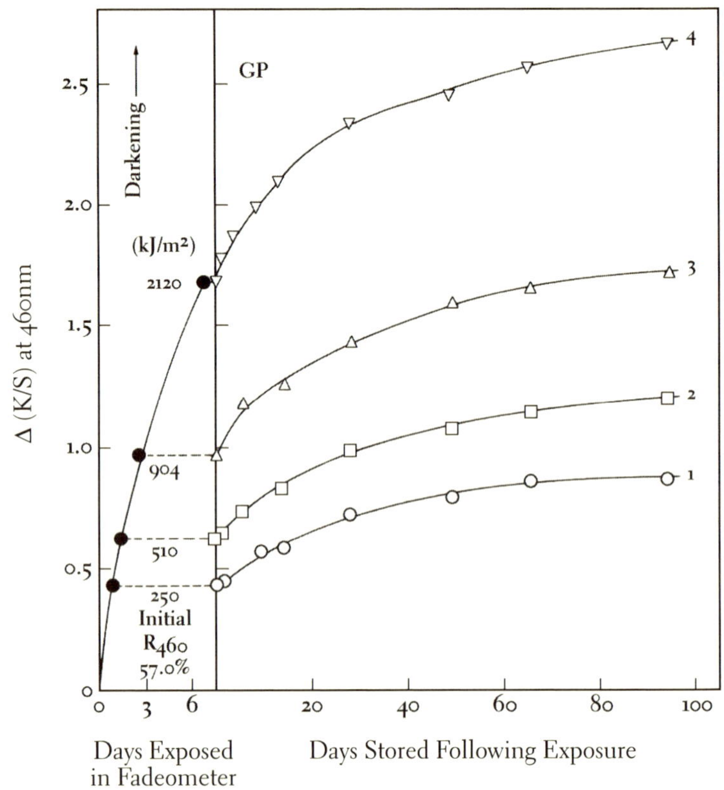

FIGURE 1
Darkening of test sheets expressed as increase in K/S values at 460 nm, Δ(K/S), of groundwood pulp (GP) during and following various exposures in the xenon-arc Fadeometer (Pyrex® Glass Filters). Exposures expressed in terms of kJ/m^2 at 420 nm. Subsequent storage at room conditions as described in text.

Three papers employed in these experiments were low in lignin. These did not change in color as much as the test sheets made with the groundwood pulp (GP). This can be seen in the much smaller changes in K/S values, Δ(K/S), in Figure 2 as compared to Figure 1. Subsequent to exposure, the filter paper (FP) darkened least of all, as one might expect in the case of a high-alpha-cellulose pulp. The particular sample of unbleached pulp (UBP) used in these tests was rather dark colored in its initial state, perhaps caused by alkalinity in the pulping treatments. In a number of laboratory experiences with this pulp we have found that it tends to bleach, at least initially, when exposed to either near-ultraviolet or visible radiation (Figure 2a). In the case of the bleached pulp (BP) sheets, the net change during exposure was very small (Figure 2c); nonetheless, the sheets darkened initially, then bleached.

Regardless of whether exposure in the Fadeometer had caused bleaching or darkening, all of the exposed samples subsequently darkened when stored in darkness in a laboratory cabinet at 50% RH and 23°C (which we shall designate as post-irradiation darkening).[10]

10. The expression "brightness reversion" has often been used for darkening that occurs after various treatments that lead to brightening. We are not in favor of this terminology because, if one component is bleached during a treatment, and another component is responsible for the subsequent discoloration, the paper may not have *reverted* to its original color or composition of chromophoric groups. It is important for the research laboratory to begin to ask: what is responsible for the color before treatment and what is responsible for the change that occurs as the immediate result of treatment, that which arises after treatment?

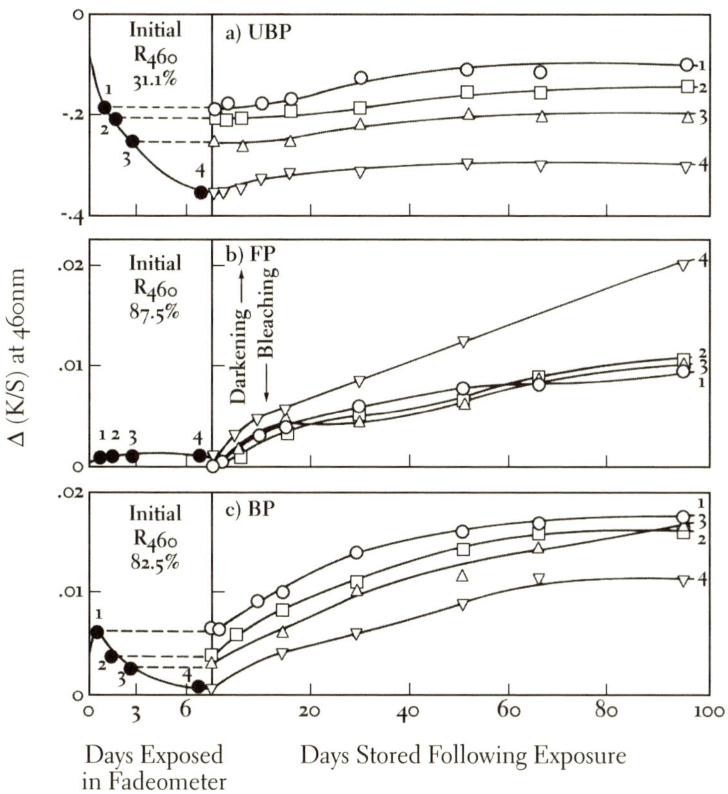

FIGURE 2
Changes in K/S value, Δ(K/S), at 460 nm for test sheets prepared with unbleached pulp (UBP), filter paper (FP) and bleached pulp (BP). Exposures 1, 2, 3, and 4 are the same conditions as in Figure 1. Note differences in vertical scale.

In our experience we have observed that test sheets containing significant amounts of lignin tend to darken when exposed to near-ultraviolet radiation and higher-than-normal temperatures. Thus, one observes that the groundwood pulp (GP) samples darkened extensively during exposure in the Fadeometer (Figure 1). Figure 3 suggests that the post-irradiation darkening [$K/S_{90\ days} - K/S_{initial}$, Δ(K/S)] of the groundwood sheets was related to the amount of exposure. This relationship was not evident, however, in the case of the three papers that contained low levels of lignin. The slight degree of post-irradiation darkening of the UBP and BP samples [only one-tenth that of the groundwood pulp, GP, sheets when expressed as Δ(K/S)] was practically the same regardless of the irradiation dosage. Any trend in post-irradiation darkening in these papers relative to the amount of exposure would have been obscured by experimental errors owing to the small changes in reflectance that took place. The longest exposed sample of filter paper (FP) seemed to have darkened more than the other samples during storage; nonetheless, the net change in all test sheets of this paper was too small to permit a firm conclusion (Figure 2b). The almost

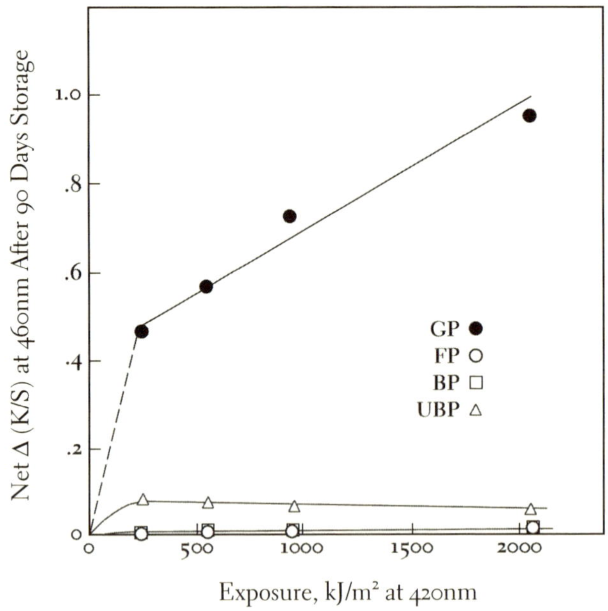

Figure 3
Post-irradiation darkening [increased Δ(K/S) at 460 nm] of various test sheets as a function of exposure. Data from Figures 1 and 2.

negligible darkening of the filter paper correlates well with the small changes observed in the low-lignin examples, BP and UBP.

On first inspection of the major differences in composition of the papers (Table 2), it appears that the degree of post-irradiation darkening following exposure in the Fadeometer might be influenced by the content of hot-alkali-soluble (HAS) matter (a measure of the oxidized and low-molecular-weight fractions of the pulp) as well as the lignin content in the paper. In this sense, the post-irradiation darkening might be considered to exhibit characteristics that one could expect from thermal-aging tests carried out on papers that have different degrees of oxidation as traditionally represented by carbonyl-group content, a property commonly reflected in the measurement of copper number or hot 1%-alkali-soluble (HAS) matter. This is essentially the conclusion reached by MacMillan using chemically-oxidized cotton linters.[7] This point of view will be further considered in a later section in which handsheets made from pulps derived from groundwood were prepared in which both lignin content and HAS matter were varied. Probably both lignin and HAS content have an influence on post-irradiation darkening, but in the experiments described here HAS matter had little or no effect.

Effect of Initial pH

Acidity greatly influences the thermally-induced deterioration of paper; how does acidity affect the post-irradiation darkening? Using handsheets prepared from two types of pulp (UBP, GP) and aged filter paper (A-FP, Whatman No. 42 filter paper thermally aged at 90°C and 50% RH for 150 hours), post-irradiation discoloration was tested as a function of initial acidity of the sheets having pH values from 3.2 to 10.2 as determined by cold-water extraction. Air-dried test sheets at different pH's were prepared by treating the sheets with various buffers or with calcium hydroxide solution (see Table 3). "Daylight" fluorescent and BLB blacklight fluorescent lamps were used in order to observe the effects of both visible radiation (having a trace of near-ultraviolet) and near-ultraviolet alone on samples, exposing them at a lower temperature than was the case with the tests carried out in the Fadeometer. The results of exposing and subsequently storing samples of the three papers having six different values of initial pH are shown in Figure 4. (Because the presentation of the data would be too crowded, not every paper treated with the buffers in Table 3 is presented in this figure.) The GP sheets and the aged filter paper (A-FP) that were on the acid side (treated with buffer solutions at low pH) tended to darken faster, not only during the period of exposure to visible or ultraviolet radiation but also during subsequent storage at room temperature in the dark (post-irradiation darkening) (Figures 4a and 4c). In contrast, sheets treated with an alkaline buffer generally were stabilized against darkening both during the initial exposure to visible and near-ultraviolet radiation as well as during the subsequent storage in the dark. Least protection by alkalinity was observed in the case of the GP samples. Thus, post-irradiation darkening took place in practically every instance, but tended to be minimized as the pH of the papers tended towards alkalinity.

The UBP sample, an initially dark-colored pulp, bleached during exposure to the "daylight" fluorescent lamps; the acid sheets bleached the least. During

TABLE 3
Buffer Solutions Used for Paper Treatment*

pH 2.8:	Potassium hydrogen phthalate + HCl[11]
pH 3.5:	Potassium hydrogen phthalate + HCl[11]
pH 4.01:	Potassium hydrogen phthalate buffer[12]
pH 5.3:	Potassium hydrogen phthalate + NaOH[11]
pH 6.86:	Phosphate buffer (sodium and potassium salts)[12]
pH 11:	Sodium bicarbonate + NaOH[11]
pH 11.7:	0. 1% $Ca(OH)_2$ solution[13]

*Test sheets dipped in the buffers and air-dried overnight did not have precisely the pH values of these impregnating solutions, but had the values as seen in Figures 4 and 5, as determined by cold-water extraction of the air-dried test sheets.

11. *Handbook of Chemistry and Physics*, 63rd ed., The Chemical Rubber Co., Cleveland, Ohio (1983), p. D-155.

12. Tappi Standard T509.

13. Hey, M., "The Washing and Aqueous Deacidification of Paper," *The Paper Conservator*, 4 (1979), p. 66-80.

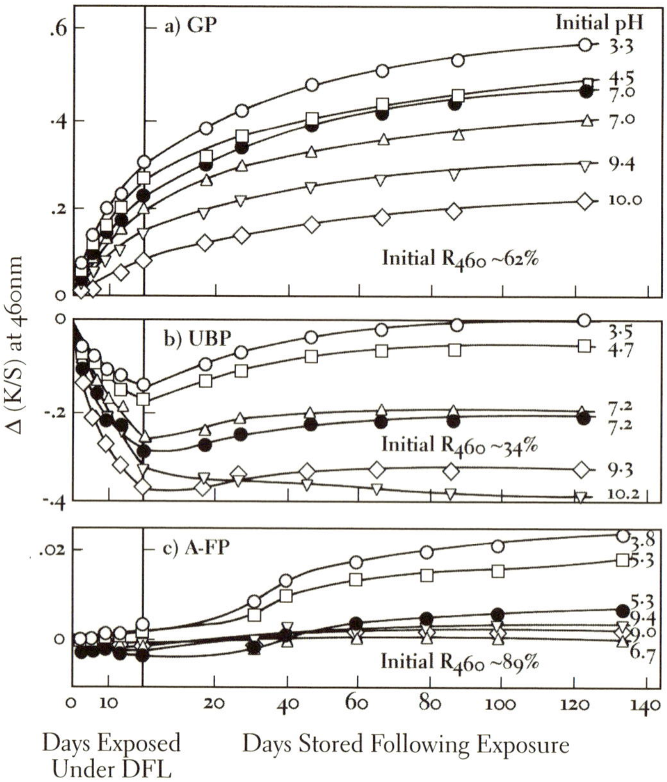

FIGURE 4
Initial changes and post-irradiation darkening [Δ(K/S) at 460 nm] of various test sheets as a function of initial pH of sheet following exposure to daylight fluorescent lamps (DFL). Note difference in vertical scale in c). The black points refer to sheets cast from water, unbuffered.

subsequent storage, the UBP sheets tended to darken, the extent likewise being a function of the pH. As seen in Figure 4b, the alkaline-buffered papers exhibited the least post-irradiation darkening; at pH 10.2 the sheet apparently continued to bleach. Much the same results were observed with the UBP sheets during and following exposure to BLB blacklight fluorescent lamps, which emit principally near-ultraviolet radiation.

The net post-irradiation change in the aged filter paper (A-FP) was an order of magnitude lower but tended to follow much the same trend (Figure 4c).

When the net change in K/S at 460 nm after 120 days' storage, Δ(K/S), taken from Figure 4, is plotted against initial pH in the paper, Figure 5a results. In the case of exposure to BLB blacklight fluorescent lamps (near-ultraviolet), the same general behavior is seen (Figure 5b). Although there are deviations in the experimental results, a general trend is indicated; post-irradiation darkening tends to be markedly reduced when the alkalinity of the sheets is above pH 9 and the degree of darkening increases when the papers have acidities greater than about pH 5. When one considers that pH is a function of the

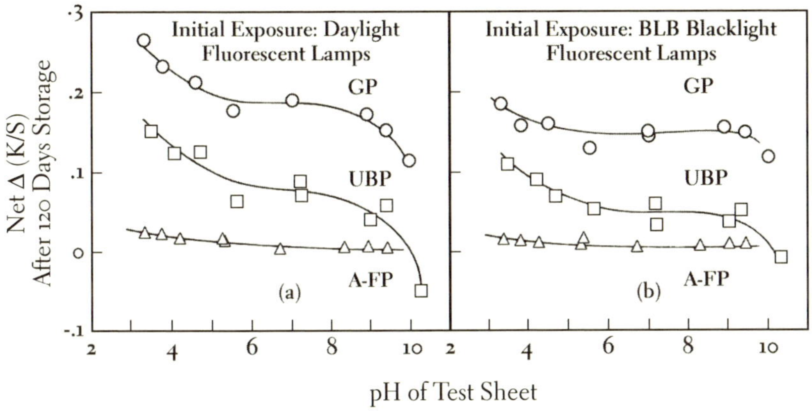

FIGURE 5
Post-irradiation darkening [represented by the change in K/S, Δ(K/S), at 460 nm] after 120 days' storage as a function of the initial pH of various test sheets exposed to daylight or BLB blacklight fluorescent lamps: groundwood pulp (GP), unbleached pulp (UBP) and aged filter paper (A-FP).

logarithm of the acidity or alkalinity, it is perhaps not unexpected that the accelerating effects at low pH would sharply curve upward and the minimizing effects of alkalinity would curve sharply downward in the linear plot of Figure 5. In the case of paper that contained little lignin, A-FP, acidity and alkalinity had scarcely any effect on the post-irradiation darkening. Again, one observes that the general level of darkening of the three types of paper, in the near-neutral condition, is in the order of the lignin content.

These results are simply technical observations regarding what happens to the papers having these values of pH when they are exposed to visible and near-ultraviolet radiation while on exhibition and are not at this stage intended to suggest changes in conservation-treatment procedures.

Effect of Gamma-Cellulose Content

Bulk material was subjected to a short period of chloriting to remove much of the original lignin content from the unbleached pulp (UBP) (about 4% initially, diminished by this treatment to less than 1%). Caustic extraction at different strengths of alkali solution was then employed to reduce the level of gamma-cellulose in individual samples. The composition of the resultant pulps is reported in Table 4; previous experiments using these same pulps have been described elsewhere.[14]

The post-irradiation effect on handsheets made from the different alkali-extracted UBP samples was investigated following exposure to the near-ultraviolet radiation from BLB blacklight fluorescent lamps. Exposures were carried out in the constant-temperature and -humidity room at 23°C and 50% RH for 19 days. Near-ultraviolet was used because it was considered that this radiation would

14. Lee, S. B., and Feller, R. L., "Influence of the Hemicellulose Fraction on Thermal and Photochemical Discoloration of Paper," in *Historic Textile and Paper Materials*, Advances in Chemistry Series, No. 212, ed. Needles, H. L., and Zeronian, S. H., American Chemical Society, Washington, DC (1986), p. 377-386.

Table 4
Characterization of Caustic-Extracted Pulps Prepared from UBP Stock
Composition of Resulting Celluloses

Treatment Conditions with Caustic Extraction	Alpha-Cellulose	Beta-Cellulose	Gamma-Cellulose	HAS Matter
Lightly Chlorited UBP	88.0	1.5 *	10.5	3.7
Same, Extracted with				
1% NaOH (80°C, 2 hrs.)	88.5	2.8**	8.8	2.8
5% NaOH (20°C, 1 hr.)	93.2	1.7**	5.1	2.2
17.5% NaOH (20°C, 1/2 hr.)	96.8	1.8**	1.5	2.0

* This figure, determined from 100 - (alpha + gamma cellulose), is high because the UBP sample contained 1.1% lignin; the content of beta-cellulose must have been about 0.4% in this sample.

** Since a substantial amount of lignin is extracted upon alkali treatment, these beta-cellulose values are only slightly in error owing to a trace of lignin remaining in the pulps.

provide a more severe exposure than daylight fluorescent lamps and therefore would enhance any detrimental effects of exposure that might be observed.

The changes in K/S values at 460 nm that occurred during exposure to the BLB lamps and during subsequent storage at room temperature of 23°C and 50% RH are shown in Figure 6. During the short exposure of about 30 hours to the near-ultraviolet radiation of the BLB blacklamps, all samples rapidly, although slightly, darkened. Thereafter during the period of exposure, samples that had been extracted with alkali, which had low gamma-cellulose content (8.8% or lower), tended to bleach, returning essentially to the K/S values at 460 nm that they had before exposure. The original sample of UBP (unextracted) also bleached extensively and, during storage, did not darken sufficiently in 90 days to return to its original K/S value.

Note that the very small change observed in these tests (the right-side curves in Figure 6) exhibited almost the same rate of post-irradiation darkening irrespective of the composition of the extracted papers or the extent of the bleaching that had taken place under the BLB lamps. The test sheets contained little lignin; the percentage probably was no more than 1%. Little net darkening occurred during 90 days of storage; $\Delta(K/S)$ for the three extracted papers was about 0.0038 (increased darkening) in this time, whereas sheets of the unextracted pulp increased only slightly more, to the extent of 0.0058, perhaps because these sheets had the slightly higher content of lignin.

The results demonstrate, once again, that the post-irradiation tendency to darken, measured at 460 nm, is largely independent of the bleaching or darkening effects that may have occurred during exposure.

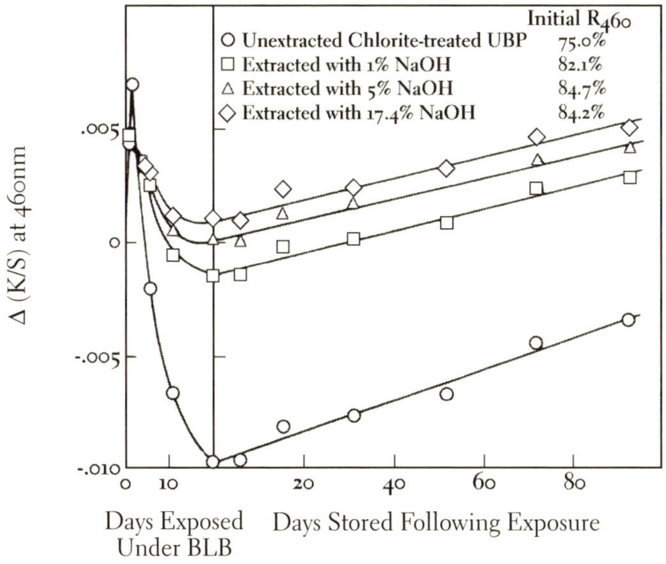

Figure 6
Initial changes and post-irradiation darkening of unbleached pulp sheets alkali-extracted to achieve decreasing levels of gamma-cellulose content. Note expanded vertical scale; the changes are small.

Effect of Lignin Content and Initial Hot-1%-Alkali-Soluble (HAS) Matter

As a final experiment, a variety of pulp samples were prepared from the groundwood pulp stock (GP) to allow the content of lignin and HAS matter to be varied as the result of a chloriting treatment followed by hot-1%-alkali extraction and by extraction with 17.5% sodium hydroxide solution. To minimize the possible effects of unintended oxidation of the cellulose, all samples were reduced with sodium borohydride prior to exposure. The values of initial lignin content and initial HAS matter in these samples are noted in Figure 7 where NC stands for not-chlorite-treated (the treatment to remove lignin); 1HC, 2HC, and 4HC are pulp stocks chlorite-treated for 1, 2, and 4 hours.

Twelve test sheets made from pulps thus variously prepared were exposed under the BLB lamps for 400 hours and then left in the dark at a room temperature of 23°C, 50% RH for up to 90 days to check post-irradiation darkening. It is clearly seen in Figure 8 that the post-irradiation darkening of these pulps is related to lignin content. This supports the general conclusion indicated by the Fadeometer exposures shown in Figure 1 versus Figure 2. When the results are compared to the initial content of HAS matter, in Figure 9, there is no indication that HAS content plays a major role in post-irradiation darkening.

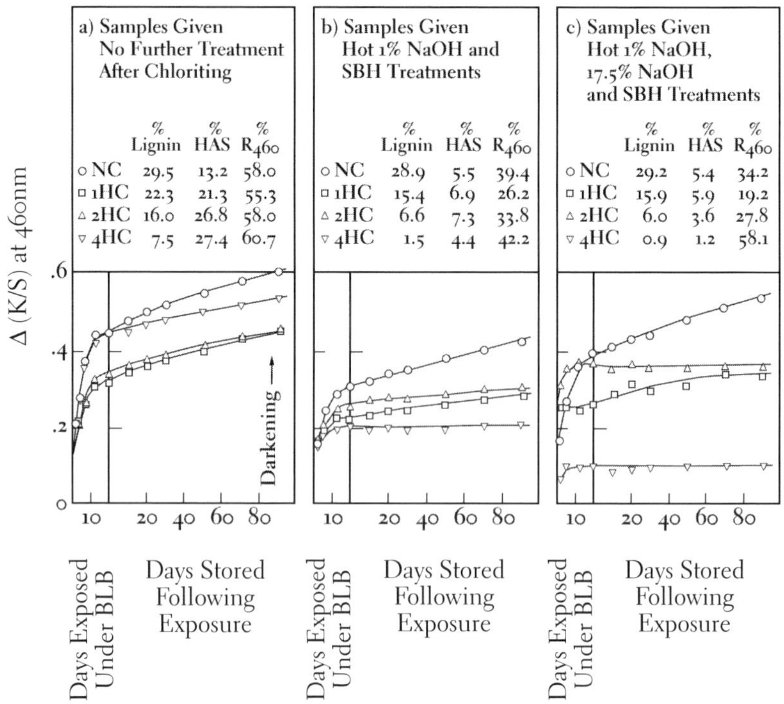

Figure 7
Initial and post-irradiation darkening of pulps prepared at various levels of lignin and hot-alkali-soluble (HAS) matter. Effects of exposure to BLB blacklight fluorescent lamps.

Conclusions

The results in Figures 1, 2, 4, 6, and 7 demonstrate that, regardless of whether darkening or bleaching took place during exposure to visible or near-ultraviolet radiation, papers darkened upon subsequent storage in the dark over three-months' time at normal room conditions. For a high-lignin-content (29.1%) groundwood pulp, the extent of post-irradiation darkening after 90 days' storage increased with increased initial exposure in a xenon-arc Fadeometer (Pyrex® filters) (Figure 3). In contrast, post-irradiation darkening was little affected by the amount of initial exposure in the Fadeometer in the case of test sheets made from pulps that contained 4.4% lignin or less.

In a series of experiments in which exposures were carried out at temperatures of about 25 to 28°C, the extent of initial darkening or bleaching and also the extent of post-irradiation darkening was found to be influenced by the pH of the test papers. Acid papers tended to darken the most upon subsequent storage and highly alkaline papers the least (Figure 4). The post-irradiation darkening of filter paper that had been previously thermally aged was scarcely affected by the pH of the test sheets (Figure 5).

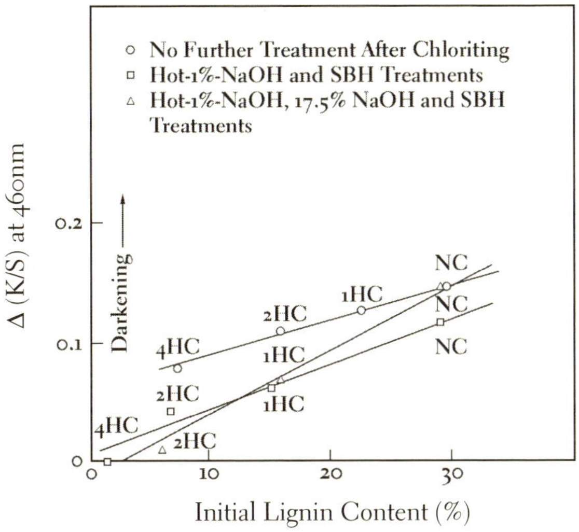

Figure 8
Post-irradiation darkening of test sheets after 90 days' storage as a function of lignin content (summary of data from Figure 7).

In an especially prepared series of pulps, post-irradiation darkening was found to be directly related to lignin content over the range between 0.9 to 29.5% lignin (Figure 8). This result confirms the trend of relative darkening noted in papers containing 29.1, 4.4 and less than 0.24% lignin which were exposed in a Fadeometer (Figures 3 and 5).

The presence of varying amounts of HAS matter was found to have no positive correlation with post-irradiation darkening (Figure 9). With a series of test sheets containing little lignin but varying amounts of gamma-cellulose (Figure 6), little darkening was induced subsequent to exposure to near-ultraviolet radiation.

Experimental

Preparation of Paper at Various pH Values

Three types of water-leaf test sheets—based on a supply of groundwood pulp (GP) and unbleached pulp (50% hardwood kraft/50% softwood kraft) (UBP), as well as unaged (FP) and aged filter paper (A-FP, Whatman No. 42 filter paper, previously aged 150 hours in an oven at 90°C, 50% RH)—were pretreated in various buffer solutions and air-dried overnight to produce test sheets having a wide range of initial pH values as measured by cold-water extract, Tappi Standard T509 Su-68.

Table 3 presents the list of the buffers used to produce the sheets at various pH values. The resultant water-extracted pH values, measured on the sheets that had been dipped into these solutions and air-dried, were not the same pH

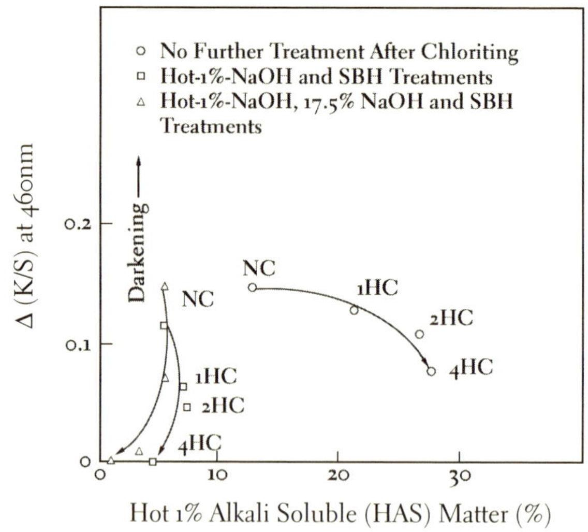

FIGURE 9
Post-irradiation darkening after 90 days' storage of test sheets plotted as a function of hot-alkali-soluble (HAS) matter (summary of data from Figure 7). Arrows indicate sequence of samples in order of chloriting treatment (to remove lignin). No positive correlation between HAS matter and post-irradiation darkening is indicated.

as that of the buffer solutions used to prepare the test sheets. The cold-water-extract pH values of the test sheets as finally achieved in the sheets are those employed in Figures 4 and 5.

Preparation of Pulps With Various Contents of Gamma-Cellulose

The unbleached pulp stock was subjected to a brief chloriting treatment followed by an extraction with different strengths of alkali solution to remove the lignin and gamma-cellulose.[15] Twenty-five grams of sodium chlorite and 12.5 ml of glacial acetic acid were reacted at a temperature of 75 to 80°C for 30 minutes at a pulp consistency of 10% using 105 g air-dried pulp. This treatment reduced the lignin content of the pulp. Caustic extractions were subsequently performed at a pulp consistency of 2%, with the concentrations of alkali, reaction temperature, and reaction time varied, as noted in Table 4.

After the caustic extractions, the pulp was filtered off (except the 17.5%-alkali treated pulps, which were first diluted with water and then decanted), 500 ml of 10% acetic acid added and the mixture allowed to soak for 1 minute before removal of the pulp. Treatment with acetic acid was repeated with a second 500 ml portion. The pulp was then washed several times until the wash water tested acid-free with pH paper. Handsheets were made from the resulting pulps at a basis weight of 75 g/m², as other types of handsheets.

Alpha-, beta-, and gamma-cellulose in the resultant pulps were determined by standard Tappi methods as noted in a previous publication.[14]

15. Wise, L. E., Murphy, M., and D'Addieco, A. A., "Chlorite Holocellulose, Its Fractionation and Bearing on Summative Wood Analysis and on Studies on the Hemicelluloses," *Paper Trade J.*, 122, No. 2 (1946), p. 35-43.

Pulp Treatments to Produce Variable Lignin Content and HAS Matter

To prepare the pulp at various contents of lignin, groundwood pulp (designated as NC, not chlorited) was subjected to chloriting, using the technique of Wise et al.[15] for 1, 2, and 4 hours, respectively (yielding stock samples 1HC, 2HC, and 4HC). Further treatments were carried out with hot-1%-alkali solution in order to remove the degradation products arising from chloriting, and with 17.5% caustic soda to remove the hemicellulose component. Thereafter, these pulps were treated with 2% sodium borohydride (SBH), at 2% pulp consistency and room temperature for 20 hours with occasional stirring, to convert aldehyde groups to hydroxyl. Lignin content and HAS matter were determined on the resultant pulps by Tappi methods T222 os-74 and T212 os-76, respectively.

Exposure to Light Sources

Three different light sources were used: (a) a bank of six General Electric High-Output (HO) "daylight" fluorescent lamps, (b) a bank of six fluorescent BLB blacklights, and (c) an Atlas 6500-W xenon-arc Fadeometer with Pyrex® filters. Data on the irradiance under the three lamps is given in Table 5, information provided primarily to indicate the percentage of ultraviolet emitted.

TABLE 5
Irradiance From Different Light Sources (mW/cm^2)

Light Source	Ultraviolet Energy (310-392 nm)	Visible Energy (400-750 nm)	Total Energy (257-750 nm)	Sample Distance from Source (inches)
G.E. Fl. Black-light F40 BLB[a]	4.13 (86%)	.70	4.90	3.5
Fl. "Daylight" High-Output[a]	.228 (3.8%)	5.70	5.928	3.5
Xenon-Arc Fadeometer® Pyrex/Pyrex Filters[b]	13.775[c] (6.3%)	203.49[d]	217.265[e]	11.6

a. Measured with an International Light IL700 Radiometer.
b. Measurements obtained from Atlas Electric Devises, Inc. catalog, multiplied by 1.9 (factor for 11.6" diameter drum).
c. Irradiance from 340-400 nm.
d. Irradiance 400-750 nm.
e. Irradiance from 340-750 nm.

Both types of fluorescent lamps were mounted about 3 1/2" above the samples in a room maintained at 50% relative humidity (RH) and 23°C. The temperature of the handsheets reached about 27°C. In the 6500-W Fadeometer the air was maintained at 31°C (±1.0°) and 27% RH; the black-panel temperature reached 65°C.

Storage Conditions

Following exposure to the various light sources, samples were kept in the closed steel cabinet in the paper-testing room maintained at 23°C and 50% RH.

Section 9: Overviews

The modern paint and protective coatings industry has had relatively little exposure to the needs and concerns of the conservation field. The occasion of a talk given before the annual meeting of the Federation of Societies for Paint Technology in 1971 was taken as an opportunity to introduce the audience to the conservator's interest in historic pigments and manufacturing methods, and to thank a number of colleagues in industry for their assistance in the past.

At the dedication of the new quarters of the Conservation Center at New York University, the Stephen Chan House, I was asked to discuss the role of science in conservation and conservation training. Mrs. Feller and I had recently returned from a conference trip to Australia and New Zealand. Shortly before, a commercial airliner had been shot down on that side of the world with all lives lost. Sober thoughts influenced the choice of remarks. Perhaps it would have been expected on such an occasion to extol the many potential contributions of science. Instead, it seemed useful to remind ourselves that the development of well-tested, innovative applications is a time-consuming task, that ready and certain success in research is not guaranteed, and that new treatment methods based on basic and applied research studies are still not extensive.

Later, an opportunity was taken to review the various types of questions that can be posed by the conservator and also the different degrees of assurance expected in the answers. Consider for example the basic question of how specific materials lose their initial properties over time: do they undergo a steady rate of decline, tend to speed up, or slow down? Under ordinary conditions of exhibition and storage, the processes of degradation usually take place so slowly that these different patterns of deterioration are not readily apparent. Thus, to date only a few materials have been recognized to deteriorate significantly in one or another of these modes, and as yet only a few appropriate methods of treatment have been devised based on such knowledge.

R. L. F.

Scientific Examination of Artistic and Decorative Colorants

ROBERT L. FELLER

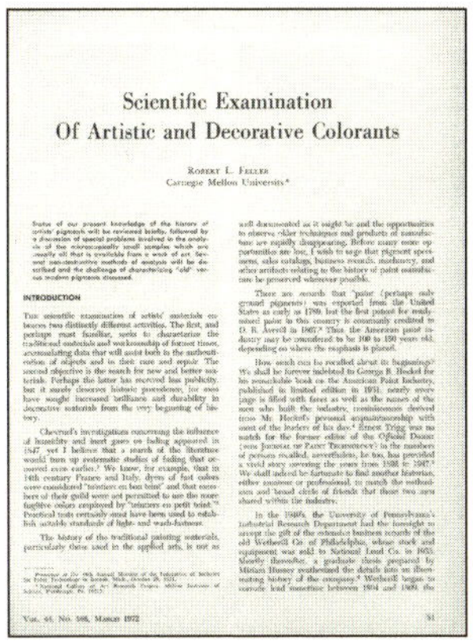
Scientific Examination
Of Artistic and Decorative Colorants
Robert L. Feller
Carnegie Mellon University*

Originally printed in:
Journal of Paint Technology
Volume 44, Number 566
(March 1972), Pages 51-58.

Introduction

The scientific examination of artists' materials embraces two distinctly different activities. The first, and perhaps most familiar, seeks to characterize the traditional materials and workmanship of former times, accumulating data that will assist both in the authentication of objects and in their care and repair. The second objective is the search for new and better materials. Perhaps the latter has received less publicity, but it surely deserves historic precedence, for men have sought increased brilliance and durability in decorative materials from the very beginning of history.

Chevruel's investigations concerning the influence of humidity and inert gases on fading appeared in 1847, yet I believe that a search of the literature would turn up systematic studies of fading that occurred even earlier.[1] We know, for example, that in 14th century France and Italy, dyers of fast colors were considered *teintiers en bon teint* and that members of their guild were not permitted to use the more fugitive colors employed by *teintiers en petit teint.*[2] Practical tests certainly must have been used to establish suitable standards of light- and wash-fastness.

The history of the traditional painting materials, particularly those used in the applied arts, is not as well documented as it might be and the opportunities to observe older techniques and products of manufacture are rapidly disappearing. Before many more opportunities are lost, I wish to urge that pigment specimens, sales catalogs, business records, machinery, and other artifacts relating to the history of paint manufacture be preserved wherever possible.

1. Chevruel, M. E., *Recherches sur la Teinture*, Martinet, Paris (1847).

2. Bancroft, E., *Experimental Researches Concerning the Philosophy of Permanent Colours*, G. Sidney, London (1813).

There are records that "paint" (perhaps only ground pigments) was exported from the United States as early as 1789, but the first patent for ready-mixed paint in this country is commonly credited to D. R. Averill in 1867.[3] Thus, the American paint industry may be considered to be 100 to 150 years old, depending on where the emphasis is placed.

3. Averill, D. R., U. S. Patent 66,773 (July 16, 1867).

How much can be recalled about its beginnings? We shall be forever indebted to George B. Heckel for his remarkable book on the American Paint Industry, published in limited edition in 1931; nearly every page is filled with faces as well as the names of the men who built the industry, reminiscences derived from Mr. Heckel's personal acquaintanceship with most of the leaders of his day.[4] Ernest Trigg was no match for the former editor of the *Official Digest* (now *Journal of Paint Technology*) in the numbers of persons recalled; nevertheless, he too has provided a vivid story covering the years from 1892 to 1947.[5] We shall indeed be fortunate to find another historian, either amateur or professional, to match the enthusiasm and broad circle of friends that these two men shared within the industry.

4. Heckel, G. B., "The American Paint Industry: Reminescences and Comments," *American Paint Journal Co.*, St. Louis (1931).

5. Trigg, E. T., *Fifty-five Colorful Years*, Pequot Press, Stonington, Conn. (1954).

In the 1940s, the University of Pennsylvania's Industrial Research Department had the foresight to accept the gift of the extensive business records of the old Wetherill Co. of Philadelphia, whose stock and equipment was sold to National Lead Co. in 1933. Shortly thereafter, a graduate thesis prepared by Miriam Hussey synthesized the details into an illuminating history of the company.[6] Wetherill began to corrode lead sometime between 1804 and 1809, the first company to do so in America; a copy of one of their labels of about the 1840s, shown in Figure 1, may be the oldest label of an American-made paint product on record. Names, places and dates of long-vanished companies may also be found in two theses recently prepared in the history department of the University of Delaware; and records of materials and methods can be found in Richard

6. Hussey, M., *From Merchants to 'Colourmen'*, University of Pennsylvania Press, Philadelphia, Pa. (1956).

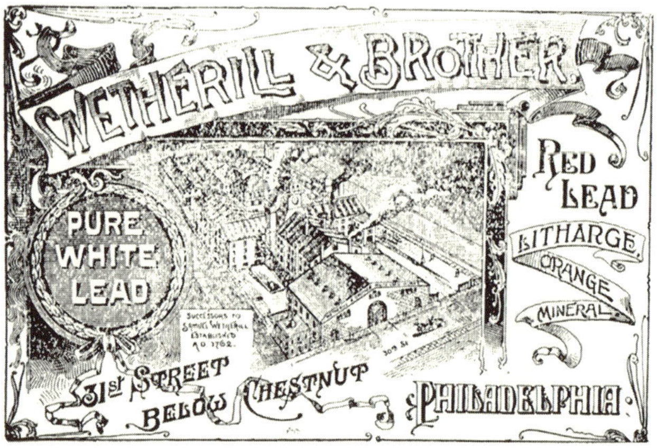

FIGURE 1
Mid-nineteenth century label of the oldest manufacturer of lead white in America (from Miriam Hussey, *From Merchants to 'Colourmen'*).

M. Candee's survey of Housepaints in Colonial America, based upon his thesis submitted to the State College of New York at Oneonta in 1966.[7-9]

These admirable studies provide a sound foundation upon which someone, someday, may construct a more complete history of the decorative coatings industry in America than we have at present.

OBSOLESCENCE OF TECHNICAL TEXTS

By the time an able and willing historian is found, many of the traditional methods of manufacturing may be lost and forgotten. The Smithsonian Institution, several years ago, made an appeal through the National Paint, Varnish and Lacquer Association for the collection of older types of machinery that would represent important facets of the industry. I understand from the curator of chemistry there, Jon Eklund, that a number of companies did make a search and managed to turn up some interesting pieces of equipment. With his permission, I shall again emphasize the urgency of his request and the desirability of continuing the search for company records, formulation books, trade catalogs, photographs, machinery, testing equipment, and textbooks which ought to be preserved. Consideration should be given to the rarity and historic interest of all such materials before they are just simply thrown away.

Technical books of scarcely a generation long ago are now quite passé. Libraries, finding that there is little call for them, often remove them from the shelves, yet such texts are of considerable historical interest for they frequently describe methods of manufacture no longer or very rarely employed. Without them, how well shall we be able to reconstruct the character of the coatings industry of 70-90 years ago (Figure 2)?[10] Books such as Louis Andes's *Oil Colours and Printers' Inks* of 1903, for example, depict a number of simple machines for producing lamp black, many of which must be completely lost.[11]

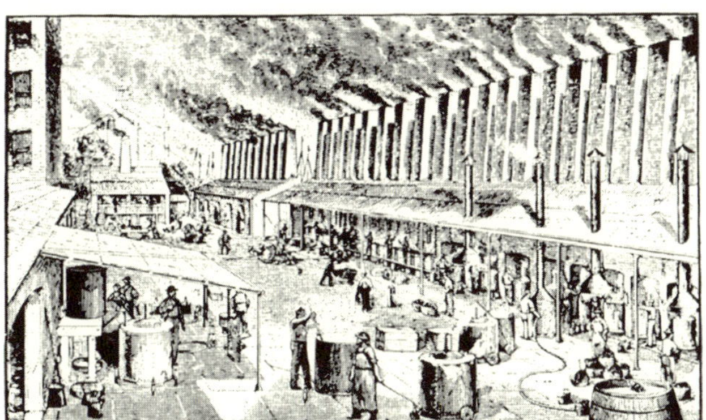

FIGURE 2
Varnish production in grandfather's day, Berry Brothers, Detroit (from Footnote 10).

7. Penn, T. Z., "Decorative and Protective Finishes, 1750-1850: Materials, Process and Craft," Thesis submitted to the faculty of the University of Delaware in partial fulfillment of the requirements for the degree of Master of Arts in History (June, 1966).

8. Green, T. O., "The Birth of the American Paint Industry," Thesis submitted to the faculty of the University of Delaware in partial fulfillment of the requirements for the degree of Master of Arts in History (June, 1965).

9. Candee, R. M., *House Paints in Colonial America*, New York Chromatic Publishing Co. (1967); Reprinted from *Color Engineering* (1966-1967).

10. Sabin, A. H., *The Industrial and Artistic Technology of Paint and Varnish*, J. Wiley, New York (1905).

11. Andes, L. E., *Oil Colours and Printers' Inks*, transl. Morris, A., and Robson, H., Scott, Greenwood and Co., London (1903).

Cruickshank Smith's *The Manufacture of Paint*, and Fluery's *The Preparation and Uses of Zinc White Paints*, describe methods and processes used just prior to World War I; George Hurst's book of the same period went through so many editions that it still may be found on some bookstore shelves for only a few dollars.[12-14] Even Zerr and Rübencamp's book in English translation of 1908, Mierzinski's of 1881, and John W. Masury's of 1868 are not yet so very scarce, but they soon will be if not appreciated and preserved.[15-17]

"Dating" Pigments

Why should the architectural historian and museum curator be particularly interested in pigments and dyes? It is commonly known, I'm sure, that if one can prove the presence of Prussian blue, it can be established that the paint containing it must have been manufactured after 1704. Cobalt blue was invented in 1802 and in 1828 a prize was granted for the development of synthetic ultramarine; titanium white was introduced at the end of World War I. These are sometimes called "dating" pigments, for their presence conveniently serves to establish the earliest possible date of application of a decorative coating. This, then, is one of the principal reasons why the analysis of paints is a valuable adjunct to the study of buildings, paintings, and decorated artifacts of various kinds.

The technical examination of artists' pigments began in the early 19th century with the examinations by Chaptal and by Humphrey Davy of pigments found in the excavations of Rome and Pompei.[18, 19] Raehlmann formally introduced microscopic examination.[20, 21] I am not sure who should be cited as the first to examine artists' pigments by emission spectroscopy or X-ray diffraction, but the latter tool still has not been fully exploited, even though museum authorities are today privileged to employ neutron activation, the electron beam microbe, Fourier-transform infrared, and other advanced techniques.

With the introduction of these increasingly sophisticated methods, we are no longer content simply to say that a pigment is lead white, ultramarine or Prussian blue; we now want to be able to specify the *kind* of lead white, ultramarine, or Prussian blue that is present. Unfortunately, however, the opportunity to assemble and learn to characterize many of the older varieties of pigment—and even some of the relatively modern types—is rapidly being lost.

Lead White

Lead white, the key pigment in easel painting for about 500 years, is a prime example. Of the dozens of processes that have been patented for the manufacture of this pigment, the traditional method used for hundreds of years was the closely related stack, Dutch, or chamber process. Plants of this type

12. Smith, J. C., *The Manufacture of Paint*, Scott, Greenwood and Co., London (1915).

13. Fleury, P., *The Preparation and Uses of Zinc White Paints*, transl. Grant, D., Scott, Greenwood and Co., London (1912).

14. Hurst, G. H., *Painters' Colours, Oils and Varnishes*, First Ed., G. Griffin and Co., London (1901).

15. Zerr, G., and Rübencamp, R., *A Treatise on Colour Manufacture*, transl. Mayer, C., Charles Griffin and Co., London (1908).

16. Mierzinski, S., *Die Erd-, Mineral- und Lack-Farben*, Fourth Ed., Voigt, Weimar (1881).

17. Masury, J. W., *A Popular Treatise on the Arts of Housepainting*, D. Appleton and Co., New York (1868).

18. Chaptal, J. A., "Sur quelques couleurs trouvées a Pompeia," *Annales de Chemie*, 70 (1809), p. 22.

19. Davy, H., "Some Experiments and Observations on the Colour Used in Painting by the Ancients," *Philosophical Transactions*, 105 (1815), p. 97.

20. Raehlmann, E., *Über die Maltechnik der Alten*, George Reiner, Berlin (1910).

21. Raehlmann, E., *Über die Farbstoffe der Malerei*, Seemann, Leipzig (1914).

have almost disappeared throughout the world. The Lead Industries Association recently informed us that there are now only five companies producing the pigment in the United States by any method whatsoever, and that the total annual production in the U. S. is a mere 6,000 tons. Fortunately, the names and dates of early U.S. manufacturers of lead white have been recorded by Holley.[22]

A survey kindly made for us by Mme. Hours, Director of the laboratory of the Musees du Louvre, Paris, failed to turn up a single primary manufacturer remaining in France or Austria and a recent inquiry to colleagues in Rome disclosed that its manufacture is prohibited in Italy. Thus, because of the precipitous decline of the manufacture of this key pigment in the history of easel painting, the National Gallery of Art Research Project is making every effort to collect samples made by known processes before it is too late.

How can one characterize lead white made at different times and by different means? It has been known for some time that two basic carbonates can be formed as well as the normal lead carbonate, but for many years there has been much confusion regarding the correct interpretation of their X-ray diffraction patterns. The latest attempt to resolve this problem has been made by Olby who considers the usual "lead white," *hydrocerussite*, $2PbCO_3 \cdot Pb(OH)_2$, to be rhombohedral in structure and a second form, *plumbonacrite*, $6PbCO_3 \cdot 3Pb(OH)_2 \cdot PbO$ to be hexagonal.[23] The compound of lower carbonate content is considered to be the type found in pearlescent lead white and indeed the current industrial pearlescent pigment is prepared in perfect hexagon-shaped plates. We have observed, however, a second morphological variety in a 1968 preparation of pearlescent lead white: six-pointed or twelve-sided "snow-flake" crystals (Figure 3d). The existence of 12-sided crystals has been reported by Katz and Lefker,[24] but we do not recall that anyone has published photomicrographs of them before. The problem of characterizing the varieties of basic lead carbonate, even by X-ray diffraction, remains a challenging one.

Work of colleagues abroad, based on emission spectroscopy and neutron activation analysis, suggests that older lead whites often have much higher impurities than those made after about 1850.[25, 26] Lead white that has been made from ore refined more recently than about 1840 will exhibit significant residual radioactivity owing to lead-210; this fact has been successfully demonstrated in research jointly sponsored by the Atomic Energy Commission and the National Gallery of Art, a finding that will enable us in many cases to confirm relatively recent manufacture.[27, 28] Research designed to characterize lead white through its lead isotope ratio has been supported by the National Science Foundation, in addition to studies on the characterization of ultramarine and vermilion by means of sulfur isotopes ratios.[29]

The success of all such investigations depends upon ones ability to obtain pigments of known source and age. When the study of lead white was first taken up a number of years ago, we had the good fortune to be in touch with

22. Holley, C. D., *The Lead and Zinc Pigments*, J. Wiley & Sons, New York (1909).

23. Olby, J. K., "The Basic Lead Carbonates," *J. Inorg. Nucl. Chem.*, 28 (1966), p. 2507.

24. Katz, G., and Lefker, R., "Basic Lead Carbonate," *Anal. Chem.*, 29 (1957), p. 1894.

25. Kühn, H., "Lead White and Its Use in Painting," *Farbe und Lacke*, 73 (1967), p. 99, 209.

26. Lux, F., and Braunstein, L., "Activation Analysis of Paintings, Determining the Trace Content in White Lead in Paintings at the Alte Pinakothek at Munich," *Zeit. Anal. Chem.*, 221 (1966), p. 235.

27. Keisch, B., Feller, R. L., Levine, A. S., and Edwards, R. R., "Dating and Authenticating Works of Art by Measurement of Natural Alpha Emitters," *Science*, 155 (1967), p. 1238.

28. Keisch, B., "Dating Works of Art Through Their Natural Radioactivity: Improvements and Applications," *Science*, 160 (1968), p. 413.

29. Keisch, B., "The Use of Isotope Mass Spectrometry in the Identification of Artists' Pigments," *Studies in Conservation*, 15 (1970), p. 1.

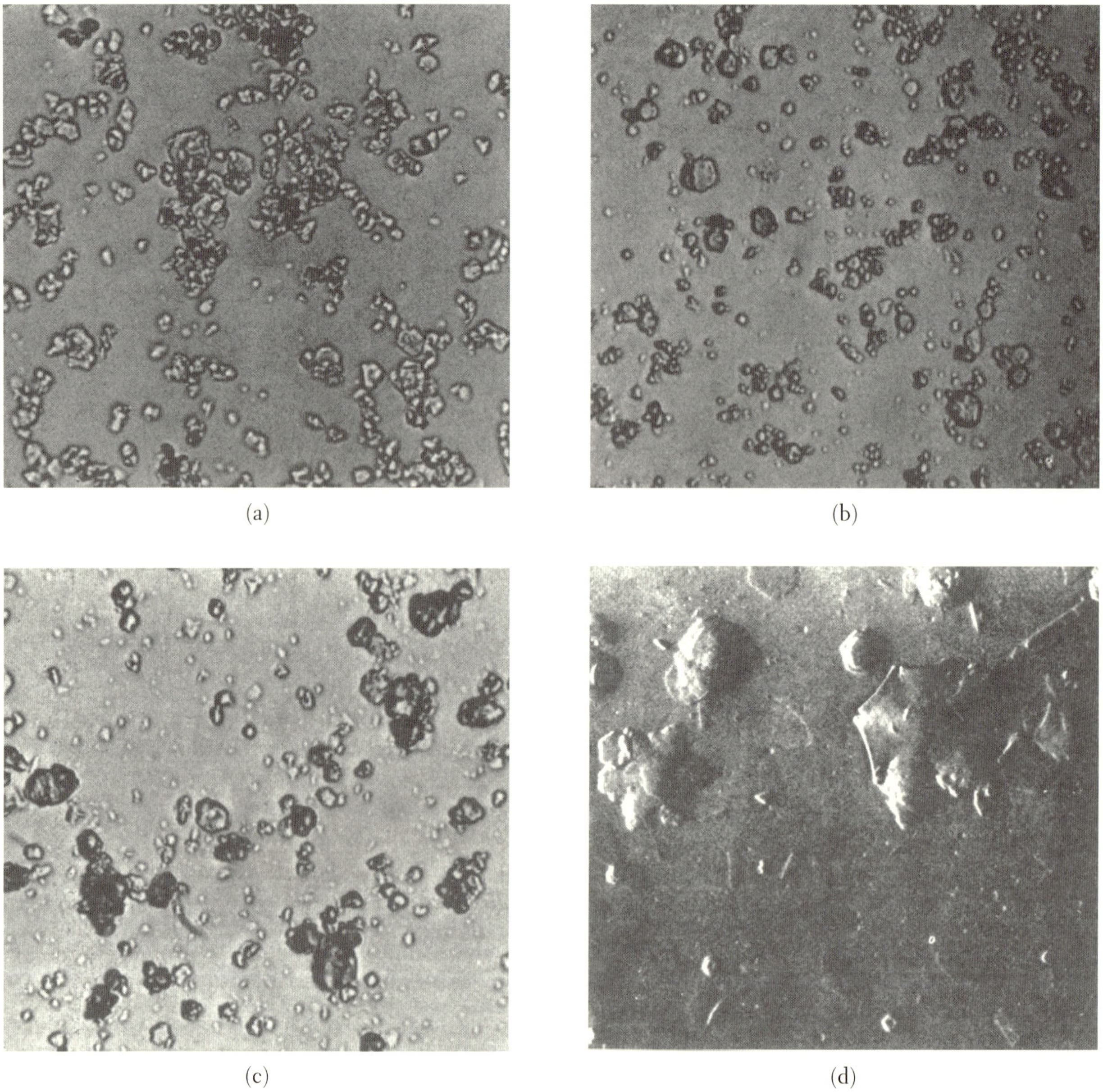

Figure 3
Various crystalline forms of basic lead carbonate; (a) Quick process from litharge; (b) Carter process; (c) Chamber process; and (d) pearlescent. Sample (a) courtesy of Lindgens and Sohne, others from E. J. Dunn Jr., NL Industries. (Magnification of sample (d) at 700x, others at 1400x).

E. J. Dunn, Jr., of NL Industries, Inc., just before his retirement. Mr. Dunn kindly supplied us with 2-1/2 oz. samples of pigment produced by different methods of manufacture; some of their striking morphological differences, which he told us we would observe, may be seen in Figure 3. Stimulated by such findings and realizing that it would be increasingly difficult to obtain samples prepared by a variety of historical methods of manufacture, we began shortly thereafter our world-wide search for specimens of known methods of production. Special thanks for well-documented samples thus far received go to NL Industries, American Cyanamid Co., Eagle-Picher Industries, Inc. and Hammond Lead Products, Inc., in this country; to the Mersey White Lead Co., Associated Lead Manufacturers and Novadel in England; to Lindgens & Söhne, Köhn-Mulheim, West Germany; and to Commonwealth Litharge & Red Lead Pty. Ltd., Australia.

Another way to obtain a chronological sequence of pigments, of course, is to sample paintings, buildings, machinery, and other artifacts that bear paint applied at a known date. Artists' canvases, for example, are often "primed" with a white base by the commercial supply house, and when the canvas is stretched prior to painting there is usually a generous half inch of "tacking edge" that remains uncovered by the artist's own paint along each side of the stretcher. Much of the success of our lead white studies was made possible through the generosity of George R. Hann, of Sewickley, Pa., and Joseph Butler, Director of the Butler Institute of American Art, Youngstown, Ohio, who kindly allowed samples to be taken from the tacking edges of paintings in their collections. Dr. Eric C. Hulmer, curator-conservator for these collections, has devoted many years to collecting minute specimens of lead white and other pigments from paintings that have come to him for documentation and treatment. Through his single-minded devotion to the needs of our research, we have been able to make measurements on milligram-sized samples of lead white taken from the edges of over a hundred 19th-century American paintings as well as a significant number of 18th-century paintings.

Ultramarine

To a degree that still may not be appreciated, the production of a number of other traditional pigments is diminishing. Ultramarine, for example, has had a long history but there is now stiff competition. In the distant past, the natural mineral, lapis lazuli, was used extensively, nearly always contaminated with accessory minerals which tend to characterize the geological source of the mineral.[30] Perhaps a forger might add the appropriate contaminants, but recent studies have shown that the sulfur-isotope ratios to be found in the principal sources of the blue mineral are highly characteristic, suggesting that this may provide an almost tamper-proof method of distinguishing the natural as well as the synthetic pigment.[29]

30. Plesters, J., "Ultramarine Blue, Natural and Artificial," *Studies in Conservation*, 11 (1966), p. 62.

The principal geological sources of lapis lazuli are conveniently limited, but the characterization of manufactured ultramarine presents a more complex problem. Several years ago, the National Gallery of Art Research Project was most fortunate in discovering that the ultramarine plant of the Holland-Suco Color Co. (a division of the Chemetron Corp.) in Huntington, W. Va., the last such plant in America, possessed a file of commercial samples dating as far back as 1914. The company very graciously permitted us to take whatever amounts of these old samples that we might need. Thus far, we have managed to survey about 3,000, sampling and cataloging some 600 of the most interesting specimens. This extraordinary collection has been of crucial value in Dr. Bernard Keisch's investigations of sulfur-isotope ratios.[31]

Ultramarine is famous for its sensitivity to acids. Its resistance to acid has been increased by several patented procedures that essentially deposit a coating of silica on the surface.[32] It occurred to me that infrared spectroscopy should turn up an extra absorption band in this variety of ultramarine; Figure 4 shows that an extra silicon-oxygen band is indeed present in the acid-resistant product. Obviously, if we should find this pigment in a paint, it may safely be characterized as a "modern" type. We are also applying electron microscopy to the study of pigment characteristics. As Figure 5 shows, the morphology of the natural, the ordinary synthetic, and the coated ultramarine is also distinctive when examined by this technique.

31. Anon., "Old Color Samples Help Scientists Detect Art Frauds," *Paint and Varnish Production*, 61 (1971), p. 40.

32. Fischer, H. C., U. S. Patent, 1,631,628, July 7, 1927, Hanahan, M. L., U. S. Patent 2,296,639, Sept. 22, 1942, Schwahl, F. J., U. S. Patent 2,628,920, Feb. 17, 1953.

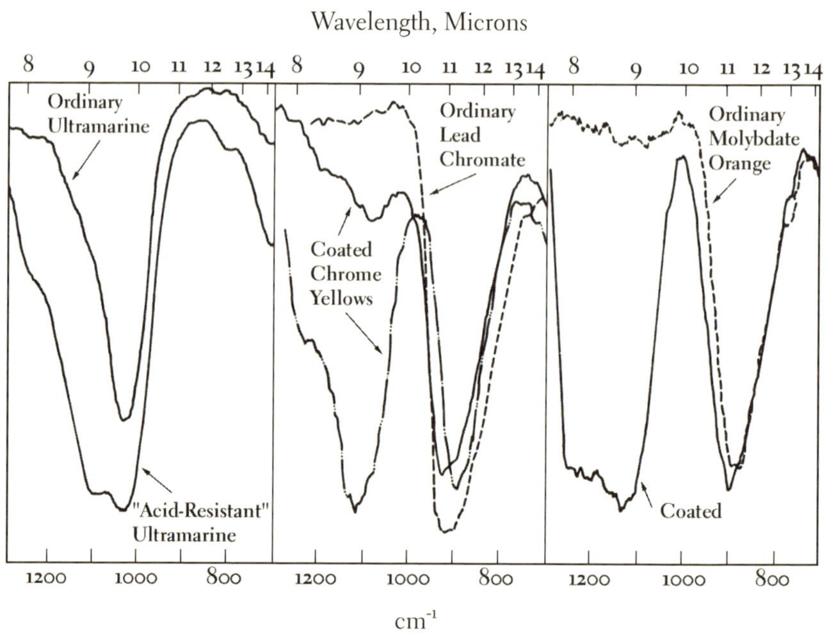

FIGURE 4
Infrared absorption in the 800-1200 cm^{-1} region in ordinary and in silica-coated ultramarine, molybdate orange and chrome yellow pigments.

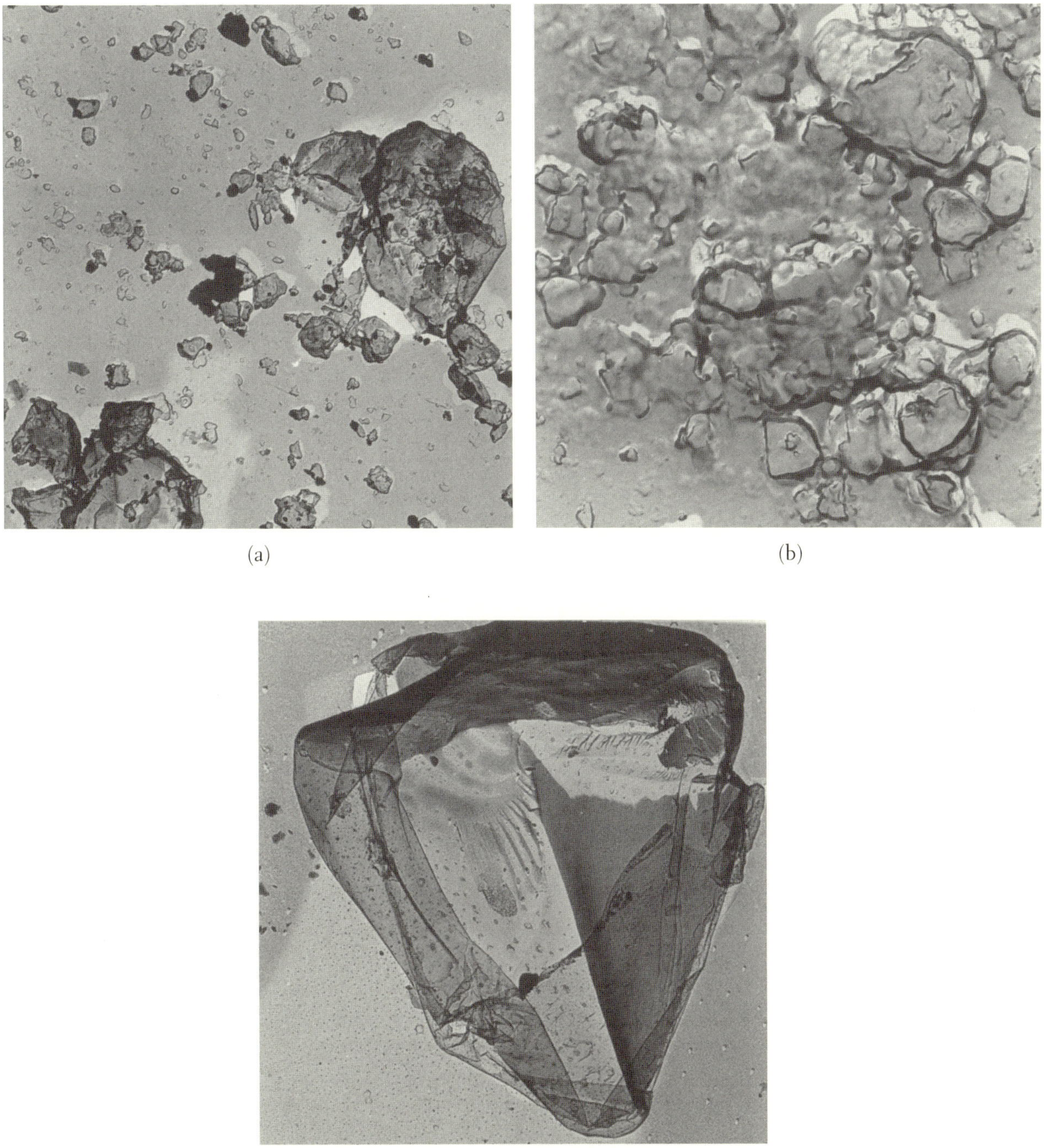

(a) (b)

(c)

FIGURE 5
Electron photomicrographs of ultramarine at 10,000x: (a) typical synthetic, (b) "acid-proof" type, (c) particle of lapis lazuli. Note the "coated" appearance of (b) and the glassy fracture in the natural mineral grain. Photomicrographs of platinum-shadowed specimens made by M. N. Haller, Mellon Institute.

OTHER PIGMENTS OF INCREASING RARITY

Vandyke Brown, introduced in the 17th century, is based on a mixture of substances found in the humus-like material of peat bogs. We have succeeded in characterizing the pigment through its infrared spectra, using reference samples from the two principal contemporary sources in Germany and a less familiar one in Czechoslovakia (the latter sample thanks to George Hawley, of W. Hawley and Son, Duffield, England). As Figure 6 shows, its infrared spectra significantly differs from two other traditional organic browns: bistre and bitumen. The presence of the latter pigment may also be indicated by the detection of vanadium and nickel in spectrographic analyses, for these are important constituents in the porphyrins that are present.[33]

33. Stewart, J. E., "Infrared Spectra of Chromatographically Fractionated Asphalts," *J. Res. Nat. Bur. Standards*, 58 (1957), p. 265.

The production of vermilion is currently diminishing owing to the competition of the cadmiums, Mercadiums®, molybdates, and other more stable colorants. In spite of its growing obsolescence, much still remains to be learned about the traditional methods of production; some of the most stable varieties of vermilion were formerly prepared by sublimation, whereas much

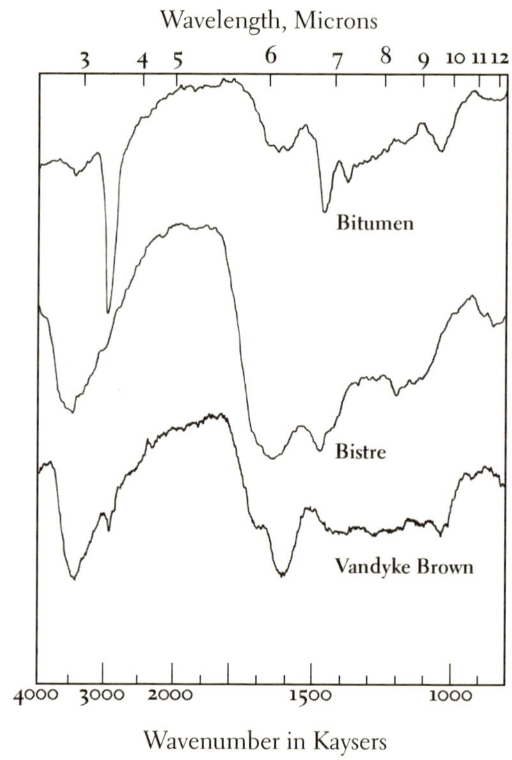

FIGURE 6
Infrared spectra of some traditional organic browns: Vandyke brown, bistre, and bitumen. Spectra shifted on vertical axis to permit convenient comparison.

of the current industrial product is made by wet process.[34] R. J. Gettens, of the Freer Gallery of Art, recently had the good fortune to discover pigment still being prepared by sublimation in more-or-less "home industries" practiced outside of Hong Kong; the samples he brought back provide us for the first time with material known to have been made in this traditional manner.

Mention of molybdate orange reminds me to point out that silica-coated varieties of this pigment and of chrome yellow are also available. They tend to be more stable to photochemically induced darkening than the untreated type; we have found, for example, that a coated molybdate orange darkened only to the extent of 6.4 MacAdam units of color difference in two summer months of exposure outdoors under glass, whereas an uncoated variety changed 20.0 MacAdam units.[35] Figure 4 indicates that the coated pigments can be detected by infrared.

34. Feller, R. L., "Studies on the Darkening of Vermilion by Light," *Report and Studies in the History of Art,* National Gallery of Art, Washington, D. C. (1967), p. 99.

35. MacAdam, D. L., "Specifications of Small Chromaticity Differences," *J. Optical Soc. Am.*, 33 (1943), p. 18; Davidson, H. R., and Friede, E., "The Size of Acceptable Color Differences," *ibid.*, 43 (1953), p. 581.

A "Pigment Bank" for Research

The examples cited illustrate just a few of the ways in which we may characterize the particular variety of a pigment or mode of manufacture. Our expertise in this area, however, will never reach its potential without collections of pigments of known date and process. Realizing the importance of this, the National Gallery of Art Research Project established a "pigment bank" several years ago which now numbers over 2,000 specimens. We will be happy to receive well-characterized pigments to add to this collection, particularly pigments of older manufacture. The Fogg Art Museum, the Freer Gallery of Art, the Intermuseum Conservation Laboratory at Oberlin College, and the Conservation Center at New York University also have excellent pigment collections, as have several museum laboratories in Europe. If you have a favorite museum, by all means present the increasingly rare materials to the institution of your choice.

Improved Colorants for the Artist

Each extension of the range in hue and purity of colorants that industry has developed has generally been accepted with considerable enthusiasm by artists and decorators. In spite of the enthusiasm, there have been disappointments, particularly with many of the fugitive early dyestuffs. By the late 19th century, the artist found himself confronted with such a variety of new and unfamiliar industrial products that he became alarmed at his lack of technical knowledge regarding their properties, a concern that continues to the present day. Just 30 years after Perkins discovery of mauve, for example, a Society for the Advancement of Rational Painting Techniques was established in Germany which sought among other things to establish a standard list of lightfast pigments.[36]

In this country a WPA art project in the late 1930s led to the establishment of

36. Munkert, A., *Die Normalfarben*, Enke, Stuttgart (1905).

a U. S. Commercial Standard of lightfast artists' oil paints. The history of this project, apparently the first such effort since the German *Normalfarbenskala* of 1886, is included in the standard, first published in 1942 and revised in 1962.[37]

37. Commercial Standard CS-98-62, "Artists Oil Paints," U. S. Department of Commerce, (1962).

The approved list of lightfast artists' pigments has remained limited. Up until just the last decade or two, it consisted of only about 20 to 40 basic chemical entities: iron, chrome, zinc, cobalt, and titanium oxides; mercury, cadmium, and selenium sulfides, etc. The pigments in Table 1 are certainly worthy of a place among the traditional highly stable pigments in the fine-arts palette, yet the latest standard still only accepts a limited number of new colorants: the phthalocyanine blues and greens; quinacradone red; and Hansa yellow 10 G.

In the last decade, the founder and retired president of the Permanent Pigments Co., Cincinnati, Ohio, Henry Levison, has evaluated many of the newer pigments and vehicles, and just in the past year, has introduced a line of fine-arts acrylic-emulsion paints that make use of modern toners to achieve high chroma and controlled value. With his guidance, the Permanent Pigments Co. continues to set high standards for the artist by revealing the essential composition of the pigments used (the Colour Index number generally is cited). In recognition of his contributions in introducing durable new painting materials, Mr. Levison was elected to the Artists' Materials Hall of Fame in 1971.[38]

38. Anon., *NAMTA News and Views*, 17, No. 6 (1971), p. 1. *The Chemist*, 47 (Sept., 1971), p. 250.

In order to advise museums concerning the most fugitive colorants that they will encounter—turnsole, Persian berries lake, gamboge, saffron, quercitron, cochineal, alizarin—our laboratory has investigated the rate at which many of these traditional colorants fade.[39, 40] We have also evaluated the lightfastness of the pigments in Table 1 on behalf of the American Artists Professional League and have called the attention to low-chalking types of titanium white.

39. Feller, R. L., "Standards of Lightfastness," *Bulletin of the American Group-IIC*, 4, No. 1 (1963), p. 10; *ibid.*, 7, No. 2 (1967), p. 8.

40. Feller, R. L., "Lightfastness of Felt-tipped Markers," *InterSociety Color Council Newsletter*, No. 192 (Jan.-Feb., 1968), p. 10.

Artists and conservators occasionally have need for soluble colorants. In seeking only those that exhibit lightfastness in the range of 5+ to 6+ on the ISO R105 scale of blue wool standards when used in pale tints, we have thus far selected CI Direct Blue 86 for water-based colorants and a green made from a mixture of this pigment and CI Solvent Yellow 13. CI Solvent Red 122, Solvent Red 130 and Solvent Yellow 13 are highly lightfast alcohol-soluble colorants; CI Solvent Red 122 and a green made from CI Solvent Blue 70 and CI Solvent Yellow 13 may be used in toluene-based formulations.[41] Toners tend to exhibit greater lightfastness than soluble colorants and we are currently considering some of them for particular artist's and conservator's applications.

41. Feller, R. L., and Curran, M., "Lightfast Soluble Colorants," *Bulletin of the American Group-IIC*, 11, No. 2 (1971), p. 15.

Our close acquaintanceship with the lightfastness problem enabled us several years ago to contribute in the development of an interesting art form which may be described as "geometric constructions in colored yarn." A Carnegie Tech-trained artist, Miss Sue Fuller, had become aware of the fact that many of the brilliantly colored modern fibers used in her early experiments were not stable upon exposure to light. With our assistance in exposure testing, she has been able to select colored fibers that will withstand as much as 800 to 1,000

TABLE 1
Some Modern Organic Colorants Having Sufficient Lightfastness to be of "Fine-Arts" Quality*

Pigment Name	*Colour Index Designation*
Carbazole dioxazine	Pigment violet 24
Thioindigo bordeaux	Pigment red 88
Quinacridone reds, violets	Pigment red 122, violet 19
Phthalocyanine blue and green	Pigment blue 15, green 7
Nickel azo yellow	Pigment green 10
Indanthrone blue (alpha)	Vat blue 4, 6
Flavanthrone yellow	Vat yellow 1
Anthrapyrimidine yellow	Vat yellow 20
Brominated anthanthrone orange	Vat orange 3
Perinone orange	Vat orange 7
Acridone red	Vat red 35
Perylene maroon	Vat red 23

* Based on Vesce, V. C., 1959 Mattiello Memorial Lecture, *Official Digest, Federation of Societies for Paint Technology*, Dec. 1959, Part 2; Gaertner, H., *J. Oil and Colour Chemists' Association*, 46 (1963), p. 13; and Levison, H., in *Painting with Synthetic Media*, Woody, R. O., Jr., Reinhold, New York (1965), p. 141.

hours in an Atlas carbon-arc Fadeometer. In her final work, striking geometric patterns of these brilliant fibers are embedded in acrylic resin; the exposure tests further demonstrated that casting in acrylics provides additional protection against photochemical deterioration, probably because oxygen is effectively excluded. The result is a successful combination of modern materials that may well outlast the finest decorative panels of ancient masters.

Advanced technology is being used to achieve many new treats for the eye: vacuum-deposited interference filters, polarization colors by transmitted light, constructions of highly polished or transparent synthetic resins, colored lights, and fluorescent pigments to name a few. Obviously, it is no small task for the chemist and engineer to provide the necessary technical assistance for the adaptation of these modern materials and only a few years ago an organization known as Experiments in Art and Technology was formed to facilitate greater interaction between artists and technical specialists.

CONCLUSION

The problems confronting the artist and museum curator involve such a range of materials that the advice of many specialists is needed. In the search for more lightfast pigments and dyes we will need much technical assistance, largely because of the variety of colorants available. However, when the particular problems of the artist and museum conservator are explained, I know of no industrial laboratory that has not given most generously of their time and knowledge. Special thanks go to a number of companies who have been particularly helpful in our investigation of pigments: Allied Chemical Co.; Holland-Suco; PPG Industries Inc.; the C. K. Williams Division of Pfizer, Inc.; Whittaker, Clark and Daniels, Inc.; Reichard-Coulston, Inc.; the lead white manufacturers mentioned earlier; and the artists' materials companies of Permanent Pigments, Talens and Son, and Winsor and Newton.

Concerning the Place of Science in the Scheme of Things

ROBERT L. FELLER

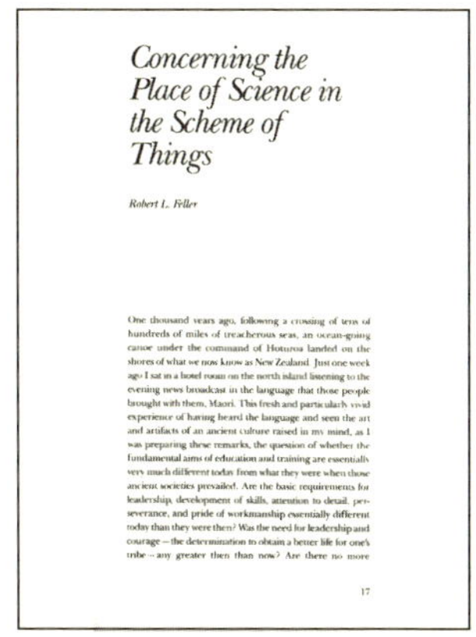

Originally printed in:
TRAINING IN CONSERVATION
ed. Baer, N. S., Institute of Fine Arts, New York University, New York (1989), Pages 17-32.

One thousand years ago, following a crossing of tens of hundreds of miles of treacherous seas, an ocean-going canoe under the command of Hoturoa landed on the shores of what we now know as New Zealand. Just one week ago I sat in a hotel room on the north island listening to the evening news broadcast in the language that those people brought with them, Maori. This fresh and particularly vivid experience of having heard the language and seen the art and artifacts of an ancient culture raised in my mind, as I was preparing these remarks, the question of whether the fundamental aims of education and training are essentially very much different today from what they were when those ancient societies prevailed. Are the basic requirements for leadership, development of skills, attention to detail, perseverance, and pride of workmanship essentially different today than they were then? Was the need for leadership and courage—the determination to obtain a better life for one's tribe—any greater then than now? Are there no more uncharted oceans or remarkable mountains to cross?

We are meeting today to discuss aspects of education and training in a rather extraordinary area of study, the conservation of the material artifacts of history and art. A time-honored cultural institution, the university, has elected to place this unprecedented subject within its purview. When the Institute of Fine Arts at New York University first contemplated undertaking such a program, it was questioned whether the university was the proper place in which to teach such a discipline, one that, among other things, necessitates the acquiring of certain craftsman-like skills of hand and eye. Today, two decades later, there should be no question that this is the place. Indeed the university is the place in which to train for a profession that demands so much knowledge and understanding of such a broad range of subjects: art, history, technology, and science. Alfred North Whitehead has said "a university imparts information... but it imparts it imaginatively... a university which fails in this respect has no reason for existence." We can be most thankful indeed for those who have had the imagination to develop the several graduate programs in conservation that have come into existence since the Center at New York University opened its doors in 1961.

I have been asked to make some remarks regarding the place of science. In fulfilling this obligation, I shall feel more comfortable being descriptive rather than prescriptive. Rather than suggest the way things ought to be, I shall attempt instead to describe the way things are and to point out certain deficiencies that remain.

Let us recognize at the very beginning that "science is here to stay." The pursuit of science and technology represents a strong subculture in our society to say the least. Both the artist and the art historian have been, and will continue for some time to be, markedly influenced by the practice and philosophy of the scientist and engineer. Given this circumstance, the humanist—as well as the technologist—must make the best of it. We must try to utilize the tools of science and technology to advantage, to assist ourselves toward the goals that society considers worthwhile. The operative word is assist.

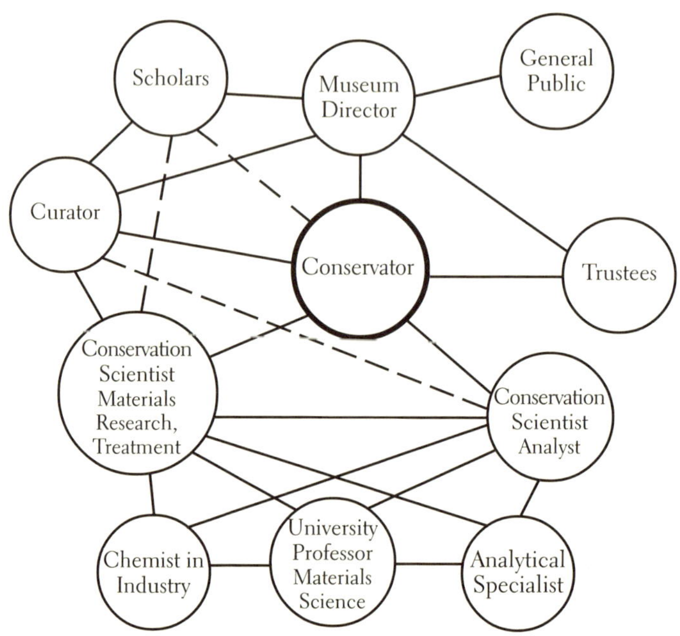

FIGURE 1
The conservator and those interacted with.

The diagram in Figure 1 was devised to emphasize the fact that this individual whom we wish to educate, the conservator, must be capable of understanding, communicating and interacting with persons having a remarkably broad range of interests, concerns and abilities. We are asking a lot of this young professional. It is not necessary to belabor the lessons that might be drawn from this doodle, but apropos of my obligations today, let me point out that the scientist is not at the center of the scheme. Instead, he finds himself slightly off to one side in a supporting role. In addition, there are several types of such persons indicated, some more directly connected to the conservator's needs than others. That is one of the points I should like to make: science is not monolithic; there are many ways of practicing the profession. Moreover, some practitioners are in a better position to be of direct assistance to the conservator than others. The scientist is but one member of a team; science, but one of an array of disciplines and skills that must be called upon.

How did science get into the field of conservation? It was not one thousand years ago that we set forth upon this voyage but practically within the memory of our own lifetime. The idea was born in the minds of a number of colleagues that the care of artistic and historic works need not continue slavishly to employ methods and materials that come into the conservator's hands primarily though tradition. Better adhesives, coatings, packaging materials, fungicides and insecticides exist today than were available to our forefathers. There are better analytical tools at hand with which to determine the composition of craftsman's mediums and the causes of their disintegration. Should not these tools be used? If society is ever to change its way of doing things, one must try. Those who set forth upon uncharted waters, must try. The vision held by such men as F. I. G. Rawlins, Edward Forbes, George L. Stout, and others of their generation was simply that, where a technical understanding of the materials of construction could help to preserve the physical evidences of the past, it was their, as it is our, duty to bring these skills and this knowledge to the task.

That's why science and technology were introduced: in the hope that useful benefits might come about. Note the use of the earlier word, assist; now hope. There is no guarantee that doing things the scientific way—whatever that may mean to you—will necessarily make things better. In considering the role of science, then, let us retain some of the sense of humility that our forebears had in these matters, emphasizing once again that this discipline has been introduced into conservation in the same spirit that it had been introduced into agriculture, forestry, animal husbandry, and medicine: in the hope—perhaps faith, today—that an objective analysis of the way the physical world seems to work will help us do a better job and reach our objectives most efficiently.

Some months ago there appeared a story in the *Wall Street Journal* which implied that there existed two schools of restoration, one of which might be called the "scientific school." I know of no such simplistically labeled school of thought or practice. The *Wall Street Journal* is a rather careful publication and when these words were first used they were placed in quotation marks, an editorial device which I take to mean that the writer considered that there was no such entity that could be properly defined and identified. The use of quotation marks implied the modifier, "so-called." Unfortunately, if the lack of a precise and accepted definition of all such fuzzy terms is not immediately pointed out, the words are likely to be used in the future without the quotation marks. Indeed, that is exactly what occurred later in the same article.

This expression may have been introduced into the newspaper columns to suggest the idea that such a school of thought, if it really did exist, would be dreadfully wrong for conservation. However, I would like to raise with you the question whether the introduction of science and technology is likely per se to be fundamentally bad. Our concept of science can be traced to Robert Boyle in the seventeenth century who introduced a process of problem-solving that has come to be known as the scientific method (Table 1). The keys to this process are (a) direct observation and (b) controlled experimentation. One assembles a collection of facts based on meticulous observation. After

TABLE 1
The Scientific Method*

1. State the objective
2. Assemble the facts
3. Organize the facts
4. Propose a likely hypothesis or solution
5. Test the hypothesis
6. If necessary, modify and repeat
7. Take action

* Robert Boyle (1627-1691) proposed steps 2, 4, 5 and 6.

preliminary organization of the facts, one then attempts to explain the findings and apparent interrelationships. A hypothesis is formed and tested. If the ideas prove to be wrong, one must have sufficient self-discipline to say—so be it, we will try something else. The essential difference between this and the earlier Aristotelian way of approaching problems is the emphasis on the direct observation of nature and on the planned experimental verification of one's hypothesis, one's proposed explanation.

The success of this way of attempting to exert some degree of control over the physical world has been astounding. Surely there is nothing fundamentally wrong with having brought this discipline to bear on some of the problems that conservators face. In "The Aims of Education," Alfred North Whitehead observes that "the particular merit of a scientific education should be that it bases thought upon first-hand observation. . . . In the teaching of science, the art of thought should be taught: namely, the art of forming clear conceptions applying to first-hand experience. . . ." Whitehead was not without bias; nonetheless, I trust that the application of principles that have been with us for two hundred years is not, in itself, taken to be potentially harmful or inappropriate to the field of conservation.

Let us keep in mind then that science is essentially a way of analyzing and attempting to solve problems posed by the physical world. We introduce this discipline in the same spirit that it has been introduced in so many other activities of our daily life: in the hope that in a rigorous effort to cultivate objectivity and to use our powers of observation we may achieve practical and efficient solutions to some of the problems nature places before us, whether in our role as conservator or as citizens of planet Earth.

What are some of the specific things that we hope science will do for the field of conservation? First of all, it can render a technical service; the tools of the chemist-analyst can be used to analyze or identify the paints, metals, and other materials of the artist and artisan. It is in the role as an analyst that the chemist's and physicist's contributions to the field of art and archeology are perhaps most familiar. Considering that the field is represented by three functions—examination, preservation, and restoration—it is fair to say that science has principally benefited the first of these through the introduction of such apparatus and techniques as the polarizing and the scanning electron microscopes, radiography, infrared sensing devices, and x-ray fluorescence analysis.

Summed up in Table 2 are a number of roles in which the scientist may serve. Individuals may be called upon for specific technical expertise in corrosion, mineralogy, entomology, the chemistry of paper and leather, the metallurgy of copper alloys. One may also find the chemist and physicist in

Table 2
Some Roles of the Scientist

Analyst
Technical Specialist
Translator-Interpreter of Technical Matters
Teacher
Product and Process Tester-Evaluator
Research and Development Investigator

the role of a translator-interpreter of technical matters; the conservator and curator frequently have need for assistance in interpreting literature that describes the results of chemical analyses or the possible uses and potential hazards of new materials. There is, of course, the role as teacher, imparting something of his methods, special understanding, and way of looking at phenomena. The scientist is also employed to test, compare, and evaluate products and processes. Last, but not least, there is the role as an investigator in research and development. For the reasons of insufficient time, money, and manpower, the field of conservation has yet to take full advantage of the scientist in this latter capacity as we shall surely see in the coming generation.

If we agree that science was brought into this field not only to help in the examination of historic and artistic works but also to assist in doing a better job of preservation and restoration, we must learn to be patient. The latter objectives are challenging and the assignment formidable. Industrially or academically trained colleagues, whom we invite from time to time to assist in this effort, are not always cognizant of just how challenging the parameters of the task really are. The conservation scientist is not interested in introducing new methods of bleaching, new methods of lining, new methods of varnishing. Not at all. Rather, he is interested in introducing new and improved methods of bleaching, lining and varnishing (a) *that will not lead to disastrously harmful results during the next generation and* (b) *that will not result in more difficulties in analysis, preservation, and restoration than are presently encountered in the use of traditional materials.* These are the constraints that make the attainment of acceptable solutions elusive.

The conservator has had high hopes that the scientist would carry out research that will lead to improved ways of preservation and restoration. Let us consider these expectations. The word "research" unfortunately has become much corrupted in the everyday language of television and advertising. When someone wants to buy a new automobile, television set, or refrigerator, many will say that they have done a lot of "research on the subject" before deciding on the final choice. Looking up something in books and catalogs is not research, certainly not as it relates to the world of science and technology. Let me define what I mean by the word: it is the search, in the laboratory-workshop, for new facts and practical applications or for the revision of accepted theories that are unknown to the contemporary scientific community. Emphasis on the *unknown*. We are not referring to one's personal ignorance of a fact or of the knowledge that resides in some published book or journal in whatever language and of whatever date that may be. To find those facts that are, or were, already known by others simply involves a diligent search of the literature, or search for the expert who has direct knowledge and experience. The process should not be confused with the far more challenging effort of inquiring into the unknown. Because what we seek is not known and because the mind of man tends to be fallible and unruly, advancement is bound to be slow.

Reading about the scientific needs of the nation, one soon encounters reports of expenditures for R&D: research and development. Studying these figures, one quickly finds that development regularly takes the largest slice of the pie (Figure 2). We may talk readily about the needs of research in the field of conservation; more rarely is the other word encountered.

What is "development"? It is all those problems that are associated with reduction to practice. If the laboratory comes up with a new product, or a new way of carrying out an old task, development refers to the myriad problems associated with making these untried ideas a practical success: engineering, fabrication, testing, packaging and foolproof instructions. If one has not thought very much about the subject, it may come as a surprise at first to learn that the major cost in the introduction of new and improved ways of doing things—in terms of time and manpower—occurs in the matter of development.

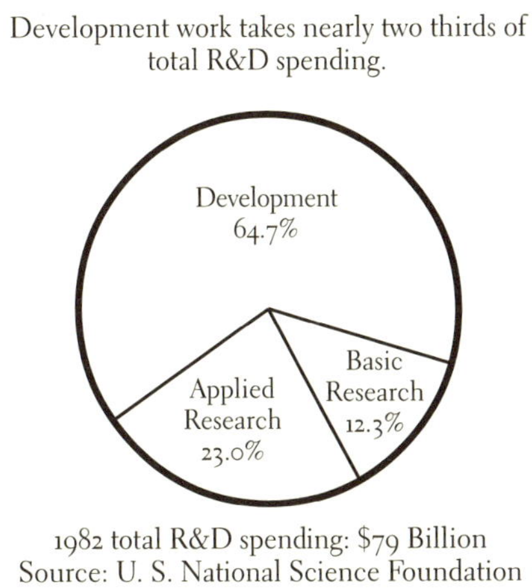

FIGURE 2
Apportionment of national expenditures for research and development, 1982.

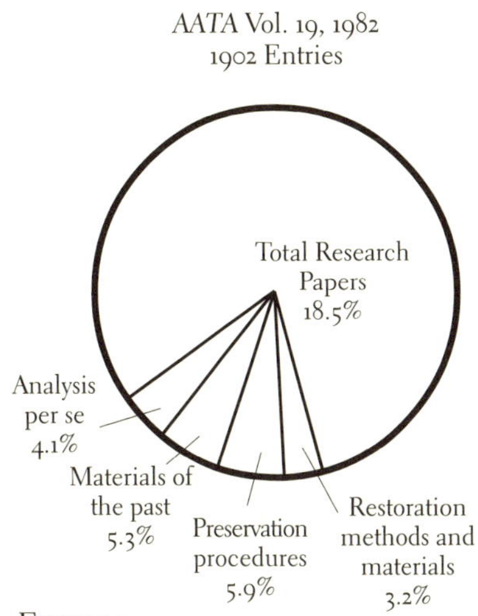

FIGURE 3
Less than twenty percent of the papers abstracted in *Art and Archaeology Technical Abstracts* in 1982 were reports of scientific research and development. Less than ten percent concerned materials and processes of preservation and restoration.

In the areas of preservation and restoration, we find ourselves still weak in R&D (Figure 3) but thus far have suffered most with respect to resources that we have been able to devote to development. One of the reasons for this is that long-standing policies and practices of funding have made it more difficult to obtain grants for development. Perhaps because an inquiry into the unknown is indeed the more fundamental task, governmental and private sources have generally tended to favor grants for "research." Moreover, in a sense, efforts to perfect and refine processes and materials can go on forever; hence, support for such activities requires restraint. A further basic reason for the benign neglect of developmental proposals by many funding agencies is that, once the point is reached where manufacturing and application techniques need to be developed, financial assistance is usually forthcoming because the opportunity to turn a profit appears imminent. Yet conservation is primarily a non-profit undertaking; new processes and procedures are not likely to lead to ready profit. In the future we will need more concerted effort towards the development of materials and processes.

Keeping in mind this relatively neglected matter, let us return to the question of conducting scientific research and how this might benefit the field of conservation. There are perhaps four principal areas in which research can make a contribution. The first is in the technical study of the materials of the past; the second, the application of advanced methods of analysis to the conservator's needs. In these two areas technical investigation has thus far contributed much to conservation. It is important to observe, however, that many of the advanced analytical techniques have not been derived from within this field but rather through the efforts of industrial and academic scientists. Thermoluminescence dating represents a notable exception, but, in general, we are the fortunate beneficiaries of numerous techniques discovered, developed, and perfected by sources of brainpower outside the field of conservation; scanning electron microscopy, x-ray fluorescence analysis, radiocarbon dating and chromatography. A third area of research is the inquiry into the causes of deterioration and means for minimizing the effects; a fourth, the investigation of new and improved methods and materials both for preservation and restoration. In these two areas, there has been precious little effort thus far and only modest success. There hasn't yet been sufficient time, money and manpower in this young field to achieve quite as much as the generation of Paul Coremans, George Stout, and John Gettens may have hoped for.

This is not surprising. Many fail to realize the length of time that it takes to introduce new ways of doing things. I have kept a file regarding the introduction of a number of successful industrial products. In the development of cavity-fighting toothpaste, for example, we find that the idea that fluorides would stop decay occurred to investigators over forty years ago. It required 12 years of effort and nearly 4 million dollars before the results of clinical tests could be published and another year or two before the product could be put on the market. The Tennessee Eastman Company needed seven years to put cyanoacrylate adhesives on the market. Dycril was one of the first plastic materials used to make printing plates which could be developed by exposure to ultraviolet radiation. With all of their technical know-how, it took the duPont Company 10 years and $6 million to bring Dycril to practical realization. Translated into terms of manpower, that represents at least 100 man-years. Conservation as we know it has not been around for 100 years; perhaps rarely will we be able to apply 100 man-years of effort directed towards one of our problems. It isn't surprising that the development of new materials and new ways of doing things in conservation has been extremely slow.

Table 3 outlines two aspects in the history of introducing solubility parameters and acrylic resins into the business of picture varnishes. The story covers several decades. The Fogg Art Museum's experiments on acrylic resins began in 1939: the idea of solubility parameters was first enunciated in 1949. Even today the conservator is only just beginning to use some of these materials and principles with confidence. These examples have not been brought up to discourage. Simply, we must be cognizant of the magnitude of the undertaking and the need for coordination and long-term support.

Table 3
Evolution of Two Important Aspects of Protective Coating Technology

	Methacrylate Polymers And Crosslinking	Solubility Parameters and Solvent Action
1949	(Fogg Art Museum experiments began 1939)	Hildebrand initial theory, 1916
1950	National Gallery of Art Research Project established, Mellon Institute	
1951		
1952	27H Varnish tested (isoamyl methacrylate)	
1953		
1954		
1955	Crosslinking note by Bovey	Simple explanation, Burrell
1956		
1957	Crosslinking under near-UV applied to picture varnish problem	Oberlin Conference, Stolow Thesis
1958		Publication of Oberlin Conference (Feller, Stolow, and Jones)
1959		
1960		
1961		
1962		
1963	Properties of Acryloid B-72 reported, IIC Rome Conference	
1964		
1965		
1966	Ratios of crosslinking to chain breaking reported	3D parameters, Crowley, Teaque and Lowe
1967		3D parameters, Hansen; Teas diagram
1968		
1969		
1970		
1971		3D included in new edition of Feller, Stolow and Jones
1972	Standard method of reporting crosslinking recommended (log-log plot)	Simple scale of solvent power IIC Lisbon Conference
1973		
1974	Standards of photochemical stability proposed	
1975		Applied to varnish removal, Feller and Curran, Torraca
1976		
1977		
1978	Standards of durability and testing proposed, ICOM	Applied to varnish removal, LaFontaine
1979		
1980		Applied to varnish removal, Raft

We have been talking dollars and manpower but there is another ingredient required to reach our destination: the establishment of feasible goals and the coordination of effort to reach them, that is, the successful administration of technical services and of research. In the last decade or two a number of major institutions in the world have successfully mastered the ability to provide efficient analytical services to assist the conservation and curatorial departments. In contrast, although there have been some notable successes such as the development of new treatments for water-logged wood and the mass deacidification of books, the field has still much to learn in the provision of effective research and development concerning processes and materials. There is a new and growing consciousness of the need, however, which bodes well for the future.

In talking to recent graduates about the applications of science, I have occasionally detected something that has made me a bit ill at ease and yet which for a long time I have not been able to define. I now believe that it is the fact that many do not understand the complexities and limitations described. They do not understand that science does not always render the ready answer, that its application does not regularly result in quick success.

Whitehead points out that the process of education proceeds in stages. Initially there is a stage of romance, in which the new ideas are found to be especially exciting; it is a process nearly all go through. Following this, there must come a stage of precision in which one begins to sharpen one's perceptions and definitions, to realize that there are distinctions and that some conclusions and generalizations are solidly based, others not. Perhaps many students in conservation are at the first stage of romance concerning the application of science. They have great enthusiasm for what science has been able to tell them thus far and for what it might do for them in the future. We should not discourage enthusiasm. Yet the conservation scientist, in his desire to be of assistance, is confronted with a formidable task. In spite of a conscious effort to foster student research projects, we may still have failed to convey very much concerning the way the research investigator must go about his work and concerning uses and limitations of his advisory role. Many may not have reached that second stage, where they perceive that scientists come in many shapes and forms, that they can flounder about with new ideas as well as anyone else, that analyses can be in error, and that the employment of highly advanced analytical techniques can at times be premature. In the few short years spent in formal programs we may not have been able to carry them to the next state of sophistication. To learn of the exciting promise of science is part of the conservator's brief period of formal training. To learn how difficult it is to change man's ways, yet all the while to take heart and persevere—above all, to wear some modicum of modesty and humility with learning—is part of a lifetime of education.

This occasion has been proclaimed a dedication program. Indeed, a most significant and marvelous turn of events has taken place upon this meeting ground: this institution, now happily joined by a few others, has begun to introduce some sense and sensibility into the conservator's task. We come to pay our respects to those who have provided the leadership and perseverance to make this achievement possible.

One builds buildings with hope. I did not find yesterday's affair quite as expected. If all this had taken place on some village green in the hinterlands, the form and content might have been very different; there would have been speeches certainly, perhaps a band, a bit of flag waving, and some prayers. Such folkways become much transformed by time and place. We may no longer feel the need to call upon the shaman to help sanctify such occasions, to help express our hopes for a successful journey as confreres embark. Yet the meaning of a dedication ceremony, in this age as in the past, remains the same: casting forth on seas of uncertainty, one hopes that one's venture will be successful. The past twenty-two years have been notably so. Congratulations. As the Conservation Center moves into handsome new facilities, one can only hope and pray that this institution may continue to find outstanding teachers and gifted students—even more successfully in the future than it has in the past. Take with you the fire of imagination.

Great Expectations: Questions and Answers on Conservation Materials and Methods

ROBERT L. FELLER

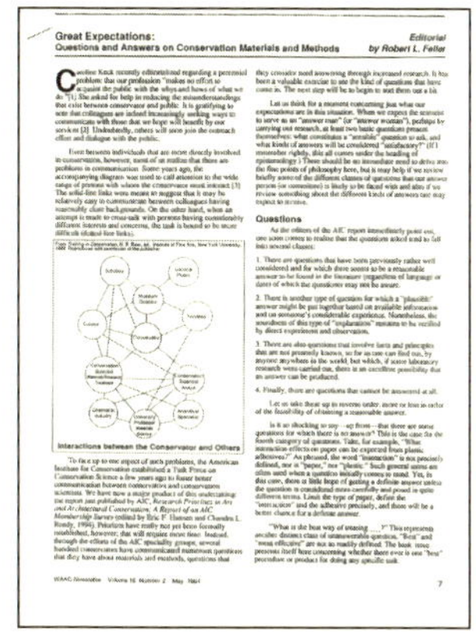

Originally printed in:
Western Association for Art Conservation Newsletter
Volume 16, No. 2 (May 1994), Pages 7-8.

Caroline Keck recently editorialized regarding a perennial problem: that our profession "makes no effort to acquaint the public with the whys and hows of what we do."[1] She asked for help in reducing the misunderstandings that exist between conservator and public. It is gratifying to note that colleagues are indeed increasingly seeking ways to communicate with those that we hope will benefit by our services.[2] Undoubtedly, others will soon join the outreach effort and dialogue with the public.

Even between individuals that are more directly involved in conservation, however, most of us realize that there are problems in communication. Some years ago, the accompanying diagram (Figure 1) was used to call attention to the wide range of persons with whom the conservator must interact.[3] The solid-line links were meant to suggest that it may be relatively easy to communicate between colleagues having reasonably close backgrounds. On the other hand, when an attempt is made to cross-talk with persons having considerably different interests and concerns, the task is bound to be more difficult (dotted-line links).

To face up to one aspect of such problems, the American Institute for Conservation established a Task Force on Conservation Science a few years ago to foster better communication between conservators and conservation scientists. We have now a major product of this undertaking: the report just published by AIC, *Research Priorities in Art and Architectural Conservation: A Report of an AIC Membership Survey* (edited by Eric F. Hansen and Chandra L. Reedy, 1994). Priorities have really not yet been formally established, however; that will require more time. Instead, through the efforts of the AIC speciality

1. Keck, C. K., "An Urgent Request," *WAAC Newsletter*, 16, No. 1 (January 1994), p. 9-10.

2. Flynn, M. C., "Public Outreach Report," *AIC News*, 19, No. 2 (March 1994), p. 5-6.

3. Feller, R. L., "Concerning the Place of Science in the Scheme of Things," in *Training in Conservation*, ed. Baer, N. S., Institute of Fine Arts, New York University, New York (1989), p. 17-32.

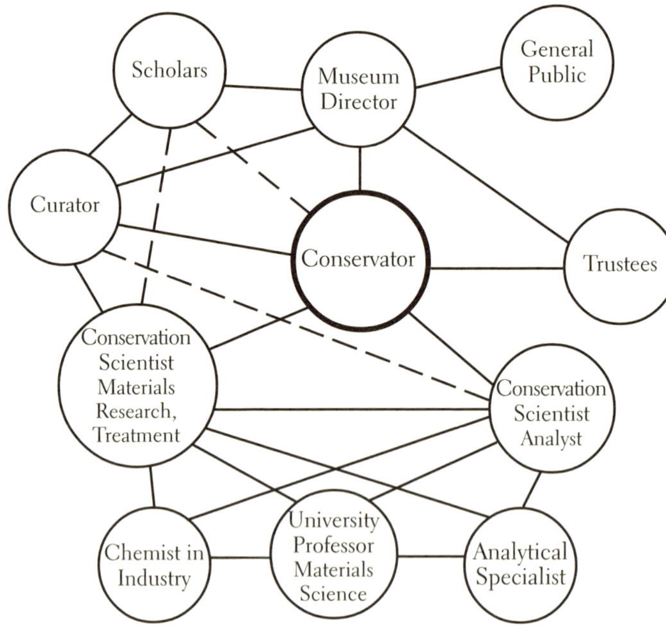

FIGURE 1
Interactions between the conservator and others. (From Ref 3. Reproduced with permission of the publisher.)

groups, several hundred conservators have communicated numerous questions that they have about materials and methods, questions that they consider need answering through increased research. It has been a valuable exercise to see the kind of questions that have come in. The next step will be to begin to sort them out a bit.

Let us think for a moment concerning just what our expectations are in this situation. When we expect the scientist to serve as an "answer man" (or "answer woman"), perhaps by carrying out research, at least two basic questions present themselves: what constitutes a "sensible" question to ask, and what kinds of answers will be considered "satisfactory?" (If I remember rightly, this all comes under the heading of epistemology.) There should be no immediate need to delve into the fine points of philosophy here, but it may help if we review briefly some of the different classes of questions that our answer person (or committee) is likely to be faced with and also if we review something about the different kinds of answers one may expect to receive.

QUESTIONS

As the editors of the AIC report immediately point out, one soon comes to realize that the questions asked tend to fall into several classes:

1. There are questions that have been previously rather well considered and for which there seems to be a reasonable answer to be found in the literature (regardless of language or date) of which the questioner may not be aware.

2. There is another type of question for which a "plausible" answer might be put together based on available information and on someone's considerable experience. Nonetheless, the soundness of this type of "explanation" remains to be verified by direct experiment and observation.

3. There are also questions that involve facts and principles that are not presently known, so far as one can find out, by anyone anywhere in the world, but which, if some laboratory research were carried out, there is an excellent possibility that an answer can be produced.

4. Finally, there are questions that cannot be answered at all.

Let us take these up in reverse order, more or less in order of the feasibility of obtaining a reasonable answer.

Is it so shocking to say—up front—that there are some questions for which there is no answer? This is the case for the fourth category of questions. Take, for example, "What interaction effects on paper can be expected from plastic adhesives?" As phrased, the word "interaction" is not precisely defined, nor is "paper," nor "plastic." Such general terms are often used when a question initially comes to mind. Yet, in this case, there is little hope of getting a definite answer unless the question is considered more carefully and posed in quite different terms. Limit the type of paper, define the "interaction" and the adhesive precisely, and there will be a better chance for a definite answer.

"What is the best way of treating ____ ?" This represents another distinct class of unanswerable question. "Best" and "most effective" are not so readily defined. The basic issue presents itself here concerning whether there ever is one "best" procedure or product for doing any specific task.

In the third category cited, there is the situation in which no one (anywhere) in fact knows the answer to the question. In some instances, a brief set of tests could, perhaps, reveal the answer (whether, for example, more rapid sun bleaching of the discolored degradation products of cellulose would occur if an ordinary glass or if a Plexiglas UF-4 filter were interspersed). Let us say that such a research investigation might be completed in a matter of a few weeks' or months' time. One can envision other questions that may require as much as five years or more of research by well-trained personnel in a well-equipped laboratory to arrive at a reasonably satisfactory answer. All parties involved must appreciate the difference in "magnitude" of the questions being asked if one is to have a reasonable expectation of receiving an answer in a convenient period of time. (We've all had the response "don't hold your breath" while waiting for an answer.)

In the second category cited, there is the question for which it appears that a reasonable answer can be synthesized on the basis of a respectable amount of information already available or out of the extensive experience of some authority on the subject. Such answers, put together out of much learning and no small measure of creative insight, are not based on direct experimental or specific analytical verification. These are "proposed" or "plausible" explanations. It is important to clearly recognize this situation and not expect the offered "explanation" to represent the final word on the matter.

In the first category in our list, there are those questions for which a well-studied and reasonably specific answer seems to be available in the literature. This, however, brings up the very basic matter as to whether the questioner will be satisfied with the answer thus available! For example, there were several questions in the AIC report regarding the stability and mechanism of deterioration of cellulose nitrate. In view of the outstanding booklet by Selwitz [4] and the research studies more recently reported by Shashoua et al.[5] and Derrick et al.,[6] one is led to ask: does the information to be found in these studies not provide a satisfactory answer? Did the questioner not know of them? Also: How difficult is it for the conservator to seek out such information? How difficult was the information to comprehend or interpret on one's own? These are some of the considerations that the AIC's Research and Technical Studies Speciality Group will be dealing with in the future with respect to questions in category one.

4. Selwitz, C., *Cellulose Nitrate Conservation*, Getty Conservation Institute, Marina del Rey, California (1988).

5. Shashoua, Y., Bradley, S. M., and Daniels, V. D., "Degradation of Cellulose Nitrate Adhesive," *Studies in Conservation*, 87 (1992), p. 113-119.

6. Derrick, M., Stulik, D., and Ordonez, E., "Deterioration of Cellulose Nitrate Sculptures Made by Gabo and Pevsner," in *Saving the 20th Century: The Conservation of Modern Materials*, ed. Grattan, D., Canadian Conservation Institute, Ottawa (1993), p. 169-182.

Answers

The last paragraph raises a very basic matter What constitutes a "satisfactory" answer? In part, this depends on the person asking, does it not? If one is asked, "How do babies come about?", a satisfying answer to a five-year-old might be different from the one given to a teenager. And it would be very much different if the question had been posed by a researcher investigating the action of X and Y chromosomes. Thus, did the question come from a practicing conservator, a scientist, a member of the public, curator, or lawyer? This brings up another matter: Can an effective mechanism be developed to deliver answers that will satisfy a variety of needs, concerns, and backgrounds? Will one tidily composed answer do? Twenty-five words or less? More? How extensive? To an agency intending to provide answers to all comers, this constitutes a basic subject that will have to be considered.

Such thoughts lead directly to another major topic: just how *certain* does the questioner expect an answer to be? (Einstein long ago said that "man has an intense desire for assured knowledge."[7] Are answers that only stand a 50-50 chance of being "correct" (reproducible, verifiable) of value to the conservator? Will one be satisfied with a 30 percent, 80 percent, 90 percent chance of the answer being correct? (Recall that there are several levels of precision in answer to the question: "What is this pigment?"[8])

7. Einstein, A., *Ideas and Opinions*, Crown Publishers, New York (1954), p. 22.

8. Feller, R. L., ed., *Artists' Pigments*, Vol. 1, Cambridge University Press, National Gallery of Art, Washington, DC (1986).

Let us not forget that, within his culture (of Snow's "two cultures" [9]), the scientist strives to be "most nearly correct" in his answers and explanations. To put it another way: "What is most important for scientists is for their results to be *right*. This is what the whole scientific endeavor is geared toward." [10] This then is one of the key reasons why a conservator may not always get a prompt and unequivocal answer. It is also for this reason that information published in the literature or in a report, however limited it may appear or how little it may satisfy, is likely to be the most reliable we have; it is "on the record" over someone's signature (authorship).

We all have our questions that we want answered. There is no limit. The task of deciding what matters deserve extensive research (category three) and what sort of questions and answers will be of benefit to the field, both in the immediate and long term, is a demanding one for any "group of experts" or "committee of concerned citizens" to undertake. Perhaps we can help each other a bit if we remind ourselves, to begin with, that questions and answers differ notably, dependent in part upon the state of prior knowledge, ease of verification, and degree of assurance required.

9. Snow, C. P., *The Two Cultures, and A Second Look: An Expanded Version of 'The Two Cultures and the Scientific Revolution,'* Cambridge University Press, London (1963).

10. Segestale, Book Review in *Science*, 263 (1994), p.837-838.

Aspects of Chemical Research in Conservation: The Deterioration Process

ROBERT L. FELLER

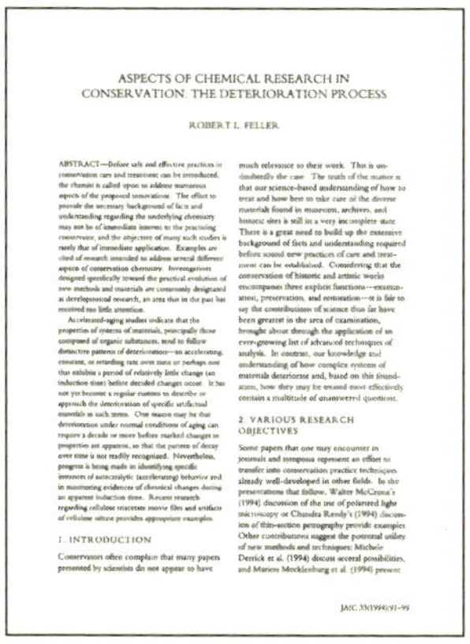
ASPECTS OF CHEMICAL RESEARCH IN CONSERVATION: THE DETERIORATION PROCESS

ROBERT L. FELLER

Originially printed in:
JOURNAL OF THE AMERICAN INSTITUTE FOR CONSERVATION
Volume 33 (1994), *Pages* 91-99.

Introduction

Conservators often complain that many papers presented by scientists do not appear to have much relevance to their work. This is undoubtedly the case. The truth of the matter is that our science-based understanding of how to treat and how best to take care of the diverse materials found in museums, archives, and historic sites is still in a very incomplete state. There is a great need to build up the extensive background of facts and understanding required before sound new practices of care and treatment can be established. Considering that the conservation of historic and artistic works encompasses three explicit functions—examination, preservation, and restoration—it is fair to say the contributions of science thus far have been greatest in the area of examination, brought about through the application of an ever-growing list of advanced techniques of analysis. In contrast, our knowledge and understanding of how complex systems of materials deteriorate and, based on this foundation, how they may be treated most effectively contain a multitude of unanswered questions.

Various Research Objectives

Some papers that one may encounter in journals and symposia represent an effort to transfer into conservation practice techniques already well-developed in other fields. In the presentations that follow, Walter McCrone's[1] discussion of the use of polarized light microscopy or Chandra Reedy's[2] discussion of

1. McCrone, W. C., "The Use of Polarized Light Microscopy in Conservation," *Journal of the American Institute for Conservation,* 33 (1994), p. 101-114.

2. Reedy, C. L., "Thin-Section Petrography in Studies of Cultural Materials," *Journal of the American Institute for Conservation,* 33 (1994), p. 115-129.

thin-section petrography provide examples. Other contributions suggest the potential utility of new methods and techniques: Michele Derrick et al.[3] discuss several possibilities, and Marion Mecklenburg et al.[4] present the potential insights offered by computerized modeling of stress development. Still others seek to establish a foundation upon which solid advances may be made; Hansen's consideration of the effect of solvents on the physical properties of thermoplastic coatings[5] and Michalski's[6] proposed model of cleaning (varnish removal by solvent action) are of this type. Other contributions seek to establish general principles that can guide our thinking about a variety of materials and situations: the concept of solubility parameter (regarding solvent action) and Pourbaix diagrams (regarding the corrosion of metals) are prime examples. In any number of fields, a time-honored service has been to identify and begin to fill in significant gaps in current knowledge.[7-9] A major area of activity aims principally to refine and enlarge upon presently available information; such would be the determination of both number average and weight average molecular weights of familiar thermoplastic resins, more precise data than had been available hitherto.[10] Table 1 summarizes just a few of the numerous objectives behind the research contributions that one will encounter.

TABLE 1

Some Research Objectives in Conservation Science

Transfer knowledge and tools from other fields
Explore potential new techniques and methods
Search for general principles
Build foundations for further advances
Fill in gaps
Refine or enlarge upon current knowledge

The conservator probably does not see much in this list likely to be of immediate assistance in his or her work, partly because a significant investigative activity is missing: development. Calls for fulfillment of national needs in research and development (R and D) are familiar. Developmental research concerns the myriad problems associated with making untried ideas a practical success. Developmental research involves mission-oriented research, "the practical application of basic knowledge to attaining a specific goal." [11] The objectives cited in Table 1 are not those immediately oriented toward the development of practical applications. They represent instead various aspects of the effort to build up the background of information needed to develop safe, sound, innovative, and practical methods of conservation.

A decade ago the effort devoted to developmental research in conservation was limited. Perhaps long-standing policies and practices of funding made it

3. Derrick, M. R., Doehne, E. F., Parker, A. E., and Stulik, D. C., "Some New Analytical Techniques for Use in Conservation," *Journal of the Amer. Inst. for Cons.*, 33 (1994), p. 171-184.

4. Mecklenberg, M. F., McCormick-Goodhart, M., and Tumosa, C. S., "Investigation into the Deterioration of Paintings and Photographs using Computerized Modeling of Stress Development," *Journal of the Amer. Inst. for Cons.*, 33 (1994), p. 153-170.

5. Hansen, E. F., Derrick, M. R., Schilling, M. R., and Garcia, R., "The Effects of Solution Application on some Mechanical and Physical Properties of Thermoplastic Amorphous Polymers used in Conservation: Poly(vinyl acetate)s," *Journal of the Amer. Inst. for Cons.*, 30 (1991), p. 203-213.

6. Michalski, S., "A Physical Model of the Cleaning of Oil Paint," in *Cleaning, Retouching, and Coatings*, ed. Mills, J. S., and Smith, P., Intl. Inst. for Cons. of Historic and Artistic Works, London (1990), p. 85-92.

7. Howie, F. M., ed., *The Care and Conservation of Geological Material: Minerals, Rocks, Meteorites and Lunar Finds*, Butterworth-Heinemann, Guildford, England (1992).

8. Williams, R. S., "Composition Implications of Plastic Artifacts: A Survey of Additives and their Effects on Longevity of Plastics," in *Saving the Twentieth Century: The Conservation of Modern Materials*, ed. Grattan, D., Can. Cons. Inst., Ottawa (1993), p. 135-153.

9. Wilhelm, H., *The Permanence and Care of Color Photographs*, Preserv. Pub. Co., Grinnell, IA (1993).

more difficult to obtain grants for development than for the acquisition of basic facts, information, and understanding. In addition, there were far fewer scientists in the conservation field than there are today.

Things have definitely changed. One finds increasing numbers of laboratories dedicated to support for developmental research in specific areas of conservation. Among others, there are now an Image Permanence Institute, a Leather Conservation Centre, a Laboratoire de Recherches des Monuments Historiques, Centro di Studio Cause di Deperimento e Metodi di Conservazione della Opere d'Arte, and a Centre for Archival Polymeric Materials. Most notable, of course, has been the major commitment of the Library of Congress and others toward developing a method of mass deacidification.[12] There is also considerable interest internationally regarding practical solutions to the consolidation of building stone.[13, 14] The Getty Conservation Institute is currently interested in ways of preserving adobe.

Among these efforts are many basic studies undertaken to obtain supporting information. Does deacidification with calcium hydroxide protect paper against mold growth?[15] Does quartz or calcite interfere with the consolidation of stone by methyltrimethoxysilane?[16] Such inquiries are rarely intended or ready for immediate application by the practicing conservator. Let us also admit that scientists often do not make their objectives clear. Nor is it always a simple matter to explain how a contribution is intended to fit into the larger picture. Perhaps it is assumed that these points usually should be rather obvious. It might be a useful exercise, nonetheless, upon being confronted with a rather involved report by a scientist, to speculate as to just what category of topic, such as those in Table 1, it is intended to address. There is a further basic problem facing authors in this field: their presentation often must satisfy an extremely wide range of interested parties extending from museum directors, curators, and practicing conservators to a peer group of conservation scientists as well as specialists in industry and academia. It is virtually impossible to accommodate all these interests in a single presentation.

There are not many occasions in which the research findings proceed directly to the practical end. Caution is certainly commendable if not mandatory. The case of soluble nylon provides a striking example in which a material was introduced directly into conservation practice. Sease[17] and Hanna and Lee[18] note that soluble nylon was first mentioned in 1958, recommended largely on the basis of some of its initial properties. Curiously, extensive aging tests were never carried out, nor was background information provided concerning whether this particular chemical substance was likely to be stable. It was more than a decade before a formal publication drew attention to potential problems.[19]

10. de la Rie, R., "The Influence of Varnishes on the Appearance of Paintings," *Studies in Conservation*, 32 (1987), p. 1-13.

11. *Report of the Study Committee on Scientific Support for Conservation of Cultural Property*, Nat. Cons. Advisory Council, Washington, DC (1979).

12. Luner, P., ed., *Paper Preservation: Current Issues and Recent Developments*, Tappi Press, Atlanta (1990).

13. Baer, N. S., Sabbioni, C., and Sors, A. I., eds., *Science, Technology and the European Cultural Heritage: Proceedings of the European Symposium, Bologna, Italy, 13-16 June 1989*, Butterworth-Heinemann, Guildford, England (1991).

14. Rodrigues, J. D., Henriques, F., and Jermias, F. T., eds., *Proceedings of the 7th International Congress on Deterioration and Conservation of Stone, Lisbon, Portugal, June 15-18, 1992* (1992), p. 1043-1052.

15. Marconi, C., "Influenza della Deacidificazione Acquosa con Idrossido di Calcio nei Riguardi della Resistenza della Carte Whatman all Attacco di Miceta Carticoli," *Bollettino dell Istituto Centrale per la Patologia del Libro Alfonso Gallo*, 43 (1989), p. 177-184.

16. Denehey, C., Wheeler, G. S., and Su, S.-C., "The Influence of Quartz and Calcite on the Polymerization of Methyltrimethoxysilane," *Proceedings of the 7th International Congress on Deterioration and Conservation of Stone, Lisbon, Portugal, June 15-18, 1992*, ed. Rodrigues, J. D., Henriques, F., Jermias, F. J., and Telmo, F. (1992), p. 1043-1052.

A Gap In Present Knowledge? Patterns of Deterioration

Conservators are fundamentally concerned with retarding the inevitable deterioration of materials. Yet, in this year of 1994, just what is it that we are able to say in answer to the basic question: How do the physical properties of specific organic substances, or systems of such materials, decline with age? If there is not a serious gap, certainly our knowledge of the subject remains limited. There is a general sense that deterioration is rarely a linear process, i.e., that the readily observed changes in properties seldom proceed at a steady rate over time. One may cite many examples, such as those in Figure 1, in which plastics and protective coatings initially increase in strength before their properties begin to decline, a decidedly nonlinear pattern of aging.[8] (This behavior is not universal but is understandable when it occurs. The initial increase in the strength of coatings can come about through crosslinking or the fact that volatile components are lost.) Perhaps Shakespeare's Seven Ages of Man can serve to remind us that it is seldom "all downhill from day one": the child first grows toward the prime of life; only thereafter does decline set in.

From what is presently known regarding the phenomenon of deterioration, it seems possible to make two statements of general principle: during aging, changes in properties usually (1) are nonlinear over time, and (2) tend to occur in stages. Figure 2, published some years ago,[20, 21] proposes four basic ways in which the chemical processes of deterioration in organic materials may take

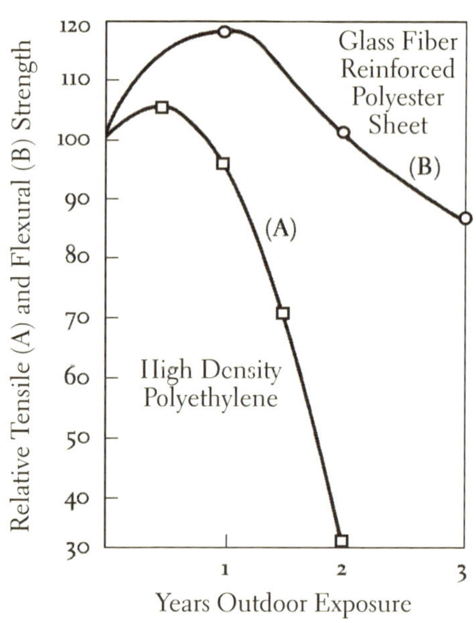

Figure 1
Change in strength of polymers during outdoor exposure: (A) data from Winslow and Hawkins;[22] (B) data from Jain et al.[23]

17. Sease, C., "The Case Against Using Soluble Nylon in Conservation Work," *Studies in Conservation*, 26 (1981), p. 102-110.

18. Hanna, S. B., and Lee, N. J., "The Consequences of Previous Adhesives and Consolidants used for Stone Conservation at the British Museum," in *Early Advances in Conservation*, British Museum Occasional Paper 65, ed. Daniels, V., British Museum, London (1988), p. 89-102.

19. DeWitte, E., "Soluble Nylon as Consolidation Agent for Stone," *Studies in Conservation*, 20 (1975), p. 30-34.

20. Kelleher, P. G., "Thermal Oxidation of Thermoplastics," *Journal of Applied Polymer Science*, 10 (1966), p. 843-857.

21. Feller, R. L., "Stages in the Deterioration of Organic Materials," in *Preservation of Paper and Textiles of Historic and Artistic Value*, Advances in Chemistry Series 164, ed. Williams, J. C., American Chemical Society, Washington, DC (1977), p. 314-335.

22. Winslow, F. H., and Hawkins, W. L., "Some Weathering Characteristics of Plastics," in *Weatherability of Plastic Materials*, Applied Polymer Symposia 4, ed. Kamal, M. R., J. Wiley, New York (1967), p. 29-39.

23. Jain, R. K., Saxena, B. K., Bansal, G. D., and Asthana, K. K., "Weathering Behaviour of Glass Fibre Reinforced Polyester Sheets," *Durability of Building Materials*, 2 (1983), p. 27-36.

place over time. In some cases, perhaps infrequently, the change really does occur in a linear manner (Fig. 2a). Far more commonly one observes processes speeding up in time (Fig. 2b) or apparently slowing down (Fig. 2c). The apparent slowing down of fading or loss of molecular weight is often simply the case of a constant fraction of colorant being lost or chemical bonds being broken in a given unit of time—"first order" decay.[24] Other aging behavior may exhibit a combination of these characteristics (Fig. 2d). Kelleher[20] referred to the manner in which oxygen might be consumed in organic materials. Nevertheless as in chemical processes, changes in physical properties—yellowness and embrittlement—often follow similar patterns.

24. Johnston-Feller, R. M., "Reflections on the Phenomenon of Fading," *Journal of Coatings Technology*, 58, No. 736 (1986), p. 32-50.

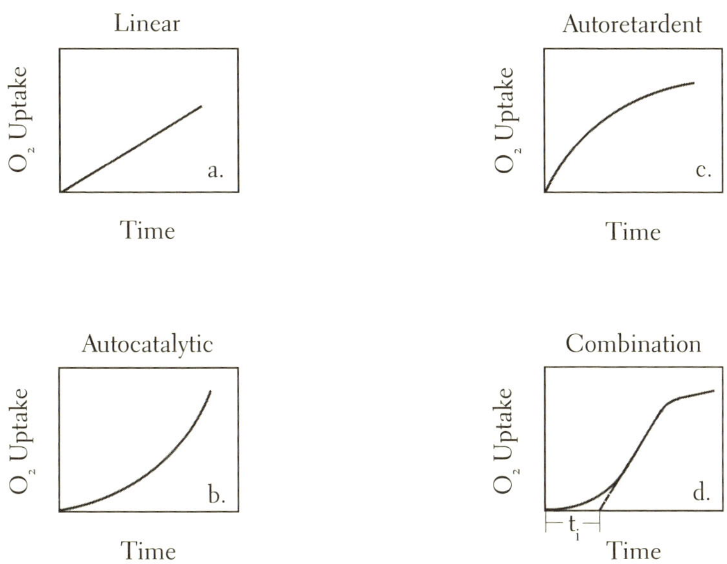

FIGURE 2
Modes of oxygen uptake versus time. After Kelleher.[20]

It is one of the present limitations in our knowledge that we are seldom in a position to state with confidence that the fading of a given dye-substrate system or the embrittlement of a particular type of textile or paper will follow one or the other of these patterns. Perhaps, with a little consideration, the field can make a few such statements, but it has not been a regular custom to look at deterioration problems from the point of view of Figure 2. Williams[8] has urged that increased attention be given to the subject.

A basic factor in our not yet being able to assign particular examples of deterioration to one of these patterns is the matter of the time scale over which deterioration takes place under the mild conditions encountered in museums and archives. In the laboratory, one can usually characterize a pattern of behavior because the full course of deterioration often can be observed, from start to finish, in a matter of hours or weeks. But under museum situations, detecting and demonstrating a change in properties may require monitoring the condition of an artifact over many decades. Naum Gabo's and Antoine Pevsner's sculptures in cellulose nitrate, done in the 1920s, only began to show clear signs of deterioration after some 25 to 60 years had passed.[25] Initial studies regarding the lamination of documents in cellulose acetate were made over the period from 1937 to 1959; it was another 16 years (about 1976), however, before observations of poor performance led the Library of Congress to recommend that the practice be discontinued.[26] Cellulose triacetate-based motion picture film reportedly showed signs of serious deterioration some 30 to 40 years after its introduction.[27] On such a time scale, it is not surprising that patterns of deterioration in terms of Figure 2 are not broadly recognized.

25. Derrick, M., Stulik, D., and Ordonez, E., "Deterioration of Cellulose Nitrate Sculptures made by Gabo and Pevsner," *Saving the Twentieth Century: The Conservation of Modern Materials*, ed. Grattan, D., Canadian Conservation Institute, Ottawa (1993), p. 169-182.

26. Stiber, L., "The Delamination of the Washington and Lee Ledger, Part 1. An Overview of Cellulose Acetate Lamination," in *Early Advances in Conservation*, British Museum Occasional Paper 65, ed. Daniels, V., British Museum, London (1988), p. 27-40.

27. Edge, M., Allen, N. S., Williams, D. A. R., Thomson, F., and Horie, V., "Methods for Predictive Stability Testing of Archival Polymers: A Preliminary Assessment of Cellulose Triacetate-based Motion Picture Film," *Polymer Degradation and Stability*, 35 (1992), p. 147-155.

28. Allen, N. S., Edge, M., Appleyard, J. H., Jewitt, T. S., Horie, C. V., and Francis, D., "Degradation of Historic Cellulose Triacetate Cinematographic Film: The Vinegar Syndrome," *Polymer Degradation and Stability*, 19 (1987), p. 379-387.

29. Shashoua, Y., Bradley, S. M., and Daniels, V. D., "Degradation of Cellulose Nitrate Adhesive," *Studies in Conservation*, 87 (1992), p. 113-119.

Basic Understanding of Patterns of Deterioration

The effort to establish general principles regarding the various ways in which the properties of materials can change over time has great significance to the practicing conservator. First of all, if one wishes to estimate the possible condition of an artifact some time in the future, it is important to understand whether the changes that have already taken place are likely to continue at a steady pace, apparently slow down, or seriously speed up in time (the latter is often labeled autocatalytic or autoaccelerating behavior). The degradation of cellulose triacetate movie film appears to be of the latter character;[28] so does the deterioration of cellulose nitrate plastics.[25, 29] Within a day of opening a vitrine in which cellulose nitrate materials have been kept in reasonably satisfactory condition for a number of years, serious fracturing can occur; the rapid rate of disintegration then becomes a matter requiring immediate attention.[25] The autocatalytic stage can be said to have begun with the opening of the vitrine. Research must attempt to answer many questions: Why do the physical changes occur? Can the behavior be successfully retarded? Can it be circumvented? Is it too late to carry out satisfactory treatment once a particular stage of deterioration has been reached?

The practical importance to the conservator in realizing that a particular deterioration process tends to speed up in time is to emphasize noticing the process and initiating treatment at the earliest possible stage, catching potential trouble at an early point in time. Moreover, when devising accelerated-aging tests to evaluate the potential stability of different types of materials, one must appreciate that it is fundamentally a difficult task to evaluate the relative merit of those that age in the manner of Figure 2b versus 2c or 2d.

Induction-Time Behavior

Of considerable interest is the phenomenon of induction time (Fig. 2d): a "time during which no observable change occurs in a chemical reaction or in a physical property, subsequently followed by such a change."[30] If conservators observe the condition of an artifact during an induction period and this period is a feature of its particular aging behavior, they may be lulled into thinking that changes are not occurring and that the artifact is in no imminent danger of change, only to find, a relatively short time thereafter, that decided changes have become apparent. Thus, as an initial reaction to a frustrating problem, Calmes[31] has been led to say of cellulose nitrate: "There appears to be no intermediate stage between the time when the film is in good condition and the time when it is obviously deteriorated." We now know that research has been able to detect a number of intermediate stages.

30. Alger, M. S. M., *Polymer Science Dictionary*, Elsevier Applied Science, London and New York (1989), p. 46.

31. Calmes, A., "Plastics Found in Archives," *Saving the Twentieth Century: The Conservation of Modern Materials*, ed. Grattan, D., Canadian Conservation Institute, Ottawa (1993), p. 95-102.

One of the earliest examples of induction time that the author encountered is in the crosslinking of butyl methacrylate polymers. When the data reported many years ago are considered (Fig. 3), one sees that there is an apparent induction period of 11 years before insoluble (crosslinked) matter was formed in the various butyl methacrylate-based coatings that had been exposed under skylights on the wall of the laboratory of the Fogg Art Museum.[32, 33] If these coatings had been tested for insoluble matter after eight years, the conservator might have concluded that they had no tendency to crosslink. Yet, a decade later, there would have been convincing evidence that insoluble matter had begun to form. Basic scientific studies and principles support these observations; the theory of how crosslinking takes place leads one to expect the existence of an induction period.

32. Feller, R. L., "The Deterioration of Organic Substances and the Analysis of Paints and Varnishes," *Preservation and Conservation: Principles and Practices*, ed. Timmons, S., Preservation Press, Washington, DC (1976), p. 287-298.

33. Feller, R. L., Stolow, N., and Jones, E. H., *On Picture Varnishes and their Solvents*, rev. and enlarged edn., National Gallery of Art, Washington, DC (1985), p. 158-159.

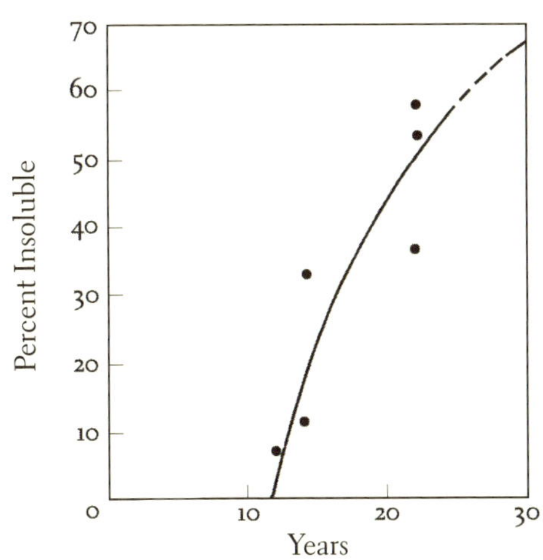

Figure 3
Insoluble matter found in n-butyl and isoamyl copolymers exposed on a well-illuminated wall of the Fogg Art Museum laboratory.[32]

The chalking of paints—i.e., the degradation of the vehicle so that the paint begins to flake from the surface—has also been observed to exhibit induction-time behavior.[34] A situation that resembles this behavior is also encountered in the corrosion of metals that have been protected by a paint or varnish. Graphs of arbitrary "corrosion ratings" often show a period of time in which no noticeable corrosion has occurred. The explanation seems plain enough: corrosion of the metal substrate tends not to occur until the organic coating has begun to fail. This situation need not fit a rigorous definition of induction-time chemistry. Yet from the practical point of view of persons responsible for the care of metal artifacts, there is a time when all goes well. Thereafter, a moment occurs at some point in time, often difficult to predict or to "catch," at which corrosion becomes evident.

34. Feller, R. L., "Fundamentals of Conservation Science: Induction Time and the Autooxidation of Organic Compounds," *Bulletin of the American Institute for Conservation*, 14, No. 2 (1974), p. 142-151.

When these concepts were initially being reviewed in 1972-77, it was possible to suggest that "most of the measures that are taken, or that should be taken, to conserve organic material are concerned with processes that occur during the induction time of oxidative deterioration and are measures essentially concerned with prolongation of the period of induction."[35] This point of view is well understood by polymer chemists. Most recently, Edge et al.[36] stated again, "It is obvious that the small physicochemical changes which occur during the early stages of aging, i.e., during the induction period prior to the onset of degradation, will be of prime importance with respect to subsequent stability." Although specific instances can be documented, the phenomenon of apparent induction time is not universally encountered in deterioration problems. When it does occur, however, it is of fundamental significance in preventive care.

35. *Ibid.*, p. 146.

36. Edge, M., Allen, N. S., Williams, D. A. R., Thomson, F., and Horie, V., *op. cit.*, p. 148.

EARLY DETECTION OF IMPENDING DETERIORATION

It is possible that the subtle processes taking place during the "apparent induction period" can be monitored successfully. In the case of rubber cement, it was shown that an increase in weight and in hydroperoxide concentration could be detected before serious discoloration took place.[37] Moreover, a "sticky" stage was also observed before serious loss of solubility took place. These observations held out the possibility that one could find and measure properties that were changing during a period when few or no visually observable changes were occurring. More recently, Shashoua et al.[29] report that they are able to follow the early loss of nitro (NO_2) groups relative to carbonyl (CO) in the infrared spectra of nitrocellulose adhesives that had been applied to objects over a period of 30 years. The loss of the camphor plasticizer can be similarly monitored.[25] There are also early stages in which odors are detectable or exudations observed. Perhaps one day these emissions can be precisely measured to monitor condition.

37. Feller, R. L., and Enke, D. B., "Stages in Deterioration: The Examples of Rubber Cement and Transparent Tape," in *Science and Technology in the Service of Conservation*, ed. Brommelle, N. S., and Thomson, G., International Institute for Conservation of Historic and Artistic Works, London (1982), p. 19-23.

Studies regarding the deterioration of cellulose triacetate motion picture film, representing more than 30 man-years of effort at the Centre for Archival Polymeric Materials, Manchester, England, have confirmed that the process is autocatalytic and that it tends to exhibit an induction period. The Centre has since demonstrated that early changes within the film can be monitored by analysis of the moisture content by means of a "novel technique of microwave spectroscopy," a nondestructive method.[38] This example and that of the cellulose nitrate studies are cited to indicate that progress is indeed currently being made in our ability to monitor the subtle changes that can take place in organic materials before serious physical deterioration sets in.

38. Edge, M., Allen, N. S., Williams, D. A. R., Thomson, F., and Horie, V., *op. cit.*, p. 147.

CONCLUSIONS

Many research contributions are not intended for immediate application by the practicing conservator. Before sound practices of prevention and treatment can be developed, extensive background information needs to be acquired regarding the way that materials and systems of materials can be characterized and how and why specific materials behave as they do. Only then can research be directed effectively toward the development of practical techniques and procedures.

Deficiencies remain in our understanding of the task of retarding deterioration, and they are the concern of all entrusted with the care of collections. One particular aspect concerns the pattern of degradation that specific classes of materials undergo. One might call this an understanding of the "pathogenesis" of degradation, or as Williams[39] calls it, "the degradation profile," as illustrated in Figure 2. Increasingly, conservation scientists are beginning to consider deterioration processes from this point of view. Thus, in spite of the long time scale over which the deterioration of museum artifacts usually takes place, the fact that certain specific materials degrade autocatalytically is now well recognized. Moreover, in situations in which an induction period seems to be part of the picture, techniques are being developed to monitor the chemical changes that occur during this deceptively quiescent period.

39. Williams, R. S., *op. cit.*, p. 140.

A Robert Feller Bibliography

"On Cyclic Intermediates in Substitution Reactions. II. Steric Effects in the Alkaline Hydrolysis of the Epimeric 3-Chloro-6//7-Chlorestandi-carboxylic Acids," *Journal of the American Chemical Society*, 73 (1951), p. 2392-2394 (with D. W. Wujciak and J. F. Lane).

"Reactions of Diazoketones. IV. Kinetics of the Decomposition of Some p-Substituted-α-diazoacetophenones in Acetic Acid," *Journal of the American Chemical Society*, 73 (1951), p. 4230-4234 (with J. F. Lane).

"The Conservation of Paintings," *Carnegie Magazine*, 26, No. 1 (1952), p. 370-373.

"Hardness and Flexibility of Natural and Synthetic-Resin Varnishes," *Museum News*, 29, No. 20 (1952), p.7-8.

"Science Serving the Fine Arts," *Carnegie Magazine*, 26, No. 2 (1952), p. 46-50.

"Color Change in Oil Paintings," *Carnegie Magazine*, 28 (1954), p. 276, 279, 281, 285.

"Dammar and Mastic Infrared Analysis," *Science*, 120 (1954), p. 1069-1070.

"Crosslinking of Methacrylate Polymers by Ultraviolet Radiation," *American Chemical Society Division of Paint, Plastics, and Printing Ink Chemistry, Papers Presented at New York Meeting*, 17, No. 2 (September 1957), p. 465-470. This volume, p. 327

"Factors Affecting the Appearance of Picture Varnish," *Science*, 125 (1957), p. 1143-1144.

"Dammar and Mastic Varnishes—Hardness, Brittleness, and Change in Weight Upon Drying," *Studies in Conservation*, 3, No. 4 (1958), p. 162-174. This volume, p. 3

"Picture Varnish," in *Encyclopedia of Chemistry*, Supplement, ed. Clark, G. L., Reinhold Publishing Co., New York (1958), p. 220-222.

"Identification and Analysis of Resins and Spirit Varnishes," in *Application of Science in Examination of Works of Art*, Museum of Fine Arts, Boston (1959), p. 51-76.

On Picture Varnishes and Their Solvents, Reports presented to the seminar on resinous surface coatings sponsored by the Intermuseum Conservation Association, Oberlin, Ohio (1959) (with E. H. Jones and N. Stolow).

"Review of *The Discoloration of Coloured Objects Under the Influence of Daylight, Incandescent Lamplight and Fluorescent Lamplight*, by J. J. Balder, Museums Association, Leiden, 1956," *Studies in Conservation*, 4 (1959), p. 155.

"The Technical Examination of Museum Objects," *The Indicator*, 40, No. 9 (1959), p. 20, 22, 24, 26, 28.

"Characteristics of Resin AW-2," *Technical Notes, Bulletin of the American Group-IIC*, 3, No. 1 (October 1962), p. 9-10.

"Considerations Regarding the Illumination of Museum Objects," in *Catalog of Exposition of Painting Conservation*, The Brooklyn Museum, Brooklyn, NY (1962).

"Levels of Illumination and Action of Light on Oil Paintings," *Intermuseum Conservation Association Information Bulletin*, No. 6 (July 1962), p. 29-32.

"Spirit Varnishes and Thermoplastic Resins," in *Catalog of Exposition of Painting Conservation*, The Brooklyn Museum, Brooklyn, NY (1962).

"Wax," in *Catalog of Exposition of Painting Conservation*, The Brooklyn Museum, Brooklyn, NY (1962).

This volume, p. 333

"New Solvent-type Varnishes," in *Recent Advances in Conservation*, ed. Thomson, G., Butterworths, London (1963), p. 171-175.

"Review of *Recent Advances in Conservation*, Butterworths, 1963," *Science*, 142 (1963), p. 1565.

"Standards of Exposure to Light," *Bulletin of the American Group-IIC*, 4, No. 1 (1963), p. 10-12.

Synthetic Materials Used in the Conservation of Cultural Property, International Centre for the Preservation and Restoration of Cultural Property, Rome (1963) (with G. Thomson and A. E. A. Werner).

"The Use of Differential Spectral Curve Analysis in the Study of Museum Objects," *Dyestuffs*, 44, No. 9 (1963), p. 1-10 (with R. M. Johnston). This volume, p. 155

"Control of Deteriorating Effects of Light Upon Museum Objects," *Museum*, 17 (1964), p. 57-98. This volume, p. 53

"Critical Pigment Volume Concentration and Chalking in Paints," *Bulletin of the American Group-IIC*, 5, No. 1 (1964), p. 25-26 (with J. J. Matous). This volume, p. 243

"Detection of an Epoxy-Resin Coating on a Seventeenth-Century Painting," *Studies in Conservation*, 9 (1964), p. 1-8 (with S. Keck).

"The Deteriorating Effect of Light on Museum Objects," *Museum News Technical Supplement*, No. 3 (1964), p. i-viii.

"A Note on the Exposure of Dammar and Mastic Varnishes to Fluorescent Lamps," *Bulletin of the American Group-IIC*, 4, No. 2 (1964), p. 12-14. This volume, p. 17

"The Use of an Electrically Conducting Glass Panel as a Heating Surface," *Bulletin of the American Group-IIC*, 5, No. 1 (1964), p. 23-24 (with J. L. Kostich).

"What's in a Name: Dammar, or Serendipity in the Library," *The Crucible*, 49, No. 8 (1964), p. 214, 216, 218. This volume, p. 21

"Combining Art and Science: The National Gallery of Art Research Project in Artists' Materials," *Rutgers Alumni Monthly* (November 1965), 3 p.

"The Early Use of Prussian Blue," *Bulletin of the American Group-IIC*, 5, No. 2 (1965), p. 30.

"On the Fading of Dyes," *Bulletin of the American Group-IIC*, 5, No. 2 (1965), p. 31-33.

"Polymer Emulsions," *Bulletin of the American Group-IIC*, 6, No. 2 (1966), p. 24-28.

"Polymer Emulsions II," *Bulletin of the American Group-IIC*, 7, No. 1 (1966), p. 18-19.

This volume, p. 247
"Problems in Retouching: Chalking of Intermediate Layers," *Bulletin of the American Group-IIC*, 7, No. 1 (1966), p. 32-34.

This volume, p. 281
"Rediscovery of the Wheel," *Color Engineering*, 4, No. 6 (1966), p. 20-23.

This volume, p. 343
"Studies of the Effect of Light on Protective Coatings Using Aluminum Foil as a Support: Determination of Ratio of Chain Breaking to Crosslinking," *Bulletin of the American Group-IIC*, 6, No. 1 (1966), 2 p. (with C. W. Bailie).

This volume, p. 27
"Translation of First Description of Dammar Picture Varnish, by F. Lucanus in 1829," National Gallery of Art Research Project (November 1966).

"Barytes Found in Blanched Paint," *Bulletin of the American Group-IIC*, 8, No. 1 (1967), p. 10.

This volume, p. 89
"Control of Deteriorating Effects of Light on Museum Objects: Heating Effects of Illumination by Incandescent Lamps," *Museum News*, 46, No. 9 (1967), p. 39-47.

"Dating and Authenticating Works of Art by Measurement of Natural Alpha Emitters," *Science*, 155, No. 3767 (1967), p. 1238-1242 (with B. Keisch, A. S. Levine and R. R. Edwards).

"Exposure to Xerox Copier Safer Than Ten Seconds of Sunlight," *Bulletin of the American Group-IIC*, 7, No. 2 (1967), p. 21.

"Felt-tipped Markers and the Need for Standards of Lightfastness for Artists' Colorants," *Bulletin of the American Group-IIC*, 8, No. 1 (1967), p. 24-26.

"Lightfastness of Felt-tipped Markers," *Inter-Society Color Council Newsletter*, 192 (January-February 1968), p. 10-11.

This volume, p. 251
"Optics of Paint Films: Glazes and Chalking," in *Application of Science in the Examination of Works of Art*, Museum of Fine Arts, Boston, MA (1967), p. 86-95 (with R. M. Johnston).

"Research on Durable Thermoplastic Polymers for the Conservation of Works of Art," in *Atti della XLIX Riunione SPIS*, Sienna, Italy (September 1967), p. 23-27; Rome, Italy (1968), p. 1099-1110.

"A Solvatochromic Dye as a Convenient Indicator of the Solubility Parameter of Petroleum Solvents," *Bulletin of the American Group-IIC*, 7, No. 2 (1967), p. 29-30 (with J. B. Page).

"Standards of Exposure to Light II," *Bulletin of the American Group-IIC*, 7, No. 2 (1967), p. 8, 32.

"Studies of the Darkening of Vermilion by Light," *Report and Studies in the History of Art*, National Gallery of Art, Washington, DC (1967), p. 99-111. This volume, p. 163

"Problems in Spectrophotometry," in *London Conference on Museum Climatology*, ed. Thomson, G., International Institute for Conservation of Historic and Artistic Works, London (1968), p. 257-269.

"Solubility Parameters," *Bulletin of the American Group-IIC*, 8, No. 2 (1968), p. 20-24.

"Polymeric Emulsions III," *Bulletin of the American Group-IIC*, 9, No. 2 (1969), p. 15-17.

"Transportation of a Panel Painting by Courier in Winter," in *Papers Given at the Annual Meeting of IIC-American Group, Los Angeles* (1969), p. 13-14.

"Identifying Works of Art: A Diagnostic Problem," *Abbottempo*, Book 4 (1970), p. 12-17 (with B. Keisch).

"On the Color Space of Sigfrid Forsius 1611," *Color Engineering*, 8, No. 3 (1970), p. 48-51 (with A. S. Stenius). This volume, p. 289

"Polymeric Emulsions IV," *Bulletin of the American Group-IIC*, 10, No. 2 (1970), p. 11-12.

"Solubility and Crosslinking Characteristics of Ethylene Vinylacetate Copolymers," *Bulletin of the American Group-IIC*, 11, No. 1 (1970), p. 42-45 (with M. Curran). This volume, p. 349

"Analysis of Pigments," in *American Painting to 1776: A Reappraisal*, ed. Quimby, I. M. G., University Press of Virginia, Charlottesville (1971), p. 327-344.

"Lightfast Soluble Colorants," *Bulletin of the American Group-IIC*, 11, No. 2 (1971), p. 15 (with M. Curran).

"Notes on the Chemistry of Bleaching," *Bulletin of the American Group-IIC*, 11, No. 2 (1971), p. 39-56.

"Notes on Modern Pigments," *Bulletin of the American Group-IIC*, 12, No. 1 (1971), p. 60-62 (with B. Keisch and M. Curran).

On Picture Varnishes and Their Solvents, Revised Edition, The Press of Case Western University, Cleveland, OH (1971) (with N. Stolow and E. H. Jones).

"Analysis of the Pigments in the Toledo Fayum Portrait," in *Conservation of Paintings and the Graphic Arts, Preprints of the Lisbon Congress*, International Institute for Conservation of Historic and Artistic Works, London (1972), p. 804-805.

"Review of *Artists' Pigments c. 1600-1835*, by R. D. Harley, Butterworths, London, 1972," *The Burlington Magazine*, 114, No. 830 (May 1972), p. 341-342.

"Rubens's The Gerbier Family: Technical Examination of the Pigments and Paint Layers," *Studies in the History of Art*, National Gallery of Art, Washington, D.C. (1972), p. 54-74.

This volume, p. 605 "Scientific Examination of Artistic and Decorative Colorants," *Journal of Paint Technology*, 44, No. 566 (March 1972), p. 51-58.

This volume, p. 31 "Solubility of Aged Coatings Based on Dammar, Mastic and Resin AW-2," *Bulletin of the American Group—IIC*, 12, No. 2 (1972), p. 72-81 (with C. W. Bailie).

"Thermochemically Activated Oxidation: Mother Nature's Book Burning," *P.L.A.* (Pennsylvania Library Association) *Bulletin, 28, No. 6* (November 1973), p. 219-251.

"Fundamentals of Conservation Science: Induction Time and the Autoxidation of Organic Compounds," *Bulletin of the American Institute for Conservation*, 14, No. 2 (1974), p. 142-151.

This volume, p. 41 "Changes in Solubility and Removability of Varnish Resins with Age," *Bulletin of the American Institute for Conservation*, 15, No. 2 (1975), p. 17-26 (with M. Curran).

"A Project to Prepare Monographs on Ten Artists' Pigments," in *Preprints of the 4th Triennial Meeting, Venice*, ICOM Committee for Conservation (1975), Paper 75/21/6.

This volume, p. 107 "Speeding Up Photochemical Deterioration," *Bulletin de l'Institut Royal du Patrimoine Artistique*, XV (1975), p. 135-150.

"Studies on Photochemical Deterioration," in *Preprints of the 4th Triennial Meeting, Venice*, ICOM Committee for Conservation (1975), Paper 75/19/4.

"Studies on the Photochemical Stability of Thermoplastic Resins," in *Preprints of the 4th Triennial Meeting, Venice*, ICOM Committee for Conservation (1975), Paper 75/22/4.

"Deterioration of Organic Substances and the Analysis of Paints and Varnishes," in *Preservation and Conservation: Principles and Practices*, ed. Timmons, S., Preservation Press, Washington, DC (1976), p. 287-299.

"On the Crystallography of Chrome Orange," *Journal of Coatings Technology*, 48, No. 618 (1976), p. 68 (with S. S. Pollack).

"Problems in the Investigation of Picture Varnishes," in *Conservation and Restoration of Pictorial Art*, ed. Brommelle, N., and Smith, P., Butterworths, London (1976), p. 137-144.

"The Relative Solvent Power Needed to Remove Various Aged Solvent-type Coatings," in *Conservation and Restoration of Pictorial Art*, ed. Brommelle, N., and Smith, P., Butterworths, London (1976), p. 158-161.

"Conservators Advise Artists (on use of polymer emulsion paints)," *Art Journal*, 37, No. 1 (1977), p. 34-39 (with L. Pomerantz and D. Goist).

"Stages in the Deterioration of Organic Materials," in *Preservation of Paper and Textiles of Historic and Artistic Value*, Advances in Chemistry Series No. 164, ed. Williams, J. C., American Chemical Society, Washington, DC (1977), p. 314-335.

This volume, p. 353

"Further Studies on the International Blue-Wool Standards for Exposure to Light," in *Preprints of the 5th Triennial Meeting, Zagreb*, ICOM Committee for Conservation (1978), Paper No. 78/18/2.

This volume, p. 127

"Standards in the Evaluation of Thermoplastic Resins," in *Preprints of the 5th Triennial Meeting, Zagreb*, ICOM Committee for Conservation (1978), Paper No. 78/16/4.

This volume, p. 143

"Use of the International Standards Organization's Blue-Wool Standards for Exposure to Light. I. Use as an Integrating Light Monitor for Illumination Under Museum Conditions," in *Preprints of the 6th Annual Meeting, Fort Worth, Texas*, American Institute for Conservation, Washington, DC (1978), p. 73-80 (with R. M. Johnston-Feller).

This volume, p. 121

This volume, p. 135

"Use of the International Standards Organization's Blue-Wool Standards for Exposure to Light. II. Instrumental Measurement of Fading," in *Preprints of the 7th Annual Meeting, Toronto, Canada*, American Institute for Conservation, Washington, DC (1979), p. 30-36 (with R. M. Johnston-Feller).

"The Importance of Research in the Conservation of Library and Archival Materials," in *A National Preservation Program: Proceedings of a Planning Conference*, Preservation Office, Library of Congress, Washington, DC (1980), p. 44-50.

"Light Sources and the Factors that Influence the Deteriorating Effects of Illumination," paper presented at Textile Preservation Symposium (1980), 16 p.

"Concerning the Exposure of Paper to Light: Discoloration of Handsheets of Known Initial Lignin and Hot-Alkali-Soluble Content," in *Preprints of the 6th Triennial Meeting, Ottawa*, ICOM Committee for Conservation (1981), Paper No. 81/14/4 (with S. Lee and J. Bogaard).

"Continued Investigations Involving the ISO Blue-Wool Standards of Exposure," in *Preprints of the 6th Triennial Meeting, Ottawa*, ICOM Committee for Conservation (1981), Paper No. 81/18/1 (with R. Johnston-Feller).

"Developments in the Testing and Application of Protective Coatings," in *Preprints of the 6th Triennial Meeting, Ottawa*, ICOM Committee for Conservation (1981), Paper No. 81/16/1.

This volume, p. 269

"The Effect of Pigment Volume Concentration on the Lightness or Darkness of Porous Paints," in *Preprints of the 9th Annual Meeting, Philadelphia, Pennsylvania*, American Institute for Conservation, Washington, DC (1981), p. 66-74 (with N. Kunz).

This volume, p. 383

"Photochemical Studies of Methacrylate Coatings for the Conservation of Museum Objects," in *Photodegradation and Photostabilization of Coatings*, ACS Symposium Series No. 151, ed. Pappas, S. P., and Winslow, F. H., American Chemical Society, Washington, DC (1981), p. 183-196 (with M. Curran and C. Bailie).

This volume, p. 529

"The Darkening and Bleaching of Paper by Various Wavelengths in the Visible and Ultraviolet," in *Postprints of the Book and Paper Group at the 10th Annual Meeting, Milwaukee, Wisconsin*, American Institute for Conservation, Washington, DC (1982), 17 p. (with S. B. Lee and J. Bogaard).

"Stages in Deterioration: The Examples of Rubber Cement and Transparent Mending Tape," in *Science and Technology in the Service of Conservation*, ed. Brommelle, N. S., and Thomson, G., International Institute for Conservation of Historic and Artistic Works, London (1982), p. 19-23 (with D. B. Encke). This volume, p. 371

"Who Invented the Color Wheel?" *Color Research and Application*, 7, No. 3 (1982), p. 217-230 (with C. Parkhurst). This volume, p. 295

"Kinetics of the Fading of Alizarin Lake," in *Preprints of the 11th Annual Meeting, Baltimore, Maryland*, American Institute for Conservation, Washington, DC (1983), p. 65-73 (with R. Johnston-Feller, C. Bailie and M. Curran).

"Identification of Traditional Organic Colorants Employed in Japanese Prints and Determination of Their Rates of Fading," in *Japanese Woodblock Prints: A Catalogue of the Mary A. Ainsworth Collection*, Keyes, R. S., Allen Memorial Art Museum, Oberlin College, Oberlin, OH (1984), p. 253-266 (with M. Curran and C. Bailie). This volume, p. 179

"The Kinetics of Fading: Opaque Paint Films Pigmented with Alizarin Lake and Titanium Dioxide," *Journal of the American Institute for Conservation*, 23, No. 2 (1984), p. 114-129 (with R. Johnston-Feller, C. Bailie and M. Curran). This volume, p. 211

"Thermoplastic Polymers Currently in Use as Protective Coatings and Potential Directions for Further Research," *Institute for the Conservation of Cultural Material (ICCM) Bulletin*, 10, No. 2 (1984), p. 5-15.

"Uses of Instrumental Color-matching Techniques for Studying Pigment Fading," *Bulletin of the Institute for the Conservation of Cultural Materials (Australia)*, 10, No. 2 (1984), p. 71-91 (with R. Johnston-Feller and C. Bailie).

"The International Standards Organization's Blue-Wool Fading Standards (ISO R105)," from *Textile and Museum Lighting*, Harpers Ferry Regional Textile Group (1985), p. 41-57.

On Picture Varnishes and Their Solvents, Reprint of 1971 Edition, National Gallery of Art, Washington, DC (1985) (with N. Stolow and E. H. Jones).

"Relation of Cellulose Chain Scission to Hot-Alkali-Soluble Content During Thermal and Photochemical Degradation of Paper," *Journal of Imaging Science*, 29, No. 2 (1985), p. 61-64 (with S. B. Lee and J. Bogaard). This volume, p. 483

This volume, p. 397 "Three Fundamental Aspects of Cellulose Deterioration," *Art and Archaeology Technical Abstracts*, Supplement, 22, No. 1 (1985), p. 277-358 (with S. B. Lee and M. Curran).

Artists' Pigments, A Handbook of their History and Characteristics, Vol. 1, ed. Feller, R. L., Cambridge University Press and National Gallery of Art, Washington, DC (1986).

"Barium Sulfate—Natural and Synthetic," in *Artists' Pigments: A Handbook of Their History and Characteristics*, Vol. 1, ed. Feller, R. L., Cambridge University Press and National Gallery of Art, Washington, DC (1986), p. 47-64.

This volume, p. 227 "Determination of the Specific Rate Constant for the Loss of a Yellow Intermediate During the Fading of Alizarin Lake," *Journal of the American Institute for Conservation*, 25, No. 2 (1986), p. 65-72 (with R. M. Johnston-Feller and C. Bailie).

"Indian Yellow," in *Artists' Pigments: A Handbook of Their History and Characteristics*, Vol. 1, ed. Feller, R. L., Cambridge University Press and National Gallery of Art, Washington, DC (1986), p. 17-36 (with N. S. Baer, A. Joel and N. Indictor).

This volume, p. 577 "Influence of the Hemicellulose Fraction on Thermal and Photochemical Discoloration of Paper," in *Historic Textile and Paper Materials*, Advances in Chemistry Series No. 212, ed. Needles, H. L., and Zeronian, S. H., American Chemical Society, Washington, DC (1986), p. 377-386 (with S. Lee).

This volume, p. 493 "The Kinetics of Cellulose Deterioration," in *Historic Textile and Paper Materials*, Advances in Chemistry Series No. 212, ed. Needles, H. L., and Zeronian, S. H., American Chemical Society, Washington, DC (1986), p. 329-346 (with S. B. Lee and J. Bogaard).

"Science at the Galleries," in *Encyclopaedia Brittanica 1987 Yearbook of Science and the Future*, ed. Calhoun, D., Encyclopaedia Brittanica, Chicago, IL (1986), p. 227-245.

"Terminology and Procedures Used in the Systematic Examination of Pigment Particles with the Polarizing Microscope," Appendix in *Artists' Pigments: A Handbook of Their History and Characteristics*, Vol. 1, ed. Feller, R. L., Cambridge University Press and National Gallery of Art, Washington, DC (1986), p. 285-298 (with M. Bayard).

"Comments on the Measurement of 'Yellowness' in Pulp and Paper," *Book and Paper Group (American Institute for Conservation) Annual*, 6 (1987), p. 40-51.

"Some Factors to be Considered in Accelerated-Aging Tests," in *Preprints of the 15th Annual Meeting, Vancouver, B.C., Canada*, American Institute for Conservation of Historic and Artistic Works, Washington, DC (1987), p. 56-67.

"The Fading of Some Traditional Pigments as a Function of Relative Humidity," in *Materials Issues in Art and Archaeology, Mat. Res. Soc. Symp. Proc.*, 123, ed. Sayre, E. V., Vandiver, P. B., Druzik, J., and Stevenson, C., Materials Research Society, Pittsburgh, PA (1988), p. 287-292 (with C. W. Bailie and R. M. Johnston-Feller). This volume, p. 235

"Accelerated Thermal Degradation of Pulp Sheets: Effect of Beating and Importance of Humidity," in *Cellulose and Wood: Chemistry and Technology*, ed. Schuerch, C., John Wiley & Sons, New York (1989), p. 863-884 (with S. B. Lee and J. Bogaard). This volume, p. 515

"Concerning the Place of Science in the Scheme of Things," in *Training in Conservation*, ed. Baer, N. S., Institute of Fine Arts, New York University, New York (1989), p. 17-32. This volume, p. 619

"Damaging Effects of Visible and Near-Ultraviolet Radiation on Paper," in *Historic Textile and Paper Materials II*, ACS Symposium Series No. 410, ed. Zeronian, S. H., and Needles, H. L., American Chemical Society, Washington, DC (1989), p. 54-62 (with S. Lee and J. Bogaard). This volume, p. 507

"Darkening of Paper Following Exposure to Visible and Near-Ultraviolet Radiation," *Journal of the American Institute for Conservation*, 28 (1989), p. 1-18 (with S. B. Lee and J. Bogaard). This volume, p. 587

Evaluation of Cellulose Ethers for Conservation, Research in Conservation Series No. 3, Getty Conservation Institute, Marina del Rey, CA (1990) (with M. Wilt).

"Calcium Carbonate Whites," in *Artists' Pigments*, Vol. 2, ed. Roy, A., National Gallery of Art, Washington, DC, and Oxford University Press, New York, NY (1993), p. 203-226 (with R. J. Gettens and E. W. FitzHugh).

"Vermilion and Cinnabar," in *Artists' Pigments*, Vol. 2, ed. Roy, A., National Gallery of Art, Washington, DC, and Oxford University Press, New York, NY (1993), p. 159-182 (with R. J. Gettens and W. T. Chase).

This volume, p. 541

"Bleaching by Light I: Effect of pH on the Bleaching or Darkening of Papers in the Dry and in the Immersed Condition under Visible and Near-Ultraviolet Radiation," in *Conservation of Historic and Artistic Works on Paper*, ed. Burgess, H., Canadian Conservation Institute, Ottawa, Ontario, Canada (1994), p. 181-190 (with S. B. Lee and J. Bogaard).

This volume, p. 561

"Bleaching by Light II: Studies of the Bleaching of Thermally Discolored Sugars and Other 'Model' Compounds," in *Conservation of Historic and Artistic Works on Paper*, ed. Burgess, H., Canadian Conservation Institute, Ottawa, Ontario, Canada (1994), p. 191-198 (with S. B. Lee and J. Bogaard).

Accelerated Aging: Photochemical and Thermal Aspects, Research in Conservation Series No. 4, The J. Paul Getty Trust, Marina del Rey, CA (1994).

This volume, p. 637

"Aspects of Chemical Research in Conservation: The Deterioration Process," *Journal of the American Institute for Conservation*, 33 (1994), p. 91-99.

This volume, p. 631

"Great Expectations: Questions and Answers on Conservation Materials and Methods," *Western Association for Art Conservation Newsletter*, 16, No. 2 (1994), p. 7-8.

"Vandyke Brown," in *Artist's Pigments*, Vol. 3, ed. FitzHugh, E. W., National Gallery of Art, Washington, DC, and Oxford University Press, New York, NY (1997), p. 157-190 (with R. M. Johnston-Feller).

Index